Springer Climate

Series Editor

John Dodson, Institute of Earth Environment, Chinese Academy of Sciences, Xian, Shaanxi, China

Springer Climate is an interdisciplinary book series dedicated to climate research. This includes climatology, climate change impacts, climate change management, climate change policy, regional climate studies, climate monitoring and modeling, palaeoclimatology etc. The series publishes high quality research for scientists, researchers, students and policy makers. An author/editor questionnaire, instructions for authors and a book proposal form can be obtained from the Publishing Editor.

Now indexed in Scopus® !

More information about this series at http://www.springer.com/series/11741

Małgorzata Falarz
Editor

Climate Change in Poland

Past, Present, Future

Springer

Editor
Małgorzata Falarz
Faculty of Natural Sciences
Institute of Earth Sciences
University of Silesia in Katowice
Katowice, Poland

ISSN 2352-0698 ISSN 2352-0701 (electronic)
Springer Climate
ISBN 978-3-030-70330-1 ISBN 978-3-030-70328-8 (eBook)
https://doi.org/10.1007/978-3-030-70328-8

This Springer imprint is published by the registered company Springer Nature Switzerland AG
The registered company address is: Gewerbestrasse 11, 6330 Cham, Switzerland

To our Masters

Authors

Preface

This monograph would not have been created without the activity and determination of its authors, 36 Polish climatologists who contributed its release. Authorship of individual book chapters was entrusted to the experienced scientists, specialists in the research of individual elements and characteristics of climate. The authors of the monograph come from eight scientific centres in Poland (Gdańsk, Katowice/Sosnowiec, Kraków, Łódź, Lublin, Poznań, Toruń, Warszawa), representing eight universities (Adam Mickiewicz University in Poznań, Jagiellonian University in Kraków, Maria Curie-Skłodowska University in Lublin, Nicolaus Copernicus University in Toruń, the University of Gdańsk, the University of Łódź, the University of Silesia in Katowice, the University of Warsaw), two technical universities (Kraków University of Technology, Agricultural University of Kraków), the Polish Academy of Science and Institute of Meteorology and Water Management—National Research Institute. The vast majority of the research presented here is new and was conducted for the purposes of this monograph. However, the book was preceded by many works, publications and studies that influenced its form.

This book would not have been created without the tireless work of many unnamed observers at the weather stations (both the Institute of Meteorology and Water Management—National Research Institute and university stations), who, day after day, have painstakingly measured, observed and documented the state of the weather for many years. In addition, the book could not have been published without the creation and availability of a meteorological database by the Institute of Meteorology and Water Management—National Research Institute and universities.

Finally, the monograph would not have been published without the active help of English native speaker, Leah Morawiec, who took care of the linguistic correctness of most chapters of the book. Special thanks are due to the scientific and technical staff of Springer International Publishing, especially for their great patience and kind help.

I express my deep gratitude to everyone for their work, commitment and effort.

The book is addressed to scientists (climatologists, geographers, etc.), academic teachers, students, journalists, and all those interested in Poland and climate change in Poland.

Climate variability and its contemporary changes are happening right in front of us. Therefore, we would appreciate any feedback to help improve the next editions of the book.

Sosnowiec, Poland Małgorzata Falarz

Acknowledgments In this monograph meteorological data of Institute of Meteorology and Water Management—National Research Institute and meteorological stations of both Jagiellonian University in Kraków (stations: Kraków Obs. UJ and Gaik-Brzezowa) and University of Silesia in Katowice (station Sosnowiec) has been used and processed.

The original version of the book was revised: In Chapter 11 figures 23 and 33 corrections have been updated and also the affiliation city name "Katowice" of the volume editor is corrected in web version. The correction to the book is available at https://doi.org/10.1007/978-3-030-70328-8_25

Contents

Part III Recent Climate Change

Part IV Future Climate Data

Part V Discussion and Conclusions

Contributors

Krzysztof Bartoszek Faculty of Earth Sciences and Spatial Management, Institute of Earth and Environmental Sciences, Department of Hydrology and Climatology, Maria Curie-Skłodowska University, Lublin, Poland

Ewa Bednorz Faculty of Geographical and Geological Sciences, Institute of Physical Geography and Environmental Planning, Department of Meteorology and Climatology, Adam Mickiewicz University, Poznań, Poland

Zuzanna Bielec-Bąkowska Faculty of Natural Sciences, Institute of Earth Sciences, University of Silesia in Katowice, Katowice, Poland

Krzysztof Błażejczyk Institute of Geography and Spatial Organization, Past Landscape Dynamic Laboratory, Polish Academy of Sciences, Warsaw, Poland

Barbara Bogdańska Institute of Meteorology and Water Management—National Research Institute, Warsaw, Poland

Anita Bokwa Faculty of Geography and Geology, Institute of Geography and Spatial Management, Jagiellonian University, Kraków, Poland

Tomasz Budzik Faculty of Natural Sciences, Institute of Earth Sciences, University of Silesia in Katowice, Katowice, Poland

Adam Choryński Institute for Agricultural and Forest Environment, Polish Academy of Sciences, Poznań, Poland

Dominika Ciaranek Faculty of Geography and Geology, Institute of Geography and Spatial Management, Jagiellonian University, Kraków, Poland

Danuta Czekierda Institute of Meteorology and Water Management—National Research Institute, Warsaw, Poland

Małgorzata Falarz Faculty of Natural Sciences, Institute of Earth Sciences, University of Silesia in Katowice, Katowice, Poland

Janusz Filipiak Faculty of Oceanography and Geography, Institute of Geography, Department of Meteorology and Climatology, University of Gdańsk, Gdańsk, Poland;
Institute of Meteorology and Water Management—National Research Institute, Warsaw, Poland

Dariusz Graczyk Institute for Agricultural and Forest Environment, Polish Academy of Sciences, Poznań, Poland

Monika J. Hajto Institute of Meteorology and Water Management—National Research Institute, Warsaw, Poland;
Faculty of Geography and Geology, Institute of Geography and Spatial Management, Jagiellonian University, Kraków, Poland

Agnieszka Halaś Institute of Geography and Spatial Organization, Past Landscape Dynamic Laboratory, Polish Academy of Sciences, Warsaw, Poland

Leszek Kolendowicz Faculty of Geographical and Geological Sciences, Institute of Physical Geography and Environmental Planning, Department of Meteorology and Climatology, Adam Mickiewicz University, Poznań, Poland

Magdalena Kuchcik Institute of Geography and Spatial Organization, Climate Impacts Laboratory, Polish Academy of Sciences, Warsaw, Poland

Monika Lisowska Faculty of Geography and Regional Studies, Chair of Physical Geography, Department of Climatology, University of Warsaw, Warsaw, Poland

Ewa Łupikasza Faculty of Natural Sciences, Institute of Earth Sciences, University of Silesia in Katowice, Katowice, Poland

Łukasz Małarzewski Faculty of Natural Sciences, Institute of Earth Sciences, University of Silesia in Katowice, Katowice, Poland

Dorota Matuszko Faculty of Geography and Geology, Institute of Geography and Spatial Management, Jagiellonian University in Kraków, Kraków, Poland

Tadeusz Niedźwiedź Faculty of Natural Sciences, Institute of Earth Sciences, University of Silesia in Katowice, Katowice, Poland

Magdalena Opała-Owczarek Faculty of Natural Sciences, Institute of Earth Sciences, University of Silesia in Katowice, Katowice, Poland

Iwona Pińskwar Institute for Agricultural and Forest Environment, Polish Academy of Sciences, Poznań, Poland

Katarzyna Piotrowicz Faculty of Geography and Geology, Institute of Geography and Spatial Management, Jagiellonian University, Kraków, Poland

Rajmund Przybylak Faculty of Earth Sciences and Spatial Management, Department of Meteorology and Climatology, Nicolaus Copernicus University, Toruń, Poland;
Centre for Climate Change Research, Nicolaus Copernicus University, Toruń, Poland

Jakub Soroka Institute of Meteorology and Water Management—National Research Institute, Hydrological and Meteorological Station, Gorzów Wielkopolski, Poland

Małgorzata Szwed Institute for Agricultural and Forest Environment, Polish Academy of Sciences, Poznań, Poland

Mateusz Taszarek Faculty of Geographical and Geological Sciences, Institute of Physical Geography and Environmental Planning, Department of Meteorology and Climatology, Adam Mickiewicz University, Poznań, Poland

Wojciech Tylmann Faculty of Oceanography and Geography, Environmental Change Reconstruction Laboratory, University of Gdansk, Gdańsk, Poland

Joanna Uscka-Kowalkowska Faculty of Earth Sciences and Spatial Management, Department of Meteorology and Climatology, Nicolaus Copernicus University, Toruń, Poland

Zbigniew Ustrnul Faculty of Geography and Geology, Institute of Geography and Spatial Management, Jagiellonian University, Kraków, Poland;
Institute of Meteorology and Water Management—National Research Institute, Warsaw, Poland

Stanisław Węglarczyk Faculty of Environmental and Power Engineering, Department of Geoengineering and Water Management, Cracow University of Technology, Kraków, Poland

Joanna Wibig Faculty of Geographical Sciences, Department of Meteorology and Climatology, University of Lodz, Lodz, Poland

Jakub Wojkowski Faculty of Environmental Engineering and Geodesy, Department of Ecology, Climatology and Air Protection, University of Agriculture, Kraków, Poland

Agnieszka Wypych Faculty of Geography and Geology, Institute of Geography and Spatial Management, Jagiellonian University in Kraków, Kraków, Poland;
Institute of Meteorology and Water Management—National Research Institute, Warsaw, Poland

Part I
Background

Chapter 1
Introduction

Małgorzata Falarz, Rajmund Przybylak, Janusz Filipiak, Agnieszka Wypych, and Małgorzata Szwed

Abstract Climate change is now one of the most important global challenges. The problem was noticed worldwide as early as the 1970s, when the first world Climate Conference was held in Stockholm (1972) under the motto "Only One Earth". Great progress, however, did not begin in climate protection until 1992, when the Rio de Janeiro Earth Summit was organised. Later on, since 1995, the Climate Summits (Conferences of the Parties, COP) of the United Nations have been held every year, bringing together representatives from almost all countries of the world and resulting in important global agreements on combating global warming (e.g. the Kyoto Protocol in 1997, Paris Agreement in 2015).

M. Falarz (✉)
Faculty of Natural Sciences, Institute of Earth Sciences, University of Silesia in Katowice, Katowice, Poland
e-mail: malgorzata.falarz@us.edu.pl

R. Przybylak
Faculty of Earth Sciences and Spatial Management, Department of Meteorology and Climatology, Nicolaus Copernicus University, Toruń, Poland

Centre for Climate Change Research, Nicolaus Copernicus University, Toruń, Poland

R. Przybylak
e-mail: rp11@umk.pl

J. Filipiak
Faculty of Oceanography and Geography, Institute of Geography, Department of Meteorology and Climatology, University of Gdańsk, Gdańsk, Poland
e-mail: janusz.filipiak@ug.edu.pl

J. Filipiak · A. Wypych
Institute of Meteorology and Water Management—National Research Institute, Warsaw, Poland
e-mail: agnieszka.wypych@uj.edu.pl

A. Wypych
Faculty of Geography and Geology, Institute of Geography and Spatial Management, Jagiellonian University, Krakow, Poland

M. Szwed
Institute for Agricultural and Forest Environment, Polish Academy of Sciences, Poznań, Poland
e-mail: mszwed@man.poznan.pl

M. Falarz (ed.), *Climate Change in Poland*, Springer Climate,
https://doi.org/10.1007/978-3-030-70328-8_1

Climate change is now one of the most important global challenges. The problem was noticed worldwide as early as the 1970s, when the first world Climate Conference was held in Stockholm (1972) under the motto "Only One Earth". Great progress, however, did not begin in climate protection until 1992, when the Rio de Janeiro Earth Summit was organised. Later on, since 1995, the Climate Summits (Conferences of the Parties, COP) of the United Nations have been held every year, bringing together representatives from almost all countries of the world and resulting in important global agreements on combating global warming (e.g. the Kyoto Protocol in 1997, Paris Agreement in 2015).

The Intergovernmental Panel on Climate Change (IPCC), the United Nations body for assessing the science related to climate change, publishes reports every 5–7 years with scientific assessments on climate change (e.g. IPCC 1992, 2007, 2013), its implications and potential future risks. In addition, special reports are published. The report appearing in 2018 for example, highlights the uneven distribution of temperature values on the globe for scenarios of 1.5 °C and 2.0 °C global warming compared to the preindustrial level. In line with expectation, the biggest changes will be in the extreme temperature in land areas, e.g. with global mean surface temperature warming of 2.0 °C, the change in average temperature of hottest days reaches 3–4 °C in some land areas, including Europe (IPCC 2018).

A number of climate change studies have been carried out worldwide, the results of which have been published in thousands of scientific articles or multifaceted monographs (e.g. Jones et al. 1996, Houghton et al. 2001, Jones et al. 2001, Hurrell et al. 2003, Knieling and Leal Filho 2013, Leal Filho et al. 2016).

Interest in climate change in Poland has a long history and tradition, for review see e.g. Przybylak et al. 2010 and Section I.2 in this volume. Here we mention only some of the more recent and most important works. They generally cover many aspects of climate change in Poland, including analyses of: (1) climate before instrumental measurements (e.g. Przybylak et al. 2010; Opała 2015; Przybylak 2016), and (2) contemporary climate: circulation conditions (e.g. Niedźwiedź and Łupikasza 2016, 2019), air pressure (e.g. Bielec-Bąkowska 2016), solar radiation (e.g. Uscka-Kowalkowska et al. 2007), sunshine (e.g. Matuszko 2014; Matuszko et al. 2020), cloudiness (e.g. Filipiak and Miętus 2009), air temperature (e.g. Ustrnul 2000), air humidity (e.g. Wypych 2010), precipitation (e.g. Łupikasza 2010, 2016), snow cover (e.g. Falarz 2008), wind (e.g. Araźny et al. 2007; Lorenc 2012), thunderstorms (e.g. Bielec-Bąkowska 2003) and tornadoes (e.g. Taszarek and Brooks 2015), hail (e.g. Bielec-Bąkowska 2013), fog (e.g. Łupikasza and Niedźwiedź 2016), urban climate (Fortuniak 2019), bioclimatic indices (e.g. Błażejczyk and Matzarakis 2007), weather types (e.g. Piotrowicz et al. 2016), temperature projections (e.g. Szwed and Graczyk 2006), precipitation projections (e.g. Pińskwar 2010) and projected changes in indices related to agriculture, water resources and human health (e.g. Szwed et al. 2010).

In the last 20 years there have been several conference publications in Polish presenting: circulation and radiation conditions of climate change and variability in Poland (Kolendowicz 2010; Bielec-Bąkowska et al. 2012; Kolendowicz et al. 2019), climate change in Poland at different spatial and time scales (Piotrowicz and Twardosz 2007), thermal and precipitation variability (Bednorz 2010), thermal, radiation and hydrological aspects of change (Chojnacka-Ożga and Lorenc 2019) and climate change and its consequences with references to the climate of Europe (Zmiany i zmienność klimatu Polski [...] 1999); Bednorz and Kolendowicz 2010).

Particular attention has been paid to extreme meteorological and hydrological phenomena in Poland (Kundzewicz and Jania 2007; Ustrnul and Czekierda 2009).

Of particular note are monographs in English, with an international composition of authors presenting: climate change in Poland in relation to European change (Przybylak et al. 2010) and climate change of the Baltic Sea Basin area covering Poland (The BACC Author Team 2008; The BACC II Author Team 2015). The first (Przybylak et al. 2010) provides both a synthesis of current knowledge about climate of Poland in recent centuries and new finding results for selected issues in selected regions of Poland or for individual meteorological stations. The book relates to the results published till 2009.

The above-mentioned works were created in different years, using different methods, with different levels of detail. None of them undertakes a systematic spatial description of the change of all climate elements throughout the area of Poland. The knowledge of Poland's climate change so far is very rich and valuable. However, it is somewhat dispersed; there is no harmonisation of methods, databases or research periods. There is not yet a book about climate change in Poland which presents new research that uses an uniform method for all the elements of climate as well as for the selected meteorological phenomena, bioclimatic conditions and weather types for past, present and future change. In this book, knowledge of these subjects is sorted out and harmonised.

The current volume consists of five main parts. Part I is an introduction. Part II presents the results of the study of climate change before the beginning of the instrumental measurements in Poland in the last millennium. In Part III the long-term trends and variability of circulation indices, climate elements, selected meteorological phenomena, bioclimatic conditions and weather types are analysed using data from the Institute of Meteorology and Water Management–National Research Institute and two university stations. A spatial diversity of climate trends and variability in Poland is presented in maps. Part IV of the book deals with projected changes up to 2100 in temperature, precipitation and thermal indices related to the agriculture and energy sectors in Poland. Part V contains summary, discussion and conclusion. More than 80% of the results presented in this monograph are the results of new studies, not yet published, which were carried out specifically for the purposes of this book.

References

Araźny A, Przybylak R, Vizi Z et al (2007) Mean and extreme wind velocities in Central Europe 1951–2005. Geogr Pol 80(2):69–78

Bednorz E (ed) (2010) *Klimat Polski na tle klimatu Europy. Warunki termiczne i opadowe*, Studia i Prace z Geografii i Geologii, 15, Bogucki Wydawnictwo Naukowe, Poznań, 214 pp
Bednorz E, Kolendowicz L (eds) (2010) *Klimat Polski na tle klimatu Europy. Zmiany i ich konsekwencje*, Studia i Prace z Geografii i Geologii, 16, Bogucki Wydawnictwo Naukowe, Poznań, 244 pp
Bielec-Bąkowska Z (2003) Long-term variability of thunderstorms occurrence in Poland in the 20th century. Atmos Res 67–68:35–52
Bielec-Bąkowska Z (2013) Burze i grady w Polsce. Prace Geograficzne Instytutu Geografii i Gospodarki Przestrzennej UJ 132:99–132
Bielec-Bąkowska Z (2016) Long-term variability of the frequency and persistence of strong highs over Poland. Environ Socio-Econ Stud 4(1):12–23
Bielec-Bąkowska Z, Łupikasza E, Widawski A (eds) (2012) Rola cyrkulacji atmosfery w kształtowaniu klimatu; Prace Wydziału Nauk o Ziemi Uniwersytetu Śląskiego, 74, Sosnowiec
Błażejczyk K, Matzarakis A (2007) Assessment of bioclimatic differentiation of Poland based on the human heat balance. Geogr Pol 80(1):63–82
Chojnacka-Ożga L, Lorenc H (eds) (2019) Współczesne problemy klimatu Polski, Institute of Meteorology and Water Management—National Research Institute, Warsaw, Poland, 260 pp
Falarz M (2008) Changes of extreme nival conditions in Poland during the second half of the 20th century. Meteorol Z 17(3):339–444
Filipiak J, Miętus M (2009) Spatial and temporal variability of cloudiness in Poland, 1971–2000. Int J Climatol 29:1294–1311. https://doi.org/10.1002/joc.1777
Fortuniak K (2019) Studies on urban climate in Poland. Przegląd Geofizyczny LXIV 1–2:73–106. https://doi.org/10.32045/PG-2019-003
Houghton JT et al (eds) (2001) Climate change 2001: the scientific basis. Cambridge University Press, Cambridge, UK
Hurrell JW, Kushnir Y, Ottersen G, Visbeck M (eds) (2003) North Atlantic oscillation: climatic significance and environmental impact. Geophysical Monograph 134. American Geophysical Union, Washington
IPCC (1992) Climate change: the 1990 and 1992 IPCC Assessments Report. JT Houghton (ed), Cambridge, UK, 212 pp
IPCC (2007) Climate change 2007, The Physical Science Basis. Contribution of working group I to the fourth assessment report of the intergovernmental panel on climate change. In: Solomon S. Qin D, Manning M, Chen Z, Marquis M, Averyt KB, Tignor M, Miller HL (eds). Cambridge University Press, Cambridge, United Kingdom and New York, NY, USA, 996 pp
IPCC (2013) Climate change 2013, The physical science basis. Contribution of working group I to the fifth assessment report of the intergovernmental panel on climate change. In: Stocker TF, Qin D, Plattner G.-K, Tignor M, Allen SK, Boschung J, Nauels A, Xia Y, Bex V, Midgley PM (eds). Cambridge University Press, Cambridge, United Kingdom and New York, NY, USA, 1535 pp, https://doi.org/10.1017/cbo9781107415324
IPCC (2018) Global Warming of 1.5 °C. An IPCC Special Report on the impacts of global warming of 1.5 °C above pre-industrial levels and related global greenhouse gas emission pathways, in the context of strengthening the global response to the threat of climate change, sustainable development, and efforts to eradicate poverty. In: Masson-Delmotte V, Zhai P, Pörtner H-O, Roberts D, Skea J, Shukla PR, Pirani A, Moufouma-Okia W, Péan C, Pidcock R, Connors S, Matthews JBR, Chen Y, Zhou X, Gomis MI, Lonnoy E, Maycock T, Tignor M, Waterfield T (eds)
Jones PD, Bradley RS, Jouzel J (eds) (1996) Climatic variations and forcing mechanisms of the last 2000 years. Springer, Berlin/Heidelberg/New York
Jones PD, Ogilvie AEJ, Davies TD, Briffa KR (eds) (2001) History and climate, memories of the future. Kluwer, New York
Knieling J, Leal Filho W (eds) (2013) Climate change governance. Springer International Publishing, 320 pp
Kolendowicz L (ed) (2010) *Klimat Polski na tle klimatu Europy. Warunki cyrkulacyjne i radiacyjne*, Studia i Prace z Geografii i Geologii, 14, Bogucki Wydawnictwo Naukowe, Poznań, 170 pp

Kolendowicz L, Bednorz E, Tomczyk AM (2019) Zmienność klimatu Polski i Europy oraz jej cyrkulacyjne uwarunkowania, Studia i Prace z Geografii, 77, Bogucki Wydawnictwo Naukowe, Poznań, 264 pp

Kundzewicz ZW, Jania J (2007) Extreme meteorological and hydrological events in Poland. Geogr Pol, 80(2)

Leal Filho W, Musa H, Cavan G, O'Hare P, Seixas J (eds) (2016) Climate change adaptation. Springer International Publishing, Resilience and Hazards, p 459

Lorenc H (2012) Maksymalne prędkości wiatru w Polsce. IMGW PIB, Warszawa

Łupikasza E (2010) Spatial and temporal variability of extreme precipitation in Poland in the period 1951–2006. Int J Climatol 30:991–1007. https://doi.org/10.1002/joc.1950

Łupikasza E (2016) The climatology of air-mass and frontal extreme precipitation. Study of meteorological data in Europe. Springer International Publishing, 313 pp

Łupikasza E, Niedźwiedź T (2016) Synoptic climatology of fog in selected locations of southern Poland (1966–2015). Bull Geogr, Phys Geogr Ser 11:5–15

Matuszko D (2014) Long-term variability in solar radiation in Krakow based on measurements of sunshine duration. Int J Climatol 34:228–234

Matuszko D, Bartoszek K, Soroka J, Węglarczyk S (2020) Sunshine duration in Poland from ground- and satellite-based data. Int J Climatol. https://doi.org/10.1002/joc.6460

Niedźwiedź T, Łupikasza E (2016) Change in atmospheric circulation patterns In: Kundzewicz ZW, Stoffel M, Niedźwiedź T, Wyżga B (eds) Flood risk in the Upper Vistula Basin. Springer International Publishing Switzerland, pp 189–208, https://doi.org/10.1007/978-3-319-41923-7_9

Niedźwiedź T, Łupikasza E (2019) Atmospheric circulation in the investigations of Polish climatologists. Przegląd Geofizyczny LXIV 1–2:107–166. https://doi.org/10.32045/PG-2019-004

Opała M (2015) The 443-year tree-ring chronology for the scots pine from Upper Silesia (Poland) as a dating tool and climate proxy. Geochronometria 42(1):41–52

Pińskwar I (2010) Projekcje zmian w ekstremach opadowych w Polsce. (Projections of changes in extreme precipitation in Poland) Monografia KGW PAN, 34 (in Polish)

Piotrowicz K, Bielec-Bąkowska Z, Domonkos P (2016) High atmospheric pressure and accompanying cold season weather types in Poland (1951 – 2010). Clim Res 67:165–177. https://doi.org/10.3354/cr01364

Piotrowicz K, Twardosz R (eds) (2007) Wahania klimatu w różnych skalach przestrzennych i czasowych. IGiGP UJ, Kraków, p 428

Przybylak R (2016) Poland's climate in the Last Millennium, Oxford Research Encyclopedia, Climate Science, Oxford University Press USA, 36 pp., https://doi.org/10.1093/acrefore/9780190228620.013.2

Przybylak R, Majorowicz J, Brázdil R, Kejna M (Eds.) (2010) The Polish climate in the European context: an historical overview, Springer, 535 pp. https://doi.org/10.1007/978-90-481-3167-9

Szwed M, Graczyk D (2006) Thermal seasons in Poland—the present and the future based on HadRM3-PRECIS results. Pap Glob Chang 13:59–77

Szwed M, Karg G, Pińskwar I, Radziejewski M, Graczyk D, Kędziora A, Kundzewicz ZW (2010) Climate change and its effect on agriculture, water resources and human health sectors in Poland. Nat Hazards Earth Syst Sci 10:1725–1737. https://doi.org/10.5194/nhess-10-1725-2010

Taszarek M, Brooks HE (2015) Tornado climatology of Poland. Mon Weather Rev 143(3):702–717

The BACC Author Team (2008) Assessment of climate change for the Baltic Sea Basin. Springer-Verlag, Berlin Heidelberg

The BACC II Author Team (2015) Second assessment of climate change for the Baltic Sea Basin, ser.: Regional Climate Studies, Springer, London

Uscka-Kowalkowska J, Przybylak R, Vízi Z, Araźny A, Kejna M, Maszewski R (2007) Variability to global solar radiation in Central Europe during the period 1951–2005 (on the basis of data from NCEP/NCAR reanalysis project). Geogr Pol 80(2):59–68

Ustrnul Z (2000) Synoptic-climatic structure of the extreme air thermal phenomena in Poland. Geogr Pol 73(2):99–109

Ustrnul Z, Czekierda D (2009) Atlas ekstremalnych zjawisk meteorologicznych oraz sytuacji synoptycznych w Polsce. IMGW, Warszawa
Wypych A (2010) Twentieth century variability of surface humidity as the climate change indicator in Kraków (Southern Poland). Theor Appl Climatol 101:475–482
Zmiany i zmienność klimatu Polski. Ich wpływ na gospodarkę, ekosystemy i człowieka (1999) Ogólnopolska konferencja naukowa, Łódź, 4–6 listopada 1999 r., 315 pp

Chapter 2
Initial Research of Climate Change in Poland

Małgorzata Falarz, Magdalena Opała-Owczarek, Tadeusz Niedźwiedź, Zuzanna Bielec-Bąkowska, Jakub Wojkowski, Dorota Matuszko, Janusz Filipiak, Zbigniew Ustrnul, Agnieszka Wypych, Monika Lisowska, Anita Bokwa, Krzysztof Błażejczyk, Katarzyna Piotrowicz, and Małgorzata Szwed

Abstract This chapter includes the results of the earliest studies on changes and variability of climatic elements, bioclimatic indices and weather types across Poland. The first pioneering works on climatic studies were presented (since 1858), even if they did not relate to climate change.

M. Falarz (✉) · M. Opała-Owczarek · T. Niedźwiedź · Z. Bielec-Bąkowska
Faculty of Natural Sciences, Institute of Earth Sciences, University of Silesia in Katowice, Katowice, Poland
e-mail: malgorzata.falarz@us.edu.pl

M. Opała-Owczarek
e-mail: magdalena.opala@us.edu.pl

T. Niedźwiedź
e-mail: tadeusz.niedzwiedz@us.edu.pl

Z. Bielec-Bąkowska
e-mail: zuzanna.bielec-bakowska@us.edu.pl

J. Wojkowski
Faculty of Environmental Engineering and Geodesy, Department of Ecology, Climatology and Air Protection, University of Agriculture, Kraków, Poland
e-mail: jakub.wojkowski@urk.edu.pl

D. Matuszko · Z. Ustrnul · A. Wypych · A. Bokwa · K. Piotrowicz
Faculty of Geography and Geology, Institute of Geography and Spatial Management, Jagiellonian University, Kraków, Poland
e-mail: d.matuszko@uj.edu.pl

Z. Ustrnul
e-mail: zbigniew.ustrnul@uj.edu.pl

A. Wypych
e-mail: agnieszka.wypych@uj.edu.pl

A. Bokwa
e-mail: anita.bokwa@uj.edu.pl

K. Piotrowicz
e-mail: k.piotrowicz@uj.edu.pl

M. Falarz (ed.), *Climate Change in Poland*, Springer Climate,
https://doi.org/10.1007/978-3-030-70328-8_2

This chapter includes the results of the earliest studies on changes and variability of climatic elements, bioclimatic indices and weather types across Poland. The first pioneering works on climatic studies were presented (since 1858), even if they did not relate to climate change. Already in the nineteenth century, the results of air temperature (e.g. Karliński 1868; Kuczyński 1884), precipitation (e.g. Kremser 1884; Wild 1887), cloudiness (Satke 1898), humidity (Wierzbicki 1878) and snow cover (Satke 1896, 1899) investigations in Polish lands were published in Polish and German.

The climatic history in Poland for the **preinstrumental period** has been reconstructed from various natural archives. Information on climate change during the Holocene period was obtained on the basis of geomorphological, sedimentological and botanical data (Stasiak 1968; Starkel 1977; Ralska-Jasiewiczowa and Starkel 1988, 1991; Starkel et al. 1996; Kotarba and Baumgart-Kotarba 1997) and the most accurate climatic reconstructions were carried out for the historical period (the last millennium) (Niedźwiedź et al. 2015). Based on a literature review, it can be concluded that information on climate change in Poland during historical times were provided mainly by four types of proxy data: documentary evidences, tree rings, lake sediments and geothermal profiles.

The earliest study on Poland's climatic conditions was conducted using documentary evidences (Semkowicz 1922; Polaczkówna 1925). A considerable amount of data can be found in the excerpts from the chronicles (e.g. Walawender 1932; Namaczyńska 1937; Szewczuk 1939; Inglot 1962, 1966, 1968, Rojecki 1965); however, the first climatological interpretation of these historical records was made no longer than 30 years ago by Maruszczak (1988), Sadowski (1991), Limanówka (2001), and Przybylak et al. (2001). Detailed information on the development of historical climatology in Poland was presented by Przybylak et al. (2001).

J. Filipiak
Faculty of Oceanography and Geography, Institute of Geography, Department of Meteorology and Climatology, University of Gdańsk, Gdańsk, Poland
e-mail: janusz.filipiak@ug.edu.pl

J. Filipiak · Z. Ustrnul · A. Wypych
Institute of Meteorology and Water Management—National Research Institute, Warsaw, Poland

M. Lisowska
Faculty of Geography and Regional Studies, Chair of Physical Geography, Department of Climatology, University of Warsaw, Warsaw, Poland
e-mail: mlisowska@uw.edu.pl

K. Błażejczyk
Polish Academy of Sciences, Institute of Geography and Spatial Organization, Climate Impacts Laboratory, Warsaw, Poland
e-mail: kblaz@twarda.pan.pl

M. Szwed
Institute for Agricultural and Forest Environment, Polish Academy of Sciences, Poznań, Poland
e-mail: mszwed@man.poznan.pl

The earliest mention of the possibility of conducting dendroclimatological research in Poland comes from 1914 (Merecki 1914). Although preliminary dendroclimatological research started after the Second World War (Zinkiewicz 1946; Ermich 1953), the first reconstruction of climate on the basis of tree rings was presented by Bednarz (1984) for the Tatra Mountains. Since these pioneer work many long tree-ring chronologies were developed for different parts of the country (see summary presented by Zielski et al. 2010). However, development of studies on past climatic changes in Poland based on dendroclimatic records has taken place in the last two decades (e.g. Krąpiec et al. 1998; Cedro 2004; Niedźwiedź 2004; Büntgen et al. 2007; Szychowska-Krąpiec 2010; Koprowski et al. 2012; Opała and Mendecki 2014; Balanzategui et al. 2017).

Another source of paleoclimatic data covering the last millennium is laminated lake sediments, whose potential for paleoenvironmental and paleoclimatic reconstructions has been thoroughly examined for the Lake Gościąż site in central Poland (Ralska-Jasiewiczowa et al. 1992, 1998; Starkel et al. 1996). Recently, high potential for quantitative paleoenvironmental reconstructions was also shown for Lake Żabińskie in north-eastern Poland (Amann et al. 2014; Hernández-Almeida et al. 2015; Larocque-Tobler et al. 2015).

The first use of geothermal profiles in thermally stabilised wells to reconstruct temperature history in Poland over the last 500 years was made by Majorowicz et al. (2001).

In the first half of the twentieth century, there was no work on the long-term variability of **atmospheric circulation** over Poland, but there were publications on the impact of atmospheric circulation on some elements of the climate (Bartnicki 1924; Arctowski 1927; Kaczorowska 1933; Milata 1935; Bartnicki and Kołodziejczyk 1935; Lisowski 1935). In some later works, the role of circulation factors in the formation of the Polish climate was emphasised (Kożuchowski 2003, 2004; Miętus 1996; Wibig 2001). In the second half of the twentieth century, several classifications of circulation types and indices were created (Osuchowska-Klein 1975; Niedźwiedz 1988; Ustrnul 1997; Piotrowski 2009). Attention was also paid to the long-term variability of circulation types, air masses and atmospheric fronts (Parczewski 1964; Niedźwiedz 1996; Kaszewski and Filipiuk 2003). Considering the Vangengeim-Girs hemispheric circulation indices for the European sector, seven circulation periods were distinguished (Degirmendžić et al. 2000).

The analyses of the distribution and variability of **atmospheric pressure** are the basis of research into atmospheric circulation. One of the first studies on the distribution of atmospheric pressure in Poland and Europe includes Gorczyński's work from (1917), in which the author presented 54 maps showing the distribution of monthly and annual isobars in Poland, Europe and all over the globe.

The Kraków series of measurements was subjected to the most detailed study of the meteorological element in question. It was used, among others, in Weisse's (1858), Hann's (1887), and Trepińska's (1988) works. The oldest studies concerned mainly general comparisons of pressure values (their daily and annual courses as well as deviations from the average) between the various stations located in Poland and in Europe. Later studies, mainly Trepińska's works also tackled the issues related to the analysis of trends of changes since 1792 (Trepińska 1988, 1997a; Bärring et al. 2002). The author also made a number of comparisons of a long-term (since 1901) series of pressure measurements in Kraków and Warszawa, as well as in other European cities. Similar studies using a long series of measurements in Warszawa were also made by Ustrnul and Czekierda (2000). In the works mentioned above, it was found, among others, that the long-term variability of average annual pressure values shows a slight increase, which is mainly caused by pressure changes in the winter season. An important characteristic of the long-term variability of pressure throughout the year and in individual months is the occurrence of clear periods of higher or lower values, which is most likely associated with the occurrence of the so-called circular epochs.

Nevertheless, most studies using the atmospheric pressure measurements in Poland are based on shorter measurement series. They usually concern the impact of pressure and its changes on the human body (for example Koźmiński and Michalska 2012) or explain the changes and variability of other meteorological elements (for example Degirmendžić et al. 2004).

The **actinometric measurements** started in Poland at the end of the nineteenth century, but they were of experimental character and were short series. The oldest measurements of solar radiation were carried out in 1894 in the meteorological observatory in Puławy (Kolomijcov 1894). The pioneer of actinometric measurements in Poland, as well as the creator of the instruments used to measure the intensity of solar radiation (Moll-Gorczyński solarimeter, later known as Kipp & Zonen pyranometer), was professor Władysław Gorczyński. He initiated the long-term measurements of global solar radiation in Warszawa starting in 1900 (Gorczyński 1913), and then in Gdynia from 1920 (Gorczyński 1951).

It is known that short-term series of measurements were carried out in Grodzisk during the interwar period (Gorczyński 1911) in Sopot (1928–1935), Gdańsk (1931–1935) (Frischmuth 1935), in Kołobrzeg (Gorczyński 1951), Wrocław (1929–1932) (Grundmann 1933), Ursynów (Gorczyński 1951), Remiszewice near Łódź (June 1932), Januszewice near Opoczno (August–October 1932) (Gorczyński and Ostrowski 1934), Racibórz (1929–1941) (Mackiewicz 1957), the Sudetes Mountains (September–November 1931 and March–November 1932) (Stenz 1959) and in the Tatra Mountains (July–August 1903, January, April and September 1924 and 1935–1939) (Stenz 1925, 1959).

The measurements of solar radiation on a larger scale began in Poland only after the end of World War II, when in 1952, the State Hydrological and Meteorological Institute (PIHM) opened several actinometric stations (Bogdańska et al. 2002). However, it was difficult to carry out the analysis of the components of the radiation budget, especially its spatial distribution aspect, because the number of the stations was insufficient. Additionally, regular measurements were carried out in meteorological stations belonging to a few universities, the Polish Academy of Sciences Institutes (PAN) and the Institute of Soil Science and Plant Cultivation (IUNG). The development of actinometry in Poland, which was observed in the 1950s, provided the source material for numerous climatological research studies. These works were published on the basis of the research on long-term variability and regional differentiation of solar radiation (Kuczmarska and Paszyński 1964a, b; Paszyński 1966; Podogrocki 1977, 1978; Miara et al. 1987; Olecki 1986, 1989; Słomka 1988; Bryś 1994) as well as the fluctuation, trends and periodicity analysis of global solar radiation (Bogdańska and Podogrocki 2000).

Until now, most of the **sunshine** studies have been based on short series of measurements, most often 10-year, and mainly includes the analysis of data from the 1960s and 1970s (Kuczmarski 1990) or earlier periods (Gorczyński 1913; Merecki 1914). Only a few stations in Poland can boast of long, homogeneous heliographic series. These include the IGiGP scientific station of the Jagiellonian University (formerly the Astronomical Observatory), where measurements have been taken continuously in the same place and using the same instrument since 1883. Based on the data from this station, numerous studies have been conducted on the subject of long-term variability of sunshine duration and its causes (among others, Morawska 1963; Matuszko 2014). In Kraków, as in other European cities (Matuszko 2016), one can notice common periods of decreases ("global dimming": 1951–1980) and increases ("global brightening": 1981–1995) in sunshine duration with extreme values occurring in the same years (maximum in 1943, 1921, minimum in 1980). Other stations with long, but partly reconstructed heliographic series, include Wroclaw (Matuszko 2016), Śnieżka, Warszawa and Puławy (Bogdańska et al. 2002).

In addition, the various aspects of **cloudiness** variability in Poland have been analysed for many years, e.g. Satke (1898), Gorczyński and Wierzbicka (1915). Stenz (1952) and Okołowicz (1962) dealt with the problem of spatial variability of cloudiness amount in Poland. Warakomski (1969) reported on the seasonal course of cloud types in Poland. However, the number of papers dealing with the analysis of long-term changes in cloudiness has been very limited. Morawska-Horawska, in her analyses (1963, 1985), described in detail the observational series of cloudiness in Kraków, the longest in Poland, covering the period since 1859.

Air temperature is the element most often analysed and has one of the longest data series. In the case of this element, there have been countless analyses carried out on a local, regional and national (Polish) scale. The oldest and most important are studies by Gorczyński (1913, 1915a, b, 1916, 1918) and Gorczyński and Kosińska (1916), and in the later period by Romer (1947b, 1948/1949). Some of these studies are of particular importance because their results refer to spatial diversity across Europe. The issue of changes and variability of this element was considered for the numerous measuring stations that were located in areas currently in Poland (e.g. in Wroclaw, Gdańsk, Poznań, Puławy, and on Śnieżka). However, most of the oldest studies presenting the variability of air temperature in different aspects concerned Warszawa and, especially, Kraków (Kowalczyk 1881; Karliński 1868, 1876; Kuczyński 1884; Merecki 1899). Nowadays, these series have been subjected to multiple detailed analyses. Studies for Kraków, edited by J. Trepińska (1997b), and those for Warszawa, the author of which is H. Lorenc (Lorenc 2000), contain their review. Owing to the complex history of measurements and observations, it is worth mentioning the temperature series for Gdansk (Miętus 2007; Filipiak 2007).

Air humidity is an element that plays an important role among meteorological processes within the atmosphere; however, the variety of humidity indices makes the global view of air moisture changes difficult to ascertain. Moreover, some methodical difficulties in measuring and computing particular air humidity parameters have been often pointed out.

Despite the above-mentioned doubts about the reliability of the material at hand, studies on trends and distribution of air humidity parameters, in both regional and local aspects, have been conducted for many years. They present diurnal and annual trends in relative air humidity, saturation deficit and, rarely, in vapour pressure. The studies are most often excerpts from greater works, generally monographs of towns or regions. The data used in them are measurement series no longer than 30 years. Attempts to analyse the long-term variability of air humidity were based only on one of the parameters, usually the relative air humidity, or on much shorter observation series. The results describe tendencies at particular locations, and, up to the present, no research concerning the whole area of Poland has been undertaken.

The oldest works, in accordance with global trends, comprised mostly measurement and terminology problems (e.g. Gorczyński 1948; Demiańczuk 1963; Janiszewski 1975). Some studies on annual and seasonal distribution of humidity parameters were conducted on the regional or local scale (e.g. Wierzbicki 1878; Gumiński 1927; Kosiba 1952; Michna 1972), also regarding urban climate conditions (e.g. Tarajkowska 1974; Młostek and Sobik 1984; Gluza and Kaszewski 1984; Kłysik et al. 1985). Air humidity variability was seldom the topic of detailed examination in Poland. Hohendorf was the first (1967, 1969) to take up this topic and described changes in saturation deficit. B. Obrębska-Starklowa and A. Grzyborowska (1997) presented the analysis of air humidity variability made for the plateau section of the Raba River (relative air humidity in the period of 1971–1992), T. Bryś (2003) described air saturation deficit variation in Wrocław over the twentieth century, and A. Wypych (e.g. 2004, 2010) focused on long-term variability of air humidity in Kraków, expressed by most of humidity parameters. Essentially, the conducted studies confirmed the drying of the atmosphere.

The oldest paper on **precipitation** variability in Poland by Kremser (1884) was based on chronological series starting in 1799 for Wroclaw. Further research used the data from Warszawa (Wild 1887; Pietkiewicz 1889). Both authors found spatial differentiation in temporal variability of precipitation between 1813 and 1887. Analysing almost 100-year long precipitation series from Warszawa (1803–1910), Gorczyński (1911) established several precipitation epoch and mentioned the periodic character of precipitation fluctuations. The Warszawa series (1811–1910) was again analysed by Rychliński (1923a, b, 1924, 1927), who discussed indices describing precipitation variability year by year and its periodicity. Moreover, Rychliński (1927) found a diminishing degree of pluvial continentality between 1811 and 1910. This result was later proved by Romer (1947a) who found increased pluvial oceanity between 1851 and 1930 at some stations in Europe, including Poland.

The oldest works on **snow cover** in the Polish area originated at the turn of the nineteenth and twentieth centuries (Satke 1899; Kamińska 1912). They were, however, concerned mostly with the main characteristics of the snow cover and its physical properties, sometimes in individual winter seasons (Satke 1896; Kosińska-Bartnicka 1924). Research on the long-term variability of snow cover began to develop only in the 1990s, with the progress of the investigations of the greenhouse effect and global warming. The first work on this subject was often published in the conference materials (e.g. Głowicki 1996; Piotrowicz 1996; Falarz et al. 1998), or even remained manuscripts (Głowicki and Jaskiewicz 1995). Detailed investigations of the long-term variability of different characteristics of the snow cover in Poznań for almost 70 winter seasons (1920/21-1989/90 with a war break) were conducted (Szustakowska 1991; Bednorz 1999a). As a result, a statistically significant positive trend of the period of potential snow cover duration in Poznań (3 days/10 years) was discovered, among others. It was launched the first research on the influence of atmospheric circulation on the snow cover long-term changes (Bednorz 1999b). A comparison of changes in snow conditions was conducted in the centre of a large city (Kraków) with the rural areas, resulting in a higher rate of decline in the values of snow-related characteristics in the city centre compared with rural areas (Falarz 1998). In addition, research was carried out in the mountainous areas of Poland (Głowicki 1996; Falarz et al. 1998). Within the 105-year snow cover series for the Śnieżka summit, a downward trend of the number of days with snow cover to the end of the 1930s, and a slight positive trend since the turn of the 1930s and 1940s has been observed (Głowicki and Jaskiewicz 1995). Polish-Slovak studies of snow cover on the southern and the northern slope of the Western Carpathians have shown a small negative trend in the number of days with snow cover depth $\geq$ 20 cm and in a maximum depth of snow in most of the investigated stations, except for the area of the Tatra mountains (Falarz et al. 1998). Research has been also done on changes and variability in snow conditions for skiing in the Polish Tatras (Falarz 1999).

Older climatological studies on spatial variability in the direction and speed of **wind** in Poland include the works of Stopa-Boryczka (1989), Niedzwiedz et al. (1985, 1995). At the end of the 1990s, research on climate change and variability in Poland began to emerge. However, in the climatological literature, studies describing the variability of temperature and precipitation in Poland dominate. Few items concern the issue of wind speed variability. Kożuchowski (2004) studied the variability of the speed of geostrophic wind and its components—meridian and zonal over Poland. Lorenc (1996, 2012) noted the increase in incidence of strong winds and whirlwinds. There are numerous publications concerning the forecasts of the change of wind field over Europe. However, the wind speed changes developed by the IPCC (2007) are not conclusive. Some models indicate an increase in the average and maximum wind speeds over Northern and Central Europe, others a decrease in wind. However, all forecasts developed by the IPCC highlight the possibility of a large seasonal wind speed variation, which is associated with a change in the pressure field over the Euro-Atlantic area.

The description of the occurrence of convective phenomena complements the climate characteristics of a given area. These phenomena include, above all, **thunderstorms**, accompanied by strong wind gusts and precipitation including hail, and tornadoes. Compared to most meteorological elements, the research of the phenomena mentioned above does not use measurement data, but mainly visual observations carried out at meteorological stations. Depending on the observation period, the records of such observations were of a different nature and were subject to various changes. Until around the mid-twentieth century, descriptive characteristics could be encountered, although symbols were more and more commonly used in compliance with the rules for describing all atmospheric phenomena. Due to the difficulty in accessing detailed descriptions of the phenomena under consideration and the low level of importance attached to their relatively rare occurrence, the first more comprehensive study of the occurrence of thunderstorms and tornadoes in Poland appeared quite late. Initially, they were concerned about the occurrence of individual phenomena or very short observation periods (Smosarski 1915). The work of Wiszniewski (1949) is considered to be the first study in which longer observational series (1891–1930) was used, followed by the work of Stopa (1962, 1965) which concerns thunderstorms occurring in Poland in 1946–1955. These studies mainly took into account the spatial, annual and long-term diversity of thunderstorm occurrences. Later on, in the 1990s interest in thunderstorms increased again and the research was a continuation of the aforementioned works. However, while conducting these studies special attention was paid to the long-term (since 1885) changes in the occurrence of the described phenomenon (Bielec-Bąkowska 2003, 2013), to the synoptic conditions conducive to the formation of thunderstorms in Poland (Kolendowicz 2005) and to forecasts of their occurrence (Grabowska 2005). An important result of the research was the identification of thunderstorm regions in Poland and the demonstration of the existence of a tendency that the number of thunderstorm days varies depending on the region of the country. The majority of stations with a drop in the number of days with a storm are located northwest of the line connecting Śnieżka and Suwałki, while the remaining stations have positive trends. Recent advancements in technology have also facilitated instrumental monitoring of thunderstorm activity within the use of lightning detection systems. Thanks to PERUN network operating operationally in Poland since 2002, it was possible to develop high-resolution thunderstorm climatologies and explore previously undiscovered aspects such as storm intensity or diurnal cycles (Taszarek et al. 2015; Czernecki et al. 2016).

The occurrence of **tornadoes** in the area of Poland is the least investigated. Most of the works, including the earliest ones, were devoted to descriptions of individual cases (Gumiński 1936; Parczewski and Kluźniak 1959). More detailed studies, often covering longer periods of observation, were only started at the turn of the twenty-first century. They concerned both the spatial and temporal diversity of tornadoes (Lorenc 1996, 2012; Taszarek and Brooks 2015), environmental conditions in which they occur most often (Taszarek and Kolendowicz 2013) and case studies (Taszarek et al. 2016; Taszarek and Gromadzki 2017; Pilguj et al. 2019). Unfortunately, the short period from which reliable reports come from and spatial and temporal inhomogenities do not allow to define reliable long-term trends in the frequency of tornadoes in Poland.

Overall, only a few studies have been devoted to the occurrence of storm precipitation, in particular **hail events**. The oldest studies include those devoted to hail tracks and areas of their most frequent occurrence (Schmuck 1959; Zinkiewicz and Michna 1955; Koźmiński and Rytel 1963). In later years, studies on hail and other storm precipitation types concerned mainly their long-term variability and diurnal cycles, as well as the circulation conducive to their occurrence (Twardosz et al. 2011; Suwała 2014). On their basis, it was found that the most characteristic feature of storm precipitation is their annual course and the occurrence of clear fluctuations in long-term variability of the number of days and the thunderstorm precipitation totals (Twardosz et al. 2011; Bielec-Bąkowska 2014). However, apart from the southern regions of Poland, it is difficult to indicate clear trends in their frequency over longer periods of time (Twardosz et al. 2011; Bielec-Bąkowska 2014). In recent years, the application of machine learning techniques in radar data and numerical weather prediction models allowed for more detailed analysis of atmospheric potential for producing hail events in Poland (Czernecki et al. 2019). The rapid development of the European Severe Weather Database also enabled the assessment of large and very large hail events over the course of a few recent years (Pilorz 2015).

Lisowski and Bartnicki (1935) analysed the mean number of **days with fog**, for particular months, and for 19 localities. Their analyses covered additionally the frequency of fog in the periods spring-summer and autumn-winter in 28 stations placed in the areas which belonged to Poland before the Second World War. Gumiński (1952a) studied the annual course of fog occurrence in central and eastern Poland using data from the period 1929–1938. Prawdzic and Sucheta (1975) presented the mean number of days with fog in particular months and for the whole year, for 15 stations in the Pomorze and Mazury regions, for the period 1951–1960, while Piwkowski (1976) used data from 18 stations across the country and demonstrated the annual course of fog occurrence and the frequency of their origin and duration in seasons in the years 1961–1970. Morawska (1966) analysed data from the period 1861–1960 from Kraków and described multi-annual variability of the number of days with fog, together with the relationship between fog frequency, wind speed and air temperature. A similar study was published for Warsaw by Janiszewski (1967), with the application of data from the period 1948–1962.

Research of long-term changes of **bioclimatic conditions** is undertaken only in the last 20 years. Different biometeorological indices are used as the measures of bioclimate. Such studies were done by few research teams, mostly in the Institute of Geography and Spatial Organization of the Polish Academy of Sciences. As the first studies of the changes of bioclimatic conditions have based on data from Jagiellonian University station in Botanic Garden (UJBG) in Cracow. The data covered the period of 1901–2000. As the bioclimatic measures, the authors (Błażejczyk et al. 2003) have used Subjective Temperature (STI) and Insulation Predicted (Iclp) indices. The changes in STI and Iclp were analysed on the background of changes in air circulation. The research was continued by Błażejczyk and Twardosz (2010) who have studied changes in Wind Chill Temperature (*WCT*), *HUMIDEX*, accepted level of physical activity (*MHR*) as well as Physiological Subjective Temperature (*PST*) and Physiological Strain (*PhS*) indices at UJBG station in Cracow in the longer period of 1826–2006. Owczarek (2009) in her PhD thesis for the northern Poland has applied data for the period 1951–2000 to analyse changes in PhS, HL, STI, WCT and Heat Stress Index (HSI). In the last decade the changes of bioclimatic conditions were studied by Bąkowska (2011) for Kołobrzeg, Poznań and Szczawno in the period 1951–2000. She has used PST, Iclp and MHR indices and furthermore—the newly developed Universal Thermal Climate Index (UTCI). UTCI was also used by Błażejczyk et al. (2015) and Kuchcik (2017) for the periods 1966–2012 and 1975–2014 as predictor of mortality risk in Poland.

The new definition of climate as a perennial weather regime (weather groups, weather types) was defined in the 1920s by Wojejkow and then disseminated by Fedorov. Nowadays, the **weather types** are understood as the result of: (1) genetic weather classification, the starting point of which is the analysis of synoptic situations (synoptic weather types) or (2) morphological classifications, in which several meteorological elements are analysed in total.

This second approach is the basis for research in complex climatology (a research stream initiated by E.E. Fiedorow). In Poland, this research trend was developed among others by Gumiński (1952b), Zinkiewicz (1953), and Wodzińska and Osuchowska (1963).

Nevertheless, the greatest contribution to learning about the Polish climate based on the classification of weather types was made by Woś (1999, 2010). While the frequency of occurrence of various weather types in Poland is well known, their long-term variability and change in trends have not been analysed in detail. Only Piotrowicz (2010) determined the variability of weather types in Kraków on the basis of a 108-year series of measurements (1901–2008).

The issue of climate change and climate **projections for the future** based on global climate models GCMs appeared in the scientific literature in the 1990s, along with increasingly visible signals of global warming and after the first IPCC report. Polish researchers have also taken up this topic. One of the first papers using global climate models was the article by Gutry-Korycka et al. (1994). In this paper, the authors used GCM models to predict changes in the spatial distribution of climate variables in hydrological models.

In 2000, the results of research of the ACACIA project (**A** **C**oncerted **A**ction Towards A Comprehensive **C**limate **I**mpacts and **A**daptations Assessment for the European Union) were published. This project assessed climate impacts and potential adaptation in Europe until the 2080 s (Parry 2000). ACACIA developed four scenarios on the basis of a combination of the UKCIP and SRES approaches. Temperature and precipitation projections for Poland can be found among the results (average value for the whole territory).

In the following years, European research projects enabled further development of studies on climate projections for Poland at a much higher resolution. Thus, projection results for Poland from the MICE and ENSEMBLES projects have become available. While in the MICE project (**M**odelling the **I**mpact of **C**limate **E**xtremes), the results from the Hadley Centre Regional Climate Model (HadRM3-P), with spatial resolution 0.44° by 0.44°, were used in the analysis of future changes in the characteristics of precipitation and temperature (e.g. Szwed et al. 2007), the ENSEMBLES project (Ensembles-Based Predictions of Climate Changes and their Impacts) produced a set of regional climate models RCMs, which cover Europe with a spatial resolution of about 25 by 25 km and outline just one possible vision of the future, corresponding to a specific SRES emission scenario, A1B. All in all, they also enabled the study of a greater number of climate variables (e.g. Van der Linden and Mitchell 2009).

Other results of the current research of climate change and variability are discussed in particular chapters of parts II–IV of the volume.

References

Amann B, Lobsinger S, Fischer D, Tylmann W, Filipiak J, Grosjean M (2014) Spring temperature variability and eutrophication history inferred sedimentary pigments in the varved sediments of Lake Żabińskie, north-eastern Poland, 1907–2008 AD. Glob Planet Chang 123:86–96

Arctowski H (1927) Wstęp do studiów nad transportem mas atmosferycznych. Kosmos 52:328–340

Bąkowska M (2011) Zmienność dobowa, sezonowa i wieloletnia warunków biotermicznych w Polsce. Doctoral dissertation, IGiPZ PAN, Warszawa

Balanzategui D, Knorr A, Heussner KU, Wazny T, Beck W, Słowiński M, Helle G, Buras A Wilmking M, Van Der Maaten E, Scharnweber T, Dorado-Liñán I and Heinrich I (2017) An 810-year history of cold season temperature variability for northern Poland. Boreas 47:443–453

Bärring L, Trepińska J, Kowanetz L (2002) Variability of the mean monthly values of air pressure based on the Cracow series of records (1792–2000). Prace Geograficzne IGiGP UJ 110:9–24

Bartnicki L (1924) O przebiegu czynników meteorologicznych w Polsce pod wpływem różnych układów barometrycznych. Prace Meteorologiczne i Hydrologiczne 1:7–32

Bartnicki L, Kołodziejczyk S (1935) Warunki synoptyczne powstawania zamieci i zawiei śnieżnych w Polsce. Prace PIM 6:3–26

Bednarz Z (1984) The comparison of dendroclimatological reconstructions of summer temperatures from the Alps and Tatra Mountains from 1741–1965. Dendrochronologia 2:63–72

Bednorz E (1999a) Pokrywa śnieżna w Poznaniu w latach 1920–1990, [w:] Zmiany i zmienność klimatu Polski i ich wpływ na ekosystemy, gospodarkę i człowieka – Ogólnopolska konferencja naukowa, materiały konferencyjne, Łódź, 4–6 listopada 1999, pp 7–10

Bednorz E (1999b) Zmiany grubości pokrywy śnieżnej i ich związek z kierunkami napływu mas powietrza w rejonie Legnicy i Kołobrzegu; Bad. Fizjogr. nad Polską Zach., ser. A – Geogr. Fiz., 50:15–26

Bielec-Bąkowska Z (2003) Long-term variability of thunderstorms occurrence in Poland in the 20th century. Atmos Res 67–68:35–52

Bielec-Bąkowska Z (2013) Burze i grady w Polsce. Prace Geograficzne Instytutu Geografii i Gospodarki Przestrzennej UJ 132:99–132

Bielec-Bąkowska Z (2014) Thunderstorms and thunderstorm precipitations in southern Poland. Environ Socio-Econ Stud 2(3):33–46

Błażejczyk K, Baranowski J, Błażejczyk A (2015) Wpływ klimatu na stan zdrowia w Polsce: stan aktualny oraz prognoza do 2100 roku. Wyd. Akademickie SEDNO

Błażejczyk K, Twardosz R (2010) Long-term changes of bioclimatic conditions in Cracow (Poland). [w:] Przybylak R, Majorowicz R, Brázdil J, Kejna M (red.), The polish climate in the european context: an historical overview. Springer, Science + Business Media B.V., 235–246

Błażejczyk K, Twardosz R, Kunert A (2003) Zmienność warunków biotermicznych w Krakowie w XX wieku na tle wahań cyrkulacji atmosferycznej. [w:] Błażejczyk K, Krawczyk B, Kuchcik M (red.) Postępy w badaniach klimatycznych i bioklimatycznych, IGiPZ PAN, Prace Geograficzne 188:233–246

Bogdańska B, Podogrocki J (2000) The variability of the global solar radiation over Poland in 1961–1965 (in Polish) [Zmienność całkowitego promieniowania słonecznego na obszarze Polski w okresie 1961–1995]. Mat. Badawcze, Meteorologia - 30, IMGW, Warszawa, p 43

Bogdańska B, Podogrocki J, Żółtowska K (2002) The polish solar radiation measurement network and actinometric database (in Polish) (in:) The scientific activity of Professor Władysław

Gorczyński and its continuation (in Polish). Symposium on Climatology at the Nicolaus Copernicus University, Toruń, 16–17 September 1993, 55–70

Bryś K (1994) The intensity of solar radiation in Wrocław in the period 1961–1990 (in Polish) [Natężenie promieniowania słonecznego całkowitego we Wrocławiu w okresie 1961–1990]. Zeszyty Naukowe AR 243:37–49

Bryś T (2003) Variability of saturation deficit in Wroclaw in the 20th century and some of its conditions [w:] Pyka JL, Dubicka M, Szczepankiewicz-Szmyrka A, Sobik M, Błaś M (red.), Man and climate in the 20th century, Studia Geograficzne, 75, Uniwersytet Wrocławski, 203–215

Büntgen U, Frank D, Kaczka RJ, Versteg A, Zwijacz-Kozica T, Esper J (2007) Growth responses to climate in a multi—species tree—ring network in the Western Carpathian Tatra Mountians, Poland and Slovakia. Tree Physiol 27:689–702

Cedro A (2004) Zmiany klimatyczne na Pomorzu Zachodnim w świetle analizy sekwencji przyrostów rocznych sosny zwyczajnej, daglezji zielonej i rodzimy gatunków dębów. In Plus Oficyna, Szczecin

Czernecki B, Taszarek M, Kolendowicz L, Konarski J (2016) Relationship between human observations of thunderstorms and the PERUN lightning detection network in Poland. Atmos Res 167:118–128

Czernecki B, Taszarek M, Marosz M, Półrolniczak M, Kolendowicz L, Wyszogrodzki A, Szturc J (2019) Application of machine learning to large hail prediction-The importance of radar reflectivity, lightning occurrence and convective parameters derived from ERA5. Atmos Res 227:249–262

Degirmendžić J, Kożuchowski K, Żmudzka E (2004) Changes of air temperature and precipitation in Poland in the period 1951–2000 and their relationship to atmospheric circulation. Int J Climatol 24:291–310

Degirmendžić J., Kożuchowski K., Wibig J (2000) Epoki cyrkulacyjne XX wieku i zmienność typów cyrkulacji atmosferycznej w Polsce. Przegląd Geofizyczny 45(3–4)

Demiańczuk P (1963) Analiza poprawności tablic psychrometrycznych używanych w polskiej sieci meteorologicznej oraz opracowań elementów wilgotności powietrza. Biuletyn PIHM 4:2–31

Ermich K (1953) Wpływ czynników klimatycznych na przyrost dębu szypułkowego (Quercus robur L) i sosny zwyczajnej (Pinus sylvestris L). Próba analizy zagadnienia, Prace Rolniczo-Leśne PAU

Falarz M (1998) Wieloletnia zmienność pokrywy śnieżnej w Krakowie na tle zmian w obszarach podmiejskich. Acta Univ. Lodz., Folia Geogr. Phys. 3:473–481

Falarz M (1999) Wieloletnia zmienność warunków śnieżnych korzystnych dla narciarstwa w polskich Tatrach; [in:] Zmiany i Zmienność Klimatu Polski i ich Wpływ na Ekosystemy, Gospodarkę i Człowieka – Ogólnopolska Konferencja Naukowa - materiały konferencyjne, Łódź, 4–6 listopada 1999, s. 65–70

Falarz M, Faško P, Lapin M (1998) Long-term variability of the snow cover in the Western Carpathians; w: Proceedings of the 2nd European Conference on Applied Climatology, ECAC 98, 19 to 23 October 1998, Vienna, Austria, „Oesterreichische Beitrage zu Meteorologie und Geophysik" Zentralanstalt fur Meteorologie und Geodynamik, 384, 19, Wien 1998, p. 19; full text on „ECAC CD-ROM", session 1, p 6

Filipiak J (2007) Rekonstrukcja warunków klimatycznych Gdańska w okresie pomiarów instrumentalnych–fakty i niepewności, [in:] 200 lat regularnych pomiarów i obserwacji meteorologicznych w Gdańsku, Monografie IMGW. Warszawa, 20–33

Frischmuth G (1935) Beitrage zum Strahlungsklima von Danzig. I. Forschungs- arbeiten des S. Observatoriums. Danzig

Głowicki B (1996) Pokrywa śnieżna w Polsce południowo-zachodniej w okresie 1888/89–1995/96; [W:] Międzynarodowa Konferencja Naukowa nt. „Klimatyczne warunki produkcji roślinnej" Puławy, 25–26 września 1996, Streszczenia prac, s. 31–32, IUNG w Puławach

Głowicki B, Jaśkiewicz H (1995) Zmienność charakterystyk pokrywy śnieżnej w Polsce (etap I), [w:] Ocena aktualnego stanu oraz zmienność klimatu Polski – raport końcowy z realizacji zadania badawczego M-2, ss. 12, maszynopis w IMGW Oddział Kraków

Gluza A, Kaszewski B (1984) Zróżnicowanie temperatury i wilgotności względnej powietrza w Lublinie [w:] Klimat i Bioklimat Miast, Materiały I Ogólnopolskiej Konferencji, Łódź, 22–24 listopada 1984, 107–114
Gorczyński W (1911) O zmienności opadu według obserwacji warszawskich od 1803 roku. Tow. Nauk. Warszaw., Spraw. z Posiedzeń, 4, 9
Gorczyński W (1913) Pyrheliometric values and heat sums in Warsaw according to measurements in the period 1901–1913 (in Polish) [Wartości pyrheliometryczne i sumy ciepła dla Warszawy według pomiarów w okresie 1901–1913] Wyd. Tow. Nauk. Warsz
Gorczyński W (1915a) O zmianach długoletnich temperatury powietrza w Polsce i Eurazyi, Warszawa, Drukarnia i Litografia Jana Cotty
Gorczyński W (1915b) O zmienności temperatury z dnia na dzień w Polsce i w Eurazyi. Warszawa, Drukarnia i Litografia Jana Cotty
Gorczyński W (1916) Badanie współzależności przebiegów temperatury metodą korelacyjną, Warszawa, Drukarnia i Litografia Jana Cotty
Gorczyński W (1917) Linie jednakowych korelacyj ciśnienia i temperatury powietrza w stosunku do Warszawy. Sprawozdanie Towarzystwa Naukowego Warszawskiego 10, 2, Warszawa
Gorczyński W (1918) Nowe izotermy Polski, Europy i kuli ziemskiej, Warszawa, Drukarnia i Litografia Jana Cotty, Warszawa
Gorczyński W (1948) O wartościach wilgotności powietrza i prostej metodzie ich wyznaczania. Gazeta Obserwatora PIHM 1(4):6–10
Gorczyński W (1951) Intensity and totals of solar radiation ol Gdańsk, Sopot and Gdynia in comparison with Warsaw (in Polish) [O insolacji Gdańska, Sopotu i Gdyni w porównaniu z Warszawą]. Studia Societatis Scientiarum Torunensis. Sectio A, Mathematica-Physica 2(4)
Gorczyński W, Kosińska S (1916) O wartościach średnich temperatury powietrza i o przebiegu izoterm w Polsce, Warszawa, Drukarnia i Litografia Jana Cotty
Gorczyński W, Ostrowski S (1934) Interdependence between the amount of diffuse sky radiation and the duration of sunshine (in Polish) [O związku między promieniowaniem rozproszonem nieba i stopniem usłonecznienia]. Sprawozdania Towarzystwa Naukowego Warszawskiego, Wydział Nauk Mat. i Przyr., 27, Warsaw
Gorczyński W, Wierzbicka W (1915) O wartościach średnich zachmurzenia w Polsce. Sprawozdania Towarzystwa Nauk Warszawskiego 8:609–649
Grabowska K (2005) Tendencje zmian i prognozy aktywności burzowej w Polsce. In: Bogdanowicz E, Kossowska-Cezak U, Szkutnicki J (eds) Ekstremalne zjawiska hydrologiczne i meteorologiczne. IMGW, Warszawa, pp 385–391
Grundmann W (1933) Über die Wärmeeinstrahlung von Sonne u. Himmel in Breslau – Krietern. Breslau – Stadt und auf der „Hohen Eule". Gerlands Beitr. Zur Geophysik, B. 40, Leipzig
Gumiński R (1927) Wilgotność powietrza w Polsce (wahania roczne i rozkład geograficzny). Prace Meteorol. i Hydrol. 3:1–71
Gumiński R (1936) Trąba powietrzna pod Lublinem w dniu 20 lipca 1931r. Wiadomości Meteorologiczne i Hydrologiczne 16:7–9
Gumiński R (1952a) Bieg roczny występowania mgły w centralnej i wschodniej Polsce, wiadomości Służby Hydrologicznej i Meteorologicznej, III(2a)
Gumiński R (1952b) Materiały do poznania genezy i struktury klimatu Polski. Przegląd Geograficzny 24(3):3–36
Gutry-Korycka M, Werner P, Jakubiak B (1994) Generation of time series of the meteorological values in changing climatic conditions. Geographia Polonica 62:23–45
Hann J (1887) Atlas der Meteorologie. Justus Perthes, Gotha
Hernández-Almeida I, Grosjean M, Przybylak R, Tylmann W (2015) A chrysophyte-based quantitative reconstruction of winter severity from varved lake sediments in NE Poland during the past millennium and its relationship to natural climate variability. Quat Sci Rev 122:74–88
Hohendorf E (1967) Opracowanie statystyczne ciągów obserwacyjnych niedosytu wilgotności powietrza, Prace i Studia Kom. Inż., i Gosp. Wodnej, 8, 263–299

Hohendorf E (1969) Niedosyty wilgotności powietrza w Polsce w latach 1881–1939, Roczniki Nauk Roln., Ser. F, 73(3):413–483
Inglot S (1962) Historyczne aspekty zjawisk klimatyczno-meteorologicznych na Śląsku od XVI do połowy XIX wieku. Spraw Wrocł Tow Nauk 17A:82–86
Inglot S (1966) Badanie zjawisk klimatyczno-meteorologicznych na Śląsku od XVI do połowy XIX w. Studia z Dziejów Gospodarstwa Wiejskiego 8:69–71
Inglot S (1968) Zjawiska klimatologiczno-meteorologiczne na Śląsku od XVI do połowy XIX w. Prace Wrocł Tow Nauk, Seria B, 139
IPCC AR4 WG1 (2007) Solomon S, Qin D, Manning M, Chen Z, Marquis M, Averyt KB, Tignor M, Miller HL (ed.), Climate change 2007: the Physical Science Basis, Contribution of Working Group I to the Fourth Assessment Report of the Intergovernmental Panel on Climate Change, Cambridge University Press
Janiszewski F (1967) Występowanie mgły w Warszawie, Prace PIHM, 91
Janiszewski F (1975) Problem tablic psychrometrycznych, Wiadomości IMGW, II(XXIII), 1, 53–59
Kaczorowska Z (1933) Przyczyny meteorologiczne letnich wezbrań Wisły. Prace PIM 2:1–33
Kamińska EW (1912) Trwałość szaty śnieżnej na północnym stoku Karpat. Rozprawy Wydz. Mat.-Przyr. Akademii Umiejętności, ser. III, 12:317–344
Karliński F (1868) Mittlere Temperatur zu Krakau nach 40-jaehrigen Beobachtungen 1826–1865, Jahrbuecher der k.k. Central-Anstalt fuer Meteorologie und Erdmagnetismus Bd. 3, Wien
Karliński F (1876) Unterschied zwischen dem Mittel der taeglichen Extreme und dem wahren Mittel in Krakau und Wien, Meteorologische Zeitschrift, 11, Wien
Kaszewski BM, Filipiuk E (2003) Variability of atmospheric circulation in Central Europe in the summer season 1881–1998 (on the basis of the Hess-Brezowski classification). Meteorologische Zeitschrift 12(3):123–130 (June 2003)
Kłysik K (1985) Wpływ struktury termiczno-wilgotnościowej przyziemnych warstw powietrza na klimat lokalny w wybranych warunkach terenowych. Acta Geographica Lodziensia 49:1–119
Kolendowicz L (2005) Wpływ cyrkulacji atmosferycznej oraz temperatury i wilgotności powietrza na występowanie dni z burzą na obszarze Polski. Wydawnictwo Naukowe UAM, Poznań
Kolomijcov NL (1894) Meteorologiceskij Biulleten Observatorii Instituta Selskogo Chozjajstva i Lesovodstva v Novoj Aleksandrii, 1–3
Koprowski M, Przybylak R, Zielski A, Pospieszyńska A (2012) Tree rings of Scots pine (Pinus sylvestris L.) as a source of information about past climate in northern Poland. Int J Biometeorol 56:1–10
Kosiba A (1952) Zagadnienie bilansu wilgoci na Śląsku w świetle zmian wilgotności względnej. Przegląd Meteorologiczny i Hydrologiczny 5(3/4):58–69
Kosińska-Bartnicka S (1924) Uśnieżenie w Polsce zimą 1922/23. Czas. Geogr. II(3/4):393–406
Kotarba A, Baumgart-Kotarba M (1997) Holocene debris-flow activity in the light of lacustrine sediment studies in the High Tatra Mountains, Poland [in:] Matthews JA, Brunsden D, Frenzel B, Gläser B, Weiß MM (eds.), Rapid mass movement as a source of climatic evidence for the Holocene. Paläoklimaforschung, vol. 19, Stuttgart: Gustav Fischer Verlag, 147–158
Kowalczyk J (1881) O spostrzeżeniach meteorologicznych w Warszawie, Pam. Fizjogr. I-II:1–46
Koźmiński C, Michalska B (2012) Międzydobowe zmiany ciśnienia atmosferycznego w Polsce niekorzystne dla organizmu człowieka. Przegląd Geograficzny 84(3):361–374
Cz Koźmiński, Rytel M (1963) Próba wykreślenia izarytm prawdopodobieństwa wystąpienia burz gradowych w Polsce na podstawie lat 1947–1960. Czasopismo Geograficzne 34(1):51–60
Kożuchowski K (2004) Cyrkulacja atmosferyczna nad Polską i jej wpływ na warunki klimatyczne. [in:] Kożuchowski K (red.) Skala, uwarunkowania i perspektywy współczesnych zmian klimatycznych w Polsce. Łódź, 69–87
Kożuchowski K (2003) Cyrkulacyjne czynniki klimatu Polski. Czasopismo Geograficzne 74(1–2):93–105
Krąpiec M, Jędrysek M, Skrzypek G, Kałużny A (1998) Carbon and hydrogen isotope ratios in cellulose from oaks tree-rings as record of palaeoclimatic conditions in Southern Poland during the last millenium. Folia Quat 69:135–150

Kremser V (1884) Über die Veränderlichkeit der niederschläge. Meteor Zeit 3–4:84
Kuchcik M (2017) Zmiany warunków termicznych w Polsce na przełomie XX i XXI wieku i ich wpływ na umieralność. Prace Geograficzne 263
Kuczmarska L, Paszyński J (1964a) The annual course of total solar radiation in Poland (in Polish) [Przebieg roczny całkowitego promieniowania słonecznego w Polsce]. Przegl Geofiz 3(4):275–279
Kuczmarska L, Paszyński J (1964b) Distribution of total solat radiation in Poland (in Polish) [Rozkład promieniowania całkowitego na obszarze Polski]. Przegl Geogr 4:691–702
Kuczmarski M (1990) Usłonecznienie Polski i jego przydatność dla helioterapii. Dokumentacja Geograficzna 4:67
Kuczyński LS (1884) Przebieg roczny ciepłoty powietrza w Krakowie, obliczony na podstawie 50-letnich spostrzeżeń (1826–1875) sposobem nowym, prostszym i ściślejszym niż dotąd używane, Pamiętnik Wydz. Matem.-Przyr. Akademii Umiejętności, IX, Kraków
Larocque-Tobler I, Filipiak J, Tylmann W, Bonk A, Grosjean M (2015) Comparison between chironomid-inferred mean-August temperature from varved Lake Żabińskie (Poland) and instrumental data sine 1896 AD. Quat Sci Rev 111:35–50
Limanówka D (2001) Rekonstrukcja warunków klimatycznych Krakowa w pierwszej połowie XVI wieku. Materiały Badawcze IMGW, Seria: Meteorologia 33, Warszawa
Lisowski K (1935) O częstotliwości i warunkach synoptycznych powstawania mgły w Polsce. Prace PIM 5:61–94
Lisowski K, Bartnicki L (1935) O częstotliwości i warunkach synoptycznych powstawania mgieł w Polsce, Prace PIHM, 5
Lorenc H (1996) Struktura i zasoby energetyczne wiatru w Polsce. Instytut Meteorologii i Gospodarki Wodnej, Państwowy Instytut Badawczy, Warszawa
Lorenc H (2000) Studia nad 220-letnią (1779–1998) serią temperatury powietrza w Warszawie oraz ocena jej wiekowej tendencji, Materiały Badawcze, Seria: Meteorologia, 31, IMGW, Warszawa
Lorenc H (2012) Maksymalne prędkości wiatru w Polsce. Instytut Meteorologii i Gospodarki Wodnej-Państwowy Instytut Badawczy
Mackiewicz M (1957) Solar radiation in Racibórz (in Polish) [Promieniowanie słoneczne w Racioborzu]. Przegl. Geofiz., Ch. II (X), 1–2:3–13
Majorowicz J, Šafanda J, Przybylak R, Wójcik G (2001) Rekonstrukcja zmian temperatury powierzchni gruntu w Polsce w ostatnim 500-leciu na podstawie profili geotermicznych. Przegląd Geofizyczny 4:305–321
Maruszczak H (1988) Zmiany środowiska przyrodniczego kraju w czasach historycznych. In: Starkel L (ed) Przemiany środowiska geograficznego, Wszechnica PAN, Ossolineum, Wrocław
Matuszko D (2014) Long-term variability in solar radiation in Krakow based on measurements of sunshine duration. Int J Climatol 34(1):228–234
Matuszko D (2016) Usłonecznienie w miastach na podstawie wybranych stacji w Europie. Acta Geographica Lodziensia 104:45–56
Merecki R (1899) O zmienności nieokresowej temperatury powietrza, Rozprawy Wydz. Matem-Przyr. Akademii Umiejętności, 35, Kraków
Merecki R (1914) Klimatologia ziem polskich, Księgarnia Gebethnera i Wolffa, Warszawa. 313 pp
Miara K, Paszyński J, Grzybowski J (1987) Spatial distribution of the radiation balance in Poland (in Polish) [Zróżnicowanie przestrzenne bilansu promieniowania na obszarze Polski]. Przegl Geogr 4:487–509
Michna E (1972) O wilgotności względnej powietrza w Polsce. Przegląd Geofizyczny 17(1):3–14
Miętus M (1996) Zmienność lokalnej cyrkulacji atmosferycznej nad Polską Północną i jej związek z elementami klimatu. Wiadomości IMGW 19/40(1):9–30
Miętus M (2007) 200 lat regularnych obserwacji i pomiarów meteorologicznych w Gdańsku – od fascynacji do praktycznego działania [in:] 200 lat regularnych pomiarów i obserwacji meteorologicznych w Gdańsku, Monografie IMGW. Warszawa, 7–19
Milata W (1935) Meteorologiczne przyczyny powodzi w lipcu 1934 r. Czasopismo Geograficzne 13(3):273–282

Młostek E, Sobik M (1984) Usłovija vłažnosti vozducha w gorode v izbrannych sinopticzeskich situacjach na primiere goroda Wrocława [w:] Mikroklimat i mezoklimat gorodskich aglomeracji, Brno, 36–43
Morawska M (1963) Zachmurzenie i usłonecznienie Krakowa w latach 1859 – 1958, Prace PIHM, 81, 3–46, Wyd. Komunikacji i Łączności, Warszawa
Morawska M (1966) Mgły w Krakowie (1861–1960). Przegląd Geofizyczny 11:3
Morawska-Horawska M (1985) Cloudiness and sunshine in Cracow, 1861–1980, and its contemporary tendencies. J Climatol 5:633–642
Namaczyńska St (ed) (1937) Kronika klęsk elementarnych w Polsce i w krajach sąsiednich w latach 1648–1696, Lwów
Niedźwiedź T (1988) Kalendarz sytuacji synoptycznych dla dorzecza górnej Wisły (1951–1985). Pr Geogr UJ 71:37–86
Niedźwiedź T (1996) Long-Term Variability of the Zonal Circulation Index above the Central Europe. W: Obrębska-Starkel B, Niedźwiedź T (eds) Proceedings of the International Conference Climate Dynamics and the Global Change Perspective, October 17–20 1995, Kraków, Zeszyty Naukowe UJ - Prace Geograficzne 102:213–219
Niedźwiedź T (2004) Rekonstrukcja warunków termicznych lata w Tatrach od 1550 roku. In: Kotarba A (ed) Rola małej epoki lodowej w przekształcaniu środowiska przyrodniczego Tatr. Prace Geograficzne 197:57–88
Niedźwiedź T, Glaser R, Hansson D, Helama S, Klimenko V, Łupikasza E, Małarzewski Ł, Nordli Ø, Przybylak R, Riemann D, Solomina O (2015) The Historical Time Frame (Past 1000 Years), In: The BACC II Author Team (eds), Second assessment of climate change for the Baltic Sea Basin. Regional Climate Studies, 51–65
Niedźwiedź T, Orlicz M, Orliczowa J (1985) Wiatr w Karpatach Polskich, Dokumentacja Geograficzna, IG i PZ PAN, z. 6, ss. 90
Niedźwiedź T, Paszyński J, Czekierda D (1995) Wiatry, plansza 31.6 (w:) Atlas Rzeczypospolitej, PPWK, Warszawa
Obrębska-Starklowa B, Grzyborowska A (1997) Tendencje zmian wilgotności względnej powietrza w pogórskim odcinku doliny Raby w latach 1971–1992. Roczniki AR w Poznaniu 291(17):15–38
Okołowicz W (1962) Zachmurzenie w Polsce. Prace Geograficzne, 34, IG PAN, Warszawa, p 107
Olecki Z (1986) On the components of the radiation balance in Cracow. Zeszyty Naukowe UJ, Prace Geograficzne 69:27–38
Olecki Z (1989) Balance of solar radiation in the Upper Vistula Basin (in Polish) [Bilans promieniowania słonecznego w dorzeczu Górnej Wisły]. Rozprawy Habilitacyjne UJ 157:126
Opała M, Mendecki MJ (2014) An attempt to dendroclimatic reconstruction of winter temperature based on multispecies tree-ring widths and extreme years chronologies (example of Upper Silesia, Southern Poland). Theor Appl Climatol 115(1–2):73–89
Osuchowska-Klein B (1975) Prognostyczne aspekty cyrkulacji atmosferycznej nad Polską. Pr IMGW 7:4–51
Owczarek M (2009) Warunki bioklimatyczne na Wybrzeżu i Pomorzu w drugiej połowie XX wieku jako element monitoringu środowiska. Doctoral dissertation, IMGW, Warszawa
Parczewski W (1964) Fronty atmosferyczne nad Polską. Wiadomości Służby Hydrologicznej i Meteorologicznej 59(4):20–35
Parczewski W, Kluźniak S (1959) Trąba powietrzna w województwie szczecińskim w dniu 25 sierpnia 1956 roku. Przegląd Geofizyczny 4:237–245
Parry ML (ed) (2000) Assessment of potential effects and adaptations for climate change in Europe: the Europe ACACIA Project. University of East Anglia, Norwich, p 320
Paszyński J (1966) Atlas of radiation balance in Poland. Materials for the heat balance of Poland (in Polish) [Atlas bilansu promieniowania w Polsce. Materiały do bilansu cieplnego Polski, IG PAN Dok. Geogr., 4
Pietkiewicz A (1889) Stosunki opadu atmosferycznego w Warszawie. Pam. Fizjogr., 9(1)
Pilguj N, Taszarek M, Pajurek Ł, Kryza M (2019) High-resolution simulation of an isolated tornadic supercell in Poland on 20 June 2016. Atmos Res 218:145–159

Pilorz W (2015) Very large hail occurrence in Poland from 2007 to 2015. Contemp Trends Geosci 4(1):46–56
Piotrowicz K (1996) Variability of the number of days with snowfall and days with snow cover against the background of air temperature changes in winter in Cracow, [w:] Obrębska-Starkel B, Niedźwiedź T (ed) Proceedings of the international conference on climate dynamics and the global change perspective, Cracow, October 17–20 1995, Zesz. Nauk. UJ, Pr. Geogr. 102:435–440
Piotrowicz K (2010) Sezonowa i wieloletnia zmienność typów pogody w Krakowie, IGiGP UJ, Kraków
Piotrowski P (2009) Obiektywna metoda klasyfikacji atmosferycznej dla Polski. Folia Geographica Physica 10:216
Piwkowski H (1976) Rozkład mgieł w Polsce i ich długotrwałość. Przegląd Geofizyczny 21:1
Podogrocki J (1977) Temporal changes of global solar radiation in Poland (in Polish) [Zmienność czasowa promieniowania całkowitego w Polsce], Zesz. Nauk. ART. Olsztyn, Rolnictwo, 21:113–125
Podogrocki J (1978) Spatial distribution of global solar radiation in Poland, Publ. Inst. Geophys. Pol. Acad. Sc., D-5, 120:17–29
Polaczkówna M (1925) Wahania klimatyczne w Polsce w wiekach średnich. Prace Geograficzne 5:65–126
Prawdzic K, Sucheta J (1975) Prawdopodobieństwo występowania mgieł w polskich portach Bałtyku w okresie 1950–1969. PWN, Warszawa-Poznań
Przybylak R, Majorowicz J, Wójcik G (2001) Zmiany temperatury powietrza i opadów atmosferycznych w Polsce w okresie XVI–XX wiek. Prace i Studia Geograficzne 29:79–92
Ralska-Jasiewiczowa M, Goslar T, Madeyska T, Starkel L (eds) (1998) Lake Gościąż, central Poland a monographic study. Szafer Institute of Botany, Polish Academy of Sciences, Kraków. p 340
Ralska-Jasiewiczowa M, Starkel L (1988) Record of the hydrological changes during the Holocene in the lake, mire and fluvial deposits of Poland, Folia Quaternaria, 57
Ralska-Jasiewiczowa M Starkel L (1991) Climate changes and water regimes in the Holocene. In: L. Starkel (ed) Geografia Polski, Środowisko Przyrodnicze, pp. 177–182. Wydawnictwo Naukowe PWN; Warszawa [In Polish]
Ralska-Jasiewiczowa M, van Geel B, Goslar T, Kuc T (1992) The record of the Late Glacial/Holocene transition in the varved sediments of Lake Gościąż, central Poland. Sveriges Geologiska Undersökning Ca 81:257–268
Rojecki A (ed) (1965) Wyjątki ze źródeł historycznych o nadzwyczajnych zjawiskach hydrologiczno-meteorologicznych hydrologiczno na ziemiach polskich w wiekach od X do XVI. Selection and Polish translation: R. Girguś and W. Strupczewski, Wydawnictwa Komunikacji i Łączności, Warszawa
Romer E (1947a) O współczesnej oceanizacji klimatu europejskiego. Przegl Geogr 21:1–2
Romer E (1947b) Rozmyślania klimatyczne, Czasopismo Geogr. XVII(3–4)
Romer E (1948/1949) Rehabilitacja wartości średniej temperatury roku, Przegl. Geogr., XXII
Rychliński JP (1923a) Wahania opadów w Polsce. Wiad. Met. 2
Rychliński JP (1923b) Teorja pluwiotermicznego kontynentalizmu. Wiad. Met. 12
Rychliński JP (1924) Stopnie kontynentalizmu pluwiotermicznego Europy. Wiad. Met. 1/2
Rychliński JP (1927) O wieloletnich średnich wysokościach rocznych opadów w Warszawie. Prace Met. i Hydr., 4
Sadowski M (1991) Variability of extreme climatic events in Central Europe since the 13th century. Meteorol Z 41:350–356
Satke W (1896) Ciepłota śniegu w zimie 1893/4 w Tarnopolu. Spraw. Kom. Fizjogr. P. A. U. 31: 248–255
Satke W (1898) Roczny przebieg stanu zachmurzenia Galicji. Rozprawy Akademii Umiejętności 34:291–432
Satke W (1899) Badania nad pokrywą śnieżną w Tarnopolu, Kosmos, 24(89–93):183–216 Lwów
Schmuck A (1959) Zarys klimatologii Polski. PWN, Warszawa
Semkowicz W (1922) Zagadnienie klimatu w czasach historycznych. Przegląd Historyczny 3:18–42

Słomka J (1988) Ultraviolet and global solar radiation at Belsk 1980–1986., Publs. Inst. Geoph. Pol. Ac. Sc., D- 30(220):163–167

Smosarski W (1915) Burze w Królestwie Polskim w roku 1912. Warszawa

Starkel L (1977) Paleogeografia holocenu. Wydawnictwo Naukowe PWN, Warszawa

Starkel L, Pazdur A, Pazdur MF, Wicik B, Więckowski K (1996) Lake-level and groundwater-level changes in the Lake Gościąż area, Poland: palaeoclimatic implications. The Holocene 6(2):213–224

Stasiak J (1968) Próba odtworzenia przemian klimatycznych w okresie subborealnym i subatlantyckim w północno-wschodniej Polsce, Folia Quaternaria 29

Stenz E (1925) The measurements of solar radiation in Zakopane in 1924 (in Polish) [Pomiary promieniowania słonecznego w Zakopanem w 1924 roku]. Drukarnia Głosu Narodu, Kraków, 26–31

Stenz E (1952) Zachmurzenie Polski. Przegląd Meteorologiczny i Hydrologiczny 1–2:69–81

Stenz E (1959) Selected data on the solar radiation in the Carpathians, the Sudetes and their foreland (in Polish) [Niektóre dane o promieniowaniu słonecznym w Karpatach, Sudetach i ich przedgórzu]. Acta Geophysica Polonica 7(1):12–24, Warsaw

Stopa M (1962) Burze w Polsce. Prace Instytutu Geografii PAN 34:109–179

Stopa M (1965) Podział Polski na regiony Burzowe. Przegląd Geograficzny 37(4):659–665

Stopa-Boryczka M (ed) (1989) Atlas współzależności parametrów meteorologicznych i geograficznych w Polsce, z. 5, Wyd. UW

Suwała K (2014) Synoptyczne uwarunkowania występowania opadów gradu w Środkowej Europie. Praca doktorska, Wydział Nauk Geograficznych i Geologicznych, Uniwersytet im. Adama Mickiewicza w Poznaniu

Szewczuk J (1939) Kronika klęsk elementarnych w Galicji w latach 1772–1848, Lwów

Szustakowska E (1991) Zmiany rocznej liczby dni z pokrywą śnieżną w okresie 1921/22–1986/87 w Poznaniu, w powiązaniu ze zmianami temperatury powietrza miesięcy zimowych, Badania Fizjogr. nad Polską Zach., ser. A. Geogr. Fiz., Poznań; 42:283–290

Szwed M, Graczyk D, Pińskwar I, Kundzewicz ZW (2007) Projections of climate extremes in Poland. Geogr Pol 80(2):191–202

Szychowska-Krąpiec E (2010) Wielowiekowe chronologie sosny (Pinus sylvestris L.) i jodły (Abies alba Mill.) z regionu Małopolski oraz ich interpretacja paleoklimatyczna. Folia Quat 79:5–124

Tarajkowska M (1974) Wpływ miasta na warunki wilgotnościowe powietrza na przykładzie Częstochowy, Zeszyty Naukowe Uniwersytetu Łódzkiego, Nauki Matematyczno-Przyrodnicze, Seria II, 63:43–56

Taszarek M, Gromadzki J (2017) Deadly tornadoes in Poland from 1820 to 2015. Mon Weather Rev 145(4):1221–1243

Taszarek M, Brooks HE (2015) Tornado climatology of Poland. Mon Weather Rev 143:702–717

Taszarek M, Czernecki B, Kozioł A (2015) A cloud-to-ground lightning climatology for Poland. Mon Weather Rev 143:4285–4304

Taszarek M, Czernecki B, Walczakiewicz S, Mazur A, Kolendowicz L (2016) An isolated tornadic supercell of 14 July 2012 in Poland—a prediction technique within the use of coarse-grid WRF simulation. Atmos Res 178:367–379

Taszarek M, Kolendowicz L (2013) Sounding-derived parameters associated with tornado occurrence in Poland and Universal Tornadic Index. Atmos Res 134:186–197

Trepińska J (1988) Wieloletni przebieg ciśnienia i temperatury powietrza w Krakowie na tle ich zmienności w Europie. Rozprawy habilitacyjne UJ 140, Kraków

Trepińska J (1997a) Wieloletni przebieg średnich miesięcznych wartości ciśnienia powietrza w Krakowie, 1792–1995. In: Trepińska J (ed) Wahania klimatu w Krakowie (1792–1995). IG UJ, Kraków, pp 131–161

Trepińska J (ed.) (1997b) Wahania klimatu w Krakowie (1792–1995), Wyd. IG UJ, Kraków

Twardosz R, Niedźwiedź T, Łupikasza E (2011), Zmienność i uwarunkowania cyrkulacyjne występowania postaci i typów opadów atmosferycznych na przykładzie Krakowa. Wydawnictwo Uniwersytetu Jagiellońskiego

Ustrnul Z (1997) Zmienność cyrkulacji atmosfery na półkuli północnej w XX wieku. IMGW, Materiały Badawcze, Meteorologia 27, Warszawa: p 208

Ustrnul Z, Czekierda D (2000) Air pressure extreme during the instrumental observation period in Warsaw. Prace Geograficzne IG UJ 108:207–213

Van der Linden P, Mitchell JFB (eds) (2009) ENSEMBLES: climate change and its impacts at seasonal, decadal and centennial time scales: summary of research and results from the ENSEMBLES project. Met Office Hadley Centre, Exeter, UK, p 160

Walawender A (1932) Kronika klęsk elementarnych w Polsce i w krajach sąsiednich w latach 1450–1586, Lwów

Warakomski W (1969) Zachmurzenie i rodzaje chmur w zależności od mas powietrznych w Polsce. Rozprawy habilitacyjne, UMCS, Lublin, p 247

Weisse M (1858) Allgemeine Übersichten der K.K. Krakauer Sternwarte von Jahre 1826 bis 1829 gemachten Meteorologischen Beobachtungen, Kraków

Wibig J (2001) Wpływ cyrkulacji atmosfery na rozkład przestrzenny anomalii temperatury i opadów w Europie. Rozprawy Habilitacyjne Uniwersytetu Łódzkiego. Wydawnictwo Uniwersytetu Łódzkiego, Łódź

Wierzbicki D (1878) Ruch dzienny prężności pary i wilgotności powietrza w Krakowie, Pamiętniki AU Wydz. Mat Przyr 4:153–173

Wild (1887) Nowyje normalnyje i pjatiletnije koliczestwa osadkow. Petersburg

Wiszniewski W (1949) O burzach w Polsce. Gazeta Obserwatora PIHM 6:9–12

Wodzińska M, Osuchowska B (1963) Miesięczne i sezonowe typy pogody, Wiadomości Służby Hydrologicznej i Meteorologicznej, 55:35–46

Woś A (1999) Klimat Polski, Wydawnictwo Naukowe PWN, Warszawa

Woś A (2010) Klimat Polski w drugiej połowie XX wieku, Wydawnictwo Naukowe UAM, Poznań

Wypych A (2010) Twentieth century variability of surface humidity as the climate change indicator in Kraków (Southern Poland). Theor Appl Climatol 101:475–482

Wypych A (2004) Zmienność wilgotności powietrza w Krakowie (1901–2000), PhD dissertation, Zakład Klimatologii IGiGP UJ, 1–243

Zielski A, Krąpiec M, Koprowski M (2010) Dendrochronological data. In: Przybylak R, Majorowicz J, Brázdil R, Kejna M (eds) The Polish climate in the European context: an historical overview. Springer, Berlin, pp 191–217

Zinkiewicz W (1946) Badania nad wartością przyrostu rocznego drzew dla studiów nad wahaniami klimatycznymi. Annales UMCS, Sec B 6:177–228

Zinkiewicz W (1953) Zagadnienie kompleksów pogodowych. Annales UMCS, Sectio B 8(6):312–341

Zinkiewicz W, Michna E (1955) Częstość występowania gradów w woj. Lubelskim w zależności od warunków fizjograficznych. Annales Universitatis Mariae Curie-Skłodowska, Lublin-Polonia, Sectio B 10 (5):224–267

Chapter 3
Data and Methods of Investigation

Małgorzata Falarz, Małgorzata Szwed, Magdalena Opała-Owczarek, Magdalena Kuchcik, Łukasz Małarzewski, Katarzyna Piotrowicz, Tadeusz Niedźwiedź, Zuzanna Bielec-Bąkowska, Anita Bokwa, Janusz Filipiak, Dorota Matuszko, Zbigniew Ustrnul, Joanna Wibig, and Agnieszka Wypych

Abstract The study covers the entire area of Poland. Part II of the book presents the results of the study of climate change before instrumental measurements in Poland in the last millennium. More than 50 proxy series including documentary evidences, dendrochronological records and varved sediment records were analysed. On this basis, reconstructions of: summer and winter air temperature, winter severity, precipitation (liquid and solid) and droughts were performed. Part III analysed the long-term

M. Falarz (✉) · M. Opała-Owczarek · Ł. Małarzewski · T. Niedźwiedź · Z. Bielec-Bąkowska
Faculty of Natural Sciences, Institute of Earth Sciences, University of Silesia in Katowice, Katowice, Poland
e-mail: malgorzata.falarz@us.edu.pl

M. Opała-Owczarek
e-mail: magdalena.opala@us.edu.pl

Ł. Małarzewski
e-mail: lukasz.malarzewski@us.edu.pl

T. Niedźwiedź
e-mail: tadeusz.niedzwiedz@us.edu.pl

Z. Bielec-Bąkowska
e-mail: zuzanna.bielec-bakowska@us.edu.pl

M. Szwed
Institute for Agricultural and Forest Environment, Polish Academy of Sciences, Poznań, Poland
e-mail: mszwed@man.poznan.pl

M. Kuchcik
Institute of Geography and Spatial Organization, Climate Impacts Laboratory, Polish Academy of Sciences, Warsaw, Poland
e-mail: mkuchcik@twarda.pan.pl

K. Piotrowicz · A. Bokwa · D. Matuszko · Z. Ustrnul · A. Wypych
Faculty of Geography and Geology, Institute of Geography and Spatial Management, Jagiellonian University, Kraków, Poland
e-mail: k.piotrowicz@uj.edu.pl

A. Bokwa
e-mail: anita.bokwa@uj.edu.pl

D. Matuszko
e-mail: d.matuszko@uj.edu.pl

M. Falarz (ed.), *Climate Change in Poland*, Springer Climate,
https://doi.org/10.1007/978-3-030-70328-8_3

changes and variability of 36 climate characteristics for 14 climate elements, indices, meteorological phenomena and weather types using data from 79 weather stations. Most of the climate elements studied were analysed for the base period 1951–2018 (68 years). For some elements and indices shorter study periods were considered. In addition, climate variability was analysed for 10 long measuring series up to 239 years. Variability of circulation indices over a period of 147 years (1873–2019) was investigated as well. Uniform research methods common to all elements and indices were used: (1) coefficient of variability (%), (2) absolute trend (unit/10y); (3) relative trend (%/10y). Part IV of the book deals with projected changes in temperature, precipitation and thermal indices related to the agriculture and energy sectors. In this study, 8 regional climate models from the EURO-CORDEX experiment were used for 2 representative concentration pathways: (1) corresponding to radiative forcing value +4.5 W.m^{-2} in 2100 (RCP4.5) and (2) corresponding to radiative forcing value +8.5 W.m^{-2} in 2100 (RCP8.5) relative to pre-industrial values. Two future time horizons were carried out for each concentration pathway: (1) near future: 2021–2050 and (2) far future: 2071–2100 with reference to the period of 1971–2000. Future projections were created for: 8 characteristics of temperature, 10 characteristics of precipitation, 5 characteristics for agriculture indices and 4 for energy demands indices.

The study covers the entire area of Poland.

Part II of the book presents the results of the study of climate change before instrumental measurements in Poland in the last millennium. Part of the chapter contains the results of new studies, not yet published, which were carried out specifically for the purposes of this book, and part of the results were published previously or updated. More than 50 proxy series including documentary evidences, dendrochronological records and varved sediment records were analysed. On this basis, reconstructions of: summer and winter air temperature, winter severity, precipitation (liquid and solid)

Z. Ustrnul
e-mail: zbigniew.ustrnul@uj.edu.pl

A. Wypych
e-mail: agnieszka.wypych@uj.edu.pl

J. Filipiak
Faculty of Oceanography and Geography, Department of Meteorology and Climatology, Institute of Geography, University of Gdańsk, Gdańsk, Poland
e-mail: janusz.filipiak@ug.edu.pl

J. Filipiak · Z. Ustrnul · A. Wypych
Institute of Meteorology and Water Management—National Research Institute, Warsaw, Poland

J. Wibig
Faculty of Geographical Sciences, Department of Meteorology and Climatology, University of Lodz, Łódź, Poland
e-mail: joanna.wibig@geo.uni.lodz.pl

and droughts were performed. A detailed description of the investigation methods is contained in Part II.

Part III is the most extensive part of the book. It analysed the long-term changes and variability of 36 climate characteristics for 14 climate elements, indices, meteorological phenomena and weather types using data from 79 weather stations, i.e. 77 stations of the Institute of Meteorology and Water Management—National Research Institute (IMWM-NRI) and 3 university stations (Tables 3.1 and 3.2). This part of the book contains mostly the results of new research, not yet published, conducted specifically for the purposes of this book. Up to 14 climate elements were studied at each meteorological station (Fig. 3.1). Most of the climate elements studied were analysed for the base period 1951–2018 (68 years). However, for some elements and indices, shorter study periods were considered. Test periods for particular elements, indices, meteorological phenomena, weather types are as follows:

- 1951 (1950)–2018, i.e. 68 years, for: cloudiness, air temperature, air humidity, precipitation, snow cover, thunderstorms and bioclimatic indices;
- 1961–2018, i.e. 58 years, for weather types;
- 1966–2018, i.e. 53 years, for air pressure, wind, hail and fog,
- 1971–2018, i.e. 48 years, for sunshine,
- different periods: 19 (2000–2018) to 36 years (1983–2018) for solar radiation (Table 3.3).

In addition, climate variability was analysed for 10 long measuring series in:

- Warszawa: from 115 years (1904–2018) to 239 years (1780–2018)—for 2 climate elements,
- Kraków Obs. UJ: from 98 years (1921–2018) to 227 years (1792–2018)—for 9 climate elements,
- Śnieżka: 118 years (1901–2018)—for 1 climate element,
- Zakopane: 105 years (1914–2018)—for 1 climate element,
- Toruń: 99 years (1920–2018)—for 1 climate element,
- Poznań: 99 years (1920–2018)—for 1 climate element,
- Puławy: 98 years (1921–2018)—for 2 climate elements,
- Kalisz: 97 years (1922–2018)—for 1 climate element,
- Gdynia: 95 years (1924–2018)—for 1 climate element,
- Kasprowy Wierch: 81 years (1938–2018)—for 1 climate element.

Moreover, variability of circulation indices over a period of 147 years (1873–2019) was investigated.

The missing data shown in column *data filled* of Table 3.3 were completed for years and seasons by methods of differences, quotients and regression analysis based on available data from neighbouring stations.

Most weather stations are located below 200 m above sea level (54 stations; Fig. 3.2). Seven stations are located at the altitude of 200 to 300 m, 5 stations at the altitude of 300 to 500 m, 4 stations at 500–1000 m and 3 stations—above 1000 m. The lowest stations are Hel and Szczecin-Dąbie in the coast (1 m above sea level),

Table 3.1 Meteorological stations used in Part III—geographical coordinates and altitude

Ordinal number	WMO number	Station	Longitude λ E	Latitude φ N	Altitude [m above sea level]
1	123902	Białowieża	23°51′	52°42′	163
2	12295	Białystok	23°10′	53°06′	148
3	122802	Biebrza	22°36′	53°39′	115
4	12600	Bielsko-Aleksandrowice	19°00′	49°48′	398
5	12235	Chojnice	17°33′	53°43′	164
6	12550	Częstochowa	19°06′	50°49′	293
7	126802	Dynów	22°14′	49°50′	260
8	12160	Elbląg-Milejewo	19°26′	54°10′	40
9	*	Gaik-Brzezowa	20°04′	49°52′	259
10	12140	Gdańsk-Port Pn.	18°42′	54°24′	2
11	121402	Gdynia	18°33′	54°31′	5
12	12300	Gorzów Wlkp.	15°17′	52°45′	72
13	129628	Hala Gąsienicowa	20°00′	49°15′	1520
14	12135	Hel	18°49′	54°36′	1
15	126518	Jabłonka	19°42′	49°28′	615
16	12500	Jelenia Góra	15°48′	50°54′	342
17	12435	Kalisz	18°05′	51°47′	138
18	12650	Kasprowy Wierch	19°59′	49°14′	1991
19	12560	Katowice	19°02′	50°14′	284
20	12185	Kętrzyn	21°22′	54°04′	108
21	12570	Kielce-Suków	20°42′	50°49′	260
22	12520	Kłodzko	16°37′	50°26′	356
23	12345	Koło	18°40′	52°12′	116
24	12100	Kołobrzeg	15°35′	54°11′	3
25	123403	Kołuda Wielka	18°09′	52°44′	85
26	12105	Koszalin	16°09′	54°12′	33
27	123305	Kórnik	17°06′	52°15′	77
28	12566	Kraków Balice	19°48′	50°05′	237
29	125511	Kraków Obs. UJ	19°58′	50°04′	206
30	12670	Krosno	21°46′	49°43′	326
31	126623	Krynica	20°58′	49°25′	585
32	123603	Legionowo	20°58′	52°24′	94
33	12415	Legnica	16°12′	51°12″	122
34	12690	Lesko	22°21′	49°28′	420
35	12418	Leszno	16°32′	51°50′	91

(continued)

Table 3.1 (continued)

Ordinal number	WMO number	Station	Longitude λ E	Latitude φ N	Altitude [m above sea level]
36	12125	Lębork	17°45′	54°33′	39
37	121601	Lidzbark Warm.	20°36′	54°08′	90
38	12495	Lublin-Radawiec	22°24′	51°13′	238
39	12120	Łeba	17°32′	54°45′	2
40	12465	Łódź	19°24′	51°44′	187
41	12280	Mikołajki	21°35′	53°47′	127
42	12270	Mława	20°21′	53°06′	147
43	12660	Nowy Sącz	20°42′	49°37′	292
44	126516	Obidowa	19°58′	49°33′	805
45	12272	Olsztyn	20°25′	53°46′	133
46	12530	Opole	17°58′	50°38′	165
47	12230	Piła (Wałcz)	16°45′	53°08′	72
48	12360	Płock	19°44′	52°35′	106
49	12330	Poznań	16°51′	52°25′	83
50	124503	Puczniew	19°05′	51°47′	140
51	124705	Puławy	21°58′	51°22′	142
52	123701	Pułtusk	21°06′	52°44′	95
53	12540	Racibórz	18°12′	50°09′	205
54	124251	Radzyń	16°02′	51°52′	60
55	12210	Resko-Smólsko	15°25′	53°46′	52
56	12580	Rzeszów	22°03′	50°06′	200
57	12585	Sandomierz	21°43′	50°42′	217
58	12385	Siedlce	22°15′	52°11′	152
59	124601	Skierniewice	20°10′	51°58′	128
60	12310	Słubice	14°36′	52°21′	21
61	**	Sosnowiec	19°08′	50°17′	262
62	12469	Sulejów	19°52′	51°21′	188
63	12195	Suwałki	22°57′	54°08′	184
64	12205	Szczecin-Dąbie	14°37′	53°24′	1
65	123801	Szepietowo	22°33′	52°51′	150
66	12510	Śnieżka	15°44′	50°44′	1603
67	12200	Świnoujście	14°14′	53°55′	6

(continued)

Table 3.1 (continued)

Ordinal number	WMO number	Station	Longitude λ E	Latitude φ N	Altitude [m above sea level]
68	12575	Tarnów	20°59′	50°02′	209
69	12399	Terespol	23°37′	52°04′	133
70	12250	Toruń	18°35′	53°02′	69
71	12115	Ustka	16°52′	54°35′	6
72	123604	Warszawa-Bielany	20°58′	52°17′	98
73	12375	Warszawa-Okęcie	20°58′	52°10′	106
74	123206	Wielichowo	16°21′	52°07′	65
75	12455	Wieluń	18°34′	51°13′	200
76	12497	Włodawa	23°32′	51°33′	177
77	12424	Wrocław	16°59′	51°08′	116
78	12625	Zakopane	19°57′	49°18′	857
79	12400	Zielona Góra	15°32′	51°56′	192

*Station of Jagiellonian University
**Station of University of Silesia in Katowice

while the highest—Kasprowy Wierch summit in the Tatra-Mountains (1991 m above sea level).

The analysis of climate change was carried out for the whole year and for four seasons: spring (Mar–May), summer (Jun–Aug), autumn (Sep–Nov) and winter (Dec–Feb). For a few elements, the trend analysis was done for both the entire baseline research period (1951–2018) and for its first (1951–1983) and the second half (1984–2018).

In Part III, uniform research methods common to all elements and indices were used:

- coefficient of variability (%), i.e. expressed as a percentage ratio of the standard deviation and arithmetic mean of the analysed characteristics;
- absolute trend (unit/10y) obtained by regression analysis with the determination of the statistical significance of the trend for the 0.05 level;
- relative trend (%/10y) expressed as a percentage ratio of linear trend value per 10 years and the average long-term value of investigated characteristic at the weather station.

The results of the studies are presented in tables, maps and figures. Detailed methods, specific to particular climate elements and indices are described in the chapters dedicated to them.

Part IV of the book deals with projected changes in temperature, precipitation and thermal indices related to the agriculture and energy sectors. This section contains only the results of new studies, not yet published, conducted for the purposes of this

Table 3.2 Elements, indices, meteorological phenomena with characteristics analysed in Part III

No.	Element, index, meteorological phenomena	Characteristic
1	Atmospheric circulation	Index of zonal westerly circulation (Wi)
2		Index of meridional southern circulation (Si)
3		Cyclonicity index (Ci)
4	Atmospheric pressure	Extreme values of atmospheric pressure
5		Number of days with pressure $\leq$990 hPa and $\geq$1030 hPa
6	Solar radiation	Totals of global solar radiation
7	Sunshine	Actual sunshine duration
8		Relative sunshine duration
9		Number of days without sunshine
10	Cloudiness	Total cloudiness
11		Number of clear days
12		Number of cloudy days
13		Frequency of cloud types
14	Air temperature	Mean air temperature
15		Extreme daily air temperatures
16		Number of days with the particular air temperature thresholds
17	Humidity	Specific humidity
18		Relative humidity
19	Precipitation	Total precipitation
20		Number of days with precipitation
21	Snow cover	Number of days with snow cover
22		Maximum depth of snow cover
23	Wind	Average wind speed
24		Frequency of calms
25		Frequency of wind in different speed classes
26		Wind direction
27		Extreme wind speed
28	Thunderstorms	Number of days with thunderstorms
29	Tornadoes	Number of days with tornadoes
30	Hail	Number of days with hail
31	Fog	Number of days with fog
32		Number of fogs of a given duration

(continued)

Table 3.2 (continued)

No.	Element, index, meteorological phenomena	Characteristic
33		Number of fogs of a given intensity
34	Bioclimatic indices	Universal Thermal Climate Index (UTCI)
35		Physiological Subjective Temperature (PST)
36	Weather types	Number of days with particular weather types

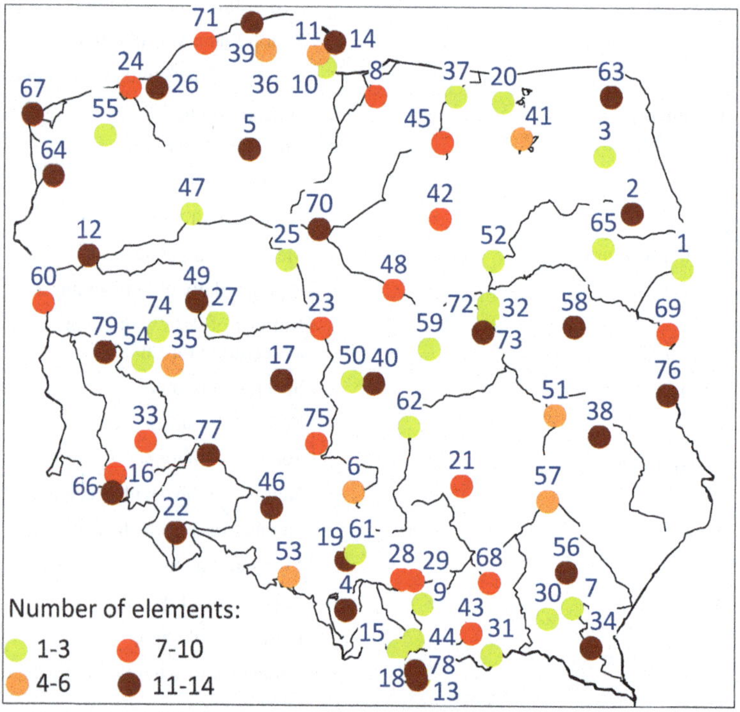

Fig. 3.1 Meteorological stations used in Part III. Number of meteorological elements, indices, phenomena, etc., considered at each station is given. The numbers correspond to the ordinal number assigned to a given station in Table 3.1

book. In this study, 8 regional climate models from the EURO-CORDEX experiment were used for 2 representative concentration pathways: (1) corresponding to radiative forcing value +4.5 $W.m^{-2}$ in 2100 (RCP4.5) and (2) corresponding to radiative forcing value +8.5 $W.m^{-2}$ in 2100 (RCP8.5) relative to pre-industrial values. Two future time horizons were carried out for each concentration pathway: (1) near future: 2021–2050 and (2) far future: 2071–2100 with reference to the period of 1971–2000.

Table 3.3 Meteorological stations, studied periods and data filled for climate elements/characteristics/indices/phenomena in Part III

No.	meteorological station	air pressure (43 stations)		solar radiation (10 stations)		sunshine (31 stations)		cloudiness (59 stations)		air temperature (59 stations)		air humidity (56 stations)		precipitation (52 stations)	
		studied period	data filled for	studied period	data filled for	studied period	data filled for	studied period	data filled for	studied period	data filled for	studied period	data filled for	studied period	data filled for
1	Białowieża	-		-		-		-		-		1951-2018	1,5%		
2	Białystok	1966-2018		-		1971-2018		1951-2018		1951-2018		1951-2018		1951-2018	
3	Biebrza	-		-		-		-		-		1951-2018	9,1%		
4	Bielsko-Aleksandrowice	1966-2018	1 day	-		-		1951-2018		1951-2018		1951-2018		1951-2018	
5	Chojnice	-		-		1971-2018		1951-2018		1951-2018		1951-2018		1951-2018	
6	Częstochowa	-		-		-		1951-2014	1994-1998	1951-2018		1951-2018	6,6%		
7	Dynów	-		-		-		-		1951-2018		-			
8	Elbląg-Milejewo	1966-2018	-	-		-		1951-2018		1951-2018		1951-2018		1951-2018	
9	Gaik-Brzezowa	-		-		1971-2018		-		-		-			
10	Gdańsk-Port Pn.	-		-		-		-		1855-2018		1951-2018			
11	Gdynia	-		1991-2015	2001	1971-2018		1951-2014		-		-		1951-2018	
12	Gorzów Wlkp.	1966-2018		-		1971-2018		1951-2018		1951-2018		1951-2018		1951-2018	
13	Hala Gąsienicowa	-		-		-		-		1951-2018		-			
14	Hel	1966-2018	3 days	-		-		1951-2018		1951-2018		1951-2018		1951-2018	
15	Jabłonka	-		-		-		1961-2018				-			
16	Jelenia Góra	1966-2018		-		1971-2018		1951-2018		1951-2018		1951-2018		1951-2018	
17	Kalisz	1966-2018	33 days	-		1971-2018		1951-2018		1951-2018		1951-2018		1951-2018	
18	Kasprowy Wierch	1966-2018		1991-2017	winter 2001	1971-2018		1951-2018		1951-2018		1951-2018		1951-2018	
19	Katowice	1966-2018	2 days	-		1971-2018		1951-2018		1951-2018		1951-2018		1951-2018	
20	Kętrzyn	-		-		-		1966-2014		-		-			
21	Kielce-Suków	1966-2018		-		-		1951-2018		1951-2018		1951-2018		1951-2018	
22	Kłodzko	1966-2018		-		1971-2018		1951-2018		1951-2018		1951-2018	1,5%	1951-2018	
23	Koło	1966-2018	25 days	-		-		1951-2014		-		1951-2018	0,1%	1951-2018	

(continued)

Table 3.3 (continued)

No.	meteorological station	air pressure (43 stations)		solar radiation (10 stations)		sunshine (31 stations)		cloudiness (59 stations)		air temperature (59 stations)		air humidity (56 stations)		precipitation (52 stations)	
		studied period	data filled for	studied period	data filled for	studied period	data filled for	studied period	data filled for	studied period	data filled for	studied period	data filled for	studied period	data filled for
24	Kołobrzeg	-		-		1971-2018		1951-2018		1951-2018		1951-2018		1951-2018	
25	Kołuda Wielka	-		-		-		-		1951-2018		-			
26	Koszalin	1966-2018		-		1971-2018		1951-2018		1951-2018		1951-2018		1951-2018	
27	Kórnik	-		-		-		-		1951-2018		-		1951-2018	
28	Kraków Balice	1966-2018	3 days	-		-		1951-2018		1951-2018		1951-2018		1951-2018	
29	Kraków Obs. UJ	1901-2018		1884-2018		1884-2018		1951-2018		1792-2018		1863-2018			
30	Krosno	-		-		-		1961-2014				-			
31	Krynica	-		-		-		1951-2018				1951-2018	10,6%		
32	Legionowo	-		-		-		-		1951-2018		-			
33	Legnica	1966-2018	6 days	1992-2016	2008	-		1951-2014		1951-2018		1951-2018		1951-2018	
34	Lesko	1968-2018		1992-2015	2003, summer 2005, winter 2006	1971-2018		1954-2018				1951-2018	5,2%		
35	Leszno	1966-2018	1 day	-		-		1958-2014		-		1951-2018	10,4%	1958-2018	
36	Lębork	-		-		-		-		1951-2018		1951-2018	0,1%	1951-2018	
37	Lidzbark Warm.	-		-		-		-		1951-2018		-		1951-2018	
38	Lublin-Radawiec	1966-2018		-		-		1951-2018		1951-2018		1951-2018		1951-2018	
39	Łeba	1966-2018	8 days	1991-2015	2012	-		1951-2018		1951-2018		1951-2018		1951-2018	
40	Łódź	1966-2018	1 day	-		1971-2018		1951-2018		1951-2018		1951-2018	0,1%	1951-2018	
41	Mikołajki			-		1971-2018		1952-2018	1992, 1993	-		1951-2018	4,0%		
42	Mława	1966-2018		-		-		1951-2018		-		-		1961-2018	
43	Nowy Sącz	1966-2018	1995	-		1971-2018	20 days in 2015-2017	1954-2014				1951-2018	4,5%		
44	Obidowa	-		-		-		-		1951-2018		-			
45	Olsztyn	1966-2018	1 day	-		-		1951-2014		1951-2018		1951-2018		1951-2018	
46	Opole	1966-2018	1992-1993	-		1971-2018		1951-2018		1951-2018		1951-2018		1951-2018	
47	Piła (Wałcz)	-		1991-2015	2011	-		1971-2014		-		-			
48	Płock	1966-2018	17 days	-		-		1951-2014		1951-2018		1951-2018		1951-2018	
49	Poznań	1966-2018		-		1971-2018		1951-2018		1951-2018		1951-2018			

(continued)

Table 3.3 (continued)

No.	meteorological station	air pressure (43 stations)		solar radiation (10 stations)		sunshine (31 stations)		cloudiness (59 stations)		air temperature (59 stations)		air humidity (56 stations)		precipitation (52 stations)	
		studied period	data filled for	studied period	data filled for	studied period	data filled for	studied period	data filled for	studied period	data filled for	studied period	data filled for	studied period	data filled for
50	Puczniew	-		-		-		-		1951-2018		-			
51	Puławy	-		-		1921-2018		1951-2018		1951-2018		1951-2018	2,7%	1951-2018	
52	Pułtusk	-		-		-		-		1951-2018		-		1951-2018	
53	Racibórz	1966-2018	17 days	-		-		1951-2018		1951-2018		1951-2018		1951-2018	
54	Radzyń	-		1991-2017	2003, 2011	-		-		-		-			
55	Resko-Smólsko	-		-		-		1951-2014	12.1995, 08. 2000	-		1951-2018	4,5%		
56	Rzeszów	1966-2018	1 day	-		-		1952-2018		1951-2018		1951-2018	1,5%	1952-2018	
57	Sandomierz	1966-2018	20 days	-		-		1951-2014		1951-2018		1951-2018		1951-2018	
58	Siedlce	1966-2018		-		-		1951-2018		1951-2018		1951-2018		1951-2018	
59	Skierniewice	-		-		-		-		1951-2018		1951-2018	0,1%	1951-2018	
60	Słubice	-		-		-		1951-2016		1951-2018		1951-2018		1951-2018	
61	Sosnowiec	-		2000-2018		-		-		-		-			
62	Sulejów	1966-2018	25 days	-		-		1961-2014		-		-			
63	Suwałki	1966-2018		-		1971-2018		1951-2018		1951-2018		1951-2018		1951-2018	
64	Szczecin-Dąbie	1966-2018		-		1971-2018		1951-2018		1951-2018		1951-2018			
65	Szepietowo	-		-		-		-		1951-2018		-		1951-2018	
66	Śnieżka	1966-2018	5 days	-		1901-2018		1951-2018		1951-2018		1951-2018		1951-2018	
67	Świnoujście	1966-2018	5 days	-		-		1951-2018		1951-2018		1951-2018	0,1%	1951-2018	
68	Tarnów	-		-		1971-2018		1951-2014		1951-2018		1951-2018		1951-2018	
69	Terespol	1966-2018		-		1971-2018		1955-2018		-		1951-2018	6,0%	1951-2018	
70	Toruń	1966-2018		1983-2018		1971-2018		1951-2018		1951-2018		1951-2018		1951-2018	
71	Ustka	-		-		-		1951-2018		1951-2018		1951-2018			
72	Warszawa-Bielany	-		-		1971-2018		-		1951-2018		-		1951-2018	
73	Warszawa-Okęcie	1966-2018		-		1904-2018		1951-2018		1780-2018		1951-2018		1951-2018	
74	Wielichowo	-		-		1971-2018		-		1951-2018		-		1951-2018	
75	Wieluń	1966-2018	8 days	-		-		1951-2014		1951-2018		1951-2018		1951-2018	
76	Włodawa	1966-2018	2 days	-		1971-2018		1951-2018		1951-2018		1951-2018		1951-2018	
77	Wrocław	1966-2018		-		1891-2018		1951-2018		1951-2018		1951-2018		1951-2018	
78	Zakopane	1966-2018		-		1971-2018		1951-2018		1951-2018		1951-2018		1951-2018	
79	Zielona Góra	1966-2018		-		1971-2018		1951-2018		1951-2018		1951-2018		1951-2018	

(continued)

Table 3.3 (continued)

No.	meteorological station	climatic element/index/characteristic/phenomena													
		snow cover (60 stations)		wind (41 stations)		thunderstorms (47 stations)		hail (40 stations)		fog (26 stations)		bioclimatic indices (24 stations)		weather types (22 stations)	
		studied period	data filled for	studied period	data filled for	studied period	data filled for	studied period	data filled for	studied period	Data filled for	studied period	data filled for	studied period	data filled for
1	Białowieża	-		-		-		-		-		-			
2	Białystok	1950-2018		1966-2018		1951-2018		1966-2018		1966-2017		1951-2018		-	
3	Biebrza	-		-		-		-		-		-		-	
4	Bielsko-Aleksandrowice	1950-2018		1966-2018		1951-2018		1966-2018		1966-2017		-		1961-2018	
5	Chojnice	1950-2018		1966-2018		1951-2018		1966-2018		-		1951-2018		1961-2018	
6	Częstochowa	1950-2018	2014-2018	-		-		-		-		-		-	
7	Dynów	-		-		-		-		-		-		-	
8	Elbląg-Milejewo	1950-2018		1966-2018		1951-2018		1966-2018		-		-		-	
9	Gaik-Brzezowa	-		-		-		-		-		-		-	
10	Gdańsk-Port Pn.	-		-		-		-		-		-		-	
11	Gdynia	1923-2018	1939-1941, 1944-1945, 2015-2016	-		-		-		-		-		-	
12	Gorzów Wlkp.	1950-2018		1966-2018		1951-2018		1966-2018		1966-2017		-		-	
13	Hala Gąsienicowa	1950-2018		-		-		-		-		-		-	
14	Hel	1950-2018		1966-2018		1951-2018		1966-2018		1966-2017		1951-2018		-	
15	Jabłonka	1950-2018		-		-		-		-		-		-	
16	Jelenia Góra	1950-2018		1966-2018		1951-2018		1966-2018		-		-		-	
17	Kalisz	1922-2018	1939-1946	1966-2018		1951-2018		1966-2018		1966-2017		-		-	
18	Kasprowy Wierch	1939-2018	1939-1940, 1943-1945	1966-2018		1951-2018		1966-2018		1966-2017		1951-2018		1961-2018	
19	Katowice	1950-2018		-		1951-2018		1966-2018		1966-2017		1951-2018		1961-2018	
20	Kętrzyn	1950-2018	2014-2017	-		-		-		-		-		-	
21	Kielce-Suków	1950-2018		1966-2018		1951-2018		1966-2018		-		1951-2018		-	
22	Kłodzko	1950-2018		1966-2018		1951-2018		1966-2018		-		1951-2018		-	
23	Koło	1950-2018	2014-2018	1966-2018		1951-2014		1966-2014		-		-		1961-2018	
24	Kołobrzeg	1950-2018		1966-2018		-		-		-		-		-	
25	Kołuda Wielka	-		-		-		-		-		-		-	
26	Koszalin	1950-2018		1966-2018		1951-2018		1966-2018		1966-2017		-		-	
27	Kórnik	-		-		-		-		-		-		-	
28	Kraków Balice	1950-2018		1966-2018		1951-2018		1966-2018		1966-2017		-		-	

(continued)

Table 3.3 (continued)

No.	meteorological station	climatic element/index/characteristic/phenomena													
		snow cover (60 stations)		wind (41 stations)		thunderstorms (47 stations)		hail (40 stations)		fog (26 stations)		bioclimatic indices (24 stations)		weather types (22 stations)	
		studied period	data filled for	studied period	data filled for	studied period	data filled for	studied period	data filled for	studied period	Data filled for	studied period	data filled for	studied period	data filled for
29	Kraków Obs. UJ	1921-2018		-		1901-2018	-	1901-2018		-		-		1901-2018	
30	Krosno	1950-2018	2014-2017	-		-		-		-		-		-	
31	Krynica	1950-2018		-		-		-		-		-		-	
32	Legionowo	-		-		-		-		-		-		-	
33	Legnica	1950-2018	2014-2017	1966-2018		1951-2014		1966-2014		-		-		-	
34	Lesko	1950-2018	1950-1954	1966-2018		1951-2018		1968-2018		-		1951-2018	1951-1954	1961-2018	
35	Leszno	1950-2018	2014-2018	-		-		-		-		-		-	
36	Lębork	-		1966-2018		-		-		-		-		-	
37	Lidzbark Warm.	-		-		-		-		-		-		-	
38	Lublin-Radawiec	1950-2018		1966-2018		1951-2018		1966-2018		1966-2017		1951-2018		-	
39	Łeba	1950-2018		1966-2018		1951-2018		1967-2018		1966-2017	1966	-		1961-2018	
40	Łódź	1950-2018		1966-2018		1951-2018		1966-2018		1966-2017		1951-2018		1961-2018	
41	Mikołajki	1950-2018	1950-1952	-		-		-		1966-2017	1992-1993	1951-2018	1951, 1992-1993	-	
42	Mława	1950-2018		-		1951-2018		1966-2018		1966-2017		-		-	
43	Nowy Sącz	1950-2018	2014-2018	1966-2018		1951-2014		1966-2014	1995-1997	-		-		-	
44	Obidowa	-		-		-		-		-		-		-	
45	Olsztyn	1950-2018	2014-2017	1966-2018		1951-2014		1966-2014		-		-		1961-2018	
46	Opole	1950-2018		1966-2018		1951-2018		1966-2018	1992-1993	1966-2017	1992-1993	-		-	
47	Piła (Wałcz)	1950-2018	2014-2018	-		-		-		-		-		-	
48	Płock	1950-2018	2014-2018	-		1951-2014		1966-2014		-		1951-2018	2014-2018	-	
49	Poznań	1920-2018	1937-1940, 1944-1946	1966-2018		1951-2018		1966-2018		1966-2017		1951-2018		1961-2018	
50	Puczniew	-		-		-		-		-		-		-	
51	Puławy	1923-2018	1944-1945	-		-		-		-		-		-	
52	Pułtusk	-		-		-		-		-		-		-	
53	Racibórz	1950-2018	2014-2017	-		-		-		-		-		-	
54	Radzyń	-		-		-		-		-		-		-	
55	Resko-Smólsko	1950-2018	2014-2018	-		-		-		-		-		-	
56	Rzeszów	1950-2018		1966-2018		1951-2018		1966-2018		1966-2017		-		1961-2018	
57	Sandomierz	1950-2018	2014-2018	-		-		-		-		-		-	

(continued)

Table 3.3 (continued)

No.	meteorological station	climatic element/index/characteristic/phenomena													
		snow cover (60 stations)		wind (41 stations)		thunderstorms (47 stations)		hail (40 stations)		fog (26 stations)		bioclimatic indices (24 stations)		weather types (22 stations)	
		studied period	data filled for	studied period	data filled for	studied period	data filled for	studied period	data filled for	studied period	Data filled for	studied period	data filled for	studied period	data filled for
58	Siedlce	1950-2018		1966-2018		1951-2018		1966-2018		1966-2017		1951-2018		-	
59	Skierniewice	-		-		-		-		-		-			
60	Słubice	1950-2018	2015-2017	1966-2018		1951-2014	2015-2017	-		-		1951-2018	2015-2017	1961-2018	
61	Sosnowiec	-		-		-		-		-		-		-	
62	Sulejów	1950-2018	1950-1956, 2014-2017	-		-		-		-		-		-	
63	Suwałki	1950-2018		1966-2018		1951-2018		1966-2018		-		1951-2018		1961-2018	
64	Szczecin-Dąbie	1950-2018		1966-2018		1951-2018		1966-2018		1966-2017		1951-2018		1961-2018	
65	Szepietowo	-		-		-		-		-		-		-	
66	Śnieżka	1950-2018	March 2009	1966-2018		1951-2018		1966-2018		1966-2017		1951-2018		1961-2018	
67	Świnoujście	1950-2018		1966-2018		1951-2018		1966-2018		1966-2017		1951-2018		-	
68	Tarnów	1950-2018	2014-2018	1966-2018		1951-2014		-		-		1951-2018	2014-2018	-	
69	Terespol	1950-2018	1950-1954	-		1951-2018	1951-1954	1966-2018		-		1951-2018	1951-1954	-	
70	Toruń	1923-2018	1939-1940, 1944-1947	1966-2018		1951-2018		1966-2018		1966-2017		-		-	
71	Ustka	1950-2018		1966-2018		1951-2014		-		-		1951-2018		-	
72	Warszawa-Bielany	-		-		-		-		-		-		-	
73	Warszawa-Okęcie	1950-2018		1966-2018		1951-2018		1966-2018		1966-2017		-		1961-2018	
74	Wielichowo	-		-		-		-		-		-		-	
75	Wieluń	1950-2018	2014-2017	1966-2018		1951-2014		1966-2014		-		-		1961-2018	
76	Włodawa	1950-2018		1966-2018		1951-2018	1953-1954, 1966	1967-2018		1966-2017	1966	-		1961-2018	
77	Wrocław	1950-2018		1966-2018		1951-2018		1966-2018		1966-2017		1951-2018		1961-2018	
78	Zakopane	1914-2018		1966-2018		1951-2018		1966-2018		1966-2017		1951-2018		1961-2018	
79	Zielona Góra	1950-2018		1966-2018		1951-2018		1966-2018		-		-		1961-2018	

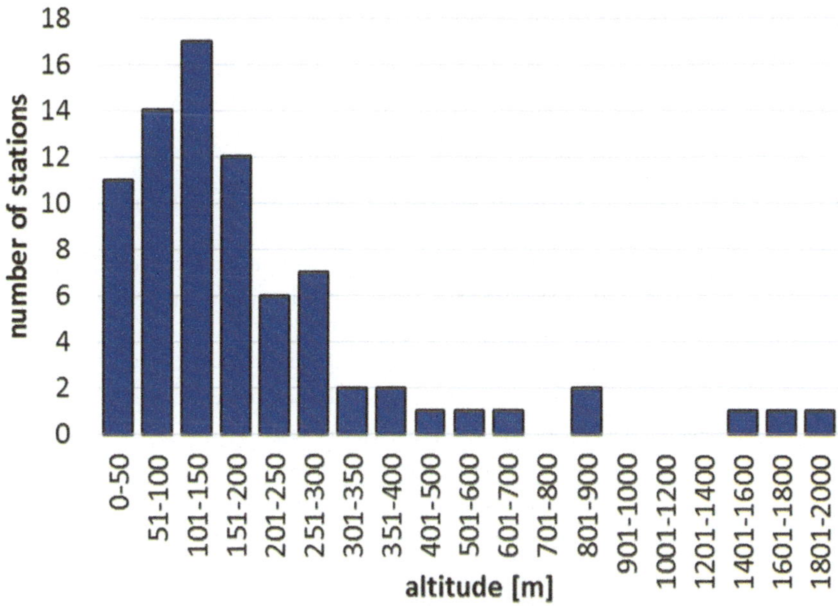

Fig. 3.2 Altitude distribution of meteorological stations used in Part III

Future projections were created for: 8 characteristics of temperature, 10 characteristics of precipitation, 5 characteristics for agriculture indices and 4 for energy demand indices (Table 3.4). A detailed description of the investigated methods for future climate changes can be found in Part IV.

Table 3.4 Elements and indices with characteristics analysed in Part IV

No.	Element, index	Characteristics
1	Temperature	Mean annual temperature
2		Absolute maximum temperature
3		Absolute minimum temperature
4		The largest number of consecutive days with maximum temperature >30 °C
5		Mean number of days with minimum temperature >20 °C
6		Mean number of days with both maximum temperature >30 °C and minimum temperature >20 °C
7		Mean number of days with minimum temperature <0 °C
8		Mean number of days with minimum temperature <−15 °C
9	Precipitation	Mean annual precipitation totals
10		Maximum 24-hour precipitation totals
11		Maximum 3-day precipitation totals
12		Number of 3-day periods with precipitation totals >50 mm for 30 years
13		Number of consecutive wet days (CWD)
14		Number of consecutive dry days (CDD)
15		Number of wet periods longer than 5 days for 30 years
16		Simple daily intensity index (SDII)
17		Number of days with intense precipitation (≥20 mm per day)
18		Period with daily precipitation amounts <1 mm and maximum daily temperature ≥30 °C
19	Agriculture indices	Growing season length (GSL)
20		Beginning of growing season (GSB)
21		Length of frost-free period (FFP)
22		Growing degree days for the threshold 5 °C (GDD5)
23		Growing degree days for the threshold 10 °C (GDD10)
24	Energy demands indices	Annual sums of heating degree days (HDD)
25		The highest daily sum of heating degree days (HDDmax)
26		Annual sums of cooling degree days (CDD)
27		The highest daily sum of cooling degree days (CDDmax)

Chapter 4
Homogeneity of Climate Series

Małgorzata Falarz, Joanna Wibig, Dorota Matuszko, Janusz Filipiak, Monika J. Hajto, Mateusz Taszarek, Zuzanna Bielec-Bąkowska, Agnieszka Wypych, and Zbigniew Ustrnul

Abstract Analyses of homogeneity of climate series and correction of heterogeneous series were undertaken in order to avoid making incorrect conclusions about climate change in Poland. The climate series homogenisation procedure was typically

M. Falarz (✉) · Z. Bielec-Bąkowska
Faculty of Natural Sciences, Institute of Earth Sciences, University of Silesia in Katowice, Katowice, Poland
e-mail: malgorzata.falarz@us.edu.pl

Z. Bielec-Bąkowska
e-mail: zuzanna.bielec-bakowska@us.edu.pl

J. Wibig
Faculty of Geographical Sciences, Department of Meteorology and Climatology, University of Lodz, Łódź, Poland
e-mail: joanna.wibig@geo.uni.lodz.pl

D. Matuszko · M. J. Hajto · A. Wypych · Z. Ustrnul
Faculty of Geography and Geology, Institute of Geography and Spatial Management, Jagiellonian University, Kraków, Poland
e-mail: d.matuszko@uj.edu.pl

M. J. Hajto
e-mail: monika.hajto@imgw.pl

A. Wypych
e-mail: agnieszka.wypych@uj.edu.pl

Z. Ustrnul
e-mail: zbigniew.ustrnul@uj.edu.pl

J. Filipiak
Faculty of Oceanography and Geography, Institute of Geography, Department of Meteorology and Climatology, University of Gdańsk, Gdańsk, Poland

Institute of Meteorology and Water Management—National Research Institute, Warsaw, Poland

J. Filipiak
e-mail: janusz.filipiak@ug.edu.pl

M. J. Hajto · A. Wypych · Z. Ustrnul
Institute of Meteorology and Water Management—National Research Institute, Warsaw, Poland

M. Falarz (ed.), *Climate Change in Poland*, Springer Climate,
https://doi.org/10.1007/978-3-030-70328-8_4

carried out in several stages: (1) collecting information on the history of meteorological stations (metadata), especially on significant changes in their location and measurements methods; (2) selecting the test for homogeneity control, calculating values of the test, selecting series with test values exceeding the specified critical level; (3) correction of non-homogeneous series; (4) recheck of the revised series. The correction of the climatic series analysed for the purposes of this book was carried out using various methods, specific to the climate element analysed. The most common was the relative and absolute Alexandersson test (Standard Normal Homogeneity Test, SNHT). The most common reasons for breaking the homogeneity of the climatic series were a significant change of station location and a change of the measuring instrument. However, in many cases, the cause of the series' heterogeneity is not clear.

Various changes during the history of meteorological stations can be the cause of loss of climate series homogeneity. The reasons for the heterogeneity of the series may be:

- a change of the location of the meteorological station (especially with a significant shift associated with a change in the height of the station above sea level),
- a change of the type of measuring instrument (in particular related to the replacement of a traditional instrument with an automatic instrument),
- a change in the environment of the measuring station (e.g. as a result of the development of the city, the creation of a water reservoir near the station),
- a change in the dates of measurements and observations,
- a change of methods of observation,
- a change of units of measured climatic elements;
- a change of the calculation method of the daily mean of any climatic element,
- a change of an observer(s) (in particular when assessing the state of the climate element observed subjectively, without the use of a measuring instrument, as in the case of cloud cover).

The monitoring of the homogeneity of the climate series and its possible correction are therefore indispensable in the analysis of climate change and variability: it makes it possible to draw appropriate conclusions on the trends of the various elements of the climate. The climate series homogenisation procedure is usually carried out in several stages: (1) collecting information on the history of meteorological stations (metadata), especially on significant changes in their location and measurements methods; (2) selecting the test for homogeneity control, calculating values of the test,

M. Taszarek
Faculty of Geographical and Geological Sciences, Institute of Physical Geography and Environmental Planning, Department of Meteorology and Climatology, Adam Mickiewicz University, Poznań, Poland
e-mail: mateusz.taszarek@amu.edu.pl

selecting series with test values exceeding the specified critical level; (3) correction of non-homogeneous series; (4) recheck of the revised series.

The correction of the climatic series analysed for the purposes of this book was carried out using various methods specific to the climate element analysed. The most common was the relative and absolute Alexandersson test (Alexandersson 1986; Alexandersson and Moberg 1997). This chapter presents the results of the analysis of the homogeneity of climate series for selected climate elements.

Average daily values of **air pressure** reduced to sea level were used for the study of pressure changes in Poland. Air pressure reduction to sea level did not affect the homogeneity of the measurement series. Only at the stations located over 500 m above sea level were the values measured at the station level. Measurement data from stations with full observation series were selected for the study. The very few cases of missing or incorrect data from one or several days were supplemented by comparing values from a few of the nearest stations and by analysing synoptic maps. In the case of three stations, the missing data concerned longer periods over one or two years. In these cases, only monthly average values were supplemented.

In addition to natural factors, i.e. astronomical, geographical and meteorological, measuring **sunshine duration** can also be influenced by instrumental and methodological elements related to the recording of sunshine duration. The magnitude of recorded sunshine duration can depend on the type of paper used in heliographic strips (colour of strips, accuracy of printing the bar scale, and paper grade), the properties of glass (transparency, colouring, scratching) of heliographic balls, as well as the professionalism of observers who manage the purity of the ball, exchange the strips, and interpret heliograms. In general, automatic measurements are more objective. They do not require the direct presence of an observer, and the results are easier to archive and send. However, in automatic measurement series, there are sometimes gaps and incorrect data due to instrument failure, power outage, or as a result of instrumental errors.

Of the initial number of 45 stations, data from 31 stations (Table 4.1) that had continuous series or slight deficiencies were used in the final analysis. At some stations, the measurement of sunshine duration began in 1966, but in recent years, due to the change of instrument, large data gaps have arisen. Examples are the meteorological stations, among others, in Lębork, Hel, Resko, Szczecinek, Ostrołęka, Leszno, Wieluń, Racibórz, Rzeszów, Bielsko Biała, and Nowy Sącz. Data verification was carried out at the level of daily totals and any possible deficiencies were supplemented using the method of similarity with the nearest station. Errors in the values of daily sunshine duration consisting in exceeding the length of the day were eliminated using the method of Forsythe et al. (1995).

For the series of data from the stations where the measuring instrument was changed in 2014, a statistical analysis (linear trends, slope coefficients, and their statistical significance (p_v values) was performed, which showed that extending the Campbell-Stokes data with converted automatic data is, in most cases, insignificant, i.e. both regression lines (i.e. of not extended and extended series) practically overlap. This is similar for monthly sunshine duration trends. At four stations (Kasprowy

Table 4.1 Homogenisation procedure for inhomogeneous series of sunshine duration

Station	Method used for the detection and correction of series inhomogeneity	The first year after homogeneity breaking	Station used for correction of inhomogeneity	The cause of the break of series homogeneity; other remarks
Białystok	1. Method of stability of a linear trend 2. method of similarity	2014	Biebrza	Change of instrument: 1 January 2014; no data for 11 days in 2016–2018
Chojnice	1. Method of stability of a linear trend 2. exceeding the duration of the day (Forsythe et al. 1995)	2014		Change of instrument: 1 July 2014; exceeding the duration of the day
Gaik—Brzezowa	1. Method of similarity 2. exceeding the duration of the day	1973	Kraków Obs. UJ	No data for 6 days in 1973, exceeding the duration of the day
Gdynia	1. Exceeding the duration of the day			Exceeding the duration of the day
Gorzów Wlkp.	1. Method of stability of a linear trend	2014		Change of instrument: 1 April 2014
Jelenia Góra	1. Method of stability of a linear trend 2. method of similarity 3. exceeding the duration of the day	2011	Radzyń	Change of instrument: 1 September 2011., no data for 44 days in 2012–2014, 2017, exceeding the duration of the day
Kalisz	1. Method of stability of a linear trend 2. exceeding the duration of the day	2014		Change of instrument: 1 April 2014., exceeding the duration of the day
Kasprowy Wierch	1. Method of similarity 2. exceeding the duration of the day	2009	Zakopane	No data for 10 days in 2009, 2015–2018, exceeding the duration of the day

(continued)

Table 4.1 (continued)

Station	Method used for the detection and correction of series inhomogeneity	The first year after homogeneity breaking	Station used for correction of inhomogeneity	The cause of the break of series homogeneity; other remarks
Katowice	1. Method of stability of a linear trend 2. method of similarity	2014	Kraków Obs. UJ	Change of instrument: 1 January 2014, no data for 1 day in 2018
Kłodzko	1. Method of stability of a linear trend	2011		Change of instrument: 1 September 2011.
Kołobrzeg	1. Method of stability of a linear trend	2014		Change of instrument: 17 September 2014
Koszalin	1. Method of stability of a linear trend 2. method of similarity 3. exceeding the duration of the day	2011	Kołobrzeg	Change of instrument: 1 September 2011., no data for 4 days in 2016, exceeding the duration of the day
Lesko	1. Method of stability of a linear trend 2. method of similarity	2014	Nowy Sącz	Change of instrument: 1 January 2014., no data for 3 days in 2018
Łódź Lublinek	1. Method of stability of a linear trend 2. method of similarity 3. exceeding the duration of the day	2013	Kalisz	Change of instrument: 1 November 2013., no data for 9 days in 2013, 2015, exceeding the duration of the day
Mikołajki	1. Method of stability of a linear trend 2. method of similarity	2014	Biebrza	Change of instrument: 1 January 2014., no data for 5 days in 2014
Nowy Sącz	1. Method of stability of a linear trend 2. exceeding the duration of the day	2014		Change of instrument: 1 November 2014., exceeding the duration of the day

(continued)

Table 4.1 (continued)

Station	Method used for the detection and correction of series inhomogeneity	The first year after homogeneity breaking	Station used for correction of inhomogeneity	The cause of the break of series homogeneity; other remarks
Opole	1. Method of stability of a linear trend 2. method of similarity	2014	Katowice	Change of instrument: 1 January 2014., no data for 4 days in 2017
Poznań	1. Exceeding the duration of the day			Exceeding the duration of the day
Puławy	1. Method of similarity 2. exceeding the duration of the day	2013	Warszawa - Bielany	No data for 4 days in 2013, 2015, exceeding the duration of the day
Suwałki	1. Method of stability of a linear trend 2. method of similarity 3. exceeding the duration of the day	2014	Biebrza	Change of instrument: 1 January 2014., no data for 1 day in 2017, exceeding the duration of the day
Szczecin Dąbie	1. Method of stability of a linear trend 2. method of similarity	2014	Kołobrzeg	Change of instrument: 1 January 2014, no data for 3 days in 2011, 2018
Śnieżka	1. Method of similarity	2009	Kłodzko	No data for 22 days in 2009
Tarnów	1. Method of stability of a linear trend	2014		Change of instrument: 19 December 2014.
Terespol	1. Method of stability of a linear trend 2. method of similarity 3. exceeding the duration of the day	2014	Warszawa - Bielany	Change of instrument: 1 January 2014., no data for 10 days in 2015, 2016, exceeding the duration of the day
Warszawa Bielany	1. Exceeding the duration of the day			Exceeding the duration of the day

(continued)

Table 4.1 (continued)

Station	Method used for the detection and correction of series inhomogeneity	The first year after homogeneity breaking	Station used for correction of inhomogeneity	The cause of the break of series homogeneity; other remarks
Wielichowo	1. Method of similarity 2. exceeding the duration of the day	2010	Poznań	No data for 39 days in 2010, exceeding the duration of the day
Włodawa	1. Method of stability of a linear trend 2. method of similarity	2014	Puławy	Change of instrument: 1 January 2014., no data for 20 days in 2015
Zakopane	1. Method of stability of a linear trend	2014		Change of instrument: 1 January 2014
Zielona Góra	1. Method of stability of a linear trend	2011		Change of instrument: 1 September 2011

Wierch, Kraków, Gorzów, Warszawa-Bielany) simultaneous comparative measurements of CSD and the Campbell-Stokes heliograph were carried out. An analysis of these data did not show the need for the application of any corrective coefficients, either.

The problem of homogenisation of the series of sunshine duration records in Poland is the subject of a separate, extensive article that is being prepared. It is worth noting that:

- the instruments have been replaced, but not at all meteorological stations;
- if the replacement occurred, it occurred in 2014, which means a 5-year period of automatic measurement in the 48-year period (1971–2018), which is 10.4% of the data length,
- the average differences between automatic and manual instrument do not have the same sign for all stations and amount only to a few percent, which does not change the sign and value of the slope of temporal trends. Such small differences are also confirmed by the results of other comparative analyses, e.g. Matuszko and Nowak (2017), Urban and Zając (2017), and Valík et al. (2019).

The analysed data series on **cloudiness** were gathered during visual observations with no use of automatic methods of cloudiness measurement. Thus, it is necessary to use caution when interpreting such research materials in comparison with those gathered using automatic measurements. Credibility of such observation data depends on numerous factors such as the place of observation, observing method used, and the way of data coding. Those aspects may be categorised as external factors affecting the process of data collection. On the other hand, there are also internal conditions that alter the observation process. They include individual skills and habits of the observer, who may have learnt some observation patterns from the former employees of a given station. On the whole, it is relatively insignificant when assessing cloud amount. However, it may affect the observation results when estimating types of clouds, which has already been confirmed by the Polish National Meteorological and Hydrological Service (Institute of Meteorology and Water Management—National Research Institute, IMGW-PIB) on the basis of some internal audits. Although it does not negatively affect the analyses concerning types of clouds observed at different levels, it does alter the appearance frequency of particular numbers of the synoptic code FM12 referring to the codes of C_L, C_M, and C_H (for instance Filipiak and Miętus 2009).

When analysing homogeneity of cloudiness data the Standard Normal Homogeneity Test (SNHT) by Alexandersson (1986; Alexandersson and Moberg 1997) was implemented in both methods of the analysis: SNHT Single Series and SNHT Alexandersson test (relative method). This procedure was implemented as there was a possibility of systemic data homogeneity disruption resulting from changes in the observation procedure. At the synoptic stations total cloud cover has been measured in oktas since 1 January 1966, while at the climatological stations oktas were introduced several dozen years later—on 1 January 1989. Moreover, the data gathered by the climatological stations had to be additionally verified in order to check whether the time change in evening observations, described in the chapter *Cloudiness change*, has influenced the data quality or not. Verifying each and every data series in both ways has to identified potential systemic errors, such as the above-mentioned change of the general cloudiness measuring method.

Nonetheless, the data homogeneity disruption resulting from the observation procedure change was not common at all analysed synoptic stations. Furthermore, it was observed only in two particular months—March and September (Table 4.2). A number of disruptions were also observed in June. However, all analysed climatological stations experienced data homogeneity disruption resulting from the change in the observation method. The change in time of the evening observation did not significantly affect the data homogeneity.

In some cases, the reasons behind the data homogeneity disruption were impossible to determine. Sometimes it was simply impossible to verify the history of observations made at the particular station. Full sets of metadata were only available for selected stations, especially those located in northern Poland.

There were few cases of data drift—systematic overestimating of the data in comparison to regional average values—for example at the climatological station in Gdynia (Fig. 4.1) and at the synoptic stations in Kielce (Fig. 4.2) and Tarnów.

The data series of **air temperature** included in the work are fully verified and provide the basis for a reliable assessment of the trends in the occurrence of temperature extremes. Of the known stations with a history of up to 200 years, which have been scientifically researched and for which data quality control and homogenisation analyses have been made, the following should be mentioned first: Warszawa, Kraków, Gdańsk, Wrocław, and Poznań.

Table 4.2 The list of detected inhomogeneity in data series concerning cloudiness

Station	Period of heterogeneity	Method used for detection and correction of data homogeneity	The first year after data homogeneity disruption and the period for which the data was corrected	Stations (regions) used when restoring data homogeneity (in the case of the relative method)	The cause of the break of series homogeneity; other remarks
Synoptic stations (first and second row stations)					
Koszalin, Łeba, Świnoujście, Resko, Gorzów Wlkp., Słubice, Zielona Góra, Legnica, Jelenia Góra, Tarnów	March	SNHT Single Series	1966, 1951–1965	–	Changes in observation practice—transition from the decile scale to the okta scale
Resko, Chojnice, Olsztyn, Gorzów Wlkp., Słubice, Leszno, Wrocław, Kalisz, Lublin, Jelenia Góra, Kłodzko, Opole, Tarnów	September	SNHT Single Series	1966, 1951–1965	–	
Lesko			1966, 1955–1965		
Koszalin, Łeba, Olsztyn, Kalisz, Lublin, Jelenia Góra, Sandomierz,	June	SNHT Single Series	1980, 1951–1979	–	Unknown

(continued)

Table 4.2 (continued)

Station	Period of heterogeneity	Method used for detection and correction of data homogeneity	The first year after data homogeneity disruption and the period for which the data was corrected	Stations (regions) used when restoring data homogeneity (in the case of the relative method)	The cause of the break of series homogeneity; other remarks
Kętrzyn			1980, 1966–1979		
Kołobrzeg	April–August	SNHT	1988, 1951–1987	Southern Baltic Coastlands Region	Unknown, data overestimated
Łeba	May		1960, 1951–1959		Changes to the order of the station, data underestimated
Hel	June		1971, 1951–1970		Unknown, data underestimated
Świnoujście	Months of cold half-year		1951–2018		
Legnica	August, October		1994, 1995–2014	Central Poland Lowlands, Podlasie-Belarus Heights and Polesie Region	Changes to the order of the station, data underestimated
Opole	June		1984, 1951–1983		Unknown, data underestimated
	July		1992, 1951–1991		
	October		1992, 1951–1991		
Wrocław	June		1984, 1951–1983		
	October		1996, 1951–1995		
Sulejów	April		1965, 1962–1964	Polish Uplands Region	
	June		1971, 1962–1970		
Kielce	All months		1951–2018		Unknown, data overestimated
Tarnów	All months		1951–2014	Northern Subcarpathians Region	Unknown, data underestimated
Lesko	June		1969, 1955–1968	Carpathians Region	

(continued)

Table 4.2 (continued)

Station	Period of heterogeneity	Method used for detection and correction of data homogeneity	The first year after data homogeneity disruption and the period for which the data was corrected	Stations (regions) used when restoring data homogeneity (in the case of the relative method)	The cause of the break of series homogeneity; other remarks
Climatological stations (third and fourth row stations)					
Gdynia	All months	SNHT Single Series	1989, 1951–1988	–	Changes in observation practice
	All months	SNHT	1951–2018	Southern Baltic Coastlands Region	Unknown
Puławy	All months	SNHT Single Series	1989, 1951–1988	–	Changes in observation practice
	All months	SNHT	1951–2018	Polish Uplands Region	Unknown
Jabłonka	All months	SNHT Single Series	1989, 1961–1988	–	Changes in observation practice
	All months	SNHT	1961–2018	Carpathians Region	Unknown

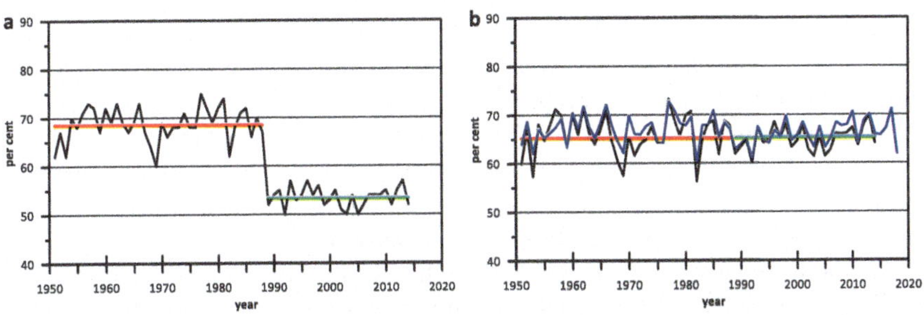

Fig. 4.1 The average annual total cloud cover at the climatological station in Gdynia in the period of 1951–2014: **a**—before homogenisation and **b**—after homogenisation. The horizontal lines depict values of the arithmetic average during the periods before (the red line) and after (the green line) the data homogeneity disruption. The blue line in Fig. 4.1b marks the data series for Southern Baltic Coastlands Region

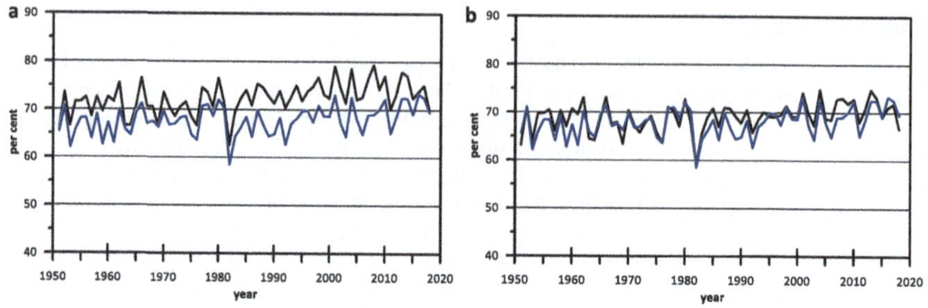

Fig. 4.2 The average annual total cloud cover at the synoptic station in Kielce in the period of 1951–2018: **a**—before homogenisation and **b**—after homogenisation. The black line marks the data series gathered in Kielce; the blue one the series for Polish Uplands Region

Monthly **relative humidity** series were examined for homogeneity using the standard normal homogeneity test (SNHT). In this study, reference series have been built from five stations, based on correlation coefficient equal to or higher than 0.7 (Tuomenvirta 2002). The critical level of the test was 95% with the critical value of the SNHT statistic T equal to 8.784 (Khaliq and Quarda 2007). At 41 stations, at least one inhomogeneity in the monthly time series was found. In total, 25% of all monthly time series were homogenised, and the adjusted time series were used in the further analyses.

Observations and measurements of **snow cover** are simple, although they can be fraught with a certain amount of subjectivity in assessing the land area occupied by the snow layer, and thus, classification of the element as: continuous snow cover (area occupied equal to 100%), snow cover breaks (100% > area ≥50%) or snow cover patches (area < 50%). This can be the reason for heterogeneity of a series of snow cover when changing the observer at the weather station. However, until the early years of the twenty-first century, the main cause of breaking the homogeneity of the snow cover series was seen in the change of station location (Sallabanda 1996; Nowosad 2000/2001; Falarz 2006). In the early decades of the twenty-first century, automatic measurements of snow cover thickness began, which could be another important cause of the heterogeneity of the tested series.

The homogeneity of snow cover series has so far been studied using various methods. Sallabanda (1996) used three non-parameter tests for series of snow cover in Albania, examining: the stability of mean, the stability of dispersion of the variations in relation to this central value, and the correlation between the consecutive terms. He showed the heterogeneity of only 3 out of the 127 snow cover series tested. Nowosad (2000/2001) used a variety of tests in his studies of many snow cover characteristics in south-eastern Poland: the Smirnov test, test of series, and the Wilcoxon test. The current work uses the Alexandersson Test (Standard Normal Homogeneity Test; Alexandersson 1986; Alexandersson and Moberg 1997). This test has already been applied by Falarz (2006) in snow cover studies until 1998 and Marcolini and coauthors (2017). In the latter research inhomogeneities have been detected in about 20% of the analysed time series.

The homogeneity of all 60 series of both seasonal number of days with snow cover and maximum snow cover depth in the winter season were examined. The AnClim of programme by Štěpánek (2008) was used for this purpose. For most of the series tested a relative test with a constancy of quotients method was used. As reference stations, a series of averaged values for snow cover regions (Falarz 2006) and additionally selected series from stations located in the same snow cover region, at similar altitudes and close to the examined station were selected. The reference station must not be a station with a significant change of location in its history. For series of more than 70 years and concerning mountain stations, both relative and absolute tests were used. The decision to consider the series to be heterogeneous was made by analysing a critical value of the SNHT test for 95% significance level. As a result of the studies, data series for 19 out of 60 weather stations (i.e. around 32% of the series) were considered heterogeneous. In 3 cases, it was shown to break the homogeneity only for maximum snow cover depth series, in 14 cases—only for series of the number of days with snow cover, while in 2 cases (Lublin, Mława)—for series of both characteristics (Table 4.3). Corrections of non-homogeneous series were made in most cases by transforming the data in the period before the break of homogeneity. The adjustment of values in the period after the break of homogeneity was made only in cases where the moment of breaking the homogeneity occurred close to the end of the entire multi-year period. The main reasons for the heterogeneity of the data were: changing the location of the measuring station and changing the measuring instrument to automatic.

Bringing a series of snow cover to homogeneity avoided drawing incorrect conclusions about the changes in this climate element. Figure 4.3 shows the changes in snow cover time and the process of elimination of series heterogeneity in the example of the Kielce station in the period 1950/51–2017/18. The station was moved in 1974. In the period 1950/51–1973/74, the values of snow cover duration in Kielce were clearly escalated in relation to the averages for the whole region (region 5), and the value of the SNHT test in 1974 was 11.99, exceeding the critical value for 95% of significance level (Fig. 4.3a,c). The correction of the series in the period before the break of homogeneity resulted in the elimination of significant differences between the values at the station in Kielce and the whole region before 1974 (Fig. 4.3d). A re-examination of the homogeneity of the corrected series showed a significant decrease in the test value and no exceedance of the critical level (Fig. 4.3b). Considering the heterogeneous series would lead to the incorrect conclusion of a greater than actual downward trend in the characteristics at issue.

The variable height of the anemometer is not the only factor that could break the homogeneity of **wind** measurements.

At the synoptic stations of IMWM-NRI changes of instruments for wind direction and speed measurements took place a number of times. Initially, until the mid-1960s, measurements were made with a Wild's anemometer. From 1966, Wild's anemometers were replaced by M-47 cup anemometers (Janiszewski 1972) at all the IMWM-NRI stations. The measurements were made manually. The observer observed the indicator oscillation for 2 min and on this basis determined the wind speed and direction: the speed with an accuracy of 1 ms^{-1} and the direction with an accuracy of

Table 4.3 The list of detected inhomogeneities in data series concerning snow cover. (D—seasonal duration of snow cover, M—seasonal maximum depth of snow cover; numbering of regions according to Falarz 2006)

Station	Inhomogeneity of series	Period of corrected data	Stations and regions used for homogenisation procedure	The cause of the break of series homogeneity; other remarks
Bielsko	M	2016/17–2017/18	Region 7a	Change the instrument to automatic (2016)
Częstochowa	D	1950/51–1969/70	Region 5, Katowice	Station moved in November 1969 by 1.5 km to the northwest; change in altitude: + 32 m
Elbląg	D	1987/88–2017/18	Region 1, 2	Unknown cause; overvaluation of recorded values after 1986 compared to neighbouring stations
Hala Gąsienicowa	M	2009/10–2017/18	Region 7b, Kasprowy Wierch	Change the instrument to automatic (2009)
Kielce	D	1950/51–1973/74	Region 5, Sandomierz	Station moved in 1974 to the south-east
Kłodzko	D	2013/14–2017/18	Region 6	Change the instrument to automatic (2013)
Koszalin	D	1950/51–1961/62	Region 1, Resko	Unknown cause; undervaluation before 1962 compared to neighbouring stations

(continued)

Table 4.3 (continued)

Station	Inhomogeneity of series	Period of corrected data	Stations and regions used for homogenisation procedure	The cause of the break of series homogeneity; other remarks
Krosno	D	1950/51–1984/85	Region 7a	Station moved to the northwest in 1984; change in altitude: + 44 m
Krynica	D	2013/14–2017/18	Region 7b	Change the instrument to automatic (2013)
Lesko	D	1950/51–1980/81	Region 7a	Station moved in November 1980 about 1 km to the east; change in altitude: + 34 m
Leszno	D	1950/51–1957/58	Region 4, Zielona Góra	Station moved in 1958 to the southwest; change of the station row from climatic to synoptic in 1958
Lublin	D,M	1950/51–1972/73	Region 5, Włodawa, Zamość	Station moved in 1973. to the southwest; change in altitude: + 66 m
Mikołajki	M	1950/51–1957/58	Region 3, Kętrzyn	Station moved in 1958
Mława	D,M	1950/51–1967/68 2000/01(2004/05)–2017/18	Region 2, Olsztyn	Station moved in 1967/1968, the second break of homogeneity in the early twenty-first century (the cause unknown)

(continued)

Table 4.3 (continued)

Station	Inhomogeneity of series	Period of corrected data	Stations and regions used for homogenisation procedure	The cause of the break of series homogeneity; other remarks
Ostrołęka	D	1980/81–2017/18	Region 3, Mikołajki	Unknown cause; undervaluation after 1980 compared to neighbouring stations
Płock	D	1950/51–1962/63	Łódź, Koło	Unknown cause; undervaluation before 1963 compared to neighbouring stations
Siedlce	D	2010/11–2017/18	Region 5	Unknown cause; breaking homogeneity in 2010
Ustka	D	1997/98–2017/18	Region 1	Unknown cause; breaking homogeneity in 1997
Warszawa	D	1991/92–2017/18, 2009/10–2017/18	Region 5, Łódź	Unknown cause; undervaluation compared to neighbouring stations in 1991 and 2009

10 degrees. Since 1 January 1976, in accordance with the recommendation of the World Meteorological Organization, the sampling interval changed. The averaging time was extended from 2 to 10 min, but still the averaging was done visually by the observer.

In the years 1966–1986 measurements of wind speed were made with anemometers of Soviet production, first type M-47, then M-63. From 1986, they were successively replaced by electric anemometers W-863 manufactured in Poland by Zootechnika Kraków (Lorenc 1996). The Vaisala cup anemometers WAA151 with wind vanes WAV151 were used from the turn of the twenty-first century. Their threshold wind speed was 0.4 ms^{-1}. In 2014, they were replaced by Vaisala sonic anemometers

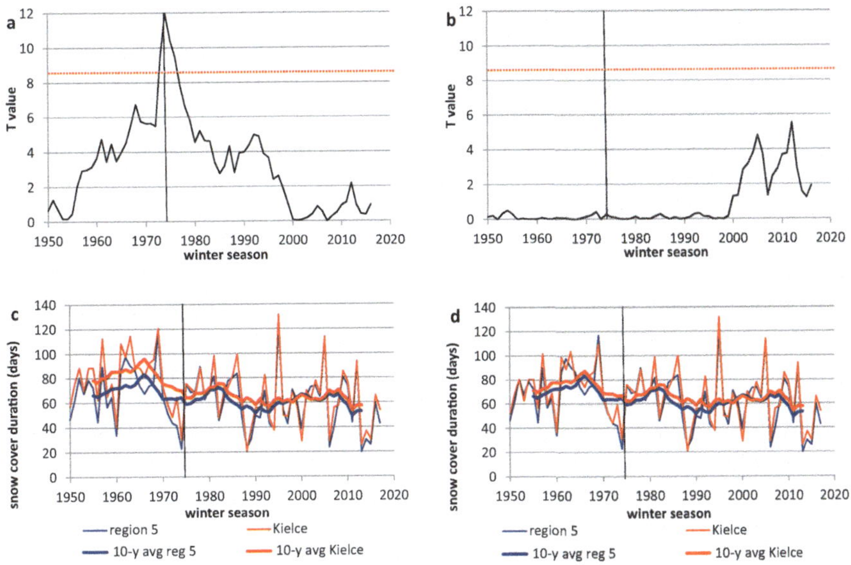

Fig. 4.3 An example of homogenisation procedure for snow cover duration in Kielce: T value for the Alexandersson test: **a** before homogenisation, **b** after homogenisation (horizontal red line indicates a critical value of the test for 95% significance level); snow cover duration in Kielce and region 5: **c** before homogenisation, **d** after homogenisation; black vertical line: the year of changing of the station location (1974)

with WS425 sensors with a threshold wind speed of virtually 0.0 ms^{-1}, but with an accuracy of 0.1 ms^{-1}. Both Vaisala anemometers automatically average wind speed and direction.

Changes of instruments often give rise to significant break-points in wind speed series. Unfortunately, changes are usually sudden, without a period of overlapping comparative measurements. The number of papers discussing this issue is very limited. Two of the most important changes are: the transition from measurements using Wild's anemometer to cup anemometers and the transition from manual to automatic measurements. In Poland, the first one took place in the mid-60s, whereas the second one at the turn of the twenty-first century.

Dziaduszyński et al. (2000) attempted to compare the of wind speed measurements by a Wild's anemometer (3 times per day) and by a YOUNG (model 05103-5) automatic weather station anemometer (every 15 min). They used data from the climatological station in Puławy from the period 1995–1998. Both anemometers were positioned 17 m above ground level. They showed that the Wild's anemometer shows a greater frequency of low- and high-speed winds and that on average the wind speed measured by the automatic anemometer was higher by 14% than that measured by the Wild's anemometer.

Brázdil et al. (2017) compared standard universal anemometers working in the Czech Republic till the mid-90s with automatic measurements from a Vaisala

WAA251 sensor (cup anemometer) and a WS425 sensor (sonic anemometer). Parallel measurements were provided at two stations only. They showed that the average wind speed measured using both Vaisala automatic sensors are slightly higher than those manually measured and the differences do not depend on wind direction. The frequencies of calms and low wind speed values are higher for standard cup anemometers than for both Vaisala sensors. These differences are more pronounced during morning and evening times because the average wind speed is lower than during the noon term. However, maximum wind speeds can be even higher for standard cup anemometers than for Vaisala sensors.

To check whether the replacement of Wild's anemometers by cup anemometers caused a break in the homogeneity of the series, the average wind speed was calculated over two 10-year periods before and after the instrument changeover (1951–1960 and 1971–1980), and it was checked how much the wind speed changed in relation to the initial value (Table 4.4). The mean increase was 8.3%, but at individual stations it varied from −23% (decrease) to 74% (increase). Decreases were mainly observed at stations with low average wind speed, whereas increases were typical at stations with high average wind speed. This is in agreement with the findings of Dziaduszyński et al. (2000). Similarly, it was done to assess whether the transition from manual to automatic measurements resulted in a break in homogeneity. This time, the averages were compared over two 5-year periods (2001–2005 and 2014–2018), as the last change was in the year 2014.

The transition from Wild's anemometer to cup anemometers resulted in an average change in wind speed of 7% (Table 4.4), which means that there was a clear break in the homogeneity of the measurement series. The transition from manual to automatic anemometers resulted in an average change in wind speed of only 1%, so it can be concluded that this is not a change that significantly disturbs the measurement series, although at individual stations the effect may be more significant. Because the step change around 1966 was very different at different stations, it was not possible to homogenise the series by simply increasing the wind speed at the beginning of the series. There are also no homogeneous series of wind speeds to which the others could be linked. For this reason, it was decided to use only data from the period 1966–2018 for further analysis.

Thunderstorms and hail events in Poland are studied using observational data from synoptic stations of IMWM-NRI and the Scientific Station of the Climatology Department of the Jagiellonian University in Kraków (Kraków Observatory). The data included in the study is qualitative and, to some extent, is influenced by the subjective assessment of the observer (Bielec and Kolendowicz 2001). An important disadvantage is that human contributions to these observations introduce errors, such as spatial and temporal inhomogeneities (Czernecki et al. 2016). For this reason, the indicator of the thunderstorm or hail occurrence used in the study was the day (24 h; 00 to 00 UTC) on which the occurrence of a given phenomenon was noted.

In addition, the spatial coverage of observing stations may be also too dispersed to capture the scale of most thunderstorms. Therefore, the homogeneity analysis consisted mainly of a manual check of observational series, metadata and comparison of monthly and annual values between stations situated next to each other. In

Table 4.4 Change (%) in the average wind speed before and after the change of instrument type. Explanations: AVE_1, AVE_2, AVE_3, AVE_4: average wind speeds over periods 1951–1960, 1971–1980, 2001–2005, and 2014–2018, respectively, $R_{2/1} = (AVE_2 - AVE_1)/AVE_1$, $R_{4/3} = (AVE_4 - AVE_3)/AVE_3$

Station	AVE_1	AVE_2	$R_{2/1}$	AVE_3	AVE_4	$R_{4/3}$
Chojnice	3.21	3.74	0.17	3.73	3.67	−0.01
Elbląg	3.97	3.84	−0.03	3.10	4.25	0.37
Koło	2.26	3.91	0.74	3.83	3.74	−0.02
Koszalin	3.85	3.93	0.02	3.20	3.35	0.05
Łeba	4.88	5.01	0.03	4.60	5.20	0.13
Lębork	2.69	2.38	−0.11	3.20	3.42	0.07
Olsztyn	3.16	3.27	0.04	3.02	3.06	0.01
Suwałki	4.04	4.83	0.19	3.75	3.46	−0.08
Świnoujście	3.79	4.18	0.10	3.42	3.25	−0.05
Szczecin	3.70	4.03	0.09	3.85	3.61	−0.06
Toruń	3.55	3.11	−0.12	2.74	2.46	−0.10
Ustka	4.32	4.14	−0.04	5.32	5.11	−0.04
Białystok	3.28	3.63	0.11	2.58	2.49	−0.04
Hel	4.15	4.99	0.20	4.04	4.08	0.01
Bielsko-Biała	3.74	3.64	−0.03	2.86	2.84	−0.01
Gorzów	3.07	3.46	0.13	2.65	2.69	0.01
Jelenia Góra	2.62	2.78	0.06	2.28	2.61	0.15
Kalisz	2.75	3.48	0.26	3.70	3.71	0.00
Kasprowy Wierch.	6.27	7.09	0.13	6.31	6.10	−0.03
Kielce	2.80	2.91	0.04	2.92	2.74	−0.06
Kłodzko	2.66	2.74	0.03	2.73	3.08	0.12
Kołobrzeg	4.03	3.64	−0.10	2.80	3.05	0.09
Kraków	2.73	2.89	0.06	2.89	3.34	0.16
Legnica	2.58	3.40	0.32	3.48	3.39	−0.03
Lesko	2.81	3.21	0.14	2.53	2.33	−0.08
Łódź	4.23	3.98	−0.06	3.16	3.50	0.11
Lublin	2.86	3.39	0.18	2.92	2.97	0.02
Nowy Sącz	1.64	1.48	−0.09	1.77	1.84	0.04
Opole	2.32	2.95	0.27	2.68	2.60	−0.03
Poznań	3.92	4.07	0.04	3.89	3.88	0.00
Rzeszów	3.61	3.75	0.04	3.91	3.81	−0.03
Siedlce	3.52	3.44	−0.02	3.26	2.88	−0.11
Słubice	2.69	2.68	0.00	2.52	3.17	0.26
Śnieżka	11.59	12.46	0.08	12.48	10.70	−0.14

(continued)

Table 4.4 (continued)

Station	AVE_1	AVE_2	$R_{2/1}$	AVE_3	AVE_4	$R_{4/3}$
Tarnów	2.80	2.28	−0.18	1.80	1.73	−0.04
Warszawa	4.08	4.29	0.05	3.98	3.56	−0.11
Wieluń	3.00	2.31	−0.23	2.99	2.92	−0.02
Włodawa	2.64	3.62	0.37	3.82	3.72	−0.03
Wrocław	3.24	3.36	0.04	3.27	3.25	−0.01
Zakopane	1.62	1.40	−0.14	1.40	1.53	0.09
Zielona Góra	2.66	3.23	0.21	2.97	2.94	−0.01
Minimum	1.62	1.40	−0.23	1.40	1.53	−0.14
Maximum	11.59	12.46	0.74	12.48	10.70	0.37
Average	3.50	3.73	0.07	3.47	3.46	0.01

addition, the homogeneity of the obtained data was inspected using the Standard Normalized Homogeneity Test (Alexandersson 1986) within The AnClim program (Štěpánek 2008). Finally, only the data series that had the longest and the most credible observations were selected for the study.

The quality of **tornado** reports in Poland varies strongly in time and space. Multiple factors are responsible for this situation. Among political, historical, and social contexts we can list changes in the borders of Poland (e.g. from 1795 to 1918 the territory of Poland was under German, Austrian, and Russian occupation), World Wars I and II and the *communist* era under the *Soviet* dominance (1945–1989). Although Poland was not an independent country in the nineteenth century, nationality and the awareness of cultural identity remained in society and thus numerous Polish newspapers operated on a regional scale. As suggested by Antonescu et al. (2016), such newspaper-type publications are the main source of information on tornado occurrence from these years. However, it is clear that only a small percentage of tornado cases were reported in these sources, unfortunately resulting in a strong quantitative underestimation (particularly for weak tornado cases). On the other hand, tornado databases are likely to be more consistent over time, especially for intense spectacular events that cause significant property damage (Brooks and Doswell 2001; Verbout et al. 2006; Rauhala et al. 2012; Taszarek and Brooks 2015). A majority of such high-end events attract more public attention and are usually better documented in media reports. Thus, they allow for a long-term estimates of strong and violent tornadoes (e.g. Taszarek et al. 2017).

In Poland, the highest number of archival newspaper editions are available for the second half of the nineteenth and first half of the twentieth century. During the socialistic period from 1945 to 1989 any information on catastrophic events is difficult to find, thus resulting in a low number of tornado reports (Taszarek and Brooks 2015). A similar situation was also observed in the Czech Republic by Setvák et al. (2003) and Brázdil et al. (2019). When Poland gained sovereignty in 1989 (transformation of political system), a small increase in tornado reporting was observed.

After a long communist era, people generally assumed that tornadoes did not occur in Poland and their occurrence was mainly limited to the Great Plains in the United States. Beginning with the twenty-first century, advances in teledetection data and communication technologies (e.g. access to Internet, mobile phones, meteorological radar network, lightning detection systems) and development of thunderstorm observer networks (e.g. Polish Stormchasing Society) resulted in a surge in tornado reporting, not only in Poland, but also in Europe as a whole (Groenemeijer and Kühne 2014). This may somehow indicate that before 2000, tornadoes were strongly underestimated in Europe. It is also worth highlighting that, after the foundation of the Polish Stormchasing Society in 2008, most tornado cases have credible documentation often accompanied by damage survey experts, witnesses, and, in some cases, by radar data. Without a doubt, regional networks of storm observers and regional damage survey experts increased the quality of tornado reports.

The tornado reports presented in this analysis derive from the European Severe Weather Database (ESWD; Dotzek et al. 2009) and cover the period from the early nineteenth century until 2018. Reports with a status of plausibility check passed (QC0+), report confirmed (QC1), and event fully verified (QC2) are involved until 2008. From 2009 only QC1, and QC2 are taken into account. This has been done because from 2009 the quality control levels were incorporated by the European Severe Storms Laboratory (ESSL). Most of the national cooperating partners joined ESWD after 2010, thus increasing the number and quality of the reports. In addition, the development of social media and the increase in those interested in severe weather that has taken place in recent years (e.g. the foundation of the Polish Stormchasing Society, Polscy Łowcy Burz) have significantly increased the number of reports (often accompanied by a photograph of the event) as well (Taszarek et al. 2017).

Data on **fog** include the results of visual observations, not measurements, and fog is an element that strongly depends on local environmental conditions. Generally, those two factors determined to a large extent the homogenisation procedure.

The homogeneity of the fog data was checked for the most basic index, i.e. annual number of days with fog. For each station, the mean annual number of days with fog in the period 1966–2018 was calculated, and then two other basic statistics were determined, i.e. the standard deviation and variability coefficient.

For four stations, the mean annual number of days with fog was distinctly higher than for other stations located nearby: Chojnice (87.2 days), Zielona Góra (82.9 days), Jelenia Góra (88.4 days), Kielce-Suków (84.8 days). In addition, the values of the variability coefficient were distinctly higher in Gdańsk (59.5%) and Terespol (43.9%) than at other stations. In the next step of the procedure, the data were standardised; for each year and each station, the mean value for the whole country was deducted from the value for a particular station and divided by the value of the standard deviation for a particular year. Standardised multi-annual courses of the yearly number of days with fog showed large increases or decreases beginning in various years at the following stations: Jelenia Góra, Kielce, Elbląg, Gdańsk, Terespol, Kłodzko, Suwałki, Zielona Góra, and Chojnice. Therefore, the stations' metadata and modifications in the local environment around the stations were studied to find out whether the changes in fog occurrence might be caused by non-climatic factors. Metadata of the stations delivered the information that the stations in Jelenia Góra, Terespol, Kłodzko, Suwałki, Zielona Góra, and Chojnice were not moved during the whole study period, and data series for Gdańsk are combined from three locations. That information was the first prerequisite to eliminate data from Gdańsk from the further analysis. Satellite images available at the time on Google Earth and data from CORINE Land Cover (CLC) data base containing CLC changes in the period 1990–2018 [https://land.copernicus.eu/pan-european/corine-land-cover] for the areas around the stations mentioned were analysed. Close to the stations in Jelenia Góra and Terespol, water reservoirs were constructed. In Kłodzko and Suwałki, the exploitation of gravel nearby the stations was connected with the rising of a few bodies of water. A similar situation can be observed in Kielce, where there was sand exploitation realised. Additionally, the station in Kielce is located in a concave land form which is the reason for the great increase in the cold air reservoir, and the occurrence of air temperature inversions and fog in comparison to nearby areas. In Zielona Góra and Chojnice, large areas around the stations were turned into transportation infrastructure or built-up areas. All the anthropogenic factors mentioned are known to significantly impact fog frequency due to major changes in atmospheric water vapour content. Water bodies deliver large amounts of water vapour into the atmosphere and can increase the fog frequency, while increases in built-up areas usually decreases water vapour content and fog occurrence. As the sudden changes in the number of days with fog matched rather well with the changes in land use/land cover in the vicinity of the stations, data from those stations were excluded from the further analysis. In the case of fog, it is not possible to correct observations from one station with observations from other stations located a few hundred kilometres away, as the phenomenon is dependent, to a large extent, on local environmental conditions. Therefore, data from the nine stations mentioned above were excluded from further analysis (Table 4.5).

Overall, the most common reasons for breaking the homogeneity of the climatic series were a significant change of station location and a change of the measuring instrument. However, in many cases the cause of the series' heterogeneity is not clear. The vast majority of the series have been brought to homogeneity and used to draw conclusions on climate change in Poland. Due to the significant spatial variability of fog occurrence, heterogeneous series of the number of days with fog were eliminated from further analysis.

Table 4.5 The list of detected inhomogeneity in data series concerning annual number of days with fog. The series were excluded from further analysis

Station	Method used for detection of data inhomogeneity	The cause of the break of series homogeneity; other remarks
Jelenia Góra	Station's metadata analysis; analysis of standardised courses of annual number of days with fog; analysis of changes in land use/land cover in the vicinity of the station	Construction of water reservoir nearby the station
Kielce		Sand exploitation was connected with arising of a few water bodies; location in a concave land form is the reason for much increased occurrence of cold air reservoir, air temperature inversions
Elbląg		Change in the station's location
Gdańsk		Three changes in station location
Terespol		Construction of water reservoir nearby the station
Kłodzko		Exploitation of gravel nearby the station was connected with arising of a few water bodies
Suwałki		
Zielona Góra		Large areas around the station were turned into transportation infrastructure or built-up areas
Chojnice		

References

Alexandersson H (1986) A homogeneity test applied to precipitation data. Int J Climatol 6:661–675

Alexandersson H, Moberg A (1997) Homogenization of Swedish temperature data. Part I: Homogeneity test for linear trends. Int J Climatol 17:25–34

Antonescu BD, Schulwtz M, Lomas F, Kühne T (2016) Tornadoes in Europe: synthesis of the observational datasets. Mon Weather Rev 144:2445–2480

Bielec Z, Kolendowicz L (2001) Problems connected with the observation and climatological elaborations of thunderstorms. Annales Universitatis Mariae Curie-Skłodowska, Lublin-Polonia, Sectio B 55/56(7): 59–65

Brázdil R, Chromá K, Dolák L, Kotyza O, Řezníčková L, Dobrovolný P, Černoch Z (2019) Spatiotemporal variability of tornadoes in the Czech Lands, 1801–2017. Theor Appl Climatol 136(3–4):1233–1248

Brázdil R, Valík A, Zahradníček P, Řezníčková L et al (2017) Wind-stilling in the light of wind speed measurements: the Czech experience. Clim Res 75:131–143

Brooks HE, Doswell CA III (2001) Some aspects of the international climatology of tornadoes by damage classification. Atmos Res 56:191–201

Czernecki B, Taszarek M, Kolendowicz L, Konarski J (2016) Relationship between human observations of thunderstorms and the PERUN lightning detection network in Poland. Atmos Res 167:118–128

Dotzek N, Groenemeijer P, Feuerstein B, Holzer AM (2009) Overview of ESSL's severe convective storms research using the European Severe Weather Database (ESWD). Atmos Res 93:575–586

Dziaduszyński K, Kozyra J, Doroszewski A (2000) Comparison of wind speed measured by wild anemometer and automatic weather station anemometer. Ann UMCS Sec B 55–56:127–131

Falarz M (2006) Wykrywanie i korekta niejednorodności wieloletnich serii niwalnych; Annales UMCS, Lublin, sec. B, LXI(18):155–163

Filipiak J, Miętus M (2009) Spatial and temporal variability of cloudiness in Poland, 1971–2000. Int J Climatol 29:1294–1311. https://doi.org/10.1002/joc.1777

Forsythe WC, Rykiel EJ Jr, Randal S, Schoolfield RM (1995) A model comparison for daylength as a function of latitude and day of year. Ecol Model 80(1):87–95. https://doi.org/10.1016/0304-3800(94)00034-F

Groenemeijer P, Kühne T (2014) A climatology of tornadoes in Europe: results from the European severe weather database. Mon Weather Rev 142:4775–4790

Janiszewski F (1972) Anemometer zdalny M-47. Gaz Obs PIHM 8:13–15

Khaliq MN, Quarda TBMJ (2007) On the critical values of the standard normal homogeneity test (SNHT). Int J Climatol 27:681–687

Lorenc H (1996) Struktura i zasoby energetyczne wiatru w Polsce. Mat Bad IMGW Meteorologia 25

Marcolini G, Bellin A, Chiogna G (2017) Performance of the standard normal homogeneity test for the homogenization of mean seasonal snow depth time series. Int J Clim 37(S1):1267–1277

Matuszko D, Nowak R (2017) Porównanie wyników pomiarów meteorologicznych wykonanych przyrządami tradycyjnymi i automatycznymi (na przykładzie danych z Krakowa), Annales UMCS, sec.: B, VOL. LXXII, z. 1, 73–89. https://doi.org/10.17951/b.2017.72.1.73

Nowosad M (2000/2001) Badanie jednorodności ciągów pomiarowych za pomocą wybranych testów nieparametrycznych na przykładzie charakterystyk pokrywy śnieżnej. Annales UMCS, sec. B, LV/LVI:259 –262, Lublin

Rauhala J, Brooks HE, Schultz DM (2012) Tornado climatology of Finland. Mon Weather Rev 140:1446–1456

Sallabanda A (1996) Homogeneity evaluation of snow cover records in Albania [w:] Proceedings of the First Seminar for Homogenization of Surface Climatological Data, Budapest, pp 6–12

Setvák M, Salek M, Munzar J (2003) Tornadoes within the Czech Republic: from early medieval chronicles to the "internet society". Atmos Res 67–68:589–605

Štěpánek P (2008) AnClim—software for time series analysis: Dept. of Geography, Faculty of Natural Sciences, Masaryk University, Brno. http://www.climahom.eu/AnClim.html

Taszarek M, Brooks HE (2015) Tornado climatology of Poland. Mon Weather Rev 143(3):702–717

Taszarek M, Brooks HE, Czernecki B (2017) Sounding-derived parameters associated with convective hazards in Europe. Mon Weather Rev 145:1511–1528

Taszarek M, Gromadzki J (2017) Deadly tornadoes in Poland from 1820 to 2015. Mon Weather Rev 145(4):1221–1243

Tuomenvirta H (2002) Homogeneity testing and adjustment of climatic time series in Finland. Geophysica 38(1–2):15–41

Urban G, Zając I (2017) Comparison of sunshine duration measurements from Campbell-Stokes sunshine recorder and CSD1 sensor. Theor Appl Climatol 129(1–2):77–87. https://doi.org/10.1007/s00704-016-1762-5

Valík A, Brázdil R, Zahradníček P, Tolasz R, Možný M, Řezníčková L (2019) Measurements of sunshine duration by automatic sensors and their effects on the homogeneity of long-term series in the Czech Republic. Clim Res 78:83–101. https://doi.org/10.3354/cr01564

Verbout SM, Brooks HE, Leslie LM, Schultz DM (2006) Evolution of the U.S. tornado database: 1954–2004. Weather Forecast 21:86–93

Part II
Long-Term Climate Change

Chapter 5
Climate Change Before Instrumental Measurements

Magdalena Opała-Owczarek, Tadeusz Niedźwiedź, Rajmund Przybylak, and Wojciech Tylmann

Abstract The chapter "Climate change before instrumental measurements" is a review and compilation of papers concerning the reconstruction of Poland's climate in the last millennium. Data for this period are gaining importance due to their comparability with modern instrumental data, which is possible thanks to their high temporal resolution (annual, seasonal). The authors compiled the research results of climate reconstruction based on the following available data sources: direct man-made observations of weather and early instrumental measurements, dendrochronological records and varved sediment records. These three types of material are so far the best-known and best-developed proxy sources of past climate information, going back centuries in Poland. Generally, the chapter is divided into three parts presenting the results of climate reconstruction made with the above-mentioned data sources. On the basis of each method, the reconstructions of both air temperature and precipitation (including extreme rainfall and drought) for the winter and summer seasons are presented.

M. Opała-Owczarek (✉) · T. Niedźwiedź
Faculty of Natural Sciences, Institute of Earth Sciences, University of Silesia in Katowice, Katowice, Poland
e-mail: magdalena.opala@us.edu.pl

T. Niedźwiedź
e-mail: tadeusz.niedzwiedz@us.edu.pl

R. Przybylak
Department of Meteorology and Climatology, Faculty of Earth Sciences and Spatial Management, Nicolaus Copernicus University, Toruń, Poland

Centre for Climate Change Research, Nicolaus Copernicus University, Toruń, Poland

R. Przybylak
e-mail: rp11@umk.pl

W. Tylmann
Environmental Change Reconstruction Laboratory, Faculty of Oceanography and Geography, University of Gdansk, Gdańsk, Poland
e-mail: wojciech.tylmann@ug.edu.pl

M. Falarz (ed.), *Climate Change in Poland*, Springer Climate,
https://doi.org/10.1007/978-3-030-70328-8_5

Introduction

The main obstacle to climatological analyses of long-term variability and change of climatic conditions in Poland in the last millennium is the shortness and scarceness of instrumental meteorological measurements. Although there are series of data dating back even to the end of the eighteenth century, this only applies to individual locations. The oldest and longest continuous instrumental series of air temperature measurements in Poland come from the following meteorological stations: Warszawa (1779), Wrocław (1791) and Kraków (1792). Measurements of precipitation started in Wrocław in 1791 and in Warszawa in 1803. Recently, Wrocław's series was prolonged until 1781 based on precipitation measurements in Żagań (formerly Sagan) within the Mannheim network of stations established for Europe and North America by the Palatine Meteorological Society in 1780 (Przybylak et al. 2020). In regional analyses of long-term climate variability, it is therefore essential to use high-resolution proxy data, as they can significantly prolong meteorological records over time. Each type of proxy data has its advantages as well as shortcomings and are characteristic to different types of environments (Bradley 2015). The high-resolution proxy archives in Poland mainly cover the period of the last millennium and comprise historical documents on extreme weather events and weather-related disasters together with tree-ring analyses. More recently, the picture of climate change in Poland before instrumental measurements has become more detailed due to the development of the network of sites with varved sediment data. These three types of material are so far best-known and best-developed proxy sources of past climate going back centuries and the last millennium in Poland (Fig. 5.1).

This chapter addresses climate change in Poland before instrumental measurements (last 1000 years). The last millennium was chosen as the period of analysis, for which we have a relatively large number of proxy data with annual resolution, in contrast to the preceding periods. Data for this period are gaining importance due to their comparability with modern instrumental data, which is possible thanks to their high temporal resolution (annual, seasonal). Furthermore, this important period in paleoclimatology is relatively well known in various parts of the world, thanks to which the results are comparable.

A great number of reconstructions of climatic conditions for Europe based on proxy data on millennia scale allow us to distinguish several characteristic climatic periods (Lamb 1977; Grove 1988; Brázdil and Kotyza 1995; Crowley 2000; Crowley and Lowry 2000; Bradley et al. 2003; Brázdil et al. 2005; Xoplaki et al. 2005; Büntgen et al. 2006, 2007; Esper and Frank 2009; Jones et al. 2009; Brázdil and Dobrovolný 2010; Büntgen and Tegel 2011; Büntgen et al. 2011c; Przybylak 2011; Ljungqvist et al. 2012). These periods are usually also visible in the data for Poland.

During the last millennium after a cold period of about 300–600 AD (Migration Period) (Büntgen et al. 2011c), four distinctly marked thermal periods occurred in the area of Central Europe and Baltic Sea Region (Niedźwiedź et al. 2015). Medieval Warm Period (MWP) or Medieval Climate Anomaly (MCA) covers the period from 900 AD to 1350 AD, with the warmest phase between 1150 AD and 1250 AD.

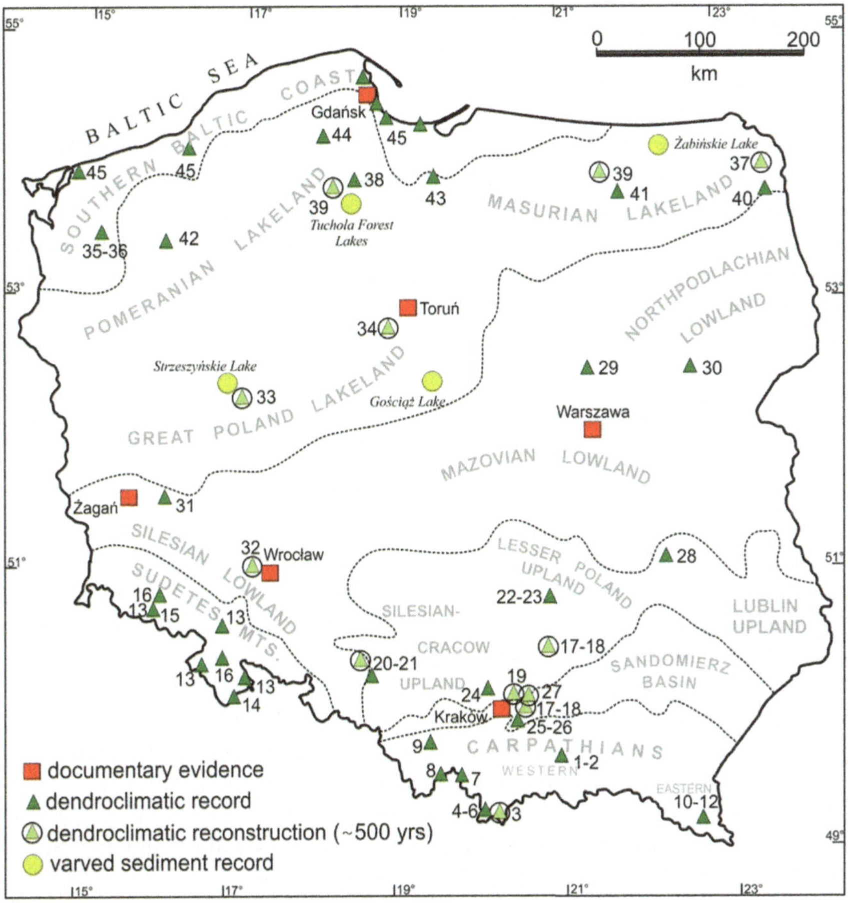

Fig. 5.1 Location of the selected proxy series described in the text: documentary evidences, dendrochronological data (numbering according to Table 5.2) and varved sediment records. Detailed information on the datasets are presented in the text

During the following 200 years (1350–1550 AD) there was great variability of intra-seasonal climatic conditions with tendency to cooling. This was the Transitional Period (TP) between MWP and the next Little Ice Age (LIA) period (Brázdil et al. 2005). The Little Ice Age (LIA 1550–1850) started with significant cooling in the period 1569–1579. In the Baltic Sea Region, there were found four cold and three relatively warm winter periods during LIA (Eriksson et al. 2007). The major warm period occurred in the years 1707–1750 with the warmest winter 1723/1724. The cool phases coincided with the Late Maunder Minimum (1675–1715) and the Dalton Minimum (1790–1840) in solar activity (Luterbacher et al. 2001; Eriksson et al. 2007). The Contemporary Warm Period (CWP) started after 1850, according to a majority of scientists (e.g. Grove 1988), or in the last decade of the nineteenth century (e.g. Lamb 1977), but the rapid increase of air temperature in Central Europe fell after 1988.

A review of paleoclimatological research results for Poland was previously discussed by Przybylak et al. (2010a). Ten years later, Polish paleoclimate science has made advances: available datasets are much improved. They are more quantitative and better spatially represented. After a brief overview of climate proxy data methods, this chapter examines the available man-made and natural proxy records, used to derive quantitative estimates of past climate in Poland (mainly temperatures and precipitation) via statistical methods calibrated against their modern distribution and associated climate.

Documentary Evidence

Documentary evidence, according to Pfister's (1999) proposition, include both direct man-made observations of weather or measurements of weather variables and indirect indications of weather obtained based on observations (plant phenology, yield of vine, etc.) or measurements (water levels, snow cover, etc.) (Table 5.1). For the reconstruction of Poland's climate in the last millennium before the nineteenth century (hereinafter called the historical period), the mentioned indirect indications of weather were not used. Thus, in this review, the most up-to-date knowledge about Poland's climate in the historical period will be based on the first group of sources, i.e. direct man-made observations and measurements. Reconstruction of climate based on direct man-made observations of weather is available for nearly the entire historical period, while using measurements of weather variables mainly for the eighteenth century. For this reason, results of climate reconstructions will be summarised and presented separately for the mentioned types of sources.

Out of all the proxy data available for the area of Poland (biological, historical, geophysical), documentary evidences were most often used to reconstruct its historical climate (for details see e.g. Przybylak et al. 2004, 2005, 2010b; Przybylak 2010, 2011, 2016). Types and quality of documentary sources were described in a detailed way by Przybylak et al. (2004, 2010b) and therefore are omitted here. In those publications, extended reviews of available literature are also given until 2009. For this reason, the review in this section is presented only for the period 2010–2019.

In the mentioned period a few new, important papers were published. A majority of them present results of air temperature instrumental measurements conducted in the eighteenth century. Przybylak and Pospieszyńska (2010) analysed air temperature changes measured by David von Grebner in Wrocław from 1710 to 1721. It is the oldest available long series of instrumental measurements in Poland. In recent years, two other air temperature series were also the subject of analysis, a shorter one (5 years: 1760–1764) from Toruń (Pospieszyńska and Przybylak 2010) and a longer one (1781–1792, Mannheim network) from Żagań (Przybylak et al. 2010c, 2014a). Three papers are devoted to the description of climate in Gdańsk. Two of them (Przybylak et al. 2014b; Filipiak et al. 2019) are based on weather notes published

Table 5.1 Types of data used for reconstructing past weather and climate (after Pfister 1999, modified)

Archives	Natural		Man-made		
Information					
Direct observation of weather and climate or instrumental measurement of meteorological parameters			Documentary	Observed • Anomalies • Natural hazards • Weather situations • Daily weather	Measured • Barometric pressure • Temperature • Precipitation • Water-gauge, etc.
Indirect references: (Proxy data) indication of controlled or affected processes through meteorological parameters	Organic • Tree rings • Fossil • Animal and plant remains • Fossil wood (trees), etc.	Non-organic • Ice-cores • Varves pollen • Terrestrial sediments • Temperature of boreholes • moraines, etc.		Organic • plant phonology • Yield of vine • Time of grain and vine harvest • Sugar content of wine etc.	Non-organic • Water levels • Snow fall • Freezing of water bodies • Snow cover, etc.
				• Cultural: • Pictorial	Rogations • Epigraphical
			Material:	• Archeological remains	

by Gottfried Reyger (1770, 1786), covering the period from December 1721 to June 1786. According to our knowledge, this is the longest series of weather observations conducted continuously by a single man. For comparison purposes in both papers, annual and seasonal mean values of air temperature (1739–1785) and totals of precipitation (1735–1772) have been presented. Air temperature data have been derived from the regular instrumental observations made in Gdańsk by Michael Christian Hanov in the period of 1739–1759 and by Johann Eilhard Reinick during the years 1760–1785. In the case of precipitation, all data comes from Hanov's observations. The third paper (Filipiak and Miętus 2010) analyses a 10-year-long (1752–1761) series of daily means of atmospheric pressure based on Reinick's measurements. Air temperature and precipitation changes in the central and north-eastern parts of the Polish–Lithuanian Commonwealth from 1656 to 1685 have been investigated by Przybylak and Marciniak (2010) using Chrapowicki's diary (Chrapowicki 1978, 1988) containing daily weather notes. More recently, documentary evidence has been utilised to reconstruct the occurrence of droughts in Poland in the period 1451–1800 (Przybylak et al. 2020).

Climate Reconstructions Based on Direct Man-Made Observations of Weather

Air Temperature

The 1001–1400 Period

For the first two centuries, the number of documentary evidence describing weather in Poland is too small to reliably reconstruct climate. Nonetheless, Maruszczak (1991) reconstructed air temperature for this period using available histories of air temperature from California, Greenland and the UK. It is important to note that the documentary evidence was only used for the reconstruction of air temperature in the UK. The reliability of Maruszczak's reconstruction is limited because the mentioned three areas are rather far from Poland and therefore often show different rhythms to changes in air temperature (see e.g. Fig. 34.2 in Bradley and Jones (eds.) (1995)), in which reconstructions of air temperature for Europe and North America are presented, or review papers written recently by Ljungqvist et al. (2012) and Pages 2k Consortium (2013).

The entire period (1001–1200) in Poland according to this reconstruction was warmer than the average millennial air temperature. Evidently, the warmest in the entire millennium was the second half of the twelfth century with the mean annual air temperature reaching 8.5 °C (Fig. 100 in Maruszczak [1991]). The first half of this century was slightly colder, while the eleventh century had air temperature slightly lower than 8 °C. For more details, see Przybylak (2016).

For the thirteenth and fourteenth centuries, a significantly greater density of historical sources exists than for the previous two centuries, but the available information is still of low reliability. They were collected mainly by Jan Długosz (1415–1480) and published in his chronicle *Annales seu cronici incliti regni Poloniae* (*Annals or Chronicles of the Famous Kingdom of Poland*). Weather descriptions available in this chronicle were used by Sadowski (1991) to calculate the frequencies of severe winters and hot summers, which indirectly enable estimation of the thermal characteristics of those seasons in the long-term perspective. Analysis of Fig. 1 in Sadowski's (1991) publication clearly indicates that the thirteenth century was a time of both the least severe winters and the least hot summers to a greater extent than at any other time from the fourteenth century onwards. That would mean in turn that the period was one of the greatest oceanicities of climate (see Fig. 6 in Sadowski [1991]). The thirteenth century evidently was colder than the fourteenth century. Maruszczak (1991) found opposite results, i.e. warmer conditions in the thirteenth century than in the fourteenth century. The results presented by Sadowski (1991) seem to be more reliable. Reconstructions of air temperature for Poland (Jan–Apr, Przybylak et al. 2005) and Scandinavia (Jun–Aug, Gouirand et al. 2008) based on tree rings confirm this conclusion, while simulated climate for the Baltic Sea Region by the regional model (Schimanke et al. 2012) do not.

The 1401–1800 Period

Since the fifteenth century, the amount of available documentary evidence is great enough to conduct quantitative reconstruction of air temperature. Such reconstructions for winter and summer seasons were conducted firstly for the period 1501–1840 (Przybylak et al. 2004, 2005) and later they were extended to include the fifteenth century (Przybylak 2011, 2016).

In addition to these long-term air temperature reconstructions, there are also available reconstructions for shorter periods (a couple of decades), but often with more detail (daily) (e.g. Bokwa et al. 2001; Limanówka 2001; Nowosad et al. 2007; Przybylak and Marciniak 2010; Przybylak et al. 2014b; Filipiak et al. 2019).

The history of winter and summer air temperature changes in the period 1401–1800 is presented in Fig. 5.2. It is clear that winters in all four centuries were markedly colder than in the twentieth century reference period 1901–1960. On average, the temperature in winter was lower by 2.2 °C. The exceptionally cold winters in the study period (anomalies oscillate between −3.5 and −4.0 °C) occurred in the decades 1451–1460, 1511–1520, 1541–1550, 1571–1580, 1701–1710 and 1741–1750. All decades (except 1521–1530) were colder than in the period 1901–1960 (black bars). The second warmest average air temperature in winter occurred in the decade 1621–1630. On average, the coldest winters were those of the sixteenth century (an anomaly then of −2.4 °C), while the warmest characterised the seventeenth century (anomaly −2.2 °C). The differences between the secular means are thus very small, which means that the long-term thermal character of winters in the entire study period was relatively stable, in spite of large year-to-year variability.

As can be concluded from Fig. 5.2, information about summer temperatures for Poland based on documentary evidence is more limited than for winter, which is also typical for other areas (see e.g. Brazdil and Kotyza 1995). This is generally a result of smaller number of weather notes available for this season in historical sources. Nonetheless, for a majority of the decades in the period 1401–1800, it was possible to reconstruct mean summer air temperature for Poland (Fig. 5.2). The summers in this period were primarily warmer than in the reference period 1901–1960. Only in seven decades this was not the case. Analysing Fig. 5.2, it is possible to distinguish three periods with the highest summer temperature, i.e. 1471–1500, the second half of the sixteenth century and 1611–1640. Positive anomalies in this time oscillate between 0.5 °C and 0.8 °C. On the other hand, the coolest summers (0.4–0.5 °C) occurred in the decades 1731–1740 and 1461–1470. Moreover, five decades (1521–1530, 1591–1600, 1671–1680, 1691–1700 and 1761–1770) were slightly colder (on average by 0.1 °C) than contemporary ones. Mean air temperature anomalies calculated for the centuries were always slightly positive, ranging from 0.1 °C in the eighteenth century to 0.3 °C in the sixteenth. Thus, we may conclude that a change of climate between the studied historical time and present time occurred mainly in winter. The observed warming of winters from historical to present times significantly reduced the thermal continentality of Poland's climate.

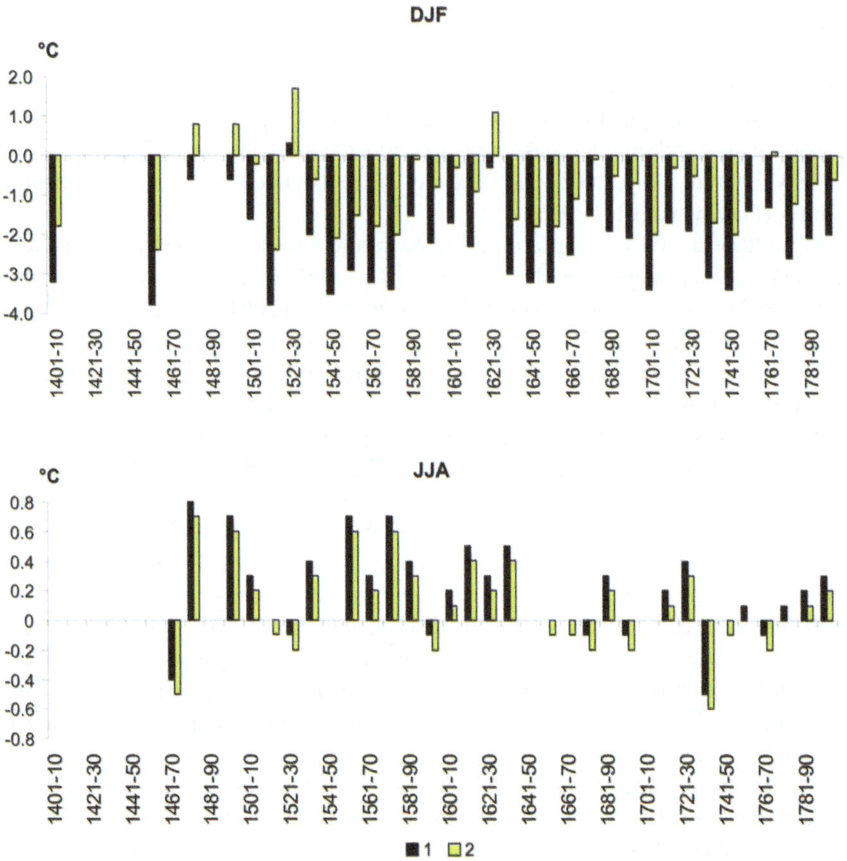

Fig. 5.2 Reconstructions of mean 10-year air temperatures (°C) in Poland from 1401 to 1800: **a** winter (DJF) and **b** summer (JJA). 1 and 2—anomalies with respect to 1901–1960 and 1789–1850 means, respectively (after Przybylak 2011)

For some periods there is information available regarding air temperature with greater time resolution. Here the example of air temperature changes and characteristics based on daily weather notes written by Jan Antoni Chrapowicki in his diary are presented for period 1656–1685. Figure 5.3 shows year-to-year changes in mean annual thermal indices for Northeastern Poland. It is clear to see that the entire period from 1656–1685 was colder than the historical norm (negative values of indices, Fig. 5.3). Five years with an index below −0.25 can be described as very cold: 1666, 1667, 1671, 1672 and 1679. Out of all years, the warmest one was evidently 1661 (index −0.03). The four years 1663, 1668, 1680 and 1681 were also relatively warm (index around −0.10). On average, spring and autumn were markedly colder, while summer and winter were near the historical norm (Fig. 5.4).

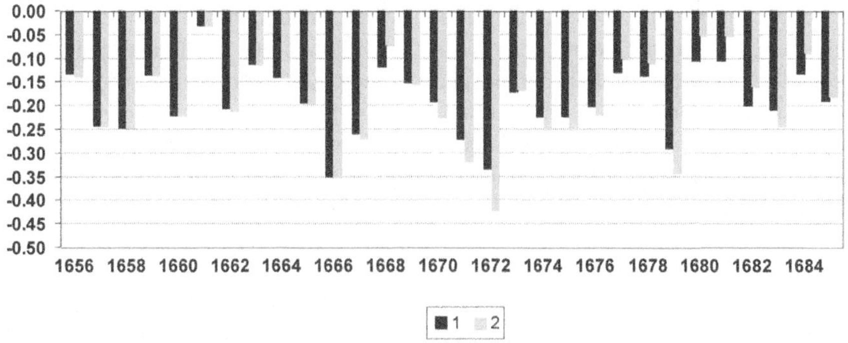

Fig. 5.3 Annual average values of thermal indices in the central and north-eastern parts of Poland from 1656 to 1685. 1—correction based on monthly data, 2—correction based on annual data (after Przybylak and Marciniak 2010)

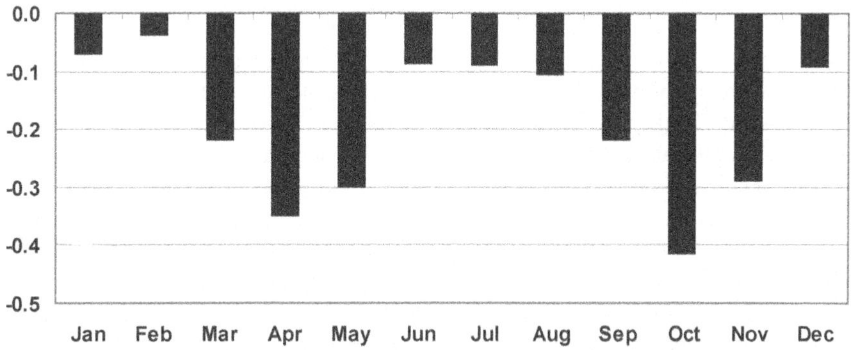

Fig. 5.4 Annual course of average values of thermal indices in the central and north-eastern parts of Poland from 1656 to 1685 (after Przybylak and Marciniak 2010)

The second very good example of similar character insight into air temperature is available for Gdańsk for the period 1722–1785, and is available in a paper recently published by Filipiak et al. (2019).

Precipitation and Droughts

The 1001–1800 Period

As it was written, the reconstruction of summer temperatures is more difficult than of winter temperatures, but reconstruction of atmospheric precipitation based on documentary evidence is certainly the most difficult out of all meteorological variables. There are two main reasons for this fact. Firstly, the influence of atmospheric precipitation on the life and activity of people in the temperate zone is smaller than e.g. of air temperature. Hence, the availability of weather notes in historical sources describing precipitation conditions is also smaller than that of thermal conditions. Secondly, to describe reliably atmospheric precipitation we need more data than for air temperature, due to fact that atmospheric precipitation is a meteorological element most variable in time and space. As a result, there is a limited number of works of reconstructed precipitation in Poland in the pre-instrumental period based on documentary evidence. Generally, there is only one work (Przybylak et al. 2004) in which a reconstruction of this kind is available for period 1501–1840. The second one (Maruszczak 1991), which presents an even longer history (back to the eleventh century), is, however, a kind of extrapolation of humidity (precipitation) conditions in Poland based on the reconstructed flow of Dniepr River (Ukraine) using both historical and sedimentological data. Thus, it is most likely not very reliable. In addition, there are also papers available which present information concerning precipitation, but only for short, isolated historical periods (Bokwa et al. 2001; Limanówka 2001; Nowosad et al. 2007; Przybylak et al. 2008; Przybylak and Marciniak 2010; Przybylak et al. 2014a; Filipiak et al. 2019).

A short description of atmospheric precipitation is described here based on the first two mentioned works. Based on Maruszczak's (1991) reconstruction of the entire history of precipitation in pre-instrumental period, the two wettest periods can be distinguished. Firstly, the wettest occurred in the years 1100–1250, while the second one in the eighteenth century, but mainly in its second half. On the other hand, the two driest 50-year periods occurred in 1301–1350 and 1551–1600. In the other 50-year periods precipitation conditions oscillated near the millennial-long norm.

For the period 1501–1840 we have more precise information, but still it is not fully satisfactory, in particular for wintertime. Extremely wet winters were noted most frequently in the first halves of the sixteenth and eighteenth centuries, while extremely dry winters mainly in the sixteenth century, in particular, however, in its first half (see Table 1 in Przybylak et al. 2004). It is possible to find significantly more information in documentary evidence about precipitation occurrence in summer than in winter. Therefore, Przybylak et al. (2004) presented the frequency of occurrence of extreme high and low precipitation in summer in the graphical form (Fig. 5.5). This figure depicts that extremely wet and very wet winters (indices 3 and 2, respectively) were most common in the sixteenth century, in particular in its first half. Most wet decades were 1501–1510 and 1561–1570 in which the respective five and four years were extremely wet and very wet. Apart from the sixteenth century, two more periods can be distinguished with highest precipitation (1641–1670 and 1721–1750). In the second period it must be stressed that in particular the last decade (1741–1750) was extremely wet and had, similarly as the decade 1501–1510, five extremely wet and very wet summers (Fig. 5.5a).

Extremely dry and very dry summers (indices −3 and −2) in Poland in the period 1501–1840 occurred most often in the sixteenth century (1531–1590) and in all the analysed decades in the nineteenth century (Fig. 5.5b). In addition, quite a lot of very dry summers were also noted in the period from the mid-seventeenth century to the mid-eighteenth century.

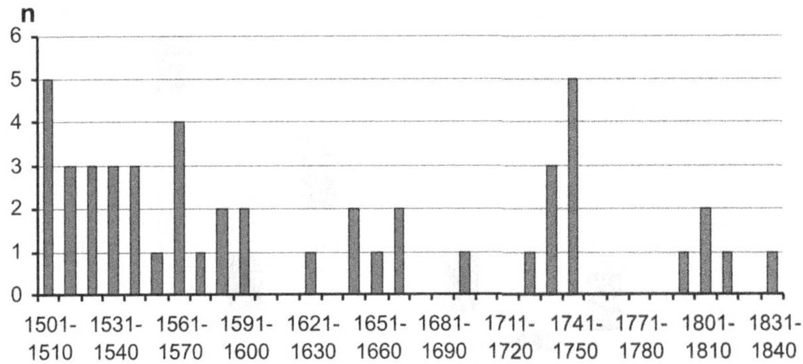

a)

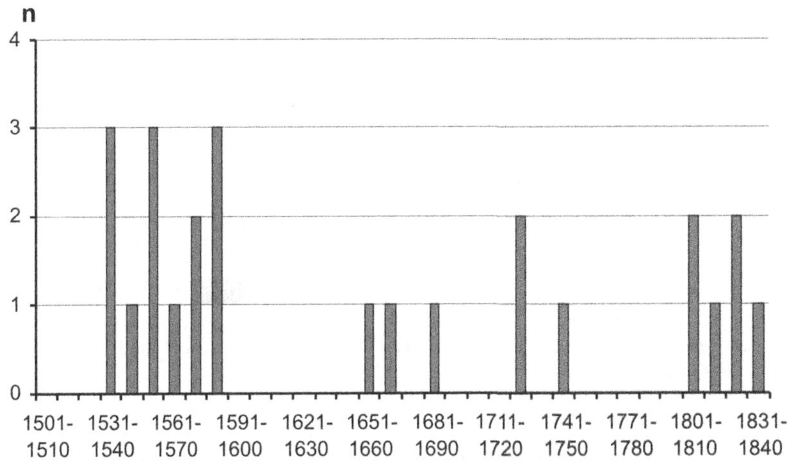

b)

Fig. 5.5 Decadal frequencies of occurrence of summers (JJA) that were: **a** extremely wet and very wet (indices 3 and 2) and **b** extremely dry and very dry (indices −3 and −2) in Poland between 1501 and 1840 (after Przybylak et al. 2004)

Recently, the occurrence of droughts in Poland in the period 1451−1800 was studied based on old (used for investigation presented by Przybylak et al. 2004) and new historical sources collected after 2004 (Przybylak et al. 2020). Their mean annual number for 50-year periods stratified into three categories is shown in Fig. 5.6. Evidently, the highest number of droughts occurred in the second halves of the eighteenth and the seventeenth centuries, respectively. What is also very interesting is that in the second half of the sixteenth century more droughts occurred than in its first half. However, the difference in drought occurrence between them is not particularly large as in the case of the analogical periods in the seventeenth and eighteenth centuries. In the entire study period, the fewest droughts were noted in

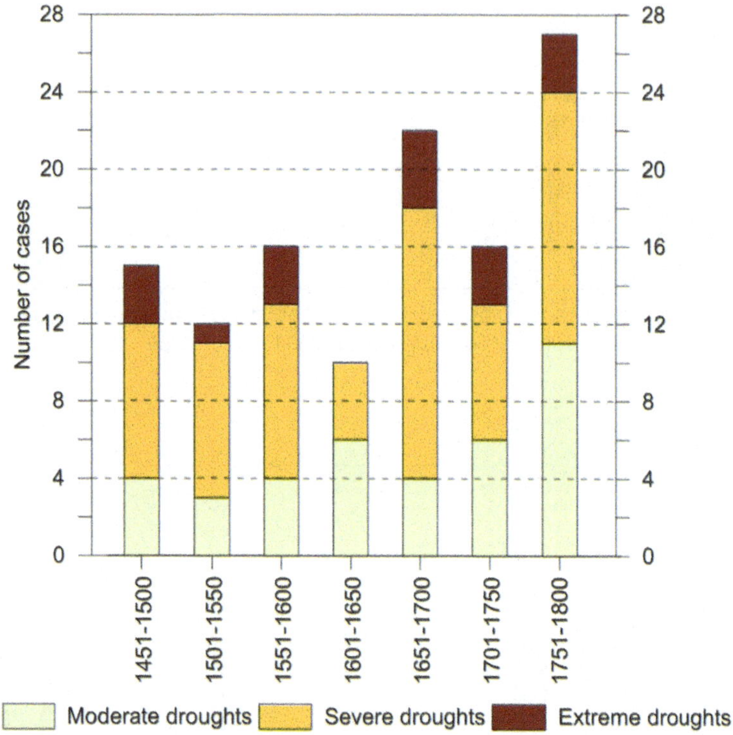

Fig. 5.6 Frequency of occurrence of three categories of droughts in Poland in 50-year periods, 1451–1800 (after Przybylak et al. 2020)

the first half of the seventeenth century, when the category of extreme droughts did not occur at all.

As in the case of air temperature, more detailed precipitation information is available for shorter periods: 1656–1685 (Chrapowicki's observations) and 1722–1785 (Reyger's observations). Precipitation is characterised for these periods by using the number of days with this phenomenon. Again, as in the case of air temperature, analysis is presented only for the first period, while the second mentioned period is described in detail in the paper published by Filipiak et al. (2019).

In the period 1656–1685, the annual number of days with precipitation in central and Northeastern Poland varied from 112 in 1676 to 206 in 1679 (Fig. 5.7). The 30-year mean number of days with precipitation was 169.6, and is very similar to the number of days with precipitation in the contemporary climate. There is also a positive correlation between mean monthly numbers of precipitation in historical (1656–1667) and present times. Differences do not exceed 5 days being positive in the warm half of the year and negative in the cold half of the year. In the annual course of the average number of days with precipitation (1656–1685), two maxima are observed, in summer and winter (from 15 to 17 days a month, except February)

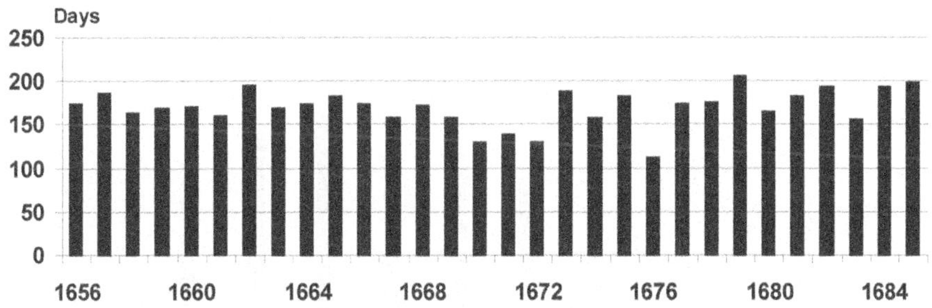

Fig. 5.7 Annual number of days with precipitation in the central and north-eastern parts of Poland, 1656–1685 (after Przybylak and Marciniak 2010)

(Fig. 5.8a). Notably the lowest number of days with precipitation (only about ten) occurred in March and April. Rainfall was noted in every month of the year, and was the dominant type of precipitation from April to November (Fig. 5.8b). The greatest average number of days with rainfall in the 10-year period (1658–1667) occurred in July (17.5), while the lowest was in January (3 days). In line with expectation, snowfall dominated from December to March with the highest number of days (>10) in December and January.

Climate Reconstruction Based on Early Instrumental Meteorological Measurements

A very detailed history of early instrumental measurements in Poland in its present and historical boundaries since the seventeenth century is presented in Chapter 5 (Przybylak 2010) of the book entitled *The Polish Climate in the European Context: An Historical Overview*. In this chapter, in Table 5.1, the reader can find 24 identified and isolated early instrumental series of meteorological observations in Poland in the seventeenth and eighteenth centuries with a wide spectrum of metadata information. For some locations, the only information preserved is that meteorological observations were conducted, but the data were lost or destroyed and are not available. The longest available series which can be used (some of them were already used) for climate analysis come from Wrocław (1710–1730), Warszawa (1725–1728, 1760–1763, 1779–1799), Gdańsk—six series covering the period from 1739 to 1812, Toruń (1760–1764) and Żagań (1781–1792). As can be seen, all of them cover mainly the eighteenth century.

A description of the climate of Wrocław in the years 1710–1721 based on series of data measured by David von Grebner was carried out by Przybylak and Pospieszyńska (2010). Various aspects of air temperature changes are presented in the paper. The greatest number of works was published using instrumental data (air temperature, atmospheric precipitation and atmospheric pressure) available for Gdańsk for the eighteenth century (Filipiak 2007a, b; Filipiak and Miętus 2010; Przybylak et al.

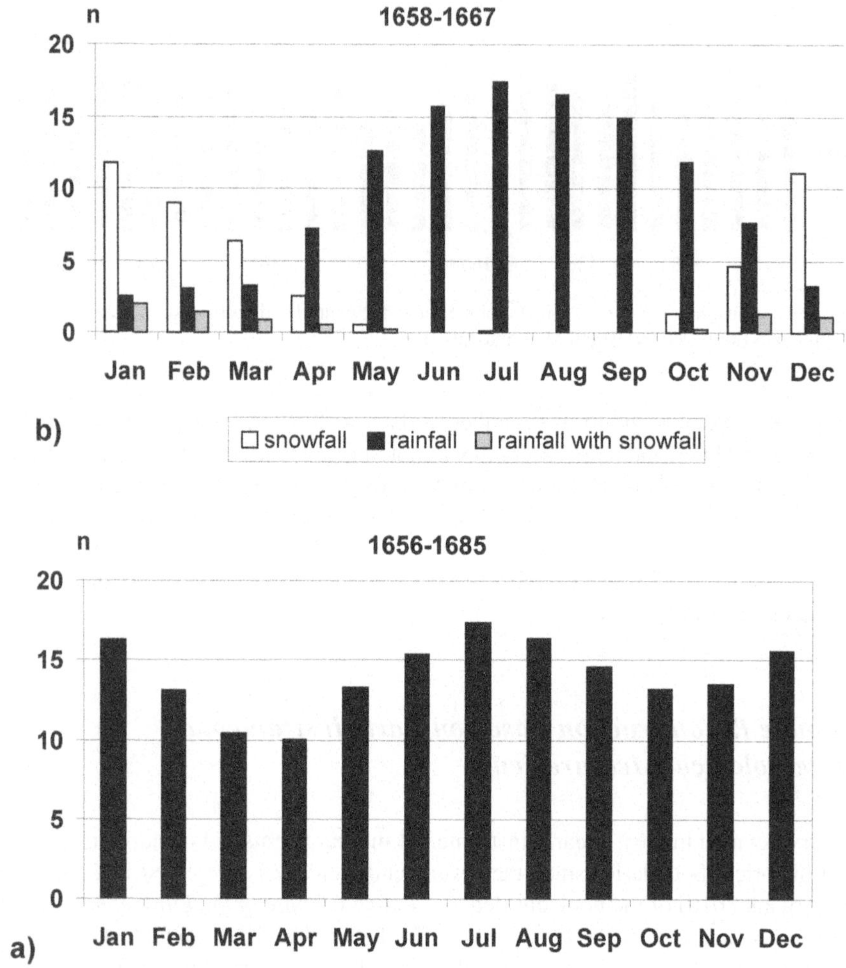

Fig. 5.8 Annual course of average number of days (n) with precipitation **a** as well as with snowfall, rainfall and rainfall with snowfall **b** in the central and north-eastern parts of Poland from 1656 to 1685 (after Przybylak and Marciniak 2010)

2014b; Filipiak et al. 2019). The mean daily air temperature in Warszawa was elaborated for the periods 1760–1763 (Michalczewski 1988) and 1779–1983 (Michalczewski 1985). In turn, Rojecki (1965) analysed instrumental observations from Toruń (1760–1762) and compared them to those available for Warszawa. Recently, Pospieszyńska and Przybylak (2010) made a comprehensive analysis of all available data from Toruń for the eighteenth century (1760–1764). The most reliable measurements of meteorological variables in the eighteenth century were likely carried out in Żagań within the network of meteorological stations called the "*Societas Meteorologica Palatina*". The results of the observations were published *in extenso* between

1783 and 1795 in volumes II–XIII of the *Ephemerides Societatis Meteorologicae Palatinae* in Mannheim. As a whole, air temperature data from Żagań were the subject of investigation by Przybylak et al. (2010b, 2014a).

To ensure the completeness of the review, it is crucial to note that in the last decade of the eighteenth century continuous meteorological instrumental measurements started in Wrocław in 1791 and in Kraków in 1792. Three series of meteorological variables covering the entire period of observations in Kraków were described in the following papers: Trepińska (1997, atmospheric pressure), Trepińska and Kowanetz (1997, air temperature), and Twardosz (1999, atmospheric precipitation). Similarly for Wrocław, air temperature and precipitation series for the periods 1791–2007 and 1791–2010 were elaborated by Bryś and Bryś (2010a) and Przybylak et al. (2013), respectively. In the first paper homogenised series of mean (total) monthly air temperature (atmospheric precipitation) values are also attached. What is more, Bryś and Bryś (2010b) present a very detailed history of air temperature measurements in the area of Wrocław in the period 1791–1890 and introduced corrections to the mean monthly data. They also attached daily mean temperatures for the period 1791–1800 after Galle (1857). Galle, in the mentioned publication published Wrocław air temperature data recorded in the years 1791–1854 at the Breslau-Sternwarte Observatory (φ = 51°06'56.5"N, λ = 17°02'10.6"E, H station = 118.0 m a.s.l, H of thermometer = 146.7 m a.s.l.). The following statistics are available for readers: mean daily, monthly and annual values, as well as extreme temperature values. In his next work (Galle 1879), the air temperature series was extended to 1875.

Dendrochronological Records

Tree Rings Definition, Formation and Climatic Potential

Annual growth rings are widely used in palaeoclimatic studies of the late Holocene, and especially of the last two thousand years. Tree rings are composed of individual cells which are the result of a complex sequence of assimilation of natural resources by the tree (Speer 2010). They constitute unique research material due to: the possibility of precise dating in calendar years, the longevity of some tree species, the possibility to construct composite tree-ring chronologies using the cross-dating procedure, and clear climatic signal recorded in specimens from sites where climate is the main growth limiting factor. According to the theoretical foundations of dendroclimatology, the growth limiting factor will be the one whose variability is recorded in annual rings and which can be reconstructed on this basis (Fritts 1976). Dendroclimatology, the science that links tree-growth patterns to climate variability, has mainly focused on trees, shrubs and dwarf shrubs growing at altitudinal or latitudinal limits, however research studies in other areas have also been conducted. Trees respond to their surroundings, thus variations in temperature, rainfall, soil moisture, wind stress and other elements can be reconstructed.

The development of dendroclimatological records usually consists of the following steps: (1) collection of samples from living trees using an increment borer or from historical buildings/archeological objects; (2) laboratory preparation including sanding the surface of samples mounted on wooden holders; (3) measurements of scanned samples or using measuring tables; (4) visual and statistical dendrochronological dating of sequences, including the procedure of cross-dating so as to link sequences from different time periods; (5) development of chronology with sufficient samples replication; (6) testing the sensitivity of the ring-width chronology to climate (usually grid point or meteorological station monthly temperature and precipitation records) via correlation and response functions; (7) calibration and verification of the transfer function based on linear regression; (8) reconstruction of climate and determination of degree of uncertainty. Statistical analysis applied in dendroclimatic reconstructions has become increasingly complex and sophisticated. Detailed descriptions of these procedures are presented by Fritts (1976), Hughes et al. (1982, 2010) and Cook and Kairiukstis (1990).

On the whole, dendroclimatic reconstructions can be considered in various spatial and temporal scales. They start with a site-level analysis of various tree species' response to climate. Trees should be sampled in climate-sensitive sites, avoiding obvious damage and visible effects of microenvironmental factors. Tree response to climate from many sites across the region can be used to identify the differentiation of climate variables that affect tree growth. A network of chronologies can be applied to reconstruct broad-scale climatic variability, also from the perspective of atmospheric circulation (Speer 2010).

Distribution of Dendroclimatological Records in Poland

Thanks to the development of dendrochronological research since the mid-twentieth century, the spatial extent of tree-ring data is quite even and covers most of the area of Poland. However, the majority of the dendrochronological records use growth-ring width sequences originating from living trees that grow nowadays, and therefore the time range of these data covers the last 100 years up to about 250 years (Table 5.2). Over the past few decades progress in dendrochronological research has led to the creation of chronologies for most native and introduced species occurring in Poland (for details see review written by Zielski et al. (2010), where all available chronologies until 2009 are listed in Table 7.1). From a large number of published papers, in this chapter we have focused only on the research that investigated the relationships of growth with climate and potentially applicable in climate reconstructions.

So far, the most attention was devoted to the study of dendroclimatology of Scots pine (*Pinus sylvestis* L.) (e.g. Zielski 1997; Feliksik et al. 2000; Cedro 2001; Wilczyński et al. 2001; Wilczyński and Skrzyszewski 2002a, 2002b; Przybylak et al. 2005; Szychowska-Krąpiec and Krąpiec 2005; Szychowska-Krąpiec 2010; Zielski et al. 2010; Koprowski et al. 2012; Opała and Mendecki 2014; Opała 2015; Balanzategui et al. 2018), which clearly indicate the influence of winter (previous year

Table 5.2 Characteristics of the selected dendroclimatological records available for Poland

No	Site/Region	Time span	Type of material	Climatic signal	References
Southern Poland (Mountains and Uplands)					
1	Polish Carpathian Mountains	1743–2015	tree-ring width of *Larix decidua*	temp. May	Danek et al. (2017)
2	Polish Carpathian Mountains	1900–1998	tree-ring width of *Pinus sylvestis*	temp. Feb–Aug prec. Mar–Apr, Jun–Aug	Wilczyński et al. (2001) and Wilczyński and Skrzyszewski (2002b)
3	Tatra Mountains	1550–2004	composite tree-ring width of *Pinus cembra, Picea excelsa, Picea abies*	temp. Jun–Aug	Ermich (unpublished) after Bednarz and Muter (2010), Feliksik (1972), Bednarz 1976, 1984), Schweingruber et al. (1979), and Niedźwiedź (2004, 2010)
4	Tatra Region	1661–2004	composite tree-ring width of *Picea abies, Pinus cembra, P. mugo*	temp. Jun–Jul	Savva et al. (2006), Büntgen et al. (2007), and Kaczka et al. (2016)
5	Tatra Region	1709–2004	maximum latewood density of *Picea abies*	temp. Apr–Sep	Büntgen et al. (2007)
6	Tatra Mountains (~1000 m a.s.l.)	1700–?	tree-ring width of *Pinus sylvestris*	summer drought, summer humidity	Büntgen et al. (2012) and Kaczka et al. (2016)
7	Żywiec Beskid, Babia Gora	1650–1910 1704–2012	tree-ring width of *Pinus cembra, Picea abies*	temp. Jun–Jul	Bednarz (1996), Bednarz et al. (1999), and Kaczka et al. (2015)
8	Żywiec Beskid, Mt Pilsko area	1641–1995 1627–2004	tree-ring width of *Picea abies*	temp. Jun–Jul	Szychowska-Krąpiec (1998) and Kaczka and Büntgen (2006)
9	Little Beskid, Żywiec Beskid, Silesian Beskid	1900–2000	tree-ring width of *Picea abies*	temp. winter/spring	Feliksik and Wilczynski (2000) and Wilczyński and Feliksik (2004, 2005)
10	Bieszczady Mountains	1840–2010	tree-ring width of *Fagus sylvatica*	temp. winter, drought Jun–Jul	Chojnacka-Ożga and Ożga (2014)
11	Bieszczady Mountains	1879–2014	tree-ring width of *Larix decidua*	temp. May	Danek and Chuchro (2016)

(continued)

Table 5.2 (continued)

No	Site/Region	Time span	Type of material	Climatic signal	References
12	Bieszczady Mountains	1926–2010	tree-ring width of *Picea abies*	temp. Jul	Kaczmarczyk and Kaczka (2014)
13	Sudetes Region	1825–2014	tree-ring width of *Picea abies*	temp. Apr–Jul	Opała-Owczarek et al. (2019)
14	Śnieżnik Massif	1827–2013	tree-ring width of *Picea abies*	temp. May–Jul	Opała and Owczarek (2016)
15	Karkonosze, Mt Szrenica	1924–2016	tree-ring width of *Pinus mugo*	temp. Jun–Aug	Migała et al. (2000) and Opała-Owczarek et al. (unpublished)
16	Klodzko and Jelenia Gora mid-mountain basins	1930–1990	tree-ring width of *Pinus sylvestis*	temp. Feb–Mar, Jul–Aug, prec. Jul	Wilczyński et al. (2001) and Wilczyński and Skrzyszewski (2002a)
17	Lesser Poland region	1091–2006	tree-ring width of *Pinus sylvestis*	temp. pDec–Mar	Szychowska-Krąpiec (2010)
18	Lesser Poland Region	1109–2004	tree-ring width of *Abies alba*	temp. pDec–Mar, Jul–Aug	Szychowska-Krąpiec (2010)
19	Lesser Poland Region	910–1997	tree-ring width of *Quercus* sp.	prec. Jun–Jul	Krąpiec (1998)
20	Upper Silesia	1568–2010	tree-ring width of *Pinus sylvestis*	temp. Feb–Mar	Opała and Mendecki (2014) and Opała (2015)
21	Upper Silesia	1739–2010	tree-ring width of *Quercus* sp.	temp. pDec–Jan	Opała and Mendecki (2014)
22	Holy Cross Mountains	1870–2005	tree-ring width of *Abies alba*	temp. Mar, drought Jun–Jul	Bronisz and Bronisz (2010)
23	Holy Cross Mountains	1931–1994	tree-ring width of *Pinus sylvestis, Abies alba, Fagus sylvatica*	temp. Feb–Mar	Feliksik et al. (2000) and Wilczyński et al. (2001)
24	Cracow Upland	1861–2008 1882–2008 1898–2008	tree-ring width of *Fagus sylvatica, Pinus sylvestis, Abies alba*	temp. Feb, prec. Jun–Jul; temp. p Dec–Mar, prec. Jun; temp. Feb–Apr, prec. Jun	Szychowska-Krąpiec and Krąpiec (2002) and Opała (2009)

(continued)

Table 5.2 (continued)

No	Site/Region	Time span	Type of material	Climatic signal	References
25	Sandomierz Basin	1900–2003	hydrogen isotope ratios in á-cellulose of *Pinus sylvestris*, *Quercus robur*	temp. summer, precip. summer	Szczepanek et al. (2006) and Treydte et al. (2007)
26	Sandomierz Basin	1881–1985	tree-ring width of *Quercus robur*	prec. May–Aug	Bednarz and Ptak (1990)
27	NE Cracow	~1100 yrs	carbon isotope ratios in á-cellulose of *Quercus*	prec. May–Jul	Krąpiec et al. (1998) and Jędrysek et al. (2003)
28	Lesser Poland, Lublin Upland	1790–2000	tree-ring width of *Picea abies*	temp. Feb–Mar, prec. Jun	Koprowski and Zielski (2006)
Central Poland (Lowlands)					
29	northern Mazovian Lowland	1796–1992	tree-ring width of *Quercus* sp.	prec. Jun–Aug	Bronisz et al. (2012)
30	Mazovian Lowland, Podlachian Lowland	1790–2000	tree-ring width of *Picea abies*	temp. May, drought Jun	Koprowski and Zielski (2006)
31	Silesian Lowland	1897–2006	tree-ring width of *Picea abies*	temp. pJun–Sep, prec. Feb, May-Jul	Koprowski (2013)
32	Silesian Lowland	780–1994	tree-ring width of *Quercus* sp.	prec. Apr–Jul	Krąpiec (1998)
Northern Poland (Lakelands and southern Baltic coast)					
33	Greater Poland	449–1994	tree-ring width of *Quercus* sp.	prec. Apr–Jul	Krąpiec (1998)
34	Kuyavian-Pomerania	1168–2000	tree-ring width of *Pinus sylvestis*	temp. Feb–Mar, prec. May–Jul	Zielski (1997) and Koprowski et al. (2012)
35	Western Pomerania	1757–2000	tree-ring width of *Pinus sylvestis*	temp. winter/spring, prec. Jun–Jul	Cedro (2001)
36	Western Pomerania	1789–2004	tree-ring width of *Quercus* sp.	prec. pOct, pDec, Feb, Jun	Cedro (2007)
37	Suwałki Region	1582–2004	tree-ring width of *Pinus sylvestis*	temp. winter	Szychowska-Krąpiec and Krąpiec (2005)
38	Tuchola Forest	1810–2010	earlywood vessel of *Quercus robur*	min. temp. winter	Pritzkow et al. (2016)
39	Masuria Lakes District, Tuchola Forest	1200–2010	tree-ring width of *Pinus sylvestis*	temp. Nov–Apr	Balanzategui et al. (2018)

(continued)

Table 5.2 (continued)

No	Site/Region	Time span	Type of material	Climatic signal	References
40	Suwalki Lakeland, Augustow Wilderness	1901–1998	carbon isotope ratios in latewood á-cellulose of *Pinus sylvestris*	temp. Jul–Aug	Pawełczyk et al. (2004) and Treydte et al. (2007)
41	Masurian Lakeland	1790–2000	tree-ring width of *Picea abies*	prec. May–Jul	Koprowski and Zielski (2006)
42	Pomerania Lakeland, Western Pomerania	1881–2004	tree-ring width of *Picea abies*	prec. Feb, May–Jul	Koprowski (2013)
43	Kuyavian-Pomerania	1893–2005	tree-ring width of *Picea abies*	prec. Feb, May–Jul, temp. Feb–Mar	Koprowski (2013)
44	Kaszubskie Lakeland	1914–2006	tree-ring width of *Abies alba*	temp. Jan–Feb	Bijak (2010)
45	Southern Baltic coast	1673–1982	tree-ring width of *Quercus*	temp. pAug–Sep, May, prec. pAug, pNov, Jun	Ważny and Eckstein (1991)

December to current year March) temperature on ring formation and also the role of summer precipitation. These climate/growth relations are similar throughout Poland. Regional differences in dendroclimatic response of pine trees have been presented by Zielski and Sygit (1998) and Wilczyński et al. (2001). Norway spruce (*Picea abies* Karst.) is also a frequently studied species, especially in the mountainous area of the Carpathian and Sudetes Mountains, where its growth-ring widths show clear temperature signal (temperature during spring/summer season, from April to September or from June to July) (e.g. Feliksik 1972; Bednarz 1976, 1984, 1996; Bednarz et al. 1999; Szychowska-Krąpiec 1998; Feliksik and Wilczynski 2000; Wilczyński and Feliksik 2004, 2005; Kaczka and Büntgen 2006; Savva et al. 2006; Büntgen et al. 2007; Zielski et al. 2010; Kaczmarczyk and Kaczka 2014; Kaczka et al. 2015, 2016; Opała and Owczarek 2016; Opała-Owczarek et al. 2019). In the lowlands and lakelands of Northern Poland, spruce tree rings mainly show dependence on February and May–July precipitation (Koprowski and Zielski 2006; Koprowski 2013). Among other coniferous species, intensive research was also carried out on silver fir (*Abies alba* Mill.) from different regions of Poland (Bijak 2010; Bronisz et al. 2010; Szychowska-Krąpiec 2010) and European larch (*Larix decidua* Mill.) (Danek et al. 2017; Danek and Chuchro 2016) proving mainly temperature signal (from previous December to March for fir, May for larch, respectively). Apart from the results presented in Table 5.2, less commonly occurring species (among others: juniper, yew, Douglas fir, plane-tree) were also investigated and presented in Polish dendroclimatological literature.

A very important species for dendrochronology and dendroarcheology is oak (*Quercus* sp.), for which the longest scales were constructed in the area of Poland. Review of these studies was included in the work of Krąpiec (1998). Existing dendroclimatological research demonstrated dependence of the annual growth width of oak upon precipitation in spring and summer (Bednarz and Ptak 1990; Ważny and Eckstein 1991; Cedro 2007; Bronisz et al. 2012; Pritzkow et al. 2016). However, the issue of climate reconstruction raises some concerns due to instability of climatic signal, unspecified local factors and diverse habitats of previous generations of oak trees (Krąpiec 1998).

Long-term dendrochronological records, exceeding the last 2–3 hundred years are the most valuable for research on climate variability and change, as they constitute unique research material for reconstructing past climatic conditions in the pre-instrumental period. Due to the lack of long-lived tree species growing in Polish territory, such unique records were constructed on the basis of historical and archaeological wood elements. For the area of Poland millennium-long composite chronologies exist for Lower Silesia (oak: Krapiec et al. 1998), Lesser Poland (oak: Krapiec et al. 1998, pine and fir: Szychowska-Krąpiec 2010), Greater Poland (oak: Krapiec et al. 1998) and Kuyavian-Pomerania region (pine: Zielski 1997) (Table 5.2). Records that go beyond the period of instrumental observations, although not covering the entire millennium, are also very valuable. Such data have been compiled for the Tatra Mountains (Bednarz 1976, 1984; Schweingruber et al. 1979; Niedźwiedź 2004), Upper Silesia (Opała 2015), the Suwalki Region (Szychowska-Krąpiec and Krąpiec 2005), the Masuria Lakes District and Tuchola Forest (Balanzategui et al. 2018).

Additionally, there is a number of floating chronologies, un-continued chronologies and chronologies which reach back into the past before the common era, but did not cover the recent time period and cannot be calibrated against recent climatic data, thus making their value for climate research limited (for details see: Zielski and Krąpiec 2004, Zielski et al. 2010).

Besides the most common type of proxy such as tree-ring width, there were attempts to use other tree-ring proxies: maximum latewood density (MXD), carbon and hydrogen isotopes ratios in latewood α-cellulose and early wood vessel parameters. Latewood density of coniferous trees is well recognised and established in dendroclimatological research (Schweingruber et al. 1979). It has been shown that density variation contains strong climatic signals, providing the best results for trees growing in extreme conditions—boreal and high elevation mountain forests. In Poland, the study made by Büntgen et al. (2007) facilitated reconstruction of April–September temperatures in the Tatra Mountains since the eighteenth century. A similar climatic signal (mean temperature from May to September), was extracted for Norway spruce growing in the Tatra Mountains from latewood Blue Intensity (BI), a new climate proxy that is less expensive in comparison to MXD (Kaczka et al. 2017). Both types of proxy, MXD and BI, correlate more strongly with climate than tree-ring width, however these differences mainly depend on the season (single months) of dendroclimatic reaction.

Isotopic variations (carbon and hydrogen isotopes ratios) in latewood α-cellulose of *Pinus sylvestris* tree ring from Augustów have been studied as a possible proxy of temperature (Pawełczyk et al. 2004; Treydte et al. 2007). While carbon isotope ratios in α-cellulose of *Quercus* sp. from the Vistula flood plain in the vicinity of Kraków and Niepołomice proved to be a sensitive indicator of summer precipitation (May–July) (Krąpiec et al. 1998; Jędrysek et al. 2003). Investigation of Szczepanek et al. (2006) demonstrates that the best climatic indicators are: hydrogen for summer temperature and oxygen for summer precipitation. It was also deduced that pine is a more sensitive indicator of environmental changes than oak.

Another attempt to improve weak climatic signal in oak species was done by Pritzkow et al. (2016), who developed wood anatomical chronologies of oak tree ring from Tuchola Forest. The authors proved that the average vessel area revealed significant positive correlations to minimum winter temperatures. The reconstruction indicates a promising direction and potential to reveal low-frequency climate information. All those other than ring-width proxies may reveal low-frequency climate information, and thus may help in further development of multicentennial dendroclimatic reconstructions in Poland.

Climate Reconstructions Based on Tree Rings

Summer Air Temperature

Dendroclimatic reconstruction of summer temperatures were provided by increment widths of Norway spruce from the highest part of the Polish mountains, in the vicinity of upper timberline. Growth-climate response analysis indicates that April–July temperatures are the main factor affecting radial growth of trees in the Sudetes Mountains and June–August temperatures in the Tatra Mountains. Reconstructions of tree-ring-based temperature variability presented here covers *circa* the last 450 years for the Tatras (Niedźwiedź 2004) and the last 200 years for the Sudetes Mountains (Opała-Owczarek et al. 2019) (Fig. 5.9). Results covering the summer temperatures in the Tatra Mountains are based on the proxy tree-ring width data of *Pinus cembra* for 1732–1969 (Bednarz 1984) and *Picea abies* for 1766–1965 (Feliksik 1972) and 1699–1978 (Schweingruber et al. 1979). Dendrochronological data for Sudetes were developed on the basis of cores from living *Picea abies* trees from leeward (eastern) slopes (less affected by pollution) (Opała-Owczarek et al. 2019), which were calibrated against instrumental data from the Mt. Śnieżka meteorological station (1603 m a.s.l.) IMGW-PIB (Institute of Meteorology and Water Management—National Research Institute).

For the Tatra Mountains it was possible the reconstruction of summer (June–August) air temperature (Niedźwiedź 2004) for the upper timberline represented by the meteorological station IMGW-PIB at Hala Gąsienicowa (1520 m.a.s.l). Direct measurements cover the period 1927–1938 and 1947–2019. Data for the years 1896–1926 and 1939–1946 were reconstructed after summer temperature in Zakopane (857 m a.s.l.). The correlation coefficient between summer temperatures of these stations is very high ($r = 0.953$, standard error of estimation se = ±0.3 K). Data for

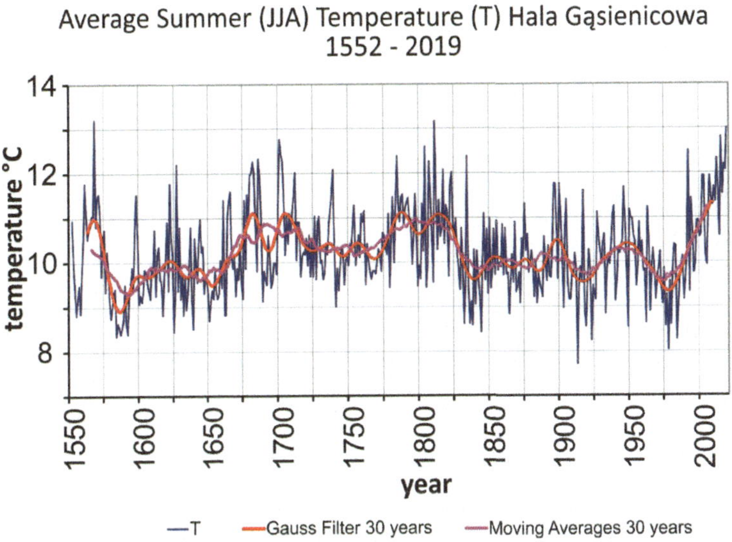

Fig. 5.9 Reconstruction of mean June–August air temperature (°C) in the Tatra Mountains at the upper timberline Hala Gąsienicowa (1520 m a.s.l.) using standardised chronologies of living Norway spruce (*Picea abies* Karst) tree-ring widths in Tatras (1700–1791) and Eastern Alps (1552–1699), and measured temperature in Kraków(1792–1895), Zakopane (1896–1926; 1939–1946), and real data for the years measured at Hala Gąsienicowa Station (1927–1938; 1947–2019) (modified after Niedźwiedź 2004; Bednarz 2015; Opała-Owczarek et al. 2019)

the period 1792–1895 were estimated from regression equation with Kraków temperatures ($r = 0.880$, se = ±0.6 K). Reconstruction of summer temperatures for the period 1700–1791 based on dendrochronological data from the Tatras (se = ±0.8 K; Bednarz 1976, 1984, 2015; Feliksik 1972; Schweingruber et al. 1979; Kaczka et al. 2016) and the Eastern Alps for the years 1552–1699 (se = ±1.0 K; Bednarz 1984; Bednarz and Niedźwiedź 2006) were less accurate. The average summer temperature in the period 1552–2019 amounted to 10.2 °C (standard deviation: std = ±0.97 K). The coldest summer was found in 1913 (7.7 °C, with deviation from average dT = −2.5 K), the warmest ones were in 1568 and 1811 (13.2 °C, dT = +3.0 K) and the last summer 2019 (13.0 °C, dT = +2.8 K).

In the course of the average summer temperature (JJA) in the period 1552–2019 (Fig. 5.9), large fluctuations are visible. There is additional information in the presentation of the occurrence of the number of cool (dt ≤ −1.0 K) and warm (dt ≥+1.0 K) summers in consecutive decades (Fig. 5.10). In the initial phase of LIA, the average summer temperature in the Tatras was marked by significant fluctuations between the cool summer of 1555 (8.8 °C, dT = − 1.4 K) to the warmest in the whole analysed series of data, the summer of 1568 (13.2 °C, dT = + 3.0 K). After this period, the first cool phase of the LIA (1576–1675) can be observed. The dense cool period covers the first 21 years (1576–1596), when 7–8 cold summers were recorded in particular decades (Fig. 5.10). In

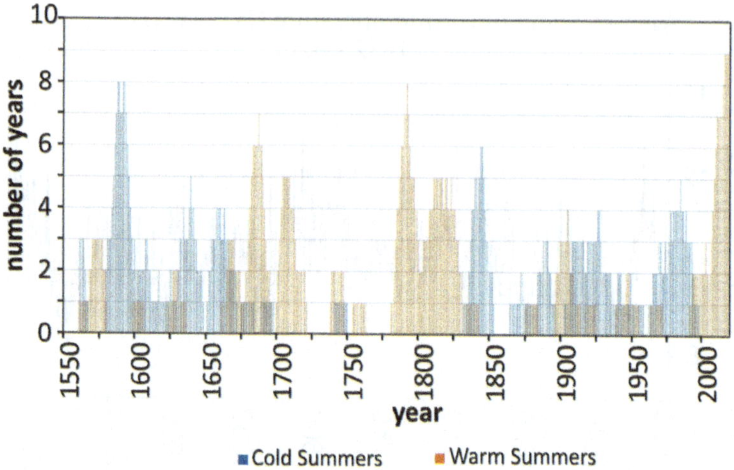

Fig. 5.10 The number of cold (≤−1 standard deviation from average) and warm summers (≥+1 standard deviation) for consecutive 10 years in the Tatra Mountains at the upper timberline Hala Gąsienicowa (1520 m a.s.l.) during the period 1552–2019. Average June–August temperature = 10.2 °C, standard deviation = ±0.97 °C

the summer of 1584, the average temperature dropped to 8.3 °C (dT = −1.9 K). In the next cool decade, 1630–1639, five cold summers were reported, with the coldest summer 1639 (8.5 °C, dT = −1.7 K). Over the next 151 years (1676–1826), relatively warm summers prevailed. This period is divided into five longer warm phases and four shorter cool phases. The warmest were two decades: 1678–1687 (with 7 warm summers) and 1783–1792 (8 warm summers). But the warmest summer, similar as in 1568, occurred in 1811 (13.2 °C, dT = +3.0 K). The last episode of LIA (1827–1895) lasted 69 years. The coolest was the period 1835–1845, with a very cold summer in 1844 (8.4 °C, dT = −1.8 K).

In the contemporary period (1896–2019), there are three warm (1896–1905, 1927–1959, 1991–2019) and two cool episodes (1906–1926 and 1960–1990). An extremely cold summer, 1913 (7.7 °C, dT = −2.5 K), was noticed in the first cool episode. It was the coldest summer in the whole data series. Such a large drop of summer temperature can be explained by the eruption of Katmai volcano in Alaska on 6 June 1912 (Bednarz 2015). The third coldest summer occurred in 1923 (8.2 °C, dT = −2.0 K). During the second cool phase, the coldest were the summers of 1978 (8.0 °C, dT = −2.2 K) and 1984 (8.3 °C, dT = −1.9 K). A period of significant warming began in the summer of 1992 (12.5 °C, dT = +2.3). After 1998, only two years (2004 and 2005) had temperatures lower than average. Since 2006, only warm summers have been recorded. During this period, as many as four summers (2019 13.0 °C, 2015 12.8 °C, 2012 12.3 °C and 2017 12.2 °C) were characterised by an air temperature two degrees higher than average.

The oldest period of the Tatra Mts. summer temperature reconstruction includes the final phase of climate warming observed in Poland in the second half of the

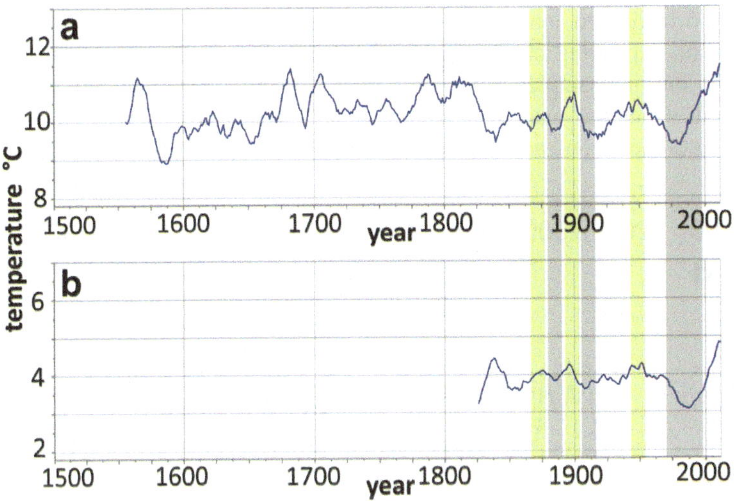

Fig. 5.11 Comparison of summer temperature dendroclimatic proxy records: reconstruction of mean June–August air temperature in the Tatra Mountains (**a**) and mean April–July air temperature in the Sudetes Mountains (**b**) using standardised chronologies of living Norway spruce (*Picea abies* Karst) tree-ring widths (data smoothed by 11-years average; modified after Niedźwiedź 2004; Opała-Owczarek et al. 2019). Common warm and cold periods are shaded in yellow and grey, respectively

sixteenth century. Warming in the mid-sixteenth century was followed by a long cool period 1576–1675. Dendroclimatological studies have identified several especially cold periods in the latter part of the LIA: 1650–1660, 1670–1680, 1720–1775. The final phase of the LIA was also marked by a sequence of exceptionally cold years at the turn of the eighteenth and nineteenth centuries. The next cold decade was reconstructed in the years 1830–1840. The course of further changes in summer temperatures in the Tatra Mountains can be compared with dendroclimatological data from Sudetes (Fig. 5.11). Except the divergent period 1825–1865, a similar course of temperature changes can be observed in both mountain massifs. The reconstructed courses of temperature data show warm periods: 1870–1880, 1890–1905, 1945–1955 and recent warming at the turn of the twentieth and twenty-first centuries. Cold decades were also identified in the years: 1881–1890, 1906–1915.

Winter Air Temperature

Tree-ring width of pines from the lowland areas were confirmed as good indicators of winter temperatures. Investigations into climate–growth relationships indicate that these records contain the signal of winter temperature variability in general, but for individual records climatic signal differs from narrow February–March window to wider November–April season. Dendroclimatic reconstructions that were developed for Poland facilitated analysis of winter temperature variability over the last 500–1000 years, depending on the region (Table 5.2, Fig. 5.12). All of these records are based on pine samples from living trees as well as historical and archaeological wood, which has made it possible to reach back over the past centuries to the beginning of the second millennium. Dendrochronological data for Northern Poland

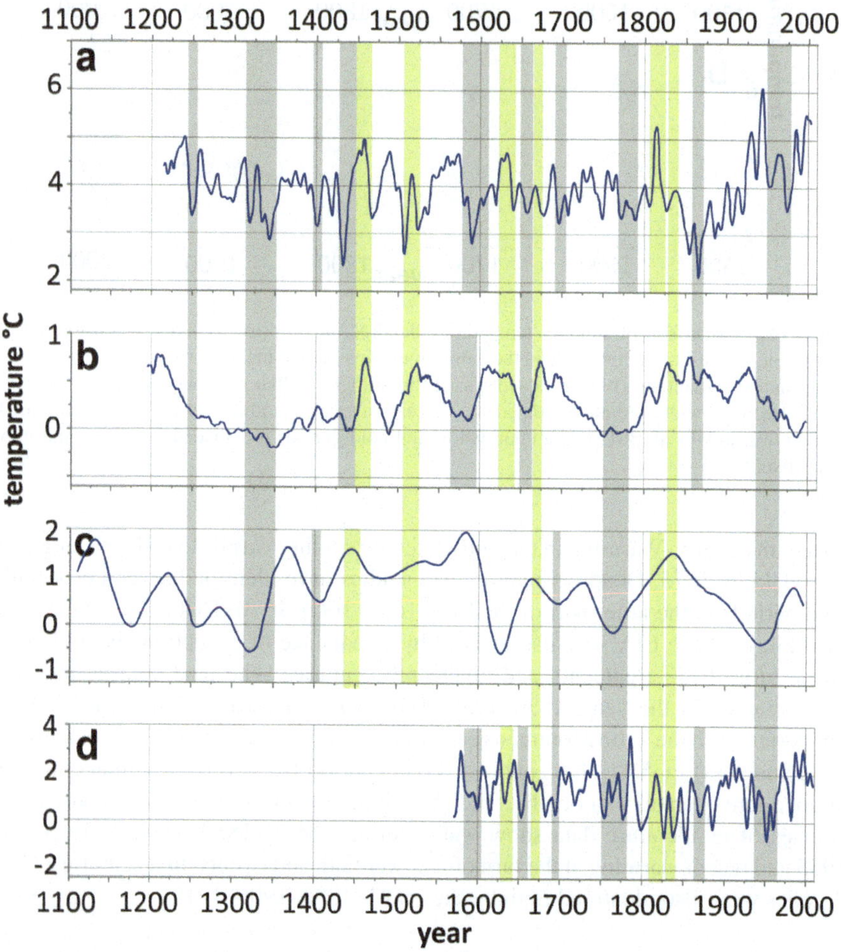

Fig. 5.12 Comparison of winter temperature dendroclimatic proxy records: reconstructions of mean November–April air temperature in Northern Poland (Bory Tucholskie and Masuria Lake District) (**a**), mean February–March air temperature in Kujawy and Pomorze region (**b**), mean February–March air temperature in Lesser Poland (**c**), mean February–March air temperature in Upper Silesia (**d**) using standardised chronologies of living and relict Scots pine (*Pinus sylvestris* L.) tree-ring widths (data smoothed using 11-years running means; modified after Szychowska-Krąpiec 2010; Zielski et al. 2010; Koprowski et al. 2012; Opała 2012, 2015; Balanzategui et al. 2018). Common warm and cold periods are shaded in yellow and grey, respectively

(1084–2010 A.D.) were developed on the basis of cores from Scots pine growing in the Masuria lakes district and Bory Tucholskie reserve (Tuchola Forest) ("MBT" record), supplemented by historic wood material collected from several town halls and churches. The sensitivity of this ring-width chronology to climate was tested against 0.5° × 0.5° grid point 53.5° N, 19.5° E from the Climate Research Unit (CRU) TS 3.23 data set (Balanzategui et al. 2018). The second dendrochronological record for Northern Poland was constructed for the Kuyavian-Pomerania region ("KP" record) (1168–2000 A.D.) using 55 samples from living trees and 111 from historical buildings, calibrated against climate data from the IMGW-PIB Bydgoszcz meteorological station (for the years 1861 to 2000) (Zielski and Krąpiec 2004; Przybylak et al. 2005; Zielski et al. 2010; Koprowski et al. 2012). For Southern Poland, millennium-long pine chronology (1091–2006) was composed for Lesser Poland region ("LP" record) utilising samples from trees growing in forests in Niepołomice and Nowy Targ and from historical wood samples originating from roof structures of churches, secular objects, wooden casings of the salt mines in Wieliczka and Bochnia and archaeological sites in Kraków. The TRW record was compared with meteorological data from Kraków and Kraków-Balice meteorological stations (Szychowska-Krąpiec 2010). The Upper Silesia chronology ("US" record), composed of samples from pines growing in six nature reserves and timber mainly from buildings located in the Open-Air Museums, covers the period 1568–2010 A.D. Direct measurements cover the periods: 1885–2012 (data from Katowice, Opole and Racibórz meteorological stations) and 1792–1884 (reconstructed after winter temperature measurements from Głubczyce, Wrocław and Kraków meteorological stations) (Opała and Mendecki 2012; Opała 2015).

Dendroclimatic reconstructions of winter temperature history over the last millennium show regional differences, however, some similarities are particularly strongly expressed during periods of strong cooling or above average conditions (Fig. 5.12). For the first two centuries of the last millennium a number of dendrochronological series reconstructing temperature conditions in Poland is too small to reliably reconstruct climate. For the thirteenth and fourteenth centuries significantly greater density of dendroclimatological series exists (>three regional records).

The earliest period of reconstructions largely agrees between tree-ring proxy series showing cooling around 1250, 1320–1350 and 1400. In all three pine records, the beginning of the thirteenth century is characterised by distinct warm conditions, similarly like the second half of the fourteenth century and decades: 1440–1460, 1510–1520. Warmer-than-average conditions were also reconstructed for the following decades: 1620–1630 (not present in "MP" data), 1660–1680, 1805–1815 (not present in "KP" data), 1825–1835.

Since the end of the sixteenth century, several extended below average cold periods prevailed in almost all pine tree-ring-based records: 1580–1600, 1650–1660, 1690–1700, 1750–1780, 1865–1875. However, most of these cool decades are not clearly marked in the "LP" data. In turn, a prolonged cold episode in the second half of the eighteenth century is particularly expressed in all but "MBT" data.

We may conclude that reconstructed temperatures over the last millennium in different regions of Poland, during winter and summer seasons, indicated similar course in dendroclimatic data. The cold periods around 1650–1660 and 1750–1780 and warm periods 1660–1675 and 1870–1880 are especially pronounced. It should be noted, however, that the most accurate reconstruction of thermal conditions was that of the Tatra Mountains, as the climate sensitivity of these proxy data is the highest, while accuracy of the dendroclimatic models for lowland areas are slightly lower.

Precipitation and Drought

For dendroclimatological data, opposite then for documentary evidence, the reconstruction of summer temperatures is more accurate than of winter temperatures, but similarly as for documentary evidence reconstruction of atmospheric precipitation based on tree-ring data is the most difficult. It is connected with the low precipitation signal present in the dendroclimatological records developed for the territory of Poland. As a result, there is a limited number of works with reconstructed precipitation in Poland in the pre-instrumental period based on tree rings in comparison to other parts of Europe (e.g. Vitas 2004; Masson-Delmotte et al. 2005; Čufar et al. 2008; Büntgen et al. 2010, 2011a, b; Dobrovolný et al. 2015). It seems that further development of multiproxy tree-ring analysis of millennium-long oak chronologies (described in the previous subchapter) will have a potential in the future dendroclimatic reconstruction of precipitation conditions in Poland.

Existing analysis of past precipitation conditions in Poland is mainly based on discontinuous dendrochronological data—analysis of pointer years. These kinds of studies were presented by different authors in regional works. Recently, frequencies of droughts derived from tree rings were presented in the work of Przybylak et al. (2020). Dendrochronological data from 22 chronologies from various locations in Poland confirmed that the greatest number of negative pointer years (proxy for droughts) occurred in the second half of the sixteenth century and eighteenth century.

Tree-ring-based reconstructions of extreme precipitation conditions can be found in dendrogeomorphological studies, however the time span of these analyses is limited, and these types of reconstruction mainly refer to extraordinary rainfall events. Zielonka et al. (2008) reconstructed flood events in a small mountain stream in the Waksmundzka Valley in the Tatra Mountains, Western Carpathians. Cross-dated flood scars found in Norway spruce trees growing along stream banks most likely formed by woody debris and stones transported during flood events indicate 17 years with flood events in the period between 1928 and 2005. It was proved that the high mid-summer rainfall (approximately 300 mm or more per month) which peaks in June, July, August—coincides with some of the flood scar formation, while high winter and spring precipitation (December–May) does not. Another paleohydrological analysis using scarred trees as paleostage indicators in the Tatra Mts. was presented by Ballesteros-Cánovas et al. (2016). Floods larger than those we experience nowadays were found to have occurred in the Strążyska and Łysa Polana valleys in the first half of the twentieth century. Tree-ring-based reconstruction of debris flows/floods history in the Sudetes was carried out by Malik and Owczarek (2009) and recently by Tichavský et al. (2017), who identified extreme hydro-geomorphic events during the period of 1889–2013, mostly triggered by extreme daily rainfall ($\geq$50 mm).

Varved Sediment Records

Varve Definition, Formation and Preservation

Varved lake sediments demonstrate characteristic horizontal bedding called lamination. The term "lamina" is understood as a single, macro- or microscopically distinguishable layer being the effect of deposition in a specific time period, e.g. one season. The term "varve", on the other hand, means a set of laminae deposited during one year. A varve may consist of a different number of laminae, in a simple case of two (Brunskil 1969), but examples of varves consisting of three (Bull and Kemp 1996), four (Dickman 1979; Renberg 1986) or even more laminae (Saarnisto et al. 1977) have been documented. The most common features that distinguish individual laminae are: sediment colour, grain-size differences and biological composition.

The history of using the term "varve" is 150 years old (Zolitschka 2007). Initially, it was used to describe rhythmically deposited clays in a proglacial environment. With the progress in lake sediment research, the meaning was extended to all annually laminated sediment records. According to Renberg (1981), the term "varve" means sediment deposited within one year, therefore it is not strictly related to any type of sedimentary environment. It is now widely accepted to use this term for lacustrine deposits in different environmental settings and climatic zones (Anderson et al. 1985; O'Sullivan 1983; Saarnisto 1986; Zolitschka 2007; Zolitschka et al. 2015), as well as marine and ocean sediments (Kemp 1996).

Varves are formed and preserved in bottom sediments only under specific conditions. Two fundamental requirements for the development of varves in lacustrine sediments are: (1) seasonal variations in biological activity, chemical processes and terrestrial input, causing changes in sediment composition, and (2) appropriate conditions for varve preservation, i.e. limited physical mixing and bioturbation. The first condition results in the fact that varved sediments are found mainly in the lakes in temperate zones characterised by strong seasonal climatic contrasts. Variability in temperature and precipitation is responsible for the annual rhythm of physical, biological and chemical processes in the lake and its catchment. Although the vast majority of lakes on Earth are located in areas with favourable climatic conditions, sites with varved sediments are quite rare. This is because a key role is played by the conditions responsible for preservation of varves, which often occur in lakes with anoxic bottom waters where bioturbation is negligible or absent. As a result, the occurrence of varved sediments is often restricted to the deepest and mostly anoxic parts of the lake basins (Petterson et al. 1993; Tylmann et al. 2012).

Varves are classified according to their composition and mechanisms of their formation (Zolitschka et al. 2015). Clastic varves consist mainly of minerogenic material and are formed as a result of seasonal variability in grain size of the allochthonous matter delivered to the lake by inflows, surficial flow or aeolian processes. In biogenic varves, the major mechanism is related to seasonal variability of biological productivity and biochemical processes in the water column. In arid climatic zone, an increase in salinity and pH as a result of intensive evaporation from the water surface leads to precipitation of minerals (e.g. gypsum, calcite, aragonite, halite) from the water column and formation of endogenic varves. However, varying climatic and environmental conditions favour the deposition of mixed varve types, e.g. biogenic varves could contain endogenic minerogenic components such as calcite. Mixed varve types are common not only in different environmental settings but are also observed in sediments deposited in different time periods at individual sites (Brauer 2004).

Potential of Varves for Climatic Reconstructions

The fundamental advantage of varved sediments is the possibility to establish precise time scales in calendar years through varve counting, which provides a solid geochronological framework that is an essential element of paleoenvironmental reconstructions. In addition to time control, varve chronology provides accurate and continuous record of changes in sedimentation rates through time. Another advantage of varves is that they provide high-resolution (annual) data. For example, the analysis of structure and composition of individual varves (thickness, size of calcite crystals, amount of minerogenic material, etc.) is a valuable tool that facilitates reconstruction of the environmental and climatic conditions during sediment deposition. High-resolution time series can be achieved using microfacies analysis (Brauer 2004) and non-destructive scanning techniques (Rothwell and Croudace 2015). For lakes with high sedimentation rates, e.g. a few millimetres per year, analyses of biological proxies may also approach the annual resolution.

As pointed out by Zolitschka et al. (2015), the nature of palaeoclimatic reconstructions based on varved sediments depends on the type of varves. For example, clastic varves may include important hydroclimatic information about changes in precipitation and fluvial transport of minerogenic matter from the catchment area into the lake basin (e.g. Cockburn and Lamoureux 2008; Ojala and Alenius 2005). In contrast, organic and mixed type varves contain a variety of fossil remains like pollen, plant macrofossils, non-pollen palynomorphs, diatoms, chrysophyte cysts, chironomids, ostracods, cladocera, etc., that enable qualitative and quantitative palaeoclimatic reconstructions (Cumming et al. 2012). A wealth of biological proxies offers possibility to reconstruct such climatic parameters as temperature, precipitation or wind intensity.

Distribution of Varved Sediments in Northern Poland

In Europe, most varved sediment profiles have been discovered as a result of systematic surveys guided by statistical analysis of lake basins and catchment data (Ojala et al. 2000; Zillén et al. 2003; Tylmann et al. 2013). Before the systematic search for varved records in northern Poland, Lake Gościąż represented the only site in East-Central Europe with a detailed study of environmental changes recorded in varved sediments (Ralska-Jasiewiczowa et al. 1998). In addition, only short laminated sections related to early stages of lake development or intermittently laminated sediment cores were occasionally reported (Więckowski 1978; Wacnik 2009). However, the postglacial landscapes of Northern Poland possess a number of lakes, which show a wide diversity of morphometric features, hydrological regimes and trophic states, making the area ideal for a survey of potential lacustrine climate archives. Also, the climate characterised by typical continental features indicated by distinct seasonal differences supports the formation of varved sediments.

Results of field surveys (Tylmann et al. 2006, 2013; Tylmann and Zawadzka 2008) as well as new findings in other parts of Northern Poland (Osadczuk 2007; Czymzik et al. 2015; Słowiński et al. 2017; Ott et al. 2018; Pleskot et al. 2018) indicate that the area is rich in lakes containing laminated sediments. Spatial distribution of lakes with laminated sediments shows most of them in Northeastern Poland. It has not been confirmed whether these laminations are annual in every case, but links between structure and composition of laminations and seasonal variability of limnological processes suggest that they can be regarded as varves (Tylmann et al. 2013). The combination of lake basin morphology, topographic position and catchment land use makes conditions favourable for varve formation and preservation in these lakes. Lakes most likely to possess varved sediments are deep relative to their surface area, are deeply incised into the catchment and surrounded by forests. These conditions minimise the influence of external factors, e.g. wind action, and lead to short water-column circulation periods, incomplete mixis or even permanent stratification, i.e. meromixis.

The large group of lakes with potentially varved sediments sheds new light on the possibilities for paleolimnological reconstructions of climatic conditions in Northern Poland. First, biogenic (calcite) varves possess a range of biological and geochemical proxies that can be used for reconstructions. Second, sedimentation rates on the order of several mm per year provide the opportunity to conduct high-resolution studies. The signal of environmental conditions preserved in these lakes theoretically allows for precise analysis of rapid climate change, extreme events and human impacts. However, quantitative reconstructions based on transfer functions require high-resolution analysis of biotic proxies sensitive to climate change, as well as establishing training sets of lakes and developing calibration models, which is a time-consuming task. Thus, so far in Poland only one attempt of quantitative reconstruction of climate conditions from varves has been made (Larocque-Tobler et al. 2015; Hernández-Almeida et al. 2017).

Reconstruction of Climatic Conditions in Northern Poland from Varves of Lake Żabińskie

Methods

Sediment cores were collected from the deepest part of Lake Żabińskie using Uwitec gravity and piston corers. The lacustrine sediment sequence was ca. 19.5 metres long and covered the entire Holocene. The two-metre-long core sections were tightly sealed and stored in a cold room until analyses. After splitting lengthwise, the core sections were photographed, described macroscopically, scanned with XRF, and correlated using diagnostic layers. Half-core A was used for the chronology, while half-core B was subsampled at annual resolution (varve-by-varve) or three-year resolution for multiproxy analyses. Chronology for the last millennium (uppermost 3.5 metres of the sediment profile) was based on varve counting and verified by radiocarbon dating, $^{137}Cs/^{210}Pb$ dating, and volcanic glass shards identified as Askja cryptotephra from 1875 CE (Bonk et al. 2015; Tylmann et al. 2016; Żarczyński et al. 2018).

The climate reconstruction presented here is based on two major biological proxies: chironomid head capsules and chrysophyte cysts. For chironomids, samples were picked at annual resolution between AD 2010 and 1939, and at 2- to 12-year resolution downcore until the beginning of the last millennium according to the varve chronology. Chironomid head capsules were extracted by placing sediment samples in KOH 10% overnight. The samples were then sieved in a 90 μm mesh and analysed under a stereomicroscope. Next, the head capsules were identified under a light microscope at 40-100X using the taxonomy followed Brooks et al. (2007) and Larocque-Tobler (2014). Details about the training set of lakes and development of transfer function for the reconstruction of August temperature were provided in Larocque-Tobler et al. (2015, 2016). The original chironomid dataset was published by Hernández-Almeida et al. (2017) in a synthesis paper summing up last millennium environmental changes recorded in varves of Lake Żabińskie.

Chrysophyte cysts were analysed on every varve subsample for the last millennium. Sediment samples were treated with H_2O_2 and HCl following the standard diatom procedure (Battarbee 1986). Samples were then washed and filtered to remove large particles. Chrysophyte cysts were analysed using scanning electron microscope (Carl-Zeiss EVO40). Identification of the cysts followed Duff et al. (1995), Wilkinson et al. (2002) and Huber et al. (2009). Details about the training set of lakes, development of transfer function and reconstruction of winter severity were published by Hernández-Almeida et al. (2015a).

Reconstruction of Summer Temperature for the Last 1000 Years

Reconstruction of chironomid-inferred August temperature variability presented here covers the last millennium. More than 200 samples representing the period 1000–2010 CE were analysed for the occurrence and species composition of chironomids. The frequency of head capsules was different, but only in some samples exceeded the required number of 50 (Heiri and Lotter 2001). Thus, the samples were merged to 83 data points in total, each represented a period of about 12 years. Finally, the number of head capsules in merged samples ranged from 31 to 450 and the low-frequency samples (<50 head capsules) constituted only 23 samples. It seems to be the optimal solution taking into account both methodological requirements and sufficient time resolution of the reconstruction.

The chironomid-inferred August temperature record shows several characteristic minima and warmer periods, but no overall long-term trend can be indicated. The record starts with a dramatic temperature increase (Fig. 5.13) and warm period resem-

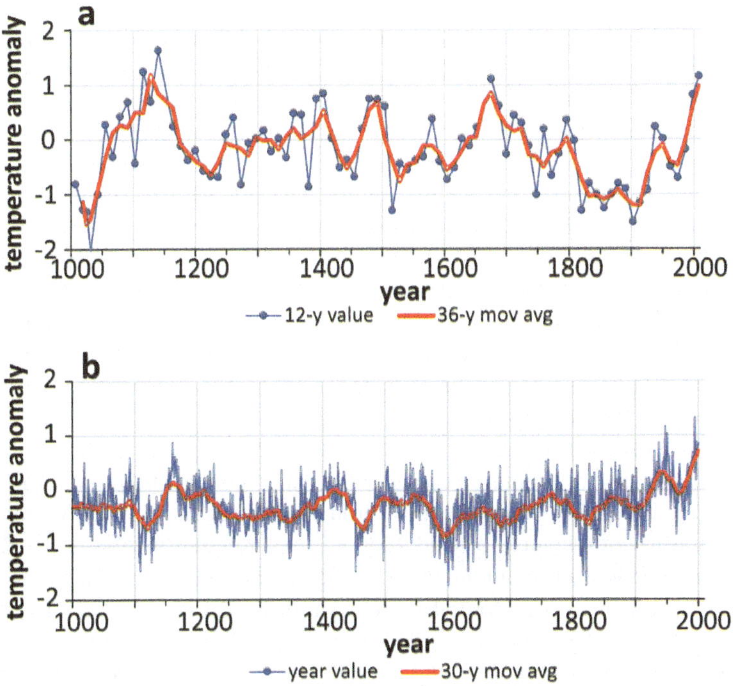

Fig. 5.13 Reconstruction of chironomid-inferred mean August temperature in Northeastern Poland (**a**) compared to PAGES 2 K reconstruction for Europe (**b**). Chironomid analyses were done by Isabelle Larocque-Tobler. Details were published in Larocque-Tobler et al. (2015, 2016) and Hernández-Almeida et al. (2017). The original dataset was transformed to obtain 12-year resolution for the whole reconstruction. Data for PAGES 2 K reconstruction from PAGES 2k Consortium (2013)

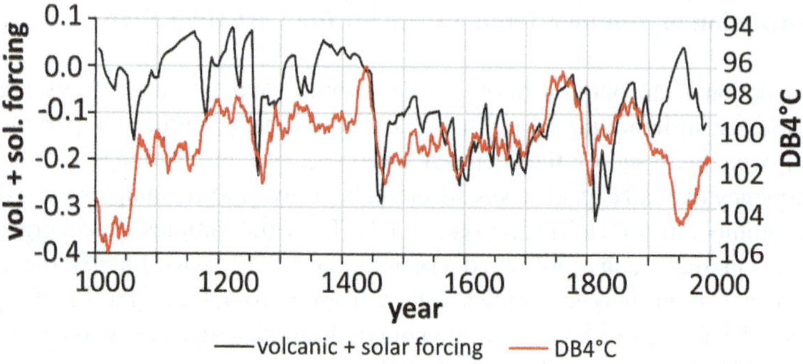

Fig. 5.14 Reconstruction of chrysophyte-based DB4 °C (red line) compared to combined solar and volcanic forcing (black line). Chrysophyte analyses were done by Ivan Hernández-Almeida. Details were published in Hernández-Almeida et al. (2015a). Data for solar and volcanic forcing from Crowley (2000)

bling Medieval Climate Anomaly (MCA), which is well pronounced between ca. 1050–1150 CE. This is inconsistent with the PAGES 2 K reconstruction for Europe, which did not record any significant warming at the beginning of the last millennium (Fig. 5.14). It must be taken into account that in this part (1000–1140 CE) the reconstruction is based on low counts of chironomid head capsules (mostly <50) which might have biased the reconstructed values of August temperatures. After rapid cooling at the end of the twelfth century, the Little Ice Age (LIA) is expressed as a long period (ca. 1200–1900 CE) with the lowest temperatures between ca. 1430–1450, 1500–1650 and 1820–1920 CE. The nineteenth century especially presents the coldest temperatures during the entire period, except the beginning of the millennium. The twentieth century shows a strong warming trend in two steps, interrupted by cooling between 1950 and 1970. Using 12-year resolution data, temperatures during the last two decades are on average 0.5 °C higher compared to the average temperature for the reconstructed MCA period.

Similarities to the PAGES 2 K reconstruction for Europe are visible (Fig. 5.13). Several characteristic minima are consistent in both reconstructions (ca. 1450, 1600, 1820, 1900, 1970 CE), while the most pronounced minimum in the chironomid-based reconstruction centred around 1030 CE has no confirmation in the PAGES 2 K reconstruction. Lower amplitude of temperature changes in PAGES 2 K reconstruction is natural as it is compiled from various individual sites and is mostly based on tree rings. Uncertainties of the chronology in both reconstructions can also cause time shifts in the short episodes. Despite some periods of divergence between the reconstructions, the reconstruction for Northeastern Poland suggests that in the long-term chironomids responded similarly with other proxies/archives to changes in temperature.

Reconstruction of Winter Severity for the Last Millennium

Chrysophytes are regarded as good indicators of winter temperatures because the number of consecutive days with low water temperatures (DB4 °C) is an ecologically important variable for them (Kamenik and Schmidt 2005; Pla and Catalan 2005). The DB4 °C affects the timing and magnitude of cyst production during an entire year and shifts the percentage of cold and warm water species. Although DB4 °C does not reflect directly cold-season temperatures, it is highly correlated to winter length and ice-cover duration. Therefore, it may be used as an indicator of "winter severity".

A quantitative reconstruction of cold-season climate from varved sediments of Lake Żabińskie covers the last millennium at a resolution of five years (Fig. 5.14). It shows pronounced decadal and multidecadal variability with two clear shifts: from colder to warmer winters in 1050 CE and from warmer to colder winter conditions from 1430–1460 CE. The first shift is consistent with the beginning of MCA recorded in the chironomid record from Lake Żabińskie. The shift to colder winters from 1430–1460 CE is consistent with winter climate indices for Poland (Przybylak et al. 2005; Przybylak 2011) and January–April air temperature for Central Poland reconstructed from tree-ring widths of Scots pine (Przybylak 2011), which show strong cooling in the period 1440–1460 CE. Higher values of DB4 °C persisted until 1700 CE, when more variable conditions occurred.

Hernández-Almeida et al. (2015a) emphasise the striking correspondences between the reconstruction of winter severity in Northeastern Poland and the combined solar and volcanic forcing (Fig. 5.14). Strong volcanic forcing could produce very long winters at sub-decadal to decadal scales. The strong evidence of this relationship may be observed in the mid-thirteenth century. The highest values of reconstructed DB4 °C are recorded between 1259–1263, i.e. after the biggest volcanic eruption during the last millennium in 1258 CE. Generally, in the period from 1200–1900 CE, the shifts towards severe winters coincide with periods of cumulative strong volcanic forcing, e.g. around 1450 CE, around 1600 CE, throughout the seventeenth century and the first two decades of the nineteenth century. During the twentieth century this relationship breaks down, which can be explained by the new combination of forcing including the anthropogenic GHG.

Discussion

Varved lake sediments are archives of past climate variability as well as other environmental changes including human impact. Varves enable precise dating and store a wealth of paleoclimatic proxies used for reconstructions of summer and winter temperature, length of ice cover, seasonal precipitation, paleofloods and wind (Brauer 2004; Zolitschka et al. 2015). However, all of these factors influence the lake system directly (e.g. water temperature, mixing regime, water column oxygenation, biological productivity) or indirectly (e.g. vegetation cover in the catchment, soil erosion, influx of minerogenic particles and nutrients) with different intensity in different time intervals. Disentangling these overlapping influences is probably one of the most difficult problems to be solved, and cause uncertainties in the interpretation of reconstructions based on proxy records from lake sediments.

The presented reconstructions of summer temperature and winter severity in Northeastern Poland are based on chironomids and chrysophyte cysts, respectively. Despite the demonstrated potential of these groups of organisms for quantitative climate reconstructions, they are also influenced by other environmental parameters, such as oxygen availability (Quinlan et al. 1998), nutrients (Lotter et al. 1997) and lake depth (Cwynar et al. 2012). Chrysophyte cysts are also sensitive to water chemistry and nutrients (Kamenik et al. 2001; Hernández-Almeida et al. 2015b). The controversial issues about temperature as the major important factor explaining the distribution of chironomids still exist (e.g. Velle et al. 2010). While there is no question that factors other than temperature can affect the chironomid assemblages, the reconstruction from Lake Żabińskie suggests that on a longer temporal scale temperature is likely to be the driving factor. Doubts will remain about the MCA which is well pronounced between ca. 1050–1150 CE, but the frequency of chironomid head capsules was below the standard in this part of the sediment core.

Information about winter temperature variability from Northern Poland may be very important because of the very strong relationship between Polish and European temperatures both at the interannual and interdecadal time scales (Luterbacher et al. 2010). Strong relationships of winter climate in Northeastern Poland with major atmospheric circulation patterns are evidenced in snow cover duration (Falarz 2007), lake ice cover (Marszelewski and Skowron 2006), documentary and early instrumental data (Przybylak et al. 2005; Luterbacher et al. 2010) for the late nineteenth and twentieth centuries. Therefore, the chrysophyte-based reconstruction of winter severity should show similar features to data available from other European regions. For example, the very cold period following 1460 CE was shown in the reconstruction from Lake Silvaplana in the Eastern Swiss Alps (de Jong et al. 2013). A similar pattern can be observed between the DB4 °C reconstruction form of Lake Żabińskie and documentary data from the Low Countries (van Engelen et al. 2001), and instrumental and documentary sources from Central Poland (Przybylak et al. 2005; Przybylak 2011).

Conclusions and Final Remarks

The density of documentary evidence was sufficient for quantitative reconstruction of some meteorological variables for the period from the fifteenth century to the end of the eighteenth century. On the other hand, prior to the fifteenth century, and in particular for the first two centuries of the last millennium, the number of historical sources is limited and their reliability is also usually not high. Therefore, the undertaken trials of climate reconstructions for this period are rather low in degree of reliability. Historians are certain that it is impossible to find more information about weather in historical sources. The only possibility to improve the knowledge about climate in the first four centuries of the last millennium is to use natural archives for this purpose, such as tree-ring widths (or density) and laminated lacustrine sediments. On the other hand, for the period 1401–1800, there is still a good deal of unused documentary evidence. We can also be sure that quite a large number of historical sources have still not been discovered. Therefore, there is still much work to be done in the future.

Dendrochronological data exceeding the last few hundred years are the most valuable for research on climate variability and change, as they cover the pre-instrumental period. Extensive studies aimed at developing chronologies from living trees in various regions of Poland proved that only utilisation of historical and archaeological wood elements can extend tree-ring records over the last 1–2 millenia. It should be recognised that the potential for further research is still very large, especially taking into account subfossil trunks and consistent collection of archaeological wood material. There is great potential for extending the scientific values of Polish dendroclimatic reconstructions by using other proxies than tree-ring width in future works and more sophisticated statistical methods of reconstruction, as they may improve the embedded dendroclimatic signal and reveal low-frequency climate information, helping in further understanding of multicentennial past climate changes in Poland.

Taking into account the occurrence and spatial distribution of lakes with varved sediments in Northern Poland, it is still possible to develop climate reconstructions at different time scales. While very interesting investigations have been conducted in the lakes of Bory Tucholskie (Czymzik et al. 2015; Słowiński et al. 2017; Ott et al. 2018), Lake Strzeszyńskie in Wielkopolska (Pleskot et al. 2018), and ongoing work on Lake Gościąż in Central Poland, none of them aimed at quantitative and high-resolution reconstruction for the last millennium. Taking into account these lakes and many sites which have not been investigated yet, the potential to increase our knowledge about past climate changes in Northern Poland based on lake sediment archives is still relatively large.

The historical period studied here was divided into two periods 1001–1400 and 1401–1800. More or less, this division is in line with observed climate change from the Medieval Warm Period (MWP) to the Littler Ice Age (LIA). The mean annual air temperature during the MWP might have been comparable to what we experience now, but data for this period are characterised by a high degree of uncertainty. Climate in this time had the greatest degree of the oceanicity in the entire millennium. On the other hand, during the LIA winters were on average about 2 °C colder than today, while summers a little warmer. As a result, in this time the highest degree of climate continentality was noted.

Knowledge about precipitation changes in Poland in historical times is significantly smaller and less reliable than in the case of air temperature. Available reconstructions depict that the most precipitation in the entire historical period occurred probably mainly in the twelfth century (especially the second half), and in the thirteenth century (in particular in the first half). Secondary maxima were noted in the first halves of the sixteenth and eighteenth centuries. Alternatively, precipitation below the norm was observed in the second halves of the fourteenth and sixteenth centuries.

For dendroclimatological data the reconstruction of summer temperatures is more accurate than of winter temperatures, but similarly as for documentary evidence, reconstruction of atmospheric precipitation based on tree-ring data is definitely most difficult. Both summer temperature and winter severity reconstructions show major features which seem to be consistent with other reconstructions and documentary or instrumental data. The cold periods in the second half of the fifteenth century, late sixteenth century and the second half of the nineteenth century are especially pronounced. However, inconsistencies remain in the details, which may be attributed to using different archives (tree rings, lake sediments, etc.), uncertainties in chronologies, varied sensitivities of the proxy used in particular reconstructions, and finally to the not fully comparable meaning of the reconstruction results. For example, the DB4 °C reconstruction is sensitive to the winter length, which is not exactly the same as low temperatures.

Acknowledgements Work presented in the section "Documentary Evidence" was supported by funds received from the NCN project DEC-2020/37/B/ST10/00710. The research presented in the subchapter *Dendrochronological records* was supported by the statutory research funds from the Faculty of Earth Sciences, University of Silesia in Katowice and within the NCN project N N306 139 638. The research presented in the subchapter *Varved sediment records* was funded by the National Science Centre grant 2014/13/B/ST10/01311. Isabelle Larocque-Tobler and Ivan Hernández-Almeida are acknowledged for their detailed analysis of chironomids and chrysophytes in the sediments of Lake Żabińskie. Special thanks go to all the CLIMPOL project members for their enthusiastic work and fruitful discussions.

References

Anderson RY, Dean WE, Bradbury JP et al (1985) Meromictic lakes and varved sediments in North America. US Geol Surv Bull 1607:1–19

Balanzategui D, Knorr A, Heussner K-U, Wazny T, Beck W, Słowiński M, Helle G, Buras A, Wilmking M, Van Der Maaten E, Scharnweber T, Doradolinán I, Heinrich I (2018) An 810-year history of cold season temperature variability for northern Poland. Boreas 47(2):443–453

Ballesteros-Cánovas JA, Stoffel M, Spyt B, Janecka K, Kaczka RJ, Lempa M (2016) Paleoflood discharge reconstruction in Tatra Mountain streams. Geomorphology 272:2–101

Battarbee RW (1986) Diatom analysis. In: Berglund BE (ed) Handbook of Holocene palaeoecology and palaeohydrology. Wiley, Chichester, pp 527–570

Bednarz Z (1976) Wpływ klimatu na zmienność szerokości słojów rocznych limby (*Pinus cembra* L) w Tatrach. Acta Agr Silv Ser Silv 16:17–34 (in Polish)

Bednarz Z (1984) The comparison of dendroclimatological reconstructions of summer temperatures from the Alps and Tatra Mountains from 1741–1965. Dendrochronologia 2:63–72

Bednarz Z (1996) June-July temperature variation for the Babia Góra National Park, southern Poland, for the period 1650-1910. In: Obrębska-Starkel B, Niedźwiedź T (eds), Proceedings of the International conference on climate dynamics and the global change perspective, Cracow, 17–20 October 1995. Institute of Geography of the Jagiellonian University, Prace Geograficzne 102:523–529

Bednarz Z (2015) Climate reconstruction. In: Dąbrowska K, Guzik M (eds) Atlas of the Tatra Mts. – Abiotic Nature. Tatrzański Park Narodowy, Zakopane, Chart II.7

Bednarz Z, Jaroszewicz B, Ptak J, Szwagrzyk J (1999) Dendrochronology of Norway spruce (*Picea abies* (L. Karst)) in the Babia Gora National Park, Poland. Dendrochronologia 16(17):45–55

Bednarz Z, Muter E (2010) Pionierska chronologia świerka Picea abies (L.) H. Karst. z Tatr Polskich autorstwa Profesora Karola Ermicha we współczesnym ujęciu metodycznym. Nauka a zarządzanie obszarem Tatr i ich otoczeniem, tom II – Zakopane, 45–51 (in Polish)

Bednarz Z, Niedźwiedź T (2006) Dendrochronologia świerka (Picea Abies (L.) Karst) z Parku Narodowego wysokie Taury (Austria). In: Trepińska J, Olecki Z (eds) Klimatyczne aspekty środowiska geograficznego. Kraków, Instytut Geografii i Gospodarki Przestrzennej Uniwersytetu Jagiellońskiego, pp 231–246 (in Polish)

Bednarz Z, Ptak J (1990) The influence of temperature and precipitation on ring widths of oak (*Quercus robur* L.) in the Niepolomice forest near Cracow. Southern Poland. Tree-ring Bull. 50:1–10

Bijak S (2010) Tree-ring chronology of silver fir and its dependence on climate of the Kaszubskie Lakeland (Northern Poland). Geochronometria 35(1):91–94

Bokwa A, Limanówka D, Wibig J. 2001. Pre-instrumental weather observations in Poland in the 16th and 17th centuries. In: Jones PD, Ogilvie AEJ, Davies TD, Briffa KR (eds) History and Climate, Memories of the Future. Kluwer Academic Publishers, pp 9–28

Bonk A, Tylmann W, Goslar T et al (2015) Comparing varve counting and C-14 chronologies in the sediments of Lake Żabińskie, Northeastern Poland: implications for accurate C-14 dating of lake sediments. Geochronometria 42:159–171

Bradley RS (2015) Paleoclimatology: Reconstructing climates of the Quaternary. Academic Press, Oxford, p 675

Bradley RS, Hughes MK, Diaz HF (2003) Climate in Medieval time. Science 302:404–405

Bradley RS, Jones PD (eds) (1995) Climate since A.D. 1500. London, Routledge, p 706

Brauer A (2004) Annually laminated lake sediments and their palaeoclimatic relevance. In: Fischer H, Kumke T, Lohmann G (eds) The climate in historical times. Towards a synthesis of Holocene proxy data and climate models. Springer Verlag, Berlin, pp 109–128

Brázdil R, Dobrovolný P (2010) Historical climate in Central Europe during the last 500 years. In: Przybylak R, Majorowicz J, Brázdil R, Kejna M (eds) The Polish climate in the European Context: An historical overview. Springer, Berlin Heidelberg, New York, pp 41–70

Brázdil R, Kotyza O (1995) History of Weather and Climate in the Czech Lands I (Period 1000–1500). Zürcher Geographische Schriften 62, Zürich, p 260

Brázdil R, Pfister C, Wanner H, von Storch H, Luterbacher J (2005) Historical climatology in Europe – The state of art. Clim Chang 70(3):363–430

Bronisz A, Bijak S, Bronisz K (2010) Dendroklimatologiczna charakterystyka jodły pospolitej (*Abies alba* Mill.) na terenie Gór Świętokrzyskich. Sylwan 154(7):463–470 (in Polish)

Bronisz A, Bijak S, Bronisz K, Zasada M (2012) Climate influence on radial increment of oak (*Quercus* sp.) in central Poland. Geochronometria 39(4):276–284

Brooks S, Langdon P, Heiri O (2007) The Identification and Use of Palaearctic Chironomidae Larvae in Palaeoecology. Quaternary Research Association, Technical Guide. 10

Brunskil GJ (1969) Fayetteville Green Lake, New York. III: Precipitation and sedimentation of calcite in a meromictic lake with laminated sediments. Limnol Oceanogr 14:830–847

Bryś K, Bryś T (2010a) Reconstruction of the 217-year (1791-2007) Wrocław air temperature and precipitation series. Bull Geogr Phys Geogr Ser 3:121–171

Bryś K, Bryś T. (2010b) The first one hundred years (1791–1890) of the Wrocław air temperature series. In: Przybylak R, Majorowicz J, Brázdil R, Kejna M (eds) The Polish climate in the European Context: An historical overview. Springer, Berlin Heidelberg New York, pp 485–524

Bull D, Kemp AES (1996) Composition and origins of laminae in late Quaternary and Holocene sediments from the Santa Barbara Basin. In: Kemp AES (ed) Palaeoclimatology and palaeoceanography from laminated sediments. Geological Society Special Publication 116, pp 143–156

Büntgen U, Tegel W (2011) European tree-ring data and the Medieval climate Anomaly. Pages News 19:14–15

Büntgen U, Frank DC, Frank DC, Nievergelt D, Esper J (2006) Summer temperature variations in European Alps, AD 755-2004. J Clim 19:5606–5623

Büntgen U, Frank DC, Kaczka RJ, Verstege A, Zwijacz-Kozica T, Esper J (2007) Growth responses to climate in a multi-species tree-ring network in the Western Carpathian Tatra Mountains, Poland and Slovakia. Tree Physiol 27(5):689–702

Büntgen U, Trouet V, Frank D, Leuschner HH, Friedrichs D, Luterbacher J, Esper J (2010) Tree-ring indicators of German summer drought over the last millennium. Quat Sci Rev 29(7–8):1005–1016

Büntgen U, Brázdil R, Dobrovolný P, Trnka M, Kyncl T (2011a) Five centuries of Southern Moravian drought variations revealed from living and historic tree rings. Theor Appl Climatol 105(1–2):167–180

Büntgen U, Brázdil R, Heussner K-U, Hofmann J, Kontic R, Kyncl TS, Pfister C, Chromá KR, Tegel W (2011b) Combined dendro-documentary evidence of Central European hydroclimatic springtime extremes over the last millennium. Quat Sci Rev 30(27–28):3947–3959

Büntgen U, Tegel W, Nicolussi K, McCormick M, Frank D, Trouet V, Kaplan JO, Herzig F, Heussner K-U, Wanner H, Luterbacher J, Esper J (2011c) 2500 years of European climate variability and human susceptibility. Science 331:578–582

Büntgen U, Kaczka RJ, Trnka M, Rigling A (2012) Ensemble estimates reveal a complex hydroclimatic sensitivity of pine growth at Carpathian cliff sites. Agric For Meteorol 160:100–109

Büntgen U, Kyncl T, Ginzler Ch, Jacks DS, Esper J, Tegel W, Heussner K-U, Kyncl J (2013) Filling the Easter European gap in millennium-long temperature reconstructions. PNAS 110(5):1773–1778

Cedro A (2001) Dependence of radial growth of *Pinus Sylvestris* L. from Western Pomerania on the rainfall and temperature conditions. Geochronometria 20:69–74

Cedro A (2007) Tree-ring chronologies of downy oak (*Quercus pubescens*), pedunculate oak (*Q. robur*) and sessile oak (*Q. petraea*) in the Bielinek Nature Reserve: comparison of the climatic determinants of tree-ring width. Geochronometria 26:39–45

Chojnacka-Ożga L, Ożga W (2014) Sygnał klimatyczny w przyrostach rocznych buka zwyczajnego (*Fagus sylvatica* L.) w Bieszczadach. Studia i Materiały CEPL w Rogowie 16(40,3):229–237 (in Polish)

Chrapowicki JA (1978) Diariusz cz. I: lata 1656-1664. Instytut Wydawniczy PAX, Warszawa (in Polish)

Chrapowicki JA (1988) Diariusz cz. II: lata 1665-1669. Instytut Wydawniczy PAX, Warszawa (in Polish)

Cockburn JMH, Lamoureux SF (2008) Inflow and lake controls on short-term mass accumulation and sedimentary particle size in a High Arctic lake: implications for interpreting varved lacustrine sedimentary records. J Paleolimnol 40:923–942

Cook ER, Kairiukstis LA (eds) (1990) Methods of dendrochronology. Applications in the environmental sciences. Springer, Netherlands, Dordrecht, The Netherlands, p 351

Crowley TJ (2000) Causes of climate change over the past 1000 years. Science 289:270–277

Crowley TJ, Lowery T (2000) How warm was the Medieval Warm Period? A comment on 'man-made versus natural climate change'. Ambio 29:51–54

Čufar C, De Luis M, Eckstein D, Kajfež-Bogataj L (2008) Reconstructing dry and wet summers in SE Slovenia from oak tree-ring series. Int J Biometeorol 52:607. https://doi.org/10.1007/s00484-008-0153-8

Cumming BF, Laird KR, Fritz SC et al (2012) Tracking holocene climate change with aquatic biota from lake sediments: Case studies of commonly used numerical techniques. In: Birks HJB, Lotter AF, Juggins S, Smol JP (eds) Tracking environmental change using lake sediments. developments in paleoenvironmental research 5, Springer, Dordrecht, pp 615–642

Cwynar LC, Rees ABH, Pedersen CR et al (2012) Depth distribution of chironomids and an evaluation of site-specific and regional lake-depth inference models: A good model gone bad? J Paleolimnol 48:517–533

Czymzik M, Muscheler R, Brauer A et al (2015) Solar cycles and depositional processes in annual ^{10}Be from two varved lake sediment records. Earth Planet Sc Lett 428:44–51

Danek M, Chuchro M (2016) Sygnał dendroklimatyczny w sekwencjach przyrostów rocznych modrzewia rosnącego we wschodniej części polskich Karpat. Studia i Materiały CEPL w Rogowie 18(48)3: 35–42 (in Polish)

Danek M, Chuchro M, Walanus A (2017) Variability in larch (*Larix Decidua* Mill.) tree-ring growth response to climate in the Polish Carpathian Mountains. Forests 8:354. https://doi.org/10.3390/f8100354

de Jong R, Kamenik C, Grosjean M (2013) Cold-season temperatures in the European Alps during the past millennium: variability, seasonality and recent trends. Quat Sci Rev 82:1–12

Dickman MD (1979) A possible varving mechanism for meromictic lakes. Quat Res 11:113–124

Dobrovolný P, Rybníček M, Kolář T, Brázdil R, Trnka M, Büntgen U (2015) A tree-ring perspective on temporal changes in the frequency and intensity of hydroclimatic extremes in the territory of the Czech Republic since 761 AD. Clim Past 11:1453–1466

Duff KE, Zeeb BA, Smol JP (1995) Atlas of chrysophycean cysts, vol 2. Kluwer Academic Publishers, Dordrecht-Boston-London

Eriksson Ch, Omsted A, Overland JE, Percival DB, Mofjeld HO (2007) Characterizing of European sub-Arctic winter climate since AD 1500 using ice, temperature and atmospheric circulation time series. J Clim 20:5316–5334

Ermich K (1951) Unpublished materials

Esper J, Frank D (2009) The IPCC on a heterogeneous medieval warm period. Clim Chang 94(3–4):267–273

Falarz M (2007) Snow cover variability in Poland in relation to the macro- and mesoscale atmospheric circulation in the twentieth century. Int J Climatol 27:2069–2081

Feliksik E (1972) Studia dendroklimatologiczne nad świerkiem (*Picea excelsa* L.). Acta Agr Silv Ser Silv 12:39–83 (in Polish)

Feliksik E, Wilczyński S (2000) Dendroclimatological analysis of the Norway spruce (*Picea abies* (L.) Karst.) from the Beskid Śląski Mountains. Zpravodaj Beskydy 13:161–170

Feliksik E, Wilczynski S, Podlaski R (2000) Wpływ warunków termiczno-pluwialnych na wielkość przyrostów radialnych sosny (*Pinus sylvestris* L.), jodły (*Abies alba* Mill.) i buka (*Fagus sylvatica* L.) ze Świętokrzyskiego Parku Narodowego. Sylwan 144(9):53–64 (in Polish)

Filipiak (2007b) Obserwacje i pomiary opadów atmosferycznych w Gdańsku w XVIII wieku. In: Piotrowicz K, Twardosz R (eds) Wahania klimatu w różnych skalach przestrzennych i czasowych, Kraków, pp 365–373 (in Polish)

Filipiak J (2007a) Rekonstrukcja warunków klimatycznych Gdańska w okresie pomiarów instrumentalnych – fakty i niepewności. In: Miętus M, Filipiak J, Wyszkowski A (eds) 200 lat regularnych pomiarów i obserwacji meteorologicznych w Gdańsku. Monografie Instytutu Meteorologii i Gospodarki Wodnej, Warszawa, pp 20–33 (in Polish)

Filipiak J, Miętus M (2010) History of the Gdańsk pre-instrumental and instrumental record of meteorological observations and analysis of selected air pressure observations. In: Przybylak R, Majorowicz J, Brázdil R, Kejna M (eds) The Polish climate in the European Context: An historical overview. Springer, Berlin Heidelberg New York, pp 267–293

Filipiak J, Przybylak R, Oliński P (2019) The longest one-man weather chronicle (1721–1786) by Gottfried Reyger for Gdańsk, Poland as a source for improved understanding of past climate variability. Int J Climatol 39:828–842. https://doi.org/10.1002/joc.5845

Fritts H (1976) Tree rings and climate. Academic Press, New York, p 582

Galle JG (1857) Grundzüge der Schlesischen Klimatologie. Josef Max & Komp. Breslau, pp XXIII + 128 (in German)

Galle JG (1879) Mittheilungen der Königlichen Universitäts-Sternwarte zu Breslau. Maruschke und Berendt, Breslau, p 128 (in German)

Glaser R (2008) Klimageschichte Mitteleuropas. 1200 Jahre Wetter, Klima Katastrophen. Primus Verlag, Wissenschaftliche Buchgesellschaft, Darmstadt, p 264 (in German)

Glaser R, Riemann D (2009) A thusand-year record of temperature variations for Germany and Central Europe based on documentary data. J Quat Sci 24:437–449

Gouirand I, Linderholm HW, Moberg A, Wohlfarth B (2008) On the spatiotemporal characteristics of Fennoscandian tree-ring based summer temperature reconstructions. Theor Appl Climatol 91:1–25

Grove JM (1988) The little ice age. Methuen & Co., London, New, York, p 498

Heiri O, Lotter AF (2001) Effect of low count sums on quantitative environmental reconstructions: an example using subfossil chironomids. J Paleolimnol 26:343–350

Hernández-Almeida I, Grosjean M, Przybylak R et al (2015a) A chrysophyte-based quantitative reconstruction of winter severity from varved lake sediments in NE Poland during the past millennium and its relationship to natural climate variability. Quat Sci Rev 122:74–88

Hernández-Almeida I, Grosjean M et al (2015b) Chrysophyte cyst-inferred variability of warm season lake water chemistry and climate in northern Poland: training set and downcore reconstruction. J Paleolimnol 53:123–138

Hernández-Almeida I, Grosjean M, Gómez Navarro JJ et al (2017) Resilience, rapid transitions and regime shifts: fingerprinting the responses of Lake Zabińskie (NE Poland) to climate variability and human disturbance since 1000 AD. Holocene 27:258–270

Huber K, Kamenik C, Weckström K et al (2009) taxonomy, stratigraphy, and paleoecology of chrysophyte cysts from Late Glacial sediment core section of Langsee, Austria. Nova Hedwigia 89:245–261

Hughes MK, Kelly PM, Pilcher JR, LaMarche VC (1982) Climate from tree rings. Cambridge University Press, Cambridge, p 223

Hughes MK, Swetnam TW, Diaz HF (eds) (2010) Dendroclimatology. Progress and prospects. Springer, Dordrecht, London, p 368

Jędrysek MO, Krąpiec M, Skrzypek G, Kałużny A (2003) Air-pollution effect and paleotemperature scale versus Δ13C records in tree rings and in a peat core (Southern Poland). Water Air Soil Poll 145:359–375

Jones PD, Briffa KR, Osborn T, Lough JM, van Ommen TD, Vinther BM, Luterbacher J, Wahl E, Zwiers F, Mann ME (2009) High-resolution palaeoclimatology of the last millennium: a review of current status and future prospects. Holocene 19(1):3–49

Kaczka RJ, Büntgen U (2006) Spatial autocorrelation and growth/climate response of a high elevation spruce network along the Carpathian arc. TRACE 6:103–112

Kaczka RJ, Czajka B, Łajczak A (2015) The tree-ring growth responses to climate in the timberline ecotone of Babia Góra Mountain. Geographia Polonica 88(2):163–176

Kaczka RJ, Spyt B, Janecka K, Niedźwiedź T, Bednarz Z (2016) Climate reconstruction from tree-rings in the Tatra Mountains. GeoPlanet Earth Planet Sci: 209–229

Kaczka RJ, Spyt B, Janecka K, Musioł R (2017) The Blue Intensity proxy for >400 years growing season temperature reconstruction from the Tatra Mountains. TRACE 15:23–30

Kaczmarczyk M, Kaczka RJ (2014) Wpływ klimatu na przyrost radialny świerka pospolitego (*Picea abies* L. Karst.) w Bieszczadach. Studia i Materiały CEPL w Rogowie 16(40)3: 238–246 (in Polish)

Kamenik C, Schmidt R (2005) Chryspohyte resting stages: a tool for reconstructing winter/spring climate from Alpine lake sediments. Boreas 34:477–489

Kamenik C, Schmidt R, Koinig K, Agustí-Panareda A, Thompson R, Psenner R (2001) The chrysophyte stomatocyst composition in a high alpine lake (Gossenkollesee, Tyrol, Austria) in relation to seasonality, temperature and land-use. Nova Hedwigia 122:1–22

Kemp AES (ed) (1996) Palaeoclimatology and palaeoceanography from laminated sediments. Geological Society Special Publication, London, p 116

Koprowski M (2013) Spatial distribution of introduced Norway spruce growth in lowland Poland: The influence of changing climate and extreme weather events. Quat Int 283:139–146

Koprowski M, Przybylak R, Zielski A, Pospieszyńska A (2012) Tree rings of Scots pine (*Pinus sylvestris* L.) as a source of information about past climate in northern Poland. Int J Biometeorol 56:1–10

Koprowski M, Zielski A (2006) Dendrochronology of Norway spruce (*Picea abies* (L.) Karst.) from two range centres in lowland Poland. Trees 20:383–390
Krąpiec M (1998) Oak dendrochronology of the Neoholocene in Poland. Folia Quatern 69:5–133
Krąpiec M, Jędrysek MO, Skrzypek G, Kałużny A (1998) Carbon and hydrogen isotope ratios in cellulose from oak tree-rings as a record of palaeoclimatic conditions in Southern Poland during the last millennium. Folia Quatern 69:135–150
Krąpiec M, Szychowska-Krąpiec E, Walanus A (2009) Rekonstrukcja termiki okresu zimowego w NE Polsce na podstawie sekwencji przyrostów rocznych sosny zwyczajnej z lat 1582-2004 AD. Prace Kom Paleog Czwart PAU 7:73–82 (in Polish)
Lamb HH (1977) Climate: Present, past and future. Vol. 2, Climate history and the future. Methuen, London, p 835
Larocque-Tobler I (2014) The Polish sub-fossil chironomids. Palaeontol Electron 17. https://doi.org/10.26879/391
Larocque-Tobler I, Filipiak J, Tylmann W, Bonk A, Grosjean M (2015) Comparison between chironomid-inferred mean-August temperature from varved Lake Żabińskie (Poland) and instrumental data since 1896AD. Quat Sci Rev 111:35–50
Larocque-Tobler I, Filipiak J, Tylmann W, Bonk A, Grosjean M (2016) Corrigendum to "Comparison between chironomid-inferred mean-August temperature from varved Lake Żabińskie (Poland) and instrumental data since 1896 AD" [Quat Sci Rev 111 (2015):35–50]. Quat Sci Rev 140:163–167
Limanówka D (2001) Rekonstrukcja warunków klimatycznych Krakowa w pierwszej połowie XVI wieku (Reconstruction of climatic conditions in Cracow in the first half of the 16th century). Materiały Badawcze IMGW, Seria: Meteorologia 33:1–176 (in Polish)
Ljungqvist FC, Krusic PJ, Brattström G, Sundqvist HS (2012) Northern hemisphere temperature patterns in the last 12 centuries. Clim Past 8:227. https://doi.org/10.5194/cp-8-227-2012
Lotter A, Birks HJ, Hofmann W, Marchetto A (1997) Modern diatom, cladocera, chironomid, and chrysophyte cyst assemblages as quantitative indicators for the reconstruction of past environmental conditions in the Alps. I. Climate. J Paleolimnol 18:395–420
Luterbacher J, Rickli R, Xoplaki E, Tinguely C, Beck C, Pfister C, Wanner H (2001) The Late Maunder Minimum (1675-1515) – a key period for studying decadal scale climatic change in Europe. Clim Change 49:441–462
Luterbacher J, Xoplaki E, Küttel M, Zorita E, Gonzalez-Rouco JF, Jones PD, Stossel M, Rutishauser T, Wanner H, Wibig J, Przybylak R (2010) Climate change in Poland in the past centuries and its relationship to European climate: Evidence from reconstructions and coupled climate models. In: Przybylak R, Majorowicz J, Brázdil R et al. (eds) The Polish climate in the European context: An historical overview. Springer, Berlin, pp 3–39
Malik I, Owczarek P (2009) Dendrochronological records of debris flow and avalanche activity in a mid-mountain forest zone (Eastern Sudetes – Central Europe). Geochronometria 34:57–66
Marszelewski W, Skowron R (2006) Ice cover as an indicator of winter temperature changes: case study of the Polish Lowland lakes. Hydrol Sci J 51:336–349
Maruszczak H (1988) Zmiany środowiska przyrodniczego kraju w czasach historycznych (Changes of the natural environment in Poland in historical times). In: Starkel L (ed) Przemiany środowiska geograficznego. Wrocław, Poland, Zakład Narodowy im. Ossolińskich, pp 109–135 (in Polish)
Maruszczak H (1991) Tendencje do zmian klimatu w ostatnim tysiącleciu (Climate change trends in the last millennium). In: Starkel L (ed) Geografia Polski- środowisko przyrodnicze. PWN Warszawa, Warszawa, Poland, pp 182–190 (in Polish)
Masson-Delmotte V, Raffalli-Delerce G, Danis PA, Yiou P, Stievenard M, Guibal F, Mestre O, Bernard V, Goosse H, Hoffmann G, Jouzel J (2005) Changes in European precipitation seasonality and in drought frequencies revealed by a four-century-long tree-ring isotopic record from Brittany, western France. Clim Dynam 24(1):57–69
Michalczewski J (1985) Temperatura średnia dobowa w Warszawie w latach 1779–1983. Zakład Klimatologii IMGW, Warszawa (in Polish)

Michalczewski J (1988) Średnia temperatura dobowa w Warszawie w latach 1760–1763. Przegl Geofiz 4:473–478 (in Polish)

Migała K, Czwartos E, Samojło S (2000) Dynamika rocznych przyrostów radialnych kosodrzewiny (*Pinus mugo* Turra) w świetle warunków klimatycznych Karkonoszy. Opera Corcontica 37:322–328 (in Polish)

Niedźwiedź T (2004) Rekonstrukcja warunków termicznych lata w Tatrach od 1550 roku. In: Kotarba A (ed) Rola małej epoki lodowej w przekształcaniu środowiska przyrodniczego Tatr. Prace Geogr IGiPZ PAN 197:57–88 (in Polish)

Niedźwiedź T (2010) Summer temperatures in the Tatra Mountains during the Maunder Minimum (1645–1715). In: Przybylak R, Majorowicz J, Brázdil R, Kejna M (eds) The Polish climate in the European Context: An historical overview. Springer Science + Business Media BV, pp 397–406

Niedźwiedź T, Glaser R, Hansson D, Helama S, Klimenko V, Łupikasza E, Małarzewski Ł, Nordli Ø, Przybylak R, Riemann D, Solomina O (2015) The historical time frame (past 1000 years). In: The BACC II Author3 Team, second assessment of climate change for the Baltic Sea Basin, regional climate studies, Springer Open, pp 51–65

Nowosad W, Przybylak R, Marciniak K, Syta K (2007) Diariusz Jana Antoniego Chrapowickiego jako źródło do badań klimatu Rzeczypospolitej w II połowie XVII w. (Diary of Jan Antoni Chrapowicki as a source for investigation of the climate of Poland in the second part of 17th century). Klio 9:21–60 (in Polish)

O'Sullivan PE (1983) Annually laminated lake sediments and the study of Quaternary environmental changes. Quat Sci Rev 1:245–313

Ojala AEK, Saarinen T, Salonen VP (2000) Preconditions for the formation of annually laminated lake sediments in southern and central Finland. Boreal Environ Res 5:243–255

Ojala AEK, Alenius T (2005) 10 000 years of interannual sedimentation recorded in the Lake Nautajarvi (Finland) clastic-organic varves. Palaeogeogr Palaeocl 219:285–302

Opała M (2009) Wpływ warunków klimatycznych na kształtowanie się szerokości przyrostu rocznego *Fagus sylvatica*, *Pinus sylvestris* i *Abies alba* z Ojcowskiego Parku Narodowego. Prądnik Prace Muz Szafera 19:219–230 (in Polish)

Opała M (2012) Rekonstrukcja warunków klimatycznych Górnego Śląska w ostatnich stuleciach na podstawie drewna historycznego (Reconstruction of the climatic conditions of the Upper Silesia in recent centuries based on historical wood). Unpublished PhD Thesis (In Polish)

Opała M (2015) The 443-year tree-ring chronology for the scots pine from Upper Silesia (Poland) as a dating tool and climate proxy. Geochronometria 42(1):41–52

Opała M, Mendecki MJ (2012) Zastosowanie skalowanych modeli dendroklimatycznych do rekonstrukcji rzeczywistych warunków termicznych i opadowych na Górnym Śląsku (Application of scaled dendroclimatological models for the reconstruction of thermal and precipitation conditions in the Upper Silesia). Studia i materiały Centrum Edukacji Przyrodniczo-Leśnej 30:55–65 (in Polish)

Opała M, Mendecki MJ (2014) An attempt to dendroclimatic reconstruction of winter temperature based on multispecies tree-ring widths and extreme years chronologies (example of Upper Silesia, Southern Poland). Theor Appl Climatol 115(1–2):73–89

Opała M, Owczarek P (2016) Zmienność warunków termicznych Masywu Śnieżnika odtworzona na podstawie słojów rocznych świerka pospolitego Picea abies Karst. i pozostałości zabudowy drewnianej (Variation of thermal conditions of the Śnieżnik Massif reconstructed based on annual growth rings of Norway spruce Picea abies Karst. and remains of wooden buildings). Przyroda Sudetów 19:211–222 (in Polish)

Opała-Owczarek M, Błaś M, Owczarek P, Sobik M, Godek M (2019) A dendroclimatological study of east- and west-facing slopes in mountainous areas subjected to strong air pollution (the Sudetes, Central Europe). Phys Geogr 40(2):186–208

Osadczuk A (2007) Co kryje dno jeziora. In: Borówka RK (ed) Jezioro Miedwie i Nizina Pyrzycka. Oficyna IN PLUS, Szczecin, pp 84–93 (in Polish)

Ott F, Kramkowski M, Wulf S, Plessen B, Serb J, Tjallingii R, Schwab M, Słowiński M, Brykała D, Tyszkowski S (2018) Site-specific sediment responses to climate change during the last 140 years in three varved lakes in Northern Poland. Holocene 28:464–477

PAGES 2 k Consortium (2013) Continental-scale temperature variability during the past two millennia. Nat Geosci 6:339–346

Pawelczyk S, Pazdur A, Halas S (2004) Stable carbon isotopic composition of tree rings from a pine tree from Augustow Wilderness, Poland, as a temperature and local environment conditions indicator. Isotopes Environ Health Stud 40(2):145–154

Petterson G, Renberg I, Geladi P, Lindberg A, Lindgren A (1993) Spatial uniformity of sediment accumulation in varved lake sediments in northern Sweden. J Paleolimnol 9:195–208

Pfister C (1992) Monthly temperature and precipitation in Central Europe 1525-1979: quantifying documentary evidence on weather and its effects. In: Bradley RS, Jones PD (eds) Climate since AD 1500. Routledge, London, pp 118–142

Pfister C (1999) Wetternachhersage. 500 Jahre Klimavariationen und Naturkatastrophen (1496-1995). Verlag Paul Haupt, Bern, Stuttgart, Wien, p 304

Pla S, Catalan J (2005) Chrysophyte cysts from lake sediments reveal the sub-millennial winter/spring climate variability in the northwestern Mediterranean region throughout the Holocene. Clim Dynam 24:263–278

Pleskot K, Tjallingii R, Makohonienko M, Nowaczyk N, Szczuciński W (2018) Holocene paleo-hydrological reconstruction of Lake Strzeszyńskie (western Poland) and its implications for the central European climatic transition zone. J Paleolim 59:443–459

Pospieszyńska A, Przybylak R (2010) Temperatura powietrza w Toruniu w okresie 1760-1764. In: Bednorz E (ed) Klimat Polski na tle klimatu Europy, Warunki termiczne i opadowe. Seria: Studia i Prace z Geografii i Geologii nr 15, Bogucki Wydawnictwo Naukowe, Poznań, pp 53–66 (in Polish)

Pritzkow C, Wazny T, Heußner KU, Słowiński M, Bieber A, Dorado Liñán I, Helle G, Heinrich I (2016) Minimum winter temperature reconstruction from average earlywood vessel area of European oak (*Quercus robur*) in N-Poland. Palaeogeogr Palaeoclimatol Palaeoecol 449:520–530

Przybylak R (2010) The climate of Poland in recent centuries: A synthesis of current knowledge: Instrumental observations. In: Przybylak R, Majorowicz J, Brázdil R, Kejna M (eds) The Polish climate in the European Context: An historical overview. Springer, Berlin Heidelberg New York, pp 129–166

Przybylak R (2011) Changes in Poland's climate during the last millennium. Czas Geogr 82(1–2):23–48

Przybylak R (2016) Poland's climate in the last millennium, Oxford research encyclopedia, climate science. Oxford University Press USA, p 36. https://doi.org/10.1093/acrefore/9780190228620.013.2

Przybylak R, Bryś K, Bryś T, Wyszyński P (2013) Ekstremalne warunki termiczno-opadowe we Wrocławiu w okresie pomiarów instrumentalnych, 1791–2010. In: Kościk E (ed) Wrocławskie spotkania z historią gospodarczą – spotkanie VIII – Gdy nadciągała wielka woda: Klęski powodzi na ziemiach polskich na przestrzeni wieków, Wrocław, pp 55–72 (in Polish)

Przybylak R, Filipiak J, Oliński P (2014b) Obserwacje meteorologiczne Gottfrieda Reygera w Gdańsku w latach 1722-1769 i ich przydatność do badań zmian klimatu. Przegląd Naukowy Inżynieria i Kształtowanie Środowiska 23(4) 66:360–375 (in Polish)

Przybylak R, Majorowicz J, Brázdil R, Kejna M (eds) (2010a) The Polish climate in the European context: An historical overview. New York, Springer, p 535

Przybylak R, Marciniak K, Dufaj E (2008) Precipitation and other atmospheric phenomena in the central and north-eastern parts of former Poland from 1658 to 1667. In: Rodriguez JS, India MB, Anfrons EA (eds) Cambio climatico regional y sus impactos M., Publicaciones de la Asociación Española de Climatologia (AEC), Serie A 6:239–248

Przybylak R, Majorowicz J, Wójcik G, Zielski A, Chorążyczewski W, Marciniak K, Nowosad W, Oliński P, Syta K (2005) Temperature changes in Poland from the 16th to the 20th centuries. Int J Climatol 25:773–791

Przybylak R, Marciniak K (2010) Climate changes in the Central and North-Eastern Parts of the Polish-Lithuanian Commonwealth from 1656 to 1685. In: Przybylak R, Majorowicz J, Brázdil R, Kejna M (eds) The Polish climate in the European context: An historical overview. Springer, Berlin Heidelberg New York, pp 423–443

Przybylak R, Oliński P, Chorążyczewski W, Nowosad W, Syta K (2010b) The climate of Poland in recent centuries: A synthesis of current knowledge: Documentary evidence. In: Przybylak R, Majorowicz J, Brázdil R, Kejna M (eds) The Polish climate in the European context: An historical overview. Springer, New York, pp 167–190

Przybylak R, Oliński P, Koprowski M, Filipiak J, Pospieszyńska A, Chorążyczewski W, Puchałka R, Dąbrowski HP (2020) Droughts in the area of Poland in recent centuries in the light of multi-proxy data, Clim Past 16(2):627–661

Przybylak R, Pospieszyńska A (2010) Air temperature in Wrocław (Breslau) in the period 1710–1721 based on measurements made by David von Grebner. Acta Agrophys, Rozprawy i Monografie 184(5):35–43

Przybylak R, Pospieszyńska A, Nowakowski M (2010c) Temperatura powietrza w Żaganiu w okresie 1781-1792. In: Bednorz E (ed) Klimat Polski na tle klimatu Europy, Warunki termiczne i opadowe. Seria: Studia i Prace z Geografii i Geologii nr 15, Bogucki Wydawnictwo Naukowe, Poznań, pp 67–78 (in Polish)

Przybylak R, Pospieszyńska A, Wyszyński P, Nowakowski M (2014a) Air temperature changes in Żagan (Poland) in the period from 1781 to 1792. Int J Climatol 34:2408–2426. https://doi.org/10.1002/joc.3847

Przybylak R, Wójcik G, Marciniak K, Chorążyczewski W, Nowosad W, Oliński P, Syta K (2004) Zmienność warunków termiczno-opadowych w Polsce w okresie 1501-1840 w świetle danych historycznych (Variability of temperature and precipitation conditions in Poland from 1501 to 1840 based on historical sources). Przegląd Geograficzny 76:5–31 (in Polish)

Quinlan R, Smol JP, Hall RI (1998) Quantitative inferences of past hypolimnetic anoxia in south-central Ontario lakes using fossil midges (Diptera: Chironomidae). Can J Fish Aquat Sci 55:587–596

Ralska-Jasiewiczowa M, Goslar T, Madeyska T, Starkel L (eds) (1998) Lake Gościąż, Central Poland. A monographic study. W. Szafer Institute of Botany, Polish Academy of Sciences, Krakow, p 340 (in Polish)

Renberg I (1981) Formation, structure and visual appearance of iron-rich, varved lake sediments. Verh Internat Verein Limnol 21:94–101

Renberg I (1986) Photographic demonstration of the annual nature of a varve type common in N. Swedish lake sediments. Hydrobiologia 140:93–95

Reyger G (1770) Die Beschaffenheit der Witterung in Danzig vom Jahre 1722–1769 beobachtet, nach ihren Veränderungen und Ursachen erwogen, und mit dem Wetter anderer Orten verglichen. Gdańsk und Leipzig: Daniel Ludwig Wedel Press, p 520 (in German)

Reyger G (1788) Die Beschaffenheit der Witterung in Danzig vom Jahre 1770–1786 beobachtet, nach ihren Veränderungen und Ursachen erwogen, und mit dem Wetter anderer Orten verglichen. Gdańsk: Daniel Ludwig Wedel Press, p 200 (in German)

Rojecki A (1965) O najdawniejszych obserwacjach meteorologicznych w Toruniu na tle wyników jednocześnie prowadzonych spostrzeżeń w Warszawie. Przegl Geofiz 2:141–151 (in Polish)

Rothwell RG, Croudace I (eds) (2015) Micro-XRF studies of sediment cores: perspective on capability and application in the environmental sciences. Developments in Paleoenvironmental Research 17, Springer, Dordrecht, pp 1–21

Saarnisto M (1986) Annually laminated lake sediments. In: Berglund BE (ed) Handbook of Holocene palaeoecology and palaeohydrology. Wiley, pp 343–370

Saarnisto M, Huttunen P, Tolonen K (1977) Annual lamination of sediments in Lake Lovojärvi, southern Finland, during the past 600 years. Annales Botanici Fennici 14:35–45

Sadowski M (1991) Variability of extreme climatic events in Central Europe since the 13th century. Z Meteorol 41:350–356

Savva Y, Oleksyn J, Reich PB, Tjoelker MG, Vaganov EA, Modrzynski J (2006) Interannual growth response of Norway spruce to climate along an altitudinal gradient in the Tatra Mountains. Poland Trees 20:735. https://doi.org/10.1007/s00468-006-0088-9

Schimanke S, Meier HEM, Kjellström E, Strandberg G, Hordoir R (2012) The climate of the Baltic Sea region during the last millennium simulated with a regional climate model. Clim Past 8:1419–1433

Schweingruber FH, Bräker OU, Schär E (1979) Dendroclimatic studies on conifers from Central Europe and Great Britain. Boreas 8:427–452

Słowiński M, Zawiska I, Ott F et al (2017) Differential proxy responses to late Allerød and early Younger Dryas climatic change recorded in varved sediments of the Trzechowskie palaeolake in Northern Poland. Quat Sci Rev 158:94–106

Speer JH (2010) Fundamentals of tree-ring research. University of Arizona Press, p 333

Szczepanek M, Pazdur A, Pawełczyk S, Boettger T, Szychowska-Krąpiec E (2006) Hydrogen, carbon and oxygen isotopes in pine and oak tree rings from Southern Poland as climatic indicators in years 1900–2003. Geochronometria 25:67–76

Szychowska-Krąpiec E (1998) Spruce chronology from Mt Pilsko area. (Żywiec Beskid Range) 1641-1995 AD. Bull Pol Acad Sci Earth Sci 46(2):75–86

Szychowska-Krąpiec E (2010) Long-term chronologies of pine (*Pinus sylvestris* L.) and fir (*Abies alba* Mill.) from the Małopolska region and their palaeoclimatic interpretation. Folia Quatern 79:1–124

Szychowska-Krąpiec E, Krąpiec M (2002) Kondycja drzewostanów sosnowych w Ojcowskim Parku Narodowym w świetle analizy dendroekologicznej. Chrońmy Przyrodę Ojczystą 58(5):5–22 (in Polish)

Szychowska-Krąpiec E, Krąpiec M (2005) The Scots Pine Chronology (1582-2004 AD) for the Suwałki region, NE Poland. Geochronometria 24:41–52

Tichavský R, Šilhán K, Tolasz R (2017) Tree ring-based chronology of hydro-geomorphic processes as a fundament for identification of hydro-meteorological triggers in the Hrubý Jeseník Mountains (Central Europe). Sci Total Environ 579:1904–1917

Trepińska J (1997) Wieloletni przebieg średnich miesięcznych wartości ciśnienia powietrza w Krakowie, 1792–1995. In: Trepińska J (ed) Wahania klimatu w Krakowie (1792–1995). Instytut Geografii UJ, Kraków, pp 131–161 (in Polish)

Trepińska J, Kowanetz L (1997) Wieloletni przebieg średnich miesięcznych wartości temperatury powietrza w Krakowie, 1792–1995. In: Trepińska J (ed) Wahania klimatu w Krakowie (1792–1995). Instytut Geografii UJ, Kraków, pp 99–130 (in Polish)

Treydte K, Frank D, Esper J, Andreu L, Bednarz Z, Berninger F, Boettger T, D'alessandro C, Etien N, Filot M (2007) Signal strength and climate calibration of a European tree-ring isotope network. Geophys Res Lett 34:L24302. https://doi.org/10.1029/2007GL031106

Twardosz R (1999) Warunki pluwialne w Krakowie w latach 1792–1998. Czasopismo Geogr 70(2):221–234 (in Polish)

Tylmann W, Bonk A, Goslar T, Wulf S, Grosjean M (2016) Calibrating ^{210}Pb dating results with varve chronology and independent chronostratigraphic markers: Problems and implications. Quat Geochronol 32:1–10

Tylmann W, Szpakowska K, Ohlendorf C, Woszczyk M, Zolitschka B (2012) Conditions for deposition of annually laminated sediments in small meromictic lakes: a case study of Lake Suminko (northern Poland). J Paleolim 47:55–70

Tylmann W, Woźniak PP, Czarnecka K, Jaźwiecka M (2006) New sites with laminated lake sediments in north-eastern Poland: preliminary results of field survey. Limnol Rev 6:283–288

Tylmann W, Zawadzka M (2008) Nowe stanowiska laminowanych osadów jeziornych na Pojezierzu Kaszubskim. Przegl Geol 56:404–408 (in Polish)

Tylmann W, Zolitschka B, Enters D, Ohlendorf C (2013) Laminated lake sediments in north-east Poland: distribution, preconditions for formation and potential for paleoenvironmental investigation. J Paleolimnol 50:487–503 van Engelen AVW, Buisman J, Ijnsen F (2001) A millennium of weather, winds and water in the low countries. In: Jones PD, Ogilvie AEJ, Davies TD et al (eds) History and climate. Springer, pp 101–124

Velle G, Brodersen K, Birks HJB, Willassen E (2010) Midges as quantitative temperature indicator species: Lessons for palaeoecology. Holocene 20:989–1002

Vitas A (2004) Tree rings of Norway spruce (*Picea abies* (L.) Karsten) in Lithuania as drought indicators: dendroecological approach. Polish J Ecol 52 (2):201–210

Wacnik A (2009) Vegetation development in the Lake Miłkowskie area, north-eastern Poland, from the Plenivistulian to the late Holocene. Acta Paleobotanica 49:287–335

Ważny T, Eckstein D (1991) The dendrochronological signal of oak (*Quercus* spp.) in Poland. Dendrochronologia 9:35–49

Więckowski K (1978) Bottom deposits in lakes of different regions of Poland (their characteristics, thickness and rates of accumulation). Pol Arch Hydrobiol 25:483–489

Wilczyński S, Feliksik E (2004) The dendrochronological monitoring of the Western Beskid mountains (Southern Poland) on the basis of radial increments of Norway spruce (*Picea abies* (L.) Karst.), EJPAU 7(2):#07. http://www.ejpau.media.pl/volume7/issue2/forestry/art-07.html

Wilczyński S, Feliksik E (2005) Disturbances in variation of the annual ring width of Norway spruce in the Polish Western Beskids Mountain. J For Sci 51(12):539–547

Wilczyński S, Krąpiec M, Szychowska-Krąpiec E, Zielski A (2001) Dendroclimatic regions of Scots pine, *Pinus sylvestris* L. in Poland. Sylwan 8:53–61

Wilczyński S, Skrzyszewski J (2002a) The climatic signal in tree-rings of Scots pine (*Pinus sylvestris* L.) from foot-hills of the Sudetic Mountains (southern Poland). Forstw Cbl 121:15–24

Wilczyński S, Skrzyszewski J (2002b) Dependence of scots pine tree-rings on climatic conditions in southern Poland (Carpatian Mts.). Electron J Pol Agric Univ, Forestry 5:2 #07. http://www.ejpau.media.pl/volume5/issue2/forestry/art-07.html

Wilkinson AN, Zeeb BA, Smol JP (2002) Atlas of chrysophycean cysts. Kluwer Academic Publishers, Dordrecht

Xoplaki E, Luterbacher J, Paeth H, Dietrich D, Steiner N, Grosjean M, Wanner H (2005) European spring and autumn temperature variability and change of extremes over the last half millennium. Geophys Res Lett 32:L15713. https://doi.org/10.1029/2005GL023424

Żarczyński M, Tylmann W, Goslar T (2018) Multiple varve chronologies for the last 2000 years from the sediments of Lake Żabińskie (northeastern Poland) – Comparison of strategies for varve counting and uncertainty estimations. Quat Geochronol 47:107–119

Zielonka T, Holeksa J, Sz Ciapała (2008) A reconstruction of flood events using scarred trees in the Tatra Mountains, Poland. Dendrochronologia 26(3):173–183

Zielski A (1997) Uwarunkowania środowiskowe przyrostów radialnych sosny zwyczajnej (*Pinus sylvestris* L.) w Polsce północnej na podstawie wielowiekowej chronologii. Wydawnictwo UMK, Toruń, p 127 (in Polish)

Zielski A, Sygit W (1998) Wpływ klimatu na przyrost radialny sosny zwyczajnej (Pinus sylvestris L.), badania wzdłuż równoleżnika 52° N i transekcie Śląsk - Białowieża. [In:] Breymeyer A, Roo-Zielińska E (red.). Reakcja borów sosnowych na zmianę klimatu wzdłuż równoleżnika 52° N (12-32° E) oraz na zmiany w depozycji zanieczyszczeń chemicznych na transekcie Śląsk - Białowieża. Dokumentacja Geograficzna IG PAN Warszawa 13:161–185 (in Polish)

Zielski A, Krąpiec M (2004) Dendrochronologia. PWN, Warszawa, p 328 (in Polish)

Zielski A, Koprowski M, Krapiec M (2010) Dendrochronological data. In: Przybylak R, Majorowicz J, Brázdil R, Kejna M (eds.) The Polish climate in the European context: An historical overview. Springer, Berlin Heidelberg New York, pp 191–217

Zillén L, Snowball I, Sandgren P et al (2003) Occurrence of varved lake sediment sequences in Värmland, west central Sweden: lake characteristics, varve chronology and AMS radiocarbon dating. Boreas 32:612–616

Zolitschka B (2007) Varved lake sediments. In: Elias SA (ed) Encyclopedia of quaternary science. Elsevier, pp 3105–3114

Zolitschka B, Francus P, Ojala AEK et al (2015) Varves in lake sediments – a review. Quat Sci Rev 117:1–41

Part III
Recent Climate Change

Chapter 6
Change of Atmospheric Circulation

Tadeusz Niedźwiedź and Zbigniew Ustrnul

Abstract The changes and variability of atmospheric circulation over Poland were analysed using several methodological approaches containing different spatial scales. The traditional manual classifications by Niedźwiedź and the well-known Grosswetterlagen were used. Simultaneously, an automatic classification was made by Ustrnul and Lityński and the NAO index was also measured for mutual comparison. All of the analyses concerned basically the main period of 1951–2018. However, thanks to the availability of the data from Niedźwiedź calendar of circulation, it was possible to extend the time series back to 1873. All of the analyses confirmed that Poland is significantly affected by the westerly circulation, which has a major influence on the formation of the climate. Throughout the entire long period from the end of the nineteenth century to the present day, its large fluctuations and a generally weak increasing trend were found. A similar situation can be also seen for the period 1951–2018, however, an increased westerly circulation was detected in the 1980s and 1990s. The turn of the 1960s and 1970s marked the weakening of this circulation pattern, as did the last 10 years. Generally, a high level of comparability in the course of circulation indices according to the classifications of Niedźwiedź, Ustrnul, Lityński and the Grosswetterlagen was found.

Introduction

Atmospheric circulation is one of the most important factors, together with solar radiation, shaping weather and climate conditions in moderate latitudes. Poland, located in Central Europe, is characterized by high weather variability associated

T. Niedźwiedź (✉)
Faculty of Natural Sciences, Institute of Earth Sciences, University of Silesia in Katowice, Katowice, Poland
e-mail: tadeusz.niedzwiedz@us.edu.pl

Z. Ustrnul
Department of Climatology, Institute of Geography and Spatial Management, Jagiellonian University, Kraków, Poland
e-mail: zbigniew.ustrnul@uj.edu.pl

M. Falarz (ed.), *Climate Change in Poland*, Springer Climate,
https://doi.org/10.1007/978-3-030-70328-8_6

with the movement of low and high-pressure systems. Most often (about 65% of days), masses of moist polar-maritime air come to Poland from the Atlantic Ocean. They cause winter thaws and, in summer, cooler periods combined with rainfall. Less often (about 29% of days) dry masses of polar-continental airflow from the east, cold in winter and hot in summer. Thus, the climatic conditions of Poland are most often affected by zonal circulation (west-east). Rapid changes in weather occur with passing atmospheric fronts (approx. 40–60% of days of the year), which are most often driven from the west and north-west.

Meridional circulation (north-south) occurs less frequently but causes large thermal anomalies. Blockage of the air supply from the Atlantic by the highlands between the British Isles and Scandinavia causes advection of Arctic air from the north (4–6% of days in a year). Such situations occur most often in spring (May and April, 18–21% of days) and in autumn (October 11%), causing frost. In the summer, the encroachment of cooler air from the northern sector, caused by Mediterranean cyclones (Degirmendžić and Kożuchowski 2017) located above western Ukraine, causes heavy rainfall for several days on the northern slopes of the Carpathians, leading to catastrophic floods (Niedźwiedź et al. 2014; Niedźwiedź and Łupikasza 2016). Advection of air from the southern sector (high pressure above Eastern Europe and low pressure in Western Europe) favours the occurrence of foehn winds in winter and transient seasons (in the foothills of the Carpathians and the Sudetes). In summer, it contributes to the occurrence of hot weather, especially if tropical air comes in. Annually, the frequency of tropical air advection does not exceed 2–3% of days.

A review of work on the circulation of the atmosphere and its impact on the climate of Poland has recently been published (Niedźwiedź and Łupikasza 2019). In Polish, the study has benefited from research in the field of synoptic climatology, the subjective circulation typology of Osuchowska-Klein (1978, 1991) and the macroscale classification of "Grosswetterlagen" for Central Europe, according to Hess and Brezowsky (1977) and Werner and Gerstengarbe (2010). The most popular quantitative classification of circulation types and weather types is that of J. Lityński (1969), with further modifications and extension by other authors (Pawłowska et al. 2000; Pianko-Kluczyńska 2007; Kulesza 2017; Nowosad 2017). An automatic classification of circulation types for the whole Northern Hemisphere was developed by Ustrnul (1997) and an objective classification for Poland by Piotrowski (2009, 2010). Regional typologies include the calendar of circulation types for southern Poland (Niedźwiedź 1981), the East-Central Europe region (Bartoszek 2017) and for the Sudetes (Ojrzyńska 2012). In addition to investigations of synoptic types, the long-term variability of deep cyclones (Bielec-Bąkowska 2010) and strong highs (Bielec-Bąkowska 2016) over Poland have been investigated.

Atmospheric circulation primarily determines the direction and speed of the wind. It has a significant impact on air temperature (e.g. Ustrnul 2000; Wibig 2001; Ustrnul et al. 2010; Wypych et al. 2017) and precipitation (e.g. Ustrnul and Czekierda 2001; Niedźwiedź 2003; Degirmendžić et al. 2004; Łupikasza 2016; Wypych et al. 2018). Circulation factors have also been shown to influence cloud formation (Żmudzka 2007; Matuszko and Węglarczyk 2018), thunderstorms (Bielec 2000; Kolendowicz 2006), snow cover (Falarz 2007; Bednorz 2011), extreme meteorological phenomena

(Ustrnul and Czekierda 2009; Ustrnul et al. 2013) and a number of other climate elements, including all types of weather (e.g. Niedźwiedź 1983; Kaszewski 1992). Air circulation also strongly influences biometeorological conditions (e.g. Błażejczyk and Twardosz 2010; Nowosad et al. 2013; Kolendowicz et al. 2018; Owczarek 2019; Tomczyk and Owczarek 2020).

In synoptic climatology, in addition to the types of circulation characterizing the synoptic conditions of each day, more synthetic circulation indices are used to determine the dynamics of the atmosphere in individual months, seasons or years. They are most often used to study the long-term variability of circulation conditions and the determination of the circulation epochs (Degirmendžić et al. 2000). In Poland, among the macroscale indices in climatological publications, the North Atlantic Oscillation Index (NAO) was most often used, defined by J. Hurrell (1995) as the difference between standardized atmospheric pressure values in the Azores High and the Icelandic Low (e.g. Wibig 2000; Marsz and Styszyńska 2002; Styszyńska et al. 2019). The macro-types of mid-tropospheric circulation developed by Russian meteorologists Wangenheim and Girs were adjusted for the European-Atlantic region by K. Kożuchowski and J. Degirmendžić (2018): the form of zonal circulation—W—and two forms of meridional circulation: C—advancement of air masses from the north to Poland, located on the east side of the high-pressure wedge; and E—air inflow from the south on the east side of the upper trough (Kożuchowski 2011). Numerical zonal Ws and meridional Wp circulation indices were developed by J. Lityński (1969) and Z. Ustrnul (1997, 2007). Three simple circulation indices, based on the frequency of circulation types, were calculated by T. Niedźwiedź (1996, 2000): zonal circulation index—Wi; meridional circulation index—Si; and cyclonicity index—Ci.

For the whole area of Poland and Central Europe, Lityński (1969) determined two circulation indices, calculated on the basis of atmospheric pressure gradients. The zonal circulation index, Ws, is the averaged parallel component of the geostrophic wind speed in $m \cdot s^{-1}$ determined on the basis of the difference in mean pressure between the parallels 40° N and 65° N (in the 0°–35° E longitudinal zone). The meridional circulation index, Wp, was calculated on the basis of the pressure difference between the 35° E and 0° E meridians (in the zone limited by latitudes 40° N and 65° N). These indices were used to create the numerical classification of circulation types described earlier. Research on them was continued by Nowosad (2017), who described the variability of the meridional circulation index and presented the multi-year zonal circulation index for the period 1948–2015 (Nowosad 2017). Geostrophic wind was also used, among others, to study the airflow over northern Poland (Miętus 1996), the whole of Poland (Marosz 2016), East-Central Europe (Bartoszek 2017) and over the entire Northern Hemisphere (Ustrnul 1997).

Data and Research Methods

Circulation Types for Southern Poland

The classification of circulation types for southern Poland (Niedźwiedź 1981) covers the period from September 1873 to today. This catalogue is available on the website of the Institute of Earth Sciences at the Faculty of Natural Sciences of the University of Silesia (Niedźwiedź 2019). The classification was modelled on the typology of atmospheric circulation developed by Lamb (1972) for the British Isles, with some modifications, especially concerning non-advection situations. On the basis of synoptic maps of Europe, the direction of air mass movement (N, NE, E, SE, S, SW, W, NW) and the type of baric system were determined (a—anticyclonic situation, c—cyclonic situation). In addition to sixteen advection situations, there are two non-directional types: Ca—high-pressure centre and Ka—anticyclonic wedge or ridge; and two cyclonic types with differentiated advection: Cc—low pressure centre and Bc—cyclonic troughs. Baric col and low gradient situations difficult to identify are marked with the letter "x". Thus, the entire classification includes twenty-one types (ten anticyclonic types, ten cyclonic types and one indefinite type). By combining adjacent types, a shortened version is also obtained for eleven situations (N + NEa or c; E + SEa or c; S + SWa or c; W + NWa or c; Ca + Ka, Cc + Bc, x), useful in studies of periods shorter than thirty years. For the years 2001–2018, the author of the classification has a version available for nine regions of Poland, bounded by parallels 51° N and 53° N and meridians 18° E and 21° E.

Circulation Indices for Southern Poland

Simple statistical indices of atmospheric circulation were developed on the basis of the number of days (frequency) of the particular circulation types (Niedźwiedź 1996, 2000). They were modelled on four indices P, M, S and C, which were proposed by Murray and Lewis (1966), using the classification developed by Lamb (1972) for circulation types for the British Isles. After some modifications, it was decided to calculate three circulation indices based on the number of days with particular circulation types: western circulation zonal index—Wi; southern circulation meridional index—Si; and cyclonicity index—Ci. Indices can be determined for months, seasons and years by adding weight points (−2, −1, 0, +1, +2) assigned to particular types of circulation. In the first version, the sum of points was the value of the index. Due to the different number of days in months, seasons and years, it was difficult to compare these indices; therefore, in the latest version, their values are expressed as a percentage of the highest possible number of points (two points times the number of days).

To calculate the circulation indices, particular scores were attributed to each circulation type, depending on air advection and summarized on a monthly, seasonal or annual basis:

- Wi index: +2 for W, +1 for NW and SW, −2 for E, −1 for NE and SE. Positive values of Wi indicate predominance of air advection from the western sector, while negative values point to a strong easterly airflow.
- Si index: +2 for S, +1 for SW and SE, −2 for N, −1 for NW and NE. Positive values of Si indicate predominance of southern advection, while negative values point to a strong northerly airflow.
- Ci index: +2 for Cc and Bc, +1 for Nc, NEc, Ec, SEc, Sc, SWc, Wc and NWc, −1 Na, NEa, Ea, SEa, Sa, SWa, Wa and NWa, −2 for Ca and Ka. Positive values of Ci indicate a predominance of cyclonic types while negative values denote anticyclonic types.

The monthly total of the scores for each index was divided by double the number of days in that particular month and multiplied by 100 to express it as a percentage. A Wi index value of 100% means that, in the entire month, throughout every day, only a western airflow was observed (conversely, −100% means that every day experienced only an eastern airflow). An Si index value of 100% means that, in the entire month, throughout every day, there was only a southern airflow (whereas −100% means that there was only a northern airflow throughout every day of the month). A Ci index value of 100% means that in the entire month, throughout every day the situation was anticyclonic Ca or Ka (while −100% means that every day was cyclonic type Cc or Bc). In reality, such values are not achieved.

The above indices were used to study the long-term variability of atmospheric circulation in southern Poland (Niedźwiedź 1996). At the Institute of Earth Sciences of the University of Silesia in Katowice, values of these circulation indices are currently available for southern Poland for the period 1873.09–2019.12, and for nine regions of Poland for the years 2000.12–2019.12. They are available to interested parties.

Other Circulation Indices

The main features of changes and variations in circulation over Poland are presented in a mesoscale approach, developed first for southern Poland. This approach was then used, for a shorter period of 2001–2018, to characterize atmospheric circulation in other regions of Poland, which allows us to conclude that the features of circulation changes throughout the entire study area are similar. Of course, the analysis confirmed some regional differences, but these, however, are not significant.

On the other hand, for a full assessment of circular changes, it is worth reaching for other measures and indices that are of a regional nature and, as a result, allow for comparative analysis. Therefore, as presented at the beginning of the chapter, other proven classifications of circulation types were used. The classification of circulation types by J. Lityński (1969), Z. Ustrnul (1997) and the Grosswetterlagen classification (Hess and Brezowsky 1977; Werner and Gerstengarbe 2010) known in Europe were taken into account. On the basis of these classifications, circulation indices were determined, giving information about the dominance of air inflow from individual major directions. Due to the importance of Europe's climate, the well-known NAO index (Hurrell 1995) for the Northern Hemisphere was also taken into consideration.

Research Results Concerning Changes of Atmospheric Circulation in Poland

Changes of Circulation Indices in Southern Poland (1873–2019)

Index of Zonal Westerly Circulation (Wi)

This index reports the intensity of parallel airflow over southern Poland. On an annual basis, circulation from the western sector prevails over the flow of air from the eastern sector. This is evidenced by the positive value of the Wi index (15.9 with standard deviation, SD = ±6.9). In the period 1874–2018, there were large fluctuations of this index from −5.7 in 1963 (intensive air inflow from the east) to 33.8 in 1990, when western circulation was most common (Fig. 6.1). Earlier, such high values of the Wi (33.0) index were recorded in 1899. Negative annual values of the Wi index, apart from 1963, occurred only three times: in 1972 (−0.9), 1996 (−2.3) and 2018 (−3.8).

Over many years, there has been a clear downward trend in the annual values of western flow intensity (Fig. 6.1), with a trend of −0.25/10 years. Changes of the Wi annual values are an effect of a decreasing trend of westerly circulation in summer. The strongest decrease in the Wi index is visible after a maximum in 1990 to modern times. The average ten-year values changed from 21.6 in the period 1985–1994 to 10.4 in the years 2009–2018. The period from 1956 to 2018 is marked by much larger fluctuations in the Wi ratio than in previous years. On an annual scale, the highest intensity of western flow was observed in three periods: 1891–1904, 1924–1932 and 1983–1994. However, the decrease in the 10-year Wi value below the long-term average was marked in 1874–1890 and the strongest in the period 1956–1982. This index is well correlated with winter temperature (Niedźwiedź 1983, 2000; Niedźwiedź et al. 1994).

In the course of a year, the Wi index changes from −3.4 in May to 26.5 in December. Extreme monthly Wi values range from −58.1 in December 1969 to 82.3 in January 1983. Among the seasons, the most intense western circulation is observed in winter (Wi = 22.7) and in autumn (20.1), while the weakest is in spring (3.7). In summer (Wi = 17.2) it is slightly higher than the annual average.

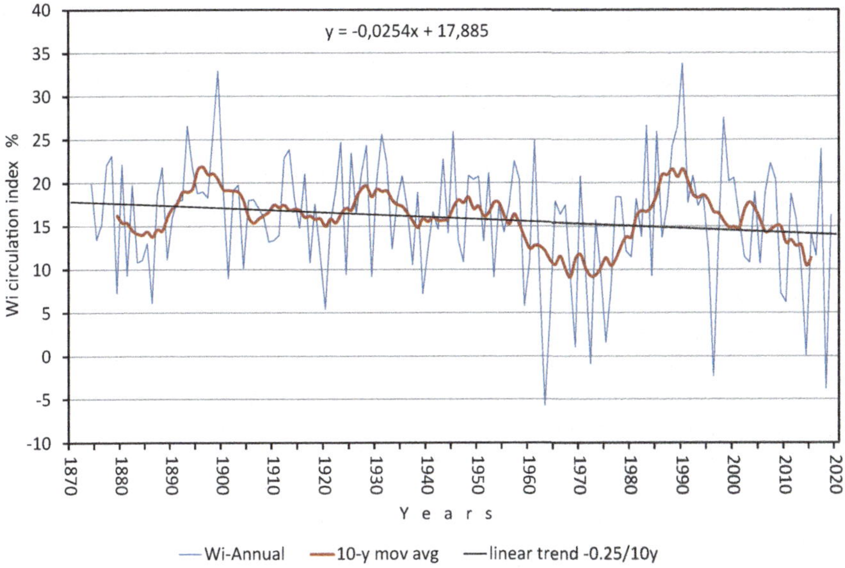

Fig. 6.1 Long-term variability of the annual westerly circulation index (Wi) values

Winter experiences the highest western circulation activity compared to other seasons. The average value of the Wi index reaches 22.7 (SD = ±17.0). Over the long term (Fig. 6.2) there is no trend, while the range of fluctuations is very wide,

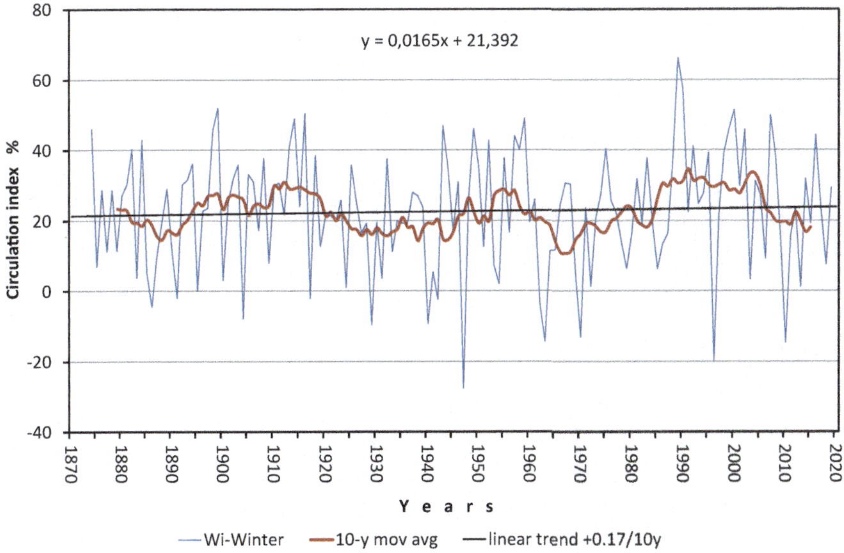

Fig. 6.2 Long-term variability of the winter (DJF) westerly circulation index (Wi) values

from −27.7 during the cold winter of 1946/1947 to +66.4 in the winter of 1988/1989. In three periods, the encroachment of air from the western sector is clearly marked: 1894–1919, with an average value of Wi = 31.1 during the winters of 1906/1907–1915/1916; 1953–1958 (Wi = +28.9 in the decade 1949/1950–1958/1959); and 1986–2005, with the highest ten-year average Wi values during the winters of 1985/1986–1994/1995 (+34.7) and 1997/1998–2006/2007 (+33.4). Nowadays, since 2006, western circulation has weakened. Earlier, three periods of reduced activity of western circulation were found during the winters before 1894 and in the years 1920–1952 and 1959–1984, with a ten-year average value of Wi = +10.7 in the winter periods 1961/1962–1970/1971.

In spring, the weakest activity of western circulation is noted in comparison with other seasons. The average value of the Wi index in spring reaches only +3.7 (SD = ±12.0). In the multiannual sequence (Fig. 6.3), as in winter, there is no statistically significant trend, while the range of fluctuations is from −25.7 in spring 1984 and −24.5 in spring 1918 to +32.5 and +30.5 in spring in 1899. The highest activity of western circulation occurred in 1925–1924, with the highest ten-year average value of the index Wi = +12.0 during spring in the years 1938–1947. Today, since 2012, a weakening of western circulation in the spring has been observed. However, the largest circulation activity from the Eastern sector appears in three ten-year periods: spring 1877–1886 (Wi = −3.4), 1916–1925 (Wi = −1.9) and 1971–1980 (Wi = −3.3).

In summer, western circulation prevails over eastern, as evidenced by the positive average value of the Wi index (Wi = +17.2, SD = ±14.0). In the multiannual

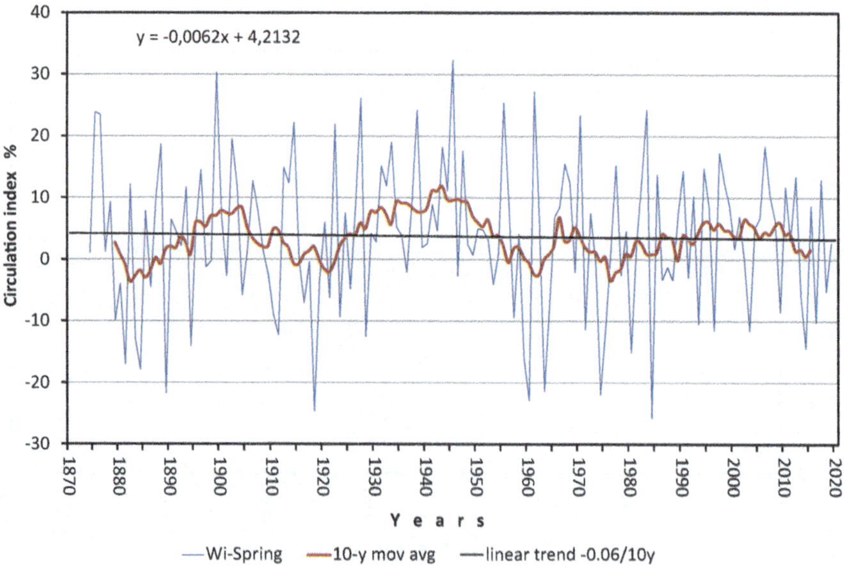

Fig. 6.3 Long-term variability of the spring (MAM) westerly circulation index (Wi) values

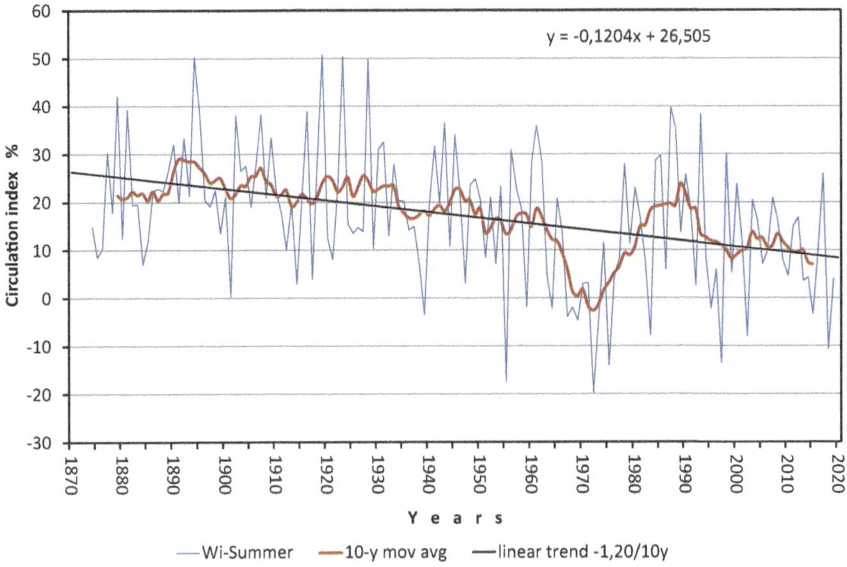

Fig. 6.4 Long-term variability of the summer (JJA) westerly circulation index (Wi) values

sequence (Fig. 6.4) there is a statistically significant downward trend (−1.20Wi/10y). The 10-year average values of the Wi index decreased from +29.2 in 1886–1895 to 7.2 in the period 2010–2019. The range of fluctuations of the Wi index is from −19.5 in the summer of 1972 and −17.2 in 1955 to +51.0 in the summer of 1919. Values above 40 occurred in four years: 1894 (+50.5), 1923 (+50.7), 1928 (+50.2) and 1987 (+40.2). The greatest summer weakening of western circulation was noted in the period of 1963–1982 with the Wi index falling to −2.4 in the decade 1967–1976.

In autumn, there is a high activity of western circulation, intermediate between summer and winter values. The average value of the Wi index reaches +20.1 (SD = ±13.4). In the multi-annual sequence (Fig. 6.5), as in winter and spring, there is no statistically significant trend, while the range of fluctuations is from −10.1 in autumn 1920 and −8.9 in autumn 1879 to an exceptionally high value of +63.7 in autumn 1899. The value +50.1 appeared in autumn in 1893 and 1919. Two periods registered consistently high levels of western circulation: 1924–1936 and 1979–1995. During these periods, in two decades (1923–1932 and 1983–1992) the average values of the Wi index reached the value of +31.8. However, the period of greatest autumn weakening of western circulation falls in 1901–1923, with an average of +8.8 in the decade 1901–1910.

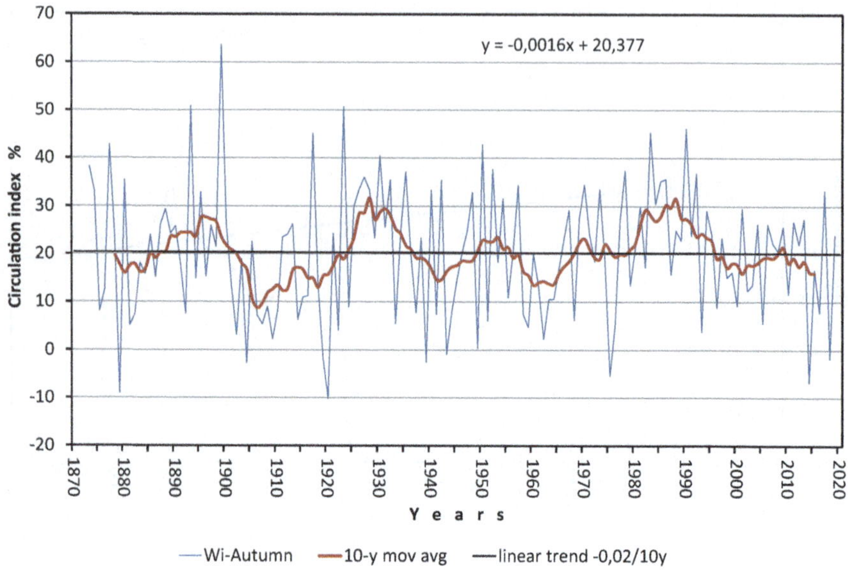

Fig. 6.5 Long-term variability of the autumn (SON) westerly circulation index (Wi) values

Index of Meridional Southern Circulation (Si)

This index reports the intensity of meridional airflow over southern Poland. On an annual basis, circulation from the southern sector slightly outweighs the airflow from the northern sector (Si = 0.9; SD = ±4.7). In the period 1874–2018, the annual average values of the Si index changed from −12.3 in 1899 (intensive air inflow from the north) to 14.1 in 2014, when circulation from the southern sector was most common (Fig. 6.6). High Si values, exceeding +10.0, were also recorded in 1951 (13.8) and 1960 (13.7). In contrast, Si lower than −8.0 was recorded in 1875 (−9.7), 1913 (−8.2) and in 1997 (−8.5).

In the multi-annual sequence (Fig. 6.6), there was a slight upward trend in the annual values of the Si index (+0.14Si/10y). Given the variability of 10-year moving averages, four periods of positive deviations from the long-term average can be distinguished (1886–1891, 1920–1942, 1951–1965 and 2010–2018) and four periods when the inflow of air from the northern sector was more intense than from the sector southern (1874–1885, 1892–1919, 1943–1950 and 1978–2009). The most consistent period of air advection from the south was noted in the years 1920–1942, when only one year, 1933, showed a negative Si index (−5.4). The highest average 10-year values (4.9) occurred in the period 1923–1932. However, the lowest negative values of the Si (−2.9) index were recorded in the first 10-year period under consideration, 1874–1883.

In the annual sequence, the Si index ranges from −14.5 in July and −13.4 in June to +13.8 in November and +10.6 in October. Extreme monthly Si values fluctuated from −50.0 in July 1934 to +75.9 in February 1972 and +75.8 in October 1907.

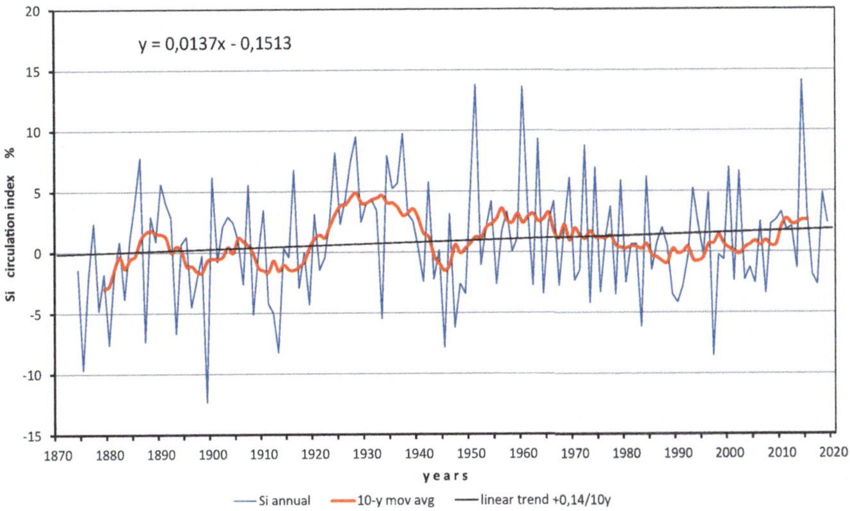

Fig. 6.6 Long-term variability of the annual southerly circulation index (Si) values

Intensive inflow of air masses from the northern sector in July 1934, combined with high cyclonal activity (Ci = +19.4), contributed to the occurrence of rainfall in the Western Carpathians, which caused one of the largest floods on record in the upper Vistula basin. Of the four seasons, autumn (Si = +8.4) and winter (Si = +6.4) are the most common times when increased advection of air occurs from the southern sector. In the summer, however, advection from the northern sector dominates (Si = −11.2). In the spring, the Si index is zero, which proves the balance between air advection from the southern and northern sectors.

In the multiannual sequence of the seasonal values of the Si index (Fig. 6.6–6.10), a slight upward trend, as for the annual values (Fig. 6.6) is marked in autumn (+0.18Si/10y). But in spring it practically ceases to exist (+0.08Si/10y). However, the strongest Si growing tendency was observed in summer (+0.59Si/10y). Undoubtedly, it has recently contributed to a significant increase in summer air temperature and the frequency of hot days. In the winter, however, there is a slight negative trend (−0.25Si/10y).

In the period of 11 winters, 1967/1968–1978/1979, there was a high level of the encroachment of air masses from the southern sector (Fig. 6.7), albeit with large fluctuations. The highest value of the Si index (+13.3) occurred in the decade covering the winters of 1968/1969–1977/1978. In a second similar, but longer, period covering 21 winters 1918/1919–1939/1940, the average value of the Si index reached +13.1 during 10 winters, 1921/1922–1930/1931. Sharp changes in Si are noteworthy, from 27.2 (winter 1929/1930 and 24.9 (winter 1930/1931) to the lowest value in the whole data series, −15.8 in the next winter, 1931/1932. In the thirty-year period, 1980–2009, the Si index weakened, including the lowest average values (−1.1) in the ten winters from 1998/1999 to 2007/2008. After this period, there were two winters with

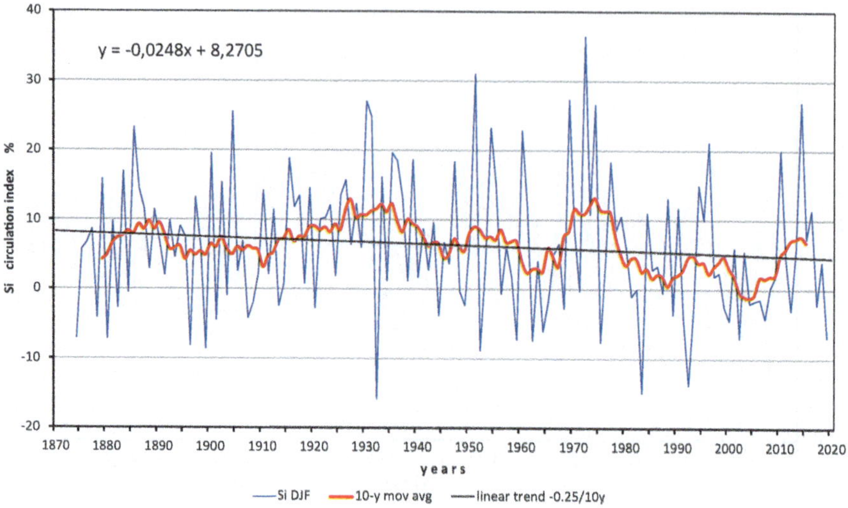

Fig. 6.7 Long-term variability of the winter (DJF) southerly circulation index (Si) values

intensive circulation from the southern sector (2009/2010 Si = +20.2 and 2013/2014 Si = +27.1), contrasting with last winter, 2018/2019 (Si = −6.8) which experienced a predominance of air advection from the northern sector.

Over the long term of the Si index (Fig. 6.8), there were large fluctuations from −25.0 in the spring of 1893 to 28.5 in the spring of 1934. Two phases with a predominance of advection from the southern sector (1919–1939 and 1967–1990) and three

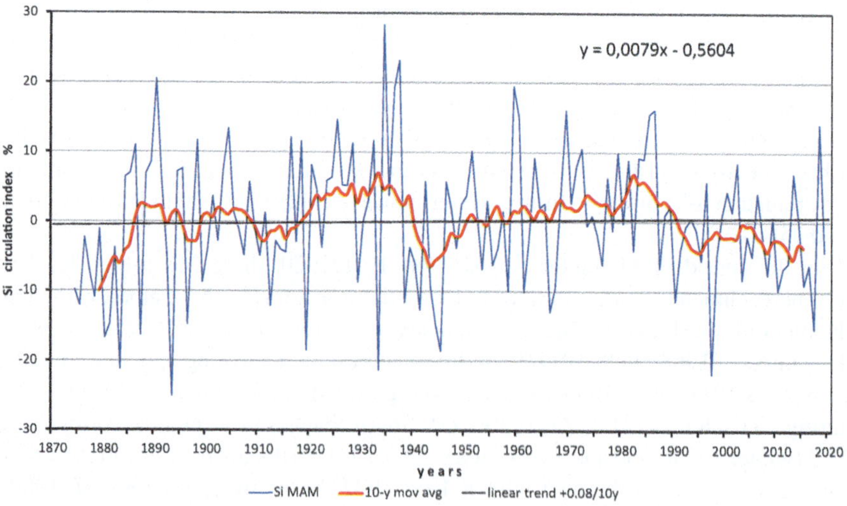

Fig. 6.8 Long-term variability of the spring (MAM) southerly circulation index (Si) values

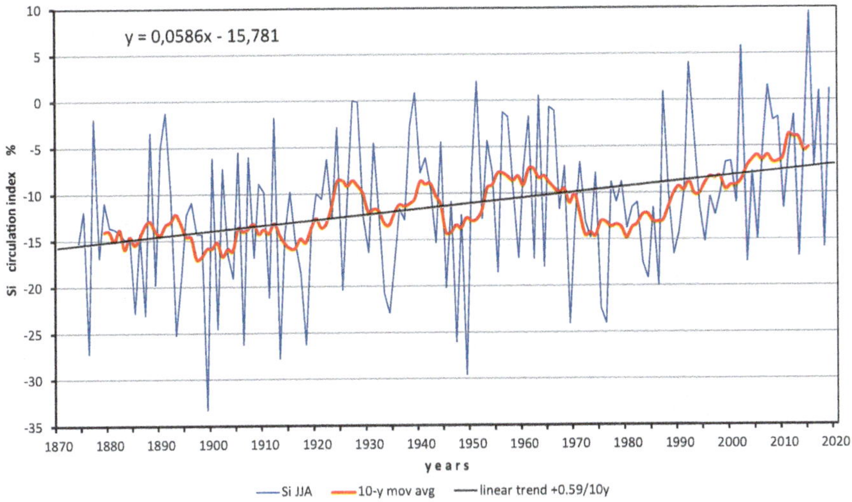

Fig. 6.9 Long-term variability of the summer (JJA) southerly circulation index (Si) values

periods with increased advection from the north are clearly marked (1874–1884, 1940–1945 and the modern period from 1991 to 2019). Rapid changes are noteworthy from year to year: a decrease from +20.7 in the spring of 1890 to −25.0 in 1893; a sharp increase from −21.3 in the spring of 1933 to +28.5 in 1934; and the most recent increase from −15.3 in spring 2017 to +14.2 in spring 2018.

For summer (Fig. 6.9) a strong upward trend of the Si index is noticeable from −14.0 in the decade 1874–1883 to −3.7 in the period 2006–2015. The average value of the Si index is negative (−11.2). Extreme values range from −33.2 in the summer of 1899 to +9.6 in the very warm summer of 2015. In the cool period of 1970–1988 there is a clear predominance of air advection from the north. The lowest average Si values were recorded in the decades 1969–1978 (−14.5) and 1975–1984 (−14.7). From 1988 to the summer of 2019, there were 7 years with the dominance of air advection from the south over advection from the north.

In autumn (Fig. 6.10), the encroachment of air from the southern sector over the entire period is maintained compared to the inflow of air from the north. The average value of the Si index is +8.4, with fluctuations from −13.2 in 1978 and −12.6 in 1972 to very high values in autumn 2000 (+33.4) and 2014 (+31.9). Two periods can be clearly distinguished with high average Si values (1924–1944 and 1953–1966) and with a weakening of southern circulation (1908–1923 and 1967–1995). After 1995, ten-year average values of the Si index remain at a level 2.0 higher (Si = +10.4) than the long-term average (8.4), but with large fluctuations.

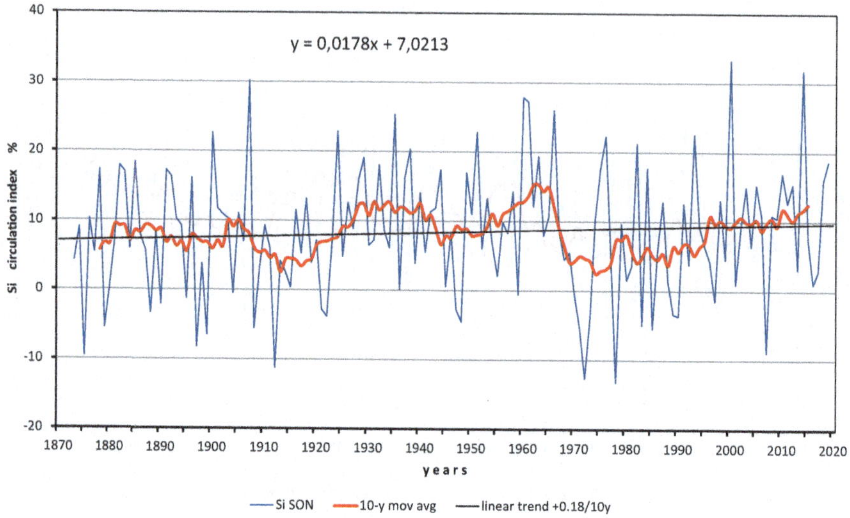

Fig. 6.10 Long-term variability of the autumn (SON) southerly circulation index (Si) values

Cyclonicity Index (Ci)

Southern Poland, unlike the northern part of the country, is more often affected by anticyclonal than cyclonal systems. This is evidenced by the negative average annual value of the cyclonicity index (Ci = −7.8; SD = ±8.0). The lowest values of the index occurred in 1921 (−31.2) and 1920 (−31.0). High anticyclonic activity was manifested during this period by the occurrence of drought and heat in the summer. On July 29, 1921, in Prószków near Opole (Lower Silesia), the highest air temperature recorded on the territory of Poland to date was 40.2 °C. However, the most frequent occurrence of cyclonal systems (Ci = 14.1) was found in 1970, which in the Upper Vistula basin was marked by the occurrence of catastrophic floods in the summer.

In the multiannual sequence (Fig. 6.11), there is a significant upward trend in the annual values of the cyclonicity index (+0.75 Ci/10y) from the average value of −13.1 in 1874–1883 to −3.7 at the end of the period under examination (2009–2018). The period of greatest cyclonal activity falls in the years 1951–1985, with the maximum in the decade of 1965–1974 (Ci = 3.4). However, the increased incidence of anticyclones falls into two periods: 1916–1935 and the more modern period 1989–1998. In these periods, the highest values of the Ci index were in the decades of 1920–1929 (Ci = −17.4) and 1989–1998 (Ci = −11.0).

In the course of a year, the Ci index changes from −16.8 in September to +4.5 in April and −0.3 in May. The extreme monthly values fluctuated from −71.7 in September 1949 to 53.3 in April 1903. Among the seasons, the most frequent occurrence of cyclonal situations is spring (Ci = +0.2), while the most anticyclonic situations occur in autumn (Ci = −13.7) and slightly less in winter (Ci = 10.7%). However, the value for summer (Ci = −6.8) is the closest to the

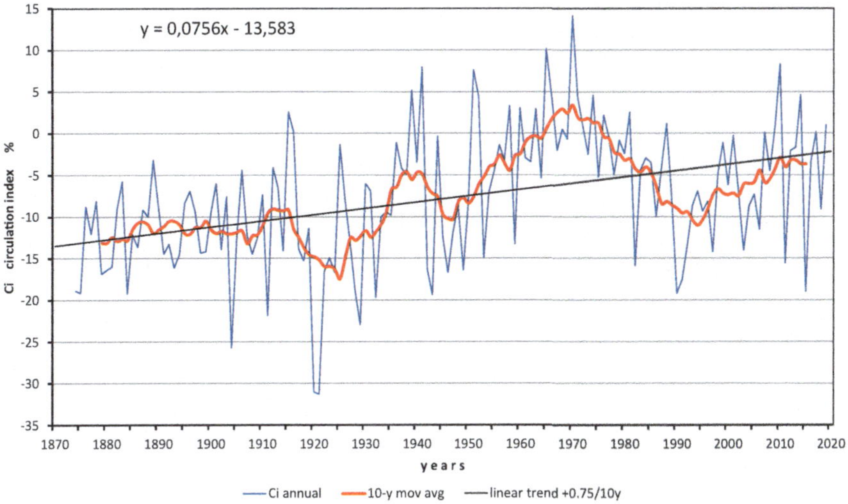

Fig. 6.11 Long-term variability of the annual cyclonicity index (Ci) values

annual average (Ci = −7.8). In all seasons, just as on an annual scale, there is an upward trend in cyclonic activity. The highest upward trend was found in winter (+1.04 Ci/10y) and in autumn (+0.87 Ci/10y), smaller in summer (+0.66 Ci/10y) and the smallest in spring (+0.43 Ci/10y). In the monthly data, the strongest upward trend was found in September (+1.36 Ci/10y). The only month that did not show a significant trend is March.

In winter, the average value of the cyclonicity index Ci is −10.7, which indicates the predominance of anticyclonic situations over cyclonal ones. The range of changes is very wide, from −46.2 in the winter of 1879/1880 to +20.6 in the winter of 1950/1951 (Fig. 6.12). Ten-year averages show an increase from −20.5 in the period 1873/1874–1882/1883 to −2.4 in the period 2009/2010–2018/2019. In the long-term sequence, two periods of low average values are visible: 1914–1934, with the average values Ci −22.7 during the winters 1924/1925–1933/1934 and 1985–1998, with the lowest mean −20.0 for the winter 1988/1989–1997/1998. A long period of increased cyclonic activity is visible in the period of 1935–1984, with the highest average of +6.1 for the winter of 1964/1965–1973/1974.

In spring, the average value of the cyclonicity index, Ci, is −10.7, which proves the dominance of anticyclonic situations over cyclonal conditions. The range of changes is from −33.1 in the spring of 1875 to +31.3 during the spring of 1970 and 31.2 in 1965 (Fig. 6.13). Ten-year averages show an increase from −3.8 in the period 1874–1883 to +2.3 in the period 2010–2019. In the long-term sequence, three periods of low average values are visible: 1874–1885, 1917–1927, with an average of −10.3 during the spring of 1917–1927, and in the contemporary period of 1990–2019. Two periods of increased cyclonic activity occurred during the spring of 1806–1903 and 1961–1989, with the highest average of +13.3 in the decade 1965–1974.

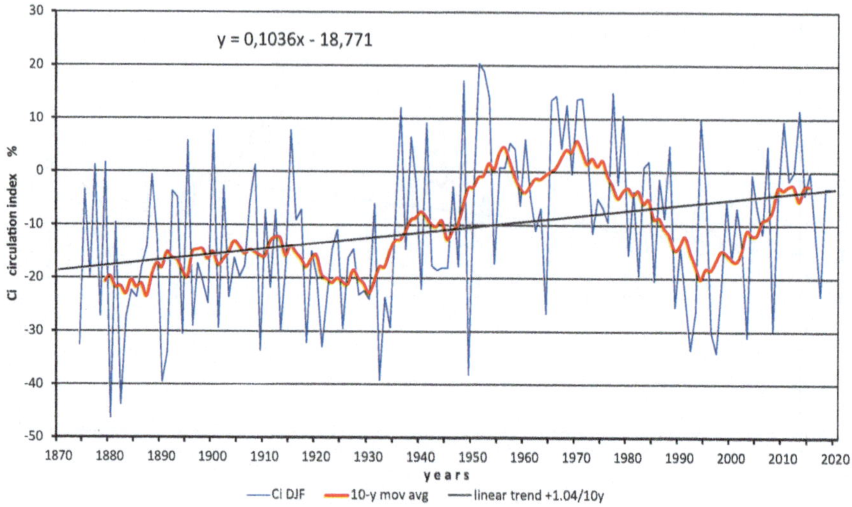

Fig. 6.12 Long-term variability of the winter (DJF) cyclonicity index (Ci) values

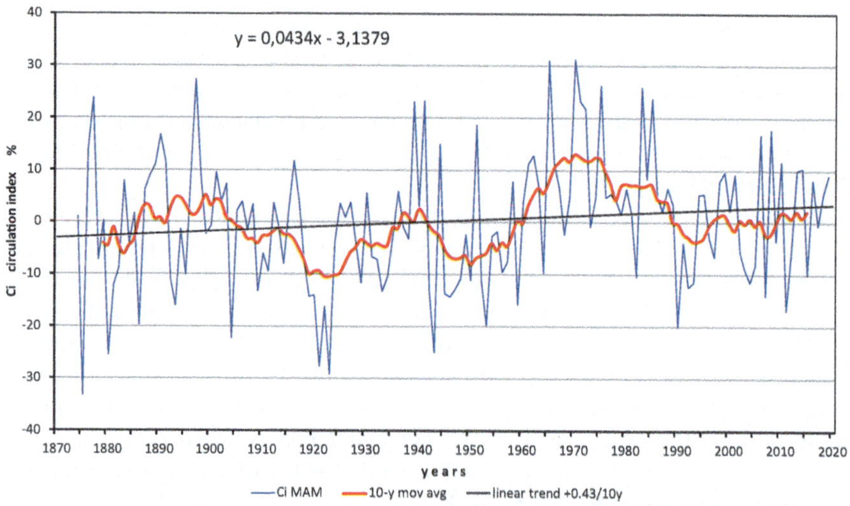

Fig. 6.13 Long-term variability of the spring (MAM) cyclonicity index (Ci) values

In the summer, the average value of the cyclonic index, Ci is −6.8 and indicates the dominance of the incidence of anticyclonic situations over cyclonal conditions. The range of changes is from −42.4 in the summer of 1904 to +19.2 in 1980, when there was a high frequency of cyclonal systems over southern Poland (Fig. 6.14). In the summer, the upward trend of the index is around +0.66 Ci/10y. The average values increase from the value of Ci −11.7 in the decade 1874–1883 to −3.1 in the last decade of 2010–2019. Over many years, large fluctuations are noted from year

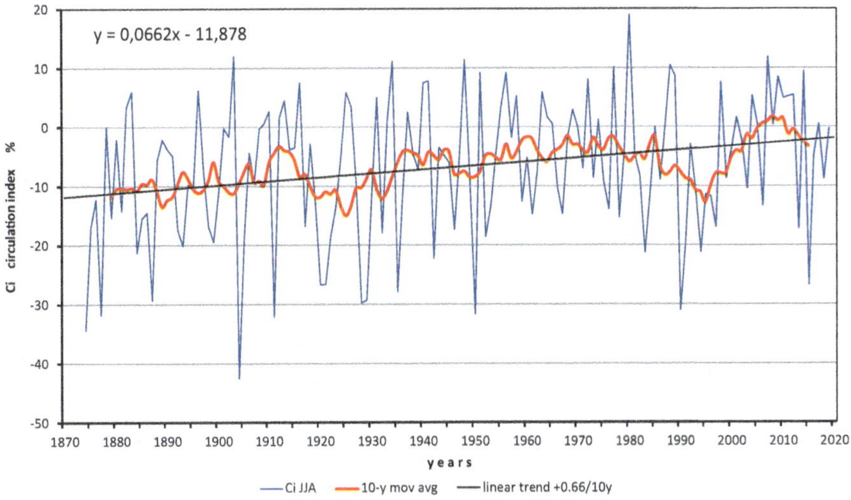

Fig. 6.14 Long-term variability of the summer (JJA) cyclonicity index (Ci) values

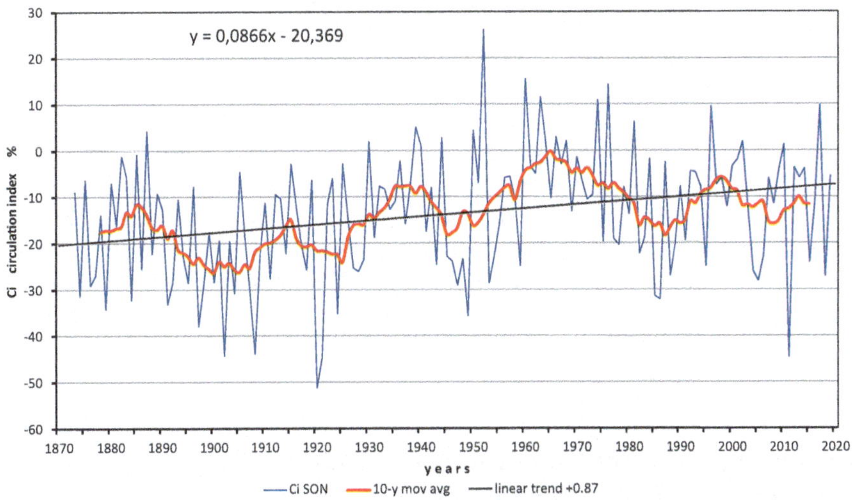

Fig. 6.15 Long-term variability of the autumn (SON) cyclonicity index (Ci) values

to year. However, one can distinguish a period of weakening cyclonal activity in the summer from 1983 to 1999.

In autumn, the average value of the cyclonicity index, Ci, is −13.7 and varies from −51.1 in 1920 and −44.6 in autumn 2011 to +26.4 in autumn 1952 (Fig. 6.15). A clear positive trend is +0.86 Ci/10y. The average 10-year values increase from −17.4 in autumn 1873–1882 to −11.4 during the last autumn of 2010–2019. A long period of weakening of anticyclonic activity, lasting up to 56 years in autumn, was recorded in 1926–1981, with a short five-year break in the period 1945–1949. The average for the decade 1960–1969 reached the value of +0.1, i.e. a balance between the frequency of cyclonic and anticyclonic situations. The highest activity of anticyclonic situations falls in the period 1888–1925, with the average value of the index Ci −26.1 in two decades, 1895–1904 and 1900–1909. Also today, in the years 1982–2019, autumn is characterized by values of the Ci index below the trend line, except for a short period of slightly higher values during the autumn of 1993–2002 (Ci = −5.5).

Circulation indices can have a significant impact on shaping or modifying other climate elements. For Katowice, correlations of average air temperature and precipitation totals with circulation indices were checked on the basis of data for 1951–2018 (Niedźwiedź and Łupikasza 2019).

Correlation coefficients significant at the 0.001 level were obtained for the average winter temperature and the Wi index ($r = 0.748$). The western circulation index increases the air temperature from December to March, with the highest correlation coefficient in February ($r = 0.738$). An increase in western circulation activity reduces the average summer temperature ($r = -0.341$, significant at 0.01), especially the average July temperature ($r = -0.586$, significant at 0.001). There is, however, no significant correlation between the Wi index and precipitation totals.

The meridional circulation index, Si, significantly affects the increase in the average autumn temperature ($r = 0.553$) and the average temperature of the months from April to June and from August to October. The average correlation coefficients show the highest temperature in May ($r = 0.573$) and September ($r = 0.571$). There is no Si correlation with precipitation.

The cyclonicity index, Ci, does not show a statistically significant correlation with air temperature. However, significant correlation coefficients were obtained with winter ($r = 0.304$), spring ($r = 0.479$) and autumn rainfall (0.647), and with monthly rainfall totals from January to May and from August to December. The highest correlation coefficients were obtained for October ($r = 0.726$) and September ($r = 0.628$) precipitation.

Comparison of Circulation Indices in Southern Poland with Other Regions of the Country

The values of circulation indices for nine regions of Poland for the period December 2000–December 2018 were compared with the average annual values of Wi, Si and Ci indices (Fig. 6.16). The highest intensity of western circulation was found in NW Poland (Wi = 20.3) and the lowest in SE Poland (Wi = 11.2). The correlation of regional annual values of the Wi index with data from S Poland is very high (in eight regions $r > 0.90$). In NW Poland, the correlation coefficient was $r = 0.89$, which is also a statistically significant value. Compared to the period 1873–2018 (Wi = 15.9), the value of the Wi index in S Poland was lower by 2.9.

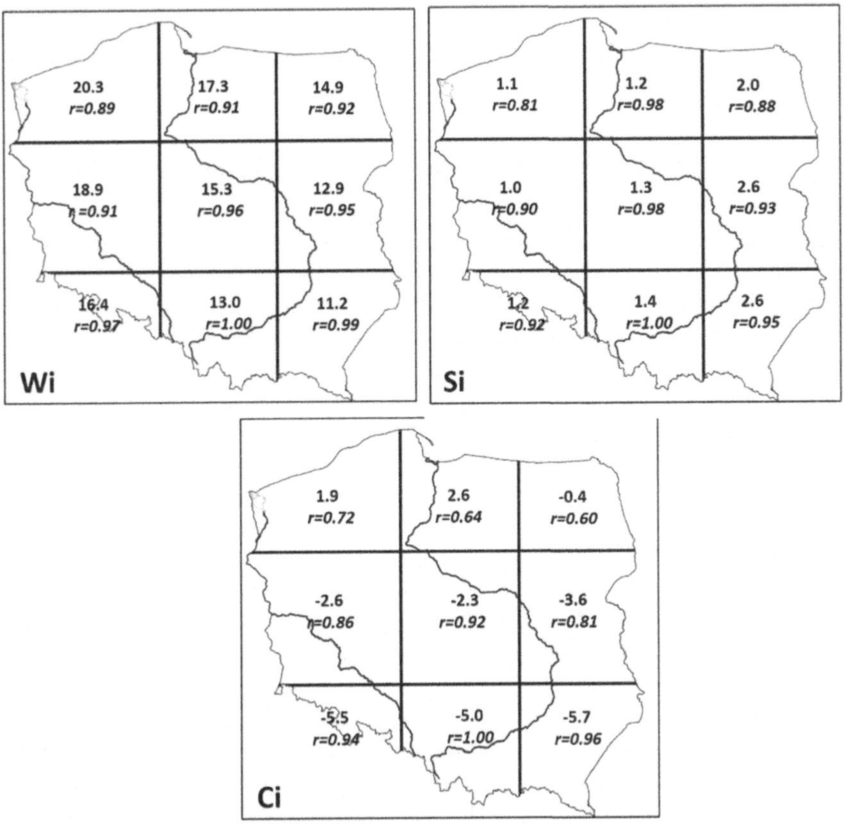

Fig. 6.16 Comparison of the mean annual circulation indices in Southern Poland with other regions; (correlation coefficient [*r*] in italic; averages for the period 2001–2018)

The value of the Si index in S Poland (Si = +1.4) was 0.5 higher than in the period 1873–2018 (Si = +0.9). Across the country, Si values varied from +1.0 in W Poland to +2.6 in E and SE Poland. The highest correlation coefficient with data from S Poland was found in central and N Poland ($r = 0.98$), while the lowest was in NW Poland ($r = 0.81$).

In S Poland, in the analysed period decreased anticyclonic activity (Ci = −5.0) was observed compared to the data for 1873–2018 (Ci = −7.8). Positive values of the Ci index indicating the dominance of the incidence of lows over high-pressure situations occurred only in two regions in N (Ci = +2.6) and NW Poland (Ci = +1.9). SE Poland has the highest anticyclonic activity (Wi = −5.7). The data from S Poland are best correlated with the Ci index from SE Poland ($r = 0.96$), SW ($r = 0.94$) and the centre of the country ($r = 0.92$). The weakest correlation was found in NE Poland ($r = 0.60$) and N ($r = 0.64$). These results indicate that, despite some differences, perennial data from S Poland quite well represent the whole country, especially the centre of Poland, and the largest differences in the values of the Wi and Si indices concern NW Poland, while for the Ci index are noticed in NE Poland.

Changes of Circulation Conditions According to Other Indices

The research focused on the analysis of those types of circulation that have the greatest impact on Poland's weather and climate conditions. Undoubtedly, the zonal type of air advection should be considered the basic type of circulation in this respect. Therefore, attention was paid to measures illustrating its intensity. Figure 6.17 shows the number of days with types of circulation with the inflow of air masses from the western sector per year, according to the manual classification of T. Niedźwiedź (1981) and the automatic approach of Z. Ustrnul (1997). As you can see, there is very good agreement in the course of this index, not only regarding the trend itself but also the values in individual years, including extreme values. Some differences exceeding 20 days per year occurred in the 70s of the twentieth century, which are probably associated with the situation of the anticyclonic wedge over Central Europe. The highest (1983, 1998, 2017) and the lowest values of the Wi index (1963, 1972, 1996) are notable, and have a significant impact on the climate conditions of the research area. Regardless of the approach, there is no clear change trend throughout the period considered.

A similar annual number of days with circulation from the western sector is visible in the light of other classifications of circulation types that can be considered macroscale. One of these is the objective classification of Lityński (1969); another is the Grosswetterlagen classification (Hess and Brezowsky 1977; Werner and Gerstengarbe 2010), well known in Europe, which, in the original version, is a subjective approach. In both cases, however, in the studied period an increasing trend is visible of the number of days with circulation from the western sector (Fig. 6.18), with

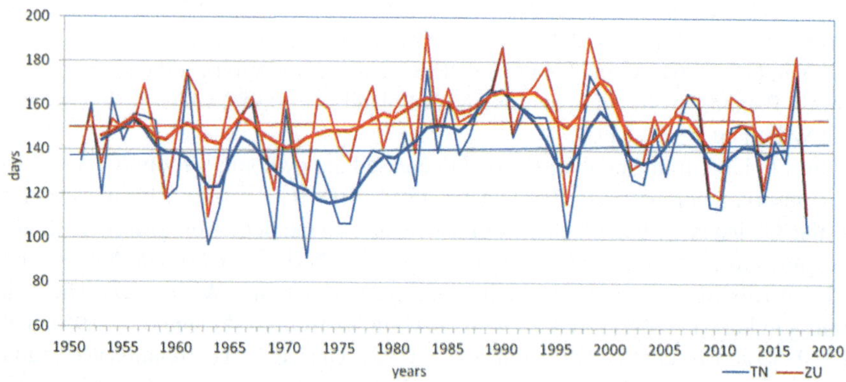

Fig. 6.17 Annual number of days with westerly circulation types according to T. Niedźwiedź (TN) and Z. Ustrnul (ZU) classifications

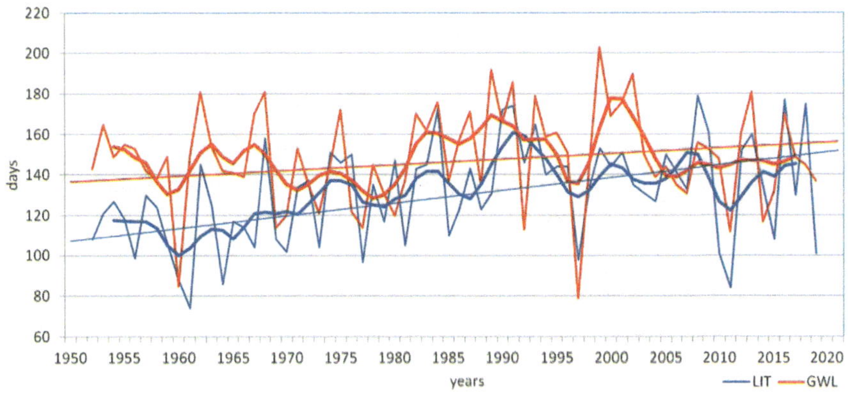

Fig. 6.18 Annual number of days with westerly circulation types according to J. Lityński (LIT) and Grosswetterlagen (GWL) classifications

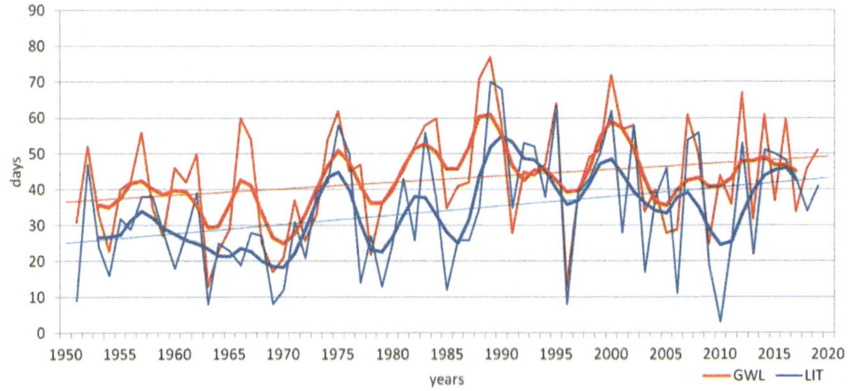

Fig. 6.19 Number of days with westerly circulation types in winter according to J. Lityński (LIT) and Grosswetterlagen (GWL) classifications

a significantly positive expression in the case of the Lityński (1969) classification. The extreme values of both indices are noteworthy, though they do not always show consistency. Analysis of synoptic maps indicates that the differences observed result from the slightly different research domains adopted.

Greater consistency in the course of the considered index is seen in the case of winter (Fig. 6.19). This applies to both the general trend and the value in individual years. This means that, in the winter season, there are less differentiated circulation conditions than during the rest of the year and the use of a given circulation index is less important than during other seasons. This high consistency of western circulation in winter is confirmed in Fig. 6.20, which shows the analogous course of days with western circulation, taking into account the classifications of T. Niedźwiedź (1981) and Z. Ustrnul (1997).

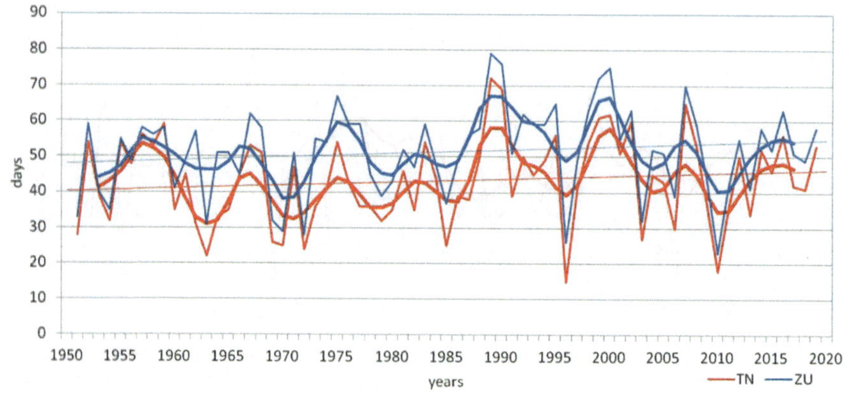

Fig. 6.20 Number of days with westerly circulation types in winter according to T. Niedźwiedź (TN) and Z. Ustrnul (ZU) classifications

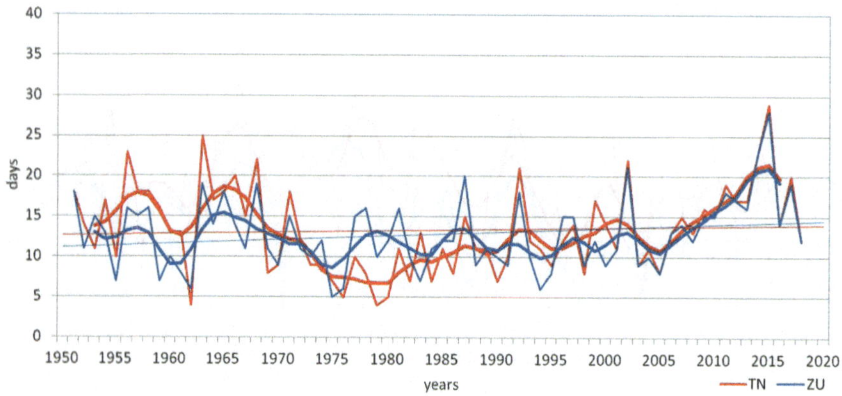

Fig. 6.21 Number of days with southerly circulation types in summer according to T. Niedźwiedź (TN) and Z. Ustrnul (ZU) classifications

Unlike the western circulation indices, measures illustrating the intensity of air inflow from the south show lower values. The importance of this index is greatest in the summer, due to its size and impact on the climate, in particular on thermal conditions. Figure 6.21 shows the pattern of days with circulation from the southern sector in the summer, which is known to significantly affect the thermal regime, especially in southern Poland. The long-term trend is slightly positive, mainly due to the extremely high value in 2015 (in the light of both considered mesoscale classifications), which affected the hot summer of this year throughout Central Europe (Hoy et al. 2017).

Western circulation in the winter period shows quite high conformity in the studied area, regardless of the measures adopted. This is also confirmed by Fig. 6.22, which shows the relationship between NAO index values (according to Hurrell [1995]) and

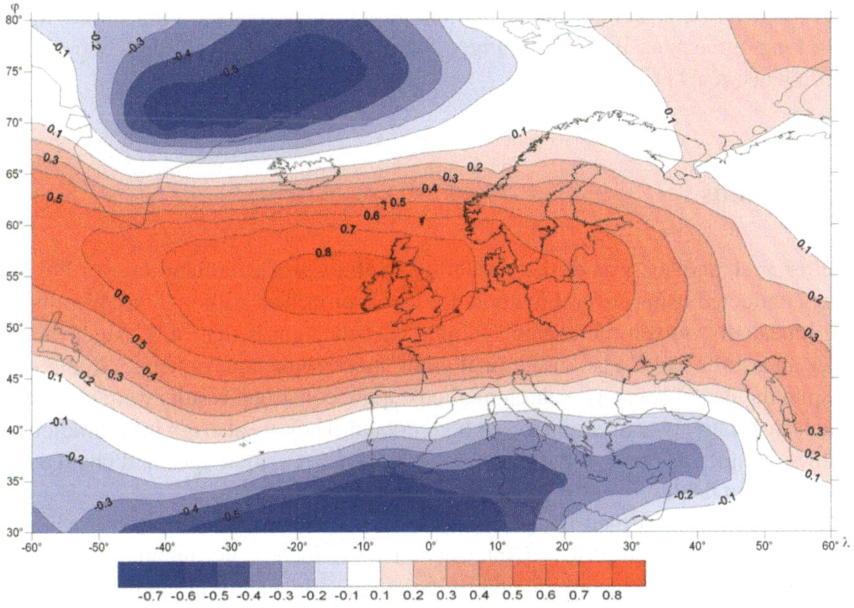

Fig. 6.22 Correlation coefficients between NAO index (according to Hurrell) and westerly circulation index over Europe—winter (1951–2018)

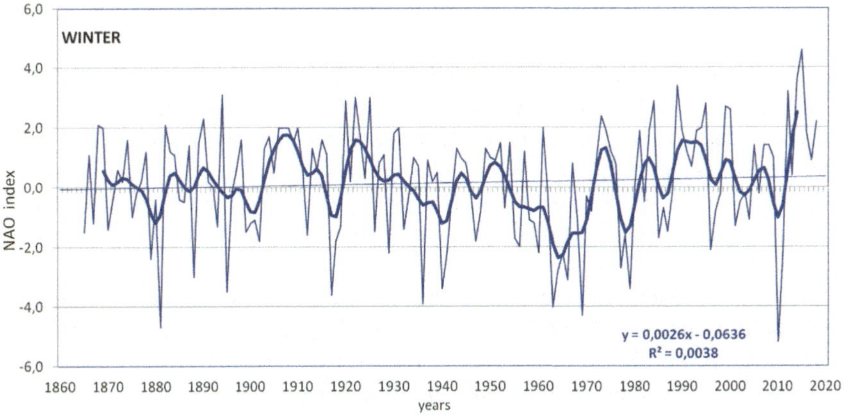

Fig. 6.23 Long-term course of the NAO index in winter (according to Hurrell)

western circulation in winter, determined by the objective method (Ustrnul 1997). At the same time, this figure confirms that the highest consistency in the course of these indices is visible in NW Poland. Due to the importance of the NAO index for shaping the climate of Central Europe, including Poland, the long-term variability in winter in the period of 1865–2018 is also presented (Fig. 6.23). Comparing (for the period

1951–2018) the course of this index with mesoscale measures (cf. Figure 6.20) there is striking agreement. In addition, Fig. 6.22 confirms the relatively high correlation coefficients, reaching about 0.60.

Conclusions

The research and analysis of the variability of atmospheric circulation over Poland were conducted using several methodological approaches. Although the most used and repeatedly proven mesoscale synoptic classification is that of T. Niedźwiedź, other measures were also used, including macroscale atmospheric circulation. Despite the use of several research methods, quite consistent results were obtained, which are relatively convergent throughout Poland. They are also not in contradiction with the results of previously published works on circulation ages and changes in circulation in the last over 100 years in Europe.

The analysis confirmed that, in Poland, circulation from the westerly sector is the most significant in terms of intensity of its influence on climatic conditions. Considering the period from the end of the nineteenth century to the present, large fluctuations and a general weak increasing trend were found. No clear trends are seen in the multi-year period, 1951–2018, although in the light of the typology of Lityński and Grosswetterlagen its increase in intensity is noticeable. This period was characterized by increased western circulation in the 1980s and 1990s. The turn of the 1960s and 1970s marked the weakening of this circulation pattern, as did the last 10 years. In view of the various indices, including a NAO index, we can conclude that the relevant multiannual period included years of very intense western circulation (years 1990 and 2015) and very poor (years 1963 and 2010). The weakening of western circulation was, however, evident during the summer. Particularly notable is that in the last several years, although it should be added that it was also observed at the turn of the 1960s and 1970s. The annual decreasing trend of the westerly circulation index is an effect of significant weakening of the Wi in summer months. For other seasons the trends are insignificant.

There is a good level of agreement in the course of circulation indices according to the classifications of Niedźwiedź, Ustrnul, Lityński and Grosswetterlagen. The best correlation between the NAO and the Ustrnul westerly circulation index determined by the objective method, was observed in NW Poland in winter.

The circulation from the south had a slightly decreasing trend throughout the year. A completely opposite trend was found in the summer, which results from relatively high values of indices connected with types of circulation with advection from the south in the last several years. Particularly noteworthy here is summer 2015, which was characterized by the most frequent advection from the southern sector, which was reflected, among other features, in the extremely warm summer season in Central Europe. Unlike the circulation from the west or the south, changes in the cyclonicity index illustrating the frequency of low pressure systems over Poland have shown increasing trends in all seasons. This is important, due to the fact that cyclonal activity contributes to greater dynamism of weather conditions.

To sum up, it should be stated that, in the light of the results obtained and the literature review, in the basic examined period of 1951–2018, no clear and constant trend of changes in atmospheric circulation was found. Such trends, although to a limited extent, could be found in the case of some circulation indices in individual seasons. Over the annual scale, however, significant fluctuations occurred in the course of the measures used, including the occurrence of their extreme values. These probably caused the occurrence of some climatic anomalies, which have been reflected in many works concerning individual elements of weather and climate.

The chapter draws attention to the main features of the variability of atmospheric circulation by assessing the intensity of air advection from major directions, which has climatological justification. The impact of air inflow from individual directions on the occurrence and course of specific elements and weather phenomena is different. It is worth mentioning, however, that for a full assessment of changes in atmospheric circulation over Poland, an analysis of the occurrence of individual air masses, as well as atmospheric fronts, would be desirable.

References

Bartoszek K (2017) The main characteristics of atmosphere circulation over East-Central Europe from 1871 to 2010. Meteorol Atmos Phys 129(2):113–129. https://doi.org/10.1007/s00703-016-0455-z

Bednorz E (2011) Synoptic conditions of snow cover occurrence in central European lowlands. Int J Climatol 31(8):1108–1118. https://doi.org/10.1002/joc.2130

Bielec Z (2000) The role of atmospheric circulation in the development of the long-term thunderstorm occurrence variability in Poland during 1949-1998. Zeszyty Naukowe UJ. Prace Geograficzne 107:157–164

Bielec-Bąkowska Z (2010) A classification of deep cyclones over Poland (1971-2000). Phys Chem Earth, Parts A/B/C 35(9–12):491–497. https://doi.org/10.1016/j.pce.2009.12.006

Bielec-Bąkowska Z (2016) Long-term variability of the frequency and persistence of strong highs over Poland. Environ Socio-econ Stud 4(1):12–23. https://doi.org/10.1515/environ-2016-0002

Błażejczyk K, Twardosz R (2010) Long-term changes of bioclimatic conditions in Cracow (Poland). In: R Przybylak, J Majorowicz, R Brazdil, M Kejna (eds) The Polish climate in the European context. An historical overview. Springer, Dordrecht, Heidelberg, London, New York, 235–246 pp

Degirmendžić J, Kożuchowski K, Wibig J (2000) Epoki cyrkulacyjne XX wieku i zmienność typów cyrkulacji atmosferycznej w Polsce (Circulation epochs of the 20th century and variability of atmospheric circulation types in Poland). Przegląd Geofizyczny 45(3–4):221–238 (in Polish)

Degirmendžić J, Kożuchowski K, Żmudzka E (2004) Changes of air temperature and precipitation in Poland in the period 1951-2000 and their relationship to atmospheric circulation. Int J Climatol 24(3):291–310. https://doi.org/10.1002/joc.l010

Degirmendžić J, Kożuchowski K (2017) Mediterranean cyclones, the atmospheric moisture content and precipitation in Poland. Geographia Pol 90(1):5–20. https://doi.org/10.7163/GPol.0075

Falarz M (2007) Snow cover variability in Poland in relations to the macro- and mesoscale atmospheric circulation in the twentieth century. Int J Climatol 27(15):2069–2081. https://doi.org/10.1002/joc.1505

Hess P, Brezowsky H (1977) Katalog der Grosswetterlagen Europas (1881-1976). 3. Verbesserte und ergänzte Auflage. Berichte des Deutschen Wetterdienstes 113:68 pp

Hoy A, Hansel S, Skalak P, Ustrnul Z, Bochnicek O (2017) The extreme European summer of 2015 in a long-term perspective. Int J Climatol 37(2):943–962. https://doi.org/10.1002/joc.4751

Hurrell JW (1995) Decadal trends in the North Atlantic oscillation: Regional temperatures and precipitation. Science 269(5224):676–679. https://doi.org/10.1126/science.269.5224.676

Kaszewski BM (1992) Typy cyrkulacji a typy pogody w Polsce (Types of circulation and types of weather in Poland). Rozprawy Wydziału Biologii i Nauk o Ziemi – Uniwersytet Marii Curie-Skłodowskiej. Lublin, Rozprawy Habilitacyjne:42, 92 pp. (in Polish)

Kolendowicz L (2006) The influence of synoptic situations on the occurrence of days with thunderstorms during a year in the territory of Poland. Int J Climatol 26(13):1803–1820. https://doi.org/10.1002/joc.1348

Kolendowicz L, Półrolniczak M, Szyga-Pluta K, Bednorz E (2018) Human-biometeorological conditions in the southern Baltic coast based on the universal thermal climate index (UTCI). Theoret Appl Climatol 134(1–2):363–379. https://doi.org/10.1007/s00704-017-2279-2

Kożuchowski K (2011) Klimat Polski, Nowe spojrzenie (Polish climate, a new look). Wydawnictwo Naukowe PWN, Warszawa, 293 pp. (in Polish)

Kożuchowski K, Degirmendžić J (2018) Zmienność form cyrkulacji środkowotroposferycznej według klasyfikacji Wangenheima-Girsa i ich relacje z polem ciśnienia na poziomie morza (Variability of mid-tropospheric circulation according to the Wangenheim-Girs classification and their relationship with the sea level pressure field). Przegląd Geofizyczny 63(1–2):89–122 (in Polish)

Kulesza K (2017) Nowe spojrzenie na klasyfikację typów cyrkulacji atmosfery J. Lityńskiego (A new look at the classification of types of J. Lityński's atmosphere circulation). Prace Geograficzne 150:79–94. https://doi.org/10.4467/20833113PG.17.018.7322 (in Polish)

Lamb HH (1972) British Isles weather types and register of the daily sequence of circulation patterns 1861-1971. Geophys Memoirs 16(116):1–85

Lityński J (1969) Liczbowa klasyfikacja typów cyrkulacji i typów pogody dla Polski (Numerical classification of circulation types and weather types for Poland). Prace PIHM 97:3–14 (in Polish)

Łupikasza E (2016) The climatology of air-mass and frontal extreme precipitation. Study of meteorological data in Europe. Springer International Publishing Switzerland, 313 pp. https://doi.org/10.1007/978-3-319-31478-5

Marosz M (2016) Variability of geostrophic airflow over Poland, 1951-2014. Bulletin of Geogr Phys Geogr Ser 10(1):5–18. https://doi.org/10.1515/bgeo-2015-0001

Marsz A, Styszyńska A (eds) (2002) Oscylacja Północnego Atlantyku i jej rola w kształtowaniu zmienności warunków klimatycznych i hydrologicznych Polski (North Atlantic Oscillation and its role in shaping the variability of climatic and hydrological conditions of Poland). Akademia Morska, Gdynia, 222 pp. (in Polish)

Matuszko D, Węglarczyk S (2018) Long-term variability of the cloud amount and cloud genera and their relationship with circulation (Kraków, Poland). Int J Climatol 38(51):1205–1220. https://doi.org/10.1002/joc.5445

Miętus M (1996) Zmienność lokalnej cyrkulacji atmosferycznej nad Polską Północną i jej związek z elementami klimatu (Variability of local atmospheric circulation over Northern Poland and its relationship with climate elements). Wiadomości IMGW 19(1):9–30 (in Polish)

Murray R, Lewis RP (1966) Some aspects of the synoptic climatology of the British Isles as measured by simple indices. Meteorol Mag 95:193–203

Niedźwiedź T (1981) Sytuacje synoptyczne i ich wpływ na zróżnicowanie przestrzenne wybranych elementów klimatu w dorzeczu górnej Wisły (Synoptic situations and their impact on spatial differentiation of selected climate elements in the upper Vistula basin). Rozprawy Habilitacyjne UJ, Kraków, 58:165 pp. (in Polish)

Niedźwiedź T (1983) The method of a synoptic-complex characterization of the climate presented upon the example of the upper Vistula River basin. Zeszyty Naukowe UJ. Prace Geograficzne 57:17–39

Niedźwiedź T (1996) Long-term variability of the zonal circulation index above the Central Europe. Zeszyty Naukowe UJ. Prace Geograficzne 102:213–219

Niedźwiedź T (2000) Variability of the atmospheric circulation above the Central Europe in the light of selected indices. Prace Geograficzne 107:379–389

Niedźwiedź T (2003) Extreme precipitation in Central Europe and its synoptic background. Pap Global Change IGBP 10:15–30

Niedźwiedź T (2019) Catalogue of circulation types for southern Poland (1873.09-2019.12). Available on website: http://www.kk.wnoz.us.edu.pl/nauka/kalendarz-typow-cyrkulacji/ or in Author by request

Niedźwiedź T, Ustrnul Z, Cebulak E, Limanówka D (1994) Long-Term climate variations in Southern Poland due to atmospheric circulation variability. In: R Heino (ed) Climate variations in Europe: Proceedings of the European Workshop on Climate Variations, Majvik, Finland, 15–18 May. SILMU, Publications of the Academy of Finland, 263–277 pp

Niedźwiedź T, Łupikasza E, Pińskwar I, Kundzewicz ZW, Stoffel M, Małarzewski Ł (2014) Variability of high rainfalls and related synoptic situations causing heavy floods at the northern foothills of the Tatra Mountains. Theoret Appl Climatol 119(1–2):273–284. https://doi.org/10.1007/s00704-014-1108-0

Niedźwiedź T, Łupikasza E (2016) Change in atmospheric circulation patterns. In Kundzewicz ZW, Stoffel M, Niedźwiedź T, Wyżga B (eds) Flood risk in the Upper Vistula Basin. Springer International Publishing Switzerland, pp 189–208. https://doi.org/10.1007/978-3-319-41923-7_9

Niedźwiedź T, Łupikasza E (2019) Cyrkulacja atmosferyczna w badaniach polskich klimatologów (Atmospheric circulation in the investigations of the Polish climatologists). Przegląd Geofizyczny 64(1–2):107–166

Nowosad M (2017) Variability of the zonal circulation index over Central Europe according to the Lityński method. Geographia Polonica 90(4):417–430

Nowosad M, Rodzik B, Wereski S, Dobek M (2013) The UTCI index in Lesko and Lublin and its circulation determinants. Geographia Polonica 86(1):29–36. https://doi.org/10.7163/GPol.2013.4

Ojrzyńska H (2012) Automatyczny kalendarz cyrkulacji dla terenu Sudetów – założenia i metody konstrukcji w oparciu o system GRASS (Automatic circulation calendar for the Sudetes - assumptions and construction methods based on the GRASS system). Przegląd Geofizyczny 57(2):17–29 (in Polish)

Osuchowska-Klein B (1978) Katalog typów cyrkulacji atmosferycznej (Catalog of atmospheric circulation types). IMGW, Wydawnictwa Komunikacji i Łączności, Warszawa, 192 pp. (in Polish)

Osuchowska-Klein B (1991) Katalog typów cyrkulacji atmosferycznej (1976-1990) (Catalog of atmospheric circulation types (1976-1990)). IMGW, Warszawa, 50 pp. (in Polish)

Owczarek M (2019) The influence of large-scale factors on the heat load on human beings in Poland in the summer months. Theoret Appl Climatol 137(1–2):855–869. https://doi.org/10.1007/s00704-018-2633-z

Pawłowska J, Jankowska A, Pindor T (2000) Kalendarz typów cyrkulacji atmosferycznej według J. Lityńskiego (1991-1999) (Calendar of atmospheric circulation types according to J. Lityński (1991-1999)). IMGW, Warszawa, 28 pp (in Polish)

Pianko-Kluczyńska K (2007) Nowy kalendarz typów cyrkulacji atmosfery według J. Lityńskiego (A new calendar of types of atmosphere circulation according to J. Lityński). Wiadomości Meteorologii, Hydrologii, Gospodarki Wodnej 51(4):65–85 (in Polish)

Piotrowski P (2009) Obiektywna metoda klasyfikacji cyrkulacji atmosferycznej dla Polski (An objective method of atmospheric circulation classification for Poland). Acta Universitatis Lodziensis. Folia Geographica Physica 10:216 pp. (in Polish)

Piotrowski P (2010) Comparison between objective and subjective method of classification of atmospheric circulation for area of Poland. Acta Agrophysica. Rozprawy i Monografie 5(184):44–54

Styszyńska A, Błaś M, Migała K (eds) (2019) NAO – jej istota, przyczyny i konsekwencje (NAO - its essence, causes and consequences). Instytut Geografii i Rozwoju Regionalnego Uniwersytetu Wrocławskiego i Stowarzyszenie Klimatologów Polskich, Wrocław, 100 pp (in Polish)

Tomczyk AM, Owczarek M (2020) Occurrence of strong and very strong heat stress in Poland and its circulation conditions. Theoret Appl Climatol 139(1–2):893–905. https://doi.org/10.1007/s00704-019-02998-3

Ustrnul Z (1997) Zmienność cyrkulacji atmosfery na półkuli północnej w XX wieku (Variability of atmosphere circulation in the Northern Hemisphere in the 20th century). Materiały Badawcze IMGW, Seria Meteorologia 27:208 pp (in Polish)

Ustrnul Z (2000) Synoptic-climatic structure of the extreme air thermal phenomena in Poland. Geographia Polonica 73(2):99–109

Ustrnul Z (2007) Zmienność wskaźnika cyrkulacji zachodniej nad Europą w XX wieku (Variability of the westerly circulation index over Europe in the 20th century). In: Piotrowicz K, Twardosz R (eds), Wahania klimatu w różnych skalach przestrzennych i czasowych (Climate fluctuations in different spatial and temporal scales). Instytut Geografii i Gospodarki Przestrzennej, Uniwersytet Jagielloński, Kraków:55–63 (in Polish, summary in English)

Ustrnul Z, Czekierda D (2001) Circulation background of the atmospheric precipitation in Central Europe (based on the Polish example). Meteorol Z 10(3):103–111. https://doi.org/10.1127/0941-2948/2001/0010-0103

Ustrnul Z, Czekierda D (2009) Atlas of extreme meteorological phenomena and synoptic situations in Poland. Institute of Meteorology and Water Management, Warszawa, p 182

Ustrnul Z, Czekierda D, Wypych A (2010) Extreme values of air temperature in Poland according to different atmospheric circulation classifications. Phys Chem Earth 35:429–436. https://doi.org/10.1016/j.pce.2009.12.012

Ustrnul Z, Wypych A, Czekierda D (2013) Composite circulation index of weather extremes (the example for Poland). Meteorol Z 22(5):551–559. https://doi.org/10.1127/0941-2948/2013/0464

Werner PC, Gerstengarbe FW (2010) Katalog der Großwetterlagen Europas (1881-2009) nach Paul Hess und Helmut Brezowsky. 7. verbesserte und ergänzte Auflage. PIK Report No. 119, Potsdam Institute for Climate Impact Research. Available online: https://www.pik-potsdam.de/research/publications/pikreports/.files/pr119.pdf. 28 Feb 2019

Wibig J (2000) Oscylacja Północnoatlantycka i jej wpływ na kształtowanie pogody i klimatu (North Atlantic Oscillation and its impact on weather and climate formation). Przegląd Geofizyczny 45(2):121–137 (in Polish)

Wibig J (2001) Wpływ cyrkulacji atmosfery na rozkład przestrzenny anomalii temperatury i opadów w Europie (Impact of atmospheric circulation on the spatial distribution of temperature and precipitation anomalies in Europe). Rozprawy Habilitacyjne Uniwersytetu Łódzkiego, Łódź, 208 pp. (in Polish)

Wypych A, Ustrnul Z, Sulikowska A, Chmielewski F-M, Bochenek B (2017) Spatial and temporal variability of the frost-free season in Central Europe and its circulation background. Int J Climatol 37(8):3340–3352. https://doi.org/10.1002/joc.4920

Wypych A, Ustrnul Z, Czekierda D, Palarz A, Sulikowska A (2018) Extreme precipitation events in the Polish Carpathians and their synoptic determinants. Időjárás 122(2):145–158. https://doi.org/10.28974/idojaras.2018.2.3

Żmudzka E (2007) Zmienność zachmurzenia nad Polską i jej uwarunkowania cyrkulacyjne (1951-2000) (Cloudiness change over Poland and its circulation determinations (1951-2000)). Wydawnictwa Uniwersytetu Warszawskiego, Warszawa, 399 pp (in Polish)

Chapter 7
Air Pressure Change

Zuzanna Bielec-Bąkowska and Katarzyna Piotrowicz

Abstract The present study takes into account the average daily values of atmospheric pressure reduced to sea level (SLP) from 43 meteorological stations for the years 1966–2018. The exceptions are the three mountain stations (Zakopane, Śnieżka and Kasprowy Wierch) for which the pressure values at the levels of these stations were used and, based on them, only the trends of changes in this meteorological element were determined. In addition, a series of measurements from the Scientific Station of the Climatology Department of the Jagiellonian University in Kraków for the years 1901–2018 was used. Annual, seasonal, and monthly pressure values, as well as the variability of the number of days with a pressure $\leq$ 990 hPa or $\geq$ 1030 hPa were characterised. The study confirmed the latitudinal distribution of pressure values and the small range of their changes in the average values, especially in the warm half-year. The highest pressure values occurred in the south and south-east of the country, while the lowest along the Baltic Sea coast. It was also determined that during analysed periods (1966–2018 and 1901–2018) no tendency was found in long-term variability of pressure values (both in annual and seasonal, as well as monthly values). The analysis revealed that strong high-pressure systems (SLP $\geq$ 1030 hPa) over Poland occur several times more often than deep lows (SLP $\leq$ 990 hPa). Although the frequency of their occurrence differs depending on the region of Poland. The conducted studies have shown that the long-term variability in the number of days with SLP $\leq$ 990 hPa and SLP $\geq$ 1030 hPa do not show a clear trend. At the same time, the occurrence of fairly regular, several-year periods of increased pressure values and number of days analysed is a characteristic feature of pressure changes in Poland.

Z. Bielec-Bąkowska (✉)
Faculty of Natural Sciences, Institute of Earth Sciences, University of Silesia in Katowice, Katowice, Poland
e-mail: zuzanna.bielec-bakowska@us.edu.pl

K. Piotrowicz
Faculty of Geography and Geology, Institute of Geography and Spatial Management, Jagiellonian University, Krakow, Poland
e-mail: k.piotrowicz@uj.edu.pl

M. Falarz (ed.), *Climate Change in Poland*, Springer Climate,
https://doi.org/10.1007/978-3-030-70328-8_7

Introduction

The location of Poland in the middle latitude zone in Central Europe determines the values and course of atmospheric pressure during the year. The Icelandic Low and the Azores High-pressure systems have a major impact on pressure changes, while the Asian Low in the summer and the Asian High in the winter (Siberian High) affect them less (Woś 2010; Kożuchowski 2011). Therefore, during the year, the atmospheric pressure field over Poland shows characteristic seasonal variability (Paszyński and Niedźwiedź 1991).

Long-term pressure changes are mainly influenced by the frequency of occurrence and changes in the position of the mentioned pressure systems, as well as by the frequency of dynamic low-pressure systems moving over Poland, and the occurrence of blocking situations. In light of the significant changes in atmospheric circulation observed in recent decades, it can be expected that these will be reflected in changes in the atmospheric pressure in Poland. The changing tracks of low-pressure systems (Zhao and Held 2012; Murakami et al. 2012), increasing in their number and intensity in northern and western Europe (Wernli et al. 2003; Rockel and Worth 2007; Haarsma et al. 2013), and decreasing below 55° N (Bartholy et al. 2006; Trigo 2006), are most important. Equally important is a change in the occurrence and an increase in the durability of high-pressure systems (Cahynová and Huth 2016; Kyselý 2008; Kyselý and Huth 2008; Bielec-Bąkowska 2014, 2016), including blocking situations (Dong et al. 2013).

Data and Methods

In the study of the spatial and temporal variability of pressure values in Poland, average daily atmospheric pressure values reduced to sea level (SLP) from 43 meteorological stations belonging to the IMGW-PIB (Polish Institute of Meteorology and Water Management—National Research Institute) network from the years 1966–2018 were used (Fig. 7.1, Table 3.3). Three of them, i.e. those on Kasprowy Wierch, Śnieżka, and in Zakopane are stations located over 850 m above sea level. Therefore, only the pressure at the level of these stations was utilised (without reduction to sea level) and it was used only in the analysis of trends of changes in this meteorological element. When considering long-term pressure changes over a longer period of time, data from the over 100-year-old Kraków observational series from the years 1901–2018 were used. These data come from the Scientific Station of the Climatology Department of the Jagiellonian University in Kraków (Kraków Observatory) and have already been used many times as an example of changes in meteorological conditions in Poland and Central Europe.

Annual, seasonal and monthly pressure values, as well as the number of days on which deep low or strong high-pressure systems occurred over Poland were characterised. Owing to the lack of available data, the maximum and minimum pressure

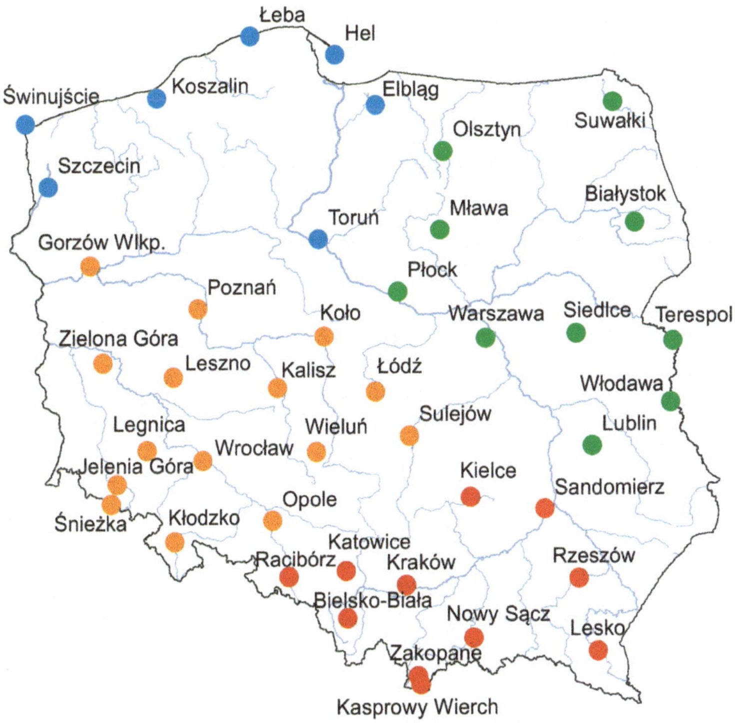

Fig. 7.1 Meteorological stations used in the study. Groups of stations with similar temporal pressure changes were distinguished by colours

values were not taken into account, but only daily averages. This means that the number of days analysed on which pressure values met the adopted criteria could be slightly higher than calculated. Adopting this way of distinguishing the described days means that the pressure systems in question occurred for a significant part of the day and usually covered a larger area. In climatological literature, there are various criteria for separating these systems. In this study, it was assumed that a day with deep low pressure would be considered one on which the value of average daily pressure is equal to or lower than 990 hPa. On the other hand, a day on which the value of average daily pressure is equal to or higher than 1030 hPa was adopted as a day on which a strong high-pressure system occurred. The criteria adopted in this way are consistent with, or similar to the values adopted by other authors (Schinke 1993; Kłysik 1995; Kożuchowski 1995; Leckebusch and Ulbrich 2004; Burt 2007; Bielec-Bąkowska and Piotrowicz 2011, 2013). At the same time, they also meet the definition determining the occurrence of extreme events (Beniston et al. 2007; IPCC 2007; Labajo et al. 2008).

Owing to the fact that spatial pressure changes are small, regions in which changes in the average daily pressure values in the studied years had a similar course were distinguished. To this end, cluster analysis was used. As a result, four groups of stations with a similar pressure pattern were distinguished (Fig. 7.1). Two stations were selected in each group to represent the singled-out area.

Spatial and Temporal Atmospheric Pressure Changes

The spatial diversification of the average annual SLP values in the years 1966–2018 was characterised by a latitudinal (belt) distribution. These values ranged from about 1014 hPa in the northern part of Poland, with the lowest value of 1014.5 hPa in Łeba, to over 1017 hPa in the south and south-east (to 1017.4 hPa in Lesko) (Fig. 7.2).

In the spring (March–May) and summer (June–August), the average SLP distribution was even less spatially diversified than annual values. It was only about 2 hPa (1014.7–1015.8 hPa in spring and 1014.3–1016.0 hPa in summer), with the pressure rising slightly from the coast of the Baltic Sea towards the south. This distribution of SLP is related to the fact that during this period Poland is often affected by the wedge of the Azores High (Woś 2010).

The average values in the autumn (September–November) and winter (December–February) also refer to the latitudinal (belt) distribution of SLP. However, the range in these seasons was definitely bigger than in the spring and summer. It amounted to

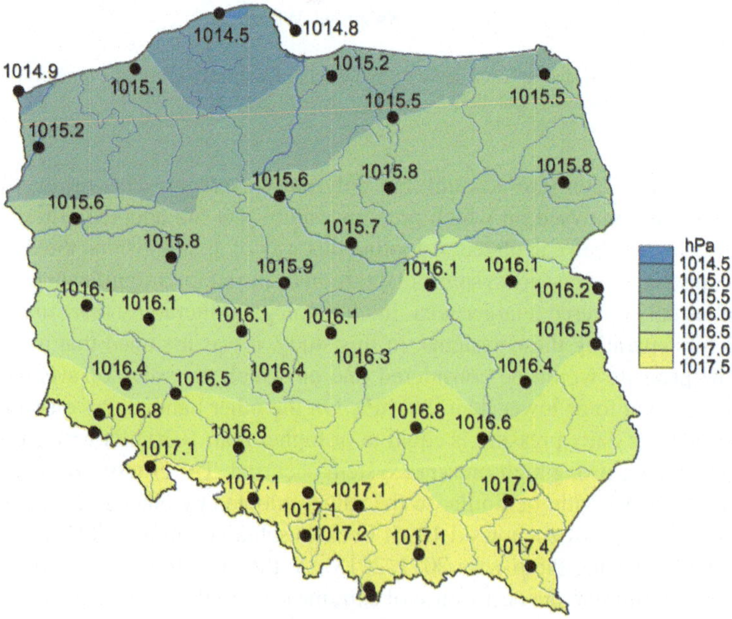

Fig. 7.2 Average annual atmospheric sea level pressure values (hPa) in the period 1966–2018

about 5 hPa between Łeba and Lesko (1014.7–1018.9 hPa in autumn and 1014.5–1019.5 hPa in winter) (Fig. 7.3). This is due to the stronger influence of the Icelandic Low in the winter, with an extensive trough reaching northern Europe. At the same time, through the centre of Europe, including on the Carpathian line, a high-pressure area (high-pressure ridge) is extended, joining the vast Azores and Asian (Siberian) Highs (Woś 2010). As a result, dynamic low-pressure systems (including deep lows) move quite often over Poland, in particular over the northern part.

In the analysed period, the average monthly and annual SLP values in Poland did not show big variability. Standard deviation values were only from about 2 to 6 hPa, with the highest in the winter (especially in January and February; from 4.3 in Zakopane to 5.8 hPa in Łeba), and the lowest in the summer (especially in June; from 1.9 in Zakopane to 2.4 hPa in Świnoujście). The coefficient of variation also

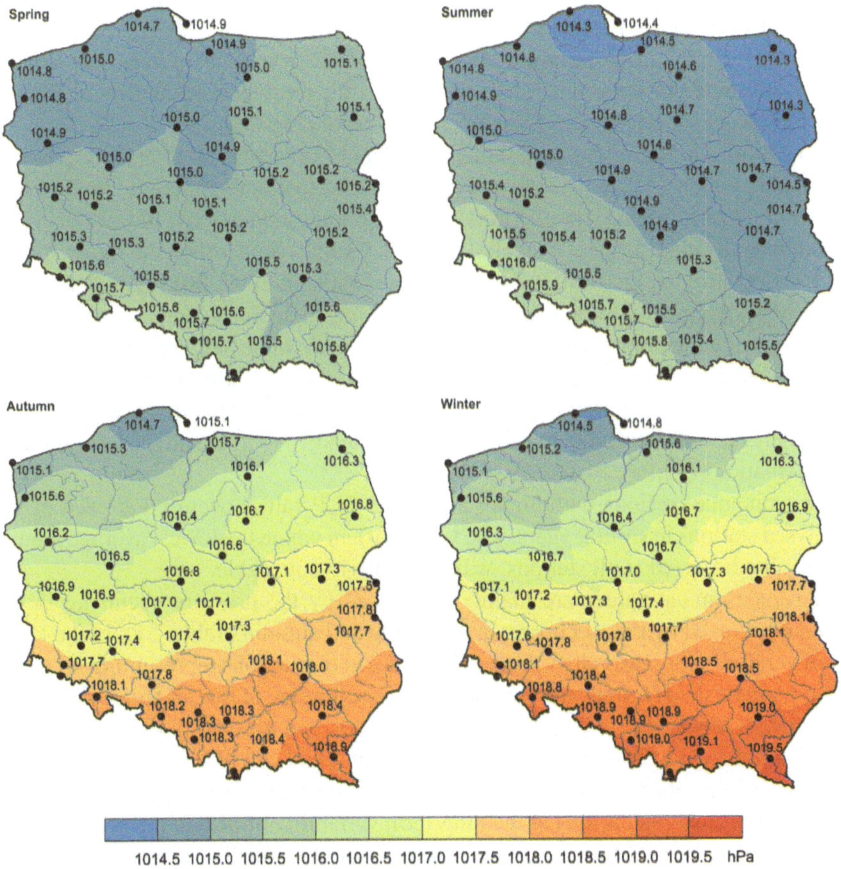

Fig. 7.3 Average annual atmospheric sea level pressure values (hPa) in particular seasons: spring (March–May), summer (June–August), autumn (September–November), winter (December–February) in the period 1966–2018

reached very small values in individual months, seasons, and in the year, from 0.1 to 0.6%.

As already mentioned in the section "Data and Methods of Investigation", four regions with similar pressure conditions were distinguished in Poland on the basis of average daily pressure values in the analysed period and using cluster analysis (Fig. 7.1).

In the period analysed, the north-western part of Poland, in which the average annual SLP was the lowest, around 1015 hPa, clearly stands out. At the turn of the 1970s and 1980s, in the second half of the 1990s, and around 2010, the pressure drop was most pronounced, especially in Świnoujście (Fig. 7.4). Values higher than the multi-annual average in this region occurred in the mid-1970s, at the turn of the 1980s and 1990s and in 2003.

The second distinguished region (mainly covering north-eastern Poland) was characterised by the highest SLP values in 1972 (even exceeding 1018 hPa), with the lowest in 1970 and 2010 (about 1014 hPa). Unlike other regions, this one was distinguished by the fact that pressure fluctuations were, from year to year, the lowest. These amounted to about 4 hPa. An example of the long-term variability of SLP in this region are the changes in Suwałki and Terespol (Fig. 7.4).

The third-largest region, represented by Jelenia Góra and Łódź (Fig. 7.4), covered the area in which SLP showed the most pronounced trend of an increase in the value over a long-term period, but the trend (as at all analysed stations) did not prove to be statistically significant at a level of 0.05. SLP values higher than the multi-annual average occurred at the turn of the 1980s and 1990s, in 2003, 2011 and 2015 (over 1018 hPa), while the lowest was in 2010, reaching about 1014 hPa.

A very similar long-term variability of SLP, with values higher and lower than the average occurring in the same years, took place in the fourth region covering the south-eastern part of Poland. This area, compared to the previous ones, is characterised by the highest SLP values (on average around 1017 hPa) and the largest range of fluctuations (5 hPa).

To conclude, periods with higher and lower SLP values occurred at each of the analysed stations in similar years. The year 2010, in which the lowest average annual SLP value occurred at all stations, was particularly noticeable.

Average annual values are mainly influenced by the pressure occurring in the winter, followed by that occurring in the spring and autumn (Table 7.1). In the long-term variability of seasonal average SLP values at the stations representing all of the four distinguished regions, the coincidence of occurrence of low and high values can be noticed, as was the case with annual values. Also, in the case of these values, the trends of change were not statistically significant at a level of 0.05, with the values of standard deviation and the coefficient of variation being very low (Table 7.1).

The described regularities of the long-term variability of SLP in the period 1966–2018, especially the absence of a clear trend of statistically significant changes, were also noticeable in the longer period of measurements. They can be traced, for example, in the pressure changes from the station at the Kraków Observatory of the Jagiellonian University in the years 1901–2018 (Fig. 7.5). Trepińska (1997) came to similar conclusions when analysing the Kraków series of pressure measurements

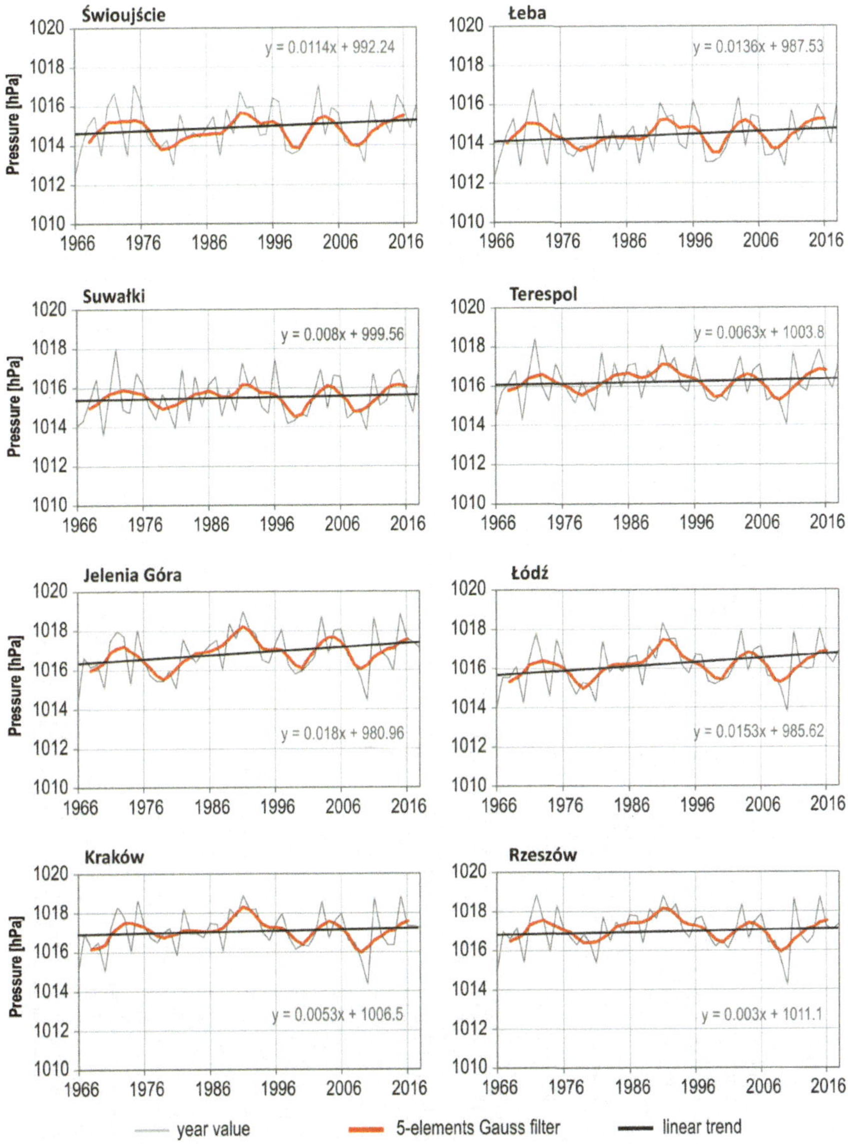

Fig. 7.4 Long-term variability of average annual atmospheric sea level pressure values (hPa) at selected stations in the period 1966–2018

Table 7.1 Average atmospheric sea level pressure values (hPa) in particular seasons: spring (March–May), summer (June–August), autumn (September–November), winter (December–February) in selected stations in the period 1966–2018

Year	Świnoujście				Suwałki				Jelenia Góra				Rzeszów			
	Spring	Summer	utumn	Winter	Spring	Summer	utumn	Winter	Spring	Summer	utumn	Winter	Spring	Summer	utumn	Winter
1966	1013.8	1012.9	1013.4	1011.7	1013.7	1013.8	1016.0	1014.6	1015.5	1014.6	1015.5	1015.9	1015.1	1014.6	1017.0	1017.6
1967	1012.4	1016.9	1013.7	1011.2	1012.5	1016.3	1015.0	1010.4	1014.7	1018.3	1016.6	1014.1	1014.5	1017.6	1018.0	1014.1
1968	1014.8	1016.1	1016.0	1014.3	1014.6	1015.1	1017.7	1019.3	1016.2	1016.1	1017.5	1015.0	1016.8	1015.4	1018.4	1018.4
1969	1015.5	1016.4	1013.9	1014.0	1016.0	1015.3	1013.5	1016.2	1015.4	1016.4	1017.2	1014.1	1016.3	1015.4	1018.0	1015.3
1970	1010.7	1015.4	1013.8	1016.1	1011.6	1014.4	1013.5	1016.6	1012.5	1015.7	1017.0	1018.6	1013.1	1015.0	1017.5	1019.1
1971	1014.6	1014.6	1018.0	1020.1	1015.1	1014.0	1016.7	1023.2	1014.0	1015.5	1020.7	1020.6	1014.7	1015.2	1020.4	1023.2
1972	1013.1	1015.8	1016.2	1018.8	1014.3	1015.3	1014.9	1020.6	1013.8	1016.8	1019.1	1022.0	1014.8	1015.5	1018.6	1023.4
1973	1016.1	1016.5	1014.7	1012.7	1016.2	1015.6	1014.0	1015.5	1017.9	1017.5	1018.2	1015.9	1017.4	1016.5	1018.5	1018.3
1974	1017.3	1013.5	1009.6	1016.7	1018.1	1012.6	1012.0	1016.0	1016.0	1015.1	1013.0	1020.8	1016.1	1014.1	1014.5	1021.1
1975	1013.9	1017.4	1018.5	1017.1	1013.9	1015.9	1020.8	1016.0	1013.7	1016.6	1019.5	1019.7	1013.8	1015.6	1021.6	1020.5
1976	1018.2	1018.0	1014.6	1010.2	1016.9	1015.8	1017.5	1013.0	1018.0	1017.8	1014.9	1012.4	1017.3	1016.9	1017.0	1014.5
1977	1015.7	1013.8	1014.3	1014.0	1015.5	1012.9	1015.1	1016.9	1016.2	1013.5	1017.2	1015.8	1016.8	1013.3	1018.4	1018.6
1978	1013.8	1013.8	1017.0	1012.8	1013.8	1012.8	1015.8	1015.7	1013.1	1015.5	1020.7	1013.4	1013.6	1015.3	1020.9	1015.4
1979	1012.3	1014.9	1016.5	1015.8	1015.2	1014.7	1018.1	1018.4	1013.6	1016.4	1018.1	1017.7	1015.4	1016.2	1019.8	1020.2
1980	1015.4	1011.5	1013.8	1014.6	1016.1	1011.0	1014.8	1014.7	1014.8	1013.3	1016.8	1018.3	1015.0	1012.7	1017.5	1018.9
1981	1013.1	1014.5	1013.5	1013.8	1015.0	1014.5	1014.4	1014.9	1013.9	1016.0	1016.6	1015.9	1015.5	1015.3	1016.3	1016.3
1982	1015.9	1014.7	1015.2	1013.7	1016.7	1014.9	1018.6	1013.1	1017.8	1015.6	1017.6	1017.8	1017.4	1014.4	1019.5	1016.6
1983	1010.2	1017.8	1014.7	1013.8	1011.8	1016.6	1014.4	1017.5	1011.9	1017.7	1018.4	1017.2	1012.0	1016.0	1019.1	1019.7
1984	1015.8	1014.4	1011.8	1018.7	1017.1	1013.9	1014.9	1019.9	1014.9	1016.4	1014.9	1020.8	1015.4	1015.9	1017.0	1021.5
1985	1011.8	1013.5	1017.5	1012.8	1014.2	1012.9	1017.3	1014.6	1013.1	1015.8	1020.8	1015.4	1014.2	1014.9	1020.7	1016.2
1986	1014.9	1014.9	1017.4	1014.8	1017.9	1014.6	1018.3	1015.3	1015.8	1016.6	1021.4	1017.7	1017.4	1015.6	1022.0	1018.2
1987	1016.1	1012.6	1015.9	1011.5	1016.9	1013.3	1019.3	1015.0	1017.2	1015.1	1018.2	1015.6	1016.7	1015.1	1020.1	1017.3
1988	1013.4	1013.0	1017.0	1018.2	1014.2	1012.4	1017.7	1016.7	1014.3	1014.9	1020.0	1023.3	1014.7	1013.9	1020.6	1023.0
1989	1014.4	1015.2	1016.4	1012.1	1015.6	1014.5	1016.3	1013.2	1015.8	1016.5	1018.9	1017.6	1015.9	1015.3	1018.4	1019.0
1990	1017.9	1014.8	1011.9	1019.1	1016.9	1014.1	1012.5	1020.4	1019.8	1016.9	1015.0	1022.0	1018.7	1015.9	1015.6	1023.0
1991	1016.4	1014.7	1014.5	1022.3	1016.9	1014.7	1016.8	1019.4	1017.1	1016.5	1017.6	1026.2	1016.5	1015.4	1018.8	1024.5
1992	1014.2	1015.1	1012.6	1021.1	1015.2	1015.7	1013.1	1020.2	1016.4	1015.6	1015.5	1024.8	1016.4	1015.5	1015.8	1024.8
1993	1016.9	1014.1	1017.9	1010.0	1017.2	1013.0	1021.4	1012.4	1017.0	1016.3	1018.6	1013.3	1017.0	1015.6	1020.5	1015.4
1994	1012.5	1015.9	1015.5	1011.4	1012.6	1015.1	1016.4	1013.1	1013.9	1016.7	1018.2	1016.3	1014.2	1016.0	1019.3	1018.0
1995	1013.2	1015.7	1015.9	1020.1	1012.5	1014.4	1016.6	1022.9	1014.5	1016.0	1018.3	1019.9	1014.3	1015.0	1018.9	1022.0
1996	1017.2	1016.4	1013.2	1018.7	1017.8	1016.7	1014.0	1018.3	1017.0	1017.8	1015.2	1022.0	1016.7	1017.2	1015.5	1022.6
1997	1017.0	1015.1	1015.3	1015.2	1014.6	1014.9	1014.4	1015.6	1018.7	1015.4	1017.7	1018.9	1017.2	1014.5	1017.7	1019.6
1998	1013.2	1012.4	1012.9	1012.3	1013.6	1011.6	1015.6	1012.1	1014.4	1014.7	1015.1	1016.3	1014.6	1014.3	1016.7	1016.9
1999	1014.0	1015.4	1016.5	1012.2	1015.1	1015.5	1018.2	1011.6	1014.9	1016.6	1018.6	1017.5	1015.6	1015.9	1019.4	1017.9
2000	1013.5	1014.5	1012.7	1013.8	1013.6	1013.8	1017.8	1015.2	1014.4	1016.2	1014.6	1016.6	1014.5	1015.5	1017.4	1017.6
2001	1012.2	1014.6	1014.0	1015.0	1012.5	1013.9	1014.3	1014.5	1012.8	1015.9	1017.1	1019.9	1012.8	1014.7	1017.1	1020.1
2002	1016.0	1014.5	1013.6	1019.9	1016.2	1015.3	1014.3	1021.2	1016.5	1015.2	1014.9	1021.0	1016.3	1014.8	1015.2	1022.1
2003	1019.7	1015.2	1016.3	1012.8	1019.5	1013.5	1017.6	1012.9	1020.3	1016.6	1018.3	1016.3	1019.4	1015.1	1018.9	1016.2
2004	1015.9	1013.7	1015.3	1015.3	1016.4	1014.2	1016.1	1015.5	1016.9	1015.8	1018.3	1019.1	1016.6	1014.9	1018.4	1018.8
2005	1015.7	1014.9	1018.6	1018.3	1015.4	1014.7	1021.5	1019.9	1016.8	1016.6	1020.8	1020.5	1015.9	1015.1	1021.7	1020.3
2006	1012.5	1015.8	1013.7	1012.9	1013.7	1015.4	1015.3	1013.5	1014.5	1016.8	1017.3	1017.7	1014.0	1015.6	1018.0	1018.1
2007	1015.2	1011.4	1017.1	1018.8	1015.2	1011.8	1016.6	1020.3	1016.7	1013.4	1019.6	1023.3	1016.2	1013.0	1018.6	1024.1
2008	1010.5	1013.3	1014.5	1014.9	1010.4	1013.3	1015.6	1016.0	1011.6	1015.0	1017.2	1016.8	1011.1	1014.4	1017.6	1017.2
2009	1015.7	1014.5	1014.0	1010.9	1016.0	1014.6	1015.5	1014.6	1016.2	1016.1	1016.4	1011.8	1016.0	1015.2	1016.8	1013.3
2010	1015.1	1014.0	1011.6	1015.6	1014.0	1013.2	1013.6	1015.9	1016.3	1014.9	1013.6	1017.9	1015.1	1013.5	1014.5	1018.0

(continued)

Table 7.1 (continued)

Year	Świnoujście				Suwałki				Jelenia Góra				Rzeszów			
	Spring	Summer	utumn	Winter	Spring	Summer	utumn	Winter	Spring	Summer	utumn	Winter	Spring	Summer	utumn	Winter
1966	1013.8	1012.9	1013.4	1011.7	1013.7	1013.8	1016.0	1014.6	1015.5	1014.6	1015.5	1015.9	1015.1	1014.6	1017.0	1017.6
1967	1012.4	1016.9	1013.7	1011.2	1012.5	1016.3	1015.0	1010.4	1014.7	1018.3	1016.6	1014.1	1014.5	1017.6	1018.0	1014.1
1968	1014.8	1016.1	1016.0	1014.3	1014.6	1015.1	1017.7	1019.3	1016.2	1016.1	1017.5	1015.0	1016.8	1015.4	1018.4	1018.4
1969	1015.5	1016.4	1013.9	1014.0	1016.0	1015.3	1013.5	1016.2	1015.4	1016.4	1017.2	1014.1	1016.3	1015.4	1018.0	1015.3
1970	1010.7	1015.4	1013.8	1016.1	1011.6	1014.4	1013.5	1016.6	1012.5	1015.7	1017.0	1018.6	1013.1	1015.0	1017.5	1019.1
1971	1014.6	1014.6	1018.0	1020.1	1015.1	1014.0	1016.7	1023.2	1014.0	1015.5	1020.7	1020.6	1014.7	1015.2	1020.4	1023.2
1972	1013.1	1015.8	1016.2	1018.8	1014.3	1015.3	1014.9	1020.6	1013.8	1016.8	1019.1	1022.0	1014.8	1015.5	1018.6	1023.4
1973	1016.1	1016.5	1014.7	1012.7	1016.2	1015.6	1014.0	1015.5	1017.9	1017.5	1018.2	1015.9	1017.4	1016.5	1018.5	1018.3
1974	1017.3	1013.5	1009.6	1016.7	1018.1	1012.6	1012.0	1016.0	1016.0	1015.1	1013.0	1020.8	1016.1	1014.1	1014.5	1021.1
1975	1013.9	1017.4	1018.5	1017.1	1013.9	1015.9	1020.8	1016.0	1013.7	1016.6	1019.5	1019.7	1013.8	1015.6	1021.6	1020.5
1976	1018.2	1018.0	1014.6	1010.2	1016.9	1015.8	1017.5	1013.0	1018.0	1017.8	1014.9	1012.4	1017.3	1016.9	1017.0	1014.5
1977	1015.7	1013.8	1014.3	1014.0	1015.5	1012.9	1015.1	1016.9	1016.2	1013.5	1017.2	1015.8	1016.8	1013.3	1018.4	1018.6
1978	1013.8	1013.8	1017.0	1012.8	1013.8	1012.8	1015.8	1015.7	1013.1	1015.5	1020.7	1013.4	1013.6	1015.3	1020.9	1015.4
1979	1012.3	1014.9	1016.5	1015.8	1015.2	1014.7	1018.1	1018.4	1013.6	1016.4	1018.1	1017.7	1015.4	1016.2	1019.8	1020.2
1980	1015.4	1011.5	1013.8	1014.6	1016.1	1011.0	1014.8	1014.7	1014.8	1013.3	1016.8	1018.3	1015.0	1012.7	1017.5	1018.9
1981	1013.1	1014.5	1013.5	1013.8	1015.0	1014.5	1014.4	1014.9	1013.9	1016.0	1016.6	1015.9	1015.5	1015.3	1016.3	1016.3
1982	1015.9	1014.7	1015.2	1013.7	1016.7	1014.9	1018.6	1013.1	1017.8	1015.6	1017.6	1017.8	1017.4	1014.4	1019.5	1016.6
1983	1010.2	1017.8	1014.7	1013.8	1011.8	1016.6	1014.4	1017.5	1011.9	1017.7	1018.4	1017.2	1012.0	1016.0	1019.1	1019.7
1984	1015.8	1014.4	1011.8	1018.7	1017.1	1013.9	1014.9	1019.9	1014.9	1016.4	1014.9	1020.8	1015.4	1015.9	1017.0	1021.5
1985	1011.8	1013.5	1017.5	1012.8	1014.2	1012.9	1017.3	1014.6	1013.1	1015.8	1020.8	1015.4	1014.2	1014.9	1020.7	1016.2
1986	1014.9	1014.9	1017.4	1014.8	1017.9	1014.6	1018.3	1015.3	1015.8	1016.6	1021.4	1017.7	1017.4	1015.6	1022.0	1018.2
1987	1016.1	1012.6	1015.9	1011.5	1016.9	1013.3	1019.3	1015.0	1017.2	1015.1	1018.2	1015.6	1016.7	1015.1	1020.1	1017.3
1988	1013.4	1013.0	1017.0	1018.2	1014.2	1012.4	1017.7	1016.7	1014.3	1014.9	1020.0	1023.3	1014.7	1013.9	1020.6	1023.0
1989	1014.4	1015.2	1016.4	1012.1	1015.6	1014.5	1016.3	1013.2	1015.8	1016.5	1018.9	1017.6	1015.9	1015.3	1018.4	1019.0
1990	1017.9	1014.8	1011.9	1019.1	1016.9	1014.1	1012.5	1020.4	1019.8	1016.9	1015.0	1022.0	1018.7	1015.9	1015.6	1023.0
1991	1016.4	1014.7	1014.5	1022.3	1016.9	1014.7	1016.8	1019.4	1017.1	1016.5	1017.6	1026.2	1016.5	1015.4	1018.8	1024.5
1992	1014.2	1015.1	1012.6	1021.1	1015.2	1015.7	1013.1	1020.2	1016.4	1015.6	1015.5	1024.8	1016.4	1015.5	1015.8	1024.8
1993	1016.9	1014.1	1017.9	1010.0	1017.2	1013.0	1021.4	1012.4	1017.0	1016.3	1018.6	1013.3	1017.0	1015.6	1020.5	1015.4
1994	1012.5	1015.9	1015.5	1011.4	1012.6	1015.1	1016.4	1013.1	1013.9	1016.7	1018.2	1016.3	1014.2	1016.0	1019.3	1018.0
1995	1013.2	1015.7	1015.9	1020.1	1012.5	1014.4	1016.6	1022.9	1014.5	1016.0	1018.3	1019.9	1014.3	1015.0	1018.9	1022.0
1996	1017.2	1016.4	1013.2	1018.7	1017.8	1016.7	1014.0	1018.3	1017.0	1017.8	1015.2	1022.0	1016.7	1017.2	1015.5	1022.6
1997	1017.0	1015.1	1015.3	1015.2	1014.6	1014.9	1014.4	1015.6	1018.7	1015.4	1017.7	1018.9	1017.2	1014.5	1017.7	1019.6
1998	1013.2	1012.4	1012.9	1012.3	1013.6	1011.6	1015.6	1012.1	1014.4	1014.7	1015.1	1016.3	1014.6	1014.3	1016.7	1016.9
1999	1014.0	1015.4	1016.5	1012.2	1015.1	1015.5	1018.2	1011.6	1014.9	1016.6	1018.6	1017.5	1015.6	1015.9	1019.4	1017.9
2000	1013.5	1014.5	1012.7	1013.8	1013.6	1013.8	1017.8	1015.2	1014.4	1016.2	1014.6	1016.6	1014.5	1015.5	1017.4	1017.6
2001	1012.2	1014.6	1014.0	1015.0	1012.5	1013.9	1014.3	1014.5	1012.8	1015.9	1017.1	1019.9	1012.8	1014.7	1017.1	1020.1
2002	1016.0	1014.5	1013.6	1019.9	1016.2	1015.3	1014.3	1021.2	1016.5	1015.2	1014.9	1021.0	1016.3	1014.8	1015.2	1022.1
2003	1019.7	1015.2	1016.3	1012.8	1019.5	1013.5	1017.6	1012.9	1020.3	1016.6	1018.3	1016.3	1019.4	1015.1	1018.9	1016.2
2004	1015.9	1013.7	1015.3	1015.3	1016.4	1014.2	1016.1	1015.5	1016.9	1015.8	1018.3	1019.1	1016.6	1014.9	1018.4	1018.8
2005	1015.7	1014.9	1018.6	1018.3	1015.4	1014.7	1021.5	1019.9	1016.8	1016.6	1020.8	1020.5	1015.9	1015.1	1021.7	1020.3
2006	1012.5	1015.8	1013.7	1012.9	1013.7	1015.4	1015.3	1013.5	1014.5	1016.8	1017.3	1017.7	1014.0	1015.6	1018.0	1018.1
2007	1015.2	1011.4	1017.1	1018.8	1015.2	1011.8	1016.6	1020.3	1016.7	1013.4	1019.6	1023.3	1016.2	1013.0	1018.6	1024.1
2008	1010.5	1013.3	1014.5	1014.9	1010.4	1013.3	1015.6	1016.0	1011.6	1015.0	1017.2	1016.8	1011.1	1014.4	1017.6	1017.2

(continued)

Table 7.1 (continued)

Year	Świnoujście				Suwałki				Jelenia Góra				Rzeszów			
	Spring	Summer	utumn	Winter	Spring	Summer	utumn	Winter	Spring	Summer	utumn	Winter	Spring	Summer	utumn	Winter
2009	1015.7	1014.5	1014.0	1010.9	1016.0	1014.6	1015.5	1014.6	1016.2	1016.1	1016.4	1011.8	1016.0	1015.2	1016.8	1013.3
2010	1015.1	1014.0	1011.6	1015.6	1014.0	1013.2	1013.6	1015.9	1016.3	1014.9	1013.6	1017.9	1015.1	1013.5	1014.5	1018.0
2011	1019.8	1012.5	1019.0	1015.2	1018.9	1012.7	1020.3	1015.8	1020.7	1014.3	1021.8	1019.5	1019.9	1013.5	1022.7	1019.6
2012	1015.5	1014.1	1012.8	1013.3	1014.3	1014.1	1014.2	1015.1	1016.5	1015.7	1015.7	1015.1	1015.5	1015.1	1016.5	1015.6
2013	1013.7	1017.7	1014.7	1012.1	1014.4	1016.4	1015.0	1016.6	1013.5	1018.4	1017.0	1015.8	1013.4	1016.8	1017.2	1018.7
2014	1015.5	1013.9	1017.1	1013.4	1015.5	1013.9	1020.9	1014.4	1016.1	1014.9	1018.0	1017.3	1015.3	1014.4	1019.8	1017.7
2015	1017.0	1016.2	1017.0	1014.4	1016.1	1016.7	1018.1	1015.7	1018.9	1017.1	1019.2	1018.4	1017.9	1016.9	1019.4	1019.6
2016	1013.9	1015.5	1018.6	1021.2	1014.1	1015.2	1019.1	1020.6	1014.8	1017.1	1019.6	1024.4	1013.8	1016.3	1019.6	1024.3
2017	1016.6	1013.8	1013.2	1013.7	1015.3	1013.6	1013.5	1015.5	1018.0	1015.4	1016.7	1016.7	1016.7	1015.2	1016.4	1017.7
2018	1013.8	1015.9	1019.3	–	1014.7	1014.4	1020.9	–	1013.4	1016.3	1020.7	–	1014.0	1014.7	1021.7	–
Avg.	1014.8	1014.8	1015.1	1015.1	1015.1	1014.3	1016.3	1016.3	1015.6	1016.0	1017.7	1018.1	1015.6	1015.2	1018.4	1019.0
S.D.	2.1	1.5	2.1	3.1	1.8	1.3	2.4	2.9	2.0	1.1	2.0	3.2	1.7	1.0	1.9	2.8
V	0.2	0.1	0.2	0.3	0.2	0.1	0.2	0.3	0.2	0.1	0.2	0.3	0.2	0.1	0.2	0.3
Trend	0.24	–0.15	0.21	0.10	0.04	–0.05	0.33	–0.05	0.28	–0.04	0.10	0.31	0.03	–0.08	0.00	0.10

Avg.—Average, S.D.—standard deviation (hPa), V—coefficient of variability (%), Trend—hPa/10 years

from the years 1792–1995. It can be seen in the twentieth century that high-pressure values in Kraków occurred mainly in the years from around 1920 to 1950. In the 1920s, high-pressure values were recorded particularly in the autumn and winter, in the 1930s in the summer, and in the 1940s in the spring and winter (Fig. 7.5). The year 2010 clearly stands out again in this multi-annual course, having the lowest pressure in the last 118 years, which is also in the spring. Comparing SLP values with those presented in Fig. 7.4, it can be stated that, especially in southern and central Poland (regions 3 and 4), and to a lesser extent in region 2, the low-pressure values occurring in 2010 were extreme in the series of measurements taken during that period of over 100 years.

Annual Course and Extreme Values of Atmospheric Pressure

The SLP annual course in Poland was traced by comparing values from stations located in the four regions distinguished. Figure 7.6 shows two stations representing each of them. It was found that in the annual course the highest SLP values occurrence mainly in cold half-year (October–March), but also at this time there are the largest differences between the stations and highlighted regions. The biggest occurring in December (6.0 hPa between Łeba and Lesko) (Fig. 7.6). In warm half-year (April–September) the values are usually the lowest in the whole year, especially in the north-east and east of Poland. A characteristic feature of this season is that the pressure differences between the stations from April to May are very small. In the analysed years, it was only 0.6 hPa (Fig. 7.6).

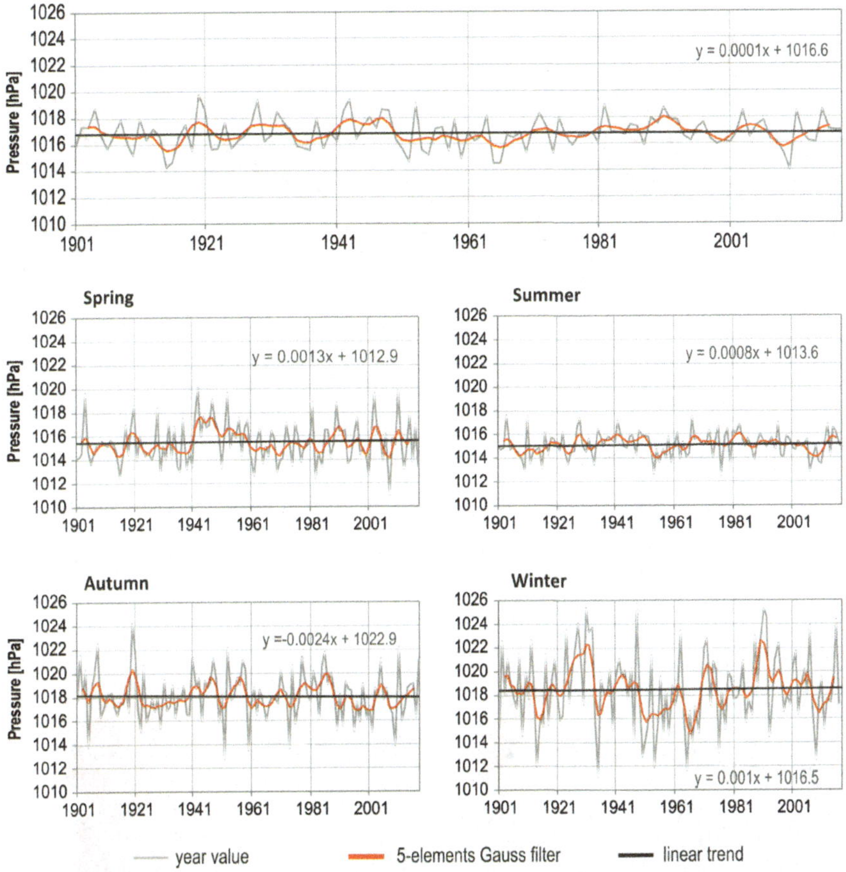

Fig. 7.5 Long-term variability of average annual atmospheric sea level pressure values (hPa) and in particular seasons in Kraków in the period 1901–2018

The average annual and seasonal SLP values do not reflect the range of fluctuations of this meteorological element in Poland. Therefore, attention was paid to the highest and lowest average daily pressure values that were recorded at individual stations in the analysed period. The absolute maxima and minima of SLP from a similar long-term period (1966–2006) can be found in Ustrnul's and Czekierda's (2009) work.

In the years 1966–2018, the highest average daily SLP values were over 1050 hPa at 57.5% of the stations located in the central and north-eastern part of Poland (Fig. 7.7). The highest value was recorded in Suwałki on 16 December 1997—1053.6 hPa. The lowest SLP values decreased almost meridionally, from south-east to north-west (Fig. 7.7) and, in extreme cases, ranged from 966.1 hPa in Świnoujście to 976.8 hPa in Lesko (26 February 1989). They were associated with the extensive lows moving from the west of Europe. All of the presented extreme cases (Fig. 7.7)

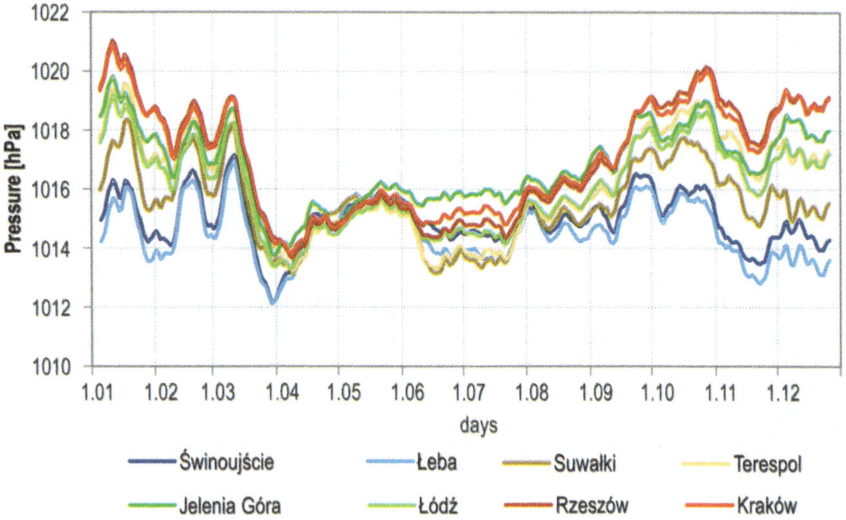

Fig. 7.6 Ten-days consecutive average daily atmospheric sea level pressure values (hPa) at selected stations in the period 1966–2018

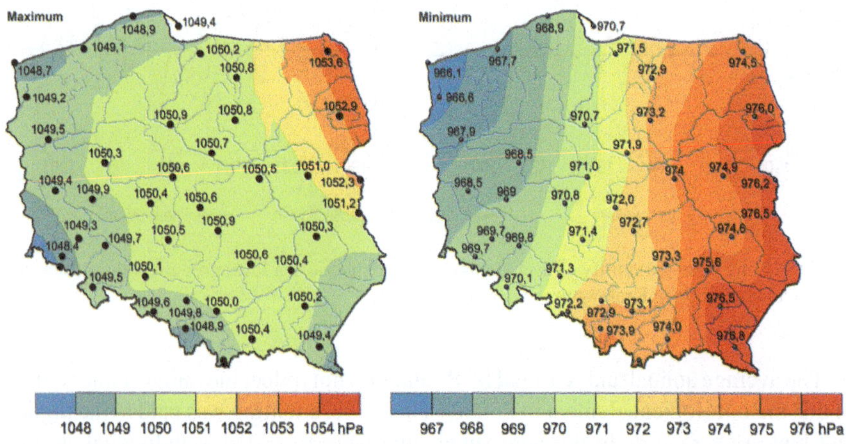

Fig. 7.7 The highest (Maximum) and lowest (Minimum) values of average daily atmospheric sea level pressure values (hPa) in the period 1966–2018

occurred over just a few days and are consistent with Ustrnul's and Czekierda's (2009) results.

In Kraków in the years 1901–2018, the highest average daily SLP value was 1050.2 hPa and occurred on 23 January 2006. It was associated with an extensive, but so-called low high-pressure system with its centre over eastern Poland (Ustrnul and Czekierda 2009). Regarding the lowest pressure value, it also occurred in Kraków in the winter, on 26 February 1989 (972.3 hPa), and was associated with the low of the centre which was over the North Sea on that day.

Number of Days with Pressure ≤ 990 hPa and ≥ 1030 hPa

Considering the spatial and temporal changes of SLP, one should mention the occurrence of exceptional pressure systems over Poland, which strongly affect the weather. These include deep lows and strong high-pressure systems. The moving of deep lows over Poland is accompanied by very strong winds, heavy rainfall and significant pressure changes. On the other hand, sunny weather is usually associated with strong highs, without precipitation and, depending on the season, with high or very low air temperature. In both cases, the occurrence of these systems may cause the development of weather conditions that are dangerous to the environment and to man.

In this part of the study, the occurrence of the pressure systems in question was characterised by analysing the occurrence of days with specific SLP values. A day with a deep low was considered one on which the value of the average daily SLP was equal to, or lower than, 990 hPa, while a day with a strong high-pressure system was one on which the value of the average daily SLP was equal to, or higher than, 1030 hPa.

In the analysed long-term period in Poland, the occurrence of strong high-pressure systems (on average 19.5–25.2 days a year) was much more frequent than of deep lows (on average 0.3–4.5 days). Most days with deep lows were recorded in the north of Poland, in Łeba (4.5 days) (Fig. 7.8). This is related to the main tracks of

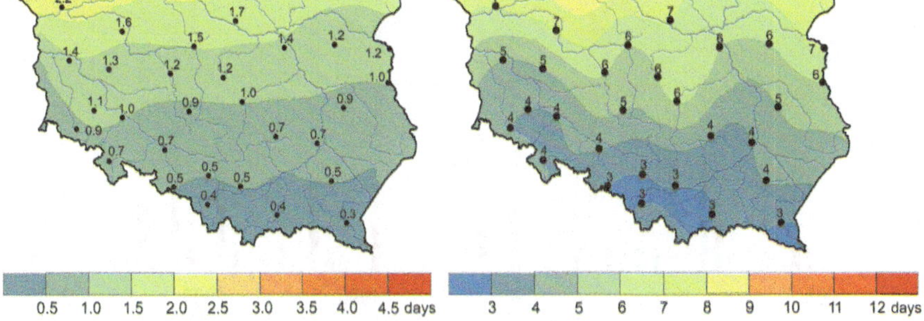

Fig. 7.8 Average and the highest (Maximum) annual number of days with atmospheric sea level pressure values ≤ 990 hPa in the period 1966–2018

movement of lows in this part of Europe (Paszyński and Niedźwiedź 1991; Bielec-Bąkowska 2010a). In this region, the studied pressure systems occur almost every year. In the south of Poland, in the analysed 53 years, such days were not recorded for 31 (Lublin, Opole) to 41 years (Bielsko-Biała). In individual years, the number of days with SLP ≤ 990 hPa can be much higher and range from three days in the south to twelve days in the north-west of the country (Fig. 7.8). At the same time, there were only nine years in the entire multi-annual period under study in which a deep low-pressure system moved at least once over each station (1974, 1976, 1981, 1983, 1989, 2008–2010 and 2015).

The long-term changes in the number of days with SLP ≤ 990 hPa indicate that the range of deep lows covered the largest area of Poland in the years 1973–1990 and 2004–2012 (Fig. 7.9). The years 1989 and 1990, in which such low SLP values during the same day were recorded by at least 30 stations for three and four days, respectively, and at 40 stations on two days, clearly stood out. However, in the last eight years of the analysed period, the number of days when deep lows moved over at least one station was very small. Within one day, the range of the analysed systems usually covered less than 10 stations. At the same time, it can be seen that in the years 1973–1990 the average number of days with SLP ≤ 990 hPa was the highest (Fig. 7.10). It was usually over four days a year. At the beginning and end of the studied period, these systems occasionally appeared over Poland.

Analysing the multi-annual changes in the number of days considered, found them to be small (from −0.29 to 0.04 days/10 years) and statistically insignificant at a level of 0.05 (Fig. 7.11). Only a slight decrease in the number of mentioned days occurred in the north of the country, in particular in the region represented by the stations in Szczecin and Gorzów Wielkopolski (from −0.22 to −0.29 days/10 years, not statistically significant). The described changes are related to annual values.

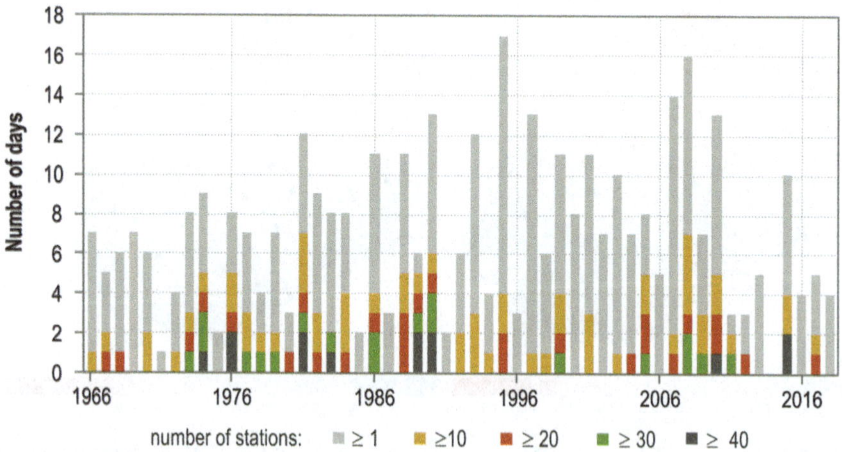

Fig. 7.9 Number of days on which the atmospheric sea level pressure values ≤ 990 hPa was recorded at least 1, 10, 20, 30, 40 stations in the period 1966–2018

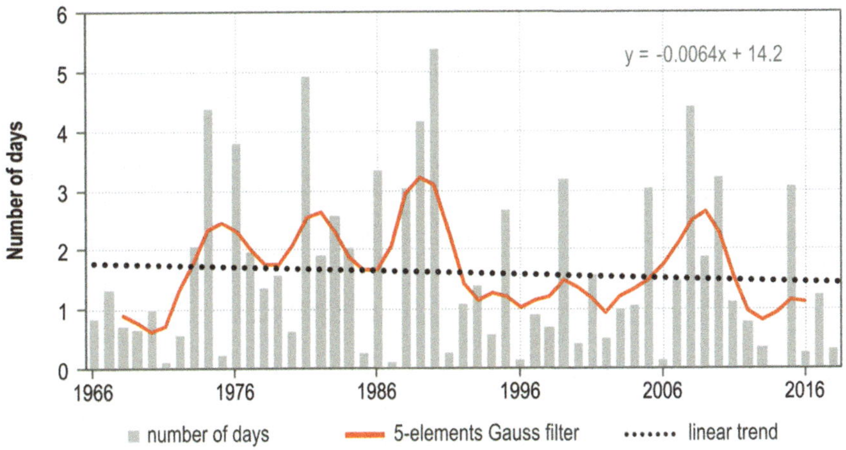

Fig. 7.10 Average number of days with the atmospheric sea level pressure values ≤ 990 hPa in Poland in the period 1966–2018

However, the small annual number of the days in question and the fact that more than 95% of all cases occur from October to March (Fig. 7.12) mean that they mainly concern this half-year period. It is also difficult to indicate the regularities of the long-term variability of days with SLP ≤ 990 hPa occurring at individual stations and for shorter periods of the year (seasons or months).

As previously mentioned, strong high-pressure systems occur over Poland several times more often than deep lows. In the years 1966–2018, the average annual number ranged from 19.5 days in Łeba to 25.2 days in Lesko (Fig. 7.13), and clearly increased from the north-west to the south and south-east of Poland. In individual years, this number could change quite significantly, with its highest values reaching 40 days in the north, and even 51 days in the south of the country. However, the spatial distribution of the maximum annual number of days considered is worth noticing. Its lowest values covered the eastern part of the Polish Baltic coast and reached central Poland in the form of a wedge. The highest maximum values were characteristic of the south-western and southern parts of the country. This distribution very well reflects the strong impact of the Azores High on the southern and western areas of Poland and that of the Asian High from the east (Martyn 1987).

In contrast to deep lows, strong highs appeared over individual regions of Poland every year. The smallest annual number ranged from three in the north to seven in southern regions. Strong highs most often occurred in the years 1982–1993, for which (except 1988 which had an average for Poland of 15.1 days) averages of 24.5–43.1 days were recorded for the analysed systems (Fig. 7.14). These occurred much less frequently over Poland at the beginning of the studied period (6.9–11.7 days) and in the years 1999–2000, as well as after 2008. In 2009, they lasted for just 5.5 days on average.

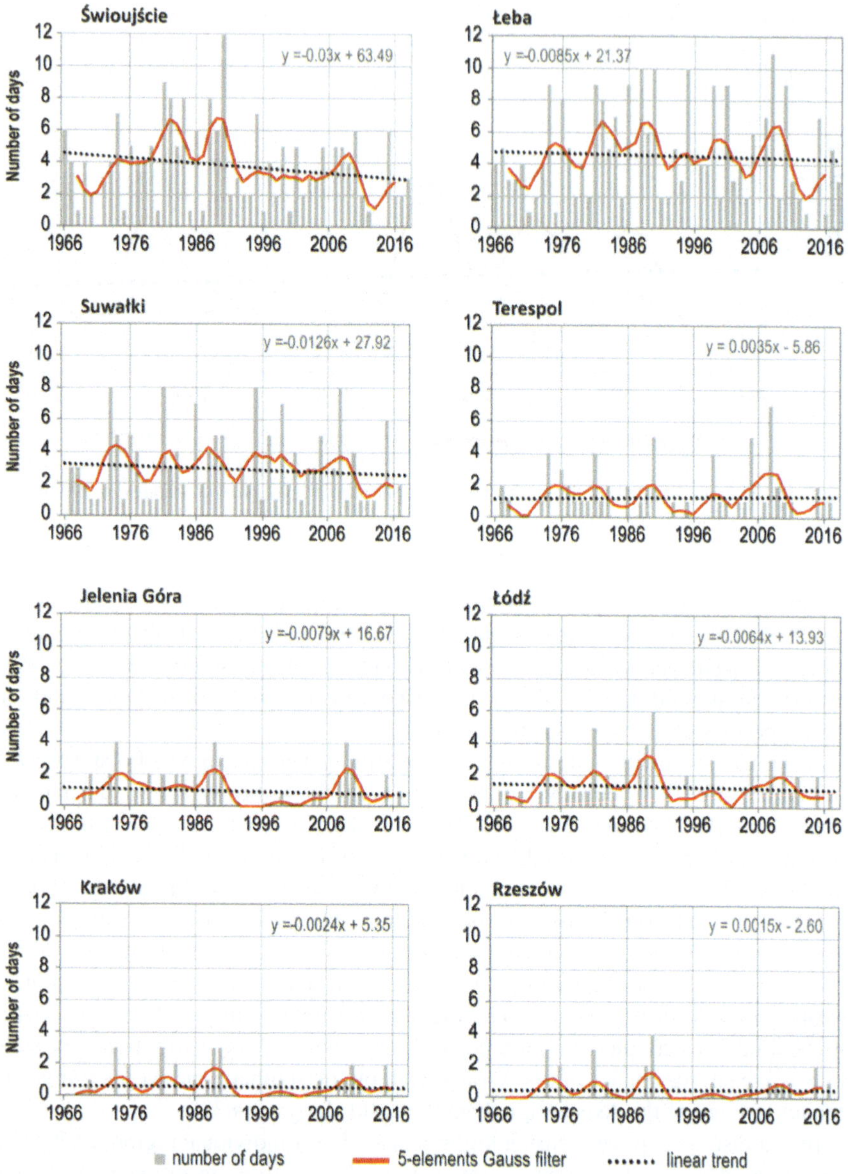

Fig. 7.11 Number of days with the atmospheric sea level pressure values ≤ 990 hPa at selected stations in the period 1966–2018

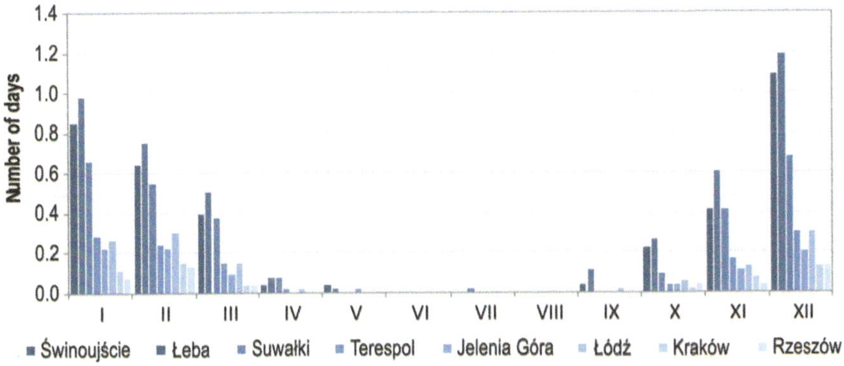

Fig. 7.12 Average monthly number of days with the atmospheric sea level pressure values ≤ 990 hPa at selected stations in Poland in the period 1966–2018

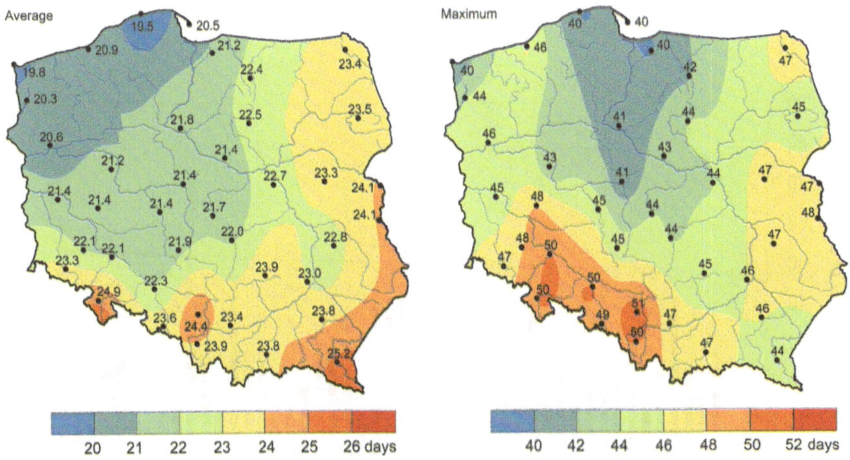

Fig. 7.13 Average and the highest (Maximum) annual number of days with atmospheric sea level pressure values ≥ 1030 hPa in the period 1966–2018

The number of days when the average daily SLP at least one station was ≥ 1030 hPa changed slightly differently. The first and final years of the multi-annual period were also characterised by a smaller number of analysed days, however, the biggest number fell in the years 1991–2006 (Fig. 7.15). At the same time, it was found that in the first half of the period studied, and especially in the years 1971–1992, strong high-pressure systems covered the entire area of the country much more often (in 1982 they even constituted 62% of all surveyed days). After 1992, situations prevailed in which the highs covering 30 or more stations during one day, usually accounted for less than 40% of the days analysed in a given year.

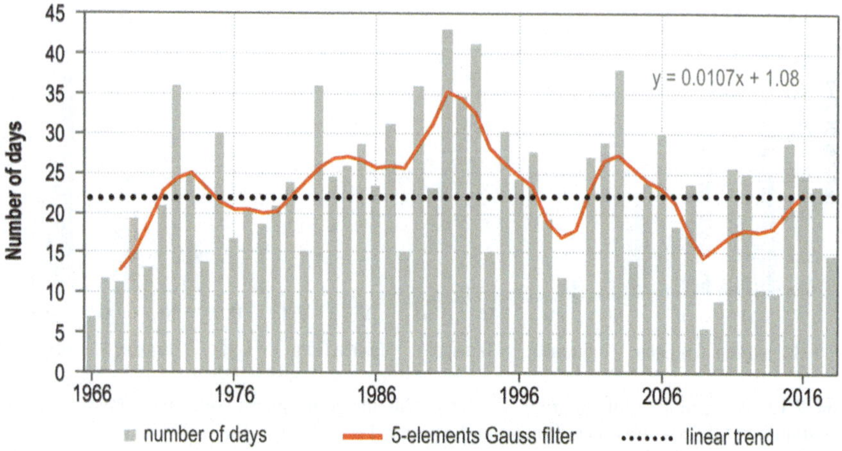

Fig. 7.14 Average number of days with the atmospheric sea level pressure values ≥ 1030 hPa in Poland in the period 1966–2018

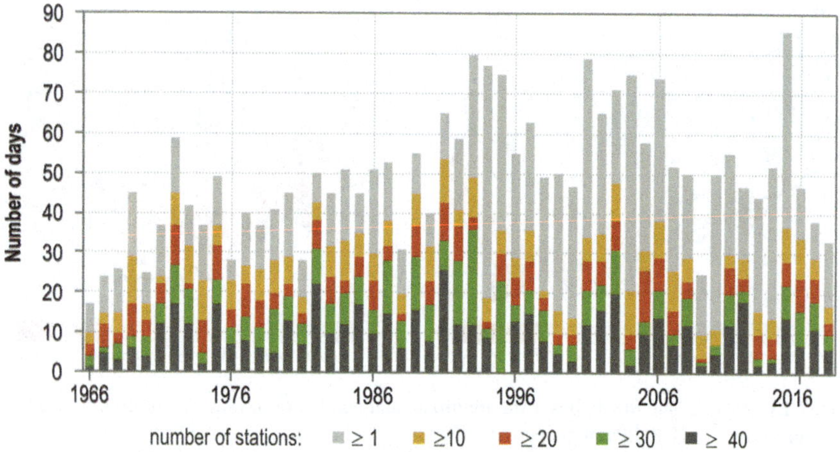

Fig. 7.15 Number of days on which the atmospheric sea level pressure values ≥ 1030 hPa was recorded at least 1, 10, 20, 30, 40 stations in the period 1966–2018

The long-term variability of the number of days with SLP ≥ 1030 hPa at individual stations is similar to the changes described for the whole of Poland (Fig. 7.16). Until around 1982, strong highs were extremely rare. In later years, the number clearly increased, reaching their highest values in the early 1990s (over 40, or even 50, days a year), to decrease again in the twenty-first century. It is for this reason that no statistically significant trend has been found in any of the regions of Poland. It is worth emphasising, however, that there are some signals of an increase in the number of mentioned systems at the Baltic coast stations, along the eastern border

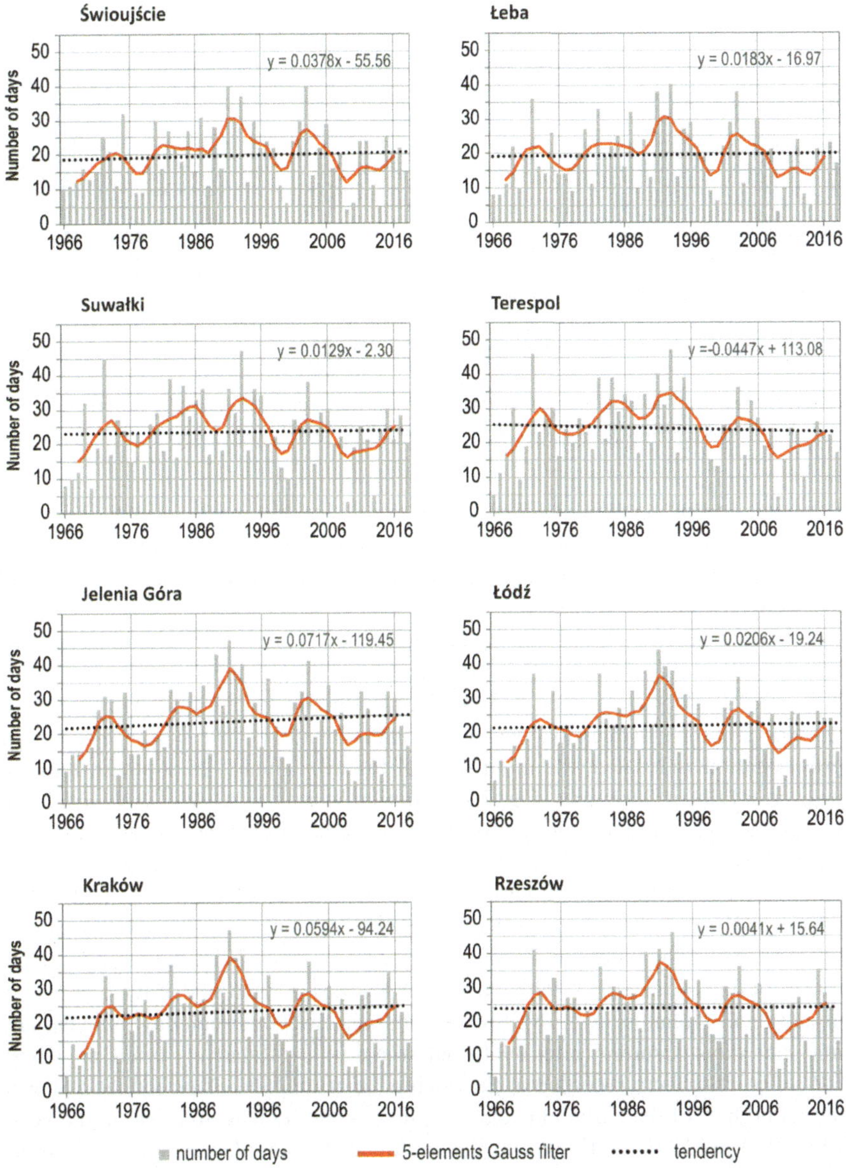

Fig. 7.16 Number of days with the atmospheric sea level pressure values ≥ 1030 hPa at selected stations in the period 1966–2018

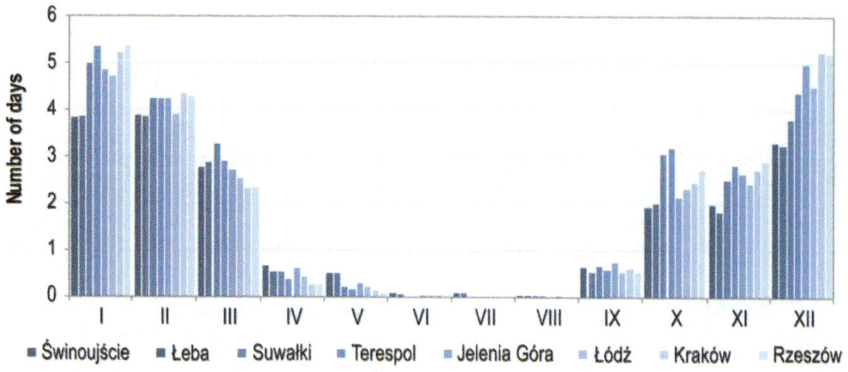

Fig. 7.17 Average monthly number of days with the atmospheric sea level pressure values ≥ 1030 hPa at selected stations in Poland in the period 1966–2018

of Poland, and in the area covering southern regions and reaching as far as Łódź. The highest estimated increase (0.81 day/10 years) occurred in Jelenia Góra, and the biggest decrease (also statistically insignificant) in Lublin and Siedlce (−0.45 and −0.44 day/10 years).

In the case of strong highs, the highest frequency is also recorded in the cool half of the year (October–March). This fluctuated from 89.6% in Świnoujście to 96.2% in Rzeszów. However, attention should be paid to the shift in the annual maximum of the number of days in question, depending on the location of the station. The farther north a station is located, the more often it falls in January. In southern regions, however, it falls in December (Fig. 7.17). Similarly, as in the case of days with deep lows, in individual months and seasons there were no clear trends of changes in the surveyed days.

Based on the long-term variability in the occurrence of deep lows and strong highs over Kraków, it was found that the presented regularities of their occurrence did not change significantly over time. As in the years 1966–2018, the average annual number of days with SLP ≤ 990 hPa since 1901 was 0.5 and such days occurred in only 45 of 118 years (Fig. 7.18). Most often, there was one such day in a year and three such days occurred (after 1960) only three times. Relatively often, days with SLP ≤ 990 hPa were recorded at the beginning of the studied period, and then in the second half of the twentieth century. These were the least frequent in the years 1918–1950 and 1993–2003. Despite the described changes, there was no statistically significant trend of changes in the number of these days in 118 years. As in the period 1966–2018, all the analysed days usually occurred in the cool half of the year and such a case was recorded only once in April and May.

Considering the occurrence of days with SLP ≥ 1030 hPa since 1901, it was found that the average annual number is similar to that from the period 1966–2018. This was 23.3 days (23.4 in the shorter period). The highest annual value—49 days (in 1932)—is two days higher than that occurring after 1966 (Fig. 7.18). The annual course of days with such high pressure is much more diversified than in the case of

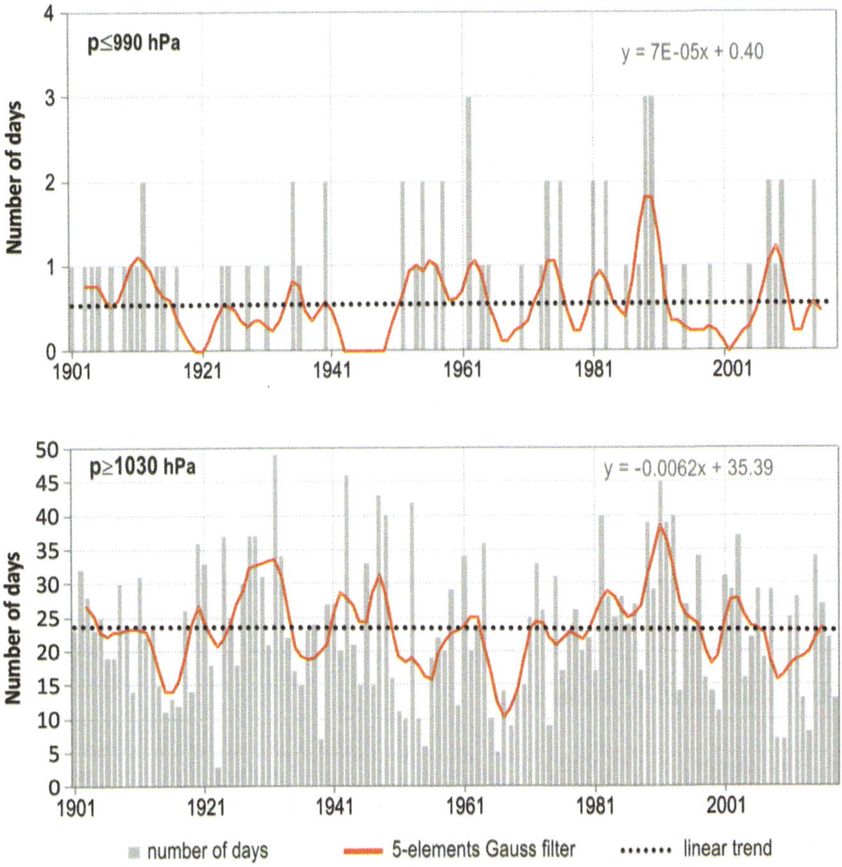

Fig. 7.18 Number of days with the atmospheric sea level pressure values ≤ 990 hPa and ≥ 1030 hPa in Kraków in the period 1901–2018

deep lows. Most of the analysed days, 94.5% on average (from 66.7% in 1989 to 100.0% in 54 years), occurred from October to March, however, they did not occur even once in July and August (Fig. 7.19). The average monthly number of such days ranged from less than 0.1 to 0.7 days in the warm half of the year and from 2.2 in October to 5.8 days in January. However, as can be seen, the annual maximum has changed over the multi-annual period. Even in the winter months, the days in question were quite often not recorded. It is also worth emphasising that in Kraków the number of days with strong highs occurring in the autumn (5.9 days on average) is more than twice as high as in the spring (2.8 days on average). In the winter, however, days with SLP ≥ 1030 hPa occur on average for 14.4, and in some years even for 20 days (1972).

The long-term variability of the number of days with SLP ≥ 1030 hPa does not indicate the occurrence of a specific trend of changes in the annual (Fig. 7.18),

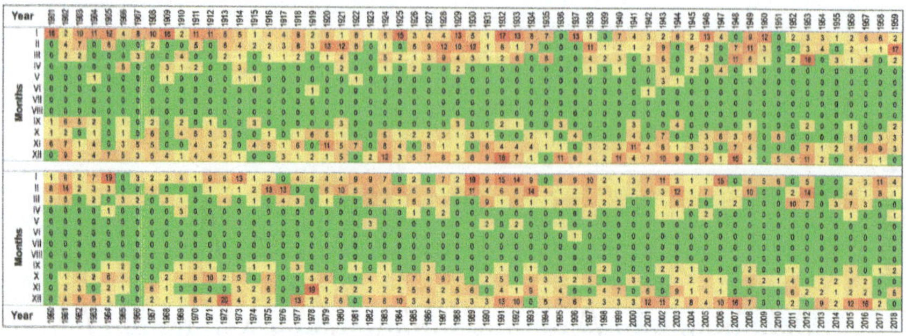

Fig. 7.19 Number of days with the atmospheric sea level pressure values ≥ 1030 hPa in Kraków in the period 1901–2018

monthly, and seasonal values. However, longer periods of an increased number of days studied are clearly visible. The most prominent were the years 1924–1933 (with an annual average of 31.9 days) and 1982–1993 (31.8 days), especially at the turn of the 1980s and 1990s. On the other hand, the least frequent number of days with SLP ≥ 1030 hPa occurred in the years 1950–1970 (on average 18.0 days a year).

Discussion and Conclusions

The conducted study confirmed the most important results of previous works on the changes in atmospheric pressure in Poland. The most characteristic feature of the spatial distribution of pressure values is its latitudinal distribution and the small range of changes in the average values. In the years 1966–2018, the highest pressure values occurred in the south and south-east of the country (up to about 1017 hPa), while the lowest along the Baltic Sea coast (below 1015 hPa). The largest range of pressure fluctuations was characteristic of the cool half of the year, especially from December to February (about 5 hPa). The smallest changes occurred in the summer and did not exceed 2 hPa. A feature of the multi-annual variability of the values of the analysed meteorological element in the period under consideration is the absence of a clear trend of changes, in the case of both annual and seasonal values. Trepińska (1988, 1997) noticed this in her research, and the aforementioned regularity concerned not only the more than 200 years of the Kraków series, but also the pressure changes at several other European stations. Burt (2009) also came to similar conclusions by examining pressure changes at three stations in Great Britain in the period 1851–2008. The absence of clear trends in pressure changes may be caused by the significant stability of the occurrence and location of the main centers controlling the circulation of the atmosphere over Poland, i.e. the Icelandic Low and the Azores High. This is confirmed by Falarz's (2009) studies, who found no trends of changes in the position of the mentioned pressure systems in the years 1901–2000. Nevertheless, in shorter periods, the occurrence of clearly increased or reduced pressure values is noticed. In the case of the long Kraków series, higher pressure values were noticeable in the years 1920–1950 and the 1990s, and were particularly visible in the cool half of the year. This may be a consequence of the pressure increase in the Azores High in January in the second half of the twentieth century, as determined by Falarz (2009). Falarz (2019) also stated that in January, in the years 1948–2018, the pressure in the centre of the Azores High showed an upward trend, statistically significant at 0.05 and 0.63 hPa/10 years, while in July the changes were not statistically significant. The author also noted the shift of the centre of this high in January from a SW direction to NE in the years 1948–1991, and its movement in a SW direction in the twenty-first century (Falarz 2019). In the studied period, the year 2010, when the lowest pressure was recorded at all stations, clearly stood out. At the same time, it was the lowest average annual pressure in the last 118 years in Kraków. Such an extremely low value of pressure (1014.0 hPa in Kraków with an average of 1016.8 hPa) was the result of a high frequency of low-pressure systems moving over Poland and the particularly long-lasting prevalence over the eastern part of the country, in August, November and December, of the low with its centre over south-eastern Europe (Bilik et al. 2014).

An important element in the development of pressure conditions over Poland is the movement of deep and very dynamic low-pressure systems (SLP $\leq$ 990 hPa) and the occurrence of strong and extensive highs (SLP $\geq$ 1030 hPa). Strong high-pressure systems occur several times more often than deep lows and have a particularly strong impact on the weather in southern Poland. On the other hand, deep lows much more often move over the northern part of the country, which is associated with the tracks of their most frequent movement over Europe. It should also be emphasised that the above-mentioned pressure systems affect meteorological conditions mainly in the cool half of the year. The conducted studies have shown that the multi-annual changes in the number of days with SLP $\leq$ 990 hPa do not show a clear trend. One can notice a slight decrease in the average number of days considered specified for the whole country, as well as at the stations located in north-western Poland, but it is not statistically significant. It is most likely related to the described and forecast changes in the incidence of low-pressure systems in the Euro-Atlantic zone. It is expected that the number and intensity of lows will be increasing mainly in the northern regions of Europe (Wernli et al. 2003) while decreasing below the latitude of 55° N (Bartholy et al. 2006; Trigo 2006). In addition, lows will more often be moving farther to the north (Knippertz et al. 2000; Leckebusch and Ulbrich 2004), and their tracks to the north-east will be shorter (Carnell and Senior 1998). Analysing the occurrence of days with SLP $\leq$ 990 hPa over a longer period of time (1901–2018), also found no clear trend in the changes. One can only indicate a few short periods of their increased incidence. Similar regularities also occur at other European stations (Sweeney 2000; Bielec-Bąkowska and Piotrowicz 2013) and in the North Atlantic region (Bärring and Fortuniak 2009). Despite the increase in the incidence of anticyclonal situations in Central Europe in the second half of the twentieth century (Kyselý and Huth 2006) and the slight increase in pressure and the number of days with strong highs, especially in the winter, in the years 1951–2015 (Bielec-Bąkowska 2014, 2016) no statistically significant changes in the number of days with SLP $\geq$ 1030 hPa were found in the periods analysed in the paper. The absence of a clear trend of changes in the number of days with high pressure in the research periods of more than 100 years, and the occurrence of fairly regular, several-year periods of an increased number, is a characteristic feature of pressure changes at many European stations (Bielec-Bąkowska 2010b). It is worth noting, however, that in the years 1966–2018 in the north and east of Poland, as well as at its south-western end, there were signals of an increasing incidence of this. Perhaps this is the result of an increase in the incidence of winter anticyclonal systems associated with the shift of the Asian High towards the north-west (Zhang et al. 2012).

References

Bärring L, Fortuniak K (2009) Multi-indices analysis of southern Scandinavian storminess 1780-2005 and links to interdecadal variations in the NW Europe-North Sea region. Int J Climatol 29(3):373–384

Bartholy J, Pongracz R, Pattantyúa-Ábrahám M (2006) European cyclone track analysis based on ECMWF ERA-40 data sets. Int J Climatol 26:1517–1527

Beniston M, Stephenson DB, Christensen OB, Ferro CAT, Frei C, Goyette S, Halsnaes K, Holt T, Jylhä K, Koffi B, Palutikof J, Schöll R, Semmler T, Woth K (2007) Future extreme events in European climate: an exploration of regional climate model projections. Clim Change 81:71–95

Bielec-Bąkowska Z (2010a) A classification of deep cyclones over Poland (1971–2000). Phys Chem Earth 35:491–497

Bielec-Bąkowska Z (2010b) Silne wyże nad Europą w XX wieku. In: Kolendowicz L (ed) Klimat Polski na tle klimatu Europy. Warunki cyrkulacyjne i radiacyjne. Bogucki Wydawnictwo Naukowe, Seria Studia i Prace z Geografii i Geologii 14:23–37

Bielec-Bąkowska Z (2014) Silne wyże nad Europą (1951–2010). Prace Naukowe Uniwersytetu Śląskiego w Katowicach, Wydawnictwo Uniwersytetu Śląskiego, Katowice, p 3135

Bielec-Bąkowska Z (2016) Long-term variability of the frequency and persistence of strong highs over Poland. Environ Socio-economic Stud 4(1):12–23

Bielec-Bąkowska Z, Piotrowicz K (2011) Weather types accompanying very high pressure in Krakow in the period 1901–2000. Int J Climatol 31:2183–2193

Bielec-Bąkowska Z, Piotrowicz K (2013) Long-term occurrence, variability and tracks of deep cyclones over Kraków (Central Europe) during the period 1900–2010. Int J Climatol 33:677–689

Blik A, Gawęda B, Gluza A, Siwek K (2014) Przebieg ciśnienia atmosferycznego w Lublinie w latach 1951-2010. Annales Universitatis Marie Curie-Skłodowska, Lublin-Polonia, Sectio B 69(2):133–142

Burt S (2007) The Highest of the Highs… Extremes of barometric pressure in the British Isles, Part 2 – the most intense anticyclones. Weather 62(2):31–41

Burt S (2009) Long-term variations in extremes of barometric pressure in the British Isles. Weather 64(7):187–189

Cahynová M, Huth R (2016) Atmospheric circulation influence on climatic trends in Europe: an analysis of circulation type classifications from the COST733 catalogue. Int J Climatol 36(7):2743–2760

Carnell RE, Senior CA (1998) Changes in mid-latitude variability due to increasing greenhouse gases and sulphate aerosols. Clim Dyn 14:369–383

Dong B, Sutton RT, Woollings T, Hodges K (2013) Variability of the North Atlantic summer storm track: mechanisms and impacts on European climate. Environ Res Lett 8:034037

Falarz M (2009) Variability of the Icelandic Low and the Azores high in January and their influence on climatic condition in Poland. Bull Geogr Phys Geogr Series 2:5–18

Falarz M (2019) Azores high and Hawaiian High: correlations, trends and shifts (1948–2018). Theoret Appl Climatol 138:417–431
Haarsma R, Hazeleger W, Severijns C, Vries H, Sterl A, Bintanja R, Oldenborgh G, Brink H (2013) More hurricanes to hit Western Europe due to global warming. Geophys Res Lett 40:1–6
IPCC (2007) In: Climate change 2007: The physical science basis. Contribution of Working Group I to the Fourth Assessment Report of the Intergovernmental Panel on Climate Change, Solomon S, Quin D, Manning M, Chen Z, Marquis M, Averyt KB, Tignor M, Miller HL (eds) Cambridge University Press: Cambridge, UK and New York, NY, USA, 996
Kłysik K (1995) Rola silnych wyżów i głębokich niżów w kształtowaniu warunków termicznych okresu zimowego w Europie Środkowej. In: Krawczyk B, Błażejczyk K (eds) Współczesne badania klimatologów polskich w kraju i za granicą. IGiPZ PAN, Warszawa, pp 19–27
Knippertz P, Ulbrich U, Speth P (2000) Changing cyclones and surface wind speeds over the North Atlantic and Europe in a transient GHG experiment. Climate Res 15:109–122
Kożuchowski K (1995) Deep cyclones, anticyclones and zonal circulation over Europe (1900-1990). Przegląd Geofizyczny 40:231–246
Kożuchowski K (2011) Klimat Polski. Nowe spojrzenie. Wydawnictwo Naukowe PWN, Warszawa
Kyselý J (2008) Influence of the persistence of circulation patterns on warm and cold temperature anomalies in Europe: Analysis over the 20th century. Global Planet Change 62:147–163
Kyselý J, Huth R (2006) Changes in atmospheric circulation over Europe detected by objective and subjective methods. Theoret Appl Climatol 85:19–36
Kyselý J, Huth R (2008) Relationships of surface air temperature anomalies over Europe to persistence of atmospheric circulation patterns conducive to heat waves. Adv Geosci 4:243–249
Labajo JL, Martín Q, Labajo AL, Piorno A, Ortega M, Morales C (2008) Recent trends in the frequencies of extreme values of daily maximum atmospheric pressure at ground level in the central zone of the Iberian Peninsula. Int J Climatol 28:1227–1238
Leckebusch GC, Ulbrich U (2004) On the relationship between cyclones and extreme windstorm events over Europe under climate change. Glob Planet Change 44:181–193
Martyn D (1987) Klimaty kuli ziemskiej. PWN, Warszawa
Murakami H, Wang Y, Yoshimura H, Mizuta R, Sugi M, Shindo E, Adachi Y, Yuki-36moto S, Hosaka M, Kusumoki S, Ose T, Kitoh A (2012) Future changes in tropical37cyclone activity projected by the new high-resolution MRI-AGCM. J Clim 25:3237–3260
Paszyński J, Niedźwiedź T (1991) Klimat. In: Starkel L (ed.) Geografia Polski. Środowisko przyrodnicze. Wydawnictwo Naukowe PWN, Warszawa, pp 296–355
Rockel B, Worth K (2007) Extremes of near-surface wind speed over Europe and their future changes as estimated from an ensemble of RCM simulations. Clim Change 81:267–280
Schinke H (1993) On the occurrence of deep cyclones over Europe and the North Atlantic in the period 1930–1991. Contrib Atmos Phys 66:223–237
Sweeney JC (2000) A three-century storm climatology for Dublin 1715–2000. Irish Geogr 33(1):1–14
Trepińska J (1988) Wieloletni przebieg ciśnienia i temperatury powietrza w Krakowie na tle ich zmienności w Europie. Rozprawy habilitacyjne UJ, 140, Kraków
Trepińska J (1997) Wieloletni przebieg średnich miesięcznych wartości ciśnienia powietrza w Krakowie, 1792-1995. In: Trepińska J (ed) Wahania klimatu w Krakowie (1792-1995). IGiGP UJ, Kraków, pp 131–161
Trigo IF (2006) Climatology and interannual variability of storm tracks in the Euro-Atlantic sector: a comparison between ERA-40 and NCEP/NCAR reanalyses. Clim Dyn 26:127–143
Ustrnul Z, Czekierda D (2009) Atlas ekstremalnych zjawisk meteorologicznych oraz sytuacji synoptycznych w Polsce. IMGW, Warszawa
Wernli H, Bader S, Hächler P (2003) Winter storms. In: Extreme events and climate change. OcCC, Bern, Switzerland, pp 81–84
Woś A (2010) Klimat Polski w drugiej połowie XX wieku. Wydawnictwo Naukowe UAM, Poznań
Zhang X, Lu Ch, Guan Z (2012) Weakened cyclones, intensified anticyclones and recent extreme cold winter weather events in Eurasia. Environ Res Lett 7:044044

Zhao M, Held I (2012) TC-Permitting GCM simulations of hurricane frequency response to sea surface temperature anomalies projected for the late-twenty-first century. J Clim 25:2995–3009

Chapter 8
Solar Radiation Change

Małgorzata Falarz, Łukasz Małarzewski, Joanna Uscka-Kowalkowska, Dorota Matuszko, and Tomasz Budzik

Abstract Global solar radiation data were analysed at 10 actinometric stations located in different regions of Poland for the period from 19 (Sosnowiec) to 125 years (Kraków; data partially reconstructed). Only series with the data gaps not exceeding 5% were considered. The most important results of the study of changes in solar radiation in Poland are as follows: (1) the average long-term totals of global solar radiation range in Poland from approximately 3750 MJ/m^2 to 4070 MJ/m^2 throughout the year; (2) the year-to-year variability of solar radiation expressed by coefficient of variability is rather small and ranges from 3.5 to 7% in Poland; the highest values of variability are observed in autumn and winter; (3) long-term trends of global solar radiation in Poland are in most cases statistically insignificant; a few significant tendencies show different trend directions; (4) relative trends of global solar radiation in the area of Poland do not exceed ±10%/10 years; (5) in Kraków, for a 125-year series of global solar radiation values, about 60 years periodicity of radiation changes can be seen, with three periods of relatively high values (1880–1900, 1940–1960, 1990–2018), separated by periods of relatively low values: 1910–1930 and 1970–1990. The global solar radiation course in Kraków largely corresponds to the periods of "global dimming" and "global brightening" described in different

M. Falarz (✉) · Ł. Małarzewski · T. Budzik
Faculty of Natural Sciences, Institute of Earth Sciences, University of Silesia in Katowice, Katowice, Poland
e-mail: malgorzata.falarz@us.edu.pl

Ł. Małarzewski
e-mail: lukasz.malarzewski@us.edu.pl

T. Budzik
e-mail: tomasz.budzik@us.edu.pl

J. Uscka-Kowalkowska
Faculty of Earth Sciences and Spatial Management, Department of Meteorology and Climatology, Nicolaus Copernicus University, Toruń, Poland
e-mail: joannauk@umk.pl

D. Matuszko
Faculty of Geography and Geology, Institute of Geography and Spatial Management, Jagiellonian University, Kraków, Poland
e-mail: d.matuszko@uj.edu.pl

M. Falarz (ed.), *Climate Change in Poland*, Springer Climate,
https://doi.org/10.1007/978-3-030-70328-8_8

parts of the world as a result of urbanisation, industrialisation and the increase in aerosols related to them. A decrease in values was observed until the end of the 1970s or 1980s, depending on the season of the year, and then there was an increase until the end of the twentieth century.

Introduction

Solar radiation is a climatic factor significantly affecting other climatic elements. Research on this subject has developed in several directions, e.g.:

- spatial diversity of solar radiation (Kuczmarska and Paszyński 1964b; Podogrocki 1978; Miara et al. 1987),
- annual variability (Kuczmarska and Paszyński 1964a) and multi-year variability of global solar radiation (Podogrocki 1977; Bryś 1994; Bogdanska and Podogrocki 2000; Uscka-Kowalkowska et al. 2007; Matuszko 2014; Kleniewska and Chojnicki 2016; Kulesza 2017, 2020),
- investigation of direct radiation (Uscka 2003; Uscka-Kowalkowska 2008, 2009, 2013, 2019),
- investigation of radiation balance (Paszyński 1966; Olecki 1986, 1989),
- variation of the radiation balance over various active surfaces (Bryś 2013),
- relationships of solar radiation with ultraviolet radiation (Słomka 1988), sunshine (Gorczyński 1934), cloudiness (Matuszko 2009) and atmospheric circulation (Kulesza 2018),
- solar radiation studies on the satellite database (Struzik et al. 2019; Kulesza 2020).

Kulesza (2018) documented the strong influence of atmospheric circulation on the radiation conditions in the area of Poland. Actinometric conditions are shaped by the North Atlantic Oscillation. In a year-course, as well as in spring, the increase in the solar radiation over Poland is mainly facilitated by a simultaneous increase in pressure in the area of the Azores High and a decrease in pressure in the area of the Icelandic Low.

Directions of changes in multi-year solar radiation in selected cities and throughout Poland in different periods of investigation are presented in Table 8.1.

Data and Methods

Global solar radiation data were analysed at 10 actinometric stations located in different regions of Poland for the period from 19 (Sosnowiec: 2000–2018) to 125 years (Kraków Observatory UJ: 1884–2018; data partly reconstructed, Matuszko 2014). The completeness of a series was the determining factor at the series selection stage: only series with the hourly data gaps not exceeding 5% were considered. Series of totals of global solar radiation in MJ/m^2 for the year and seasons were created

Table 8.1 Trends of global solar radiation in selected cities and throughout Poland

Author(s)	Location	Period analysed	Trends
Bryś (2013)	Wrocław	1881–2012	The period 1961–2012 is characterised by a positive trend of global and reflected radiation and short-wave radiation balance, after a period of strong declines in values beginning in the mid-1920s. During the period 1881–2012, there was observed a negative trend in the value of the short-term radiation balance and both its components. It is accompanied by a marked increase in the downward atmospheric radiation and long-wave radiation
Kleniewska and Chojnicki (2016)	Warszawa	1964–2013	During the period considered, an increase of 11.4 MJ/m2 per year in global solar radiation totals was observed. There were opposing trends in annual radiation totals during the period considered: a decrease until 1981 and recorded since the mid-1980s. increase in global radiation totals. Such changes were also observed during the seasons: in summer, autumn and winter, while in spring there was a steady increase in global solar radiation totals
Żmudzka and Kulesza (2019)	Zakopane	1986–2015	For the period 1986–2015, there was an increase of 0.03 MJ/m^2/day/year in the annual average daily totals of global solar radiation, despite a significant increase in the amount of cloud cover in the warm half of the year
Kulesza (2020)	Poland	1986–2015	The average annual radiation sum over Poland increased by 7.16 MJ/m^2 per year on average. A wavelet analysis showed a several-year cycle (of 12–13 years) of annual fluctuations in global solar radiation totals

on the basis of the hourly values of actinometric measurements. The data gaps were completed using the quotients method on the basis of neighbouring stations data except for the peak station Kasprowy Wierch, where a very small number of data breaks were completed by the multi-year mean values for a given season. With the exception of the university stations in Kraków and Sosnowiec, the data was taken from the IMGW-PIB database available at: https://dane.imgw.pl/data/dane_pomiarowo_obserwacyjne/dane_aktynometryczne. Standard characteristics of climate change and variability have been identified and analysed.

Results

The average of the long-term totals of global solar radiation in Poland ranges from approximately 3750 MJ/m^2 to 4070 MJ/m^2 throughout the year, about 1200 MJ/m^2 to 1350 MJ/m^2 in spring, about 1380 MJ/m^2 (with a minimum at the top of Kasprowy Wierch) to 1750 MJ/m^2 in summer, about 560 MJ/m^2 on the coast to 710 MJ/m^2 on the highest mountain peaks in autumn and about 210 MJ/m^2 on the coast to 465 MJ/m^2 in the high mountains in winter (Tables 8.2, 8.3, 8.4, 8.5, and 8.6). The spatial diversity of radiation values depends to a large extent on the latitude of the observation site. However, the cloud coverage and cloud type, changing with altitude above sea level, also play a significant role: in the high parts of the Tatra Mountains (about 2000 m above the sea level) in autumn and winter, the highest solar radiation values are recorded. In these seasons, low-level layer clouds (Stratocumulus, Stratus, Nimbostratus) dominate, and these clouds often occur at the level below the Tatra peaks, leaving them in the zone of direct solar radiation influence. In summer,

Table 8.2 Change and variability of global solar radiation in a whole year

Actinometric station	Analysed period	Mean value [MJ.m^{-2}]	Coefficient of variability [%]	Absolute trend [MJ.m^{-2}/10y]	Trend statistically significant (0.05) yes/no	Relative trend [%/10y]
Gdynia	1991–2015	3779.1	3.9	28.6	No	0.8
Kasprowy W.	1991–2017	3900.0	5.7	−77.8	No	−2.0
Kraków Obs.	1884–2018	3734.4	3.7	−7.0	Yes	−0.2
Łeba	1991–2015	3797.0	6.9	70.8	No	1.9
Legnica	1992–2016	4070.0	4.3	119.7	No	2.9
Lesko	1992–2015	3966.6	7.2	216.9	No	5.5
Piła	1991–2015	3748.4	3.5	42.1	No	1.1
Radzyń Ląd	1991–2017	3800.9	5.6	58.2	No	1.5
Sosnowiec	2000–2018	3748.0	4.9	83.0	No	2.2
Toruń	1983–2018	3749.8	5.6	−33.4	No	−0.9

Table 8.3 Change and variability of global solar radiation in spring

Actinometric station	Analysed period	Mean value [MJ.m^{-2}]	Coefficient of variability [%]	Absolute trend [MJ.m^{-2}/10y]	Trend statistically significant (0.05) yes/no	Relative trend [%/10y]
Gdynia	1991–2015	1292.4	5.9	29.7	No	2.3
Kasprowy W.	1991–2017	1346.3	7.7	−33.5	No	−2.5
Kraków Obs.	1884–2018	1196.4	6.3	−1.7	No	−0.1
Łeba	1991–2015	1315.9	7.8	39.9	No	3.0
Legnica	1992–2016	1345.0	7.1	25.8	No	1.9
Lesko	1992–2015	1276.3	8.7	59.1	No	4.6
Piła	1991–2015	1278.1	6.4	25.3	No	2.0
Radzyń Ląd	1991–2017	1271.7	7.9	27.0	No	2.1
Sosnowiec	2000–2018	1235.2	7.7	0.7	No	0.1
Toruń	1983–2018	1276.9	8.6	0.2	No	0.2

Table 8.4 Change and variability of global solar radiation in summer

Actinometric station	Analysed period	Mean value [MJ.m^{-2}]	Coefficient of variability [%]	Absolute trend [MJ.m^{-2}/10y]	Trend statistically significant (0.05) yes/no	Relative trend [%/10y]
Gdynia	1991–2015	1691.8	8.1	−31.1	No	−1.8
Kasprowy W.	1991–2017	1382.0	9.4	−36.5	No	−2.6
Kraków Obs.	1884–2018	1606.8	5.3	−0.8	No	0.0
Łeba	1991–2015	1710.8	10.3	−0.1	No	0.0
Legnica	1992–2016	1752.5	4.6	47.8	Yes	2.7
Lesko	1992–2015	1694.0	8.5	117.8	Yes	7.0
Piła	1991–2015	1659.1	5.7	−12.7	No	−0.8
Radzyń Ląd	1991–2017	1659.5	6.3	13.8	No	0.8
Sosnowiec	2000–2018	1602.7	5.8	41.8	No	2.6
Toruń	1983–2018	1640.8	6.2	−18.0	No	−1.1

however, when convective vertical clouds dominate, the high mountain areas are very often cloudier than the areas below. In autumn and winter, coastal areas are in the zone of low-floor clouds forming in very humid coastal air. These areas therefore have the lowest radiation values in Poland in autumn and winter.

The variability of year-to-year solar radiation expressed by coefficient of variability in Poland ranges from 3.5 to 7% for the whole year, 6 to 9% in spring, 5 to 10% in summer (with a maximum for coastal and mountainous areas), 7 to 14% in

Table 8.5 Change and variability of global solar radiation in autumn

Actinometric station	Analysed period	Mean value [MJ.m^{-2}]	Coefficient of variability [%]	Absolute trend [MJ.m^{-2}/10y]	Trend statistically significant (0.05) yes/no	Relative trend [%/10y]
Gdynia	1991–2015	576.8	9.5	32.3	Yes	5.6
Kasprowy W.	1991–2017	711.1	10.2	−3.8	No	−0.5
Kraków Obs.	1884–2018	625.4	6.9	−2.9	Yes	−0.5
Łeba	1991–2015	564.0	10.3	26.6	No	4.7
Legnica	1992–2016	661.8	9.9	38.2	Yes	5.8
Lesko	1992–2015	676.0	12.4	59.6	Yes	8.8
Piła	1991–2015	579.8	9.1	30.9	Yes	5.3
Radzyń Ląd	1991–2017	608.1	11.4	15.5	No	2.5
Sosnowiec	2000–2018	626.8	14.0	45.9	No	7.3
Toruń	1983–2018	585.6	8.8	−4.3	No	−0.7

Table 8.6 Change and variability of global solar radiation in winter

Actinometric station	Analysed period	Mean value [MJ.m^{-2}]	Coefficient of variability [%]	Absolute trend [MJ.m^{-2}/10y]	Trend statistically significant (0.05) yes/no	Relative trend [%/10y]
Gdynia	1991–2015	218.1	8.8	−2.3	No	−1.1
Kasprowy W.	1991–2017	465.0	9.1	−6.3	No	−1.4
Kraków Obs.	1884–2018	305.8	6.9	−1.7	Yes	−0.6
Łeba	1991–2015	206.1	11.0	4.1	No	2.0
Legnica	1992–2016	310.6	8.1	7.8	Yes	2.5
Lesko	1992–2015	345.0	9.2	11.9	Yes	3.4
Piła	1991–2015	230.9	10.5	−2.1	No	−0.9
Radzyń Ląd	1991–2017	261.7	8.6	2.0	No	0.8
Sosnowiec	2000–2018	285.9	8.4	−5.4	No	−1.9
Toruń	1983–2018	246.6	13.1	−13.6	Yes	−5.5

autumn and 7 to 11% in winter (Tables 8.2–8.6). Radiation variability values should be therefore considered small at all seasons of the year.

Long-term trends of global solar radiation are, in most cases, statistically insignificant (Tables 8.2–8.6, Figs. 8.1 and 8.2). The only significant are the trends: for the whole year in Kraków (negative; −7 MJ/m^2 per 10 years), in the summer in Legnica and Lesko (positive; relatively: 48 and 118 MJ/m^2 per 10 years), in autumn for 5 out of 10 actinometric stations (positive trend in: Gdynia: 32, Legnica: 38, Lesko: 60,

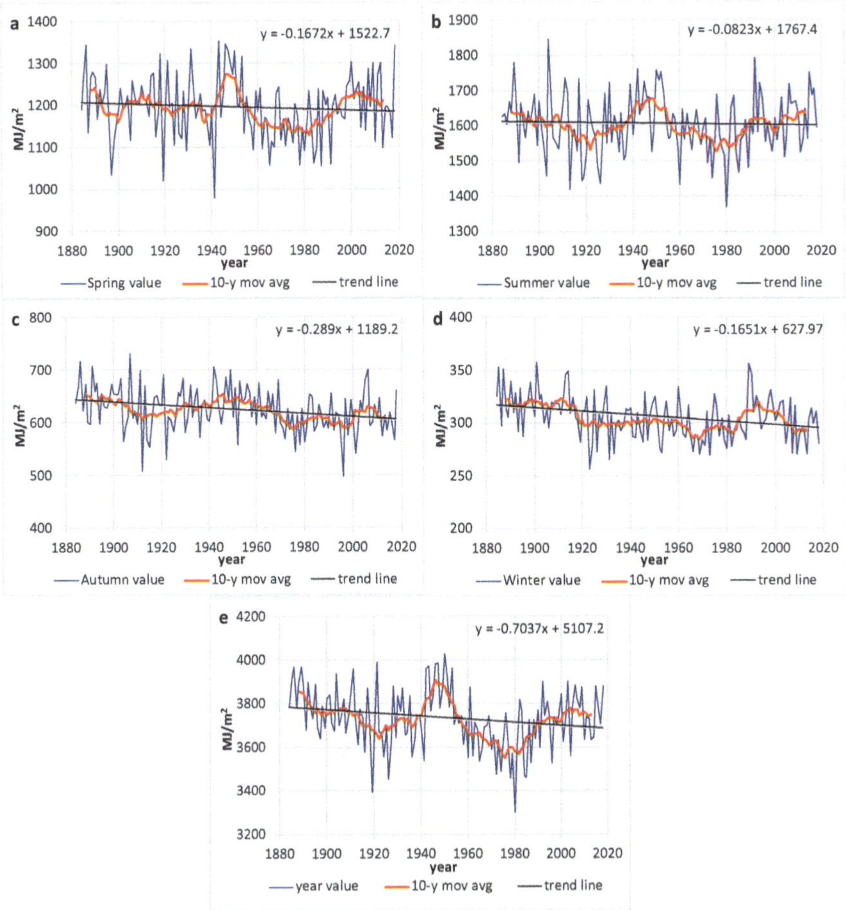

Fig. 8.1 Totals of global solar radiation [MJ/m^2] in the Kraków-Observatory in 1884–2018 (data partly reconstructed by Matuszko in 2014) in **a** spring, **b** summer, **c** autumn, **d** winter, **e** whole year. Trend line and 10-y moving average line were added

Piła: 31 MJ/m^2 per 10 years; negative trend in Kraków: –3 MJ/m^2 per 10 years), in winter at 4 out of 10 stations (negative trend in Kraków: –2 MJ/m^2 per 10 years and Toruń: –14 MJ/m^2 per 10 years, positive trend in Legnica: 8 MJ/m^2 per 10 years and Lesko: 12 MJ/m^2 per 10 years). In the spring, none of the investigated solar radiation series showed statistically significant changes at the 0.05 level. What is noteworthy is the statistically significant positive trend of solar radiation in all seasons except spring in southeastern Poland (Lesko) and in Silesian Lowland (Legnica). In the southern part of Poland (Kraków) global radiation is characterised by a negative statistically significant trend in autumn, winter and in calculations for the whole year. Essentially, no significant changes in the radiation value in the high mountains (Kasprowy Wierch station) have been shown for any season. The significant diversity

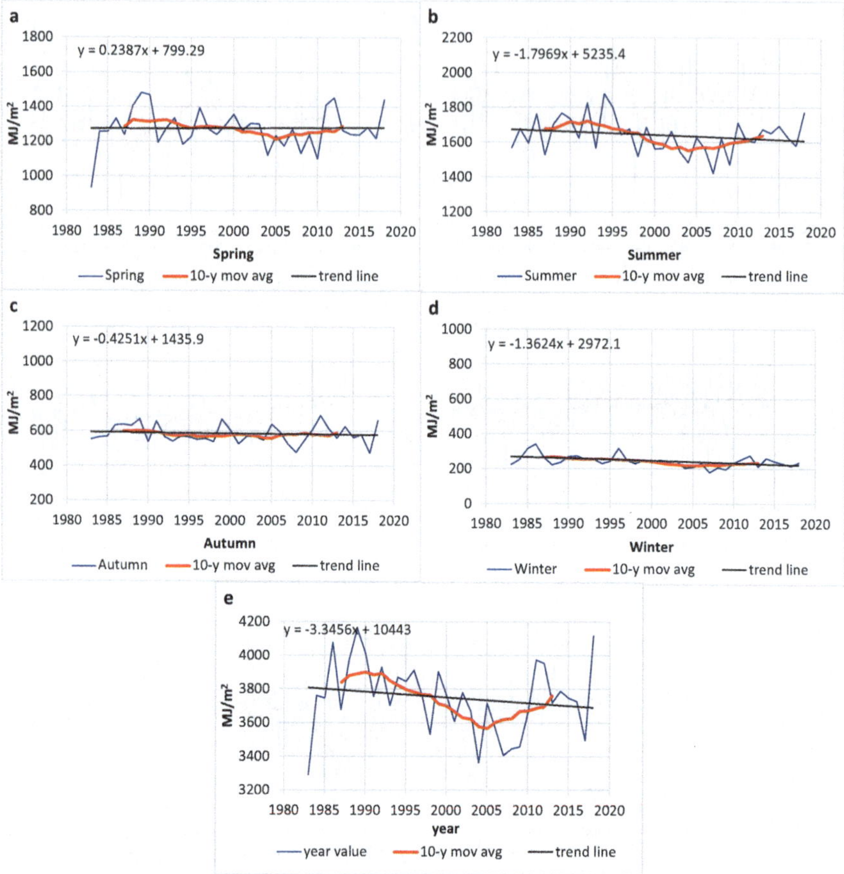

Fig. 8.2 Totals of global solar radiation [MJ/m^2] in Toruń in 1983–2018 in **a** spring, **b** summer, **c** autumn, **d** winter, **e** whole year. Trend line and 10-y moving average line were added

of directions and values of solar radiation trends is due to the climatic distinctness of the various regions of Poland, but it is also the result of analysing study periods of quite different lengths. This is especially true for the station in Kraków, whose series of radiation values is at least 100 years longer than the other series and shows only negative or almost zero changes over the 125-year period under consideration.

Relative trends, comparable for locations with different absolute values of solar radiation, do not exceed ±10%/10 years and in Poland are in the range of −2% to about 6%/10 years in calculations for the whole year, from −3% to 5%/10 years in spring, −3% to 7%/10 years in summer, −1% to 9%/10 years in autumn and −6% to about 3%/10 years in winter (Tables 8.2–8.6).

On the example of stations in Kraków and Toruń, one can trace the exact long-term course of changes in the radiation value (Figs. 8.1 and 8.2). In Kraków, about

60 years of periodicity of changes in global radiation can be noticed: in the multi-annual course of all seasons and in annual values, three periods of relatively high values are noted: in the last two decades of the nineteenth century, in the 1940s and 1950s (less noticeable in winter), followed by the early 1990s until the end of the analysis period, with a maximum in the first decade of the twenty-first century (Fig. 8.1). In all seasons except winter, the highest values occurred in the 1940s and 1950s. In winter, the values in the first and third periods are comparable. These periods of high values are separated by periods of relatively low values: in the 1910s and 1920s (less visible in spring and winter) and in the 1970s and 1980s.

In Toruń in the period 1983–2018, marked in all seasons of the year and throughout the year, there was a decrease in the value of global solar radiation up the first years of the twenty-first century, followed by a fairly rapid increase by the end of the study period (Fig. 8.2). The absence of significant differences in the course of solar radiation at different seasons of the year was noted: changes are very similar, and in the whole period, with the exception of spring, a negative trend is marked, statistically significant in winter.

Conclusion and Discussion

The most important results of the study of changes in solar radiation in Poland are as follows:

- the average long-term totals of global solar radiation range in Poland from approximately 3750 MJ/m^2 to 4070 MJ/m^2 throughout the year; the spatial differentiation of radiation values depends to a large extent on the latitude of the observation site, but the cloud coverage and cloud type, as well as the altitude above sea level also play a significant role;
- the year-to-year variability of solar radiation expressed by the coefficient of variability is rather small and ranges from 3.5 to 7% in Poland; the highest values of variability are observed in autumn and winter;
- long-term trends of global solar radiation in Poland are, in most cases, statistically insignificant; a few significant tendencies show different trend directions; a statistically significant positive trend of solar radiation has been shown in all seasons except for spring in southeastern Poland and in Silesian Lowland, while a negative one in autumn, winter and year-round in the southern part of Poland; no significant changes in the solar radiation value in the high mountains have been demonstrated at any time of the year; also during the spring period, none of the radiation series analysed showed statistically significant changes;
- relative trends of global solar radiation in the area of Poland do not exceed ±10%/10 years;
- in Kraków, for a 125-year series of global solar radiation values (partly reconstructed) one can see about 60 years of periodicity of radiation changes with three periods of relatively high values (1880–1900, 1940–1960, 1990–2018);

these periods of high values are separated by periods of relatively low values: 1910–1930 and 1970–1990.

The time trends of global solar radiation examined in this chapter correspond in part to the results of the cloud studies presented in the section *Change of cloudiness* of this book by Filipiak 2021. He divided the period of cloud studies in Poland into two sub-periods: 1951–1984 and 1985–2018. In the first sub-period Filipiak found a positive trend in cloudiness in spring and autumn, and in the second—a decrease in cloudiness in summer and autumn. These trends are opposite to the trends detected in the radiation series in Kraków and confirm the significant effect of increase/decrease in cloud cover on the decrease/increase in the value of global solar radiation already described by Matuszko (2009). In addition, the limit years of these two sub-periods (1984–1985) correspond in some cases (especially in spring, summer and throughout the year) with periods of low global radiation values in Kraków. The negative trend of annual solar radiation totals in Kraków over the entire 125-year period studied is accompanied by a positive trend in cloudiness in the city over the period 1826–2005, as proven by Lewik and co-authors (2010). An additional explanation of the radiation-cloud cover relationship is the conclusion about the positive trend of the most observed Stratocumulus and Cumulus clouds since the 1950s (Matuszko and Węglarczyk 2018).

The global solar radiation course in Kraków largely corresponds to the periods of "global dimming" and "global brightening", described by researchers in different parts of the world as a result of urbanisation, industrialisation and the increase in aerosols related to them (Liepert 2002; Alpert et al. 2005; Wild et al. 2005, 2007; Norris and Wild 2007; Ruckstuhl and Norris 2009; Wild 2009). On the whole, there is an observed decrease in values until the end of the 1970s or 1980s depending on the season of a year, and then an increase until the end of the twentieth century.

Overall, the positive trend in solar radiation since the 1980s can be a significant factor affecting the positive trend of thermal conditions (Ustrnul et al. 2021) and the negative trend for snow cover (Falarz and Bednorz 2021).

References

Alpert P, Kishcha P, Kaufman YJ, Schwarzbard R (2005) Global dimming or local dimming? Effect of urbanization on sunlight availability. Geophys Res Lett 32:L17802. https://doi.org/10.1029/2005GL023320

Bogdańska B, Podogrocki J (2000) Zmienność całkowitego promieniowania słonecznego na obszarze Polski w okresie 1961-1995. Mat. Badawcze, Meteorologia - 30, IMGW, Warszawa, 43 pp

Bryś K (1994) Natężenie promieniowania słonecznego całkowitego we Wrocławiu w okresie 1961–1990. Zeszyty Naukowe AR 243:37–49

Bryś K (2013) Dynamika bilansu radiacyjnego murawy oraz powierzchni nieporośniętej. Wrocław, Wydawnictwo Uniwersytetu Przyrodniczego we Wrocławiu, p 288

Falarz M, Bednorz E (2021) Snow Cover Change. In: Falarz M. [ed.] Climate Change in Poland - Past, Present, Future, Springer Climate. https://doi.org/10.1007/978-3-030-70328-8_14

Filipiak J (2021) Change of Cloudiness. In: Falarz M. [ed.] Climate Change in Poland - Past, Present, Future, Springer Climate. https://doi.org/10.1007/978-3-030-70328-8_10
Gorczyński W (1934) O związku między promieniowaniem rozproszonem nieba i stopniem usłonecznienia. Sprawozdania Towarzystwa Naukowego Warszawskiego, Wydział Nauk Mat. i Przyr., No. 27, Warsaw
Kleniewska M, Chojnicki BH (2016) Zmienność całkowitego promieniowania słonecznego w Warszawie w latach 1964-2013. Acta Geogr Lodziensia 104:67–74
Kuczmarska L, Paszyński J (1964a) Przebieg roczny całkowitego promieniowania słonecznego w Polsce. Przegl Geofiz 3, 4:275–279
Kuczmarska L, Paszyński J (1964b) Rozkład promieniowania całkowitego na obszarze Polski. Przegl Geogr 4:691–702
Kulesza K (2017) Comparison of typical meteorological year and multi-year time series of solar conditions for Belsk, central Poland. Renew Energy 113:1135–1140
Kulesza K (2018) Wpływ cyrkulacji atmosferycznej na ilość promieniowania słonecznego docierającego do powierzchni ziemi w Polsce. Phd dissertation, University of Warsaw, Warszawa, p. 149
Kulesza K (2020) Spatiotemporal variability and trends in global solar radiation over Poland based on satellite-derived data (1986-2015). Int J Climatol. https://doi.org/10.1002/joc.6596
Lewik P, Matuszko D, Morawska-Horawska M (2010) Multi-annual variability of cloudiness and sunshine duration in Krakow between 1826 and 2005. In: The Polish climate in the European Context: An historical overview. In: Przybylak R, Majorowicz J, Brazdil R, Kejna M (eds) Advances in global change research, vol XVIII. Springer, Dordrecht, The Netherlands, Heidelberg, Germany, London and New York, NY, pp 341–353
Liepert B (2002) Observed reductions of surface solar radiation at sites in the United States and worldwide from 1961 to 1990. Geophys Res Lett 29(10):1421. https://doi.org/10.1029/2002GL014910
Matuszko D (2009) Wpływ zachmurzenia na usłonecznienie i całkowite promieniowanie słoneczne (na przykładzie krakowskiej serii pomiarów). Wydawnictwo Uniwersytetu Jagiellońskiego, Kraków
Matuszko D, Węglarczyk S (2018) Long-term variability of the cloud amount and cloud genera and their relationship with circulation (Kraków, Poland). Int J Climatol 38:1205–1220. https://doi.org/10.1002/joc.5445
Matuszko D (2014) Long-term variability in solar radiation in Krakow based on measurements of sunshine duration. Int J Climatol 34:228–234. https://doi.org/10.1002/joc.3681
Miara K, Paszyński J, Grzybowski J (1987) Zróżnicowanie przestrzenne bilansu promieniowania na obszarze Polski. Przegl Geogr 4:487–509
Norris JR, Wild M (2007) Trends in aerosolradiative effects over Europe inferred from observed cloud cover, solar 'dimming' and solar 'brightening'. J Geophys Res 112: D08214, 1–13. https://doi.org/10.1029/2006JD007794
Olecki Z (1986) On the components of the radiation balance in Cracow, Zeszyty Naukowe UJ. Prace Geograficzne 69:27–38
Olecki Z (1989) Bilans promieniowania słonecznego w dorzeczu Górnej Wisły, Rozprawy Habilitacyjne UJ, 157, p 126, Krakow
Paszyński J (1966) Atlas bilansu promieniowania w Polsce. Materiały do bilansu cieplnego Polski, IG PAN Dok. Geogr. 4
Podogrocki J (1977) Zmienność czasowa promieniowania całkowitego w Polsce, Zesz. Nauk. ART. Olsztyn, Rolnictwo 21:113–125
Podogrocki J (1978) Spatial distribution of global solar radiation in Poland. Publ Inst Geophys Pol Acad Sci D-5(120):17–29
Ruckstuhl C, Norris JR (2009) How do aerosol histories affect solar "dimming" and "brightening" over Europe? IPCC-AR4 models versus observations. J Geophys Res 114: D00D04, 1–12. https://doi.org/1029/2008JD011066

Słomka J (1988) Ultraviolet and global solar radiation at Belsk 1980-1986. Publs Inst Geoph Pol Acad Sci D 30(220):163–167

Struzik P, Kępińska-Kasprzak M, Limanówka D, Pajek M (2019) Promieniowanie słoneczne na obszarze Polski w świetle danych satelitarnych 2004-2018. In: Chojnacka-Ożga L, Lorenc H (eds) Współczesne problemy klimatu Polski. Institute of Meteorology and Water Management - National Research Institute, Warsaw, Poland, pp 67–82

Uscka J (2003) Direct solar radiation and its attenuation by the atmosphere with different air masses in the suburban area of Toruń. Acta Universitatis Wratislaviensis, Studia Geograficzne 75:268–281

Uscka-Kowalkowska J (2008) Bezpośrednie promieniowanie słoneczne i jego ekstynkcja w atmosferze w Kołobrzegu w latach 1960-2000. Acta Agrophysica 12:221–233

Uscka-Kowalkowska J (2009) Porównanie bezpośredniego promieniowania słonecznego i jego ekstynkcji w atmosferze w Warszawie i w Mikołajkach. Acta Agrophysica 14, 2(171):501–514

Uscka-Kowalkowska J (2013) An analysis of the extinction of direct solar radiation on Mt. Kasprowy Wierch, Poland. Atmos Res 134:175–185

Uscka-Kowalkowska J (2019) Bezpośrednie promieniowanie słoneczne w Polsce w latach 1966-2010. Toruń, Wydawnictwo Naukowe Uniwersytetu Mikołaja Kopernika, p 257

Uscka-Kowalkowska J, Przybylak R, Vízi Z, Araźny A, Kejna M, Maszewski R (2007) Variability to global solar radiation in Central Europe during the period 1951-2005 (on the basis of data from NCEP/NCAR reanalysis project). Geographia Polonica 80(2):59–68

Ustrnul Z, Wypych A, Czekierda D (2021) Air Temperature Change. In: Falarz M. [ed.] Climate Change in Poland - Past, Present, Future, Springer Climate. https://doi.org/10.1007/978-3-030-70328-8_11

Wild M (2009) Global dimming and brightening: A review. J Geophys Res 114(D10): D00D16. https://doi.org/10.1029/2008JD011470

Wild M, Gilgen H, Roesch A, Ohmura A, Long CN, Dutton E, Forgan B, Kallis A, Russak V, Tsvetkov A (2005) From dimming to brightening: Decadal changes in solar radiation at earth's surface. Science 308(5723):847–850

Wild M, Ohmura A, Makowski K (2007) Impact of global dimming and brightening on global warming. Geophys Res Lett 34:L04702. https://doi.org/10.1029/2006GL028031

Żmudzka E, Kulesza K (2019) Total solar radiation in Zakopane and at mount Kasprowy Wierch (Western Carpathians, Poland) in 1986-2015. Geographia Polonica 92(2):211–231

Chapter 9
Change of Sunshine

Dorota Matuszko, Stanisław Węglarczyk, Krzysztof Bartoszek, Jakub Soroka, and Barbara Bogdańska

Abstract Sunshine duration is an important characteristic of the inflow of solar radiation, one of a multitude of factors shaping the climate conditions of a given area. Based on the data from 31 stations from the years 1971–2018, the spatial diversity of the totals of annual and seasonal sunshine duration is presented. In order to analyse the long-term variability of this element, an additional analysis of long heliographic series from five stations in Poland is used. Furthermore, actual and relative sunshine duration and the number of days without sunshine are characterised. The mean annual sunshine duration from all stations was 1,647 h, while the relative one amounted to 37%. The maximum occurred in 2018 and was, respectively, 2,069 h and 46%; the minimum was found in 1980 (1,275 h and 28%). As the latitude increases, the annual totals of sunshine duration, as well as the monthly totals in the summer and the warm half of the year increase. In the winter and the cold half of the year, sunshine duration decreases from the south to the north. The course of the annual totals of sunshine

D. Matuszko (✉)
Faculty of Geography and Geology, Institute of Geography and Spatial Management, Jagiellonian University, Kraków, Poland
e-mail: d.matuszko@uj.edu.pl

S. Węglarczyk
Faculty of Environmental and Power Engineering, Department of Geoengineering and Water Management, Cracow University of Technology, Kraków, Poland
e-mail: sweglar@pk.edu.pl

K. Bartoszek
Faculty of Earth Sciences and Spatial Management, Department of Hydrology and Climatology, Institute of Earth and Environmental Sciences, Maria Curie-Skłodowska University, Lublin, Poland
e-mail: k.bartoszek@umcs.pl

J. Soroka
Institute of Meteorology and Water Management—National Research Institute, Hydrological and Meteorological Station, Gorzów Wielkopolski, Poland
e-mail: jakub.soroka@imgw.pl

B. Bogdańska
Institute of Meteorology and Water Management—National Research Institute, Warszawa, Poland
e-mail: Barbara.Bogdanska@imgw.pl

M. Falarz (ed.), *Climate Change in Poland*, Springer Climate,
https://doi.org/10.1007/978-3-030-70328-8_9

duration in Poland shows changes similar to those occurring in other parts of Europe, with pronounced periods of "global dimming" and "global brightening". In the years 1971–2018, at all stations there is a statistically significant growing trend in actual and relative sunshine duration, more strongly expressed in the western than the eastern part of the country, and a decrease in the number of days without sunshine. An analysis of long heliographic series confirmed the high multi-annual variability of sunshine duration in Poland and showed the periods of decreases and increases in sunshine duration also at the turn of the nineteenth and twentieth centuries, and in the first half of the twentieth century, although not as pronounced as the periods of "global dimming" and "global brightening".

Introduction

Sunshine duration is the longest measured solar radiation characteristic in the world (since 1880) (Pallee and Butler 2002) until recently determined using the commonly used, simple instrument recommended by WMO, i.e. Campbell–Stokes heliograph (WMO 2008). The results of studies carried out in many places on the globe (e.g. Brázdil 1991; Manara et al. 2015; Sanchez-Lorenzo et al. 2015; Stanhill and Cohen 2005) show similar trends over many years of sunshine duration records. They relate not only to the well-researched periods of "global dimming" and "global brightening" (e.g. Norris and Wild 2007; Sanchez-Lorenzo et al. 2009), but also the decreases and increases in sunshine duration occurring at the turn of the nineteenth and twentieth centuries, and in the first half of the twentieth century (i.a., Matuszko 2016).

Terminology

In this chapter, the following characteristics were analysed: actual sunshine duration, relative sunshine duration and the number of days without sunshine.

Actual sunshine duration is the basic indicator of the heliographic conditions of an area. It is defined as the time expressed in hours or minutes in which the inflow of solar radiation reaching the Earth's surface is registered (using a heliograph or another recorder).

Relative sunshine duration is a complement of the characteristics of heliographic conditions, which reflects the solar conditions of a given area well. This is especially true in the comparison of different seasons by eliminating the impact of day length changing during the year. The relative sunshine duration is defined as the ratio of actual sunshine duration to astronomically possible sunshine duration, expressed as a fraction or in percentage, while possible sunshine duration is understood as the time from sunrise to sunset (length of the day).

The number of days without sunshine is a characteristic of heliographic conditions, which does not depend on the value of actual sunshine duration (such as relative sunshine duration) and is just an additional source of information confirming the trend in sunshine duration. A day without sunshine is understood as a day on which the daily total of sunshine duration is zero.

Results

Actual Sunshine Duration

The mean annual sunshine duration in Poland for the years 1971–2018 at all stations in the study amounts to 1,647 h, with the smallest one observed on Śnieżka (1,430 h), and the largest in Gdynia (1,782 h). Annual totals range from around 1,460 h in the south and south-west of the country (mountainous areas) to more than 1,740 h in the north, i.e. in the region of the Gdańsk Coastland (Fig. 9.1a). The area of long sunshine duration (over 1,700 h) stretches along a wedge to the south of the central part of the Koszalin Coastland to the Southern Wielkopolska Lowland. The second area with the highest sunshine duration (over 1,740) is located in the central-eastern part of Poland, including a fragment of the Northern Podlasie Plain and the Southern Podlasie Lowland, as well as the Western Polesie. In the prevailing area of the country, in the central part of Poland, sunshine duration ranges from 1,660 to 1,700 h per year, and gradually decreases from the centre towards the south-west, south, and north-east. From the mountainous areas to the central part of the country, there is a wedge of reduced sunshine duration called by Gorczyński (1913) and Merecki (1914) "the depression of sunshine duration on the central Vistula".

In the studied period, the greatest variability of annual sunshine duration (14%) occurred in the south-west (Zielona Góra, Jelenia Góra, Opole, Katowice) and decreased in the east and north-east to the value of 8%, which also occurred in mountainous areas in the south of Poland (Fig. 9.1b). At all the stations surveyed in the years 1971–2018, there was a statistically significant linear growing trend in the annual sunshine duration, with the highest intensity in the west and south-west of the country, expressed in both absolute and relative trend slope values (Fig. 9.1c, d).

In the recent period (1985–2018), the increase in sunshine duration at all the stations was particularly significant, up to maximum values above 100 h/10y in Western and Central Poland (Fig. 9.1e, f).

The magnitude of sunshine duration in individual seasons depends on the length of the day, cloud cover and the transparency of the atmosphere, so the sunshine duration is the longest in the summer (June–August), and the shortest in the winter (December–February). The contrasts in the sunshine duration in these seasons are enhanced not only by the varying degree of cloud cover, but also by cloud cover type, convective in the summer and layer in the winter, as well as the high content of aerosols of anthropogenic origin during the heating period.

In spring, sunshine duration increases from the south and south-west of Poland to the north (Fig. 9.2a). The lowest values (410 h) occur in mountainous areas, the

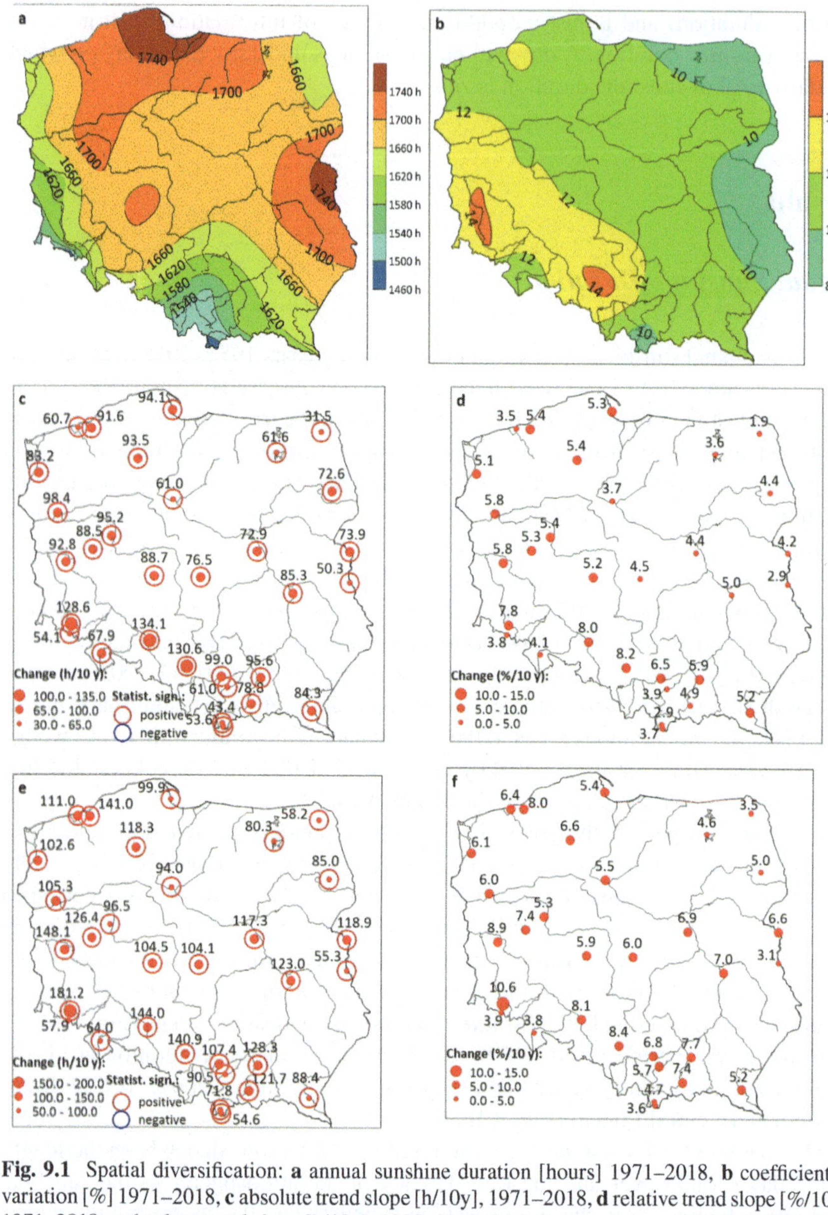

Fig. 9.1 Spatial diversification: **a** annual sunshine duration [hours] 1971–2018, **b** coefficient of variation [%] 1971–2018, **c** absolute trend slope [h/10y], 1971–2018, **d** relative trend slope [%/10y], 1971–2018, **e** absolute trend slope [h/10y], 1985–2018, **f** relative trend slope [%/10y], 1985–2018

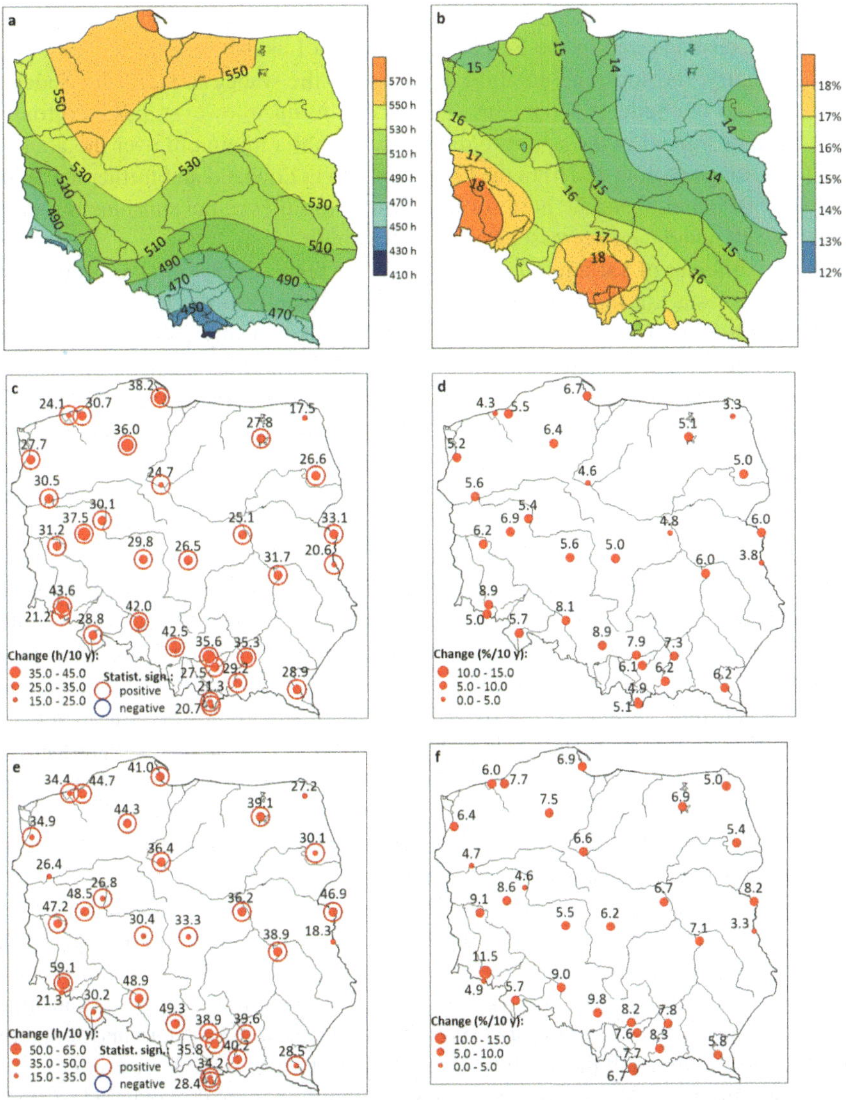

Fig. 9.2 Spatial diversification: **a** seasonal actual sunshine duration [hours]—spring, 1971–2018, **b** coefficient of variation [%] 1971–2018, **c** absolute trend slope [h/10y], 1971–2018, **d** relative trend slope [%/10y], 1971–2018, **e** absolute trend slope [h/10y], 1985–2018, **f** relative trend slope [%/10y], 1985–2018

highest (over 570 h) on the Gdańsk Coastland, and they are slightly lower on the Koszalin Coastland and in the Southern Pomerania Lakeland. In the western part of the country, the sunshine duration decreases in the wedge towards the Sudetes Foreland and the Sudetes Mountains. The second wedge, however, passes through the centre of Poland in the opposite direction, from the minimum over the area of the Tatra Range (below 420 h) to the Central Masovia Lowland and further north. In the east of Poland, from the north of the country to Polesie, the sunshine duration exceeds 530 h and then decreases further to the south.

The coefficient of variation decreases from the highest (18%) in the south-west (from the Silesia-Lusatia Lowland to the Silesia Upland) to the lowest (12–14%) in the north-eastern part of Poland (Fig. 9.2b). A statistically significant positive trend is noticeable at all stations (except for Suwałki), with the highest slopes (above 40 h/10y) in the south-west of Poland (Fig. 9.2c). The highest relative trend slopes for the years 1971–2018 (8.9%/10y) occurred at the Jelenia Góra and Katowice stations, even bigger in recent years (1985–2018), and also at the Zielona Góra and Opole stations (Fig. 9.2d, f). The smallest increase in sunshine duration occurred in Eastern Poland, especially at the stations in Włodawa and Suwałki (Fig. 9.2d, f).

Due to the length of the day in the summer, the privileged (in terms of sunshine duration) position of the northern half of Poland and the decrease in sunshine duration from the north to the south, and from the east to the west, which is associated with the predominance of continental climate features in Eastern Poland, are noticeable (Fig. 9.3a). In the Southern Podlasie Lowland and in Western Polesie, maximum values of sunshine duration occur (over 740 h), with slightly lower values over the Koszalin Coastland, and in the north-east (except for the Lithuanian Lakeland) and in the central and eastern part of Poland. The shortest sunshine duration (below 500 h) is characteristic of mountainous areas, the Tatra Range and the Sudetes Mountains, owing to the shorter length of the day than in the north of Poland and high convective cloud cover even in high-pressure situations.

The highest coefficient of variation (18%) was found for the south-western edge of Poland; it decreases from the west towards the east and north-east (Fig. 9.3b). The absolute and relative trend slopes are positive and statistically significant at all stations, except for Suwałki, Toruń and Kołobrzeg. The highest increases in sunshine duration, especially in recent years (1985–2018), occur in the western and southern parts of the country, and the smallest ones in the north-east (Fig. 9.3c, d, e, f).

In autumn, the spatial variability of sunshine duration is small compared to other seasons, with the values of less than half of the others. The total of sunshine hours varies from a minimum of 270 in the Lithuanian Lakeland to 340 in the south-eastern part of the country, and, sporadically, in the Sudetes Foreland and in the Gdańsk Coastland (Fig. 9.4a). The southern half of Poland (except for the belt in the central part) is sunnier than its north-eastern and central part.

The greatest variability of sunshine duration occurs at the station in Białystok and in the belt from the east to the west through the centre of the country, stretching as far as the Southern Wielkopolska and Silesia-Lusatia Lowlands (Fig. 9.4b). The absolute and relative trend slopes at all stations are statistically significant, except for Suwałki, Kłodzko, Zakopane and Kasprowy Wierch in the years 1971–2018. The

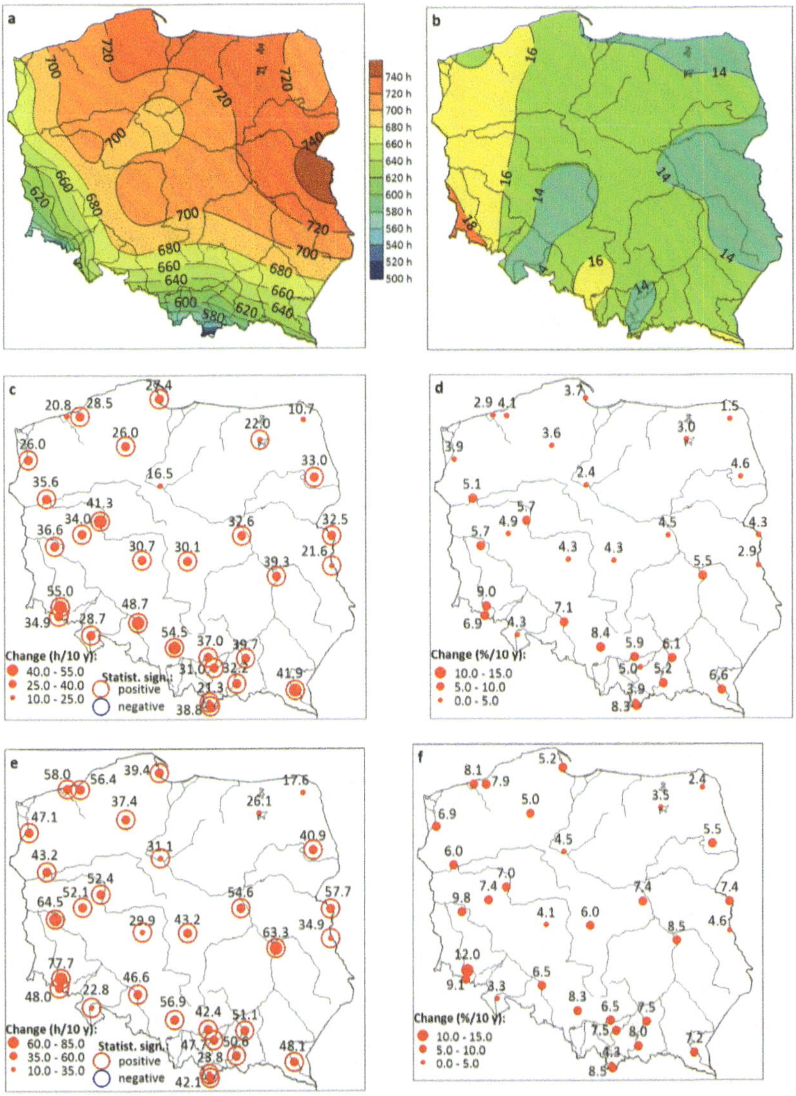

Fig. 9.3 Spatial diversification: **a** seasonal actual sunshine duration [hours]—summer, 1971–2018, **b** coefficient of variation [%] 1971–2018, **c** absolute trend slope [h/10y], 1971–2018, **d** relative trend slope [%/10y], 1971–2018, **e** absolute trend slope [h/10y], 1985–2018, **f** relative trend slope [%/10y], 1985–2018

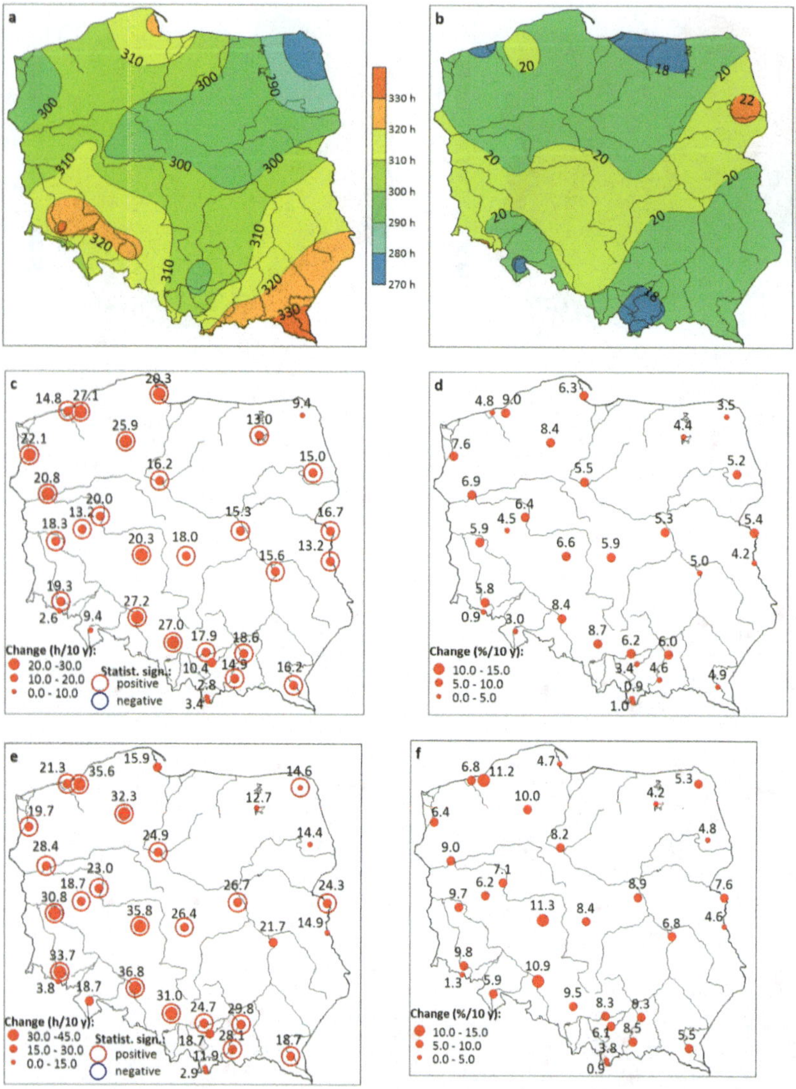

Fig. 9.4 Spatial diversification: **a** seasonal totals of actual sunshine [hours]—autumn, 1971–2018, **b** coefficient of variation [%], 1971–2018, **c** absolute trend slope [h/10y], 1971–2018, **d** relative trend slope [%/10y], 1971–2018, **e** absolute trend slope [h/10y], 1985–2018, **f** relative trend slope [%/10y], 1985–2018

highest increase in sunshine duration occurs in the north-west of Poland and the belt from Chojnice to the south, as far as Opole and Katowice, while the smallest increase is in the north-eastern part of the country (Fig. 9.4c, d, e, f).

In winter, the sunshine duration over a large area of the country does not exceed 150 h and increases towards the south and south-west (Fig. 9.5a). The shortest

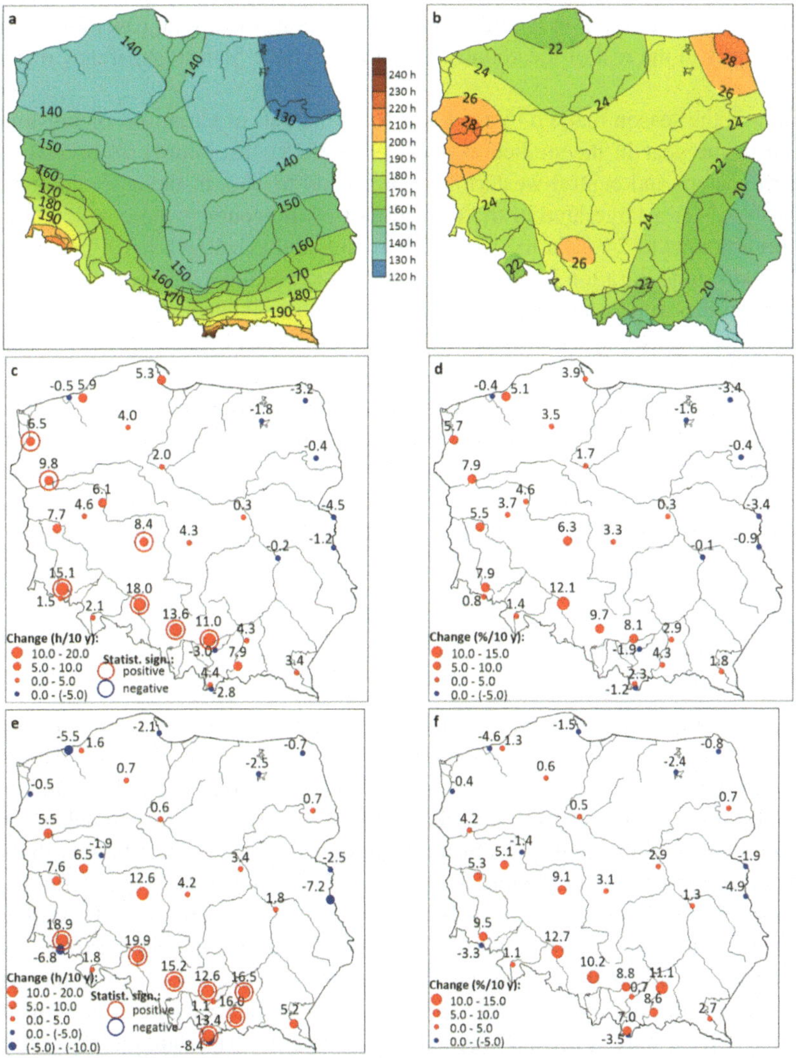

Fig. 9.5 Spatial diversification: **a** seasonal actual sunshine [hours]—winter, 1971–2018, **b** coefficient of variation [%] 1971–2018, **c** absolute trend slope [h/10y] 1971–2018, **d** relative trend slope [%/10y], 1971–2018, **e** absolute trend slope [h/10y], 1985–2018, **f** relative trend slope [%/10y], 1985–2018

sunshine duration (below 120 h) occurs in the north-eastern part of Poland due to the shorter length of the day than in the south and greater, stratiform cloud cover. Values recorded in the mountainous and foothill areas in southern Poland are more than twice as high as those in the northern part of the country. In the winter, especially during high-pressure situations, cold air stagnations occur in the valleys and the depressions of the area. Temperature inversion favours the fog and low layer clouds formation. On the elevations above the inversion layer, there is cloudless weather, which causes that the sunshine duration on mountain peaks reaches its maximum (over 240 h) in comparison with the rest of the country, also owing to the high transparency of the air.

Winter is the season marked by the highest coefficients of variations of sunshine duration in the year in the studied period (1971–2018). They range from 28% in the north-eastern and central-western part of Poland to 16% in the east and south (Fig. 9.5b). Over a large area in the central part of the country, the coefficient of variation amounts to 24%. In the winter, there are slight (below 5 h/10y) negative slopes of sunshine duration trends, both relative and absolute. They are marked at most stations in north-eastern and sporadically in Southern Poland, with simultaneous large increases (above 10 h/10y) in the west and south-west of the country (Fig. 9.5c, d).

In the recent period (1985–2018), negative trends occurred in the north-eastern half of Poland, while positive ones in the south-west, except for Śnieżka. The contrasts in the trends of sunshine duration in various parts of the country may be associated with a change in the size and structure of cloud cover and with a decrease in pollution in heavily industrialised areas in south-western Poland.

Multi-annual Course

At all stations, the course of annual sunshine duration in the years 1971–2018 showed a growing trend, statistically significant at the level of 5% (Fig. 9.6). The largest increase in sunshine hours (on average 130 h/10 years) occurred in Opole, Katowice and Jelenia Góra, and the smallest (less than 50 h/10 years) in Suwałki and Zakopane. At all stations, the shortest annual sunshine duration occurred in the 1980s, with the minimum (at 17 stations) occurring most often in 1980, and at another 5 stations in 1977, 4 stations in 1981, 2 stations in 1984 and 1987 each, and in 1985 at 1 station. In the years following the 1980s, until 2018, sunshine duration increased irregularly, particularly evidently in the first decade of the twenty-first century. At most stations, the annual sunshine duration values exceeded 1,600 h at that time and, in subsequent years, 1,700 h or more. The maximum sunshine duration was found at the end of the period under review, with record levels in 2018. That year, the mean annual sunshine duration of all the stations under consideration amounted to 2,069 h, which was by 421 h higher than in the multi-annual period on average. The highest annual actual sunshine duration values occurred in the twenty-first century, i.e. in 2018 (19 stations), 2015 (5 stations), 2003 (4 stations), 2011 (2 stations) and 2006 (1 station).

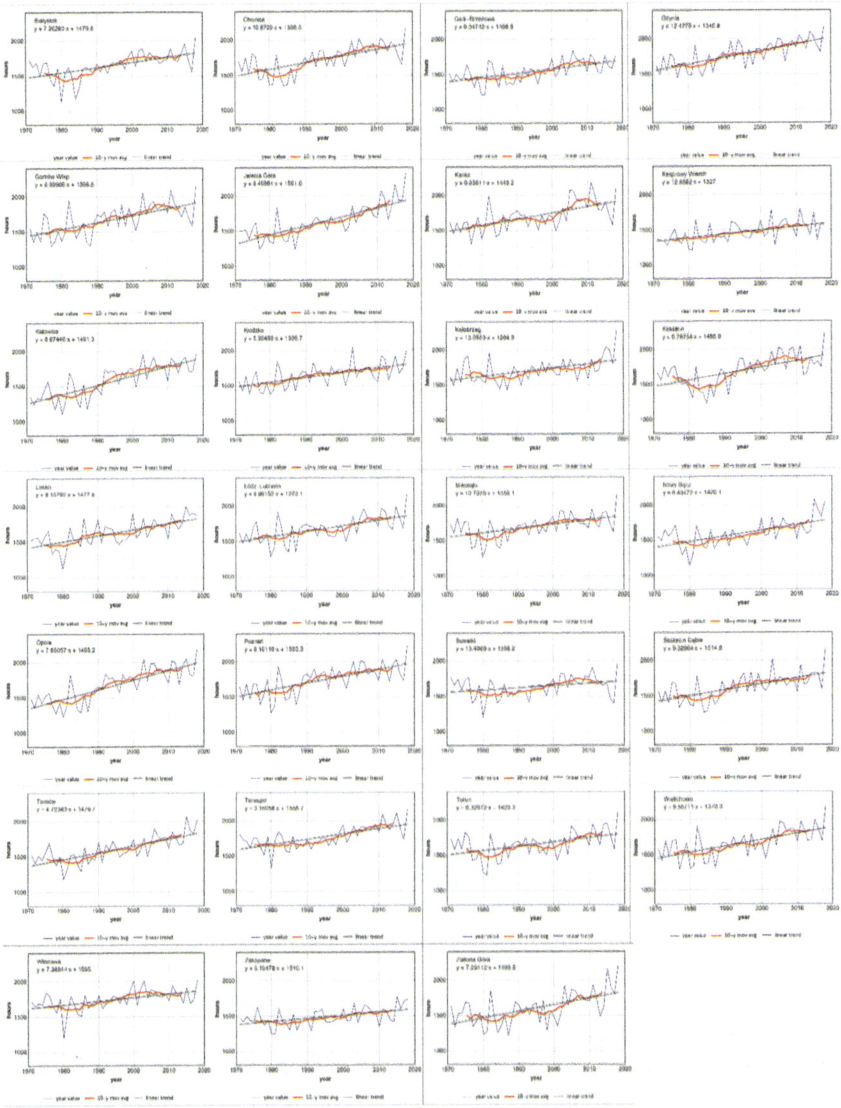

Fig. 9.6 The multi-annual course of annual sunshine duration [hours] at the selected stations included in the study (1971–2018)

Long Series Analysis

The multi-annual period of 1971–2018 is only a fragment of long heliographic series in selected Polish cities, i.e. in Kraków, Warszawa, Puławy and Wrocław, as well as on Śnieżka. It is for this reason that only an analysis of the course of annual totals from these stations gives a proper view on the long-term variability of sunshine duration. From among the aforementioned heliographic series, only the Kraków series is fully homogeneous due to the unity of the place, continuity and credibility, since its metadate is known (Matuszko 2014). At the other stations (Warszawa, Puławy, Śnieżka), data were supplemented due to deficiencies related to the interruptions in the operation of meteorological stations during World War II and changes in their location (Dubicka 1998; Podogrocki 1998; Górski and Górska 2000) or reconstructed (Wrocław) based on various stations located in Wrocław (Bryś 2013).

In Kraków (Fig. 9.7), in the beginning of the measurement period (1884–1900), high sunshine totals (over 1,600 h) were recorded followed by a downward trend from 1,860 h in 1886 to 1,222 h in 1919. In subsequent years, sunshine duration remained at the level of 1,500 h per year, except for 1921, when the annual total was 1,892 h. In the 1940s, sunshine duration increased again, exceeding 1,800 h in 1942, 1943, 1946, 1947 and 1950. Then, the annual totals decreased from 1951 to a minimum of 1,067 h in 1980. In subsequent years, until the end of the studied period, sunshine duration increased irregularly, particularly clearly in the years 1981–1995 and in the first decade of the twenty-first century. The increase in annual sunshine duration was mainly the result of increased sunshine duration in the summer and—to a lesser extent—in the winter half of the year. The years following 2006, until the end of the studied multi-annual period, were marked by annual values exceeding 1,600 h (well above 1,700 h in the years 2006, 2009, 2011 and 2012). The year 2010 was an exception: the sunshine duration was only 1,500 h. Very high sunshine durations occurred in 2015 (1,904 h) and 2018 (1,874 h), but they did not exceed the record value (1,920 h) of the year 1943. In Fig. 9.7, the periods of "global dimming" (1951–1980) and "global brightening" (from 1981), observed in many places around the world (among others Norris and Wild 2007; Sanchez-Lorenzo et al. 2009), can be seen clearly. Not so pronounced, but similar increasing and decreasing trends in sunshine duration also occurred at the turn of the nineteenth and twentieth centuries, and in the first half of the twentieth century (Fig. 9.7).

The long-term heliographic series in Warszawa, Puławy, Wrocław, and on Śnieżka also show the periods of "global dimming" (the years ca. 1951–1980) and "global brightening" (from around 1981), as well as irregular periods of increasing or decreasing sunshine duration over several years (Fig. 9.7).

In Warszawa at the beginning of the twentieth century, the annual sunshine duration ranged on average from 1,600 h to almost 1,900 h (in 1921, 1,898 h); only the 1912-value was smaller and amounted to 1,354 h. The subsequent years of the third decade of the twentieth century were characterised by annual values below 1,600 h, reaching 1,800 h again in the 1940s. The years with the shortest annual sunshine duration were 1952 (1,245 h) and 1980 (1,289 h). The maximum (2,262 h) occurred in 2018, and it was close to 2,000 h in 2006 (1,981 h).

The regular Puławy series began in 1923, although the first measurements of sunshine duration were made there at the end of the nineteenth century (Górski and Górska 2000). At the beginning of the heliographic series, the values of sunshine duration were below 1,600 h, increasing then to 1,899 h (1932), and in subsequent years, in line with the "global dimming" trend, they fell, particularly evident in the

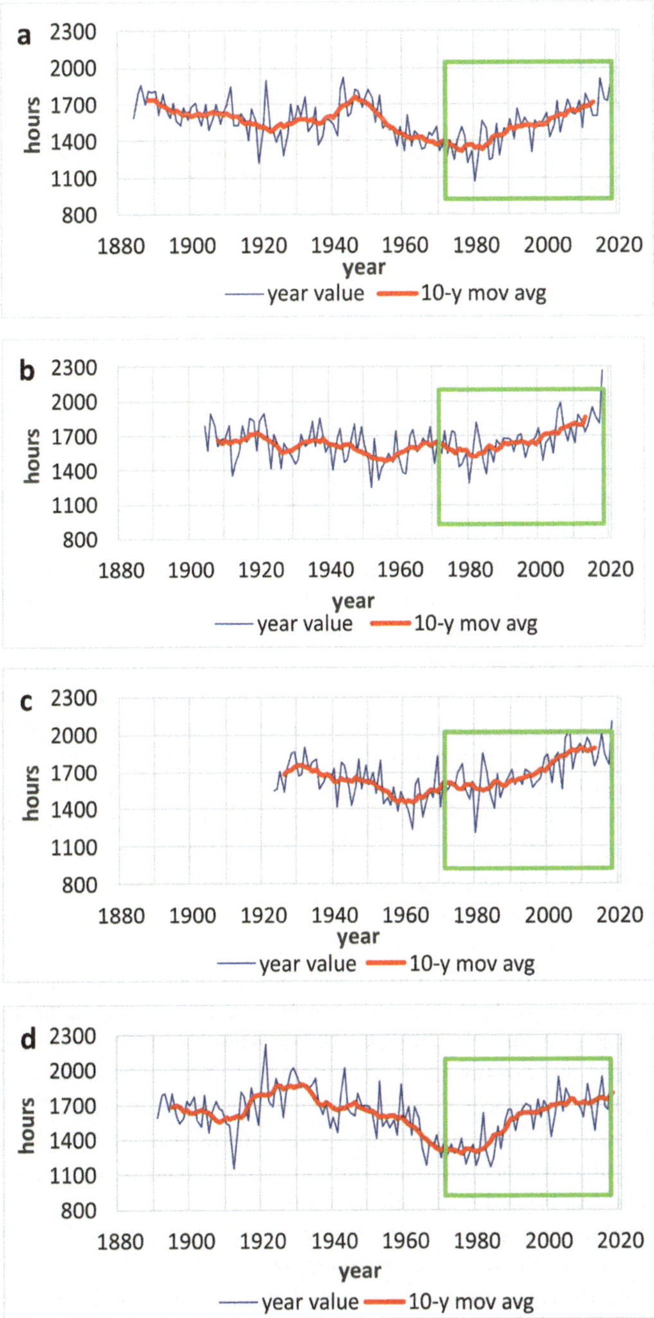

Fig. 9.7 Multi-annual course of annual sunshine duration [hours] in Kraków (1884–2018), Warszawa (1904–2018), Puławy (1923–2018), Wrocław (1891–2018), and Śnieżka (1901–2018)

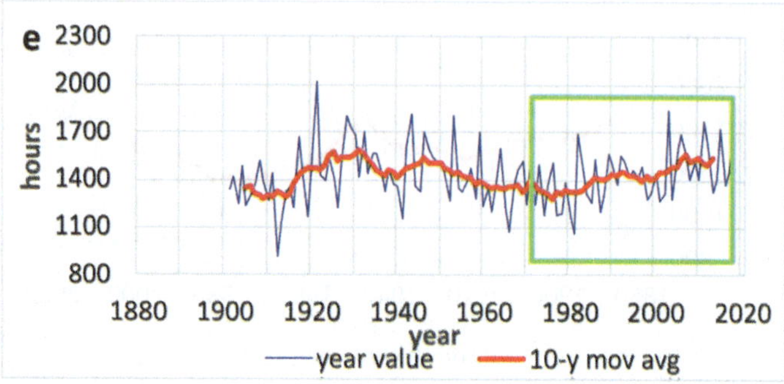

Fig. 9.7 (continued)

1950s (Fig. 9.7). Since 1962 (1,240 h), annual sunshine duration has been increasing irregularly, except for 1980, when, just like at the other stations, a minimum value was observed (1,209 h). In Puławy, like in Warszawa, the highest values of sunshine duration, exceeding 2,000 h, occurred in 2018 (2,110 h) and 2006 (2,038 h).

In Wrocław, the minimum annual sunshine duration occurred in 1912 (1,138 h), while the maximum (2,119 h) was reached in 1921. The low values of sunshine duration in 1912 were also recorded in Warszawa and on Śnieżka and can be explained by the eruption of the Katmai/Novarupta volcano in Alaska (June 6, 1912). High totals of sunshine duration in 1921 also occurred at other stations in Poland (e.g. in Kraków, Warszawa, Bydgoszcz, Kołobrzeg, and on Śnieżka) and are associated with circulation factors, i.e. the minimum cyclonicity index and minimal cloud cover in the multi-annual period (Lewik et al. 2010). Bryś (2013) divides the heliographic series in Wrocław (1875–2010) into 4 basic periods of alternating trends: decreasing (1875–1912), increasing (1912–1921), decreasing (1921–1980), and increasing (1980–2010). After the first decade of the twenty-first century, further irregular increase in annual sunshine duration was observed, up to the value of 2,035 h in 2018 (Fig. 9.7).

The 1901–2018 mean annual sunshine duration on Śnieżka is 1,429 h. This value indicates that the climatic subregion of the summit zone of the Karkonosze Mts. belongs to areas with the lowest sunshine duration in Poland, due to the heavy cloud cover in the summer months (Dubicka 1998). Also, the 1971–2018 data confirm that, compared to other stations in Poland, Śnieżka is distinguished by a short annual sunshine duration and a large number of days without sunshine (Fig. 9.1; Fig. 9.10). The multi-annual course of annual totals shows great variability from year to year (Fig. 9.7). Similar to other stations, there are several-year trends of increases and decreases, both associated with “global dimming” and “global brightening”, and also noticed in the earlier period. Record values include the minimum of 921 h and the maximum of 2,015 h, occurring at the beginning of the series, respectively, in the years 1912 and 1921, i.e., in the same years that stand out in multi-annual course at other stations, e.g. in Wrocław.

Relative Sunshine Duration

Relative sunshine duration, the ratio of actual to possible sunshine duration, eliminates the variability of the day length in the study area. It is a characteristic that indicates the impact of meteorological conditions (cloud cover, transparency of the atmosphere) on sunshine duration. The course of mean annual relative sunshine duration in Poland (Fig. 9.8a) refers to the spatial variability of actual sunshine duration (Fig. 9.1a). The highest values occur in the Gdańsk Coastland and Western Polesie (>39%). The lowest relative sunshine duration (<35%) is found in mountain and foothill regions.

In the studied period, at each meteorological station, there was a statistically significant increasing trend in relative sunshine duration, more strongly pronounced in the western than the eastern part of the country (Fig. 9.8b–c).

The highest relative sunshine duration values occurred in 2018 (46%), the lowest in 1980 (28%). At all stations, the maximum values fell in the twenty-first century,

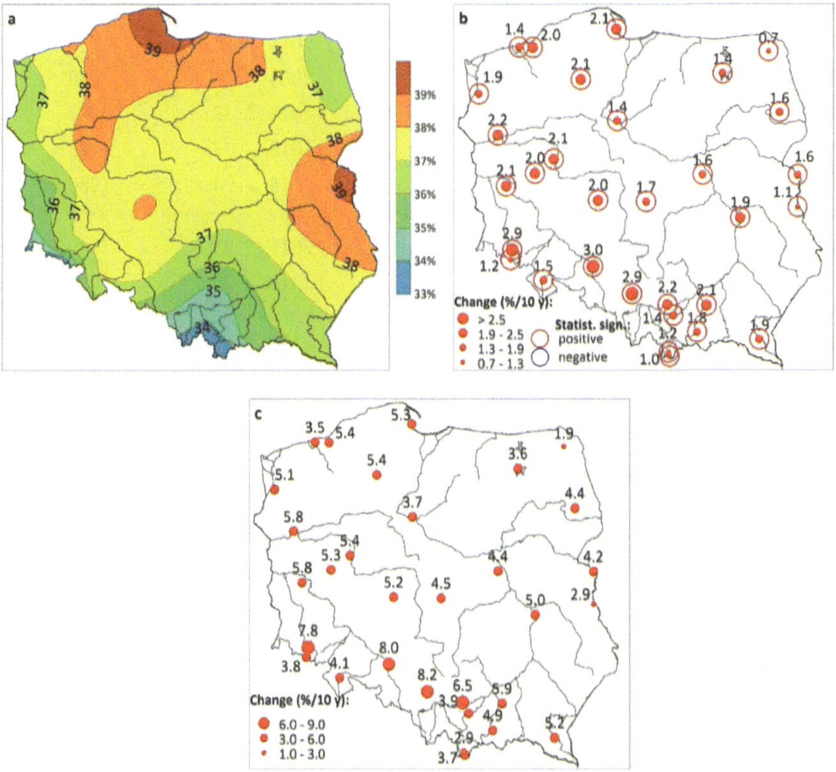

Fig. 9.8 Spatial diversification: **a** annual relative sunshine duration [%] 1971–2018, **b** absolute trend slope [%/10y], 1971–2018, **c** relative trend slope [%/10y], 1971–2018

with the most (20 stations) in 2018 and 5 stations in the south of Poland in 2015. The minimum values occurred before 2000, most of which were (20 stations) in 1980 (Table 9.1). The stations with the highest (≥50%) annual relative sunshine duration are Jelenia Góra, Kołobrzeg, Koszalin, Poznań and Warszawa (Table 9.1). The highest

Table 9.1 Mean, maximum, and minimum annual relative sunshine duration [%] at the stations included in the study (1971–2018)

Station	Mean	Maximum	Year	Minimum	Year
Białystok	37	46	2018	25	1980
Chojnice	38	48	2018	30	1980
Gaik-Brzezowa	35	41	2006	27	1980
Gdynia	40	49	2018	32	1977
Gorzów Wlkp.	38	48	2018	29	1977, 1988
Jelenia Góra	37	52	2018	28	1987
Kalisz	38	49	2012	29	1977
Kasprowy Wierch	33	40	2011	27	1980, 1981
Katowice	35	44	2018	25	1980
Kłodzko	37	46	2003	31	1980
Kołobrzeg	38	51	2018	32	1987
Koszalin	38	50	2018	27	1985
Kraków Jagiellonian University Observatory	34	43	2015	24	1980
Lesko	36	45	2015	25	1980
Łódź-Lublinek	38	48	2018	29	1980
Mikołajki	38	48	2018	28	1980
Nowy Sącz	36	46	2015	26	1980
Opole	38	49	2018	27	1980
Poznań	39	50	2018	28	1980
Puławy	38	47	2018	27	1980
Suwałki	36	44	2018	27	1980
Szczecin Dąbie	36	49	2018	28	1984
Śnieżka	32	41	2003	24	1981
Tarnów	36	46	2015	26	1980
Terespol	40	48	2018	30	1980
Toruń	37	47	2018	27	1981
Warszawa Bielany	37	50	2018	29	1980
Wielichowo	37	49	2018	29	1977
Włodawa	39	45	2003	27	1980
Zakopane	33	40	2015	28	1980
Zielona Góra	36	49	2018	27	1987

monthly values amounted to 82% and occurred in April 2009 in Chojnice and in July 1994 in Gdynia. In the warm half of the year, the maximum monthly values were recorded on the coast (Gdynia, Kołobrzeg), while in the cold half of the year the maxima occurred in the mountains (on the Kasprowy Wierch and Śnieżka Mts.). The lowest relative annual sunshine duration (24%) occurred in Kraków in 1980 and on Śnieżka in 1981 (Table 9.1). Extremely low monthly relative sunshine duration (c.a. 1%) was recorded in the cold part of the year at the stations in north-eastern Poland (Suwałki in January 1994 and November 1997, and Białystok in December 1984).

The years with the largest positive anomalies were 2003, 2006, 2011, 2015 and 2018, while the largest negative anomalies occurred in the years 1980/1981 and 1985/1989 (Fig. 9.9). Since the beginning of the twenty-first century, individual months have had a large number of positive anomalies in relative sunshine duration, compared to the 1970s and 1980s when relative sunshine duration was clearly lower (Fig. 9.9). It is important to note that high values of relative sunshine duration can occur every month, both in the warm and cold halves of the year. Most often, the largest positive anomalies occurred in March (7 years), and the smallest ones frequently in June and July (2 years for each month). The most maximum monthly anomalies occurred in 2018 (April, May, September, October, November). The largest cluster of extremely low negative anomalies occurred in July and August in the years 1977–1981 (Fig. 9.9), which indicates high cloudiness and low air transparency during that period.

Number of Days Without Sunshine

The number of days without sunshine is an important characteristic of solar radiation, independent of the value of actual sunshine duration (not like with e.g. relative sunshine duration) and less sensitive to changes of measuring instruments than actual sunshine duration. It is an important, independent source of information confirming the trend of sunshine duration. The number of days without sunshine is a good indicator in assessing the bioclimatic conditions of a given area. Longer periods with no direct inflow of solar radiation affect people's health and well-being. Sunshine deficiency is one of the most important factors that burden the human body, especially in the cold season.

The biggest number of days without sunshine occurs in the north-eastern part of Poland (>102 days) and decreases in the south-west (Fig. 9.10a). The lowest values are recorded in the Silesia Lowland, Sudetes Foreland (<77 days), and in the Outer Western Carpathians, as well as in the Gdańsk Coastland (about 80 days). In the analysed multi-annual period, there was a statistically significant decrease in the number of days without sunshine at almost all meteorological stations (Fig. 9.10b–c). The biggest negative trends, both relative and absolute, are observed at stations in the north and west of Poland.

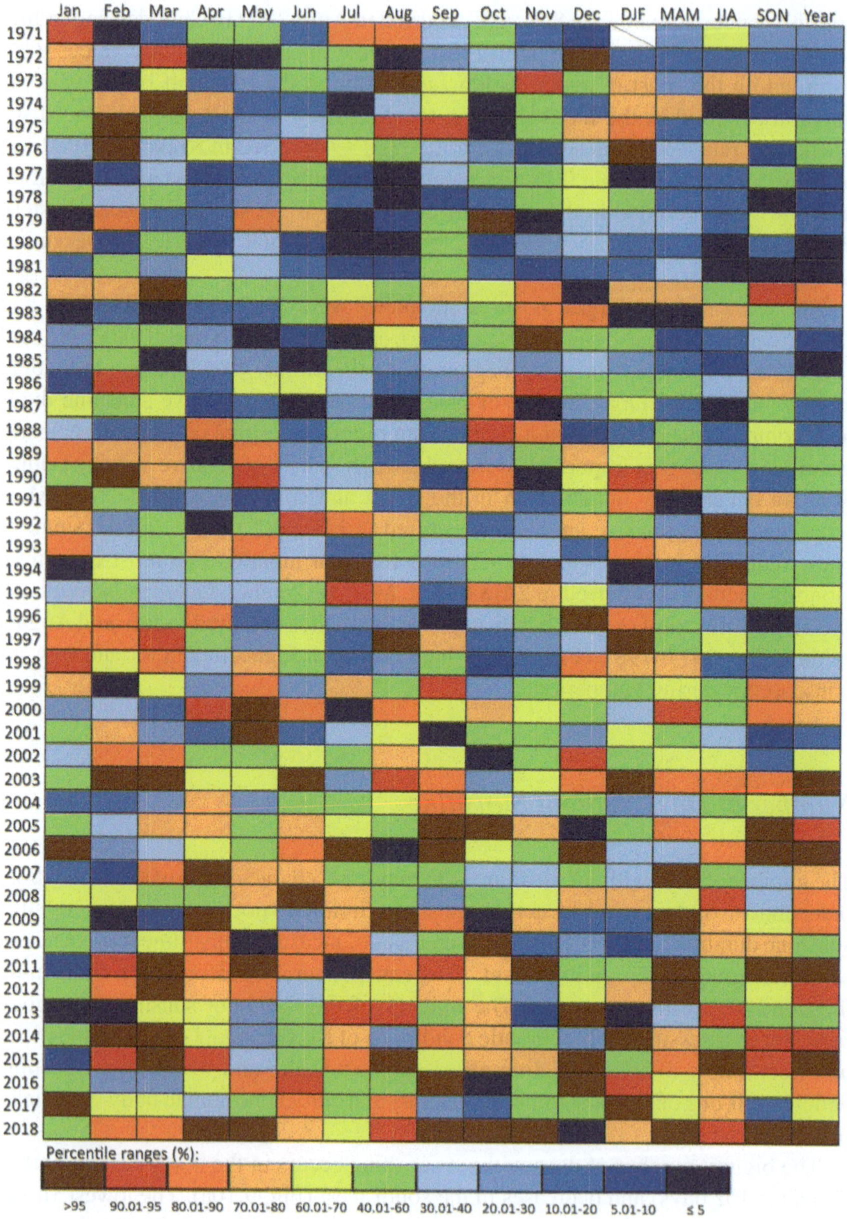

Fig. 9.9 Quantile classification of relative sunshine duration in Poland for individual months in the years 1971–2018

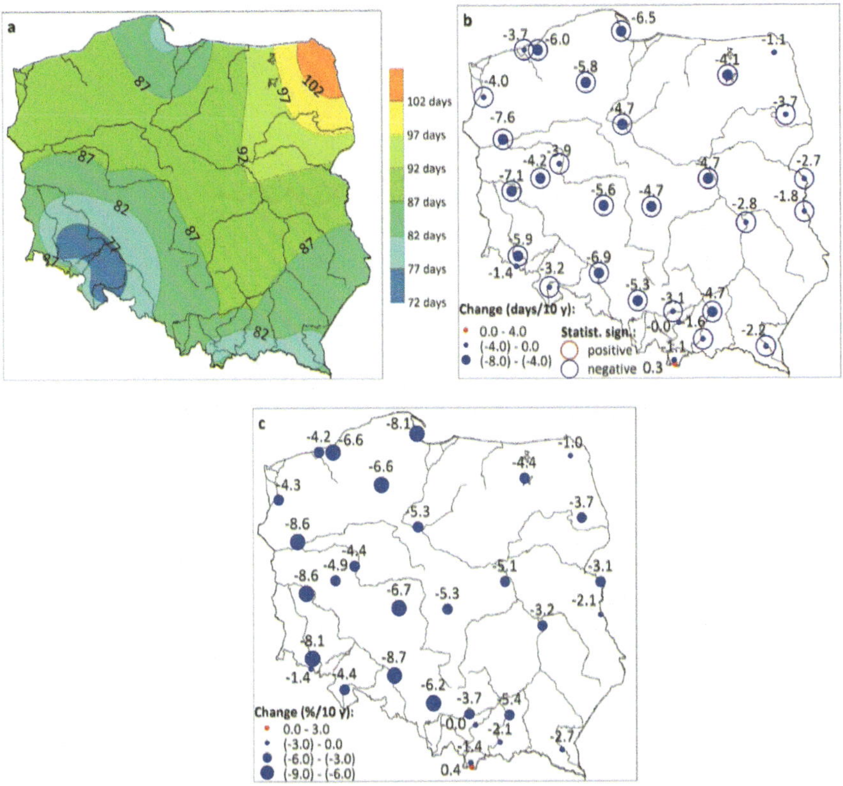

Fig. 9.10 Spatial diversification: **a** number of days without sunshine in Poland 1971–2018, **b** absolute trend slope [days/10y], **c** relative trend slope [%/10y]

The multi-annual course of the annual totals of the number of days without sunshine shows a downward trend and high year-to-year variability. Until the mid-1970s, there were on average less than 100 such days a year, and more in the 1980s, except for 1982 (78 days). In the following years, the number of days without sunshine varied from 80 to less than 100 a year (Fig. 9.11). On average, there were 87 such days in a year, with the most (105) in 1980, and the least (66) in 2015 (Table 9.2).

An analysis of the distribution of days without sunshine in individual seasons shows that half of these days occur in winter, while summer covers only 10% of their annual total, and spring and autumn about 20% each. The multi-annual course in each season (Fig. 9.11) is decreasing, more pronounced in the warm than in the cold half of the year. In spring, in the last decade of the studied multi-annual period, there were both the most (22 days in 2013) and the least (9 days in 2007, 2011 and 2015) days without sunshine, which indicates the high dynamics of weather conditions in that period. In the summer, the maximum number of days without sunshine occurred in 1980 (on average >12 days), while the minimum in 2015 (on average 1.5 days). In autumn, the number of days without sunshine ranged from 34 (1976) to 15 (2006).

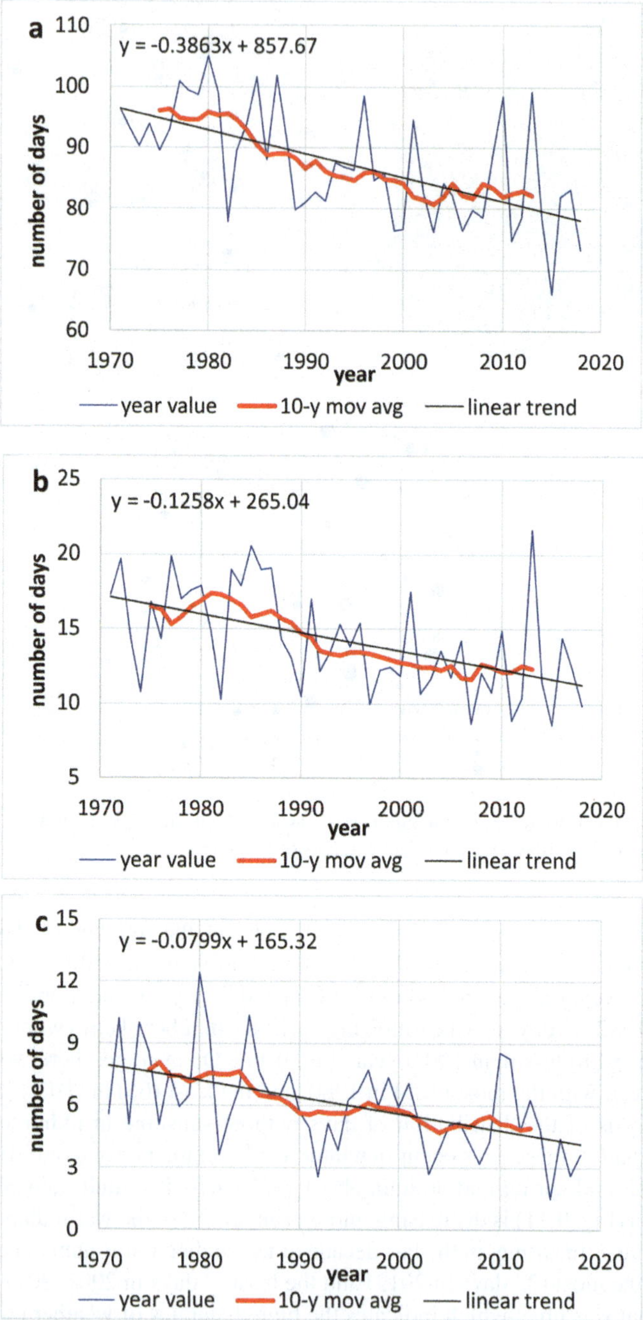

Fig. 9.11 Multi-annual course of the number of days without sunshine in Poland **a** year, **b** spring, **c** summer, **d** autumn, **e** winter

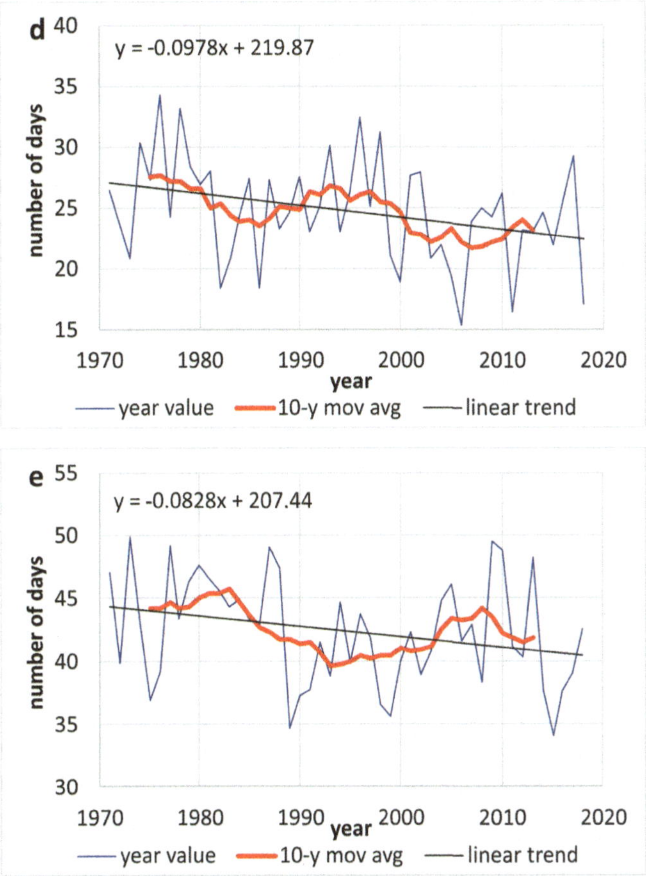

Fig. 9.11 (continued)

The winter is characterised by the greatest multi-annual variability in individual years. The highest (close to 50 days) and the lowest (34 days) values occur both in the first half of the studied multi-annual period and in recent years (Fig. 9.11).

On average, the most (>100) days without sunshine occur in Suwałki and on Śnieżka (Table 9.2). Additionally, at these stations (and in Białystok), the largest number of such days occurred. The year 1980 had the largest number of days without sunshine at 9 stations, 2010 at 4 stations, and 1985 and 2013 at 3 stations each (Table 9.2). The fewest days without sunshine (<50) were observed in Jelenia Góra and Opole, and 2015 was a record year in this respect, since the lowest number of days without sunshine was observed at 15 stations (Table 9.2). It is worth noting that the relationship between sunshine duration, both actual and relative, and the number of days without sunshine is loose, e.g. in the record-breaking for sunshine duration year 2018, the lowest number of such days in the multi-annual period was observed only at 4 stations.

Table 9.2 Mean, maximum and minimum annual number of days without sunshine at the stations included in the study (1971–2018)

Stations	Mean	Maximum	Year	Minimum	Year
Białystok	100	132	1984	68	2015
Chojnice	88	118	1984	66	2018
Gaik-Brzezowa	89	124	2013	66	2000
Gdynia	80	103	1974	52	1992
Gorzów Wlkp.	88	118	1977	63	2003
Jelenia Góra	73	97	1987	47	2011
Kalisz	84	113	1980	58	2015
Kasprowy Wierch	94	118	2010	69	1986
Katowice	85	127	1987	61	2015
Kłodzko	74	91	1980	50	2007
Kołobrzeg	87	113	2010	64	2015
Koszalin	92	117	1985	65	2018
Kraków Jagiellonian University Observatory	85	108	2010	61	1989, 2000
Lesko	81	106	1995	50	1986
Łódź-Lublinek	89	114	1985	62	2018
Mikołajki	95	120	2013	65	2015
Nowy Sącz	78	99	1979	56	2015
Opole	79	115	1971	49	2015
Poznań	87	119	1980	63	2003
Puławy	89	112	1980	69	2015
Suwałki	106	128	1980	84	2015
Szczecin Dąbie	92	122	2010	66	2015
Śnieżka	101	129	2013	67	2003
Tarnów	87	112	1985	54	2015
Terespol	88	117	1978	67	2015
Toruń	88	125	1981	62	2015
Warszawa Bielany	92	115	1980	69	2012
Wielichowo	87	119	1980	68	2015
Włodawa	87	108	1980	70	2000, 2015
Zakopane	79	101	1981	60	2018
Zielona Góra	82	109	1980	54	2017

Discussion and Conclusion

The annual sunshine duration in Poland based on data from 31 stations from the years 1971–2018 ranges from a minimum of 1,275 h in 1980 to a maximum value of 2,069 h in 2018 with the mean equal to 1,647 h. These values do not differ from the mean annual totals occurring in the neighbouring countries of Central Europe (Czech Republic—Bednar 1990; Central Europe—Brázdil 1991; Slovakia—Horecka 1990; Czech Republic and Slovakia—Vaniček 1990; Germany—Weber 1990; Austria—Dobesch 1992). In the entire study period, a statistically significant increasing trend in the annual actual and relative sunshine duration and a decreasing trend in the number of days without sunshine were observed at all stations.

The annual variability of sunshine duration and its spatial diversification depend on the length of the day, the size and type of cloud cover, and the transparency of the atmosphere. In Poland, an increase in sunshine duration in the summer, in the warm half of the year (March–August), and the whole year, occurs together with increasing latitude. The northern half of Poland is sunnier in the spring and summer than the southern part. In the spring, increased values of sunshine duration progress from the west of the country to the east. In the east, especially in the north-east, the increase in sunshine duration in the spring is delayed compared to the rest of the country. Spring is sunnier than autumn, especially in the northern half of Poland. In the summer, large convective cloud cover prevails in the mountains (south of Poland), resulting in shorter sunshine duration than in the lowlands. In the winter, and in the cold half of the year (September–February), sunshine duration decreases from the south to the north due to the longer day at lower latitudes. In addition, low layer clouds prevail in this part of the year, which results in the fact that in the mountainous areas in the south of Poland, at elevations of above 1,000 m a.s.l., there is a high amount of sunshine while below, in the valleys, fog and low *Stratus* clouds occur in greater amounts.

Comparing the maps of annual actual sunshine duration (Fig. 9.1) and days without sunshine (Fig. 9.10), an inverse relationship between these characteristics can be seen. This is clearly visible especially on the Gdańsk Coastland and in the Suwałki Region. The situation is different in the south-west and the foothill areas. There, the lowest annual number of days without sunshine coincides with lower annual sunshine duration.

The mean annual value of relative sunshine duration in Poland is 37%, with the maximum occurring in 2018 (46%) and the minimum in 1980 (28%). The transitional nature of the Polish climate, shaped under the influence of moving low and high barometric pressures and the related frontal systems in different years, with different frequency and activity, results in a very big difference in monthly relative sunshine values in individual years. However, since the beginning of the twenty-first century, there have been a much larger number of positive relative sunshine duration anomalies than in the 1970s and 1980s, and a statistically significant growing trend was observed at all stations throughout the entire study period.

The course of the multi-annual totals of sunshine duration in Poland shows trends similar to the trends in other parts of Europe (Norris and Wild 2007; Sanchez-Lorenzo et al. 2007, 2008; Manara et al. 2015; Matuszko 2014, 2016; Sanchez-Lorenzo 2015). This is indicated both by the data from the multi-annual period of 1971–2018 and the long-term heliographic series in Warszawa, Puławy, Wrocław, and on Śnieżka. The multi-annual variability of sunshine duration is most often explained by changes in cloud cover, which are conditioned by circulation (including Houghton et al. 2001; Herber et al. 2002; Sanchez-Lorenzo et al. 2008; Stjern et al. 2009; Matuszko 2014), or changes in the content of aerosols in the air (among others, Stanhill and Kalma 1995; Stanhill and Cohen 2001; Liepert 2002; Wild et al. 2005; Norris and Wild 2007; Ruckstuhl and Norris 2009; Sanchez-Lorenzo et al. 2009; Stjern et al. 2009; Folini and Wild 2011; Wild 2009, 2012; Wang et al. 2013; Vetter and Wechsung 2015). The periods of "global dimming" and "global brightening" widely presented in the climatological literature are very clearly visible in the course of sunshine duration in Poland. The years 1971–1980 mark the ending of "global dimming", which was manifested in very low values of actual and relative sunshine duration. In 1980, record low values of these characteristics were observed at many stations in Poland (Koźmiński and Michalska 2005; Gluza and Filipiuk 1995).

The decrease in sunshine duration in the 1980s was followed by an increase ("global brightening"), reaching maximum values in 2018. The occurrence of anomalously high totals of sunshine duration that year, among others in Central Europe, is also indicated by the studies based on the measurements performed by geostationary satellites (Kothe et al. 2019). At all meteorological stations, there was a statistically significant growing trend in actual and relative sunshine duration, more strongly expressed in the western than the eastern part of the country. These trends can be explained by both circulation factors, i.e. an increase in the frequency of occurrence of convective clouds in recent years, and a decrease in the frequency of stratiform clouds, as well as anthropogenic factors, i.e. the improvement of air quality in Poland. The end of the 1980s was a period of intensified international projects aimed at reducing the emission of sulphur and nitrogen compounds into the atmosphere. The measures taken yielded positive results, as evidenced by the courses of the multi-annual variability of transparency of the atmosphere published by different authors and confirmed by a decrease in the aerosol optical depth (AOD) of the atmosphere, which has been marked in Central and Eastern Europe since the mid-1980s (Weller and Gericke 2005; Ruckstuhl et al. 2008; Ohvril et al. 2009; Markowicz and Uscka-Kowalkowska 2015).

Similar tendencies in sunshine duration over a larger area indicate causes of global nature, only modified by local factors. Macro-scale conditions can be of both natural (circulation) origin and anthropogenic, as a result of industrial development and urbanisation. The impact of volcanic eruptions cannot be ruled out (Mount St. Helens in May 1980, Eyjafjallajökull in Iceland in April 2010). Floating in the air, volcanic aerosols could, directly and indirectly, by increasing cloud cover, contribute to the reduction of the inflow of solar radiation, which in those years resulted in decreases in sunshine duration at many stations. The relationship between volcanic eruptions and sunshine duration was demonstrated by, among others, Stanhill and Cohen (2005).

References

Brázdil R (1991) Kolísání vybraných meteorologických prvků ve střední Evropě v období přístrojových pozorování. Národní klimatický program České a Slovenské Federativní Republiky, vol 2. Český hydrometeorologický ústav, Praha, p 56

Bednar J (1990) Recent changes of the sunshine in Prague Region. In: Brázdil R (ed) Climatic change in the historical and the instrumental periods. Masaryk University, Brno, pp 240–242

Bryś K (2013) Dynamika bilansu radiacyjnego murawy oraz powierzchni nieporośniętej.Wyd. Uniwersytetu Przyrodniczego we Wrocławiu, Wrocław, p 288

Dobesch H (1992) On the variations of Sunshine Duration in Austria. Theor Appl Climatol 46:33–38

Dubicka M (1998) 95-letnia seria rejestracji usłonecznienia na Śnieżce. In: Sarosiek J, Štrusa J (eds) Geoekologiczne Problemy Karkonoszy, materiały z sesji naukowej, Przysieka, 15-18 X 1997. Acatus, Poznań, pp 133–143

Folini D, Wild M (2011) Aerosol emissions and dimming/brightening in Europe: Sensitivity studies with ECHAM5-HAM. J Geophys Res 116:D21104. https://doi.org/10.1029/2011JD016227

Gluza AF, Filipiuk E (1995) Usłonecznienie w Lublinie w latach 1952-1991. Wiadomości IMGW 1:91–101

Gorczyński E (1913) O insolacji ziem polskich. Encyklopedia Polska, vol 1. Akademia Umiejętności, Kraków, pp 161–170

Górski T, Górska K (2000) Usłonecznienie w Puławach w ciągu lat 1923–1992. Acta Univer N Copernici – Geografia 31(106):141–156

Herber A, Thomason LW, Gernandt H, Leiterer U, Nagel D, Schulz KH, Kaptur J, Albrecht T, Notholt J (2002) Continuous day and night aerosol optical depth observations in the Arctic between 1991 and 1999. J Geophys Res 107:D10. https://doi.org/10.1029/2001JD000536

Houghton JT, Ding Y, Griggs DJ, Noguer M, van der Linden PJ, Dai X, Maskell K, Johnson CA (eds) (2001) Climate change 2001: The scientific basis. Cambridge University Press, Cambridge

Horecka V (1990) Trend of sunshine in Slovakia. In: Brázdil R (ed) Climatic change in the historical and the instrumental periods. Masaryk University, Brno, pp 246–248

Kothe S, Hollmann R, Pfeifroth U, Träger-Chatterjee C, Trentmann J (2019) The CM SAF R toolbox – A tool for the easy usage of satellite-based climate data in NetCDF format. ISPRS Int. J. Geo-Information 8(3):109

Koźmiński C, Michalska B (2005) Usłonecznienie w Polsce, Sunshine in Poland. Akademia Rolnicza w Szczecinie, Uniwersytet Szczeciński, Szczecin, p 110

Lewik P, Matuszko D, Morawska-Horawska M (2010) Multi-annual variability of cloudiness and sunshine duration in Krakow between 1826 and 2005. In: Przybylak R, Majorowicz J, Brázdil R, Kejna M (eds) The Polish climate in the European context: An historical overview. Springer, Netherlands, Dordrecht Heidelberg London New York, pp 341–353

Liepert BG (2002) Observed reductions of surface solar radiation at sites in the United States and worldwide from 1961 to 1990. Geophys Res Lett 29(10):1421. https://doi.org/10.1029/2002GL014910

Manara V, Beltrano MC, Brunetti M, Maugeri M, Sanchez-Lorenzo A, Simolo C, Sorrenti S (2015) Sunshine duration variability and trends in Italy from homogenized instrumental time series (1936–2013). J Geophys Res 120:3622–3641. https://doi.org/10.1002/2014JD022560

Markowicz KM, Uscka-Kowalkowska J (2015) Long-term and seasonal variability of the aerosol optical depth at Mount Kasprowy Wierch (Poland). J Geophys Res Atmos 120(5):1865–1879. https://doi.org/10.1002/2014JD022580

Matuszko D (2014) Long-term variability in solar radiation in Krakow based on measurements of sunshine duration. Int J Climatol 34(228–234)

Matuszko D (2016) Usłonecznienie w miastach na podstawie wybranych stacji w Europie. Acta Geogr Lodziensia 104:45–56

Merecki R (1914) Klimatologia ziem polskich (Climatology of Poland). Druk i Litogr Jana Cotty, Warszawa, p 313

Norris JR, Wild M (2007) Trends in aerosol radiative effects over Europe inferred from observed cloud cover, solar "dimming" and solar "brightening". J Geophys Res 112(D08):D08214. https://doi.org/10.1029/2006JD007794

Ohvril H, Teral H, Neiman L, Kannel M, Uustare M, Tee M, Russak V, Okulov O, Joeveer A, Kallis A, Ohvril T, Terez EI, Terez GA, Gushchin GK, Abakumova GM, Gorbarenko EV, Tsvetkov AV, Laulainen N (2009) Global dimming and brightening versus atmospheric column transparency, Europe, 1906–2007. J Geophys Res 114(D10):D00D12. https://doi.org/10.1029/2008JD010644

Pallee E, Butler CJ (2002) Comparison of sunshine records and synoptic cloud observations: a case study for Ireland. Phys Chem Earth 27(6–8):405–414

Podogrocki J (1998) Warunki klimatyczne i meteorologiczne do wykorzystania energii promieniowania słonecznego w warunkach Polski, Konferencja Netmark Dom Ekologiczny. http://solis.pl/index.php/pliki_do_pobrania/publikacje_i_artykuly/warunki_do_wykorzystania_energii_promieniowania_slonecznego_w_polsce. Accessed 22 Aug 2019

Ruckstuhl C, Philipona R, Behrens K, Coen MC, Dűrr B, Heimo A, Mầtzler C, Nyeki S, Ohmura A, Vuilleumier L, Weller M, Wehrli C, Zelenka A (2008) Aerosol and cloud effects on solar brightening and the recent rapid warming. Geophys Res Lett 35(12):L12708. https://doi.org/10.1029/2008GL034228

Ruckstuhl C, Norris JR (2009) How do aerosol histories affect solar "dimming" and "brightening" over Europe?: IPCC-AR4 models versus observations. J Geophys Res 114(D10):D00D04. https://doi.org/10.1029/2008JD011066

Sanchez-Lorenzo A (2015) Reassessment and update of long-term trends in downward surface shortwave radiation over Europe (1939–2012). J Geophys Res 120(18):9555–9569. https://doi.org/10.1002/2015jd023321

Sanchez-Lorenzo A, Brunetti B, Calbo J, Martin-Vide J (2007) Recent spatial and temporal variability and trends of sunshine duration over the Iberian Peninsula from a homogenized data set. J Geophys Res 112(D20), 1–18. https://doi.org/10.1029/2007JD008677

Sanchez-Lorenzo A, Calbo J, Martin-Vide J (2008) Spatial and temporal trends in sunshine duration over Western Europe (1938–2004). J Clim 21(22):6089–6098

Sanchez-Lorenzo A, Wild M, Brunetti M, Guijarro JA, Hakuba MZ, Calbó J, Mystakidis S, Bartok B (2015) Reassessment and update of long-term trends in downward surface shortwave radiation over Europe (1939–2012). J Geophys Res Atmos 120(18):9555–9569. https://doi.org/10.1002/2015JD023321

Sanchez-Lorenzo A, Calbó J, Brunetti M, Deser C (2009) Dimming/brightening over the Iberian Peninsula: Trends in sunshine duration and cloud cover and their relations with atmospheric circulation. J Geophys Res 114(D10):D00D09. https://doi.org/10.1029/2008JD011394

Stanhill G, Kalma JD (1995) Solar dimming and urban heating at Hong Kong. Int J Climatol 15(8):933–941

Stanhill G, Cohen S (2001) Global dimming: A review of the evidence for a widespread and significant reduction in global radiation with discussion of its probable causes and possible agricultural consequences. Agr. Forest Meteorol 107(4):255–278

Stanhill G, Cohen S (2005) Solar radiation changes in the United States during the Twentieth Century, Evidence from Sunshine Duration Measurements. J Clim 18:1503–1512

Stjern CW, Kristjansson JE, Hansen AW (2009) Global dimming and global brightening – an analysis of surface radiation and cloud cover data in northern Europe. Int J Climatol 29(5):643–653

Vaniček K (1990) Trendy slunično svitu na územi ČSR v obdobi 1956-1985. In: Sbornik prac Českoho Hydrometeorol. Ustavu, vol. 37, Praha

Vetter T, Wechsung T (2015) Direct aerosol effects during periods of solar dimming and brightening hidden in the regression residuals: Evidence from Potsdam measurements. J Geophys Res Atmos 120(21):11299–11305. https://doi.org/10.1002/2015JD023669

Wang Y, Yang Y, Han S, Wang Q, Zhang J (2013) Sunshine dimming and brightening in Chinese cities (1955–2011) was driven by air pollution rather than clouds. Clim Res 56:11–20

Weber GR (1990) Spatial and temporal variation of sunshine in the Federal Republic of Germany. Theor Appl Climatol 41(1–2):1–9

Weller M, Gericke K (2005) Long-term observations of aerosol optical depths at the Meteorological Observatory Lindenberg. Meteorol Z 14:651–662. https://doi.org/10.1127/0941-2948/2005/0070
Wild M (2009) Global dimming and brightening: a review. J Geophys Res 114(D10):D00D16. https://doi.org/10.1029/2008JD011470
Wild M (2012) Enlightening global dimming and brightening. Bull Am Meteorol Soc 93:27–37
Wild M, Gilgen H, Roesch A, Ohmura A, Long CL, Dutton EG, Forgan B, Kallis A, Russak V, Tsvetkov A (2005) From dimming to brightening: Decadal changes in solar radiation at Earth's surface. Science 308:847–850
WMO (World Meteorological Organization) (2008) Guide to meteorological instruments and methods of observation (CIMO guide), vol. WMO-No. 8. Secretariat of the WMO, Geneva

Chapter 10
Change of Cloudiness

Janusz Filipiak

Abstract Cloudiness is an important element of the climate system as it strongly influences the global energy balance. On the basis of data from 57 stations, the most important features of spatial diversity and temporal variability of cloudiness, i.e. annual and seasonal mean and extreme values, number of clear and cloudy days and cloud type, were analysed. The mean annual cloud cover in Poland in the years 1951–2018 was 68.2%. The maximum value, equal to 73%, was recorded in 1966, whereas the minimum value of nearly 61% occurred in 1982. The lowest level of cloudiness occurs in summer, while the largest in winter. Generally, cloudiness in spring is lower than in autumn. In spring and summer, cloudiness decreases eastwards, while in autumn and winter it increases towards the north of the country. Additionally, the western part of Poland is cloudier than the east of the country. In the north, a negative tendency in annual cloudiness in the period 1951–2018 prevails, while in the south positive statistically significant changes occur. The amount of cloudiness in particular seasons also changes, however, stronger trends can be observed in subperiods: 1951–1984 and 1985–2018. In winter, in the later period, a statistically significant increase in cloudiness occurs with the coefficient increasing towards the south. Strong positive trends in spring were in the period of 1951–1984. In both remaining seasons, recorded changes of cloudiness are smaller, and, in the period of 1985–2018, a decrease of cloudiness was observed across the country. Overall, the frequency of cloudy days is much higher in Poland than clear ones. The number of clear days is decreasing considerably across the country, and, in case of cloudy days, negative changes concern only some regions. Atmospheric circulation plays a key role in shaping nephological conditions, however, processes of a local scale are also important. *Stratocumulus*, *Altocumulus* and *Cirrus* are the most frequent clouds among the low-level, middle-level and high-level clouds, respectively, in Poland, but its percentage ranges significantly in various regions. Coastlands are most favourable

J. Filipiak (✉)
Faculty of Oceanography and Geography, Institute of Geography, Department of Meteorology and Climatology, University of Gdańsk, Gdańsk, Poland
e-mail: janusz.filipiak@ug.edu.pl

Institute of Meteorology and Water Management—National Research Institute, Warszawa, Poland

M. Falarz (ed.), *Climate Change in Poland*, Springer Climate,
https://doi.org/10.1007/978-3-030-70328-8_10

in the country for the occurrence of both convective clouds and *Stratus* clouds. *Stratocumulus* clouds are most common in the whole country, particularly in areas with a diverse relief, and *Altocumulus* clouds are most observed in southwestern Poland. The frequency of occurrence of *Altostratus* decreases slightly from west to east. *Cirrostratus* clouds are the most frequent in western Poland, while *Cirrocumulus* clouds are very rare. In case of low-level clouds, excluding *Stratus*, a strong, statistically significant positive trend occurs. In addition, the number of *Stratus* observations is declining substantially. The frequency of occurrence of *Altostratus* and *Nimbostratus* is systematically decreasing, while in the case of *Altocumulus*, an increase may be observed. On the whole, high-level clouds can be observed more and more frequently in Poland.

Introduction

Cloudiness seems to be one of the most interesting among the meteorological elements. Clouds strongly influence the global energy balance: on the one hand, reflecting a large part of shortwave radiation, and, on the other, absorbing long-wave radiation and emitting the additional considerable own flux of energy (Norris and Slingo 2009; Siebesma et al. 2009; Boucher et al. 2013; Zhou et al. 2016). What is crucial is that their role in many regions of the Earth, particularly the most fragile ones, is still increasing—e.g. Vavrus et al. (2011) reported on the growing feedback on climate driven by clouds observed in the Arctic, and Schneider et al. (2019) presented the mechanism concerning how clouds can respond to greenhouse warming in the subtropical region.

Moreover, clouds greatly interact with other meteorological variables. There is a relation between the observed worldwide, long-term decrease in daily temperature range and the changes in cloud cover (Henderson-Sellers 1986), seasonal and regional variability of relative humidity and cloud cover (Cox et al. 2015), and sea ice and cloudiness (Jun et al. 2016).

Though visual observations at synoptic and climatological stations still remain the main source of information on the spatial and temporal variability of cloud cover and cloud type, the role of new methods, such as satellite observations (Hollmann et al. 2013) detecting clouds from a position above the atmosphere, is increasing. Such developments help in the discovery of new patterns of cloud changes as, e.g. increasing height of the highest cloud tops at all latitudes or poleward retreat of mid-latitude storm tracks (Norris et al. 2016). The drivers of all current changes in cloud amount and type appear to be related mostly to anthropogenic activity, specifically increasing greenhouse gas concentrations or human-induced aerosol optical depth (Charlson et al. 1992; Boers et al. 2017). The increasing analytical possibilities and growing volume of data constitute an important contribution to the enhanced analyses of long-term variability of cloud cover and cloud type (Eastman et al. 2011; Eastman and Warren 2013) as well as climate model simulations (Webb et al. 2017).

The increase in the spatial resolution of research brought to light important features of cloudiness variability in the smaller areas of special interest, as is presented in the analysis below.

Data and Methods

The structure of spatial and temporal variability of cloudiness in Poland was analysed for the period 1951–2018 using data derived from 57 observation stations located throughout Poland. The varied locations of the stations presents the variability of cloudiness in all geomorphological types of Polish landscapes (shores, lowlands, uplands, mountain foothill basins and mountains—low and high) (Solon et al. 2018).

The research material was derived mainly from manned synoptic stations, where measurements of cloud characteristics were made by visual observation. Initially, it included data on total cloud cover gathered during eight main and intermediate standard times (starting from 00 to 21 UTC).

Taking the specificity of the materials provided by the Polish National Hydrological and Meteorological Service (IMGW-PIB), the majority of data was available for the periods starting not earlier than in 1966. As for the period 1951–1965, the data available covers only three climatological times—morning, afternoon and evening. Moreover, during the 1990s, at some stations, 24/7 observation shifts were changed into a 12-h system of work, starting at 6 and finishing at 18 UTC. In 2000, some of the synoptic stations (more precisely the changes affected five stations) were fully automated. It was decided to delete these stations from the analysis due to insufficient temporal data coverage. In 2014, some more changes to the IMGW-PIB observation network were introduced, resulting in withdrawal of cloud observations from the observing practices at almost 20 stations (of total number of 63 synoptic stations in Poland). At the beginning of 2018, cloud observations were brought back in practice at approximately half of them. Currently (the end of 2019), observations of clouds are made at 48 Polish synoptic stations.

Additionally, three climatological stations were also added to the list of those analysed. The data on total cloud cover enriched the analysis concerning several physico-geographical regions. In Poland, climatological observations are carried out at times corresponding to the main morning, afternoon and evening synoptic observation times (i.e. 6, 12 and 18 UTC). Until 1970, meteorological observations in Poland had been made at 7, 13 and 21 of the local solar time.

As the analysis covered data obtained from the stations of different order (synoptic and climatological) and taking the above-mentioned changes in the observing practices under consideration, it was decided to use only the data gathered at three main standard times (06, 12 and 18 UTC), referring to the climatological ones.

Subsequently, it was necessary to standardise the unit used when assessing the total cloud cover. The one currently used—okta (covering the figures from 0 to 8 and, additionally 9 for an obscured sky), was introduced into observation practice on 1 January 1966 at the synoptic stations and on 1 January 1989 at the climatological ones. Before that time, cloud amount had been recorded in deciles (on a scale 0–10). Both units were converted into percentages in accordance with the WMO guidelines (WMO 2018). The cases with 9 oktas have been classified as situations of overcast sky (8/8). The number of such observations was very small and they constituted approximately 1% of all analysed cases. Exceptionally, at some stations located in the lakelands (Kętrzyn, Resko, Chojnice and Gorzów Wlkp.) and in the mountain regions (Zakopane—intramountain basin, Lesko—low mountains) such observations reached the value of approximately 2%. However, at two Polish high-mountain observatories, the percentage of full cloudiness cases was relatively high (Kasprowy Wierch—34%, Śnieżka—40%).

On the basis of the observation data, the mean daily total cloud cover was calculated for each analysed station. Then, the most important features of spatial diversity of mean annual and seasonal values of the element in the period 1951–2018 were determined. The aspects of temporal variability were analysed for the whole selected research period as well as its two subperiods: 1951–1984 and 1985–2018. Next, the annual frequency of clear and cloudy days was calculated. A clear day is one with a daily mean cloudiness lower than 20% of the cloud amount. Analogically, a cloudy day is the one with the daily mean cloudiness of more than 80% of the cloud amount.

Additionally, cloud type, i.e. estimate of sky fraction covered by a variety of ten cloud genera in the period of 1966–2018 for several dozen stations was analysed. The stations were selected using the criteria of observation material completeness, i.e. only from series with eight daily observations at main and intermediate standard times of cloud genera at all three levels were taken into account, which finally resulted in the number of 23 series. According to the regulations of FM-12 synoptic code (WMO 2018), results of cloud type observations are classified using a scale given in synoptic code: C_L, C_M or C_H. However, this specific way of coding does not always enable the assignment of a single genus of cloud to a particular code element, as Matuszko and Węglarczyk (2018) mentioned in their analysis. The regulations concerning the coding of C_L, C_M or C_H clouds in a situation when there is more than one genus of clouds in the sky which correspond to more than one code of clouds at particular level, put an obligation to select a proper code in accordance with the fixed priority hierarchy.

An easily noticeable example concerns low clouds—the appearance of the sky coded $C_L = 3$ and 9 means that *Cumulonimbus* is present, yet it may be accompanied by *Cumulus*, *Stratocumulus* or *Stratus*. Thus, it can be assumed that the accompanying clouds appeared in the sky in some part of the cases; in the remaining cases a huge *Cumulonimbus* completely obscured the other genera of clouds. A code of $C_L =$ 8 means coappearance of *Cumulus* and *Stratocumulus*, with their bases at different heights. Thus, all the cases were assigned to both clouds. In the case of middle clouds, a code of $C_M = 2$ means *Nimbostratus* or *Altostratus* sufficiently dense to hide the sun. The ratio of the occurrence of both clouds has to be, however, assigned. A code of $C_M = 7$ concerns both *Altocumulus* and *Altostratus*/*Nimbostratus*. Usually, observers code two genera of clouds—Ac As or Ac Ns. That is why, all observations coded that way were added to the number of *Altocumulus* cases, but part of them have to be added to the number of *Altostratus* and *Nimbostratus* ones. In the case of high clouds, a similar approach was implemented. The numbers 5 and 6 added to code of C_H mean not only *Cirrostratus*, but also the possibility of coappearance of *Cirrus* clouds. Similarly, in the case of $C_H = 9$, domination of *Cirrocumulus* does not exclude the existence of *Cirrus* and *Cirrostratus* clouds in the sky.

The above-mentioned short description explains why analysis of cloud type shall always be preceded by a detailed comparative analysis of data contained in synoptic reports and journal of synoptic observations. Nonetheless, the results of methodological analyses initially begun by the Polish National Meteorological and Hydrological Service, with the use of observation practices implemented at the Polish synoptic stations (IMGW-PIB 2013), further developed by the author for the purposes of the current paper, helped to indicate a simplified method of assigning a particular genus or genera of clouds to the code of C_L, C_M or C_H. The observation material for the years 2000–2018 collected for the current analysis helped to justify and improve the methodology. Its updated results are displayed in Table 10.1.

It is also worth emphasising that when analysing observation data concerning genera of clouds at levels higher than the low level it is important to keep in mind the fact that in some cases, the sky was completely obscured by the lower level clouds. As for the middle clouds, such a situation concerned approximately 30–40% of cases. However, the high clouds were not visible during approximately 50% of the observations. At the stations located in the north and in the mountain areas, this percentage was even higher.

Results

Total Cloudiness

The average annual total cloud cover in Poland during the period of 1951–2018 is 68.2%. The value of the variable recorded at particular stations varies from nearly 65% to more than 75% (Fig. 10.1a). Western and eastern peripheries of the Southern Baltic Coastlands belt (Świnoujście, Hel and Gdynia stations: 64, 65 and 65%, respectively) as well as the area of the Northern Subcarpathians (with value recorded at the Sandomierz station—64%) are the least clouded are in the country. The largest degree of cloud coverage is observed in the mountainous areas—the Sudety Mts. and the Carpathians, yet the average annual total cloud cover in the Karkonosze Mts. (the highest range in Sudety Mts.) is higher than in the Tatra Mts. (culmination of the Carpathians). The average annual total cloud cover at the top of Śnieżka Mountain in the Karkonosze Mts. reached almost 76%, while at Kasprowy Wierch it was 73%. Although cloudiness is predominantly dependent on the macro-scale situation (Ustrnul and Niedźwiedź 1994), there are a number of local factors affecting the forming clouds in the mountains (Trepińska 2002; Szyga-Pluta 2017). Among these, intensive dynamic and thermal turbulence can be listed as well the local convection

Table 10.1 Relation between coding of CL, CM or CH clouds as reported in synoptic reports and corresponding cloud genera reported in journals of synoptic observations[a]

Code of cloud reported in synoptic report		Corresponding cloud genus reported in journal of synoptic observations
C_L	1	*Cumulus*
	2	*Cumulus*, *Stratocumulus* (20% of cases)
	3	*Cumulonimbus*, *Cumulus* (75% of cases), *Stratocumulus* (50% of cases), *Stratus* (20% of cases)
	4	*Stratocumulus*
	5	*Stratocumulus*, *Stratus* (25% of cases)
	6	*Stratus*
	7	*Stratus*
	8	*Cumulus*, *Stratocumulus*
	9	*Cumulonimbus*, *Cumulus* (50% of cases), *Stratocumulus* (50% of cases), *Stratus* (10% of cases)
C_M	1	*Altostratus*
	2	*Nimbostratus* (80% of cases), *Altostratus* (20% of cases)
	3	*Altocumulus*
	4	*Altocumulus*
	5	*Altocumulus*
	6	*Altocumulus*
	7	*Altocumulus*, *Altostratus* (90% of cases)
	8	*Altocumulus*
	9	*Altocumulus*
C_H	1	*Cirrus*
	2	*Cirrus*
	3	*Cirrus*
	4	*Cirrus*
	5	*Cirrostratus*, *Cirrus* (75% of cases)
	6	*Cirrostratus*, *Cirrus* (75% of cases)
	7	*Cirrostratus*
	8	*Cirrostratus*, *Cirrus* (50% of cases), *Cirrocumulus* (10% of cases)
	9	*Cirrocumulus*, *Cirrostratus* (25% of cases), *Cirrus* (25% of cases)

[a]Numbers in parentheses indicate the percentage of cases when particular genera of clouds occur in relevant situation of coding. Lack of information in parentheses means that particular cloud occurs in each individual case

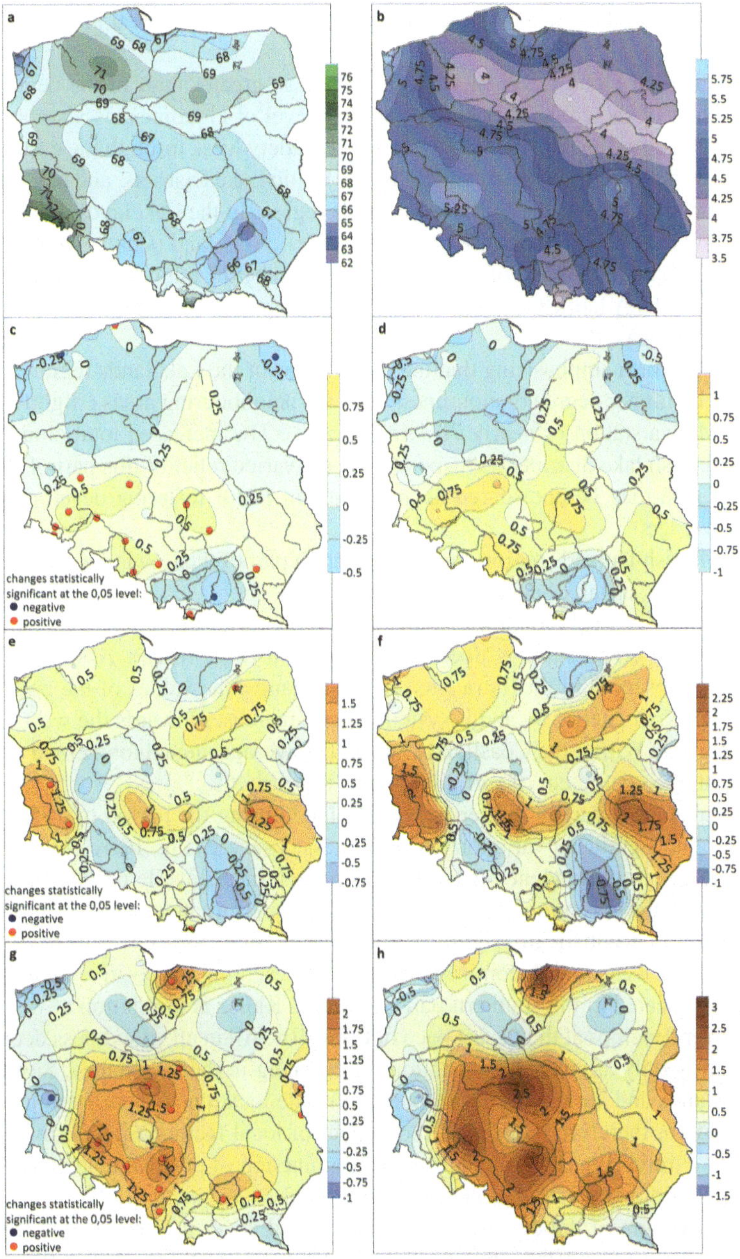

Fig. 10.1 Variability and change of annual total cloudiness: mean values (1951–2018; %; **a**), coefficient of variability (%; **b**), absolute trend for the whole period (1951–2018; %/10y; **c**), relative trend for the whole period (%/10y; **d**), absolute trend for the first half of the period (1951–1984; %/10y; **e**), relative trend for the first half of the period (%/10y; **f**), absolute trend for the second half of the period (1985–2018; %/10y; **g**), relative trend for the second half of the period (%/10y; **h**). Statistically significant (0.05) trends were shown in (**c, e, g**) as circles

movements. Cloud condensation level depends also on the vertical profile of water vapour.

Overall, total cloud cover in western Poland is greater than in the central and eastern parts of the country. A slightly higher level of cloudiness is also observed within the longitudinal belt stretching from the Sudety Mts., through the eastern part of the Central Poland Lowlands and Southern Baltic Lake Districts, and reaching the central part of the Southern Baltic Coastlands region. In the Lakelands, cloud cover exceeds 70% (at Chojnice station, it is 72%). Western Poland is more frequently, in comparison with other parts of the country, affected by the humid polar maritime air masses from the Atlantic Ocean. Their advection results in an increase in cloudiness, no matter the season.

It is also worth emphasising that the eastern part of the Lakelands region is characterised by higher degree of cloudiness than the neighbouring areas (slightly above 70% at Mława station or nearly 70% in Suwałki). The increased total cloud cover over the Polish lake districts correspond with the varied relief of this part of Poland, which in turn may be considered as an important factor fostering development of cloud coverage.

In the Polish Uplands region and in the eastern part of the Lowlands, cloud levels are below 70%, usually between 67 and 68%.

The average annual cloudiness in the Lakelands varies the least—the coefficient of variation at some stations does not exceed 4% (Fig. 10.1b). A similar stable situation is observed in the Tatra Mts. In the Uplands, the values of the coefficient reaches approximately 5%, and it systematically increases towards the west, finally exceeding 5%. Within the Coastlands, where the average annual total cloud cover is the lowest, the coefficient of variation comes to nearly 6%.

In a substantial part of the territory of Poland during the period of 1951–2018, there were no significant variations in cloudiness observed (Fig. 10.1c). However, in the north, a negative tendency in cloudiness predominates, while in the south—it is positive, with the exception of the Subcarpathians and Carpathians and excluding the Tatra Mts. Some statistically significant changes in total cloud cover were recorded in several dozen cases with domination of series presenting the positive trend in case of stations located in the southern part of Poland. Kalisz is the station at which the cloud cover develops at the fastest pace of 0.7%/10y. All in all, cloudiness decreased significantly only at three Polish stations (Kołobrzeg in the Coastlands, Suwałki in the Lakelands and Tarnów in Subcarpathians). The location of these stations seems to be random, therefore it suggests the impact of local factors. Spatial diversity in the relative trend (Fig. 10.1d) confirms the domination of positive tendencies in the development of annual total cloud cover in Poland.

The maximum of the average annual total cloud cover in Poland equalled 73% and was observed in 1966 (Table 10.2). At particular stations, the occurrence of maximum values was differentiated. It was observed most frequently in 1966 (11 stations, mostly in the northern part of Poland), 1980 (8 stations in eastern and southeastern Poland), 2013 (9 stations in central, southern and western parts of the country) and 2017 (6 stations located in central Poland). The cloudiest sky was in 2001 on Śnieżka Mt. (81%).

Table 10.2 Highest and lowest values of mean annual cloudiness [%] at stations taken into consideration (1951–2018)

Station	Highest		Lowest	
	Value	Year	Value	Year
Białystok	76.1	1980	62.3	1969
Bielsko-Biała	73.4	1952	59.8	1982
Chojnice	78.5	1966	65.8	1982
Częstochowa[a]	72.9	1952	55.8	1982
Elbląg	76.6	1962	60.7	1992
Gdynia[a]	73.3	1977	56.3	1982
Gorzów Wlkp.	77.4	1966	61.0	1982
Hel	71.9	1966	57.8	1989
Jabłonka[a]	72.6	1970	60.1	1982
Jelenia Góra	77.6	2013	62.6	1953
Kalisz	75.9	2013	60.8	1982
Kasprowy Wierch	78.8	2004	62.4	1953
Katowice	73.4	2016	58.6	1982
Kętrzyn[a]	72.8	2013	59.1	1982
Kielce	74.9	2013	59.0	1982
Kłodzko	76.5	2013	60.7	1982
Koło[a]	73.7	2013	57.8	1982
Kołobrzeg	74.6	1966	60.6	2018
Koszalin	76.8	1966	61.4	1982
Kraków Balice	71.0	1980	58.3	1982
Krosno[a]	77.7	1966	60.2	1982
Legnica[a]	78.9	2001	58.2	1953
Lesko[a]	74.8	1970	58.8	1982
Leszno[a]	76.8	2001	59.3	1982
Lublin	77.3	1980	59.0	1982
Łeba	74.3	1977	59.9	1953
Łódź	74.8	2017	59.1	1982
Mikołajki[a]	74.4	2017	60.6	1953
Mława[a]	75.4	2017	64.1	1964
Nowy Sącz[a]	72.8	1958	59.8	1982
Olsztyn[a]	73.1	1966	61.4	1982
Opole	74.5	2001	59.9	1982
Piła[a]	73.8	2001	61.4	1982
Płock[a]	74.1	1952	61.8	1956

(continued)

Table 10.2 (continued)

Station	Highest		Lowest	
	Value	Year	Value	Year
Poznań	74.0	2017	56.7	1982
Puławy	76.9	1980	58.4	1953
Racibórz[a]	71.9	2010	56.8	1982
Resko[a]	77.3	1977	61.9	2003
Rzeszów[a]	72.3	1980	56.8	1982
Sandomierz[a]	70.9	1980	56.1	1982
Siedlce	74.3	1980	62.1	1964
Słubice[a]	75.9	2013	59.8	1982
Sulejów[a]	75.0	2013	60.3	1982
Suwałki	76.6	1966	63.5	1951
Szczecin	74.2	1966	58.8	1992
Śnieżka	81.0	2001	66.0	1959
Świnoujście	72.7	1966	55.7	1982
Tarnów[a]	71.2	1958	56.4	1982
Terespol[a]	74.7	2017	61.8	1982
Toruń	74.3	1952	62.7	1982
Ustka	72.5	1966	59.3	1982
Warszawa	70.9	2017	57.8	1982
Wieluń[a]	75.1	1981	60.0	1953
Włodawa	76.5	1980	59.0	1953
Wrocław	75.1	2013	55.5	1982
Zakopane	74.6	2014	63.0	1982
Zielona Góra	75.9	1981	61.4	1953
Poland	73.0	1966	60.7	1982

[a]Period shorter than 1951–2018 (see details in Chap. 3 *Data and Methods of investigation*)

The minimum annual cloudiness, equalling 60.7%, was recorded in Poland in 1982 (Table 10.2). This year was characterised by very low value of cloud cover in the majority of stations in the country (37 stations of 56 analysed). In addition, very low value of cloud cover occurred also in 1953. The least cloudy sky was in 1982 in Wrocław (55.5%).

Periods of lower cloudiness levels were observed during the first half of the 1970s and at the turn of the 1980s and 1990s (Fig. 10.2). In fact, in the Coastlands, the minimum recorded at the turn of the 1980s and 1990s was even lower than in 1982. The sky over Poland was covered with the thickest layer of clouds at the beginning of the 1960s, in the second half of the 1970s and at the end of the twentieth century.

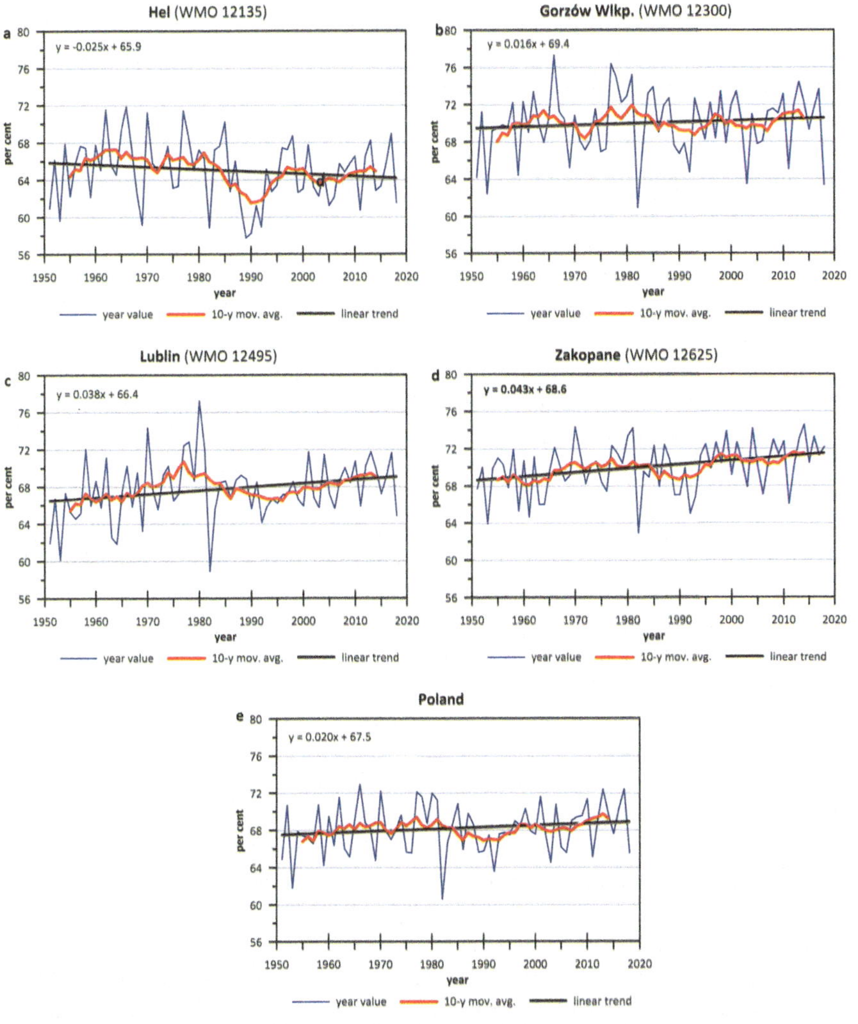

Fig. 10.2 Long-term variability of annual total cloudiness in the selected stations of: northern Poland (station Hel; 1951–2018; **a**), western Poland (station Gorzów Wlkp.; 1951–2018; **b**), eastern Poland (station Lublin; 1951–2018; **c**), southern Poland (station Zakopane; 1951–2018; **d**) and in Poland as a whole (the average, **e**). Mean annual value (*blue*), 10-year consecutive average (*red*) and line trend (*black*) are shown. The trend is statistically significant (0.05, in *bold*) in (**d**) and not significant in (**a**–**c**)

The last period of increased cloudiness began at the end of the first decade of the twenty-first century.

Temporal variability of cloudiness in Poland underwent a noticeable change in the analysed period. During the subperiod of 1951–1984, a statistically significant increase in cloudiness was observed at a relatively small number of stations (seven stations), mainly in the peripheral areas of the country (Fig. 10.1e). The increase of cloud cover was observed particularly in two pairs of stations—one located in western Poland (Legnica and Zielona Góra) and the second one in eastern Poland (Puławy and Lublin). During the later subperiod—in the years 1985–2018—the increase in cloudiness was already observed at a greater number of stations and, what is the most important, intensified and emerged as statistically significant at twice as many stations than in the previous subperiod (Fig. 10.1g). Positive trends are noticeable particularly at the stations located in the central and southern parts of the country. The values of relative trends revealed that cloudiness developed most intensively in Elbląg, Koło, Wrocław and Częstochowa, while areas with decreasing cloud cover were still in decline during the period of 1985–2018. Therefore, it can be stated that atmospheric conditions fostering cloudiness development were observed over the majority of Poland's territory. However, though the negative trends were not observed during the first subperiod, in the second they were recorded at one station (Zielona Góra, where in the years 1951–1984 cloudiness increased).

Winter is the season characterised by the highest value of total cloud cover (Fig. 10.3a). The average winter cloudiness in Poland reaches 75.7%. The values of total cloud cover decrease from north and northeast to the south. An exceptionally high level of cloudiness is observed over the Lakelands, especially those located in northeastern Poland. There, the level of cloudiness reaches approximately 80%. The lowest level of cloudiness, approximately 70%, is observed over the Carpathian Foreland and Carpathian basins (Zakopane, Nowy Sącz). Similarly, a relatively low level of cloudiness is observed in the Karkonosze Mountain Basins (Jelenia Góra station, located in a basin).

The coefficient of variation in winter is higher than the annual one. Its values increase towards the south, running almost parallel in the southern part of the country (Fig. 10.3b). The cloudiness varies the least in the lakeland areas of western Poland and in the central coast. In the Tatra Mts. the coefficient of variation is almost twice as high.

Largely, winter cloudiness gradually increases at the majority of considered stations (Fig. 10.3c, d). It is statistically significant at five stations located mostly in the southern part of the country (the western part of the Lowlands and the Uplands). Negative tendencies characterise the coastal region, the Lakelands in the northeastern part of Poland and the Subcarpathians. However, a statistically important decrease in cloudiness was recorded only in Kołobrzeg in the Coastlands.

While analysing the long-term variability of values of average winter cloudiness, a minimum is clearly visible in all data series recorded at the beginning of the 1990s (selected examples presented in Fig. 10.4). At particular stations it was 1990–1991 or 1993. In some cases, especially at stations located in the Lakelands, the reduced cloudiness was observed until the end of the twentieth century and at the very beginning of the twenty-first century as well. Lower levels of cloudiness were also observed in the 1970s, yet this anomaly was visible mainly at the southern stations. At the upland stations deep minimums and periods with high levels of cloudiness were observed alternately in the 1960s. At the Lakeland stations, the lowest cloudiness level was observed during the winters of 1954 and 2003. The level of cloudiness

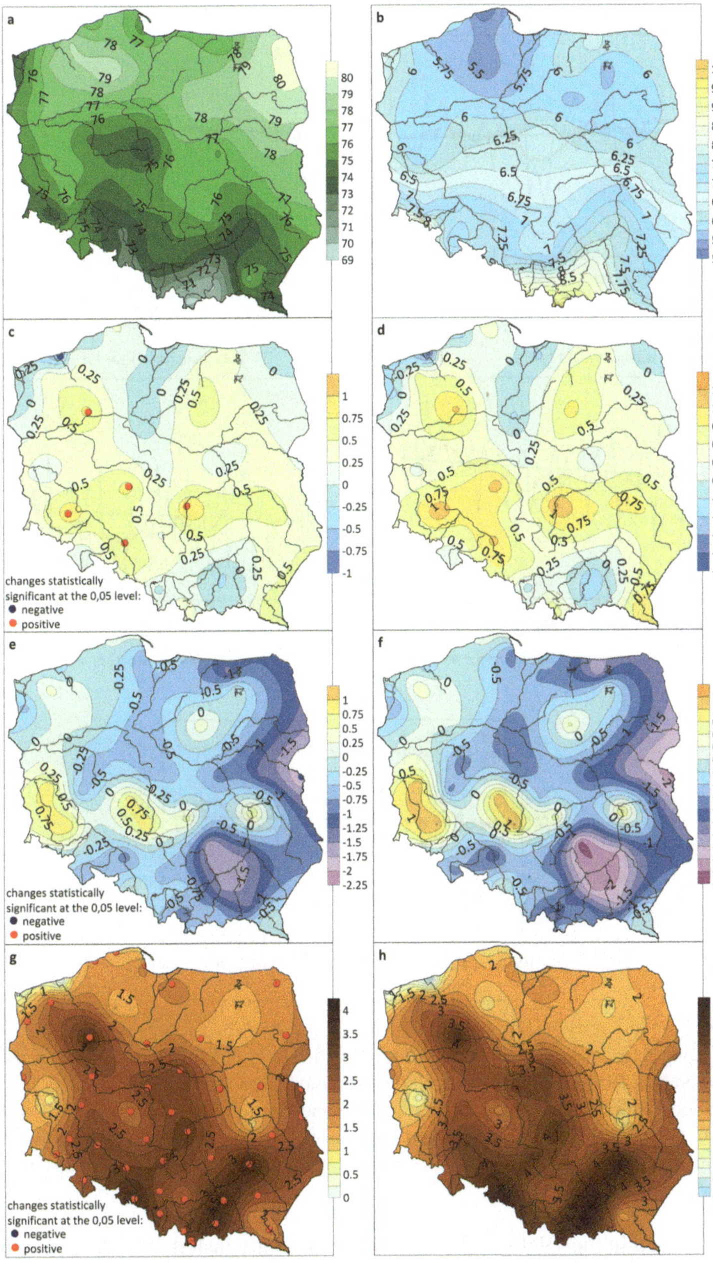

Fig. 10.3 Variability and change of total winter cloudiness: mean values (1951–2018; %; **a**), coefficient of variability (%; **b**), absolute trend for the whole period (1951–2018; %/10y; **c**), relative trend for the whole period (%/10y; **d**), absolute trend for the first half of the period (1951–1984; %/10y; **e**), relative trend for the first half of the period (%/10y; **f**), absolute trend for the second half of the period (1985–2018; %/10y; **g**), relative trend for the second half of the period (%/10y; **h**). Statistically significant (0.05) trends were shown in (**c**, **e**, **g**) as circles

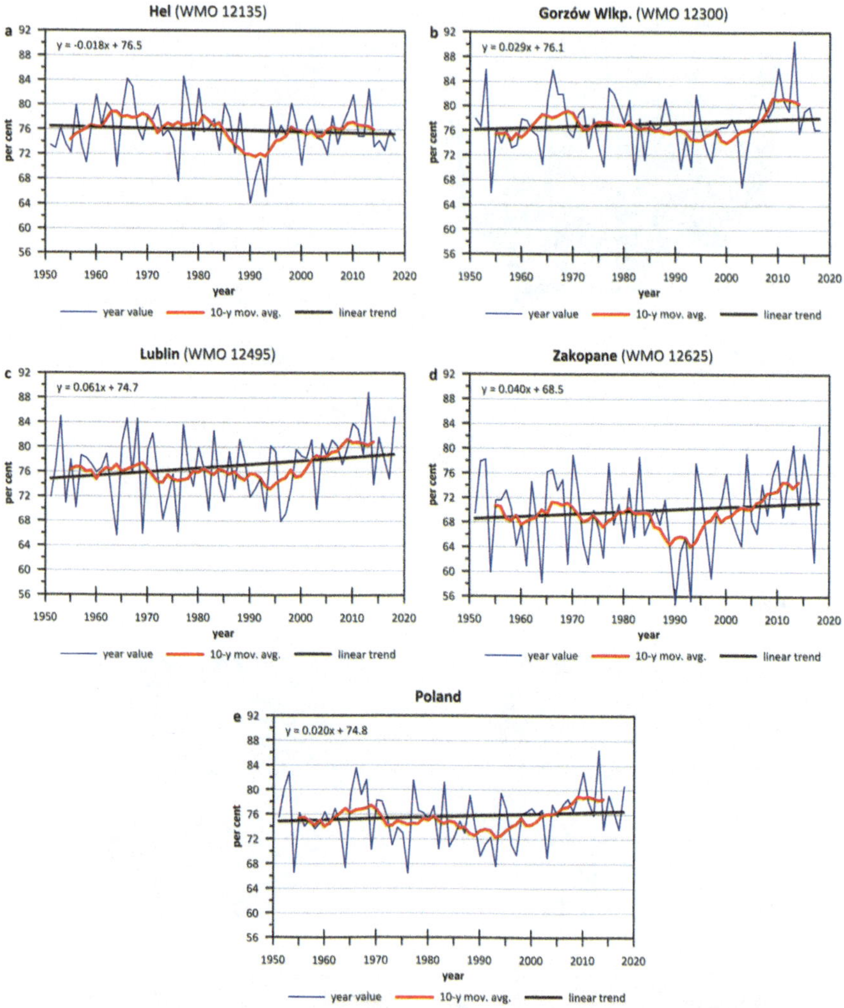

Fig. 10.4 Long-term variability of total winter cloudiness in the selected stations of: northern Poland (station Hel; 1951–2018; **a**), western Poland (station Gorzów Wlkp.; 1951–2018; **b**), eastern Poland (station Lublin; 1951–2018; **c**), southern Poland (station Zakopane; 1951–2018; **d**) and in Poland as a whole (the average, **e**). Mean seasonal value (*blue*), 10-year consecutive average (*red*) and line trend (*black*) were shown. The trend is statistically not significant (0.05) in (**a–d**)

visibly increased in the 1960s and this positive anomaly lasted for a decade. Another period of increased cloudiness started at the beginning of the present century and lasted until the end of the research period.

The highest winter cloudiness level in Poland occurred in 2013, when its value reached 86.5% (Table 10.3). At particular analysed stations it was most frequently also in 2013 (35 stations), 1953 (eight stations) and 1977 (four stations). At the

Table 10.3 Highest and lowest values of mean seasonal cloudiness [%] at stations taken into consideration (1951–2018)

Station	Winter				Spring				Summer				Autumn			
	Highest		Lowest		Highest		Lowest		Highest		Lowest		Highest		Lowest	
	Value	Year	Value	Year	Value	Year	Value	Year	Value	Year	Value	Year	Value	Year	Value	Year
Białystok	88.3	1955	65.7	1976	88.3	1955	65.7	1976	88.3	1955	65.7	1976	88.3	1955	65.7	1976
Bielsko-Biała	86.3	2013	58.3	1990	86.3	2013	58.3	1990	86.3	2013	58.3	1990	86.3	2013	58.3	1990
Chojnice	88.3	2013	69.7	1976	88.3	2013	69.7	1976	88.3	2013	69.7	1976	88.3	2013	69.7	1976
Częstochowa[a]	87.3	2013	63.7	1954	87.3	2013	63.7	1954	87.3	2013	63.7	1954	87.3	2013	63.7	1954
Elbląg	86.3	1977	67.0	1976	86.3	1977	67.0	1976	86.3	1977	67.0	1976	86.3	1977	67.0	1976
Gdynia[a]	85.5	2013	64.4	1976	85.5	2013	64.4	1976	85.5	2013	64.4	1976	85.5	2013	64.4	1976
Gorzów Wlkp.	90.7	2013	66.0	1954	90.7	2013	66.0	1954	90.7	2013	66.0	1954	90.7	2013	66.0	1954
Hel	84.7	1977	64.3	1990	84.7	1977	64.3	1990	84.7	1977	64.3	1990	84.7	1977	64.3	1990
Jabłonka[a]	84.3	1970	54.3	1990	84.3	1970	54.3	1990	84.3	1970	54.3	1990	84.3	1970	54.3	1990
Jelenia Góra	85.7	2013	58.7	1993	85.7	2013	58.7	1993	85.7	2013	58.7	1993	85.7	2013	58.7	1993
Kalisz	89.3	2013	63.3	1954	89.3	2013	63.3	1954	89.3	2013	63.3	1954	89.3	2013	63.3	1954
Kasprowy Wierch	80.3	2018	53.7	1964	80.3	2018	53.7	1964	80.3	2018	53.7	1964	80.3	2018	53.7	1964
Katowice	87.0	2013	63.7	1954,1990	87.0	2013	63.7	1954,1990	87.0	2013	63.7	1954,1990	87.0	2013	63.7	1954, 1990
Kętrzyn[a]	86.7	1988, 2013	61.3	1976	86.7	1988,2013	61.3	1976	86.7	1988,2013	61.3	1976	86.7	1988, 2013	61.3	1976
Kielce	88.8	2013	64.2	1976	88.8	2013	64.2	1976	88.8	2013	64.2	1976	88.8	2013	64.2	1976
Kłodzko	88.3	2013	62.7	1990	88.3	2013	62.7	1990	88.3	2013	62.7	1990	88.3	2013	62.7	1990
Koło[a]	87.3	2013	62.3	1996	87.3	2013	62.3	1996	87.3	2013	62.3	1996	87.3	2013	62.3	1996

(continued)

Table 10.3 (continued)

Station	Winter				Spring				Summer				Autumn			
	Highest		Lowest		Highest		Lowest		Highest		Lowest		Highest		Lowest	
	Value	Year	Value	Year	Value	Year	Value	Year	Value	Year	Value	Year	Value	Year	Value	Year
Kołobrzeg	86.7	1966	68.0	1976	86.7	1966	68.0	1976	86.7	1966	68.0	1976	86.7	1966	68.0	1976
Koszalin	89.4	2013	67.5	1976	89.4	2013	67.5	1976	89.4	2013	67.5	1976	89.4	2013	67.5	1976
Kraków Balice	86.3	1953	60.0	1990	86.3	1953	60.0	1990	86.3	1953	60.0	1990	86.3	1953	60.0	1990
Krosno[a]	88.1	1967	65.0	1990	88.1	1967	65.0	1990	88.1	1967	65.0	1990	88.1	1967	65.0	1990
Legnica[a]	88.7	2013	61.7	1954	88.7	2013	61.7	1954	88.7	2013	61.7	1954	88.7	2013	61.7	1954
Lesko[a]	84.7	1968	63.0	1976	84.7	1968	63.0	1976	84.7	1968	63.0	1976	84.7	1968	63.0	1976
Leszno[a]	87.9	2013	64.0	1984	87.9	2013	64.0	1984	87.9	2013	64.0	1984	87.9	2013	64.0	1984
Lublin	89.0	2013	65.7	1964	89.0	2013	65.7	1964	89.0	2013	65.7	1964	89.0	2013	65.7	1964
Łeba	88.7	2013	67.3	1954	88.7	2013	67.3	1954	88.7	2013	67.3	1954	88.7	2013	67.3	1954
Łódź	88.0	2013	66.3	1954	88.0	2013	66.3	1954	88.0	2013	66.3	1954	88.0	2013	66.3	1954
Mikołajki[a]	88.0	1988	65.0	1976	88.0	1988	65.0	1976	88.0	1988	65.0	1976	88.0	1988	65.0	1976
Mława[a]	87.0	1977	66.3	1976	87.0	1977	66.3	1976	87.0	1977	66.3	1976	87.0	1977	66.3	1976
Nowy Sącz[a]	85.3	2013	56.3	1990	76.7	1958, 1966	58.0	2000	74.3	1980	49.7	1992	76.3	1996	51.7	2011
Olsztyn[a]	88.3	2013	63.7	1976	75.3	1970	50.0	1953	72.2	1961	47.7	1992	84.2	1952	57.9	1951
Opole	88.0	2013	60.7	1954	75.0	2013	49.0	1953	71.8	1980	53.1	1973	84.6	1952	55.4	1982
Piła[a]	90.3	2013	68.3	1976	72.7	1983	54.3	2011	74.0	1980	50.0	1983	82.7	1978	57.3	2005

(continued)

Table 10.3 (continued)

Station	Winter				Spring				Summer				Autumn			
	Highest		Lowest		Highest		Lowest		Highest		Lowest		Highest		Lowest	
	Value	Year	Value	Year	Value	Year	Value	Year	Value	Year	Value	Year	Value	Year	Value	Year
Płock[a]	87.7	1953	65.7	1976	76.3	1970	54.3	1953	73.3	1980	46.3	1992	88.0	1952	57.3	1953
Poznań	88.7	2013	64.3	1954	73.7	1958	47.7	1953	74.7	1980	46.3	1983	83.3	1952	53.0	2005
Puławy	89.9	1977	63.3	1954	78.6	1958	47.1	1953	74.1	1977	46.8	1992	87.1	1952	50.6	1953
Racibórz[a]	85.3	2013	60.7	1964	73.7	1991	46.0	1953	72.3	1980	47.3	1983	78.7	1952	50.3	1982
Resko[a]	88.3	2013	67.3	1954	79.0	1970	54.3	1953,2007	76.3	1980	53.7	1983	85.1	1952	51.0	2005
Rzeszów[a]	87.0	1953	62.3	1976	75.0	1970	53.3	1953	71.0	1980	48.0	1992	78.0	1952	50.7	1982
Sandomierz[a]	87.0	1953	59.7	1976	71.7	2013	49.3	1953	73.8	1980	46.1	1971	80.0	1952	48.3	1982
Siedlce	88.3	2013	63.7	1976	77.0	1970	50.3	1953	77.7	1980	45.3	1992	86.0	1952	57.3	1953
Słubice[a]	88.7	2013	64.3	1954	77.0	2013	49.9	1953	72.0	1977	52.0	1983	82.9	1952	51.0	2005
Sulejów[a]	88.0	2013	62.0	1964	75.0	1970	57.4	1963	75.3	1980	51.3	1992	79.0	1996	55.0	1982
Suwałki	88.3	1967	65.0	1976	79.7	1966	51.3	1953	75.6	1980	45.0	1992	87.0	1952	60.7	2005
Szczecin	86.3	1953	65.3	1954	76.3	1970	48.3	1953	73.0	1987	48.0	1992	80.7	1952	56.3	2005
Śnieżka	88.0	1994	57.0	1964	86.0	1970	56.0	1953	83.3	1980	63.3	1983	90.3	1952	57.3	1959
Świnoujście	84.9	1953	61.5	1976	75.7	1970	46.1	1959	68.0	1987	43.3	2018	76.6	1976	53.6	2005
Tarnów[a]	88.2	1953	58.6	1990	73.5	1966	52.4	2000	71.9	1980	45.5	1992	79.8	1952	45.1	1982
Terespol[a]	88.7	2013	65.0	1976	75.7	1958	55.7	1986	76.7	1980	47.7	1992	83.7	2017	57.7	2011
Toruń	87.0	1953	67.0	1976	78.0	1970	51.3	1959	74.7	1980	49.3	1983	86.0	1952	55.3	2005
Ustka	87.7	2013	65.3	1954	78.3	1970	49.0	1990	67.3	1987	45.0	1992	82.0	1978	55.3	2005

(continued)

Table 10.3 (continued)

Station	Winter				Spring				Summer				Autumn			
	Highest		Lowest		Highest		Lowest		Highest		Lowest		Highest		Lowest	
	Value	Year	Value	Year	Value	Year	Value	Year	Value	Year	Value	Year	Value	Year	Value	Year
Warszawa	85.7	2013	65.3	1976	70.6	1962	47.9	1953	70.3	1980	44.7	1992	84.7	1952	52.7	1953
Wieluń[a]	89.0	2013	64.3	1954	76.0	1970,1977	47.0	1953	77.0	1980	49.7	1992	86.0	1952	52.3	1982
Włodawa	86.0	2013	66.0	1976	79.0	1970	46.3	1953	78.3	1980	45.7	1992	83.0	1952	56.3	2005
Wrocław	86.7	2013	60.7	1954	77.0	2013	48.0	1953	72.4	1980	49.7	1992	86.8	1952	51.5	1982
Zakopane	83.7	2018	55.0	1990	79.0	2017	55.7	1953	79.0	1980	55.0	1952	82.2	1952	53.3	2011
Zielona Góra	88.0	2013	63.7	1954	78.7	1983	49.9	1953	75.7	1980	53.3	1983	83.3	1952	52.3	2005
Poland	86.5	2013	66.6	1976	75.1	1970	50.4	1953	73.8	1980	50.5	1992	83.6	1952	57.2	1982

[a]Period shorter than 1951–2018 (see details in Chap. 3 *Data and Methods of investigation*)

remaining stations, the cloudiest sky in winter was observed in other years, i.e. in 1955, 1966 and 1967, 1970, 1988, 1994 and 2018. The highest total cloud cover at particular stations equalling 90.7% occurred in Gorzów Wlkp. in 2013.

The lowest total cloud cover in winter, 66.6%, in Poland occurred in 1976 (Table 10.3). At particular stations the years with the lowest cloudiness level in winter were: mentioned 1976 (23 stations), 1954 (17 stations) and 1990 (10 stations in the southern and southeastern part of the country). Furthermore, there were some additional years marked with very low levels of cloudiness: 1964 (minimum at five stations), 1984, 1993 and 1996. The clearest sky in winter occurred in 1964 at Kasprowy Wierch in the Tatra Mts; its value did not exceed 54% (53.7%). It should be underlined that the minima occurring in winter at stations located in mountains or mountain forelands were characterised by exceptionally low values, not exceeding 60% (e.g. Jabłonka and Zakopane—54.3 and 55%, respectively in 1990, Śnieżka Mt.—57% in 1964, Nowy Sącz—56.3% and Bielsko-Biała—58.3%, also in 1990, Jelenia Góra—58.7% in 1993). The minima recorded at the remaining stations did not drop below 60%, in the case of stations located in northern Poland these values were even close to 70%.

Both selected subperiods differ from each other in terms of sign of trend coefficients of cloudiness occurring in the country. During the first one, 1951–1984, decreases in the average seasonal cloudiness were common at stations all over Poland (Fig. 10.3e, f). The pace of decrease was higher than −1.5%/10y, and the value of the relative trend was even higher.

In the second analysed subperiod, 1985–2018, the trend completely reversed. At most stations, a strong, statistically significant increase in cloudiness was observed. The value of the trend increased towards the south, reaching 3%/10y at the Subcarpathian stations. The weakest observed changes were at the stations located along the western part of the Coastlands.

In spring, mean seasonal total cloud cover equals 65.7%, and cloudiness visibly decreases after the winter season, which is characterised by very high cloudiness levels (Fig. 10.5a). This change is the most apparent in the northern part of the country, where the cloudiness decreases from approximately 80% in winter to 65–67% in spring and even less in the coastal areas (Świnoujście and Hel—less than 60%, Ustka—61%, Łeba and Kołobrzeg—62%). In the Lakelands of northeastern Poland, which is usually mostly covered with a layer of clouds in winter in comparison with the rest of the country, clouds cover less than 65% of the sky in spring. Springs are also quite sunny in the Lowland area of central Poland. There the level of cloudiness is also less than 65% in spring. Both the Lowlands and Lakelands, located in the west, are characterised by the cloudiness reaching approximately 66–68%. In the Uplands, the total cloud cover is similar. In the Subcarpathian region, cloudiness drops to less than 65%, while in the mountains it is considerably higher. In the highest parts of the mountains—the Karkonosze Mts. and the Tatra Mts.—clouds cover more than 75% of the sky.

Spring variation of the cloudiness is also higher than in the winter (Fig. 10.5b). In the coastal areas and in the Lakelands of northeastern Poland, values of the coefficient of variation reach 9–10% and then slightly drop towards the south. The variability

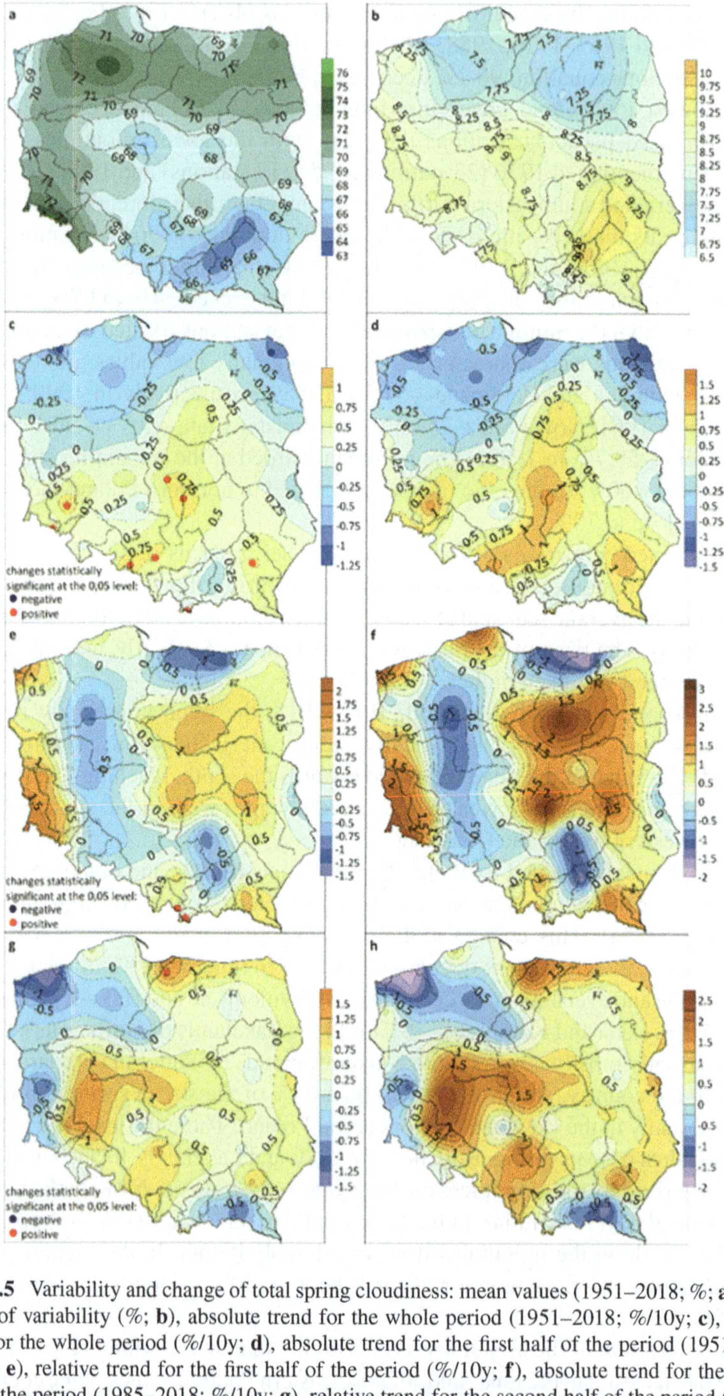

Fig. 10.5 Variability and change of total spring cloudiness: mean values (1951–2018; %; **a**), coefficient of variability (%; **b**), absolute trend for the whole period (1951–2018; %/10y; **c**), relative trend for the whole period (%/10y; **d**), absolute trend for the first half of the period (1951–1984; %/10y; **e**), relative trend for the first half of the period (%/10y; **f**), absolute trend for the second half of the period (1985–2018; %/10y; **g**), relative trend for the second half of the period (%/10y; **h**). Statistically significant (0.05) trends were shown in (**c**, **e**, **g**) as circles

is higher in the Lakelands and Lowlands of the western part of Poland and in the Upland areas in the east. In the submountainous areas and in the mountains, the cloud cover again varies slightly less.

By and large, there are not many stations in Poland which experienced statistically significant changes in spring cloudiness in the period 1951–2018 (Fig. 10.5c). In the Lowland areas of western Poland, the value of cloudiness rose; statistically significant values were recorded in Legnica and Kalisz. In the Subcarpathian region and in the Outer Western Carpathians (Beskidy Mts.), the cloudiness systematically decreased, which has been proved by the negatives values of the trend coefficient for stations in Tarnów and Nowy Sącz. In the remaining area of Poland—in the central and northern parts of the country—changes were insignificant, mostly negative and do not exceed 0.5%/10y. The strongest relative trend characterised the previously mentioned data series recorded in Nowy Sącz (approx. −1.5%) (Fig. 10.5d).

The beginning of the 1950s was particularly clear, especially the spring of 1953 when the absolute minimum cloudiness was recorded across the country (Fig. 10.6). After that, the value of total cloud cover gradually began increasing. From the 1960s to the very beginning of the 1980s, a clear positive anomaly in spring cloudiness was observed. At the turn of the 1980s and 1990s, the cloud coverage level did not change considerably. Then, in the north, a decrease in the average seasonal cloudiness was recorded. Subsequently, at the end of the last century, its value started to grow, and during the first decade of the twenty-first century, the average seasonal cloudiness dropped again. In the south, the turn of the centuries brought a negative anomaly of the element value, and, in the last few years of the research period, a slight increase in the cloud coverage was observed.

The cloudiest sky in spring in Poland, amounting to 75.1%, occurred in 1970 (Table 10.3). The minimum at particular stations was observed irregularly, however at many of them (23 stations) it was recorded in 1970. At nine stations, the minimum of spring cloudiness was observed in 1958, at another four in 1962 and at five in 1966. Additionally, the very low values of cloudiness were also recorded at the remaining stations in 1977, 1983, 1991, 2013 and 2017. The aforementioned minimum in spring cloudiness in Poland in 1953 was only 50.4% (Table 10.3). This absolute minimum was reflected in the course of series of 40 stations. The lowest cloudiness among all stations, equalling 44.8%, was then observed in Gdynia. At other stations the minimum cloudiness occurred in 1959, 1963 and 1974. In the Bieszczady Mts. (southeastern Poland), the lowest level of cloudiness was observed in 1982. At several individual stations, minima occurred at the turn of the century (at two stations in 2000) and in the twenty-first century (2007 and 2011).

Positive oscillation in the cloudiness of the 1960s resulted in the existence of strong positive trends throughout the country (Fig. 10.5e, f). More than ten stations recorded a statistically significant change in cloudiness in the period of 1951–1984, including the highest growths—approx. 3%/10y—recorded in the eastern Uplands. Although the observed trends in the south were stronger than in the north, there were a few stations in the Subcarpathian region at which slight decreases in the cloudiness were observed.

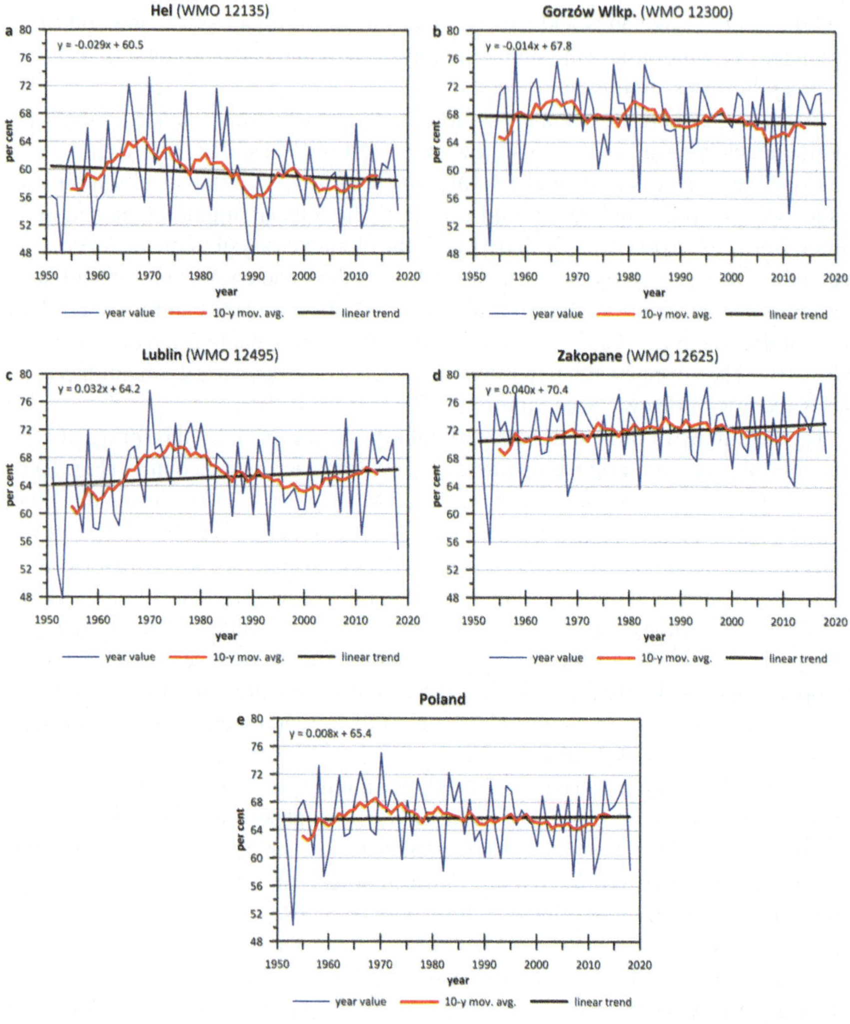

Fig. 10.6 Long-term variability of total spring cloudiness in the selected stations of: northern Poland (station Hel; 1951–2018; **a**), western Poland (station Gorzów Wlkp.; 1951–2018; **b**), eastern Poland (station Lublin; 1951–2018; **c**), southern Poland (station Zakopane; 1951–2018; **d**) and in Poland as a whole (the average, **e**). Mean seasonal value (*blue*), 10-year consecutive average (*red*) and line trend (*black*) were shown. The trend is not statistically significant (0.05) in (**a**–**d**)

In the subperiod of 1985–2018, the cloud coverage underwent changes in the opposite direction (Fig. 10.5g, h). In the northern and western regions, the cloudiness started to decrease and the station in Kętrzyn recorded statistically significant growth. A higher drop was observed in the whole Carpathian region along with the submountainous areas, and the decrease in cloudiness was even higher in Krosno than in the north. Some insignificant increases were observed in the central areas of

Lowlands. At the station located in Elbląg (northern Poland) the increase in cloudiness was extremely high during the analysed period, possibly due to the relocation of the station in the beginning of the second decade of the current century (the exposure of the station rose more than 100 m above sea level compared with the former location).

In comparison to spring, cloudiness in Poland continues to decrease in summer (Fig. 10.7a), even in the mountains where it has reduced by 2–3%. Less than 60% of the summer sky over the coastal areas is covered by clouds: 55% in Świnoujście, 56% in Hel and 58% in Gdynia and Ustka. There is a visible difference between the value of total summer cloud cover over the east and west parts of Poland; usually it is higher by 5% in the west. There is also a belt of low cloudiness across the Subcarpathian region and the Uplands of eastern Poland, along the Vistula River—the average cloudiness does not exceed 60% there; in Sandomierz it drops to 58%.

Summer is the season with the least average cloudiness, equalling 62.1%, and with the highest total cloudiness variability (Fig. 10.7b) when comparing to the other seasons. The coefficient of variation in the north exceeds 10%, which indicates great dynamics of cloud conditions. However, the value decreases towards the south reaching 6–7% in the submountainous areas and less than 6% at Zakopane station, located at the foot of the Tatra Mts. High values of the coefficient of variation were also observed in Uplands of the eastern areas of the country.

The only statistically significant changes in the summer cloudiness during the period of 1951–2018 were observed at a few stations in the south of Poland (Fig. 10.7c). An interesting situation can be observed in the mountains. In the Tatra Mts. and the Sudety Mts., cloudiness increased systematically, while at stations located in the lower parts of the mountains, in the Outer Western Carpathians (Beskidy Mts.), the cloudiness gradually decreased. However, when compared to the previously described season, in summer there are more regions where a positive tendency in cloudiness development over the long-term period of research can be observed. Taking the decrease of the average seasonal cloudiness into consideration, the relative trend is higher in summer than in spring; it exceeds 1% in the southern belt and in the northernmost edge of the country (Fig. 10.7d).

When analysing the whole long-term period 1951–2018, some significant year-to-year anomalies in total summer cloudiness are observed (Fig. 10.8). Most of the analysed data series are characterised by the existence of alternate positive and negative anomalies of cloudiness value. Total cloud cover increased at the turn of the 1950s and 1960s as well as at the turn of the 1970s and 1980s, at the end of the last century and the beginning of the present one. In 1980, clearly visible cloudiness maximums were observed at most of the stations throughout the country. However, it is noticeable that it is not observed for the data series gathered at the stations located in the north. In the case of coastal stations, the highest cloudiness was recorded in 1987 as well as in 1961 and 1963. A period of decreased cloudiness was observed in the whole country in the last decade of the twentieth century. This was the period when most stations recorded minimum cloudiness levels (1992). The former negative oscillation of cloudiness in northern Poland was observed at the turn of the 1960s and 1970s, while in the south it had been observed a few years earlier. The last several

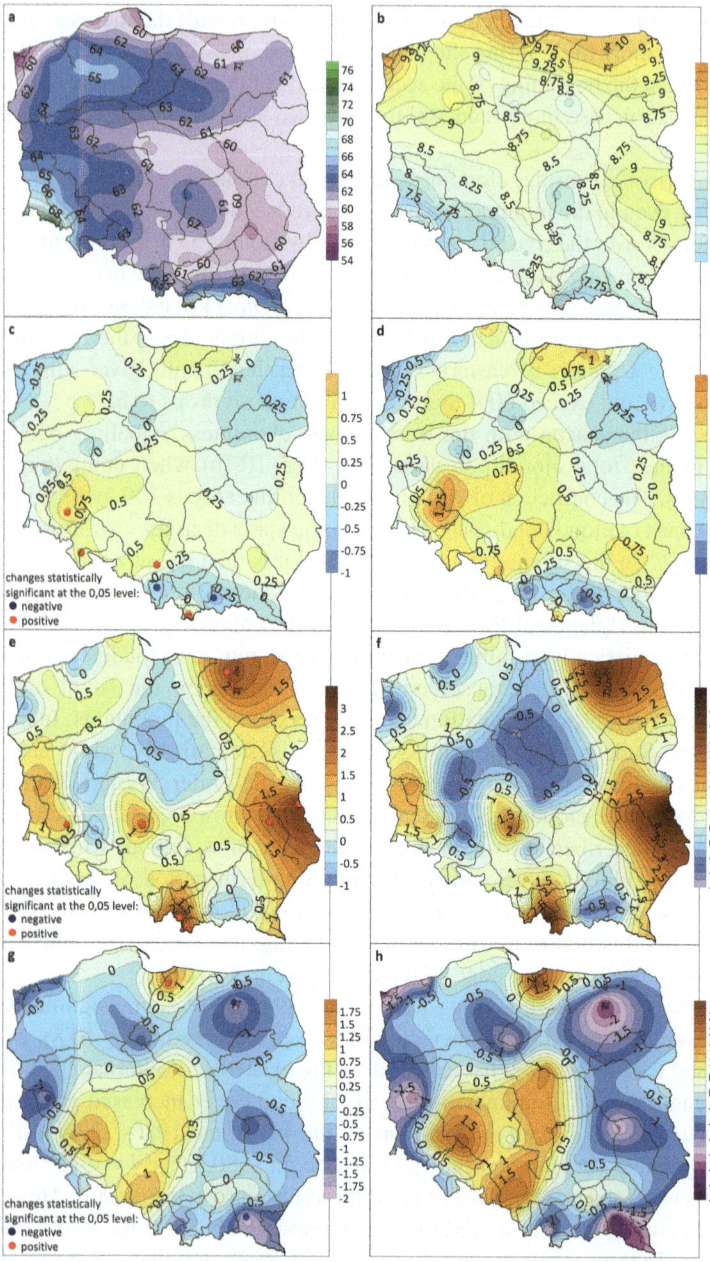

Fig. 10.7 Variability and change of summer total cloudiness: mean values (1951–2018; %; **a**), coefficient of variability (%; **b**), absolute trend for the whole period (1951–2018; %/10y; **c**), relative trend for the whole period (%/10y; **d**), absolute trend for the first half of the period (1951–1984; %/10y; **e**), relative trend for the first half of the period (%/10y; **f**), absolute trend for the second half of the period (1985–2018; %/10y; **g**), relative trend for the second half of the period (%/10y; **h**). Statistically significant (0.05) trends were shown in (**c**, **e**, **g**) as circles

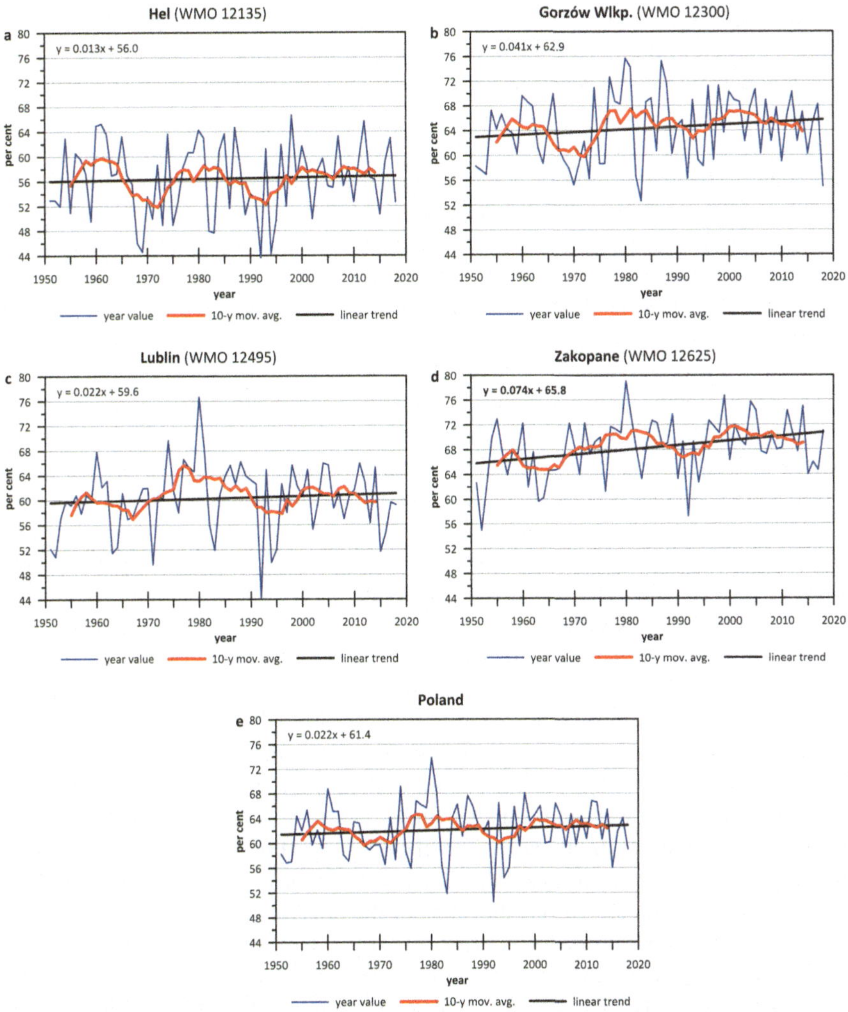

Fig. 10.8 Long-term variability of total summer cloudiness in the selected stations of: northern Poland (station Hel; 1951–2018; **a**), western Poland (station Gorzów Wlkp.; 1951–2018; **b**), eastern Poland (station Lublin; 1951–2018; **c**), southern Poland (station Zakopane; 1951–2018; **d**) and in Poland as a whole (the average, **e**). Mean seasonal value (*blue*), 10-year consecutive average (*red*) and line trend (*black*) were shown. The trend is statistically significant (0.05, in *bold*) in (**d**) and not significant in (**a**–**c**)

dozen years of the analysed period were characterised by a slight, yet continuous, drop in cloudiness.

The cloudiest sky in summer in Poland was observed in 1980, with a mean total cloud cover equal to 73.8% (Table 10.3). The indicated year was also marked with the occurrence of maximum value at 43 stations. However, particularly at coastal stations, the highest values did not occur in summer of said year. This is the region where maxima occurred irregularly, depending on the station. In western part of the Coastlands (from Świnoujście and Szczecin to Ustka) the highest level of cloudiness was observed in 1987. In Łeba and Koszalin, it was in 1961, and in Gdynia, located in western part of the Coastlands, in 1981. In Hel, the maximum occurred in 1998. At a few stations located in interior of the country, the highest values of cloudiness were also recorded irregularly—in 1961 (Olsztyn in Masurian Lakeland), 1974 (Kłodzko in the Sudety Mts.), 1977 (Słubice am Oder in the western part of Lakelands and Puławy in the Masovian Lowland), 1982 (Legnica in the Silesian Lowland) and 2012 (Kętrzyn in the Masurian Lakeland). The absolute maximum cloudiness in summer in Poland occurred on Kasprowy Wierch Mt. in 1980 and equalled 84%, at Śnieżka Mt. it was only slightly smaller and amounted to 83.3%.

The least cloudy sky in summer in Poland occurred in 1992, with the value amounting to 50.5% (Table 10.3). This absolute minimum was marked also in the course of cloudiness series of 30 stations. In the Lakelands and Lowlands of western Poland and in the Sudety Mts., the lowest level of cloudiness in summer was recorded in 1983. In addition, there are a few stations where the minimum occurred at another time, e.g. in Legnica and in the Tatra Mts. in 1952, in Łeba in 1959, in Gdynia in 1969, in Krosno (Outer Western Carpathians) and Sandomierz (Northern Subcarpathians) in 1971, in Opole (Silesian Lowland) in 1973, in Lesko (Outer Western Carpathians) in 2015 and in Kołobrzeg and Świnoujście in 2018. Exceptionally low total cloud cover of only 42.3% occurred in Cracow in the summer of 1992, and in Gdynia in 1969, the value of mean summer cloudiness was only higher by 0.2%.

During both subperiods, only a few stations recorded statistically significant changes of cloudiness. In the case of the first subperiod, beginning in 1951 and ending in 1984, there were more such stations (nearly ten) than in the latter one and all the data regard cases of increases in total cloudiness (Fig. 10.7e, f). Those stations are most often located in eastern Poland and in the highest parts of the Beskidy Mts. and Tatra Mts. For those stations the value of the relative trend reached 5%. On the whole, some drops in cloudiness were recorded only in the central part of the country.

The decrease in cloudiness dominated in the period from 1985 to 2018 (Fig. 10.7g, h). Yet, there are only four cases of statistically significant cloudiness changes—at three stations (Legnica, Mikołajki and Krosno) the cloud cover showed a tendency to decline, while in Elbląg the value of total cloud cover increased—this is the second time this station has stood out in comparison to other stations located in the same region. However, it is worth mentioning that some weak positive tendencies were also observed in Hel and Łeba in the coastal region. All in all, the values of the relative trends usually do not exceed ±1.5%.

The mean seasonal total cloud cover in autumn amounts to 69.2%, and the cloud cover over Poland visibly transforms (Fig. 10.9a) comparing to spring and summer. Clouds cover more than 70% of the sky over the northern part of the country, with culmination reaching 74% over the Lakelands (Chojnice station). However, the highest level of autumn cloudiness is observed in the high-mountain ranges: the Karkonosze Mts. and Tatra Mts., reaching approximately 76%. In general, the central areas of the country are not particularly cloudy as of yet. Within the region as a whole,

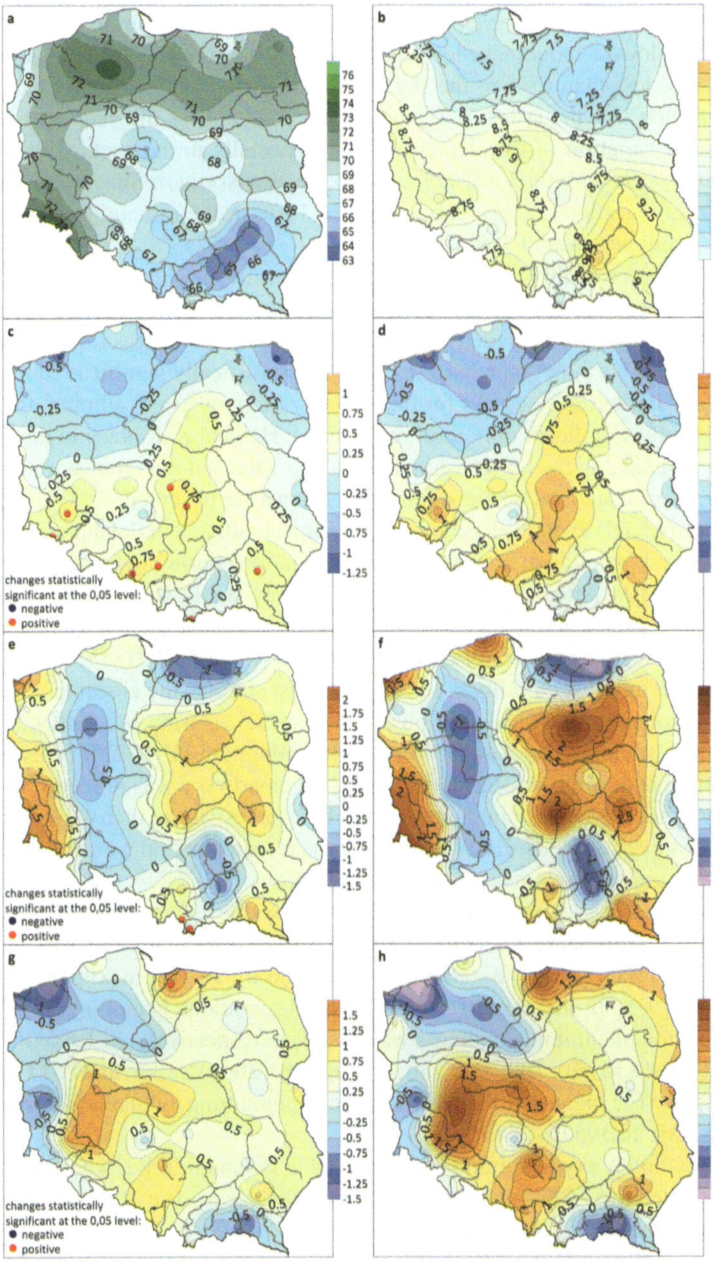

Fig. 10.9 Variability and change of autumn total cloudiness: mean values (1951–2018; %; **a**), coefficient of variability (%; **b**), absolute trend for the whole period (1951–2018; %/10y; **c**), relative trend for the whole period (%/10y; **d**), absolute trend for the first half of the period (1951–1984; %/10y; **e**), relative trend for the first half of the period (%/10y; **f**), absolute trend for the second half of the period (1985–2018; %/10y; **g**), relative trend for the second half of the period (%/10y; **h**). Statistically significant (0.05) trends were shown in (**c**, **e**, **g**) as circles

an increase in cloudiness is observed towards the west, just as during the rest of the seasons. The lowest cloudiness level in autumn is observed in the Subcarpathian region, where less than 65% of the sky is covered with clouds. Transformations of cloud cover towards the structure typical for the cold half of the year also include a decrease in variation of the element, visible especially in the north (Fig. 10.9b). Only in the region with the lowest cloudiness, i.e. the Northern Subcarpathians, the coefficient of variation reaches approximately 10%, in the remaining areas of the country it oscillates around 7–9%.

The country is visibly divided into areas opposite in signs of the long-term tendency of total cloudiness during the period from 1951 to 2018. This division is more or less latitudinal (Fig. 10.9c). Generally, total cloud cover systematically decreased in the north but increased in the south; statistically significant data series were gathered at ten stations. As for the decline of cloudiness in autumn, the trend was statistically significant only for three stations: Hel and Kołobrzeg in the coastal region and Suwałki in the northeastern Lakelands. Values of the relative trend rarely reach 1%, even at stations where significant changes are observed.

The positive and negative phases of total autumn cloudiness in Poland were consistently observed in series of almost all analysed stations during the period of 1951–2018 (Fig. 10.10). After two decades of a relatively stable year-to-year values of autumn cloud cover, a period of increased cloudiness started in the 1970s. However, it should be emphasised that there were three consecutive years (1951–1953) with alternate very high and low cloudiness seasons (at most stations the absolute maximum of average autumn cloudiness was observed in 1952). During the 1980s, a negative anomaly of cloudiness was recorded with a minimum in 1982. It is also worth mentioning that at some coastal stations the least cloudy sky was observed in 1988. Generally, in the 1990s the cloudiness was higher than usual and then during the next decade it dropped again. The last few years of the research period were characterised by relatively high year-to-year cloudiness variability.

As mentioned above, the highest autumn cloudiness level was observed in Poland in 1952. Its value amounted to 83.6%. That year the absolute maximum was recorded at a majority of stations. At the stations found along the coast, the maximum typically occurred in 1978. In the case of stations with observations beginning after 1952, the culmination of autumn cloudiness was observed, respectively: in 1996 (Sulejów, Nowy Sącz and Krosno), 1998 (Leszno and Legnica), 2002 (Lesko) and 2017 (Mława and Terespol). The highest cloudiness in autumn, amounting to 90.3%, was recorded on Śnieżka Mt. in 1952.

The least cloudy sky in autumn was observed in Poland in 1982, when total cloud cover amounted to 57.2%. At particular stations, the occurrence of minimum values was differentiated. Most frequently it was observed in 1982 (17 stations, mostly in central and southern Poland) and 2005 (16 stations in northern and western Poland). There was also a number of stations where the least cloudy sky in the described season was observed in 1950s (1951, 1953 and 1959, summarising 13 stations) and in 2005 (six stations in southern and southeastern Poland). The remaining cases concerned the singular stations with minima in other years. The least cloudy sky was in 1982 in Tarnów in the Northern Carpathians, where it was recorded as 45.1%.

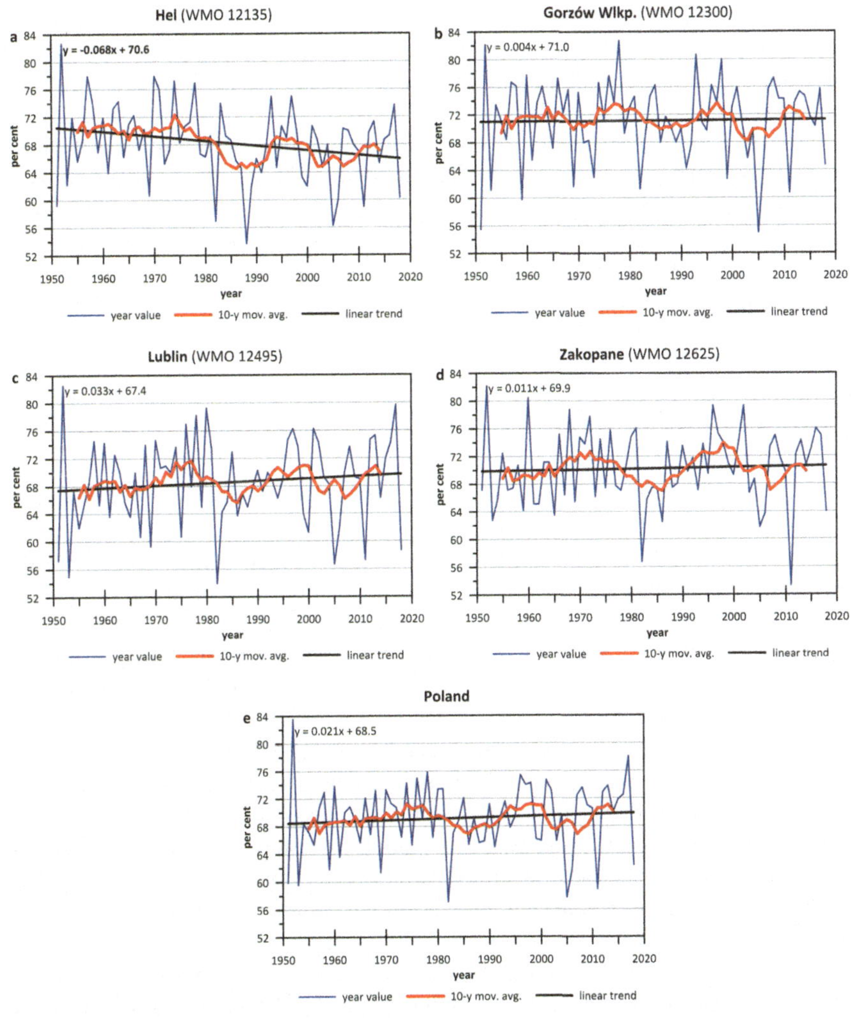

Fig. 10.10 Long-term variability of total autumn cloudiness in the selected stations of: northern Poland (station Hel; 1951–2018; **a**), western Poland (station Gorzów Wlkp.; 1951–2018; **b**), eastern Poland (station Lublin; 1951–2018; **c**), southern Poland (station Zakopane; 1951–2018; **d**) and in Poland as a whole (the average, **e**). Mean seasonal value (*blue*), 10-year consecutive average (*red*) and line trend (*black*) were shown. The trend is statistically significant (0.05, in *bold*) in (**a**) and not significant in (**b**–**d**)

Overall, there were very few significant changes in cloudiness during both subperiods. In the years 1951–1984, cloudiness slightly increased on the western edges of the country as well as in the central and northeastern parts of Poland (Fig. 10.9e, f). The region with the most intensively developing cloud cover was the Carpathians. however, statistically significant variability of the element was observed only at three stations located in this region: Jabłonka (Beskidy Mts.), Zakopane and Kasprowy Wierch Mt. in the Tatra Mts. Some insignificant declines in cloudiness were observed at less than half of the stations. The relative trend reached approximately 2% at a few stations (in western Poland).

After 1985, the structure of long-term autumn cloudiness variability significantly changed (Fig. 10.9g, h). At a greater part of the stations, increases in cloudiness were observed, but statistically significant data series come from only one station located in Elbląg. In western and central Poland, a decrease in cloudiness is observed as well as in the Karpaty Mts., while in the northern and eastern parts of the country values of the element increased.

Clear Days

Certainly the sky over Poland is more often cloudy than cloudless, as confirmed by data on the number of clear and cloudy days.

In Poland, clear days were recorded at least several dozen times a year during the period of 1951–2018 (Fig. 10.11a). On average, the number of clear days in Poland amounts to 32 days. The highest number of clear days occurred in the Subcarpathian region: in Tarnów there were 43 such days a year and a slightly lower result was noted in Sandomierz—nearly 40. Both the Polish Uplands and the Carpathians Mts. are also regions where the sky without clouds or with low levels of cloudiness can often be observed: 32–34 clear days a year, on average. More than 30 such days are recorded in the western part of the Lakelands and Coastlands (Świnoujście and Słubice: 33–34 clear days annually) as well as in the central part of the Lowlands (Wieluń—35 cases, Opole—32 cases). Generally, western Poland, except for the aforementioned regions, is characterised by a lower average annual number of clear days (less than 30 cases a year) than the rest of the country. The Karkonosze Mts. is the region with the lowest annual number of clear days—25, on average.

On the whole, the sky over Poland usually becomes clear in spring. However, this is true especially for areas by the Baltic Sea and at lower altitudes, in general. In the Uplands and in the upper part of Odra River basin, the maximum number of clear days is usually recorded in autumn. In the Subcarpathians, the number of clear spring days equals the number of such days observed in autumn. In the high-mountain regions, the highest number of clear days is recorded in winter.

In spring, the number of clear days amounts to 9, on average. Clear days are usually the most frequently noted in the coastal region (from 10 days in Kołobrzeg to 12 days in Hel and Świnoujście). The eastern part of the Lakelands is also clear—on average 10 such days a year are recorded there in spring. From 9 to 10 clear days occur in the Uplands and Lowlands of eastern Poland while in the Subcarpathians—11. In the low-mountain regions (the Beskidy and Sudety Mts.; excluding the Karkonosze Mts.) from 6 to 8 clear days are noted in spring, while in the highest mountains (the Tatra Mts. and above-mentioned Karkonosze Mts) only 5.

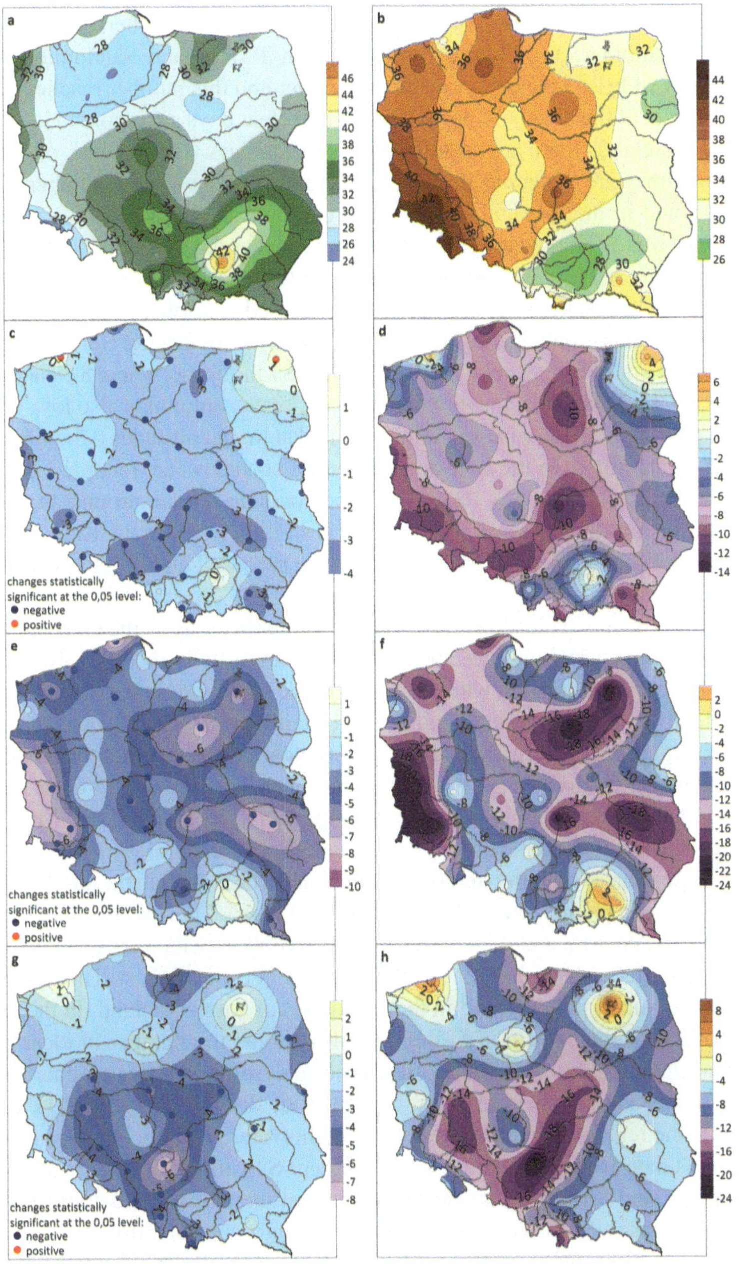

Fig. 10.11 Variability and change of annual number of clear days: mean values (1951–2018; %; **a**), coefficient of variability (%; **b**), absolute trend for the whole period (1951–2018; %/10y; **c**), relative trend for the whole period (%/10y; **d**), absolute trend for the first half of the period (1951–1984; %/10y; **e**), relative trend for the first half of the period (%/10y; **f**), absolute trend for the second half of the period (1985–2018; %/10y; **g**), relative trend for the second half of the period (%/10y; **h**). Statistically significant (0.05) trends were shown in (**c**, **e**, **g**) as circles

The Subcarpathian region is the one where the largest number of clear days is recorded in summer: 11–12 cases on average in Tarnów and Sandomierz. In the coastal areas, at least 9 clear days occur in summer and this number reaches up to 12 in Hel. In most of the country, approximately 8–9 clear days are recorded in summer, yet in the Lakelands there are usually only 6–8 such days (for instance, only 6 clear days in summer are observed in Piła and Chojnice). The lowest number of clear days are noted in the mountains (approx. 5 cases in the Tatra Mts., 3 cases in the Karkonosze Mts.). The mean seasonal number of clear days amounts to 8 days.

The numbers of clear days in autumn and winter amount to 8 and 6, respectively. In autumn, the lowest number of clear days is recorded in northern Poland, especially in the coastal areas—only 6 cases by the Gdańsk Bay (Gdynia and Hel), 6–7 cases in the remaining coastal areas, with the exception of Świnoujście, where 8 clear days are recorded. In the Lakelands, the number of clear days in autumn is also low, no more than 7–8 days, and, in the case of stations located at higher altitudes (Chojnice), only 6 clear days are noted. In the Lowlands, at least 8 such days are recorded in autumn and in the Uplands at least 9–10. In the Subcarpathian region, around 11–12 clear days occur in autumn. However, this number is slightly lower in the Karpaty Mts. In the Sudety Mts., only 7–8 autumn clear days are recorded and 6 in the Karkonosze Mts.

Furthermore, only 3 clear days are noted in winter in areas by the Gulf of Gdańsk. In the remaining coastal regions, about 4–6 cases are usually recorded. In the Lakelands, 5–6 clear days occur in winter. Further to the south, this number increases—in the Lowlands of western Poland and in the Uplands from 6 and 7 cases are noted. In the Lowlands of eastern Poland, only 5–6 clear days are noted in winter. The Carpathians can be divided into several parts: in the eastern one, 7 clear days are recorded. There are 9 cases in the western part and 12 in the Tatra Mts. In the Karkonosze Mts., 9 winter clear days are usually recorded.

Both large year-to-year fluctuations and long-term decreases in the number of clear days are common phenomena in Poland (Fig. 10.11b, c). The coefficient of variation is between 30 and 40% and it increases from east to west, which is especially visible in the Sudety Mts., where the value of this coefficient is the highest. This value is the lowest in the Subcarpathians and in the Lakelands of eastern Poland. In the country as a whole, with the exception of three small regions, the number of clear days is decreasing considerably. A vast majority of the analysed data series are characterised by statistically significant decreases. The pace of decline is 3 days/10y, and the Uplands is the region with the highest value of this rate. The only exceptions include data series taken from the stations in Kołobrzeg, Suwałki and Tarnów. In the case of Kołobrzeg and Suwałki, statistically significant increases in the number of clear days can be observed. Taking the relative scale into consideration, the reductions reach up to a dozen or so percentage points (Fig. 10.11d). The frequency of clear days decreased at majority of the analysed stations during the earlier of the two established subperiods (Fig. 10.11e, f). Statistically significant drops in the variable were a distinctive feature of the data series collected from the stations located in the central and northern parts of the country. The relative drops in the number of clear days were also considerable, reaching more than 20%. During the subperiod of 1985–2018, statistically significant decreases in the number of clear days were recorded for the data series obtained from the stations located in the central and southern parts of Poland (Fig. 10.11g, h). However, in this case, the relative scale of the decreases is smaller than the one for the earlier subperiod.

The largest annual number of clear days in the period 1951–2018 occurred in Poland, on average, in 1953, as an effect of a spring with an exceptionally low level of cloudiness (particularly in the south of the country) as it was mentioned in preceding Section Total Cloudiness. It is worth noting that summer and autumn 1953 were also among seasons with low cloudiness levels. Other years characterised by great number of clear days were 1964 and 1982, in northern Poland also 1959. The absolute maximum number of clear days was 89 and occurred in 1964 in Sulejów in the central part of Poland. In the same year, 79 clear days were recorded in Sandomierz, and in 1959, another 79 clear days occurred in Świnoujście.

The fewest clear days in Poland were recorded, on average, in 1998. It was the smallest value of the variable at 14 stations. Additionally, at particular stations, minima also occurred in 2001, 1977 and 2008 at 9 stations, 5 stations and another 5 stations, respectively. The smallest annual number of clear days came to only 6 cases at Śnieżka Mt. in 1998 and in Resko (Lakelands in western Poland) in 1978.

In the scale of seasons, the reduction in the number of clear days is strongly visible in spring and summer. The spring maxima were most often recorded at the analysed stations in 1953, 1959, 1982, 2007 and 2018. The most frequently clear days in spring occurred in Świnoujście in 1959 (31 cases) and Włodawa and Puławy in eastern Poland in 1953 (29 cases). Clear days were the least frequent in 1970 and 1983 (at five stations there were no clear days at all in this year). In the case of summer, the following years were especially clear: 1959, 1964, 1982–1983 (particularly 1983 as a summer season with the clearest sky in the analysed period 1951–2018) and 1992. In the Coastlands, clear days were also frequent in 1969. In the Uplands in Częstochowa, in the summer of 1983, there were 31 clear days, which is the absolute maximum value of the considered variable. Furthermore, in Puławy there were 30 such days noted in the same season, and 28 clear days occurred in Hel in summer of 1968. It is worth noting that at the same station 27 clear days were recorded also in 1959 and 1969, which underlines the exceptional position of Hel as the station with clearest sky in the summer in Poland, most probably due to its location—at peninsula, in the indirect vicinity of deep sea basin of Gulf of Gdańsk, which effectively reduces the likelihood of development of clouds in summer. At more than half the stations, at least one summer season with a complete lack of clear days occurred in the period 1951–2018. Among such seasons were the summers of 1980 and 2014, and in northern Poland in 1998. At high-mountain stations: Kasprowy Wierch Mt. and Śnieżka Mt., the number of such seasons reached 14 and 9 cases, respectively.

The number of clear days also decreases in autumn, yet the trend rate is lower. This was mainly caused by the high frequency of autumn seasons with frequent clear days in the beginning of the research period—in the 1950s (1951, 1953 and 1959) and in 1961. What is more, a high frequency of clear days was noticeable in Poland in 1982 and 2005–2006. The maximum numbers of clear days noted at particular stations increases from 16 to 22 cases (in 1951 in Gdynia and in 2005 in Ustka) in the north to 25 cases or more in the south (27 cases in 1959 in Kłodzko, 26 cases in Zakopane and at Kasprowy Wierch Mt., both in 1962). Again, at all 30 stations, at least one season with the lack of clear days occurred. This situation was most frequently recorded in autumn 1997, when at 12 stations no clear days occurred. In winter, the number of clear days remains relatively stable throughout the research period. However, a strong positive anomaly was observed in the 1960s and 1970s. The high numbers of clear days occurred, respectively in 1964 and 1969 and in 1973 and 1976. Other positive, but slightly weaker anomalies of the analysed variable were noted in 1993 and 2003. The highest maxima of the number of clear days at particular stations were recorded at high-mountain stations—at Kasprowy Wierch Mt. and Śnieżka Mt. In the Tatra Mts. in 1964, 30 clear days occurred during winter and in the Karkonosze Mts. 26 clear days were noted in 1993. The lowest number of clear days occurred in the winters of 1966 and 2013. In 1966, at 17 stations, there were no clear days recorded. Some strictly local minima also occurred. In 1968, the central area of the country experienced very small number of clear days, whereas in 2007 such situation occurred in the Lakelands of western part of Poland.

The presented long-term course of seasonal data series is partially reflected in the course of annual values (Fig. 10.12). In the 1950s, there were numerous clear days in Poland, and their annual number decreased over the next decade. Despite the aforementioned significant increase in the frequency of winter clear days in the 1970s, a strong negative anomaly in the variable was a distinctive feature of this decade across the country. Another visible increase in the annual number of clear days took place in the late 1980s and early 1990s. Subsequently, the number of clear days started to decrease systematically.

Cloudy Days

All in all, the mean number of cloudy days in Poland amounts to 157 days. Spatial diversity of the annual number of cloudy days recorded at selected Polish stations during the period of 1951–2018 was considerable (Fig. 10.13a). It can generally be stated that the structure of the variable changes latitudinally. In the north, especially in the coastal regions, the lowest values of this variable are usually recorded. In Świnoujście, only 110 cloudy days occur, which is the lowest value recorded in Poland. At the remaining coastal stations, the frequency of cloudy days varies from approx. 130 days in Hel and 140 days in Ustka and Łeba to approx. 160 cloudy days in Kołobrzeg. In the Lakelands, the number of cloudy days is higher, varying usually from 155 to 165, with the culmination in the areas located at the highest altitude: Chojnice—173 days and Suwałki—166 days. In the Lowlands of central and eastern Poland, and in the areas of the Lakelands located at a lower altitude and at a greater distance from the Baltic Sea coast, the number of cloudy days is lower—150 cases or less. Moving south, the number of cloudy days increases again. In the Lowlands of western Poland and in the Uplands, there are typically 160 cloudy days a year or more. In the Subcarpathian region, the number of cloudy days is much

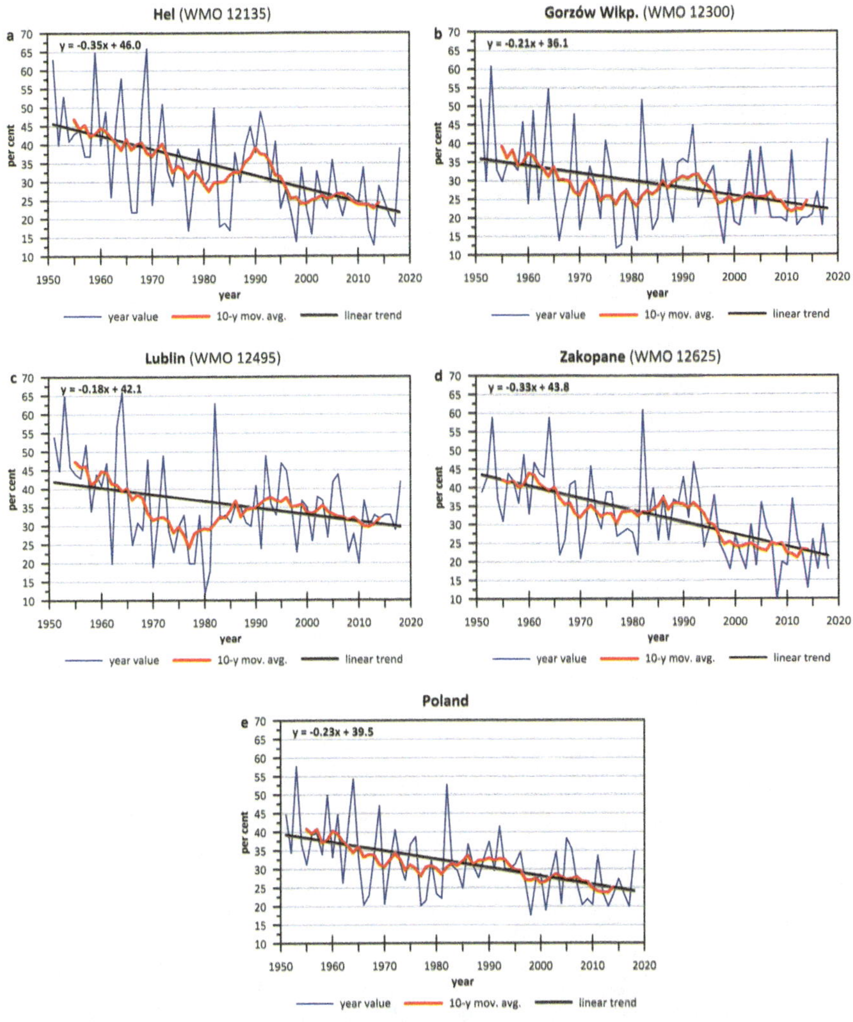

Fig. 10.12 Long-term variability of the annual number of clear days in selected stations of: northern Poland (station Hel; 1951–2018; **a**), western Poland (station Gorzów Wlkp.; 1951–2018; **b**), eastern Poland (station Lublin; 1951–2018; **c**), southern Poland (station Zakopane; 1951–2018; **d**) and in Poland as a whole (the average, **e**). Mean seasonal value (*blue*), 10-year consecutive average (*red*) and line trend (*black*) were shown. The trend is statistically significant (0.05, in *bold*) in all cases

lower, reaching 140–145 cases. In the mountains, the variable value increases again by a dozen cases or so. In the highest parts of the mountains, the number of cloudy days reaches nearly 200 (Kasprowy Wierch—192 days) or even more (Śnieżka—207 days). The variability in the number of cloudy days in Poland during the period 1951–2018 is significantly smaller than the fluctuations in the number of clear days (Fig. 10.13b). The coefficient of variation value is less than 10% in eastern Poland

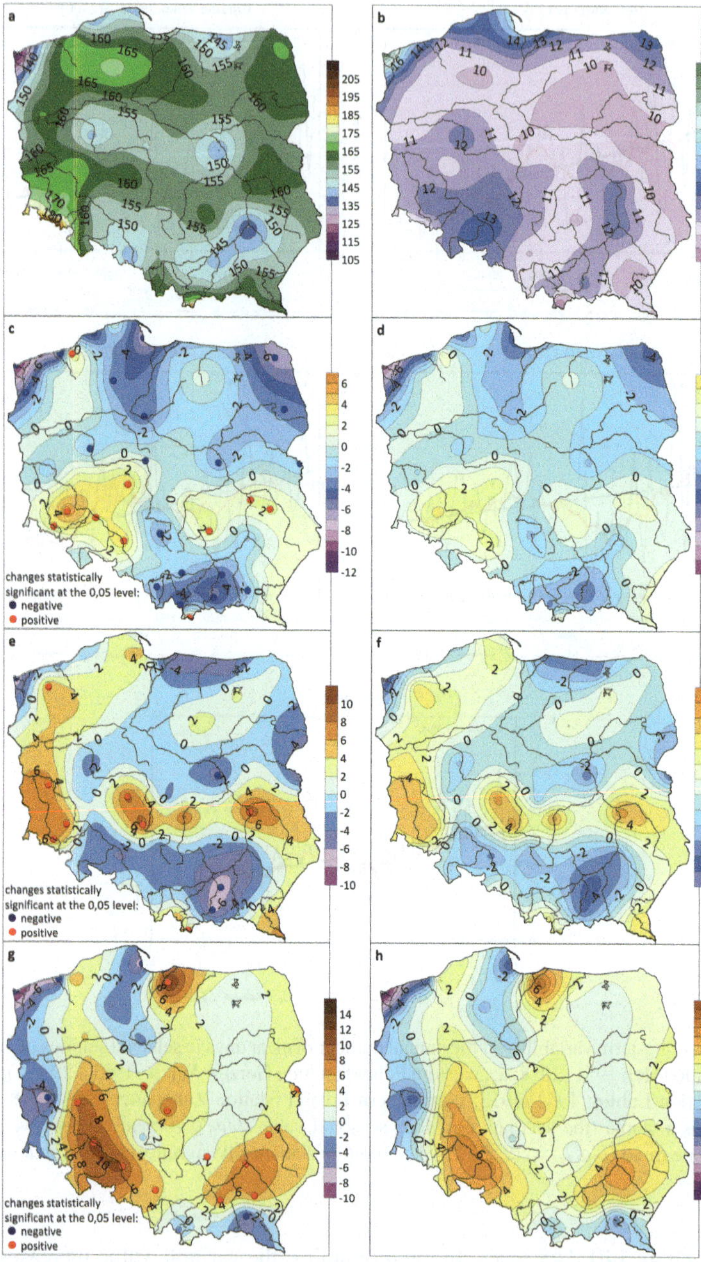

Fig. 10.13 Variability and change of annual number of overcast days: mean values (1951–2018; %; **a**), coefficient of variability (%; **b**), absolute trend for the whole period (1951–2018; %/10y; **c**), relative trend for the whole period (%/10y; **d**), absolute trend for the first half of the period (1951–1984; %/10y; **e**), relative trend for the first half of the period (%/10y; **f**), absolute trend for the second half of the period (1985–2018; %/10y; **g**), relative trend for the second half of the period (%/10y; **h**). Statistically significant (0.05) trends were shown in (**c**, **e**, **g**) as circles

and almost 20% in Świnoujście. The rate for Świnoujście is exceptionally high as a typical range of values for this variable in Poland is between 10 and 13%.

The seasonal frequency of cloudy days follows a simple pattern: summer is the season with the lowest number of cloudy days (28 cases), then spring (36 cases), autumn (41 cases) and winter when the number is the highest (52 cases).

In summer, the number of cloudy days varies from almost 20 in Świnoujście and Hel, 21 in Ustka and Łeba to more than 30 days in the lower mountains, e.g. the Beskidy Mts. Generally, the number of cloudy days is higher in the western parts of the country than in the eastern regions. In the highest mountains, there are approx. 47 cloudy days observed. Kasprowy Wierch is the only station where there are more cloudy days in summer than in autumn.

While in summer the lowest frequency of cloudy days is observed (with the one already-mentioned exception), an interesting relation has been discovered between the number of cloudy days in spring and autumn.

In the north, especially in the coastal areas, there are far more cloudy days in autumn than in spring—at least 20% (30–35 cases in spring and 8–10 cases more in autumn). The greater the distance from the sea, the lower the disproportion: the number of spring cloudy days starts to grow, while the number of autumn cloudy days remains unchanged or only slightly increases. In the Lowlands of western Poland and in the Uplands, the number of spring cloudy days (35–40 cases) is only slightly lower than the number of cloudy days observed in autumn (38–42). Fewer cloudy days, both in spring and autumn, are observed in the Subcarpathians: 30–35 and 35–37, respectively. In the Carpathians and Sudety Mts. almost the same numbers of spring and autumn cloudy days are recorded by all stations (38–41 in the Carpathians; 38–42 in the Sudety Mts.); except for Śnieżka—52 cloudy days in spring as well as in autumn and Kasprowy Wierch—52 cloudy spring days and 45 cloudy autumn days. At several stations located in the Carpathians the number of cloudy days in spring is larger than the number in autumn. Thus, in autumn the Karkonosze Mts. are the region with the largest number of cloudy days in Poland.

By and large, in winter the number of cloudy days in Poland is higher than in the remaining seasons of the year. The lowest number of cloudy days (less than 50) is recorded in the western part of the Coastlands and at some selected stations located in western Poland in the Subcarpathians and the Carpathians. The highest number of cloudy winter days is observed in the Lakelands: from 55 to almost 60 cases.

Similarly to the case of clear days, the annual number of cloudy days in Poland during the period of 1951–2018 decreased at majority of the stations (Fig. 10.13c). The decrease is especially visible in the north: in Suwałki the pace of the decrease exceeds 7 cases/10y; in Świnoujście it is even faster—12 cases/10y. The drops are also observed in the Subcarpathians and in the lower ranges of the Carpathians (the Beskidy Mts.) and the fastest pace of decrease is observed in Nowy Sącz—5 cases/10y. Yet, in many analysed stations located within this region, the negative trend is statistically significant. However, there are two regions in Poland where considerable increases are observed. The first one is the Lowlands of western Poland and the Sudety Mts. and the second is the central and eastern parts of the Uplands. The pace of increase reaches 4 cases/10y in Legnica. The number of cloudy days also increases in the Tatra Mts. The relative decreases or increases in the number of cloudy days range from −10 to +3% (Fig. 10.13d).

There were a few years in the period 1951–2018 with very high annual numbers of cloudy days. Its temporal distribution is somewhat interesting. First of all, a high frequency of such cases was characteristic for the first three decades of the analysed period (with maxima in years: 1952, 1958, 1962, 1966—the highest value in the whole research period, 1970, 1977 and 1980—the second highest one). During the next three decades, negative anomalies of the considered variable were recorded, which means the occurrence of potentially more sunny years with relatively low total cloud cover: 1982, 1989, 1992, 2003, 2006 and 2011. The minimum number of cloudy days occurred in 1982. In the last decade of the analysed period, again years with large amount of cloudy days occurred. The long-term course of the number of days without sun (Chap. 6) confirms that 1980 was characterised by a very high number of days without sun and, on the contrary, during 1982 the fewest number of such days was noted.

The greatest numbers of cloudy days were noted at mountain stations: 243 days at Śnieżka Mt. in 2016 and 223 days at Kasprowy Wierch Mt. in 2003. The smallest number of the considered days, equalling only 49, was recorded in Świnoujście in 2003.

In winter, the drops in the number of cloudy days are observed only in the north, yet some disparities were observed within the regions. The winter seasons of 1953, 1966, 2010, 2013 and 2018 were the ones with an exceptionally large number of cloudy days. Particularly frequent cloudy days were in 2013, when, at greater part of analysed stations, the maximum number of cases was recorded. In Leszno (Lowlands in western areas of Poland) there was 76 cloudy days noted then. In Białystok and Suwałki, 74 and 73 cloudy days, respectively occurred in 1959—in north and north-eastern Poland, an increase in winter cloudiness was observed that year. In 1954, 1976 and 1984, the number of cloudy days was extremely low. Furthermore, between 1990 and 2000, a considerable decrease in the number of cloudy days occurred in Poland with minima in the years: 1990–1991, 1993, 1997 and 2000. The fewest cloudy days (22 cases) were recorded in Świnoujście in 2000. Overall, spring is the season when the number of cloudy days usually considerably decreases in Poland and there are only a few stations that do not match this pattern. Thus, the spring seasons exceptionally abundant in cloudy days in spring occurred in the first two decades of the research period: 1958, 1962, 1970 and 1972. A strong maximum was recorded in the Masurian Lakeland in the northeastern part of Poland in 1966, and the largest amount of cloudy days was noted at mountain stations. At both locations, 71 such days occurred—at Śnieżka Mt. in 1958 and at Kasprowy Wierch Mt. in 2017. The lowest minimum occurred in 1953 and the number of cloudy days noted at more than 20 stations did not decrease in the subsequent years. However, it is worth noting that cloudy days were not frequent after the beginning of the current century. Minima occurred in 2007, 2011 and 2018, which means that the spring seasons of the twenty-first century have been rather sunny. Among the considered stations, the lowest values of the variable are characteristic at coastal stations. The cold water of the Baltic Sea in the spring effectively limits the development of clouds in this region. The minima recorded at coastal stations did not exceed 20 cases in spring: seven cloudy days in Świnoujście in 2011, 13 such days in Hel in 1990 and 14 days in Łeba, also in 1990. In Ustka in 1974, only 15 cloudy days occurred. In summer, the number of cloudy days decreases in Poland. The coast and Lakelands are among the regions where the drops are considerable. Thus, during the first half of the research period, the high frequency of the seasons with a large number of cloudy days occurred in Poland: 1960–1961, 1974, 1980–1981 and 1985. In 1980, the lowest number of cloudy days occurred at 34 stations. At Kasprowy Wierch Mt., 74 cloudy days were then recorded. However, in the summer of 1974 at Śnieżka Mt., an even larger amount of cloudy days was noted—79 cases. Since the 1980s, the negative anomalies of the variable frequently occurred: 1982–1983, 1992, 1994, 2003 and 2015. The summer of 1992 was characterised by a particularly low number of cloudy days at half the stations. At stations located in the Coastlands and Lakelands of eastern Poland, less than 10 cloudy days were then noted. In 2003 in Świnoujście, only four such days occurred. In autumn, the number of cloudy days decreases only in northern Poland. The decade of 1950s was characterised by a great multi-year variability in the number of cloudy days in Poland. In 1951, 1953 and 1959 a very low number of cloudy days occurred in Poland, whereas in 1952 and 1958 this number rose considerably. Absolute minima and maxima occurred year by year—in 1951 (minimum) and in 1952 (maximum). In autumn of 1952 at Śnieżka Mt., 73 cloudy days, 69 such days in Płock and 68 in Warsaw were recorded. Other negative anomalies in the amount of cloudy days occurred in 1982, 2005–2006 and 2011. In autumn of 2003, only 8 cloudy days occurred in Świnoujście. On the whole, substantial increases were noted in 1978, 1996 and 2017. Additionally, in the Coastlands the autumn of 1993 was also very cloudy.

The first subperiod (1951–1984) was characterised by a relatively low number of drops in the variable (including some statistically significant changes recorded at four stations) and by some more increases (statistically significant at 10 stations) (Fig. 10.13e). The number of cloudy days decreased mainly in the south, including the Subcarpathians. In turn, in western Poland and at several stations located in the Uplands, the analysed variable increased. Nonetheless, the relative changes in the variable were rather unnoticeable (Fig. 10.13f).

During the second subperiod (1985–2018), the pace of the increase in the number of cloudy days was visibly faster in the whole country (Fig. 10.13g). At the majority of stations, the number of cloudy days systematically increased. It is particularly visible in the upper part of Odra River basin: more than 10 cases/10y. The number of cloudy days also increased in the Subcarpathians and in Elbląg in northern Poland. In western Poland, the variable decreased, which was especially visible in Świnoujście. The range of relative changes strongly relates to the highly diversified absolute values and it varies from −12 to 8% (Fig. 10.13h).

The long-term analysis of changes in the number of cloudy days recorded by the selected stations in Poland showed that this variable is diversified (Fig. 10.14). In the north, the number of cloudy days was rather high from the beginning of the research period until the 1980s. However, there was a short-term drop in the variable recorded at the stations located in the Lakelands in the late 1960s and early 1970s. In the central

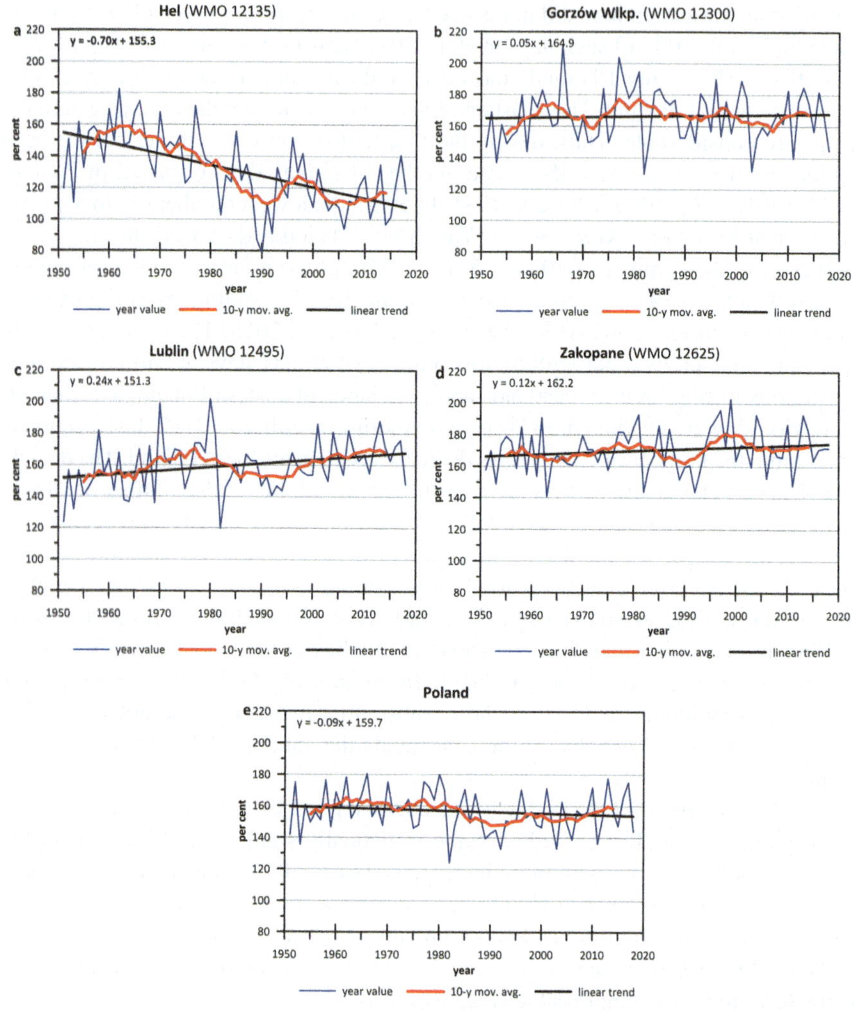

Fig. 10.14 Long-term variability of the annual number of overcast days in: northern Poland (station Hel; 1951–2018; **a**), western Poland (station Gorzów Wlkp.; 1951–2018; **b**), eastern Poland (station Lublin; 1951–2018; **c**), southern Poland (station Zakopane; 1951–2018; **d**) and in Poland as a whole (the average, **e**). Mean seasonal value (*blue*), 10-year consecutive average (*red*) and line trend (*black*) were shown. The trend is statistically significant (0.05, in *bold*) in (**a**) and not significant in (**b–d**)

part of the country and in the south, there were short-term increases in the 1970s. In the late 1980s and early 1990s a strong negative anomaly of the variable value was observed. At the beginning of the twenty-first century, the number of cloudy days in Poland increased again and now it is relatively stable at the majority of the stations.

Cloud Type

While analysing spatial and temporal variability of cloud type, i.e. cloudiness divided into 10 cloud genera, it is important to be aware of the fact that data series obtained from visual observations conducted at synoptic stations are imperfect. Occurrence frequency of a particular genera of clouds is largely affected by actual possibilities of its observation. It is a significant hindrance in the case of medium- and high-level clouds, as they may be obscured by some low-level clouds. Any clouds may be obscured by a thick layer of fog. Moreover, the specific coding requirements for synoptic observations, however, designed to give an optimum characteristic of cloudiness, may partially limit the possibility to register all genera of clouds that are present in the sky at a particular moment. Another impediment is the way the data are coded at some stations. Although National Hydrological and Meteorological Services provide trainings on proper methods of observing and recording cloudiness, there are still cases when some observers use techniques learnt from their predecessors, frequently their teachers. The above-mentioned situation does not apply to the way particular genera of clouds are recognised, but rather to the way of their subdivision into species and varieties, as well as classification in terms of coding them to report the specific state of the sky.

The data series obtained from 25 stations were analysed according to the methodology described in Chap. 3. The analysis made it possible to calculate the frequency of all ten cloud genera with a reference to the number of observations at particular levels.

On the assumption that has already been mentioned, frequency (in per cent) of low-level clouds was determined with regard to the number of their observations and the fact that in some cases the weather conditions made it impossible to carry out observations. The percentage of such cases varies from almost 0.4% in Ustka and Warszawa, 0.6% in Hel, Świnoujście, Białystok, Wrocław, Lublin and Rzeszów to 2% in Koszalin, Gorzów Wielkopolski, Kielce and Zakopane. At two other stations—in Zielona Góra and Chojnice, the weather conditions made it impossible to identify low-level clouds in 4% (Zielona Góra) and 5% (Chojnice) of all observations. Typically, results of low-level clouds observations are unsatisfactory in 1.4% of cases.

Medium-level clouds are not visible in less than 30% of all observations made at stations located in the Coastlands, Lowlands and Uplands (e.g. Ustka, Świnoujście, Warszawa, Wrocław and Kraków), while at the stations located in the Lakelands (Suwałki and Chojnice) and in the Carpathians (Zakopane and Lesko) these genera of clouds are invisible in more than 40% of all observations. The average for all the stations is 35%.

Świnoujście is a station where high-level clouds are well observable—in only 39% of cases these genera of clouds were not visible. At the remaining stations, except for ones located near the large civil airports (Warszawa and Kraków), the percentage of observations during which it was impossible to record any high-level clouds varies from 47 to 52%. In Warsaw and Kraków, it was approximately 42%. The average for all the station is 49%.

Stratocumulus is the most frequently observed cloud in Poland (Fig. 10.15a), especially at stations located in the regions with diverse orography and at high altitudes (in the Lakelands, Uplands and in the mountains). There, the *Stratocumulus* frequency exceeds 40% of all observed low-level clouds. On the coastline, this genus of clouds is observed with a frequency lower than 33%. In the remaining regions of the country, the frequency of *Stratocumulus* reaches approximately 33–35% or more. By and large, the frequency of *Stratocumulus* in Poland is 40%.

Cumulus is the second most frequently observed low-level cloud in Poland (Fig. 10.15b). On average, *Cumulus* is observed in one of every five observations of the sky in Poland. The highest frequency is recorded in the Coastlands—from nearly 20% in Kołobrzeg to more than 30%, with a maximum in Hel (33%). However, at several stations located inland (Gorzów Wlkp., Suwałki and Kraków), the frequency of *Cumulus* decreases to 15% of the observable low-level clouds and then towards the south it increases again reaching 20% in Lesko (the Carpathians). In Zakopane,

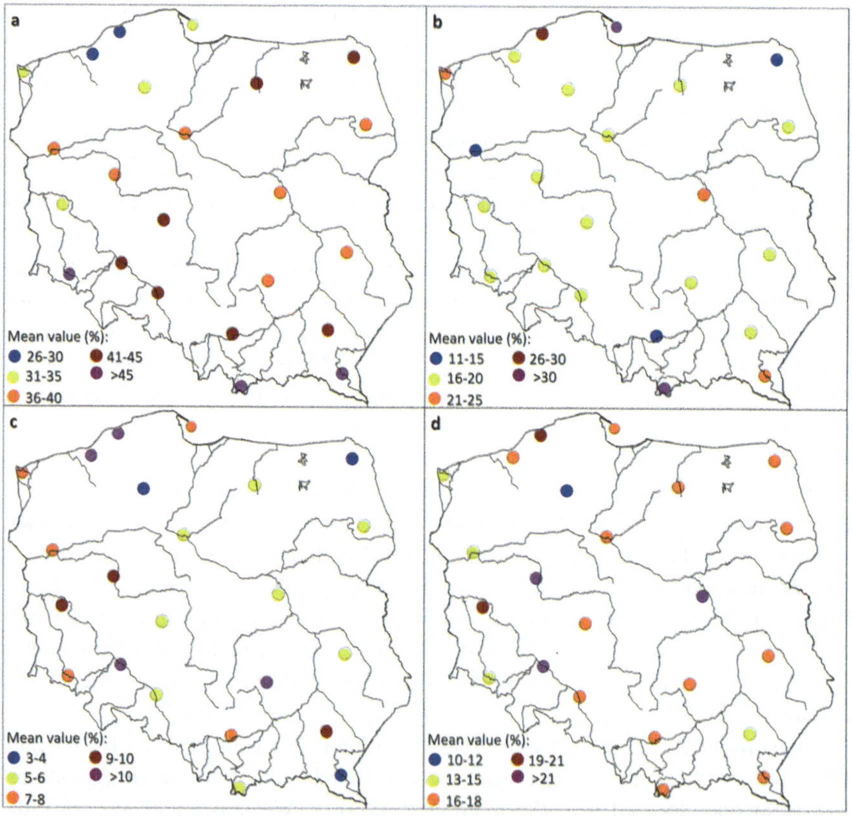

Fig. 10.15 Variability of annual average daily per cent of observations in Poland (1966–2018) for selected cloud genera: *Stratocumulus* (**a**), *Cumulus* (**b**), *Altostratus* (**c**) and *Cirrus* (**d**)

at the foot of the Tatra Mts., the frequency of *Cumulus* within the observations of low-level clouds is 30%.

When it comes to the structure of low-level cloudiness, both aforementioned stations share another distinctive feature—the lowest frequency of another low-level cloud, *Stratus*, across Poland: only 8% in Zakopane and 12% in Lesko. At the stations located at a lower altitude, this variable fluctuates between 13 and 17%, while in the Coastlands it reaches nearly 20%. The average for Poland is 15%.

Cumulonimbus, characterised by extensive vertical development, is the last analysed genus of low-level clouds. Spatial distribution of *Cumulonimbus* occurrence is highly diverse. In the Coastlands, it is observed in 6% of all cases in Hel and 11% in Ustka. In the Lakelands and Lowlands, the frequency varies from 5 to 7% (Poznań, Warszawa, Wrocław, Suwałki) to 13–14% (Olsztyn and Chojnice). In the Uplands, *Cumulonimbus* is observed once in every 9–10 cases, yet its frequency in Kraków is only 4%. At the stations located in Subcarpathians and the Carpathians, its frequency is highly diversified. In Rzeszów, it is 7%, in Zakopane—4%, while in Lesko it reaches 15%.

The analysis of long-term occurrence frequency of particular genera of low-level clouds in Poland during the period of 1966–2018 revealed a strong, statistically significant increase in the occurrence of *Stratocumulus*, *Cumulus* and *Cumulonimbus* (Fig. 10.16). As for the first listed genus, the increase is relatively stable and evenly spread over time. Both convective clouds were particularly often observed at the end of the previous century and during the first decade of the twenty-first century. Nowadays, the above-mentioned clouds are observed less frequently. The number of *Stratus* observations in terms of the whole research period also declined and the pace of this decline was rather uneven. After a period of a higher *Stratus* frequency lasting until 1980, there was a fast decline observed with minima in the 1980s and the 1990s. During the last few years, *Stratus* has been observed more frequently.

The average annual prevalence of mentioned genera of clouds in individual regions is very consistent for the *Cumulus* cloud. In the case of other low-cloud genera, some regional discrepancies can be observed. The reported dramatic increase in the frequency of the *Cumulonimbus* in the late 1990s was particularly evident in the coastal mountain foothill regions and in the mountains, whereas the other regions were characterised by a more stable course of cloud occurrence. In the case of the *Stratocumulus*, only a slight increase in its frequency was found at mountain stations. At stations located in the Coastlands, the minimum frequency of *Stratus* was shifted to the second half of the 1980s and the beginning of the 1990s, whereas at stations located in the Lakelands, there was relative stabilisation of the occurrence of this cloud.

The structure of medium-level cloudiness is dominated by *Altocumulus* clouds. They are observed in 21% of all cases, yet their frequency is substantially diversified. *Altocumulus* is most frequently observed in the sky over stations located in the Lowlands and in the south of Poland, with the exception of the Carpathians. Thus, the maximum frequency was recorded in Wrocław (36%), Zielona Góra and Warszawa (31%), and Kielce (28%). At the remaining stations of those regions, the frequency does not exceed 25%. In the mountain region, *Altocumulus* is usually

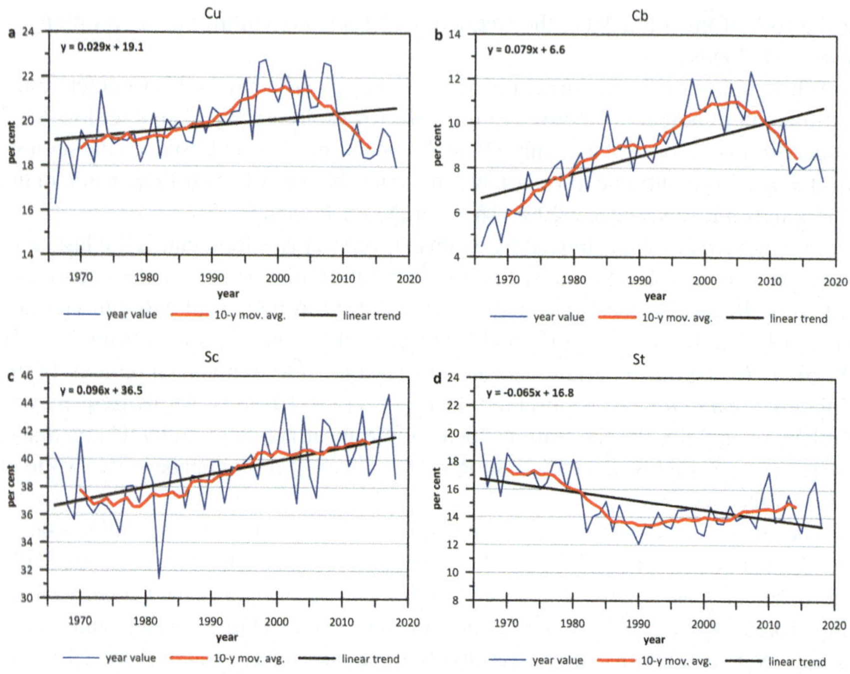

Fig. 10.16 Long-term variability of annual average daily per cent of observations in Poland (1966–2018) for low-cloud genera: *Cumulus* (Cu, **a**), *Cumulonimbus* (Cb, **b**), *Stratocumulus* (Sc, **c**) and *Stratus* (St, **d**). Mean annual value (*blue*), 10-year consecutive average (*red*) and line trend (*black*) were shown. The trend is statistically significant (0.05, in *bold*) in all cases

observed less often (Zakopane—nearly 17% and Lesko—13%, however, in Jelenia Góra in the foothills of the Karkonosze Mts., this value reaches 25%). In the Coastlands, *Altocumulus* represents from 11% (Hel) to 23% (Ustka) of all medium-level clouds observations. However, the lowest frequency is observed in the Lakelands: from almost 10% in Suwałki and 12% in Olsztyn to 20% in Toruń and 24% in Koszalin (located in the Lakeland region but very close to the coast).

The two remaining genera of medium-level clouds, *Altostratus* and *Nimbostratus*, are not frequently observed in Poland (their aggregated frequency of occurrence does not exceed the value for Ac).

On the whole, the number of the *Altostratus* observations seems to be declining from west to east (Fig. 10.15c). Nevertheless, its occurrence frequency is spatially diversified. In most cases, in the west of the country, the frequency of *Altostratus* is at least 7%, and in the east it is usually approx. 5–6%. *Altostratus* is often observable at the coastal stations (nearly 11% in Ustka, more than 6% in Świnoujście and Hel), but there are other three stations where the frequency of *Altostratus* exceeds 10%. They are: Koszalin, Wrocław (located in the Lowlands of western Poland) and Kielce in the Uplands. At a few stations located in the east, less than 7% of *Altostratus* frequency occurred, e.g. in Lesko and Suwałki it was less than 4%. The average value for the whole country is 7%.

The last analysed medium-level cloud with extensive horizontal and vertical development is *Nimbostratus*. Its frequency is even lower than the one calculated for *Altostratus*—approximately 4%. When it comes to spatial diversification, the *Nimbostratus* frequency varies from 2% in Zakopane and Jelenia Góra (both located in the foothills of the highest mountain ranges in Poland: the Tatra Mountains and Karkonosze Mountains) to almost 6% at some stations located in the north: Koszalin, Białystok and Suwałki. In the Coastlands, the frequency reaches 5% (Ustka and Świnoujście). The same value of this variable was recorded in the Lowlands, in Warszawa and Opole. In the remaining regions, observations of this cloud genus represent approx. 4–5% of all medium-level clouds observations and even less (3%) in the Uplands.

The structure of medium-level cloudiness in Poland during the period of 1966–2018 was affected by some statistically significant, long-term changes. The number of observed *Altostratus* and *Nimbostratus* systematically decreased overtime (Fig. 10.17a, b, respectively). They were observed much more often in the late 1970s and early 1980s, and then the number of observations dropped, though, since

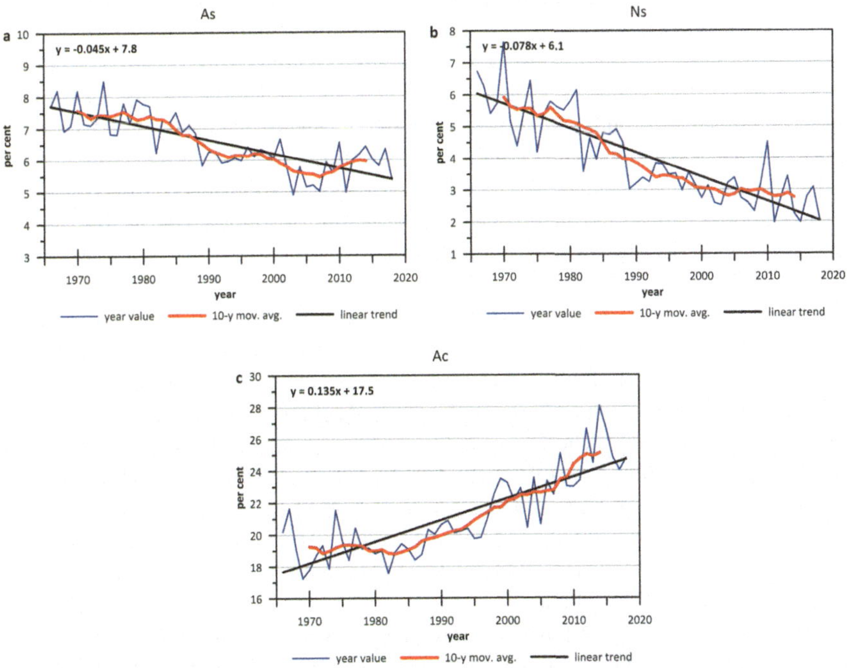

Fig. 10.17 Long-term variability of annual average daily per cent of observations in Poland (1966–2018) for middle cloud genera: *Altostratus* (As, **a**), *Nimbostratus* (Ns, **b**) and *Altocumulus* (Ac, **c**). Mean annual value (*blue*), 10-year consecutive average (*red*) and line trend (*black*) were shown. The trend is statistically significant (0.05, in *bold*) in all cases

the beginning of the twenty-first century, this negative trend has not been as strong as it used to be. *Altocumulus* was observed more frequently in the research period, but not systematically (Fig. 10.17c) and not during the first part of the analysed period. Its frequency began increasing rapidly in the last decade of the twentieth century.

Regional trends do not usually follow the average pattern determined for the whole country. In the case of *Altostratus*, a positive trend was observed in the Coastlands after a minimum had been recorded in the second half of the last decade of the former century. Moreover, the frequency of the same cloud decreased much faster at the stations located in the Subcarpathian region than at the others. In the mountains and in the Subcarpathians, *Altocumulus* formed differently than in the rest of the country. The *Altocumulus* frequency in the mountains was almost the same at the beginning and at the end of the analysed period. However, a clear minimum was recorded in the 1980s and 1990s. At the stations located in the Subcarpathian region, there was no strong positive trend observed. The regional frequency of *Nimbostratus* more or less reflects the national pattern.

The most often observed high-level cloud in Poland is *Cirrus*, which represents 17% of all observations of high-level clouds in the country. Spatial diversification of the *Cirrus* frequency does not present any distinctive features (Fig. 10.15d). In central and eastern Poland, it is observed as often as the average for the whole country. In the west and north, the variable is much more diversified, even at neighbouring stations. The lowest frequency was recorded in Chojnice, the most elevated station in the Lakelands, where *Cirrus* observations represented only 10% of all cases. Other stations where this genus of cloud is observed relatively seldom (13–15% of all observed high-level clouds) are as follows: Świnoujście (in the Coastlands), Gorzów Wlkp. (in the Lakelands) and Jelenia Góra (foothills of Karkonosze Mts.), all of which are located in western Poland. In Ustka (in the Coastlands), the frequency reaches approx. 20%, and in Poznań and Wrocław (the Lowlands in the west of Poland), the *Cirrus* observations represent almost a quarter of all cases (22.5% in Wrocław).

Cirrostratus is the second most frequently observed high-level cloud with the average frequency of almost 7% across the country, although this value decreases from the west to the east of Poland. The frequency exceeds 10% at the stations located more westward: in Koszalin, Chojnice, Kalisz and Jelenia Góra (e.g. in Zielona Góra it reaches approx. 15%) whereas in the east there is only 2–4% (Warszawa—2%, Białystok and Suwałki—3%, Olsztyn—4%) However, in Kielce and Lublin, located in the Uplands, this variable reaches the value of 11–12%. *Cirrostratus* is often observed also in the mountains—11% of all observations of high-level clouds are recorded in Zakopane. In Rzeszów, in the Subcarpathians, it is a rarely observed genus of clouds (0.7% of all cases).

Cirrocumulus is most certainly the rarest high-level cloud observed in Poland. Its average frequency in Poland is less than 1% and it is the lowest on the coast; this value increases at stations located at higher altitudes.

The occurrence frequency of all three genera of high-level clouds has been systematically increasing in Poland (Fig. 10.18). The analysis showed that the minimum frequency of high-level clouds was recorded in the late 1980s and early 1990s (a bit

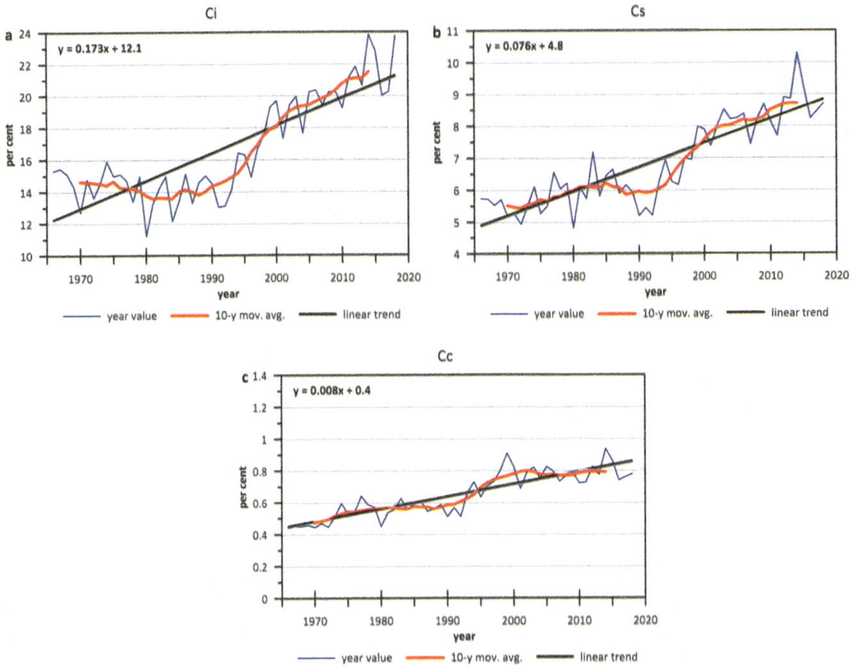

Fig. 10.18 Long-term variability of annual average daily per cent of observations in Poland (1966–2018) for high cloud genera: *Cirrus* (Ci, **a**), *Cirrostratus* (Cs, **b**) and *Cirrocumulus* (Cc, **c**). Mean annual value (*blue*), 10-year consecutive average (*red*) and line trend (*black*) were shown. The trend is statistically significant (0.05, in *bold*) in all cases

faster in the case of *Cirrus*). Then, after 1995, both *Cirrus* and *Cirrostratus* started to be observed more often. In the case of *Cirrocumulus*, the positive trend has slowed down since the beginning of the current century.

The average annual frequency of *Cirrus* is more or less the same in all analysed regions. In the case of *Cirrostratus*, its frequency different from the average is found on the coast. At the stations located by the sea, the *Cirrostratus* frequency was in line with the average only at the beginning of the analysed period, then the frequency dropped (while it increased in other regions) and increased again until the late 1990s. Afterwards, it continued to decrease during the last ten years of the analysed period. As for the *Cirrocumulus* frequency, a substantial growth was recorded at the stations located in the Lowlands in the 1990s. In the Lakelands, there was a stable increase recorded, whereas in the mountains the variable was highly diversified.

Discussion and Conclusions

The mean annual cloud cover in Poland in the years 1951–2018, calculated on the basis of data from 57 stations distributed relatively evenly throughout the country, was 68.2%. The maximum value equal to 73% was recorded in 1966, while the minimum value of nearly 61% occurred in 1982. There is a noticeable spatial variation in the value of the element at individual stations and regions. For example, the mean annual total cloudiness amount ranges from 64% in the Coastlands to 76% in the Karkonosze Mts.

The course of cloud cover throughout the year in Poland is also varied. The lowest level of cloudiness occurs in summer and the largest in winter. Between both seasons there is a difference in cloud cover of about 15–20%. Cloudiness in spring is lower than in autumn. In general, in spring and summer, cloudiness amount decreases eastwards. In autumn and winter, the total cloudiness amount increases towards the north of the country. However, the western part of Poland is more cloudy than the east of the country.

Based on the results of this study, it can be concluded that at the greater part of analysed stations the cloud cover did not change considerably and such observations on the long-term cloudiness variability at least up to the year 2000 are consistent with the results of other authors. In the analyses by Warren et al. (2007) and Eastman and Warren (2013) covering the period from 1971 to the first decade of the twenty-first century, Poland is an area of a continuing decline in cloudiness. Wibig (2008) indicated a significant decrease in cloudiness in the city of Łódź in central Poland in the second half of the twentieth century. Similar conclusions were reached by Żmudzka (2005), also in relation to the cloudiness over Poland in the period 1951–2000, and Filipiak and Miętus (2009), characterising total cloud cover in Poland in the years 1971–2000. However, in the most recent period, up to 2018, a decrease in mean annual cloudiness is still proceeding, but only in the northern part of the country. In the south, the increase of the cloudiness is observed and the process has been advanced in the later part of subperiods analysed in this study, 1985–2018. According to Matuszko (2003) and Lewik et al. (2010), cloudiness in Kraków increase from 1983 onwards.

Statistically significant changes in the cloudiness amount in particular seasons in the period 1951–2018 are present only at a few stations in Poland, mostly located in the southern and southwestern parts Poland. More extreme changes are found by applying the analysis of the long-term variability of cloudiness in subperiods: 1951–1984 and 1985–2018. In case of winter, it can be observed the significant differentiation of both periods in terms of signs of trend coefficients of cloudiness in the country. While during the first subperiod cloudiness decreased, in the second, the trend completely reversed and a substantial, statistically significant increase in cloudiness was observed with the coefficient growing towards the south. More than 10 stations recorded a statistically significant change in cloudiness in spring in the period of 1951–1984, and the observed trends in the south were stronger than in the north. In the later subperiod, the observed trend is reversed. A few stations located in eastern Poland and in the highest parts of the Carpathians recorded statistically significant changes of cloudiness in summer in the years 1951–1984. In the period of 1985–2018, the decrease in cloudiness dominated in the country. The cloudiness in autumn slightly increased at the majority of stations in Poland in the years 1951–1984. After 1985, at a majority of stations, increases in cloudiness were observed.

Another important feature of the cloudiness conditions in Poland is related to the variability of the characteristic nephological days. In general, cloudy days are much more frequent in Poland than clear ones. They are most often observed in the Lakelands, where they constitute up to 160 days or more a year. There is a particularly strong contrast between the Lakelands and the Coastlands, where the discussed cases constitute on the average about 140 days or less. The number of clear days is decreasing considerably across the country, while in case of cloudy days, such negative changes concern only selected stations in the north, the Subcarpathians and lower ranges of the Carpathians. According to Żmudzka (2012), the number of both types of analysed characteristic nephological days declined in the period 1966–2000. Moreover, the pace of the decrease in the annual number of cloudy days was greater than in case of clear days, and this rate was particularly considerable in the case of winter. Matuszko (2007) commented on the course of the number of characteristic nephological days in Kraków in the twentieth century, which is closely related to the course of the cloud amount.

The above-mentioned situation is influenced by the fact that cloudiness is a meteorological element of complex origin and strongly related to other meteorological variables. However, many authors underline the key role of atmospheric circulation in shaping nephological conditions (Żmudzka 2003, 2004a, b, 2007; Matuszko 2003, 2009; Filipiak and Miętus 2009; Matuszko and Pluta 2012; Matuszko and Węglarczyk 2018). As with other elements (such as temperature or precipitation, Miętus and Filipiak 2001, 2002), its role increases in the cold season. As Filipiak (2012) reported, the results of the analyses of the influence of regional atmospheric circulation on the variability of the average monthly cloudiness demonstrated a strong relationship between the development of clouds and the properties of the inflowing air masses, as well as the characteristic weather features in baric systems. It was also clearly proved that characteristic properties of cloudiness in the months of the cool half of the year are largely determined by the character of the atmospheric circulation. However, in the warm half of the year, a more important role is played by the processes of thermal convection. Żmudzka (2004a, b, 2007, 2012) emphasised that the decrease in cloudiness is caused by anticyclonic systems with meridional air inflow—in the cold half of the year from the south, and in the warm mainly from the north—while the high frequency of cyclonic systems is conducive to the increase in cloudiness. Moreover, the trends of changes in the amount of cloudiness over Poland correspond with changes in macro-scale circulation conditions (so-called circulation epochs). Even up to 80% of changes resulting in variance in cloudiness results directly from atmospheric circulation. Her findings were partially proved by Szyga-Pluta (2015) in the study describing circulation influence on cloudiness in the city of Poznań. According to Henderson-Sellers (1986) the mentioned long-term decrease in cloudiness in Central Europe up to the beginning of 1980s was directly dependent on the circulational conditions and particularly in the higher than usual frequency of radiative weather associated with baric highs. The evidence on the strong relationship between cloudiness in Kraków and North Atlantic Oscillation was presented by Lewik et al. (2010).

The key role of regional atmospheric circulation in shaping the described element means that the mean annual cloud amount in Poland does not differ significantly from the values recorded in neighbouring European countries and in the entire Baltic Sea basin (Henderson-Sellers 1986; Miętus 1998; Tuomenvirta et al. 2000; BACC 2008). This regional climatic factor may also explain why relatively high amount of clouds in Poland can be observed. It is perhaps related to the effect of the Atlantic Ocean, the source of humid masses of maritime polar air, prevailing throughout almost the whole year in Poland. When this type of air mass flows into the country, there is a characteristic increase in cloudiness.

Despite this fact, it should be highlighted that regional atmospheric circulation is not the only factor shaping the variability of the cloudiness in Poland. An important role in this case is also played by processes on a local scale. The diverse relief of the Lakelands of western Poland, situated on the route of the inflow of mentioned humid maritime polar air masses into Poland, may be a good example of how the local conditions may modify the level of cloudiness. It constitutes a factor favourable for the development of cloudiness. Mountains are another example which illustrates the influence of local factors affecting the formation of clouds. Orography greatly influences the dynamic and thermal conditions in this area, causing the intensification of turbulent fluxes and the development of local dynamic convective movements (Trepińska 2002). According to Szyga-Pluta (2017), who described the variability of cloud cover in the Sudety Mts the increase of cloudiness is, in general, observed with an altitude as it can be seen at Śnieżka Mt. However, there are some specific conditions related to the occurrence of anticyclonic circulation in autumn and winter, when the thermal inversion may develop in the valleys, leading to the reverse of the cloudiness field over the mountains and its foothills. Thus, cloud cover may develop at a lower altitude, below high parts of the mountains, which remain very clear. This situation, observed mostly in the cold half of the year, with the maximum in January and least frequently from May to July, explains why mountain stations are the most clouded places in Poland in summer and spring and not in winter. Moreover, the number of clear days in winter recorded at Śnieżka Mt., and especially at Kasprowy Wierch Mt., are the greatest in Poland.

It is worth noting that observations of cloudiness are conducted at stations usually located in cities. Thus, the variability of the analysed element is also influenced by factors related to the development of the city—its territorial and industrial development, anthropogenic heat emission, change in land use, or increased presence of aerosols above the urban area. Morawska-Horawska (1985) and Matuszko (2003) in the analyses of the variability of cloudiness over Kraków, drew attention to the impact of the urban environment in shaping the variability of nephological conditions. One of the aspects of the city's impact may be, in their opinion, the fact that economic activity is conducive to, inter alia, drying the atmosphere over urban areas and the associated decrease in air humidity, which in turn hinders the development of cloud cover.

Filipiak (2003) and Limanówka et al. (2012) studied the character and rate of changes of selected hygric elements at 10 selected stations in northwestern Poland in the years 1951–2000 and more than 50 stations in Poland in 1966–2008. Both analyses revealed that long-term relative air humidity decreases dominate both annually and seasonally in the analysed periods. Overall, the air dried out the most in spring. A less dynamic decrease in the value of the element is also observed in summer. In autumn and winter, situations of long-term growth in the value of the element are markedly more frequent than in the above discussed two seasons. In winter, in 10 series of more than the 50 analysed, the changes are statistically significant. In the case of water vapour pressure, the value increases during 1966–2008, particularly in winter. In spring and summer, the changes are not so dramatic, but above the majority of the country, an increase in the value of the element can be observed. In autumn the changes are the weakest, but positive trends also predominate in long-term water vapour pressure changes. Therefore, long-term changes of hygric elements favour the increase in precipitable water in the atmosphere, but the decreasing relative humidity may limit the development of clouds.

Another interesting aspect of cloudiness analysis concerns cloud type. Its spatial variability and temporal changes have been analysed by numerous Polish authors over the last few decades. Warakomski (1962) found that the most frequently recorded cloud genera in Poland are *Stratocumulus* and *Altocumulus*. Both clouds show the largest regional differences in the frequency of occurrence. They are most frequent in southwestern Poland, whereas convective clouds *Cumulus* and *Cumulonimbus* occur most often over the northeastern part of Poland. Nowak (1971) compared the incidence of *Cumulus* and *Cumulonimbus* clouds in Gdańsk and Kraków in 1961–1965, finding convective clouds more frequent in Gdańsk due to the effect of breezes, and in Kraków the relation of occurrence of both clouds to the thermal convection. Szyga-Pluta (2002) concluded that *Stratocumulus, Altocumulus* and *Cirrus* clouds were the most common in northwestern Poland in 1971–1990, while the other high-level clouds were the rarest. A long-term increase in the number of cases of *Cirrus, Altocumulus* and *Cumulus* clouds, and a decrease in the number of *Cirrocumulus, Cirrostratus* and *Cumulonimbus* clouds was also detected. According to Miętus et al. (2003), in Hel in the second half of the twentieth century, there is a rise in convective clouds, accompanied by a decrease in the incidence of *Stratocumulus, Stratus, Altostratus* and *Nimbostratus* clouds. Similar changes, expressed by a decrease in the frequency of stratiform clouds *Stratus* and *Nimbostratus*, associated with a slightly smaller decline of *Stratocumulus* and a rapid increase in cloudiness by *Cumulonimbus* and *Cumulus* clouds, also occur in Łódź (Wibig 2008). Moreover, it is associated with the more frequent occurrence of *Altocumulus* and high-level clouds and the fall in *Altostratus*. The study by Filipiak and Miętus (2009), related to the years 1971–2000, confirmed that *Stratocumulus, Altocumulus* and *Cirrus* are the most frequent clouds among, respectively the low-level, middle-level and high-level clouds in Poland, but its percentage ranges considerably in various regions. Coastlands are more favourable than the interior of the country for the occurrence of convective clouds and *Stratus* clouds. *Stratocumulus* clouds are the most frequent in areas with a diverse relief. Essentially, the frequency of *Altostratus* and *Nimbostratus* clouds is similar all over the country. On the whole, *Cirrostratus* clouds are the most frequent in western Poland, while *Cirrocumulus* clouds are the rarest.

The observations made in this paper on the spatial cloud type variability and its long-term changes are consistent with the findings of other authors presented above, and constitute a logical extension of information on the variability of cloud genera from the first decade of the twenty-first century. Some important discrepancies do exist, however.

Stratocumulus is still the most frequently observed of all low-level clouds in Poland, especially at stations located in the regions with diverse orography and at high altitudes (in the Lakelands, Uplands and in the mountains). The highest frequency of *Cumulus* is recorded in the Coastlands, while in case of *Cumulonimbus*, it is highly diversified, from 4% in Zakopane at the foothills of the Tatra Mts. to 14% in Chojnice in the Lakelands. *Stratus* occurs most frequently in the Coastlands. The first three genera of low-level clouds were observed more and more frequently in Poland during the period of 1966–2018. In the case of *Stratocumulus*, such a conclusion is distinct from the findings related to the twentieth century by the mentioned authors, e.g. Szyga-Pluta (2002), Miętus et al. (2003) and Wibig (2008). However, it is Matuszko and Węglarczyk (2018), who, in their recent paper on the long-term variability of amount and genera of clouds in Kraków in the years 1906–2015, presented evidence of the noteworthy increase of *Stratocumulus* since the 1960s. The finding on the decrease in frequency of both convective clouds since the beginning of the current century, after the earlier few decades of significant increase in their occurrence, is also consistent with the conclusions presented by Matuszko and Węglarczyk (2018). The number of *Stratus* observations also declined in the research period in Poland, which was indicated by mentioned authors (Szyga-Pluta 2002; Miętus et al. 2003; Wibig 2008; Matuszko and Węglarczyk 2018).

Altocumulus clouds dominate in the structure of medium-level cloudiness. As Warakomski (1962) observed, it is mostly observed at the stations in the southwestern part of Poland. Again, dependence on the diverse relief can be observed, like in case of *Stratocumulus*. *Altostratus* and *Nimbostratus*, are not frequently observed in Poland. The frequency of occurrence of *Altostratus* slightly decreases from west to east. The more rare *Nimbostratus* is observed in 5–6% of cases maximum. The number of observed *Altostratus* and *Nimbostratus* systematically decreased, while in case of *Altocumulus*, an increase in its occurrence can be observed. These conclusions are consistent with the results related to the earlier periods in mentioned papers. Matuszko and Węglarczyk (2018) also reported on the considerable increase in both middle-level stratiform clouds, but it worth noting that, in contrast to this study, the occurrence of *Altocumulus* in Kraków has been declining since the beginning of the twenty-first century.

Among high-level clouds, *Cirrus* is most frequently observed in Poland with a frequency of occurrence between 10 and 22% at particular stations, but with quite chaotic spatial distribution. *Cirrostratus* is observed with an average frequency of almost 7% in the country, however this value decreases from the west to the east of

Poland. The average frequency of *Cirrocumulus* in Poland is less than 1%, with the lowest value in the Coastlands, increasing with altitude. The occurrence frequency of all three genera of high-level clouds has been systematically increasing in Poland. The positive trend is most pronounced in case of *Cirrus*. Wibig (2008) reported on the proliferation of all genera of high-level clouds in Łódź in the second half of the twentieth century. Alternatively, Matuszko and Węglarczyk (2018) found a decreasing tendency for *Cirrostratus* in Kraków.

It is the matter of discussion why such diversified trends in the occurrence of various genera of clouds are present in the analysed series. In the case of *Cirrus*, one of the most probable explanations is related to the intensification of the air traffic resulting in the increase in contrails transforming into *Cirrus* clouds (Henderson-Sellers 1986). The occurrence of other cloud genera is associated with the specific circulation conditions. According to Warakomski (1962) and Matuszko and Węglarczyk (2018), the great frequency of occurrence of *Stratocumulus* and *Altocumulus* clouds results from their linkage to maritime polar mass inflow, prevailing throughout nearly the whole year in Poland, especially in the cold half of the year. Though very high values of North Atlantic Oscillation index occurred in the 1990s (Hurrell et al. 2003) resulting in the rise in advection of humid maritime polar masses, a positive anomaly in the long-term variability of *Stratocumulus* and *Altocumulus* occurrence corresponding to it can be observed. However, in the course of *Altocumulus* occurrence, a positive anomaly can be observed in the 2010s when very high values of NAO index occurred again (Deser et al. 2017). Current global climate change manifested in the rise of surface temperature may create conditions favourable to the more frequent occurrence of convective clouds. On the contrary, a decrease in relative humidity may limit such development. It may prove that the coupling between the occurrence of the particular cloud genera and circulation patterns is non-linear. Additionally, the opinion by Matuszko and Węglarczyk (2018) on the correctness of visual observations which constitute the research material is sensible. For a group of observers, it is still challenging to differentiate between some genera of clouds, which may result in misleading results of any analyses of the occurrence of cloud type.

There is also an important question about the future long-term development of cloud cover. The currently observed climate change, manifested in many ways, e.g. in the ongoing warming of the troposphere, have significant implications for the energy and mass fluxes between the atmosphere and the ground. Systematic changes of the total cloud amount are associated with long-term declines or increases in the values of a number of other climatic elements. For the area of Europe, and especially the part located within the Baltic Sea basin, there is a positive correlation between changes in air temperature and atmospheric precipitation (BACC 2008; BACC II 2015), at least for winter, spring and autumn. This phenomenon is consistent with the Clausius-Clapeyron equation, whereby the mean air temperature increase of 1 °C entails about 7% increase in the potential quantity of water in the atmosphere, including, in particular, water vapour content, the most crucial long-wave radiation absorbent. The consistency of the nature of temperature and precipitation changes is primarily of advection origin. In summer, the presence of water in the atmosphere is to a lesser degree determined by the advection factor in favour of local processes, which also significantly influence cloudiness (Numaguti 1999). A decrease in daily temperature amplitude, observed in the majority of areas on the planet, is connected with an increase in total cloudiness degree (Karl et al. 1993; Zhou et al. 2016). Changes in cloudiness are in turn the result of a number of meteorological processes related to the quantity of water vapour, its condensability and thermal balance of the atmosphere, and the associated development or inhibition of atmospheric instability (Dai et al. 1999; Willett et al. 2010; Tobin et al. 2012). Hence, the question arises to what extent the expected changes in cloudiness amount will be concurrent with changes in other climatic elements.

Given the results the statistical empirical projections of climate elements of Poland for the selected period of the twenty-first century obtained by Filipiak (2012) and Miętus et al. (2012), it can be concluded that the most noticeable changes in cloudiness, on the scale of the twenty-first century, reaching nearly 10% compared to the years 1971–2000, will take place in summer. In light of the research of the above-mentioned authors, it can be assumed that the progressive decrease in cloudiness will also find its implications in the variability of other climatic elements. Although in summer, the importance of local-scale processes, such as the effect of the ground, or local convection, grows significantly in comparison to other seasons (Namaguti 1999), an increase in air temperature during this century, forecast by a majority of climatologists, should be kept in mind (Boucher et al. 2013). The warming of the lower troposphere is accompanied by changes in the value of evaporation resulting, on the one hand, from the acceleration of the very evaporation rate, or on the other, from ground desiccation. Sensible and latent heat fluxes intensify. Although the water vapour pressure increases, relative humidity decreases quite rapidly, which means less favourable conditions for the development of cloudiness. The decrease in cloudiness during the day may lead to an increase in direct solar radiation. Considering also strong negative correlation between cloudiness degree and maximum temperature, an increase in the value of the thermal index should also be expected.

The expected increase in cloudiness in winter will also be conducive to systematic warming projected by various simulations performed with selected RCMs (BACC 2008; Filipiak 2012; Miętus et al. 2012; BACC II 2015). The same simulations also indicate an increase in precipitation totals in the coldest months of the year. The probable reason for the changes is a combination of the process of the air warming with the intensification of moisture flux from lower latitudes and the Atlantic Ocean into the northern and central regions of the continent, where Poland is located. The cloudiness increase will be consistent with the projected minimum air temperature increase, and the presence of clouds due to the increased downward thermal radiation will mitigate the nighttime cooling of the ground.

The reduction in cloudiness in spring is in contradiction with the increase in temperature in the light of the suggested schemes. However, it is likely that a further decline in stratiform cloudiness, with a simultaneous increase in convective cloudiness, will continue (Wibig 2008; Miętus et al. 2003; Matuszko and Węglarczyk 2018), formed as a result of increasing insolation in spring when the sky is clear.

The autumn changes in all the climatic elements, from temperature, to hygric indicators, to cloudiness are, in turn, small enough that a certain stability of the situation, characteristic of the described reference period, can be assumed.

References

BACC Author Team (2008) Assessment of climate change for the Baltic Sea basin. Regional climate studies, Springer Verlag, Berlin, Heidelberg

BACC II Author Team (2015) Second assessment of climate change for the Baltic Sea basin. Regional climate studies, Springer, Cham, Heidelberg New York, Dordrecht and London

Boers R, Brandsma T, Siebesma AP (2017) Impact of aerosols and clouds on decadal trends in all-sky solar radiation over the Netherlands (1966–2015). Atmos Chem Phys 17:8081–8100. https://doi.org/10.5194/acp-17-8081-2017

Boucher OD, Randall P, Artaxo C, Bretherton G, Feingold P, Forster V-M, Kerminen Y, Kondo H, Liao U, Lohmann P, Rasch SK, Satheesh S, Sherwood B, Stevens B, Zhang XY (2013) Clouds and aerosols. In: Stocker TF, Qin D, Plattner G-K, Tignor M, Allen SK, Boschung J, Nauels A, Xia Y, Bex V, Midgley PM (eds) Climate change 2013: the physical science basis. Contribution of working group I to the fifth assessment report of the intergovernmental panel on climate change. Cambridge University Press, Cambridge, UK and New York, NY

Charlson RJ, Schwartz SE, Hales JM, Cess RD, Coakley JA Jr, Hansen JE, Hofmann DJ (1992) Climate forcing by anthropogenic aerosols. Science 225:423–430

Cox C, Walden V, Rowe P, Shupe MD (2015) Humidity trends imply increased sensitivity to clouds in a warming Arctic. Nat Commun 6:10117. https://doi.org/10.1038/ncomms10117

Dai A, Trenberth KE, Karl TR (1999) Effects of clouds, soil moisture, precipitation and water vapour on diurnal temperature range. J Clim 12:2451–2473. https://doi.org/10.1175/1520-0442(1999)012%3c2451:EOCSMP%3e2.0.CO;2

Deser C, Hurrell JW, Phillips AS (2017) The role of the North Atlantic Oscillation in European climate projections. Clim Dyn 49:3141–3157. https://doi.org/10.1007/s00382-016-3502-z

Eastman R, Warren SG (2013) A 39-Yr survey of cloud changes from land stations worldwide 1971–2009: long-term trends, relation to aerosols, and expansion of the Tropical Belt. J Clim 26(4):1286–1303. https://doi.org/10.1175/jcli-d-12-00280.1

Eastman R, Warren SG, Hahn CJ (2011) Variations in Cloud Cover and Cloud Types over the Ocean from Surface Observations, 1954–2008. J Clim 24(22): 5914-5934. https://doi.org/10.1175/2011jcli3972.1

Filipiak J (2003) Zmienność zachmurzenia na wybrzeżu Morza Bałtyckiego i Pojezierzu Pomorskim w drugiej połowie XX wieku. Sprawozdanie z Tematu Badawczego KBN-5. IMGW Oddział Morski w Gdyni (manuscript)

Filipiak J, Miętus M (2009) Spatial and temporal variability of cloudiness in Poland, 1971-2000. Int J Climatol 29:1294–1311. https://doi.org/10.1002/joc.1777

Filipiak J (2012) Scenariusze zmian klimatu Polski. Warunki nefologiczne. Raport końcowy z wykonania prac w zakresie podzadania 1.2. Sprawozdanie z Projektu KLIMAT Wpływ zmian klimatu na środowisko, gospodarkę i społeczeństwo. IMGW-PIB, Warszawa (manuscript)

Henderson-Sellers A (1986) Cloud changes in a warmer Europe. Clim Change 8:25–52

Hollmann R, Merchant CJ, Saunders R, Downy C, Buchwitz M, Cazenave A, Chuvieco E, Defourny P, de Leeuw G, Forsberg R, Holzer-Popp T, Paul F, Sandven S, Sathyendranath S, van Roozendael M, Wagner W (2013) The ESA climate change initiative: satellite data records for essential climate variables. Bull Am Meteorol Soc 94:1541–1552. https://doi.org/10.1175/bams-d-11-00254.1

Hurrell JW, Kushnir Y, Ottersen G, Visbeck M (2003) An overview of the North Atlantic Oscillation. In: Hurrell JW, Kushnir Y, Ottersen G, Visbeck M (eds), The North Atlantic Oscillation: climatic significance and environmental impact. Geophys Monograph 134:1–35. https://doi.org/10.1029/134GM01

IMGW-PIB (2013) Metody określania rzetelności obserwacji meteorologicznych oraz eliminacji lub redukcji zmian tej rzetelności. Temat Badawczy DS-M Doskonalenie jakości procesu pozyskiwania i weryfikacji danych meteorologicznych z naziemnego systemu stacji synoptycznych (manuscript)

Jun S-Y, Ho CH, Jeong J-H, Choi Y-S, Kim B-M (2016) Recent changes in winter Arctic clouds and their relationships with sea ice and atmospheric conditions. Tellus A 68. https://doi.org/10.3402/tellusa.v68.29130

Karl TR, Knight RW, Gallo KP, Peterson TC, Jones PD, Kukla G, Plummer N, Razuvayev V, Lindseay J, Charlson RJ (1993) A new perspective on recent global warming: asymmetric trends of daily maximum and minimum temperature. Bull Am Meteorol Soc 74:1007–1023. https://doi.org/10.1175/1520-0477(1993)074%3c1007:ANPORG%3e2.0.CO;2

Lewik P, Matuszko D, Morawska-Horawska M (2010) Multi-annual variability of cloudiness and sunshine duration in Krakow between 1826 and 2005. In: Przybylak R, Majorowicz J, Brazdil R, Kejna M (eds) The Polish climate in the European context: an historical overview. Advances in global change research, vol XVIII. Springer. Dordrecht, The Netherlands, Heidelberg, Germany, London and New York, NY, pp 341–353

Limanówka D, Biernacik D, Czernecki B, Farat R, Filipiak J, Kasprowicz T, Pyrc R, Urban G, Wójcik R (2012) Zmiany i zmienność klimatu od połowy XX w. In: Wibig J, Jakusik E (eds) Warunki klimatyczne i oceanograficzne w Polsce i na Bałtyku Południowym. IMGW-PIB, Warszawa, pp 7–32

Matuszko D (2003) Cloudiness changes in Cracow in the 20th century. Int J Climatol 23(8):975–984. https://doi.org/10.1002/joc.887

Matuszko D (2007) Zachmurzenie. In: Matuszko D (ed) Klimat Krakowa w XX wieku. IGiGP UJ, Kraków

Matuszko D (2009) Wpływ zachmurzenia na usłonecznienie i całkowite promieniowanie słoneczne (na przykładzie krakowskiej serii pomiarów). Wydawnictwo Uniwersytetu Jagiellońskiego, Kraków

Matuszko D, Pluta G (2012) Rola cyrkulacji atmosfery w kształtowaniu zachmurzenia (na przykładzie Krakowa). In: Bielec-Bąkowska Z, Łupikasza E, Widawski A (eds) Rola cyrkulacji atmosfery w kształtowaniu klimatu. Uniwersytet Śląski, Sosnowiec, pp 147–156

Matuszko D, Węglarczyk S (2018) Long-term variability of the cloud amount and cloud genera and their relationship with circulation (Kraków, Poland). Int J Climatol 38:1205–1220. https://doi.org/10.1002/joc.5445

Miętus M (ed) (1998) The climate of the Baltic Sea basin. World Meteorological Organization, Marine Meteorology and Related Oceanographic Activities, Rep. No. 41, WMO/TD-No. 933

Miętus M, Filipiak J (2001) Struktura czasowo-przestrzennej zmienności warunków termicznych w rejonie Zatoki Gdańskiej, Materiały Badawcze IMGW, Seria Meteorologia, 32, 53 pp

Miętus M, Filipiak J (2002) Struktura czasowo-przestrzennej zmienności warunków opadowych w rejonie Zatoki Gdańskiej, Materiały Badawcze IMGW, Seria Meteorologia, 34, 79 pp

Miętus M, Filipiak J, Owczarek M (2003) Czasowa i przestrzenna zmienność opadów atmosferycznych w rejonie wybrzeża Zatoki Gdańskiej oraz jej spodziewane zmiany w skali XXI wieku. In: Cyberski J (ed) The Flood in Gdansk. GTN, Gdansk, pp 35–55

Miętus M, Biernacik D, Czernecki B, Filipiak J, Marosz M, Owczarek M, Pilarski M, Wójcik R (2012) Statystyczno-empiryczne projekcje wybranych elementów klimatu Polski na lata 2011–2030. In: Wibig J, Jakusik E (eds) Warunki klimatyczne i oceanograficzne w Polsce i na Bałtyku Południowym. IMGW-PIB, Warszawa, pp 34–91

Morawska-Horawska M (1985) Cloudiness and sunshine in Cracow, 1861–1980, and its contemporary tendencies. J Climatol 5:633–642. https://doi.org/10.1002/joc.3370050605

Norris JR, Slingo A (2009) Trends in observed cloudiness and Earth's radiation budget: what do we not know and what do we need to know? In: Clouds in the perturbed climate system. Heintzenberg J, Charlson RJ (eds.). MIT Press, Cambridge, MA, pp 17–36

Norris JR, Allen R, Evan A, Zelinka MD, O'Dell CW, Klein SA (2016) Evidence for climate change in the satellite cloud record. Nature 536:72–75. https://doi.org/10.1038/nature18273

Nowak M (1971) Częstość występowania chmur Cumulus i Cumulonimbus w Gdańsku i Krakowie, 1961-1965. Zesz. Nauk. Wydz. BiNoZ UG 1

Numaguti A (1999) Origin and recycling processes of precipitating water over the Eurasian continent: Experiments using an atmospheric general circulation model. J Geophys Res 104: 1957–1972 https://doi.org/10.1029/1998JD200026

Schneider T, Kaul CM, Pressel KG (2019) Possible climate transitions from breakup of stratocumulus decks under greenhouse warming. Nat Geosci 12:163–167. https://doi.org/10.1038/s41561-019-0310-1

Siebesma AP, Brenguier J-L, Bretherton CS, Grabowski WW, Heintzenberg J, Kärcher B, Lehmann K, Petch JC, Spichtinger P, Stevens B, Stratmann F (2009) Cloud-controlling factors. In: Heintzenberg J, Charlson RJ (eds) Clouds in the perturbed climate system: their relationship to energy balance, atmospheric dynamics, and precipitation. MIT Press, Cambridge, pp 269–290

Solon J, Borzyszkowski J, Bidłasik M, Richling A, Badora K, Balon J, Brzezińska-Wójcik T, Chabudziński Ł, Dobrowolski R, Grzegorczyk I, Jodłowski M, Kistowski M, Kot R, Krąż P, Lechnio J, Macias A, Majchrowska A, Malinowska E, Migoń P, Myga-Piątek U, Nita J, Papińska E, Rodzik J, Strzyż M, Terpiłowski S, Ziaja W (2018) Physico–geographical mesoregions of Poland: verification and adjustment of boundaries on the basis of contemporary spatial data. Geographia Polonica 91:143–170. https://doi.org/10.7163/gpol.0115

Stjern CW, Kristjànsson JE, Hansen AW (2009) Global dimming and global brightening—an analysis of surface radiation and cloud cover data in northern Europe. Int J Climatol 29:643–653. https://doi.org/10.1002/joc.1735

Szyga-Pluta K (2002) Częstość występowania rodzajów chmur w Polsce Północno-Zachodniej. UAM, Poznań

Szyga-Pluta K (2015) Circulation influence on cloudiness in Poznań. Quaestiones Geographicae 34:141–149. https://doi.org/10.1515/quageo-2015-0021

Szyga-Pluta K (2017) Variation of cloudiness in the mountain region on the example of the Sudetes. Badania Fizjograficzne Seria A Geografia Fizyczna A 68: 205–221. https://doi.org/10.14746/bfg.2017.8.15

Tobin I, Bony S, Roca R (2012) Observational evidence for relationships between the degree of aggregation of deep convection, water vapor, surface fluxes, and radiation. J Clim 25(20):6885–6904. https://doi.org/10.1175/jcli-d-11-00258.1

Trepińska J (2002) Górskie klimaty. Wyd. IGiGP UJ, Kraków

Tuomenvirta H, Alexandersson H, Drebs A, Frich P, Nordli PØ (2000) Trends in Nordic and Arctic temperature extremes and ranges. J Clim 13:977–990. https://doi.org/10.1175/1520-0442(2000)013%3c0977:TINAAT%3e2.0.CO;2

Ustrnul Z, Niedźwiedź T (1994) Wstępna analiza zachmurzenia Polski. Sprawozdanie z Tematu Badawczego M-2. IMGW Oddział w Krakowie (manuscript)

Vavrus SJ, Bhatt US, Alexeev VA (2011) Factors influencing simulated changes in future Arctic cloudiness. J Clim 24(18):4817–4830. https://doi.org/10.1175/2011JCLI4029.1

Warakomski W (1962) O częstości występowania poszczególnych rodzajów chmur w Polsce. Prz Geof 3:185–192

Warren SG, Eastman RM, Hahn CJ (2007) A survey of changes in cloud cover and cloud types over land from surface observations, 1971–1996. J Clim 20:717–738. https://doi.org/10.1175/JCLI4031.1

Webb MJ, Andrews T, Bodas-Salcedo A, Bony S, Bretherton CS, Chadwick R, Chepfer H, Douville H, Good P, Kay JE, Klein SA, Marchand R, Medeiros B, Siebesma AP, Skinner CB, Stevens B, Tselioudis G, Tsushima Y, Watanabe M (2017) The Cloud Feedback Model Intercomparison Project (CFMIP) contribution to CMIP6. Geosci Model Dev 10:359–384. https://doi.org/10.5194/gmd-10-359-2017

Wibig J (2008) Cloudiness variations in Łódź in the second half of the 20th century. Int J Climatol 28:479–491. https://doi.org/10.1002/joc.1544

Willett KM, Jones PD, Thorne PW (2010) A comparison of large scale changes in surface humidity over land in observations and CMIP3 general circulation models. Environ Res Lett 5:1–13. https://doi.org/10.1088/1748-9326/5/2/025210

WMO (2018) Manual on codes—international codes. Volume I.1 Annex to the WMO Technical Regulations: part A - Alphanumeric Codes. WMO TD No. 306

Zhou C, Zelinka MD, Klein SA (2016) Impact of decadal cloud variations on the Earth's energy budget. Nat Geosci 9:871–874. https://doi.org/10.1038/ngeo2828

Żmudzka E (2003) Wielkość zachmurzenia w Polsce w drugiej połowie XX wieku. Prz Geofizyczny 3–4:159–185

Żmudzka E (2004a) Epoki cyrkulacyjne a zmienność zachmurzenia w Polsce. Prz Geofizyczny 1–2:25–41

Żmudzka E (2004b) Uwarunkowania cyrkulacyjne zmienności zachmurzenia w Polsce. Czas Geogr 75:1–2

Żmudzka E (2007) Zmienność zachmurzenia nad Polską i jej uwarunkowania cyrkulacyjne (1951–2000). Wydawnictwo UW, Warszawa

Żmudzka E (2012) Nefologiczne dni charakterystyczne – różne kryteria wydzielania i ich wpływ na wyniki badań. Prz Geofizyczny 1: 21-33.

Chapter 11
Air Temperature Change

Zbigniew Ustrnul, Agnieszka Wypych, and Danuta Czekierda

Abstract The chapter presents the results of analyses made on the basis of daily homogeneous series of maximum, minimum, and average temperature data originating from 58 meteorological stations. Standard measures of climate variability based on absolute values as well as on their relative measures were used. Attention was paid to long-term trends and extreme values, including the number of days with specific air temperature thresholds. All characteristics have been compiled for the period 1951–2018. Changes in the average air temperature since the beginning of their operation were assessed for three stations (Krakow, Warszawa, and Gdansk).

Introduction

In the widespread opinion of most researchers, air temperature is considered to be a guiding element of the climate. Climate change is sometimes equated with and related only to changes in air temperature. It is indeed difficult to identify any comprehensive study on climate change where this element would not be considered. Actually, there is nothing strange in this because it is crucial in many natural and socio-economic dimensions. As the basic factor conditioning the functioning of the environment and human activity, it is considered in various temporal and spatial scales using many measures and indicators as well as research methods.

Given the above, it is difficult to point out most of the most important papers on changes in air temperature that can be found in the world literature. Individual

The original version of this chapter was revised: In this chapter figures 23 and 33 corrections have been updated. The correction to this chapter is available at https://doi.org/10.1007/978-3-030-70328-8_25

Z. Ustrnul (✉) · A. Wypych
Faculty of Geography and Geology, Institute of Geography and Spatial Management, Jagiellonian University in Krakow, Kraków, Poland
e-mail: zbigniew.ustrnul@uj.edu.pl

Z. Ustrnul · A. Wypych · D. Czekierda
Institute of Meteorology and Water Management—National Research Institute, Warsaw, Poland

M. Falarz (ed.), *Climate Change in Poland*, Springer Climate,
https://doi.org/10.1007/978-3-030-70328-8_11

publications include research on various spatial scales from global, through regional approaches, to those presenting the variability of this element on a local scale. Owing to the scale of the issue and the year of publication, H. Lamb's monumental work *Climate: Past, Present, and Future* (Lamb 1972, 1977) should be considered one of the most important studies on air temperature changes in an actually broader context. In its second volume *Climatic History and the Future* (Lamb 1977), the author presents, on examples from different parts of the world, information on changes in air temperature. In addition to data from previous eras based on indirect data, he also cites the results of instrumental measurements of air temperature and its changes. The secular changes are presented on the basis of two known data series for central England (Manley 1974) and the eastern United States based on a study by Landsberg et al. (1968). He also makes a preliminary assessment of changes in the average global temperature since 1870, using, among others Mitchell's calculations (1961). These data clearly show an increase in the average global air temperature since the mid-nineteenth century. Another important study addressing climate change, including primarily air temperature, is the monograph *History and climate* (Jones et al. 2001). In this work, the authors of individual chapters, in addition to intermediate data, present the results of analyses of instrumental temperature measurements in selected regions of the world, in particular Europe, presenting results obtained on the basis of new homogeneous data series. The extensive monograph by Broenimann (2015), in which the author synthesizes knowledge about climate change and its causes, taking into account the period of instrumental measurements, is of particular scientific importance. In addition to the above monographs, the issues of changes in air temperature have been the goal of many hundreds of scientific articles and reports. They dealt with both large regions and individual climate series, which were the basis for the reconstruction of thermal conditions of air. Examples of this type of studies include the works of Bradley (1988), Bradley and Jones (1995), Camuffo and Zardini (1997), Jones and Hulme (1997), Moberg and Bergstrom (1997), Camuffo and Jones (2002), Luterbacher et al. (2004), Boehm et al. (2010), Camuffo et al. (2010).

In addition to the above oldest works, which form the basis of research on global temperature changes, one should mention contemporary studies that are based on a large number of instrumental data. Some of them are based only on data *in extenso*, others use data already processed into grid data. A significant number of them are based on data from reanalyses, which are currently becoming the basis for many climatological studies. Undoubtedly, such work may include studies by, among others, Vose et al. (2005), Jones et al. (2012), or Osborn and Jones (2014). In all of these works, apart from the methodological issues related to "data collection", their processing and generation of data series, there are synthetic conclusions regarding temperature changes in large spatial scales. However, they are usually based on average monthly temperature values and it is difficult to assess daily fluctuations, including also to assess, for example, the occurrence of characteristic days.

The issue of changes in air temperature can also be considered depending on the period covered by the analyses. Certainly, the research in these studies is based on instrumental measurements, although in many of them one can also find the use of various proxy data (e.g. Bradley and Jones 1995, Brazdil 1994, 1996; Brazdil et al. 1996, 2010). There are also numerous such works relating to Poland. Examples include studies by Bednarz (1996), Wójcik et al. (2000), Niedźwiedź (2004), or by Przybylak et al. (2005). Among the numerous studies on changes in air temperature, mention should be made of those that reach the longest measuring series in Europe (e.g. Balling et al. 1998; Luterbacher et al. 2004), and in its individual regions (Tuomenvirta et al. 2000, Auer et al. 2007, Bokwa et al. 2013). Studies devoted to individual data series are also valuable from the methodological point of view. Such studies include, among others, those by Camuffo and Zardini (1997), Demaree et al. (2002), Jurkovič et al. (2011) or, about Poland, by Trepińska (1997), Lorenc (2000), Filipiak (2007), Miętus (1998, 2007), and Bryś and Bryś (2010). Works based on shorter sequences would be difficult to count, especially if they relate to specific geographical regions. Possible examples of such studies, on a Central European scale, can include those by Brazdil et al. (1996), Trepińska et al. (1997), Hoy et al. (2017), Wypych et al. (2017a), or Haensel et al. (2019).

There are also numerous works devoted to the area of Poland and contemporary thermal changes there. The oldest and most important studies for the area of Poland include those by Merecki (1899), Gorczyński (1915a, b, 1916), and later by Romer (1947, 1948/1949), which deal with specific issues of air temperature variability. Among contemporary studies on a Polish scale, works of Ustrnul (2000), Kożuchowski and Żmudzka (2003), Cebulak and Limanówka (2007), Ustrnul and Wypych (2011), Ustrnul et al. (2011, 2014), or Marsz and Styszyńska (2018) are worth mentioning. When reviewing the literature, we sometimes fail to mention here very interesting studies covering a smaller scale, i.e. individual regions or all the more specific measurement series. Undoubtedly, some of them are important even from the methodical point of view.

At the end of the literature review, one should also mention a number of papers devoted to various climatological characteristics based on daily air temperature values, which are based on average or extreme values. Analyses of the temporal variability of the occurrence of so-called characteristic days, including temperature extremes determined taking into account various threshold values, are increasingly becoming the subject of studies on a regional and local scale (Wypych et al. 2017b, c). In addition to threshold values, the analyses also use relative measures (e.g. percentiles) and indicators the purpose of which is a comprehensive approach (Sulikowska and Wypych 2020).

The chapter presents the results of analyses using the most commonly used air temperature characteristics. Owing to editorial restrictions, no analysis was performed using other measures such as day-to-day variability, amplitude, degree-days, or the relationship with air circulation, which is beyond the purpose of this part of the monograph. However, the used measures reflect the main features of variability and, in principle, the changes in air temperature in Poland. They coincide with the trends observed across the entire continent and the globe.

In accordance with the assumptions of this monograph, the current chapter focuses on the analysis of changes in air temperature in the period 1951–2018. However, wherever possible, a short background of changes based on a longer time series was presented.

Owing to the great interest in and importance of knowledge about weather and climate extremes that has been observed in science around the world in recent years, it was decided to extend the analysis to include this issue. A professional approach to it is often not easy owing to the absence of reliable and homogeneous data sequences. The data series included in the work are fully verified and provide the basis for a reliable assessment of the trends in the occurrence of temperature extremes. From a number of methods used in this type of research, the probabilistic approach to exceed certain critical values was used. Owing to the relatively large variation in climatic conditions in Poland, the probability of occurrence of less than 10%, defined by the 10th and 90th percentile values, respectively, for the lowest and highest values of the considered element, was adopted as the main method of determining extremes. This means that extreme values were those found in the population of 10% of the lowest (below the 10th percentile) and 10% of the highest (above the 90th percentile). This approach is in accordance with the recommendations of the IPCC (2007), widely used in numerous works in which the probabilistic method is applied.

Air Temperature Climatological Background

At the beginning, it is worth paying attention to the features of spatial diversification of thermal conditions in Poland based on the period 1951–2018. Figures 11.1, 11.2, and 11.3 show this diversification presenting them on the basis of average air temperatures as well as maximum and minimum air temperatures. The average annual air temperature in Poland varies from below 7 °C at its north-eastern ends, to over 8.5 °C in the south-western part of the country (Fig. 11.1). Certainly, much lower temperatures, which results from their vertical distribution, are recorded in the mountain areas of the Sudetes and the Carpathians. In principle, similar spatial relationships can be found at different times of the year. However, the meridian layout of isotherms in the winter should be noted, which means large thermal differences between the eastern and western parts of Poland. They are conditioned by circulation factors and a much greater influence of polar maritime air masses over the western part of the country, while the eastern part of Poland is in the area of influence of the seasonal high pressure centre, which causes the inflow of frosty continental air masses. It is worth noting that the average winter temperature there is slightly above 0 °C, while it drops to below −3 °C in north-eastern Poland.

The diversification of the average maximum temperature confirms the fact that the western and south-western parts of Poland are privileged in thermal terms, while its north-eastern ends are characterized by thermal severity (Fig. 11.2). It is particularly visible in the winter when, in north-eastern Poland, the average maximum temperature drops below 0 °C. The spatial distribution of the average minimum temperature shows the greatest diversification among all thermal characteristics (Fig. 11.3). In addition to the previously stated general regularities, significant regional differences are visible. They result from hypsometric determinants and the significant impact of the terrain on the minimum air temperature.

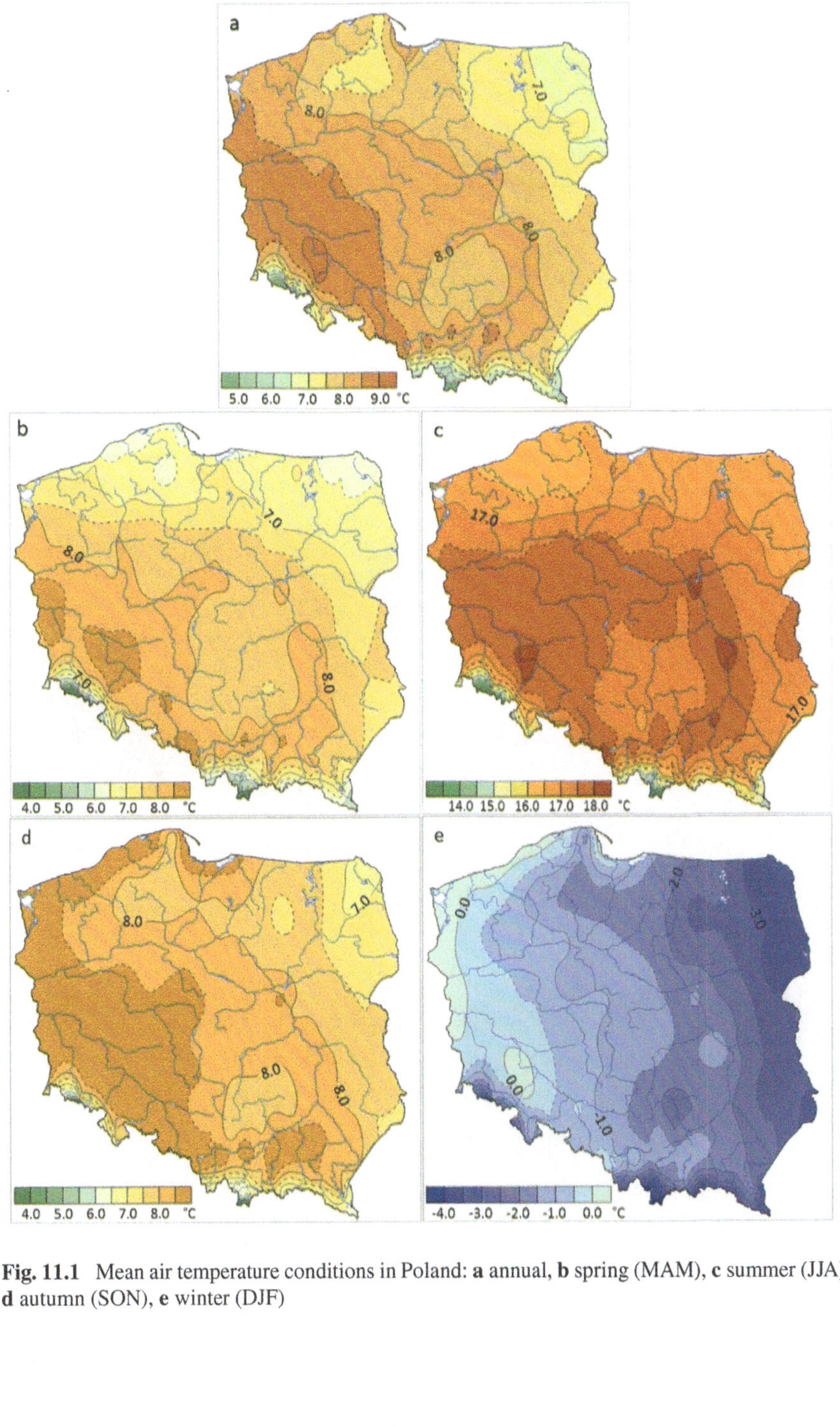

Fig. 11.1 Mean air temperature conditions in Poland: **a** annual, **b** spring (MAM), **c** summer (JJA), **d** autumn (SON), **e** winter (DJF)

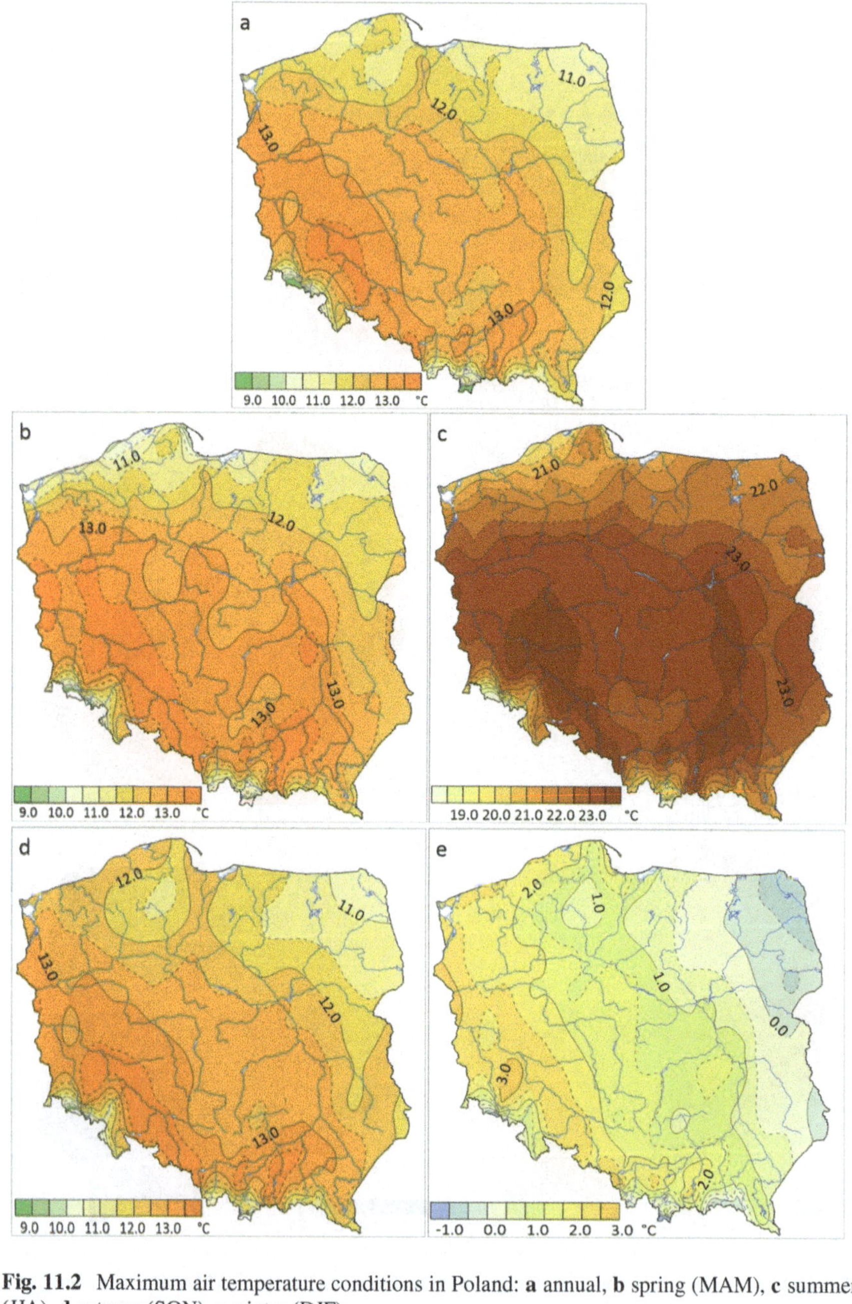

Fig. 11.2 Maximum air temperature conditions in Poland: **a** annual, **b** spring (MAM), **c** summer (JJA), **d** autumn (SON), **e** winter (DJF)

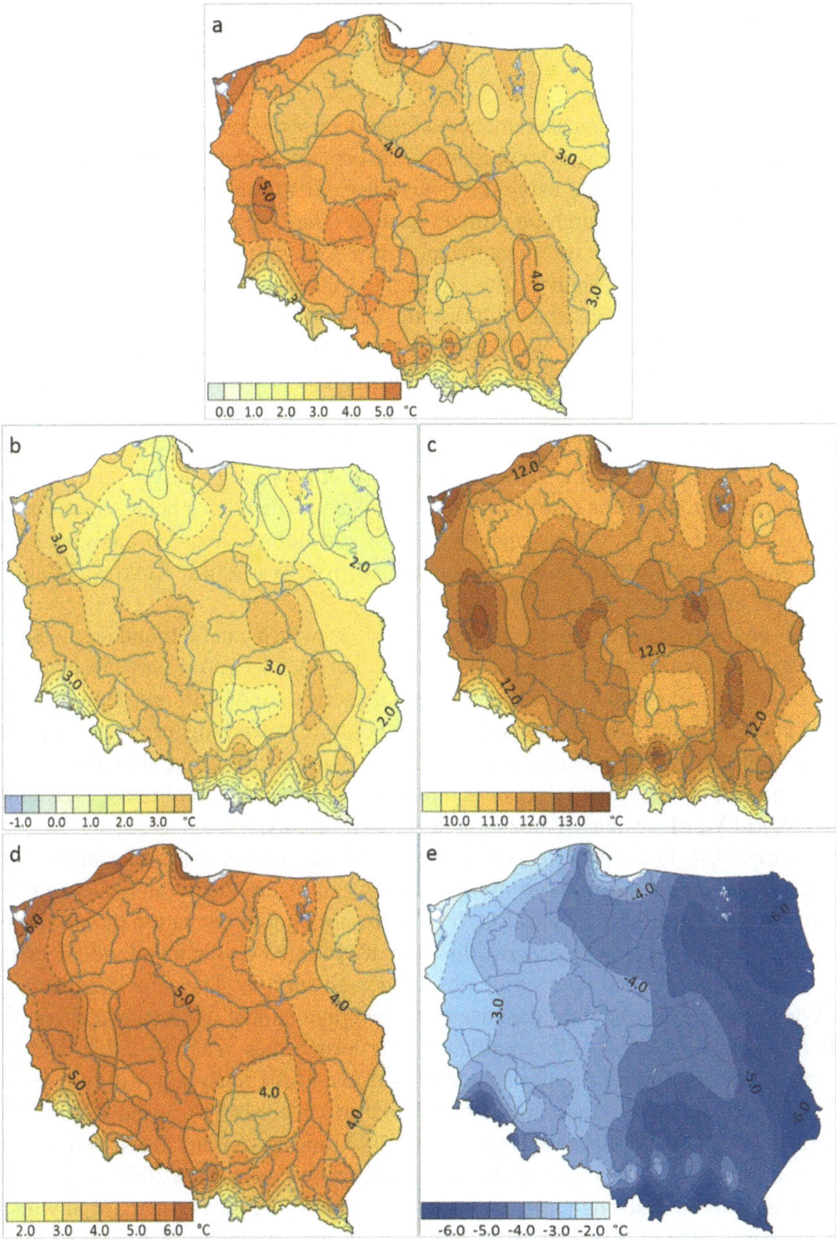

Fig. 11.3 Minimum air temperature conditions in Poland: **a** annual, **b** spring (MAM), **c** summer (JJA), **d** autumn (SON), **e** winter (DJF)

Air Temperature Changes During the Instrumental Measurement Period

The first meteorological measurements in Poland were made in Warszawa (Warsaw) and Gdansk in the mid-seventeenth century. Unfortunately, these measurements were very short: they were made using unverified instruments and hence it is difficult to use them in modern research on climate change. The beginnings of measurements that make the assessment of climate change possible, in particular air temperature, date back to the second half of the eighteenth century (Warszawa 1779, Krakow 1792), although, as Miętus states (2007), the beginnings of the long-term series for Gdansk date back to as early as 1739. Analysing all the information collected so far, it can be concluded that there are several meteorological stations in the current area of Poland, the history of which dates back to the nineteenth century. To this day, new meteorological records are still being found which, after painstaking analyses, make it possible to reconstruct the climatic conditions at least in the last 2 centuries. Of the known stations with a history of up to 200 years, which have been scientifically researched and for which data quality control and homogenization analyses have been made, the following should be mentioned first: Warszawa, Krakow, Gdansk, Wroclaw, and Poznan.

This chapter presents changes in air temperature which can be found on the basis of 3 stations representing northern Poland (Gdansk), its central part (Warszawa), and southern part (Krakow). Although the data series are of different lengths, they reflect the characteristics of changes in air temperature over a long term. Owing to the location of all three stations, especially Warszawa and Krakow, one should be aware of the impact of the urban effect on the formation of thermal conditions, especially after World War II, when urbanization developed rapidly.

Figures 11.4, 11.5, and 11.6 show successively the courses of average air temperatures on an annual scale, the summer (June–August), and the winter (December–February) at the stations in question. It is worth noting that the nature of variability is similar in all 3 regions. In general, this similarity is also visible in individual years. Extremely warm or cold seasons are observed at all stations. This applies especially to Warszawa and Krakow. In the case of the average annual temperature, there are clearly increasing, statistically significant, long-term trends, reaching 2.2 °C and 1.9 °C, respectively, during the measurement periods. Particularly noteworthy are the very high, previously unobserved air temperatures recorded since the 1980s. The average annual temperature values exceeded 10 °C in Krakow and Warszawa, which had been sporadic earlier, and in the case of Warszawa only since 2000 (Fig. 11.4). The annual trend is strongly affected by a statistically significant increase in winter air temperature (Fig. 11.6). In over 200 years of measurements, it reached 3.4 °C in both cities. The temperature variability in the summer is slightly less spectacular. Although its growing trend is also noticed, it is statistically insignificant (Fig. 11.5). Summer seasons in the last few years in Warszawa and Krakow are characterized by average temperature values over 20 °C, although it is worth adding that such an average in the last 30 years had already appeared in 1992. Looking back at history,

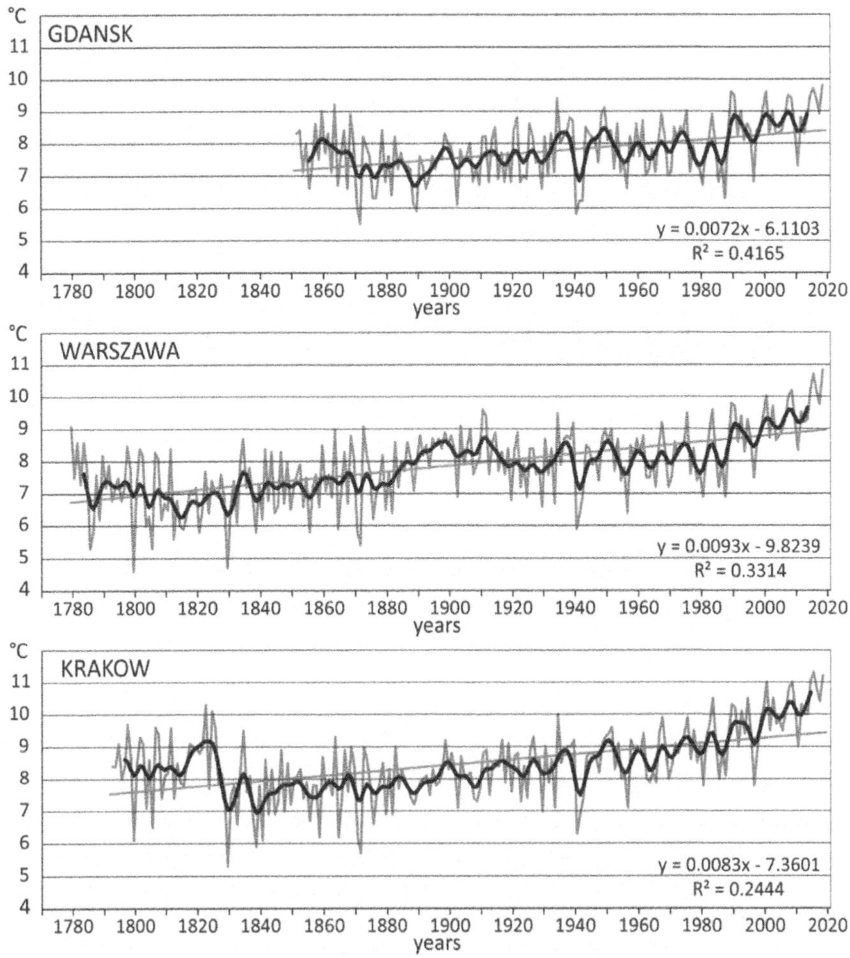

Fig. 11.4 Long-term variability of the annual mean temperature in Gdansk, Warszawa, and Krakow

it should be noted that in Warszawa and Krakow exceptionally warm summers were recorded much earlier, at the beginning of the nineteenth century. The Gdansk series shows slightly smaller fluctuations in air temperature in individual seasons of the year, which is associated with the mitigating influence of the Baltic Sea. This is especially evident in the winter, when it is in Gdansk that the smallest range of variations in its values are recorded, and the long-term trend reaches 1.6 °C.

To sum up, the secular sequences of air temperature from the 3 included stations confirm its growing trends. In their case, the last 30 years, during which extremely high temperature values have occurred, taking into account the year as well as summer and winter, have had a significant impact. The role of these last years is clearly

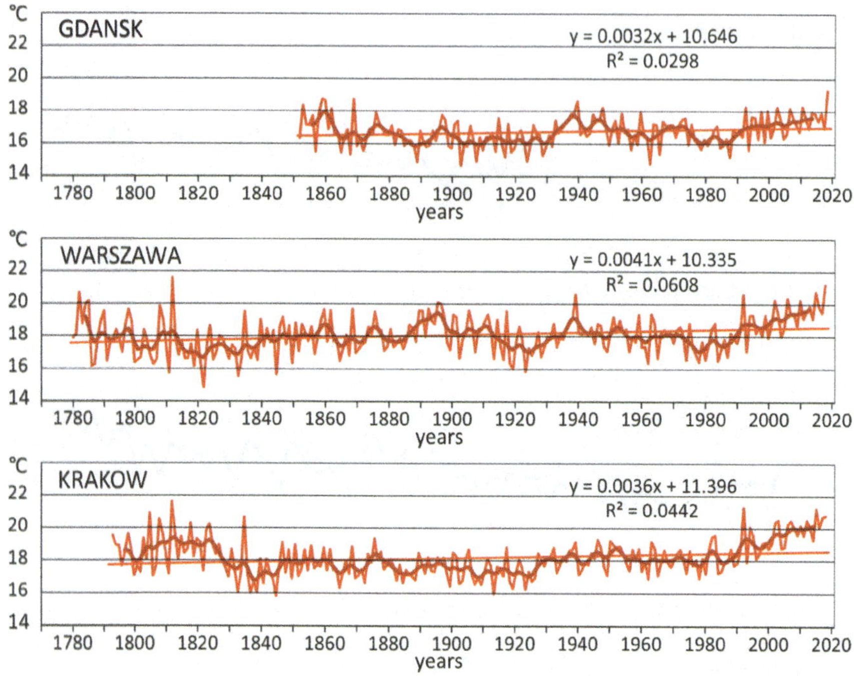

Fig. 11.5 Long-term variability of the seasonal mean temperature in Gdansk, Warszawa, and Krakow—summer (JJA)

expressed in the number of the warmest and coldest seasons that have been determined for the 1851–2018 long-term period common for all 3 stations. Tables 11.1 and 11.2 show the five warmest and coldest seasons. As can be seen, among the highest average annual and seasonal values, the vast majority occurred after 2000. Only in Gdansk there were 4 warm summer seasons, which were recorded in the mid-nineteenth century. Among the coldest seasons, on the other hand, it is no use looking for the last 30–50 years at the first 5 places (Table 11.2). The year 1871, when the average annual temperature reached 5.4–5.7 °C, should be considered the coldest year in the entire history of all 3 measurement series.

In accordance with the assumptions of this monograph, special attention was devoted to the variability of air temperature observed in the years 1951–2018. Figures 11.7, 11.8, 11.9, 11.10, and 11.11 depicts annual and seasonal temperature deviations from the long-term average calculated on the basis of values from the entire analysed period for the three secular stations under consideration. As we can see, average air temperature trends for all seasons are generally positive and mostly statistically significant (Fig. 11.7). The largest growing trends are observed for the annual average, when the average annual increase for Warszawa and Krakow is 0.4 °C per 10 years. This trend is really exceptionally pronounced even when taking into account the additional warming influence of the urban effect. It is worth noting

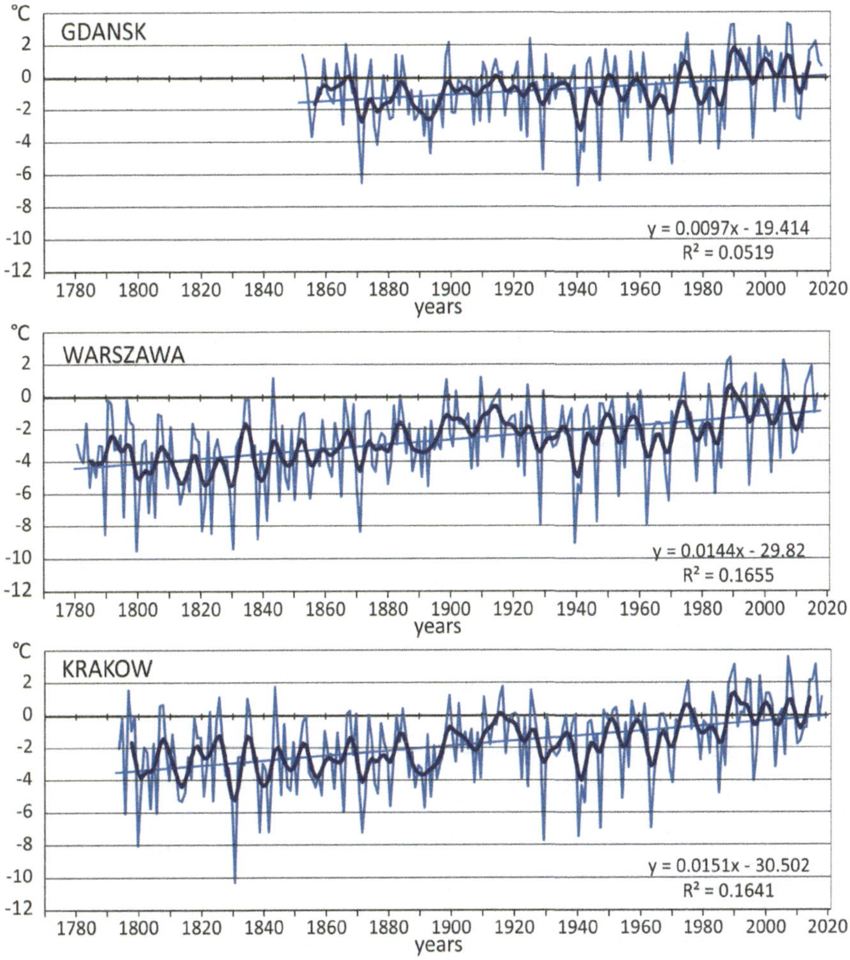

Fig. 11.6 Long-term variability of the seasonal mean temperature in Gdansk, Warszawa, and Krakow—winter (DJF)

that it is so significant mainly owing to positive deviations often exceeding +1.0 °C in the last several years. The years 2014–2018, with a slightly smaller deviation in 2017, showed one of the highest deviations at all 3 stations, especially in Warszawa and Krakow. It should be added that negative deviations prevailed up to the 1980s. A similar nature of the deviations, although slightly less pronounced, is visible in all 4 seasons (Figs. 11.8–11.11). However, large positive trends in the spring and summer, and much smaller ones in the autumn and winter, should be noted here.

Since the occurrence of extreme values is an important indicator of changing climatic conditions, an analysis of the variability of thermal extremes was also carried out in addition to the described changes. To this end, it was decided to check

Tab. 11.1 The highest annual and seasonal air temperatures (°C) in Gdansk, Warszawa, and Krakow and their occurrence year (1851–2019)

No.	Gdansk			Warszawa			Krakow		
	Year	Summer	Winter	Year	Summer	Winter	Year	Summer	Winter
1	9.8 2018	19.3 2018	3.4 2006/07	10.8 2018	21.2 2018	2.7 1989/90	11.3 2015	21.4 1992	3.7 2006/07
2	9.7 2015	18.8 1858 1868	3.3 1988/89 1989/90	10.7 2015	20.8 2015	2.5 2006/07	11.2 2018	21.3 2015	3.2 1989/90 2015/16
3	9.6 1989 2000	18.7 1859	3.2 2007/08	10.2 2008 2014 2016	20.6 1939 1992	2.4 1988/89	11.1 2014	20.9 2018	2.6 1988/89
4	9.5 1990 2007 2014	18.6 1839	2.8 1974/75	10.0 2000 2007	20.4 2006	2.1 2015/16	11.0 2008 2000	20.8 2017	2.5 1997/98
5	9.4 1934 2008 2016	18.4 1852	2.6 1997/98	9.8 1989 2017	20.3 2002 2010	1.8 2007/08	10.8 2007 2016	20.6 2003 2007 2012	2.3 1993/94 2007/08

Tab. 11.2 The lowest annual and seasonal air temperatures (°C) in Gdansk, Warszawa, and Krakow and their occurrence year (1851–2019)

No.	Gdansk			Warszawa			Krakow		
	Year	Summer	Winter	Year	Summer	Winter	Year	Summer	Winter
1	5.5 1871	14.7 1902	−6.7 1939/40	5.4 1871	15.9 1923	−8.8 1939/40	5.7 1871	16.1 1913	−7.7 1928/29
2	5.8 1940	14.8 1962	−6.5 1870/71	5.8 1855 1870	16.0 1919	−8.1 1870/71	6.2 1858 1864 1870	16.4 1864 1919 1923	−7.4 1939/40
3	5.9 1889	14.9 1888	−6.3 1946/47	5.9 1864 1940	16.4 1849 1918 1962	−7.7 1962/63 1928/29	6.3 1940	16.5 1899	−7.2 1870/71
4	6.1 1870 1888 1902	15.0 1907 1923	−5.7 1928/29	6.2 1875	16.5 1864 1978 1984	−7.5 1946/47	6.6 1875	16.6 1909	−6.9 1946/47 1962/63
5	6.2 1941 1942	15.2 1916 1987	−5.3 1969/70	6.4 1941 1881 1956	16.6 1916 1974 1980	−6.2 1969/70	6.7 1855	16.7 1872 1884 1902 1907	−6.0 1864/65

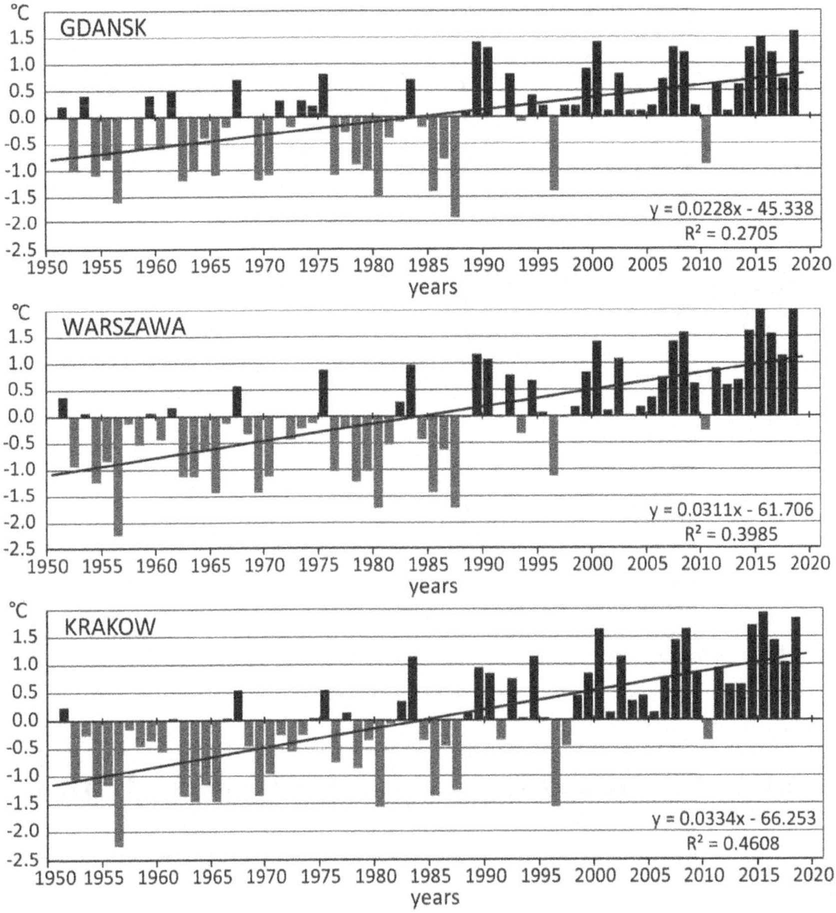

Fig. 11.7 Annual mean temperature anomalies (with respect to the period 1951–2018)

the behaviour of the maximum and minimum air temperature trends for an almost 140-year homogeneous data series from Krakow—the Observatory—from the years 1881–2018. Although the station represents the city's climate, the presentation of changes in the number of days with extreme values determined by the 10th and 90th percentile thresholds well illustrates the long-term variability of this element.

Long-term trends in the number of days with a maximum temperature below the 10th percentile in all seasons are negative. Their magnitudes are small and, as a result, statistically insignificant. The decrease in value reaches 0.3–0.4 per day per 10 years (Fig. 11.12). Over the past several years, the number of days with these temperature peaks has been clearly smaller than in previous decades. Noteworthy is their extremely high winter variability, which is also characterized by their high number, exceeding 30 days in the years 1925–1965. Opposite trends in the number of extremes, because they are increasing, are evident for the long-term course of

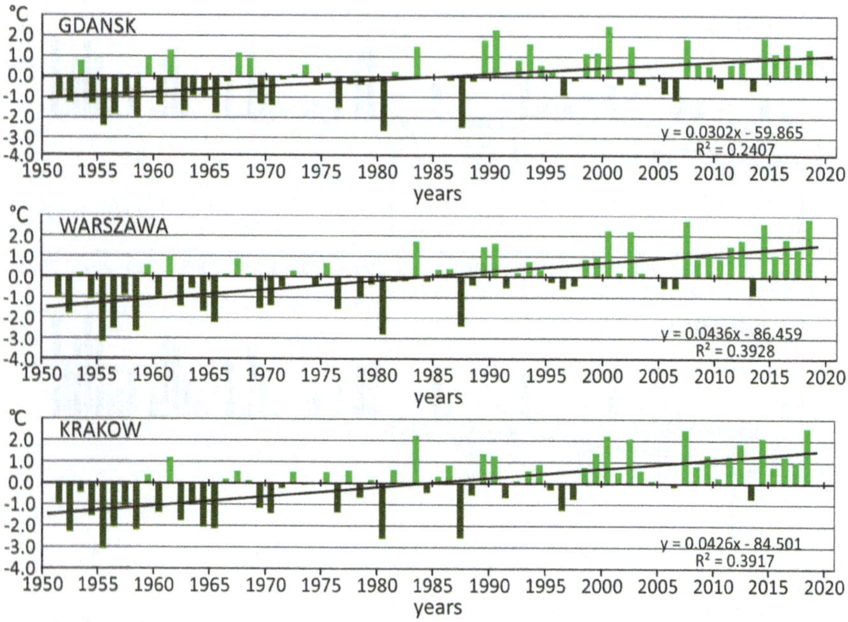

Fig. 11.8 Seasonal mean temperature anomalies (with respect to the period 1951–2018)—spring (MAM)

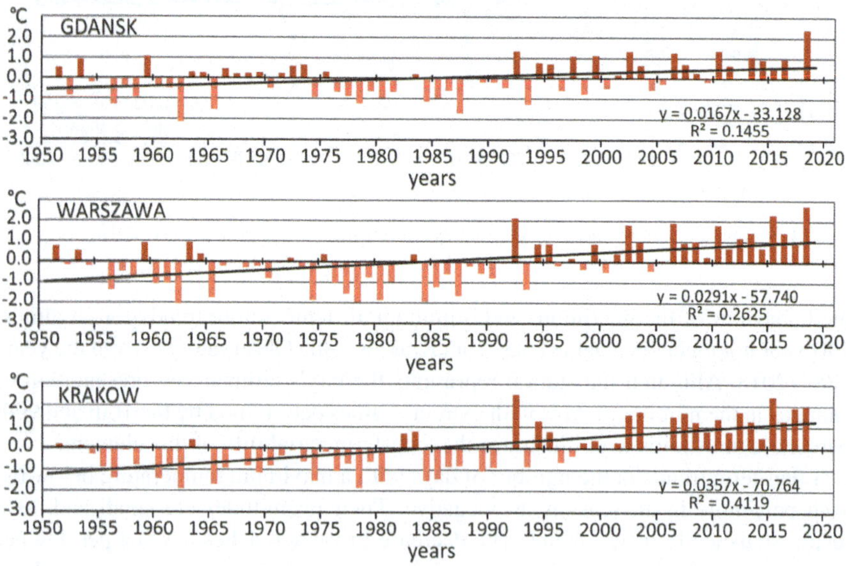

Fig. 11.9 Seasonal mean temperature anomalies (with respect to the period 1951–2018)—summer (JJA)

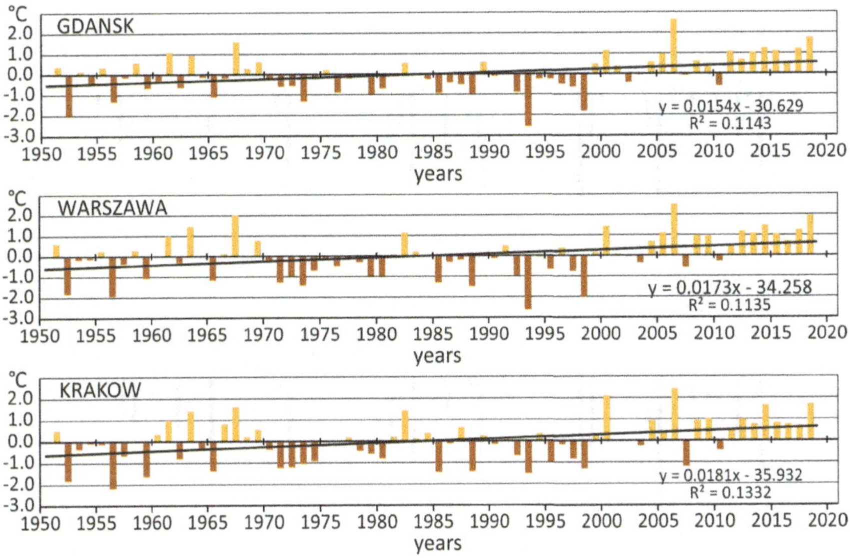

Fig. 11.10 Seasonal mean temperature anomalies (with respect to the period 1951–2018)—autumn (SON)

extremes determined by the value of the 90th percentile (Fig. 11.13). There is a clearly increased number of days in the last few years, especially in the summer and winter. This is the result of a series of extremely hot summer seasons and warm winters. Trends for these seasons reach 0.9 and 0.8 per day per 10 years, respectively.

The course of the number of days with extreme minimum temperature values is quite similar. In all seasons, trends in the number of days with values below the 10th percentile are negative, while the positive ones are above the 90th percentile (Figs. 11.14 and 11.15, respectively). This is owing to the higher incidence of days with higher minimum temperatures. As in the case of maxima, this applies especially to the winter and summer. It was in the summer that the trend value reached 1.5 days per 10 years, which was caused by a significant number of days with high minimum temperature values.

Long-Term Variability of the Daily Mean Air Temperature

The features of air temperature variability presented on the example of the above 3 stations, having one of the longest data sequences in Poland, were analysed in more detail for the period 1951–2018 on the basis of data from all 58 available measurement points. At the beginning, the assessment of the average annual temperature variability was made on the example of 5 stations representing different regions and having complete data sequences. At the same time, these stations were selected taking into

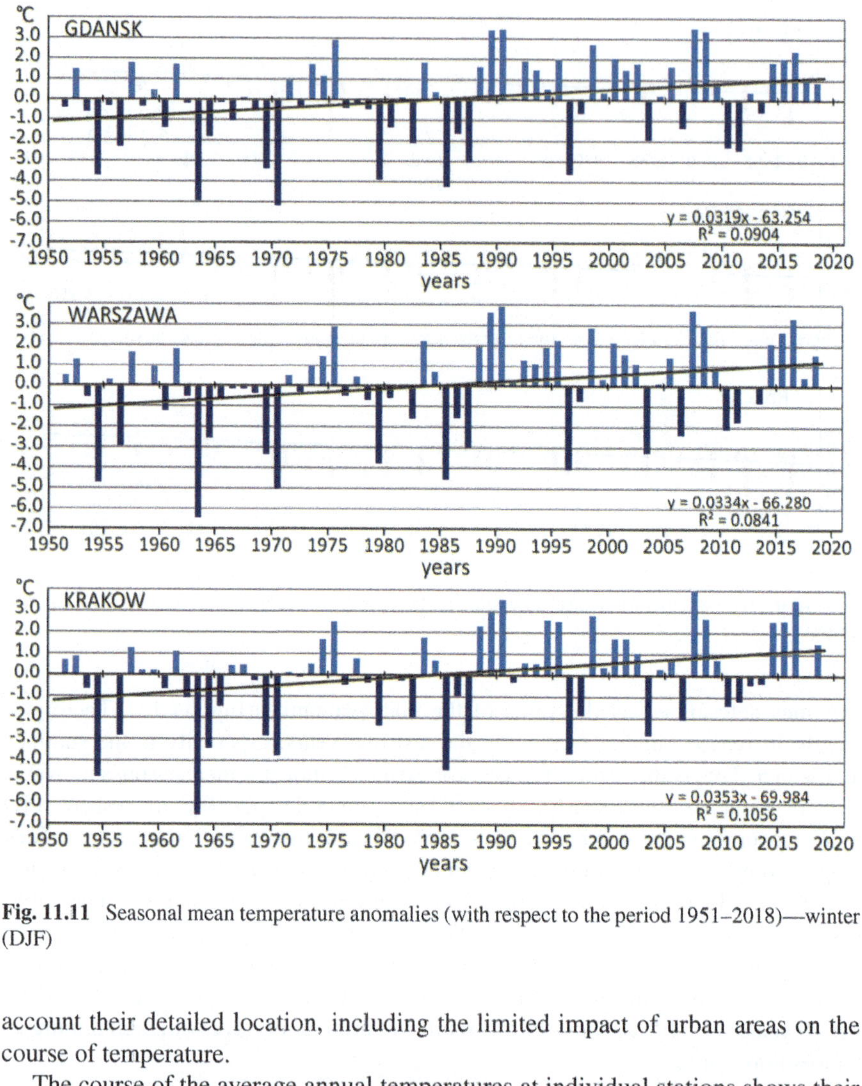

Fig. 11.11 Seasonal mean temperature anomalies (with respect to the period 1951–2018)—winter (DJF)

account their detailed location, including the limited impact of urban areas on the course of temperature.

The course of the average annual temperatures at individual stations shows their strong growing trend (Fig. 11.16). It is the result of much higher values in the last 20–25 years, especially in the period 2014–2018. Throughout the entire multi-year period, the increase in average values can be estimated at even as much as 2 °C. This trend is highly statistically significant and observed throughout the country. Its highest values, exceeding 0.3 °C/10 years, were found in central-western Poland, and slightly lower below 0.2 °C/10 years in its south-eastern part.

The nature of the changes, similar to the average for the year, is also visible in the case of seasonal values (Figs. 11.17, 11.18, 11.19, 11.20, 11.21). The trends of changes are positive in all seasons, with the spring showing the highest temperature

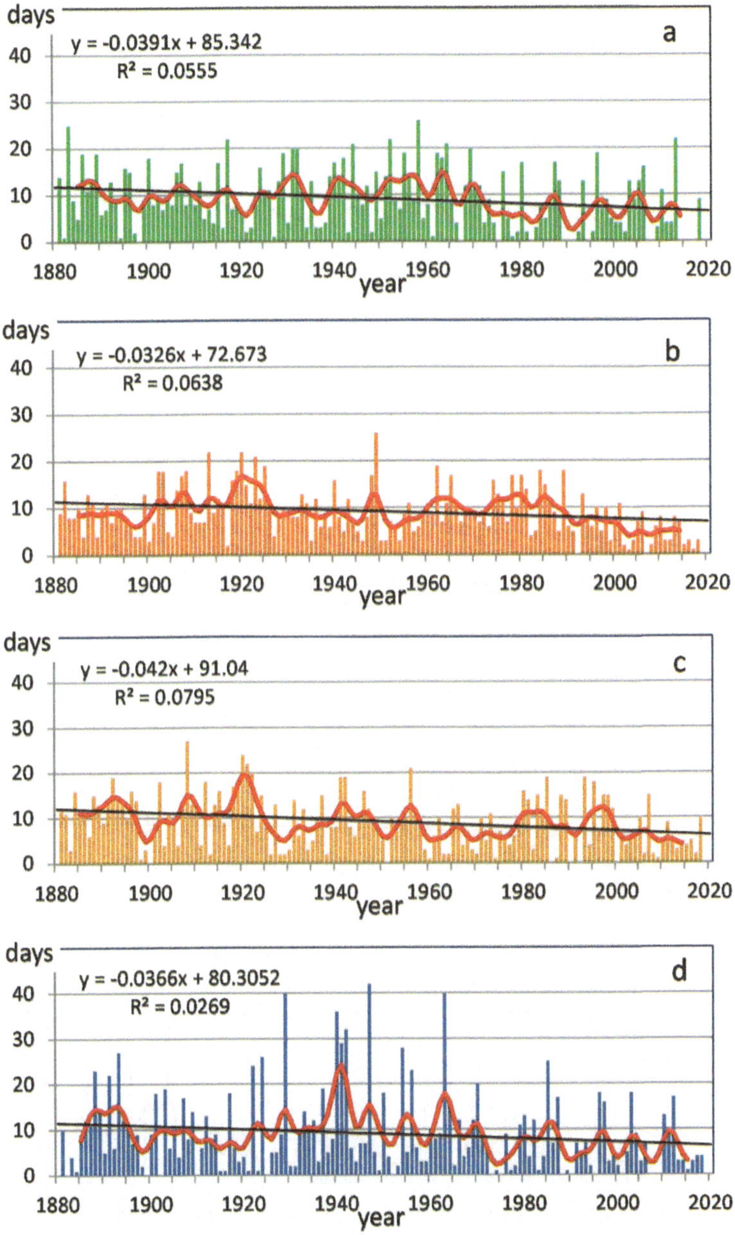

Fig. 11.12 Variability of the number of days with maximum air temperature below 10 percentile values in particular seasons in Krakow (1881–2018); **a** spring (MAM), **b** summer (JJA), **c** autumn (SON), **d** winter (DJF)

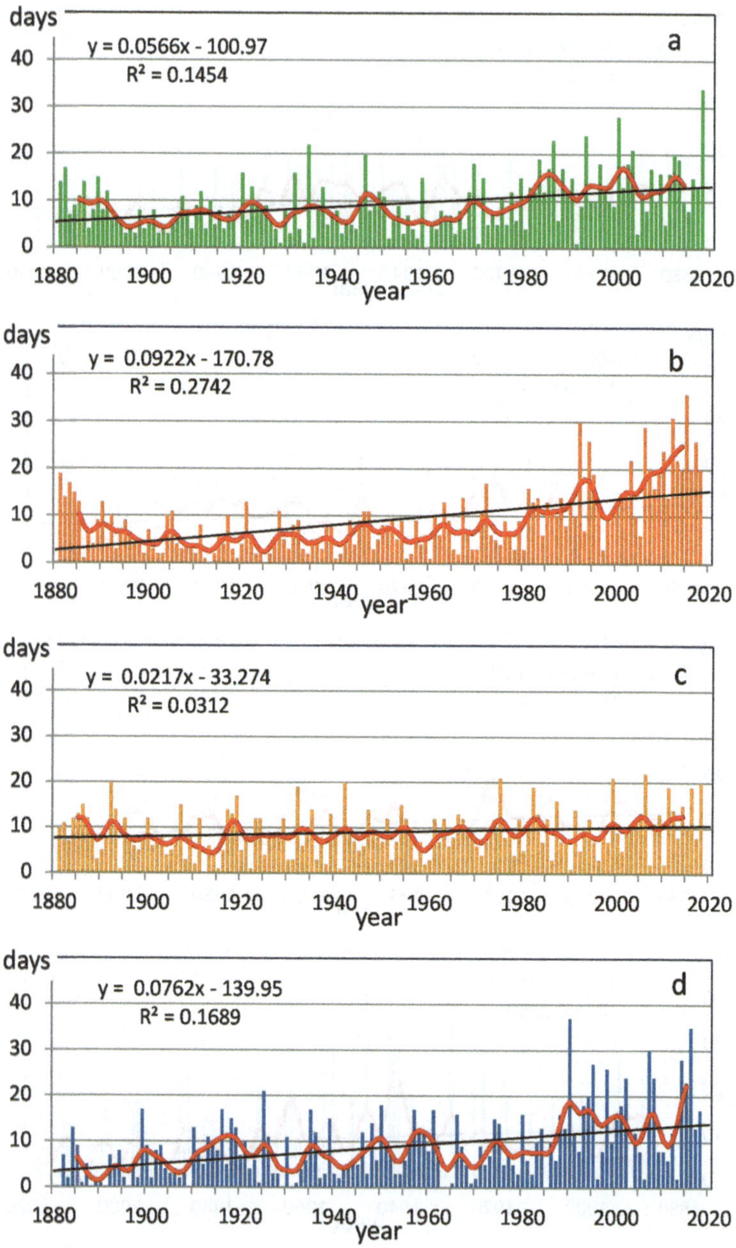

Fig. 11.13 Variability of the number of days with maximum air temperature exceeding 90 percentile values in particular seasons in Krakow (1881–2018); **a** spring (MAM), **b** summer (JJA), **c** autumn (SON), **d** winter (DJF)

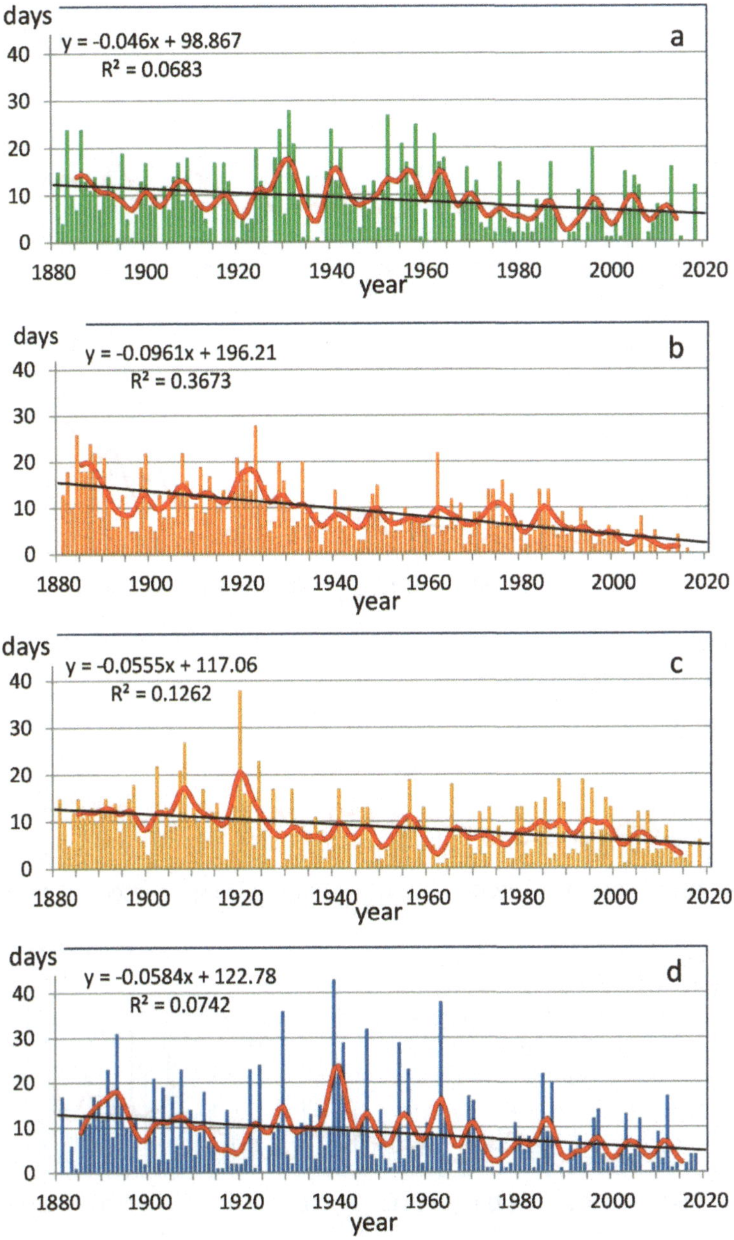

Fig. 11.14 Variability of the number of days with minimum air temperature below 10 percentile values in particular seasons in Krakow (1881–2018); **a** spring (MAM), **b** summer (JJA), **c** autumn (SON), **d** winter (DJF)

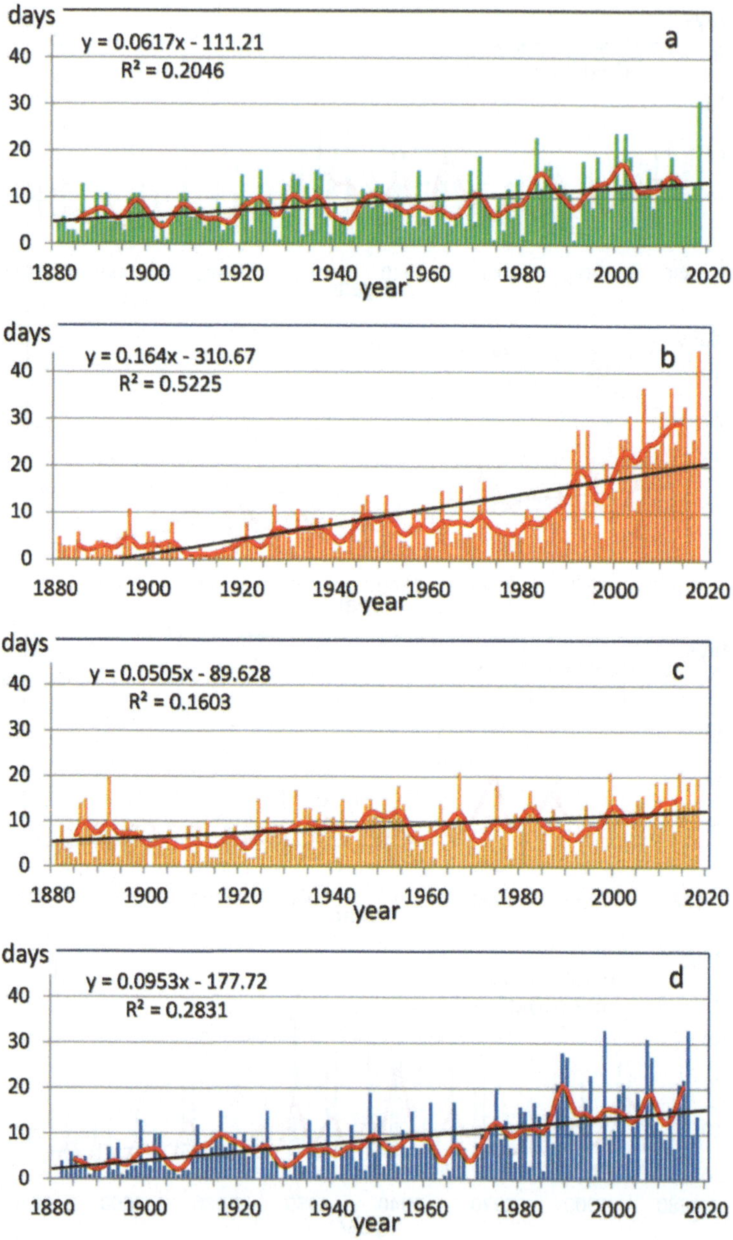

Fig. 11.15 Variability of the number of days with minimum air temperature exceeding 90 percentile values in particular seasons in Krakow (1881–2018); **a** spring (MAM), **b** summer (JJA), **c** autumn (SON), **d** winter (DJF)

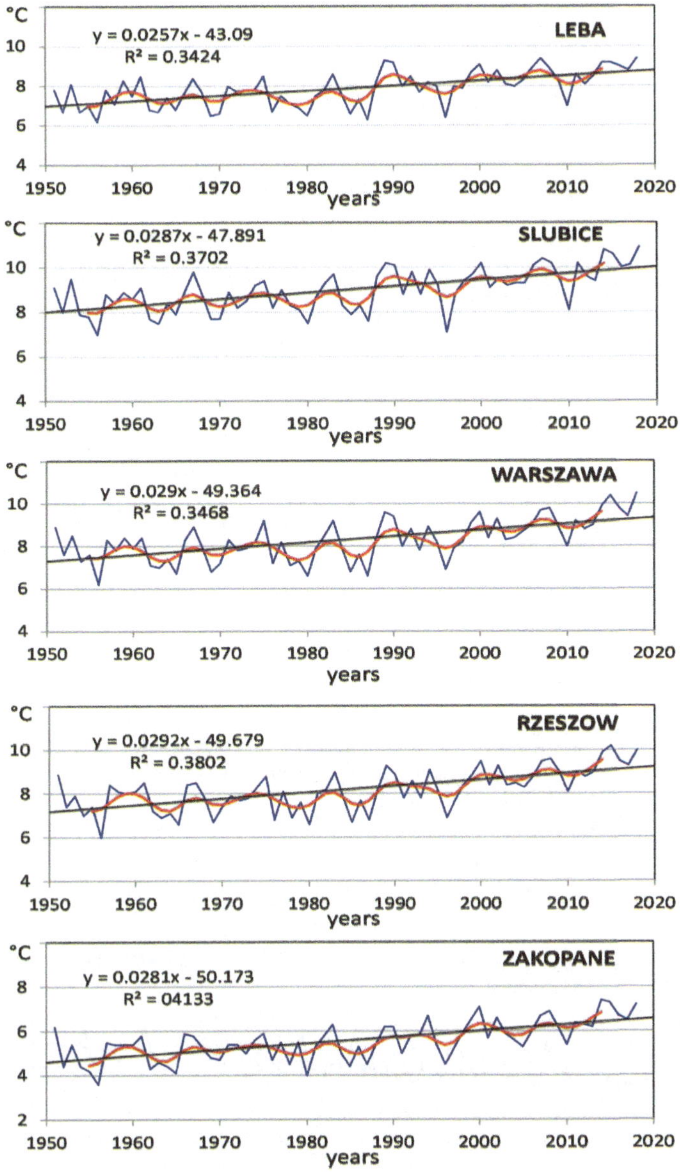

Fig. 11.16 Variability of the annual mean air temperature at selected stations; *red line—Gaussian filter; straight line—linear trend*

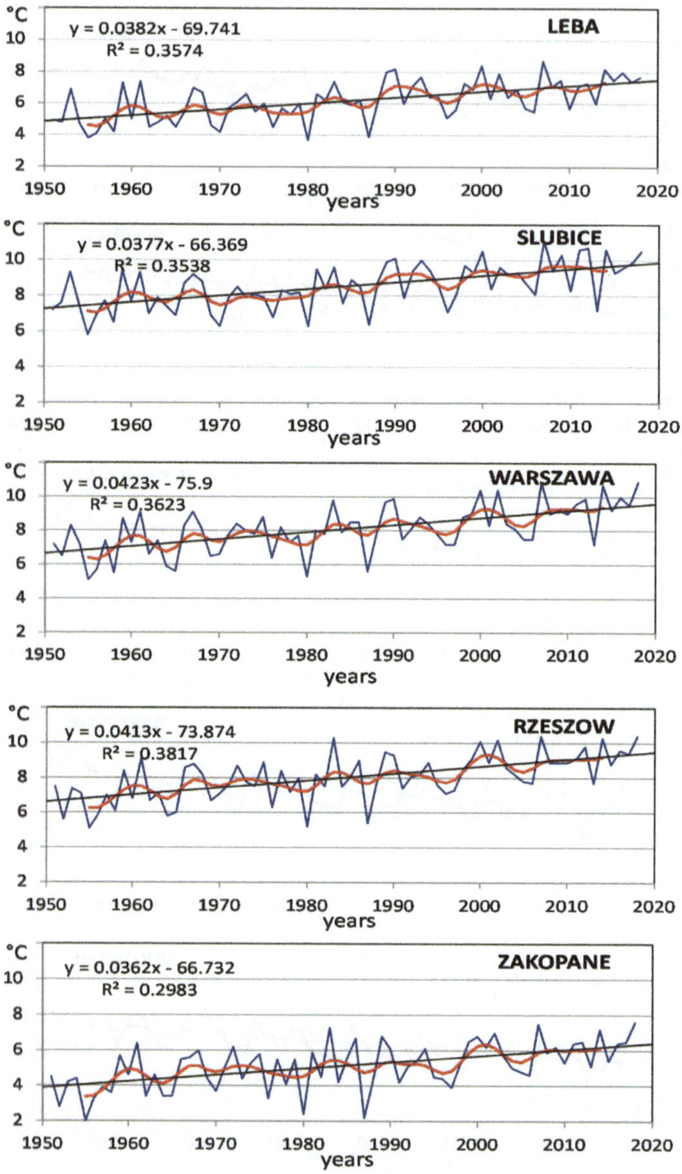

Fig. 11.17 Variability of the seasonal mean air temperature at selected stations—spring (MAM); *red line—Gaussian filter; straight line—linear trend*

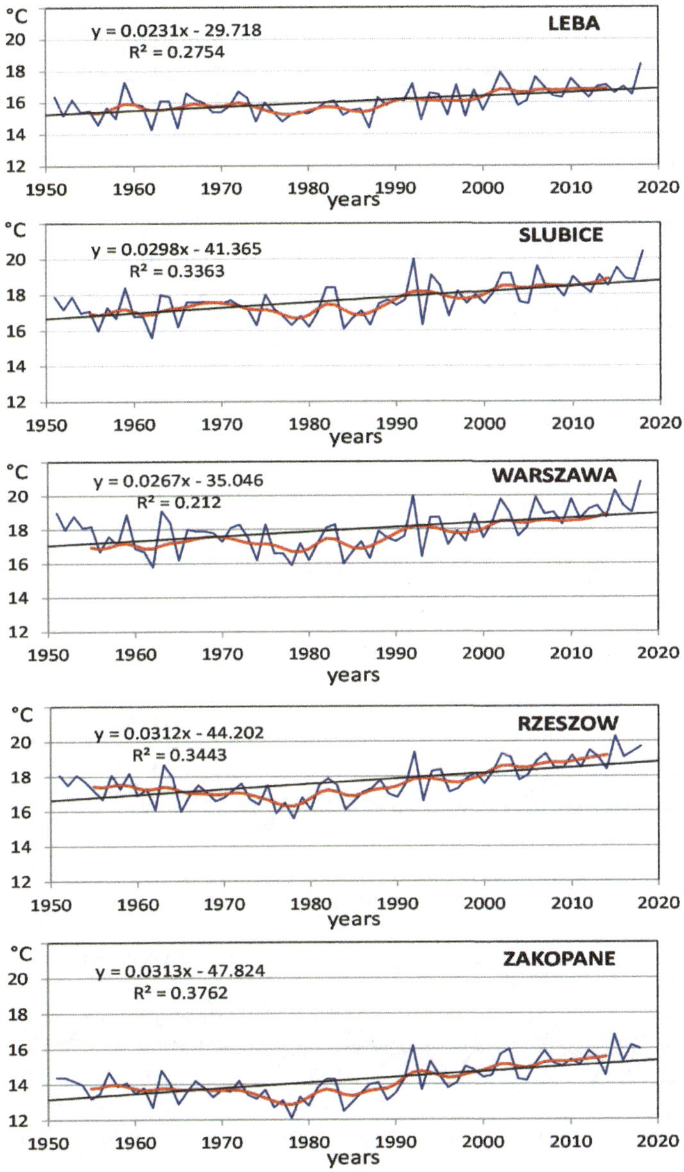

Fig. 11.18 Variability of the seasonal mean air temperature at selected stations—summer (JJA); *red line—Gaussian filter; straight line—linear trend*

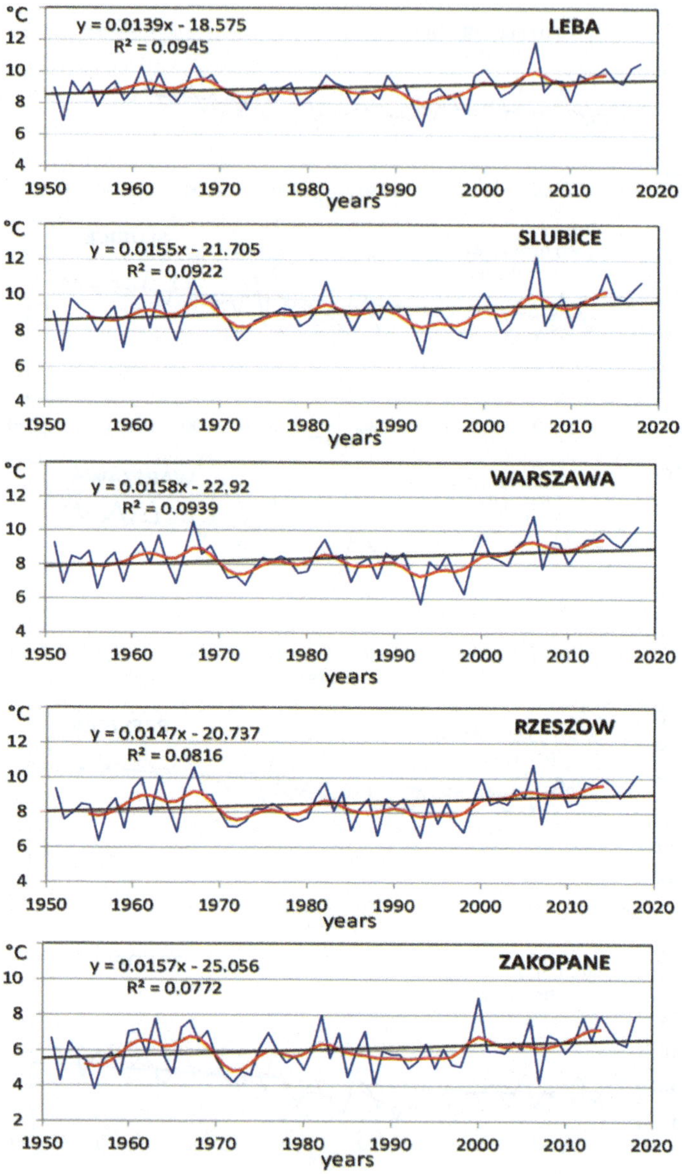

Fig. 11.19 Variability of the seasonal mean air temperature at selected stations—autumn (SON); *red line—Gaussian filter; straight line—linear trend*

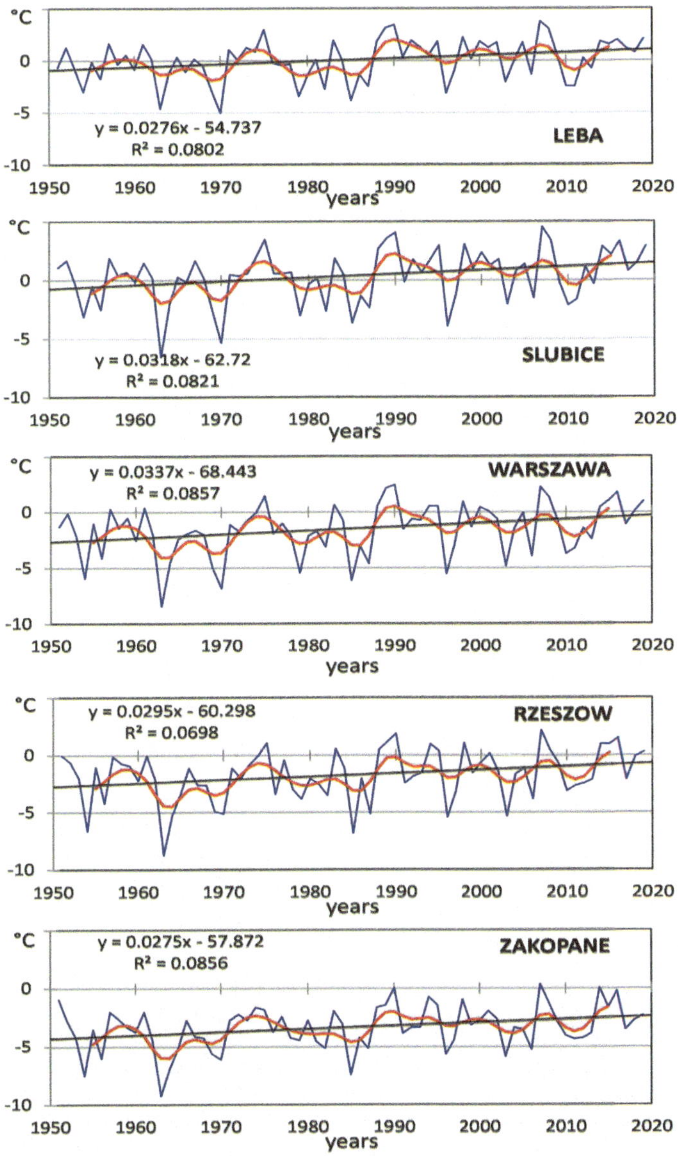

Fig. 11.20 Variability of the seasonal mean air temperature at selected stations—winter (DJF); *red line—Gaussian filter; straight line—linear trend*

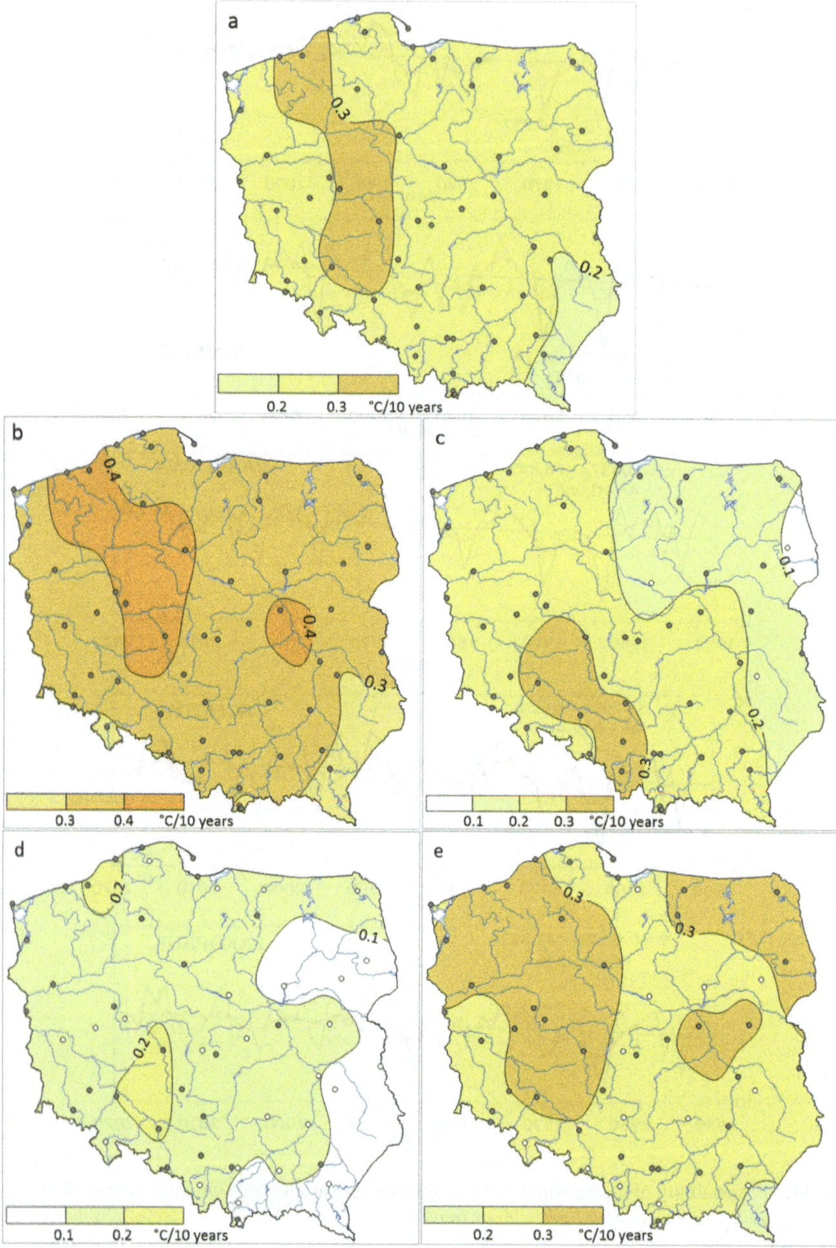

Fig. 11.21 Linear trends of the annual and seasonal mean air temperature in Poland (°C/10 years, (dark dots denote significant trends at the particular stations at the significance level of 0.05): **a** annual, **b** spring (MAM), **c** summer (JJA), **d** autumn (SON), **e** winter (DJF)

rise. In turn, it is the smallest in the autumn, when in eastern Poland it is practically difficult to observe (Fig. 11.21).

Analysing in detail the graphs in Figs. 11.16, 11.17, 11.18, 11.19, and 11.20, it is also possible to indicate particularly outstanding years. Considering the average annual temperature value, the year 1956 turned out to be the coldest at all 5 stations, and the years 2014–2015 the warmest (Fig. 11.16). The only season in which it is difficult to indicate years with the highest or lowest temperature values occurring simultaneously throughout the country was the spring (Fig. 11.17). In the summer, the lowest average temperature occurred in 1962 and 1978. In turn, its highest values fell in 2018 and 1992 (Fig. 11.18). It is worth noting that these years do not correspond to the extreme values recorded in Western Europe, where an extremely hot summer occurred in 2003 and 2015. In a significant area of the country, the coldest autumn occurred in 1993, and the warmest in 2006 (Fig. 11.19). In the entire 68-year period, the harshest winter was in the 1962/1963 season, and the mildest in 2006/2007. In that season, even in Zakopane, the average air temperature was positive at that time (Fig. 11.20).

In order to make a synthetic presentation of air temperature variability in Poland, its average area value was determined. It was calculated for the annual average and individual seasons, which made it possible, at the next stage to construct a diagram illustrating the average area thermal characteristics of Poland. This average was determined on the basis of air temperature values from all stations included. Owing to their relatively even location in Poland, the presented values represent the average temperature throughout the country. Although this is a simple measure, it characterizes long-term temperature variability very well and makes the hierarchical assessment of individual seasons possible. Figure 11.22 clearly shows that the last 20 years have been distinctly warmer than the beginning of the analysed period. In the 1951–2018 period under review, the coldest year was 1956, while the coldest summer was in 1962 and the coldest winter was in 1962/1963. Since 1999, practically every year at least one of the seasons has been classified as anomalous or even extremely warm, i.e. in the range above the 90th or even 95th percentile, which is reflected in the assessment of the annual temperature. Extremely warm and, in the light of the area average, the warmest years in the whole period under consideration were the years 2014, 2015, and 2018, in which most or all seasons were characterized by a temperature higher than that assumed as normal for the period 1951–2018. This general unambiguous trend indicating warming in all seasons is slightly disturbed by the thermal conditions of the winter. Despite the media opinion about increasingly mild winters, in the last 10 years up to 3 winters (2010, 2011, and 2013) should be considered very cold or anomalously cool according to the adopted classification.

Long-Term Variability of Extreme Daily Air Temperatures

Average air temperature is the thermal variable most frequently considered. It is undoubtedly of great importance as a basic indicator of climatic conditions. It is considered by many researchers to be the most important synthetic indicator of

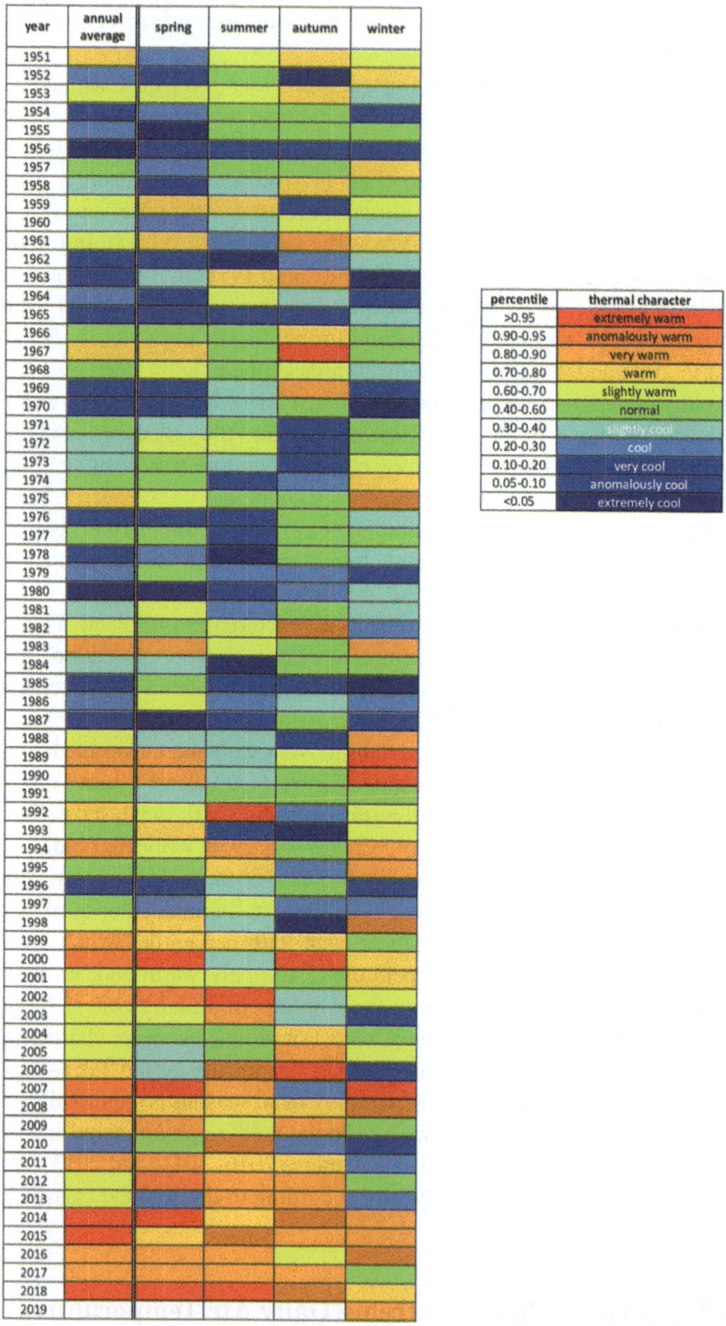

Fig. 11.22 Annual and seasonal air temperature classification—areal mean for Poland

climate change. Sometimes these changes are assessed only from the perspective of average air temperature. However, more in-depth analyses require reaching for extreme values based on daily maximum and minimum temperature values, which, incidentally, are also used in Anglo-Saxon countries to determine mean values. Without going into the various opinions on the importance of mean values (referring, for instance, to Romer 1948/1949), taking into account temperature values based on its extremes becomes a necessity. Climate change is often understood through changes in extreme temperatures (IPCC 2007, 2013).

On the basis of homogeneous, daily series of maximum and minimum air temperatures, the changes in their occurrence in Poland in the analysed period were determined. In addition to the standard analysis, given the importance of the occurrence of weather extremes in the analysis of changing climatic conditions, an assessment was made of the long-term variability of values considered extreme for both the maximum and minimum air temperature. These extremes were determined both on a seasonal basis, determining on the basis of the considered period 1951–2018 the values of the 10th and 90th percentiles, daily minimum temperature (Tmin) and daily maximum temperature (Tmax), in subsequent seasons of the year, followed by the long-term variability of the number of days with the above extremes, and on an annual basis, where the 10th and 90th percentiles were calculated for each year separately, which makes the analysis of the long-term variability of their values possible.

The average values of the maximum temperature trends in Poland show similar, increasing trends as in the case of the average temperature. The average annual maximum temperature shows positive trends reaching 0.2–0.3 °C/10 years in a significant part of the country (Fig. 11.23). They are the highest in the spring, when they exceed 0.4 or even 0.5 °C/10 years. These values should be considered alarming, because with such high trends over 100 years, the average maximum temperature of the spring can increase by as much as 5 °C. Significantly, only slightly smaller trends are observed in the summer. They reach the highest values of 0.4 °C/10 years in the southern part of Poland. In the winter, as was to be expected, they are also growing and in most of the country they exceed 0.2 °C/10 years. The only season of the year where there are no such trends, similar to the average daily air temperature, is the autumn.

Trends in changes in the minimum temperature do not differ significantly from those for the average and maximum temperatures. The average annual value of the minimum temperature in a significant part of the country shows positive trends reaching 0.2–0.3 °C/10 years (Fig. 11.24). They are the highest in the spring and winter, when in northern and eastern Poland they exceed even 0.4 °C/10 years. The lowest (below 0.3 °C/10 years) occur in mountain areas, indicating the greater thermal and climatic stability of these areas. As could be expected, the autumn is the season with relatively lowest long-term trends, when they do not exceed 0.3 °C/10 years.

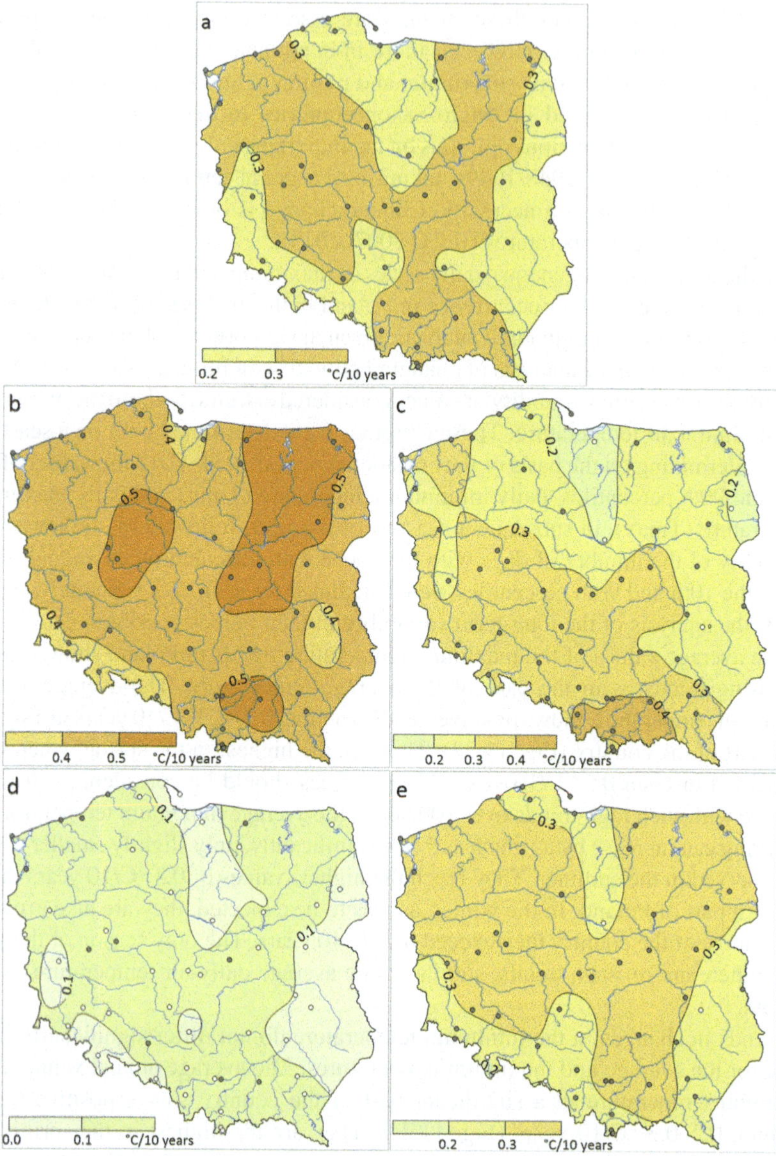

Fig. 11.23 Linear trends of the annual and seasonal maximum air temperature in Poland (°C/10 years (dark dots denote significant trends at the particular stations at the significance level of 0.05): **a** annual, **b** spring (MAM), **c** summer (JJA), **d** autumn (SON), **e** winter (DJF)

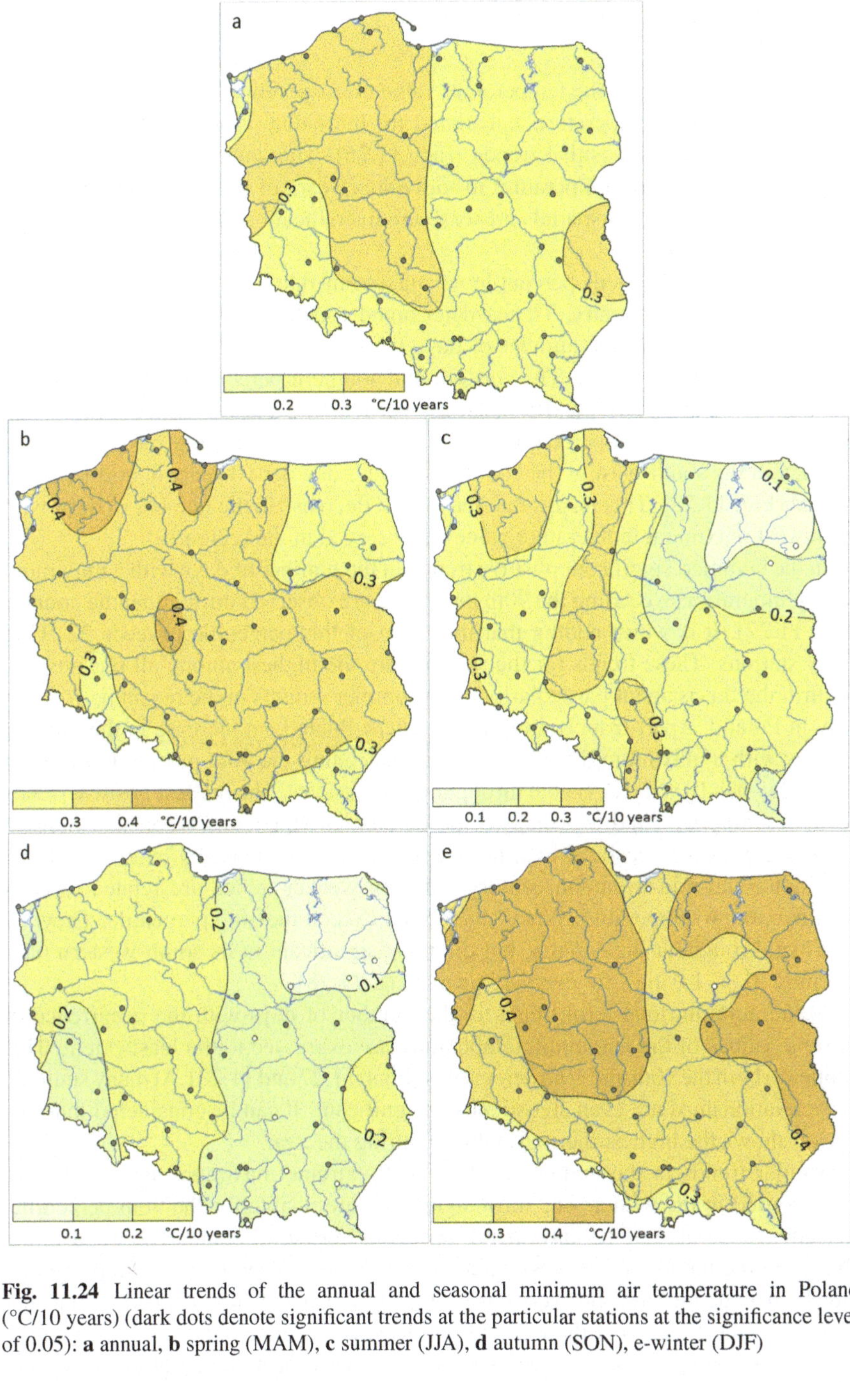

Fig. 11.24 Linear trends of the annual and seasonal minimum air temperature in Poland (°C/10 years) (dark dots denote significant trends at the particular stations at the significance level of 0.05): **a** annual, **b** spring (MAM), **c** summer (JJA), **d** autumn (SON), e-winter (DJF)

Maximum Daily Temperature

The long-term course of the temperature value corresponding to the 90th percentile of daily maximum temperature, calculated for individual years and the 5 selected stations, shows weak positive trends (Fig. 11.25). The course of extreme values of the daily maximum temperature in individual years of the multi-year period is similar (Fig. 11.26). The spatial and seasonal diversification of change trends is not surprising.

In the spring, the 90th percentile of maximum daily air temperature is positive throughout the country. Trends determined for the examined period indicate an average increase in the number of days with these values by over 0.5 days for every 10 years (Fig. 11.27). The western and southern parts of the country, where this trend exceeds 1 day per 10 years, are particularly noteworthy. A similar map constructed for the 10th percentile shows a negative trend ranging from just above -0.6 day in the south-west of the country to over 1 day per 10 years in north-eastern Poland (Fig. 11.28). This means that the largest decrease in the number of days with the lowest maximum temperature values is observed in the region.

In the summer, significant positive trends in the number of days with a maximum daily temperature exceeding the 90th percentile are observed throughout the country (Fig. 11.27). It is worth adding that the value of this percentile exceeds 32 °C at many stations. These trends for the summer are the highest among all seasons and confirm the thesis about particularly warm summer months in the recent period, not only in Poland. Apart from central and northern Poland, an increase in the number of days with a maximum daily temperature exceeding the 90th percentile reaching over 1 day per 10 years is visible throughout the country. The highest positive trends occur in southern Poland, especially in the Carpathian Mountains, where this increase reaches 2 days per decade, which in turn results in an increase by over 12 extreme days since the mid-twentieth century. Slightly weaker trends are related to days with maximum temperature values that do not exceed the 10th percentile. They are negative throughout the country, but the most significant in its south-western part, reaching -0.8/–1.0 days for every 10 years (Fig. 11.28).

In the autumn, long-term trends in the number of days with the occurrence of extreme values of the maximum temperature are expressed to the least extent. This applies to both the 90th and 10th percentiles (Figs. 11.27 and 11.28). At many stations, these relationships are even statistically insignificant. It can be considered that this season shows the highest thermal stability during the year.

Similar trends in the number of days with extreme maximum air temperature are recorded in the winter. The most significant are visible for the 90th percentile, exceeding the value of 1.5 days per 10 years in northern Poland. The trend in the change of the number of days with a temperature not exceeding the 10th percentile threshold is decreasing, but very insignificant, and it nowhere exceeds 1 day per 10 years. This means that there is a slight decrease in the number of days with the temperature being in the range of 10% of the lowest peak temperatures recorded, which is associated with the fact that the majority of days reach values within the

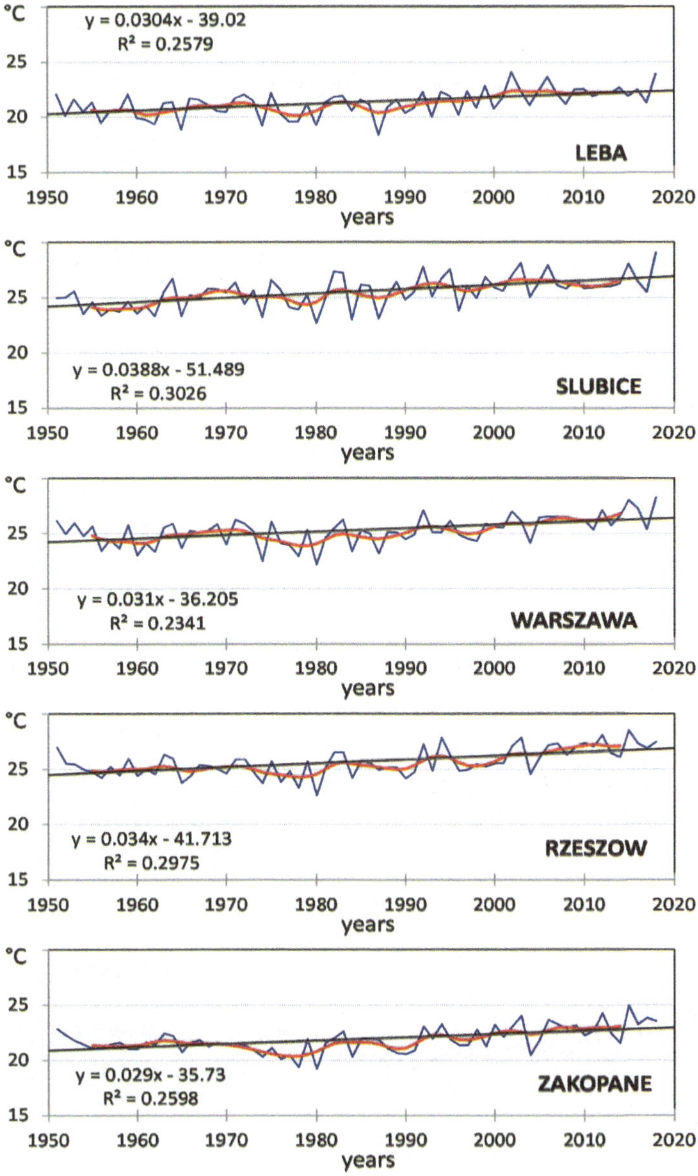

Fig. 11.25 Variability of the 90th percentile air temperature values at selected stations

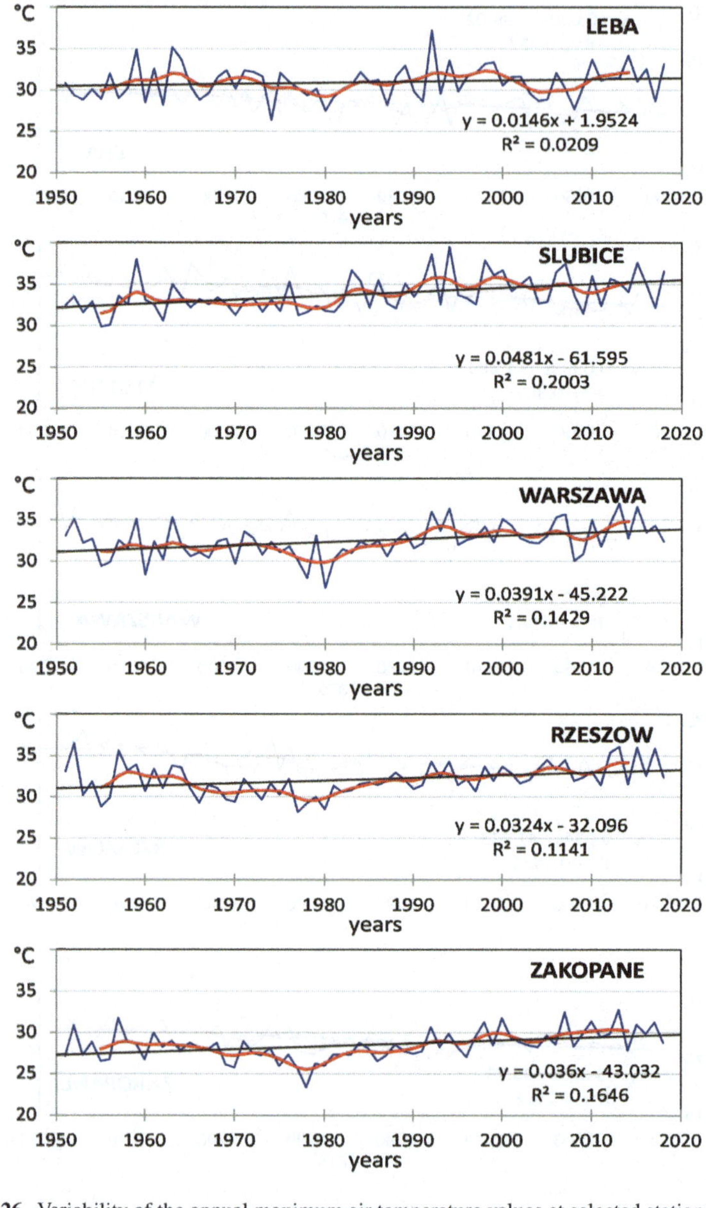

Fig. 11.26 Variability of the annual maximum air temperature values at selected stations

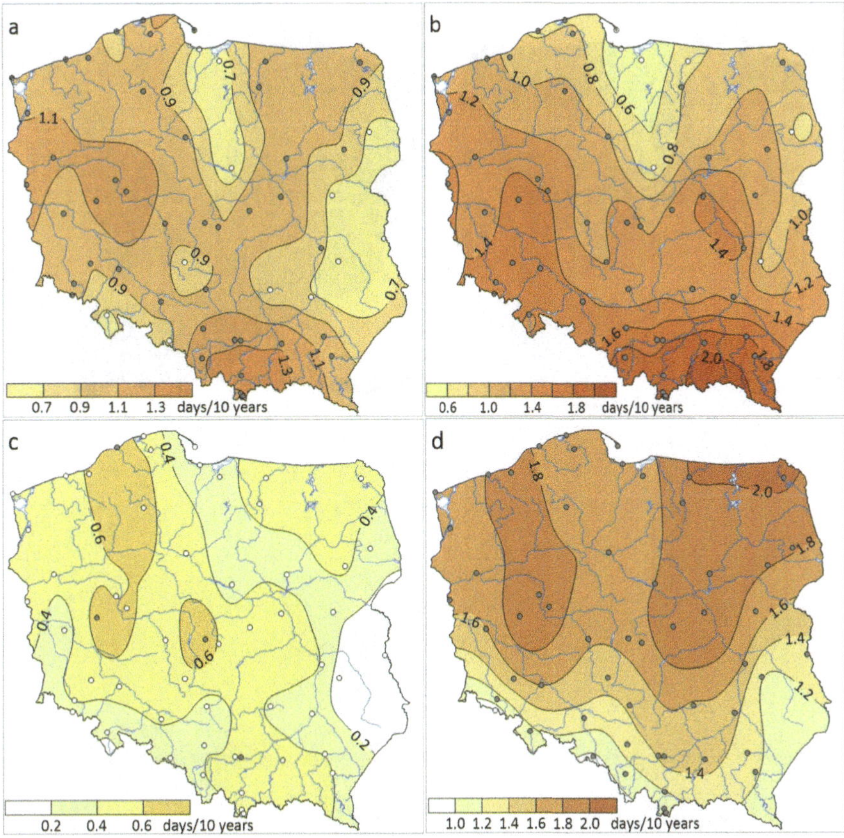

Fig. 11.27 Linear trends of the number of days with maximum seasonal air temperature values exceeding 90 percentile in Poland (days/10 years, dark dots denote significant trends at the particular stations at the significance level of 0.05): **a** spring (MAM), **b** summer (JJA), **c** autumn (SON), **d** winter (DJF)

normal range, or have high maximum temperature values. This trend, like in the autumn, is not statistically significant.

Minimum Daily Temperature

The attached graphs for the selected 5 stations showing the long-term course of the 10th percentile of the daily minimum temperature, show a positive trend at all 5 analysed stations (Figs. 11.29 and 11.30). In the multi-year period under consideration, the minimum temperature limit value of the 10th percentile increases, primarily influenced by the temperature in recent years. Smaller, though positive, trend values

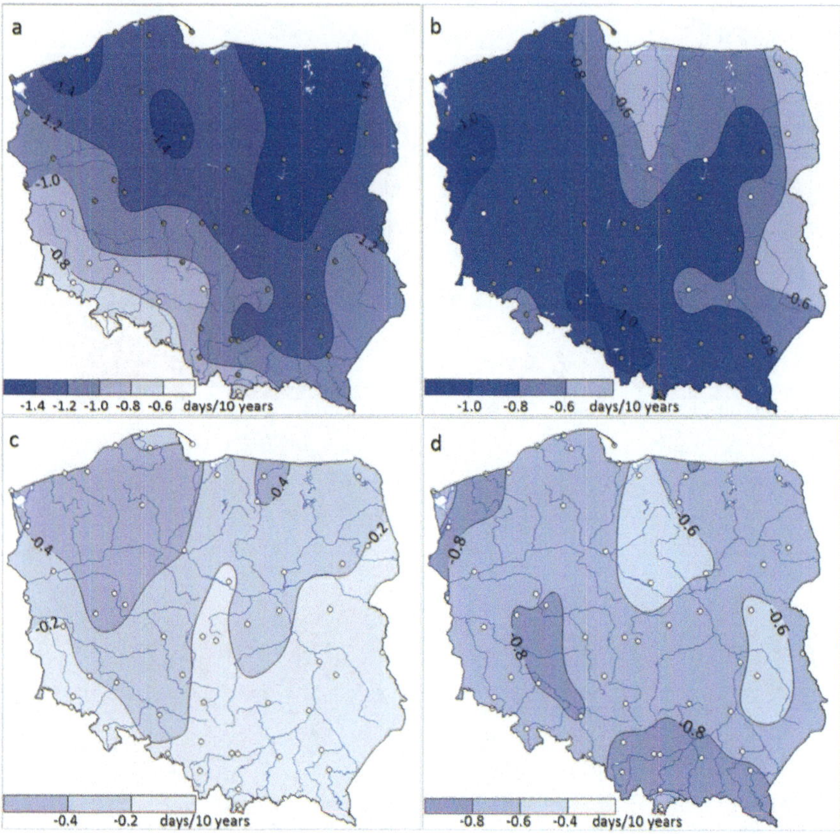

Fig. 11.28 Linear trends of the number of days with maximum seasonal air temperature values below 10 percentile in Poland (days/10 years, dark dots denote significant trends at the particular stations at the significance level of 0.05): **a** spring (MAM), **b** summer (JJA), **c** autumn (SON), **d** winter (DJF)

can also be found in the course of the extreme values of the daily minimum temperature in the considered multi-year period (Figs. 11.29 and 11.30). It is worth noting, however, that even in recent, undoubtedly warmer years, including certainly winters, the lowest values of the minimum temperature can be as low as in previous decades.

An analysis of the temporal and spatial diversification of trends in the number of days with a minimum temperature below the 10th percentile showed their negative values in all seasons (Fig. 11.31). The largest negative trends can be seen in the winter, when in all of Poland, except its north-eastern part, the magnitude of the trend reaches at least 1 day per 10 years. This means a decrease in the number of days with the minimum temperature in the range of the lowest 10% of all recorded values over the multi-year period. Significantly negative trends can also be found in

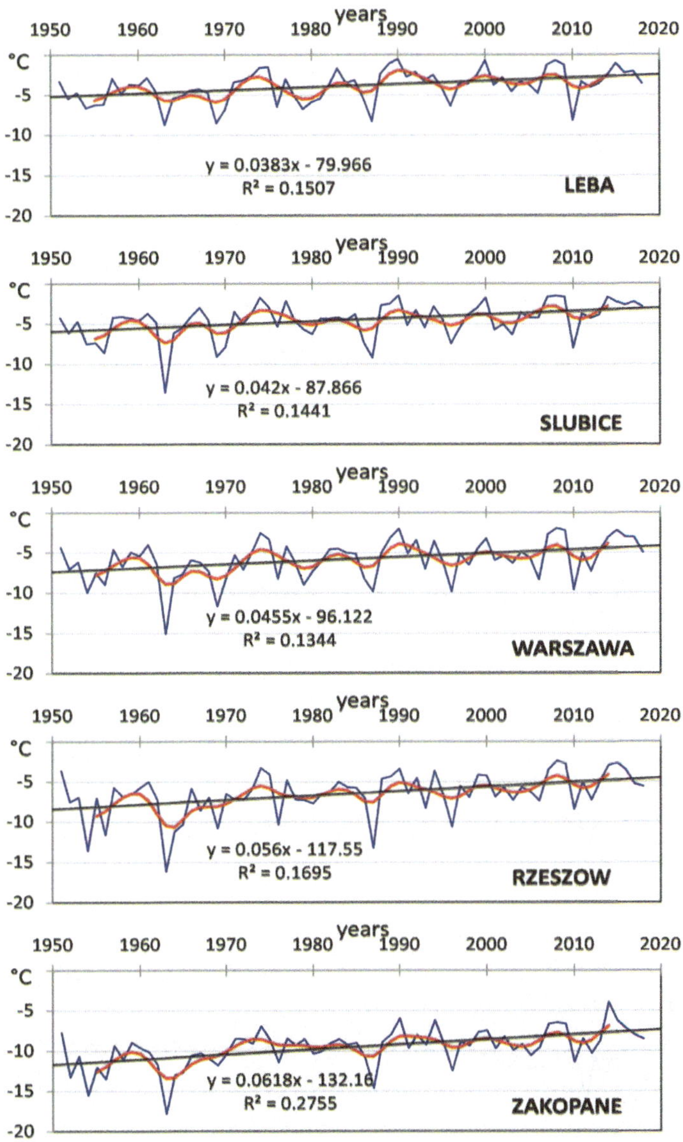

Fig. 11.29 Variability of the 10 percentile air temperature values at selected stations in Poland

the spring and summer. The exception, as in the case of the maximum temperature, is the autumn, when these trends are small and generally not statistically significant.

The image of trends in the number of days with a minimum air temperature above the 90th percentile is different. In all seasons, it shows pronounced positive trends (Fig. 11.32). Considering the already described changes in the average minimum air temperature and the number of days with values below the 10th percentile, it

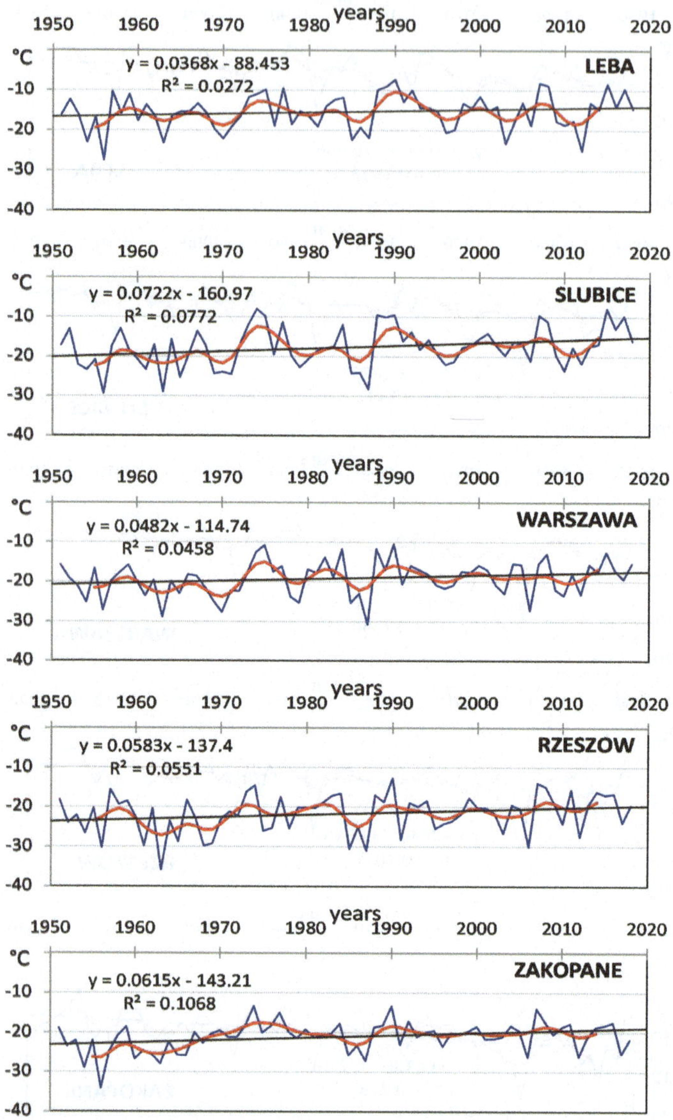

Fig. 11.30 Variability of the annual minimum air temperature values at selected stations in Poland

can be stated that an increase in the value of the minimum daily temperature has been observed in recent years. The highest values of the examined trends are visible in the summer, when the increase in the south-west of Poland is 2 per 10 years. Slightly lower values are observed in the winter and spring. However, large regional differences can be found in both these seasons (Fig. 11.32). Similarly to the previously analysed characteristics, the smallest trends are visible in the autumn.

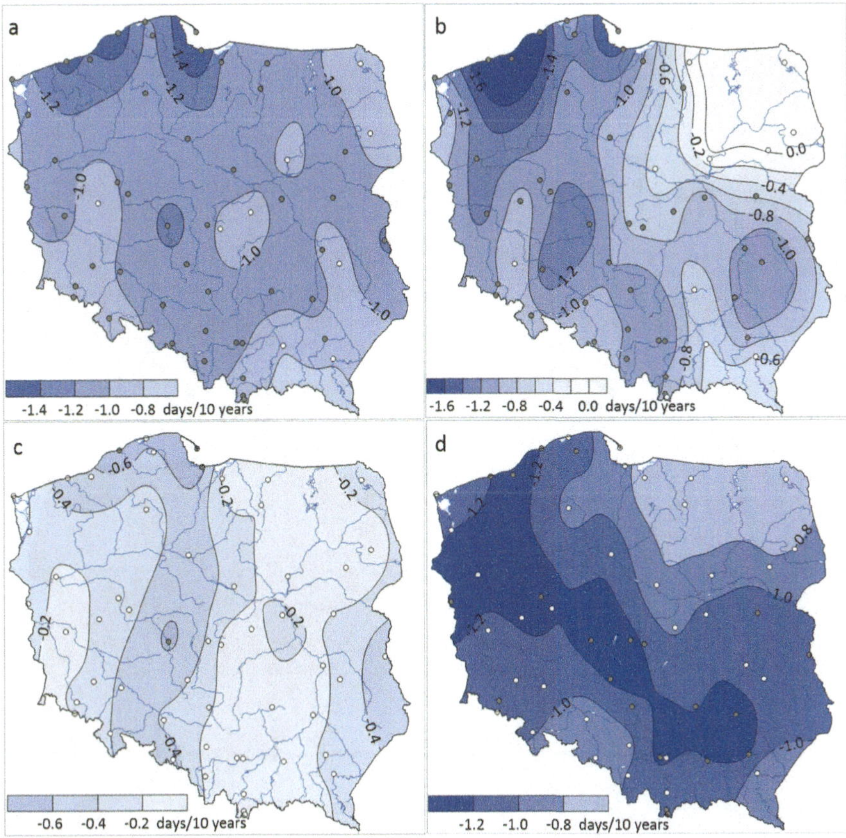

Fig. 11.31 Linear trends of the number of days with minimum seasonal air temperature values below 10 percentile in Poland (days/10 years, dark dots denote significant trends at the particular stations at the significance level of 0.05): **a** spring (MAM), **b** summer (JJA), **c** autumn (SON), **d** winter (DJF)

Number of Days with the Particular Air Temperature Thresholds

Changes in the maximum and minimum temperatures have been presented above using average values and percentile values, which can be considered a relative measure of extreme values. Modern climatology has also used simple characteristics based on specific extreme temperatures for many years. The so-called characteristic days, determined on the basis of conventional, generally accepted maximum and/or minimum temperature thresholds, are considered to be applied most frequently. They make an absolute assessment of the thermal air regime of the area possible, and thus analysing the long-term variability of their occurrence is an important indicator of

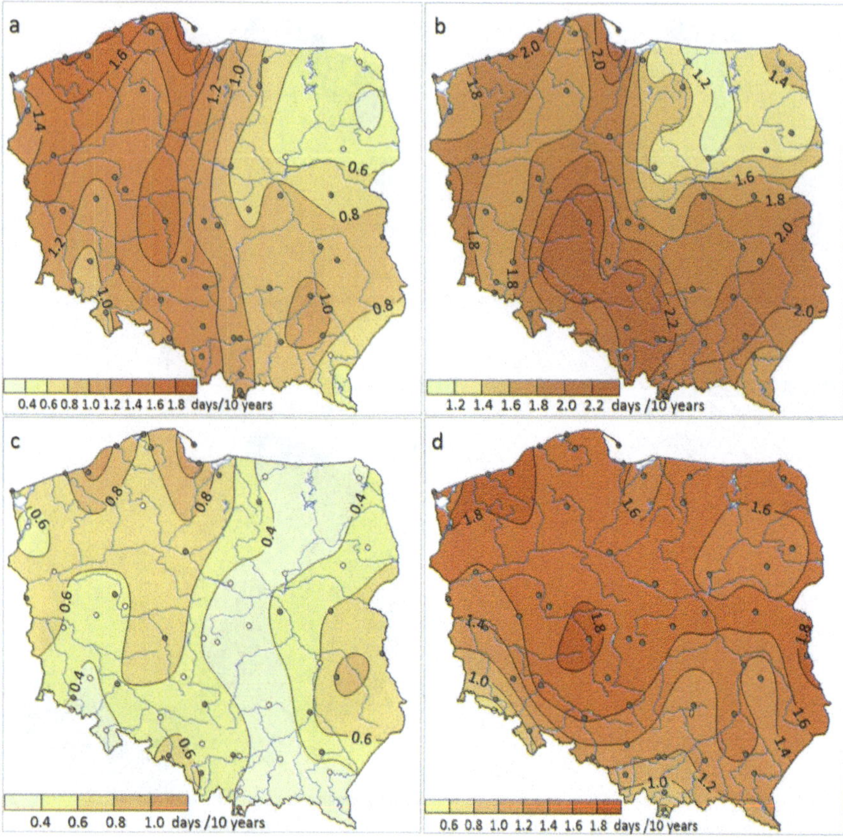

Fig. 11.32 Linear trends of the number of days with minimum seasonal air temperature values exceeding 90 percentile in Poland (days/10 years, dark dots denote significant trends at the particular stations at the significance level of 0.05): **a** spring (MAM), **b** summer (JJA), **c** autumn (SON), **d** winter (DJF)

changing climatic conditions. The characteristics used most often for the area of Central Europe, including Poland, are briefly described below.

The most spectacular effect of the increase in air temperature in recent years is the change in the number of hot days (maximum temperature >30 °C). A clear rising trend in the number of these days is visible at selected stations (Fig. 11.33), while their incidence has been definitely higher in recent years. If, at the beginning of the multi-year period, there were on average several such days in a year, there have been even more than 20 hot days in the past 30 years. Most were recorded in 2015 and 2018, and earlier in 1992 and 1994. It is worth adding that especially few of these days were recorded in the 1970s.

The number of warm days (maximum temperature >25 °C) is also a characteristic of the recent global warming of the climate (Fig. 11.34). The constructed map of

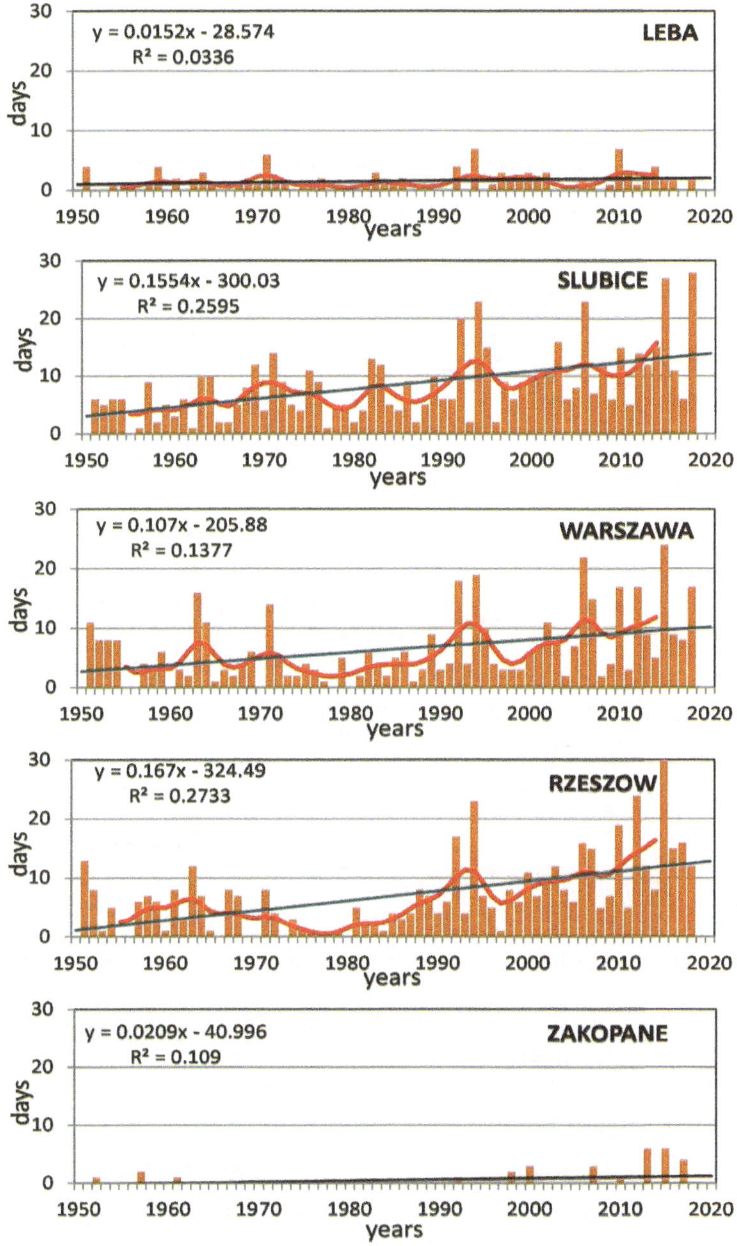

Fig. 11.33 Number of hot days (Tmax > 30 °C) at selected stations in Poland

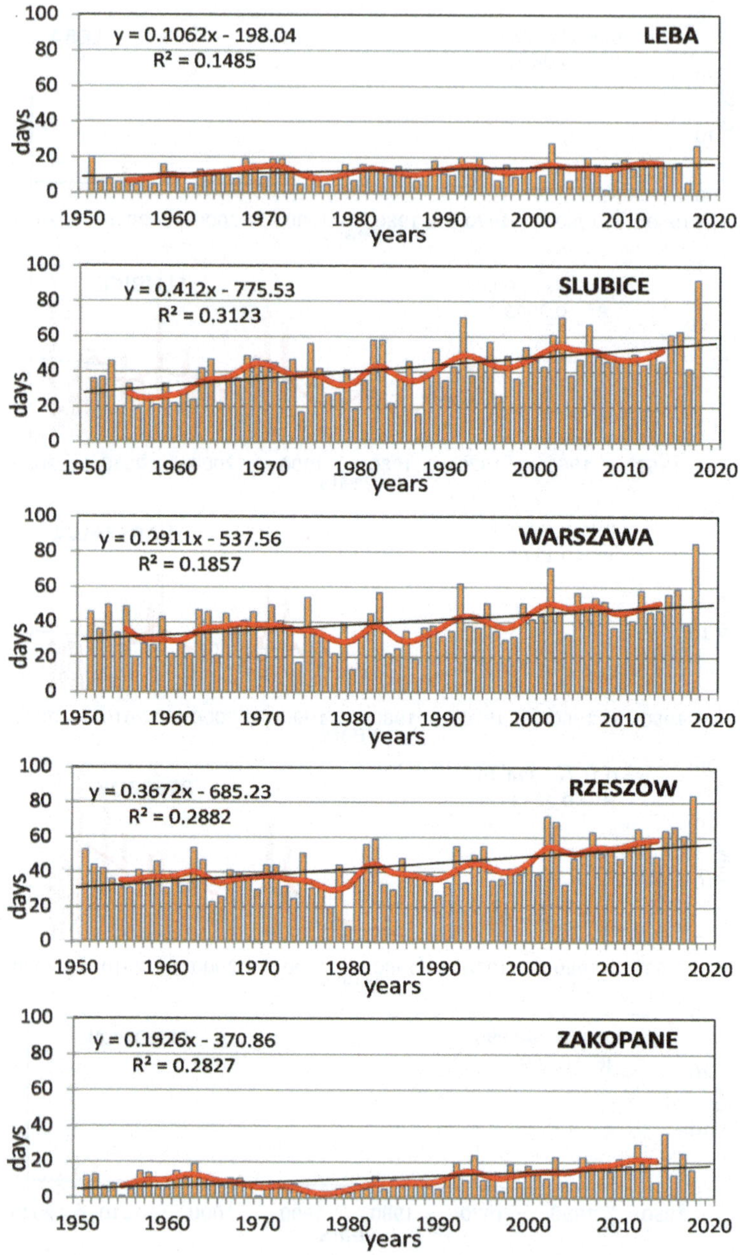

Fig. 11.34 Number of warm days (Tmax > 25 °C) at selected stations in Poland

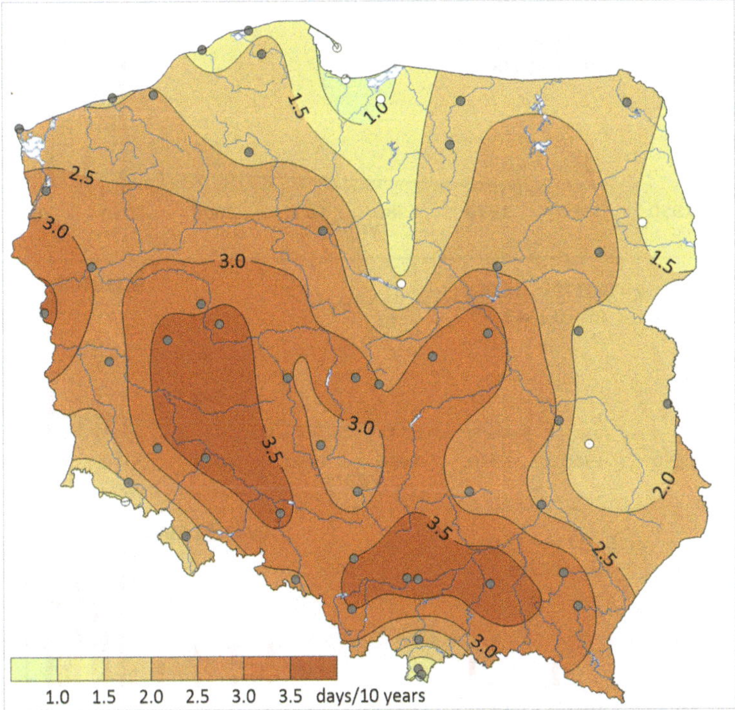

Fig. 11.35 Linear trend of the number of warm days (Tmax > 25 °C) in Poland (days/10 years, dark dots denote significant trends at the particular stations at the significance level of 0.05)

the long-term trend of the number of these days indicates their significant increase in the south and south-west of Poland, where it reaches over 3 days per 10 years (Fig. 11.35). The smallest changes are visible in northern Poland, with a minimum of about 1 day per 10 years in the Gulf of Gdansk.

The number of days with so-called tropical nights (minimum temperature >20 °C) complements and, at the same time, highlights the extremity of thermal conditions. These days, which had previously occurred sporadically at various stations, i.e. once every few or even a dozen or so years, in the last decade appeared much more often and they were no longer single episodes (Fig. 11.36). Owing to their small number, it is difficult to perform a spatial analysis. It is worth adding, however, that they can occur, not only in highly urbanized areas, but also in mountainous areas (e.g. Zakopane).

Since the warming of climatic conditions is visible not only in the summer, but also in the cool half of the year, the occurrence of characteristic days from November to April is an important part of the analysis.

The number of cold days (minimum temperature <−15 °C) attests to the thermal severity of the climate, especially in the winter. On the example of individual stations, it should be noted that they have decreased significantly in the last 3 decades.

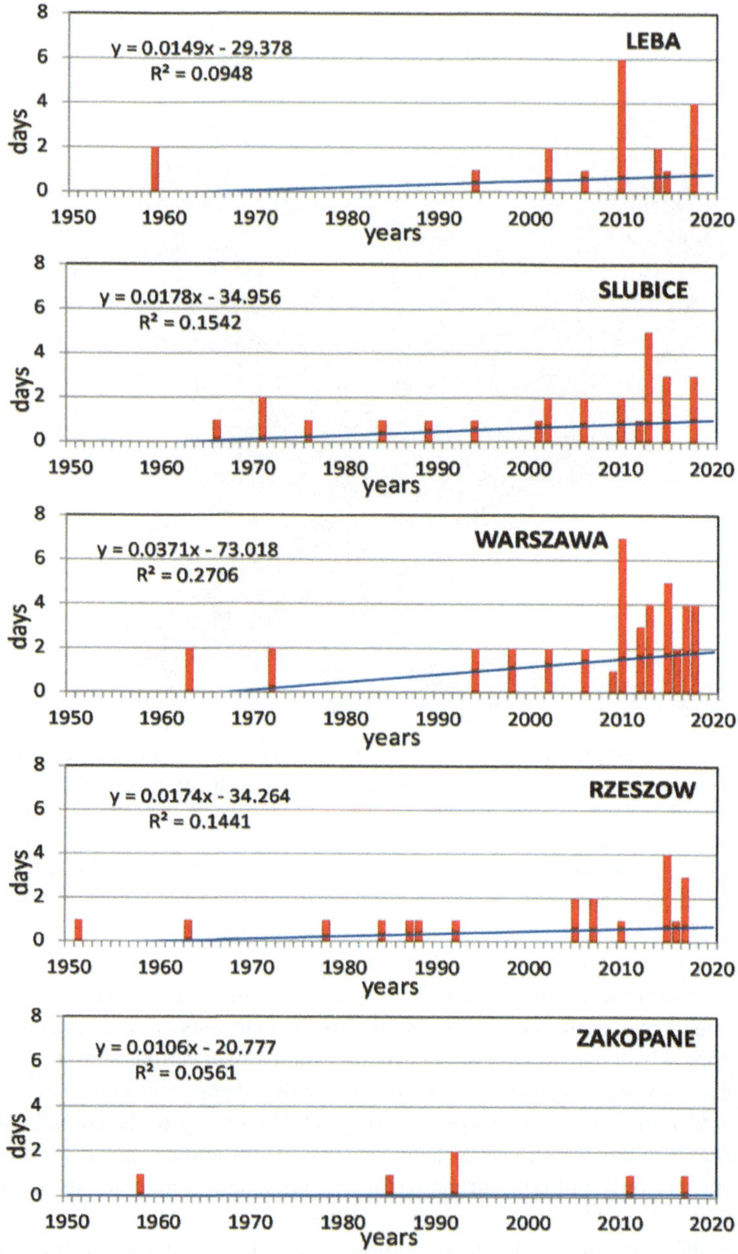

Fig. 11.36 Number of tropical nights (Tmin > 20 °C) at selected stations in Poland

Regardless of the downward trend, it should be emphasized that such days have also occurred in recent years, although their number generally is only a few (Fig. 11.37). In the surveyed multi-year period, the largest number of these days at most stations, exceeding 30, occurred in 1963.

The number of days with a minimum temperature below <0 °C (frost days) is another indicator on the basis of which one can draw conclusions about global warming. The performed analyses clearly confirm the suppositions that the number of these days has decreased. In the last 2–3 decades, there have been years when this number has fallen to fewer than 90 days or fewer than 3 months (Fig. 11.38). Certainly, at some stations, such a small number of days has also occurred earlier. The spatial distribution of linear trends indicates that the largest decrease in the number of days with a minimum temperature below 0 °C occurs in north-western Poland, where the values of the described changes reach over 5 days per 10 years. This is undoubtedly conditioned by the mitigating effect of the Baltic Sea (Fig. 11.39). For the record, it should be added that significant negative trends were also found in the east of Poland, which may be associated with circulation factors. A slightly smaller decrease in the number of considered days (about 3–4 days/10 years) is visible in southern Poland, especially in the Carpathians and their foreground. Most likely, the terrain, which contributes to the occurrence of slight frost in concave landforms, has some effect on the number of these days.

An interesting characteristic, significant from both the environmental and practical point of view, is the number of days with the passage through 0 °C (maximum temperature >0 °C and minimum temperature <0 °C), according to some approaches identified with frost days. The clearly decreasing number of days with a minimum temperature <0 °C is, certainly, a direct cause of the decrease in the number of days with maximum temperature >0 °C and minimum temperature <0 °C (Fig. 11.40). The spatial distribution of the value of trends in the number of days with freezing point (0 °C) transition indicates that the most significant changes (a decrease of about 4 days/10 years) occur in north-western Poland, especially along the coast, which undoubtedly should be associated with the mitigating role of the Baltic Sea in the winter and thus its effect on the simultaneous reduction of the number of days with a minimum temperature below 0 °C (Fig. 11.41). These relationships can be easily contrasted with each other by comparing the spatial distribution of the discussed trends in Figs. 11.39 and 11.41).

This analysis is complemented by the characteristics of the variability of the number of frost-free days (minimum temperature >0 °C) considered here as the length of the frost-free period (between the last spring frost and the first autumn frost). This is an important climate indicator that is of great practical importance especially in the field of agriculture around the world. As expected in the light of other research results, the trends of length of this period are clearly positive, i.e. this length increases significantly. Their values reach 4–6 days per 10 years, which in turn causes a very large extension of the frostless period (Fig. 11.42). On the Polish scale, this is particularly visible in its south-eastern part (Fig. 11.43). Although, on the one hand, this general tendency seems to be a very positive agrometeorological

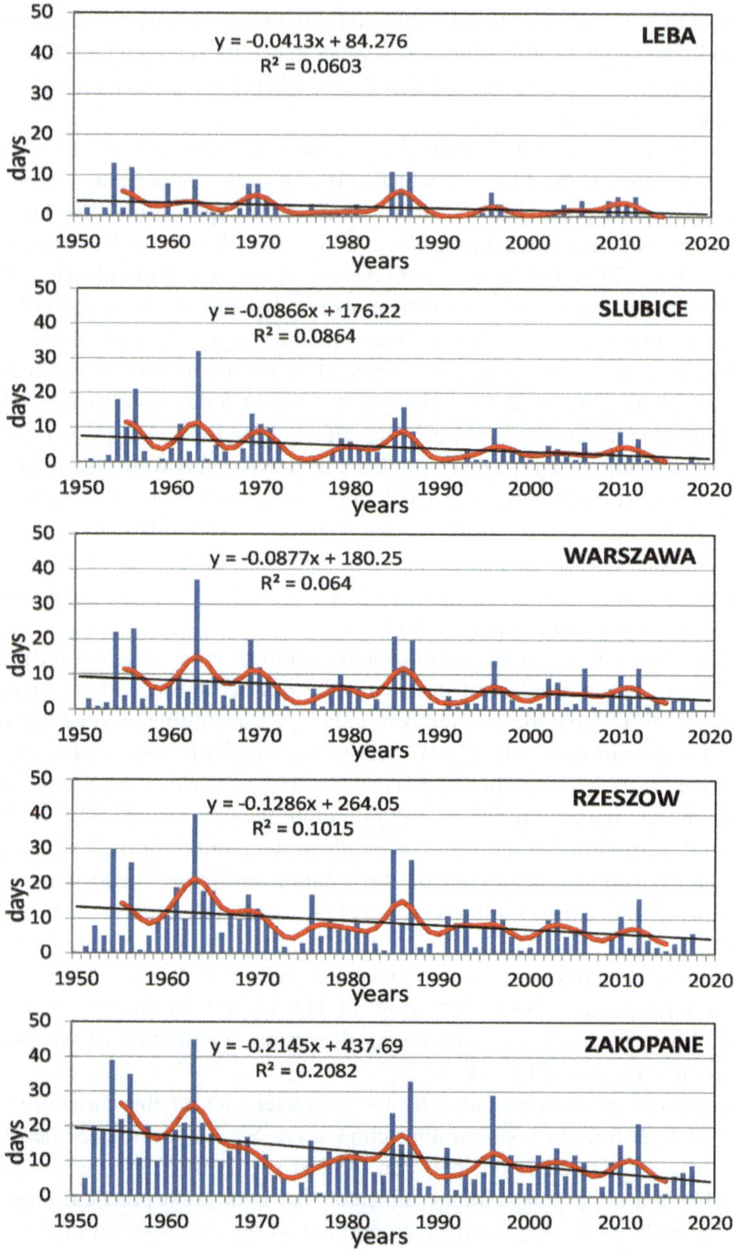

Fig. 11.37 Number of cold days (Tmin <−15 °C) at selected stations in Poland

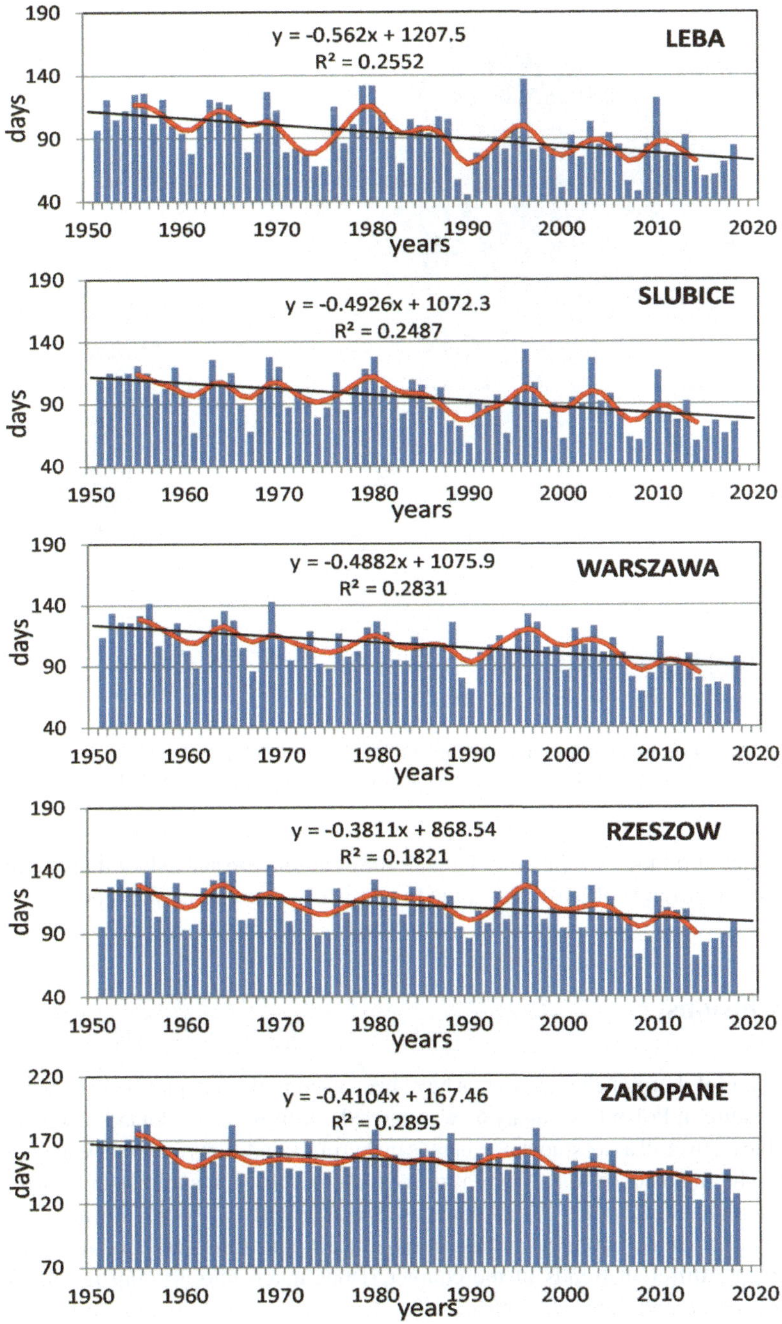

Fig. 11.38 Number of frost days (Tmin < 0 °C) at selected stations in Poland

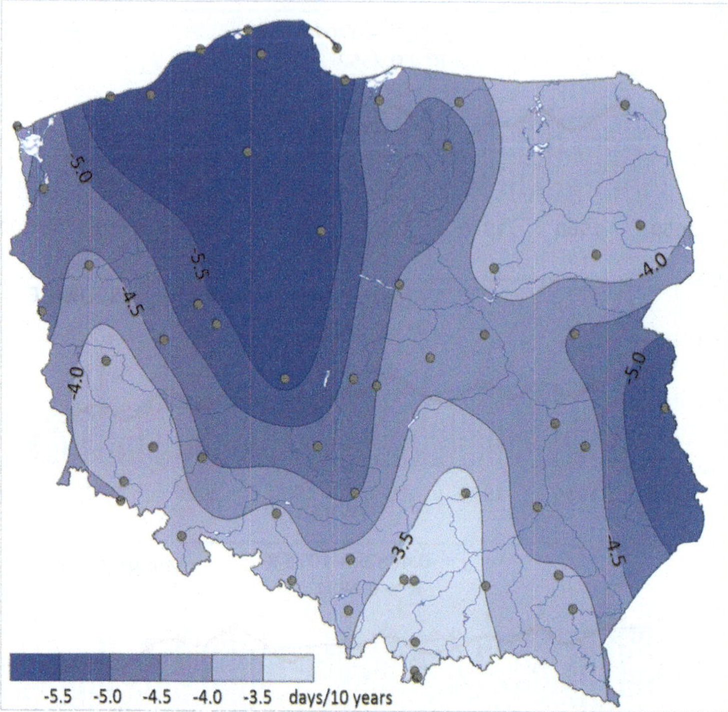

Fig. 11.39 Linear trend of the number of frost days (Tmin <0 °C) in Poland (days/10 years, dark dots denote significant trends at the particular stations at the significance level of 0.05)

and agricultural fact, on the other hand, it may cause a greater risk of damage in the event of occurrence of frost during the theoretically ongoing frostless period.

Conclusions

A number of thermal characteristics have been used in this chapter on changes in air temperature in Poland. An analysis was carried out throughout Poland, which made possible an overall assessment of changes in thermal conditions over the period 1951–2018. The inclusion of data from almost 60 meteorological stations and posts also made a regional approach to the issue possible. Certainly, although it is not possible to fully exclude certain local peculiarities, in the light of the data used it is difficult to indicate different trends of the characteristics taken into account in individual regions of Poland. This is confirmed by the cited selected studies performed on smaller scales. Possibly finding such differences in the trends considered would be very doubtful and, in the first place, would suggest the existence of heterogeneity in the data used. As a continuous meteorological element, from the physical point of

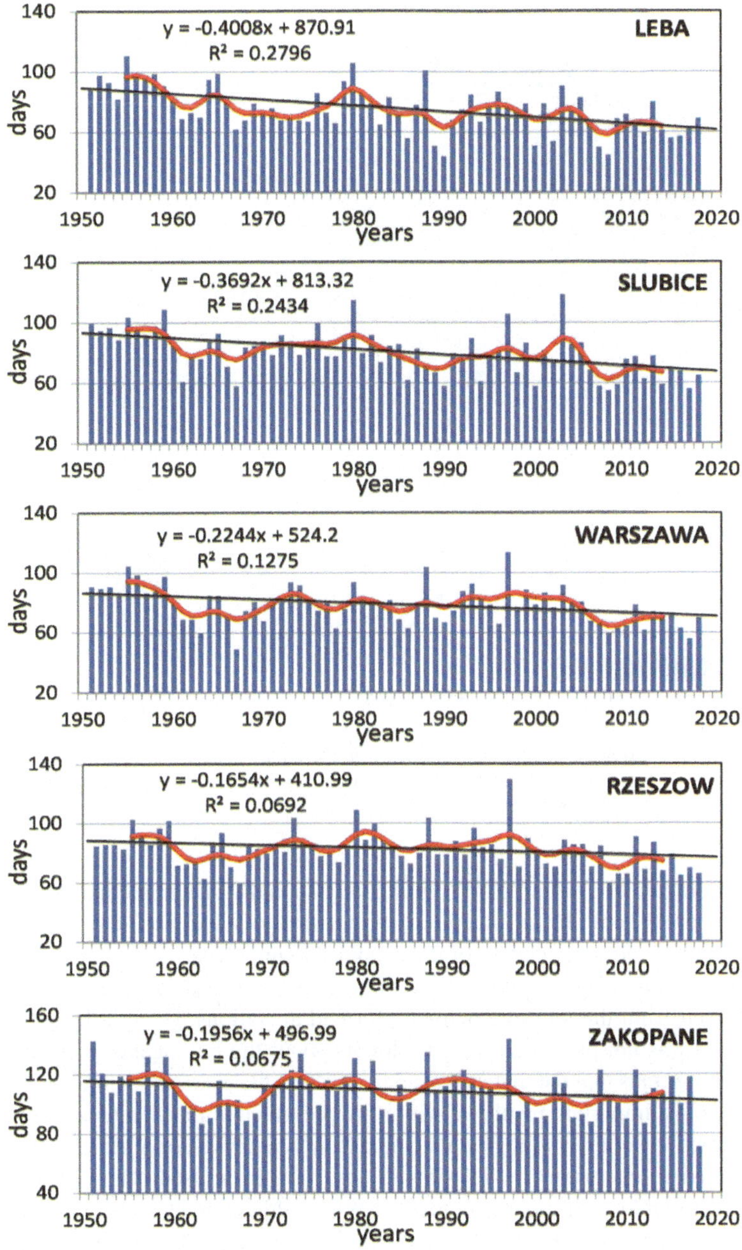

Fig. 11.40 Number of days with the temperature crossing 0 °C (Tmax > 0 °C and Tmin < 0 °C) at selected stations in Poland

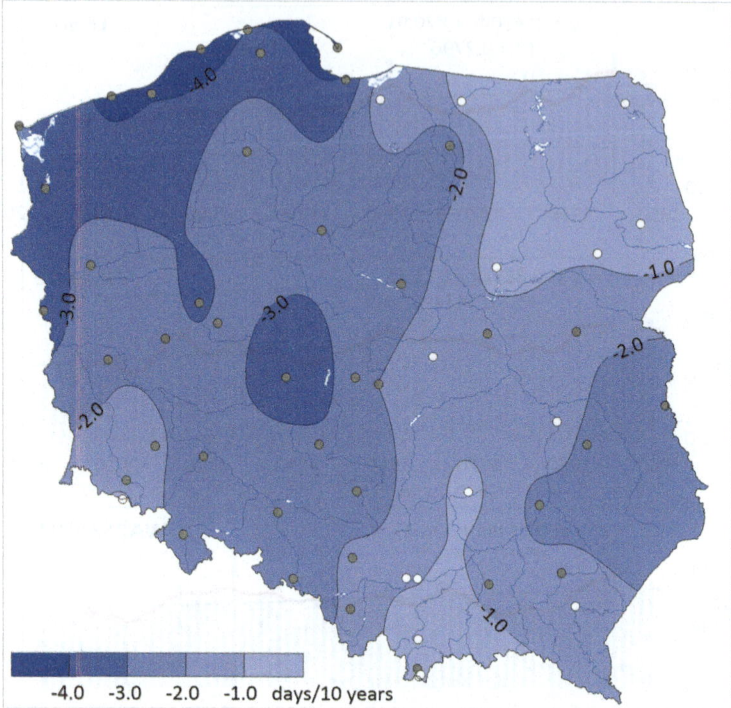

Fig. 11.41 Linear trend of the number of days with the temperature crossing 0 °C (Tmax > 0 °C and Tmin <0 °C) in Poland (days/10 years, dark dots denote significant trends at the particular stations at the significance level of 0.05)

view air temperature cannot show significantly spatially different changes in trends. Mountain areas could be an exception, but the analysis and work carried out so far do not confirm the separateness of the changes taking place.

As stated, all analyses were carried out on the basis of data with a daily resolution derived from the measurements performed at stations. Using a larger number of stations was not possible owing to doubts about the homogeneity of the series. However, it seems that the material used from 58 points was sufficient to perform the tests, the most important results of which are presented in the chapter. According to the assumptions of this monograph, no data from reanalyses, or other data that could be obtained from various model simulations, have been used. The use of such is currently possible and probably in many cases would facilitate some of the analyses, including in particular the construction of individual maps. This applies especially to elements and characteristics very sensitive to local environmental conditions, which sometimes disturb the continuity of their distribution. However, there is a concern that an uncritical use of such data may lead to somewhat idealized results, which are not reliable enough to illustrate the features of the spatial diversification and temporal variability of air temperature.

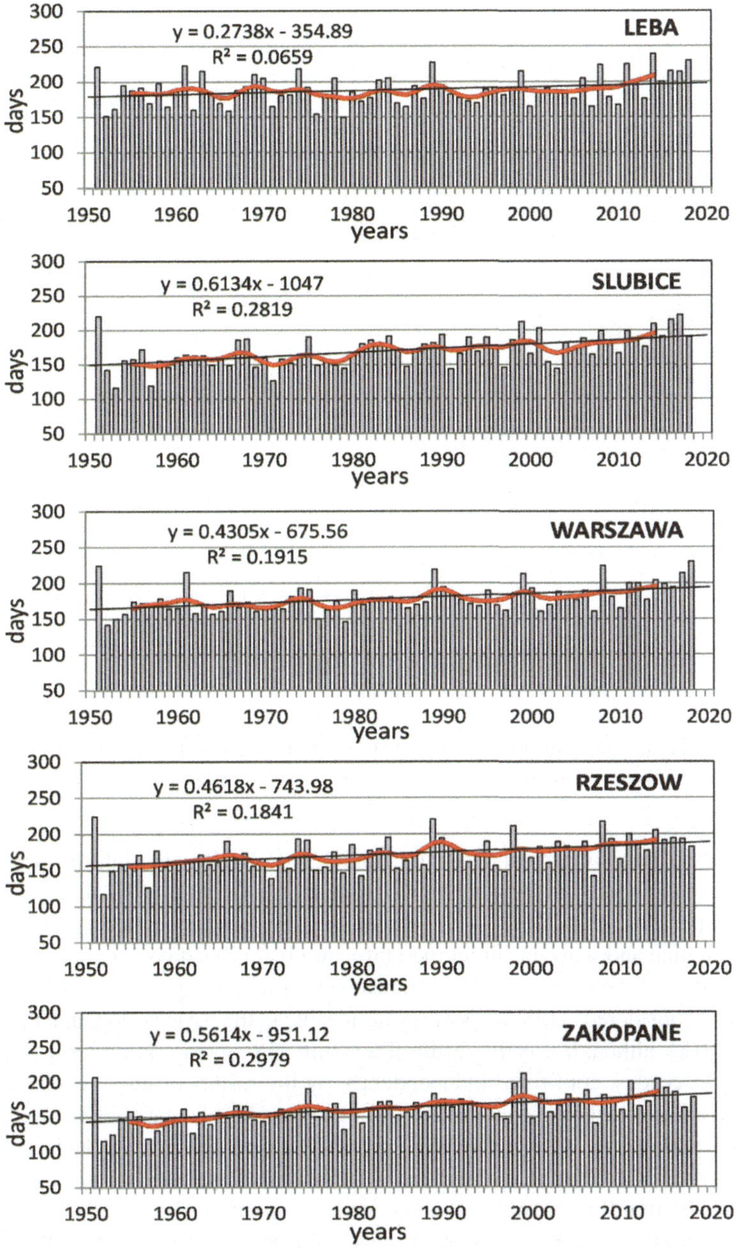

Fig. 11.42 Variability of the frostless period (Tmin > 0 °C) length at selected stations in Poland

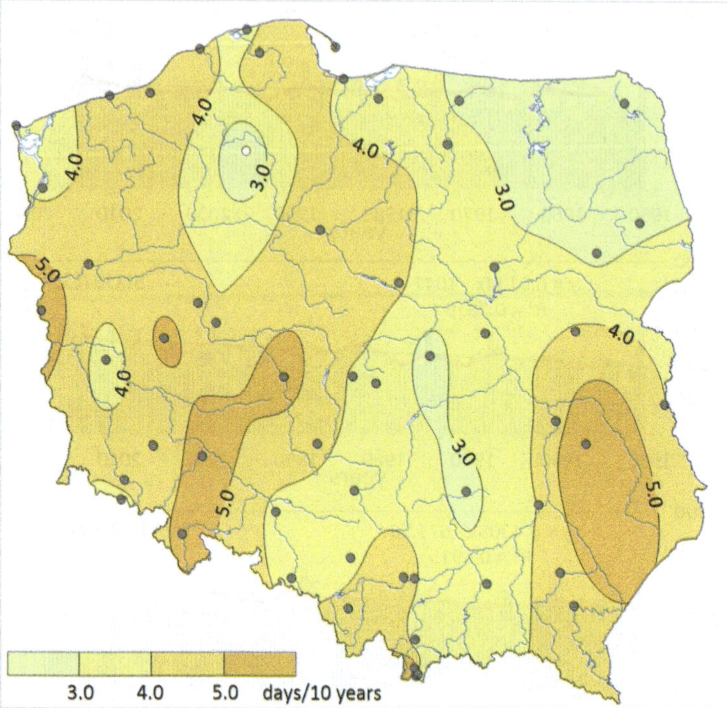

Fig. 11.43 Linear trend of the frostless period (Tmin >0 °C) length in Poland (days/10 years, dark dots denote significant trends at the particular stations at the significance level of 0.05)

The assessment of air temperature changes in Poland in the light of most characteristics and indicators showed their significant trends resulting from the increase in air temperature. Depending on the characteristics, these are either increasing (e.g. average annual and seasonal air temperature, number of hot days) or decreasing (e.g. number of frost days) trends. These trends were found irrespective of the length of the period considered. This is especially noticeable in the basic period under analysis (1951–2018), although it is the result of a significant increase in air temperature in the last 3 decades, confirmed independently by the results of analyses for various thermal indicators. These changes are in most cases highly statistically significant, however, particular attention should be paid to the trends in average and maximum air temperatures during the year and in the summer and winter. They are also associated with the trends of changes in the occurrence of various so-called characteristic days. Although the coming years are going to confirm to what extent these trends will continue, they give cause for pessimism about the pace of climate change represented by its guiding element, i.e. air temperature.

The demonstrated variability of thermal conditions in Poland, which according to the suggestions of some researchers (Mitchel et al. 1966) can already be considered as symptoms of change, is fully in line with the trends observed in recent decades in Europe and elsewhere in the world. This applies in particular to the courses of average air temperature values for which there are a number of key studies. The most important ones were cited at the beginning of this chapter.

References

Auer I, Böhm R, Jurkovic A, Lipa W, Orlik A, Potzmann R, Schoener W, Ungersboeck M, Matulla C, Briffa K, Jones PD, Efthymiadis D, Brunetti M, Nanni T, Maugeri M, Mercalli L, Mestre O, Moisselin J-M, Begert M, Müller-Westermeier G, Kveton V, Bochnicek O, Stastny P, Lapin M, Szalai S, Szentimrey T, Cegnar T, Dolinar M, Gajic-Capka M, Zaninovic K, Majstorovic Z, Nieplova E (2007) HISTALP–Historical instrumental climatological surface time series of the greater Alpine region 1760-2003. Int J Climatol 27:17–46

Balling RC, Michaels PJ, Knappenberger PC (1998) Analysis of winter and summer warming rates in gridded temperature time series. Clim Res 9:175–181

Bednarz Z (1996) June-July temperature variation for the Babia Góra National Park, Southern Poland, for the period 1650–1910. Prace Geogr 102:523–529

Boehm R, Jones PD, Hiebl J, Frank D, Brunetti M, Maugeri M (2010) The early instrumental warm-bias: a solution for long central European temperature series 1760–2007. Clim Change 101:41–67. https://doi.org/10.1007/s10584-009-9649-4

Bokwa A, Wypych A, Ustrnul Z (2013) Climate changes in the vertical zones of the Polish Carpathians in the last 50 years. The Carpathians: Integrating nature and society towards sustainability, Springer, Heidelberg, New York, Dordrecht, and London, pp 89–110

Bradley RS (1988) The explosive volcanic eruption signal in Northern Hemisphere continental temperature records. Clim Change 12(3):221–243

Bradley RS, Jones PD (eds.) (1995) Climate since A.D. 1500, Routledge, London and New York

Brazdil R (1994) Climatic fluctuation in the Czech Lands during the last millennium. GeoJournal 32:199–205. https://doi.org/10.1007/BF01122109

Brazdil R (1996) Reconstructions of past climate from historical sources in the Czech Lands. Climatic variations and forcing mechanisms of the last 2000 years, Springer, Berlin pp 409–431

Brazdil R, Budikova M, Auer I, Bohm R, Cegnar T, Fasko P, Lapin M, Gajic-Capka M, Zaninovic K, Koleva E, Niedzwiedz T, Ustrnul Z, Szalai S, Weber RO (1996) Trends of maximum and minimum daily temperatures in central and southeastern Europe. Int J Climatol 16(7):765–782. https://doi.org/10.1002/(SICI)1097-0088(199607)16:7%3c765:AID-JOC46%3e3.0.CO;2-O

Brazdil R, Dobrovolny P, Luterbacher J, Moberg A, Pfister C, Wheeler D, Zorita E (2010) European climate of the past 500 years: new challenges for historical climatology. Clim Change 101:7–40

Broenimann S (2015) Climatic changes since 1700, Series: Advances in Global Change Research, Springer. 10.1007/978-3-319-19042-6_4

Bryś K, Bryś T (2010) Reconstruction of the 217-year (1791–2007) Wrocław air temperature and precipitation series. Bull Geogr, Phys Geogr Ser 3(1):121–171

Camuffo D, Bertolin C, Barriendos M, Dominguez-Castro F, Cocheo C, Enzi S, Sghedoni M, della Valle A, Garnier E, Alcoforado MJ, Xoplaki E, Luterbacher J, Diodato N, Maugeri M, Nunes MF, Rodriguez R (2010) 500-year temperature reconstruction in the Mediterranean Basin by means of documentary data and instrumental observations. Climatic Change, 101:169–199

Camuffo D, Jones P (2002) Improved understanding of past climatic variability from early daily European instrumental sources. Kluwer Academic, Dordrecht, Boston, and London

Camuffo D, Zardini F (1997) Controlling the homogeneity of a long meteorological series: The series of Padova (1725–today). Applications of Time Series for Meteorology and Astronomy, Chapman & Hall, London, pp 441–459

Cebulak E, Limanówka D (2007) Dni z ekstremalnymi temperaturami powietrza w Polsce. Wahania klimatu w różnych skalach przestrzennych i czasowych, IGiGP UJ, Kraków, pp 185–194

Demaree GR, Lachaert P-J, Verhoeve T, Thoen E (2002) The long-term daily Central Belgium Temperature (CBT) series (1767–1998) and the early instrumental meteorological observations in Belgium. Clim Change 53(1):269–293. https://doi.org/10.1023/A:1014931211466
Filipiak J (2007) Rekonstrukcja warunków klimatycznych Gdańska w okresie pomiarów instrumentalnych – fakty i niepewności. In: 200 lat regularnych pomiarów i obserwacji meteorologicznych w Gdańsku, Monografie IMGW. Warszawa, pp 20–33
Gorczyński W (1915a) O zmienności temperatury z dnia na dzień w Polsce i w Eurazyi. Warszawa, Drukarnia i Litografia Jana Cotty
Gorczyński W (1915b) O zmianach długoletnich temperatury powietrza w Polsce i Eurazyi. Warszawa, Drukarnia i Litografia Jana Cotty
Gorczyński W (1916) Badanie współzależności przebiegów temperatury metodą korelacyjną. Warszawa, Drukarnia i Litografia Jana Cotty
Haensel S, Ustrnul Z, Lupikasza E, Skalak P (2019) Assessing seasonal drought variations and trends over Central Europe. Adv Water Resour 127:53–75. https://doi.org/10.1016/j.advwatres.2019.03.005
Hoy A, Haensel S, Skalak P, Ustrnul Z, Bochnicek O (2017) The extreme European summer of 2015 in a long-term perspective. Int J Climatol 37(2):943–962. https://doi.org/10.1002/joc.4751
IPCC (2007) Climate change 2007, the physical science basis. Contribution of Working Group I to the Fourth Assessment Report of the Intergovernmental Panel on Climate Change [Solomon, S., D. Qin, M. Manning, Z. Chen, M. Marquis, KB. Averyt, M. Tignor and H.L. Miller (eds.)]. Cambridge University Press, Cambridge, UK and New York, NY, USA, 996 pp
IPCC (2013) Climate change 2013, the physical science basis. Contribution of Working Group I to the Fifth Assessment Report of the Intergovernmental Panel on Climate Change [Stocker, T.F., D. Qin, G.-K. Plattner, M. Tignor, S.K. Allen, J. Boschung, A. Nauels, Y. Xia, V. Bex and P.M. Midgley (eds.)], Cambridge University Press, Cambridge, UK and New York, NY, USA, 1535 pp, https://doi.org/10.1017/cbo9781107415324
Jones PD, Hulme M (1997) The changing temperature of Central England. In: Climates of the British Isles: present, past and future, Routledge, London, pp 173–196
Jones PD, Ogilvie AEJ, Davies TD, Briffa KR (Eds.) (2001) History and climate. Memories of the future? Springer, New York
Jones PD, Lister DH, Osborn TJ, Harpham C, Salmon M, Morice CP (2012) Hemispheric and large-scale land surface air temperature variations: an extensive revision and an update to 2010. J Geophys Res 117:D05127. https://doi.org/10.1029/2011JD017139
Jurkovic A, Boehm R, Auer I, Gruber Ch, Orlik A, Majstorovic Z, Hodzic S, Zulum D (2011) The Mountain Observatory Bjelašnica-History, analysis, homogenization and interpretation of a more than 100 years long temperature data set. Meteorologische Zeitshrift 20:291–303. https://doi.org/10.1127/0941-2948/2011/0205
Kożuchowski K, Żmudzka E (2003) 100-year series of areally averaged temperatures and precipitation totals in Poland, Acta Universitatis Wratislaviensis, No. 2542. Studia Geograficzne 75:116–122
Lamb HH (1972) Climate: present, past and future, vol. 1: Fundamentals and climate now, Methuen, London
Lamb HH (1977) Climate: present, past and future, vol. 2: Climatic history and the future, Methuen, London
Landsberg HE, Yu CS, Huang L (1968) Preliminary reconstruction of a long time series of climatic data for the Eastern United States, Technological Note B14-571. University of Maryland, College Park
Lorenc H (2000) Studia nad 220-letnią (1779–1998) serią temperatury powietrza w Warszawie oraz ocena jej wiekowej tendencji, Materiały Badawcze, Seria: Meteorologia, 31. IMGW, Warszawa
Luterbacher J, Dietrich D, Xoplaki E, Grosjean M, Wanner H (2004) European seasonal and annual temperature variability, trends and extremes since 1500. Science 303:1499–1503. https://doi.org/10.1126/science.1093877

Manley G (1974) Central England temperatures: monthly means 1659 to 1973. Q J R Meteorol Soc 100:389–405
Marsz A, Styszyńska A (2018) Przebieg temperatury zim na obszarze Polski w latach 1720–2015. Prace Geograficzne 155:85–138. https://doi.org/10.4467/20833113PG.18.018.9541
Merecki R (1899) O zmienności nieokresowej temperatury powietrza, Rozprawy Wydz. Matem-Przyr. Akademii Umiejętności, t. 35, Kraków
Miętus M (1998) O rekonstrukcji i homogenizacji wieloletnich serii średniej miesięcznej temperatury ze stacji w Gdańsku-Wrzeszczu 1851–1995. Wiadomości IMGW 21(2):41–63
Miętus M (2007) 200 lat regularnych obserwacji i pomiarów meteorologicznych w Gdańsku – od fascynacji do praktycznego działania. In: 200 lat regularnych pomiarów i obserwacji meteorologicznych w Gdańsku, Monografie IMGW. Warszawa, pp 7–19
Mitchell JM (1961) Recent secular changes of global temperature, Annals NY Acad Sciences, 95(1):235–250
Mitchell JM, Dzerdzeevskii B, Flohn H, Hofmeyr WL, Lamb HH, Rao KN, Walleen CC (1966) Climatic change, report of a working group of the Commission for Climatology, WMO No. 195, Technical Note 79, Geneva, 79 pp
Moberg A, Bergstrom H (1997) Homogenization of Swedish temperature data. Part III: the long temperature records from Uppsala and Stockholm. Int J Climatol 17:667–699
Niedźwiedź T (2004) Rekonstrukcja warunków termicznych lata w Tatrach od 1550 roku. In: Rola małej epoki lodowej w przekształcaniu środowiska przyrodniczego Tatr, Prace Geograficzne IGiPZ PAN, 197, Warszawa, pp 57–88
Osborn TJ, Jones PD (2014) The CRUTEM4 land-surface air temperature data set: construction, previous versions and dissemination via Google Earth. Earth Syst Sci Data 6:61–68. https://doi.org/10.5194/essd-6-61-2014
Przybylak R, Majorowicz J, Wójcik G, Zielski A, Chorążyczewski W, Marciniak K, Nowosad W, Oliński P, Syta K (2005) Temperature changes in Poland from the 16th to the 20th centuries. Int J Climatol 25:773–791. https://doi.org/10.1002/joc.1149
Romer E (1947) Rozmyślania klimatyczne, Czasopismo Geogr. XVII:3–4
Romer E (1948/49) Rehabilitacja wartości średniej temperatury roku, Przegl. Geogr., XXII
Sulikowska A, Wypych A (2020) Summer temperature extremes in Europe: how does the definition affect the results? Theor Appl Climatol 141:19–30
Trepińska J (ed) (1997) Wahania klimatu w Krakowie (1792–1995). Wyd, IG UJ, Kraków
Trepińska J, Ustrnul Z, Kowanetz L (1997) Variability of air temperature in the Central Europe in the period 1792–1995. Geographia Polonica 70:43–52
Tuomenvirta H, Alexandersson H, Drebs A, Frich P, Nordli O (2000) Trends in Nordic and Arctic Temperature Extremes and Ranges. J Clim 13:977–990. https://doi.org/10.1175/1520-0442(2000)013%3c0977:TINAAT%3e2.0.CO;2
Ustrnul Z (2000) Synoptic-climatic structure of the extreme air thermal phenomena in Poland. Geogr Pol 73(2):99–109
Ustrnul Z, Wypych A (2011) Ekstremalne temperatury powietrza w Polsce w świetle różnych klasyfikacji typów cyrkulacji. Prace i Studia Geograficzne, Wydawnictwa Wydziału Geografii i Studiów Regionalnych UW 47:87–95
Ustrnul Z, Wypych A, Kosowski M (2011) Ekstrema termiczne i opadowe w Polsce w świetle różnych metod wyróżniania wartości ekstremalnych, Wiadomości Meteorologii Hydrologii Gospodarki Wodnej, V(1–2):55–68
Ustrnul Z, Wypych A, Winkler J, Czekierda D (2014) Late spring freezes in Poland in relation to atmospheric circulation. Quaest Geogr 33(3):165–172
Vose RS, Easterling DR, Gleason B (2005) Maximum and minimum temperature trends for the globe: An update through 2004. Geophys Res Lett 32:L23822. https://doi.org/10.1029/2005GL024379
Wójcik G, Majorowicz J, Marciniak K, Przybylak R, Šafanda J, Zielski A (2000) The last millennium climate change in Northern Poland derived from well temperature profiles, tree – rings and instrumental data. Prace Geogr 107:137–148

Wypych A, Ustrnul Z, Sulikowska A, Chmielewski F-M, Bochenek B (2017a) Spatial and temporal variability of the frost-free season in Central Europe and its circulation background. Int J Climatol 37(8):3340–3352. https://doi.org/10.1002/joc.4920

Wypych A, Sulikowska A, Ustrnul Z, Czekierda D (2017b) Temporal variability of summer temperature extremes in Poland. Atmosphere 8:51. https://doi.org/10.3390/atmos8030051

Wypych A, Sulikowska A, Ustrnul Z, Czekierda D (2017c) Variability of growing degree days in Poland in response to ongoing climate changes in Europe. Int J Biometeorol 61(9):49–59. https://doi.org/10.1007/s00484-016-1190-3

Chapter 12
Air Humidity Change

Agnieszka Wypych

Abstract Water vapour is a major component of the Earth's atmosphere, believed to be the most important trace gas found there. The observed phenomena of global warming suggest that temperature-dependent atmospheric moisture should also demonstrate visible long-term fluctuations. Long-term variability of humidity conditions in Poland was analysed using three data sources: surface measurements of relative humidity (RH) from the period 1951–2018, historical measurements of relative humidity conducted in Kraków since the year 1863, and ERA-5 specific humidity (SH) reanalyses data for the available period of 1981–2018. Without unlimited water supplies, Poland experiences only a slight increase in moisture, which is significant particularly in summer (up to 0.2 g·kg^{-1} per 10 years) and in autumn. The relative trend reaches approximately 3% of mean values. Together with increasing temperature, it brings a statistically significant decrease in RH, observed since the second half of the twentieth century. The trend value of almost 1% per decade is observed in the southeastern part of the country and much less at the coast. The relative trend is the highest, about 5% of the mean, in central Poland. The most intensive decreasing trends all around the country have been found for spring (March–May) and summer (June–August).

State of the Art

Water vapour is a major component of the Earth's atmosphere. Given its role in processes such as radiation, cloud formation and energy exchange in the ocean-atmosphere system, water vapour is believed to be the most important trace gas in the atmosphere. It has a positive feedback, based on an exponential increase in the equilibrium of water vapour pressure with rising temperatures estimated at about

A. Wypych (✉)
Faculty of Geography and Geology, Institute of Geography and Spatial Management, Jagiellonian University, Krakow, Poland
e-mail: agnieszka.wypych@uj.edu.pl

Institute of Meteorology and Water Management—National Research Institute, Warsaw, Poland

M. Falarz (ed.), *Climate Change in Poland*, Springer Climate,
https://doi.org/10.1007/978-3-030-70328-8_12

7% per 1 K (according to Clausius-Clapeyron equation), and a quasi-exponential increase in maximum water vapour holding capacity of the Earth's atmosphere (Manabe and Wetherald 1967). The observed phenomena of global warming suggest that temperature-dependent atmospheric moisture should also demonstrate visible long-term fluctuations. Over recent decades, increasing specific humidity has been observed over most of the ocean and land surface (Dai 2006; Willett et al. 2013). Nevertheless, the trend is more pronounced with the presence of unlimited water supplies, whereas over many land areas where the moisture is restricted, there should be less of an increase in specific humidity, thereby allowing temperatures to increase by higher amounts and RH to decrease (Willet 2007).

Changes in water vapour content in the Earth's atmosphere remain a popular subject of research due to the significance of the subject matter. Nevertheless, an insignificant number of research studies have examined its course and variability. This is caused, among others, by the lack of long-term hygrometric or psychrometric measurements as well as the methodical difficulties related to the analysis of numerical material (Heino 1994). Long-term variability and change in water vapour was described by Elliott (1995), Peixoto and Oort (1996), Ross and Elliot (1996) as well as Gaffen and Ross (1999), Dai (2006) and Vincent et al. (2007). Atmospheric moisture content has also been analysed in the context of contemporary climate change (Schneider et al. 1999; Hall and Manabe 1999; Mieruch et al. 2008; Allan and Zveryaev 2011; Mattar et al. 2011), particularly on a regional scale (Groisman et al. 2004; Morland et al. 2009; Ye and Fetzer 2010; Ortiz de Galisteo et al. 2014) and also towards the parametrisation of climate models (Allan et al. 2003; Ingram 2010; Kahn et al. 2011; Ning et al. 2013). The newest climate models indicate that any trends may be exacerbated by high moisture mobility associated with changes in circulation patterns. Insightful analyses of data quality as well as temporal and spatial variability of water vapour (different variables) were conducted by Willet (2007) and Willet et al. (2008, 2013, 2014). To create a global database of atmospheric humidity (HadCRUH, http://hadobs.metoffice.com/hadcruh/), the authors examined available moisture content data sources with detailed data quality control and homogenisation procedures, demonstrating a variety of problems related to atmospheric humidity measurements.

Despite the above-mentioned doubts about the reliability of the material on hand, studies on the trends and distribution of air humidity parameters, in both regional and local aspects, have been conducted in Poland for many years. They present diurnal and annual trends in relative air humidity, saturation deficit and, rarely, in vapour pressure. The studies are most often excerpts from greater works, generally monographs of towns or regions. The data used in them are measurement series no longer than thirty years. Michna (1972) described annual and seasonal distribution of relative air humidity in Poland on the basis of data for the period of 1946–1965 received from 53 stations, using daily mean values and the midday measurement. Gumiński (1927), among others, carried out similar research earlier on analysing spatial variability of vapour pressure and relative air humidity. In the 1960s, Wierzbicki (1959a, b, 1960) and Hohendorf (1955, 1960) studied spatial distribution of relative air humidity in Poland, using air saturation deficit as the primary air humidity parameter.

Among regional studies, the most noteworthy are the works by Obrębska-Starklowa et al. (1986a, b) and works by Niedźwiedź (1973), in which thermal-humidity conditions in the Carpathian Mountains and the Sub-Carpathian region (Beskid Mountains, Podhale, the Carpathian Plateau) are described in detail. The authors stress the influence of morphometry and morphography of landforms on diurnal variations of humidity parameters and the great effect the weather types have on changes in humidity conditions. In the 1950s, Kosiba (1952) was engaged in studies on humidity balance in Silesia. He discovered that the observed decrease in relative air humidity that caused a decrease in precipitation and an increase in evaporation, and, consequently, in saturation deficit, was the result of changes in atmospheric circulation. The work by Wierzbicki, published in 1878, was one of the earliest publications characterising local humidity conditions. Wierzbicki had inhomogeneous material (missing data) at his disposal, therefore he interpolated the missing values using Bessel's formula. Satke (1904) pointed out errors in records, missing data, frequent changes of observers as well as incorrect location of measurement stations. Describing air humidity in Tarnopol, he nevertheless emphasised the regional importance of his work because of specific environmental conditions of psychrometer location. In the publication "The Climate of Rabka", Trybowskis' (1967) described humidity conditions in this town from the angle of health resort treatment. Głowicki (1970a, b) analysed the annual variations of saturation deficit and the relative humidity (in classes), connecting variability of described parameters with frequent changes in air masses. He called attention to the great importance of air humidity research for agrometeorological and hydrological applications, especially for preparing forest fire danger forecast. Kotońska (1974), based on a 10-year observation series, characterised humidity conditions in Poznań using relative air humidity and vapour pressure as well as saturation deficit. She analysed the annual variation of these parameters and also the occurrence frequency of days with characteristic humidity and thermal conditions. Furthermore, in her dissertation, Tarajkowska (1974) focused on the influence of changed subsoil on humidity conditions in Częstochowa. Describing long-term (1924–1965) variations in air humidity parameters, the author did not prove their direct connection with the city's development, but she pointed out diversified humidity conditions in urban and non-urban areas that provided valuable clues for works concerning city climates. For many years, studies on air humidity distribution in the aspect of city climate changes have also been conducted in Wrocław's research centre (Młostek and Sobik 1984; Dubicka et al. 2003), Łódź's research centre (Dubaniewicz 1977; Kłysik 1985), and more and more frequently also in other cities in Poland (e.g. in Lublin—Gluza and Kaszewski 1984). The aim of research works is to establish the influence of urban environment (pattern of development) on diurnal and annual variations in air humidity, with particular regard to the impact of urban heat islands. The results corroborate other authors' findings (Lewińska 2000), which highlight the fact that values of vapour pressure and saturation deficit are higher in towns than outside them, while values of relative air humidity are considerably lower.

As already mentioned, air humidity variation was seldom the topic of detailed examination. In Poland, Hohendorf (1967, 1969) was the first who took up this topic in his research and described the variability of saturation deficit. Obrębska-Starklowa and Grzyborowska (1997) presented the analysis of air humidity variability made for the plateau section of the Raba River (relative air humidity in the period of 1971–1992), whereas Bryś (2003) described air saturation deficit variation in Wrocław and Wypych (e.g. 2008, 2010) the variability of humidity conditions in Kraków, both for the period of the twentieth century. The conducted research confirmed the drying out of the atmosphere above Poland on the basis of point location studies.

Slovak (Kveták 1985, 1993; Brázdil and Budiková 1994) and Finnish (Heino 1994) researchers made a considerable contribution to the studies of long-term variability of this meteorological element. The results of conducted analyses confirm those obtained in Poland. Heino, despite breaks in series homogeneity, documented a decrease in the value of relative humidity in Helsinki in the twentieth century. In winter months, the relative humidity dropped by 5–8%, while in May and June by as much as 18–19%. In Hurbanovo, Slovakia, in the twentieth century, there was a decrease in both relative air humidity, whose value dropped by 3–9%, and in vapour pressure. All authors stressed unanimously that, mainly because of the scarcity of material, the obtained results did not provide an unequivocal answer to the question of air humidity variability and its causes, thus confirming the necessity of conducting more detailed studies and analyses.

Data and Methods

To analyse long-term variability of humidity conditions in Poland, three data sources were used. Surface measurements of relative humidity (RH) were obtained at meteorological stations evenly distributed all over the country using a psychrometer with both a dry-bulb and wet-bulb thermometer, covering the period of 1951–2018. Historical measurements of relative humidity (RH) were conducted in Kraków starting in the year 1863, consequently in the same place (psychrometer located in Stevenson screen by the NNW window, 12 m above the ground level), and ERA-5 reanalyses data (with spatial resolution of 0.25°) describing the amount of water vapour in the troposphere (specific humidity, SH) were also obtained for the available period of 1981–2018. To reduce the problem of changes in observing times within the analysed period for all the data sources, only subdaily measurements performed at 12 UTC were taken into consideration.

In modern reanalysis, including ERA-5 the parametrisation of the representation of the hydrologic cycle is strongly improved, which leads to a reduction in the magnitude of errors for humidity-sensitive variables. Many comparative studies have confirmed the usefulness of reanalysis data in the examination of water vapour fields (Schröder et al. 2016). On the contrary, psychrometric measurements are associated both with instrument handling and reading errors. All the available data (more than 150 stations of the Institute of Meteorology and Water Management—National Research Institute) were thoroughly inspected and checked by quality control and weather dependent analyses. Finally, the 56 selected stations have data completeness of at least 90% for 1951–2018.

The non-parametric Mann-Kendall test was applied to assess the trends for the entire period and subperiods and to measure the statistical significance of the changes, while the Theil-Sen estimator was used to calculate the trend slopes.

Moisture Content (Specific Humidity)

Specific humidity (SH, g·kg^{-1}) is sometimes regarded as a measure of 'actual' humidity, relating the mass of water vapour in the atmosphere to the total mass of the moist atmosphere. Increasing SH directly implies increasing absolute atmospheric moisture content as opposed to the relative measure of RH.

Analysis of spatial and temporal differences in water vapour content in the troposphere shows large seasonal differences in the distribution of water vapour. Although mean annual values of specific humidity in Poland demonstrate almost no spatial differentiation, reaching around 5.5 g·kg^{-1}, the standard deviation and coefficient of variation indicate climate continentalism in southeastern part of the country, showing significant seasonality from 3.0 g·kg^{-1} in winter to more than 7.0 g·kg^{-1} in the summer months (Wypych 2018).

Regarding rising temperatures and the increase in maximum water vapour holding capacity of the atmosphere for recent decades, increasing specific humidity has also been observed in Central Europe (Fig. 12.1). Without unlimited water supplies,

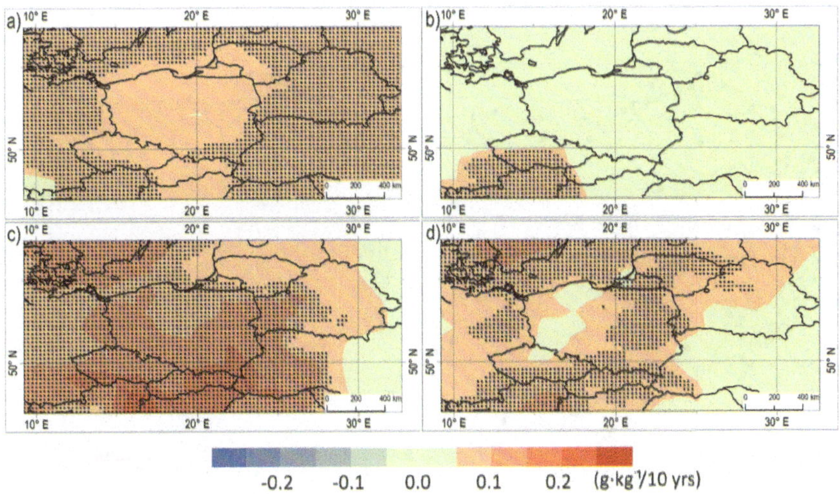

Fig. 12.1 Specific humidity change over Central Europe—absolute trend for the period 1981–2018 (g·kg^{-1}/10 years): **a** winter (DJF), **b** spring (MAM), **c** summer (JJA), **d** autumn (SON). Black dots—statistically significant trends ($\alpha = 0.05$)

the region experiences only a slight increase in moisture, significant particularly in summer (up to 0.2 g·kg^{-1} per 10 years) and in autumn (Fig. 12.1). The relative trend reaches approximately 3% of mean values.

In comparison to spring time, greater water vapour content in autumn as well as a substantial increase in its amount confirms the significance of evaporation surfaces. In terms of water heat capacity, water bodies and moist surfaces are less active in spring, while in summer or autumn play a significant role, which confirms the significance of active surfaces in the formation of air humidity.

Relative Humidity

Relative humidity, usually expressed in per cent (RH, %), describes the ratio of the vapour pressure to the saturation vapour pressure. With increasing temperature and no external source of water vapour, the decreasing RH indicates drying of the atmosphere.

Although temporal and spatial differentiation of relative humidity in Poland, regarding many factors, is more definite than in the case of SH, the annual range reaches less than 25% and the areal about 6–10%, depending on the season. The annual distribution of RH confirms the highest values at the seaside (>70%) decrescent towards the south (app. 66%), outside of the mountains, where the amount of water vapour is mostly the same as in the north (Fig. 12.2a).

In Poland, a statistically significant decrease in RH has been observed since the second half of twentieth century, with the trend value of almost 1% per 10 years in southeastern part of the country and much less at the sea coast (Fig. 12.2a). The

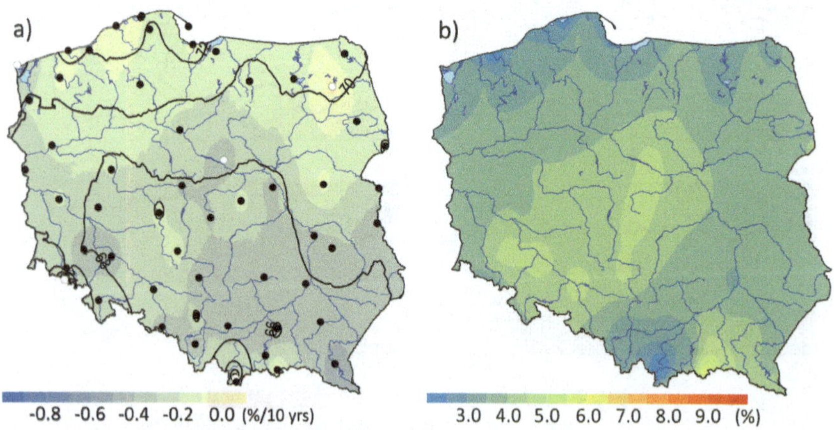

Fig. 12.2 Relative humidity annual variability and change: **a** mean values (isolines, %) and absolute trend for the whole period 1951–2018 (%/10 years), **b** relative trend the whole period 1951–2018 (%/10 years). Black dots—statistically significant trends ($\alpha = 0.05$)

relative trend is the highest in central Poland, about 5% of the mean, while in the north and south edges, it does not even reach 2.5%, which confirms the low magnitude of the change (Fig. 12.2b).

With a general declining trend, long-term variability demonstrates different change directions in succeeding halves (Fig. 12.3). The example of stations representing selected PL regions, i.e. Łeba (seaside), Słubice (western PL), Warszawa

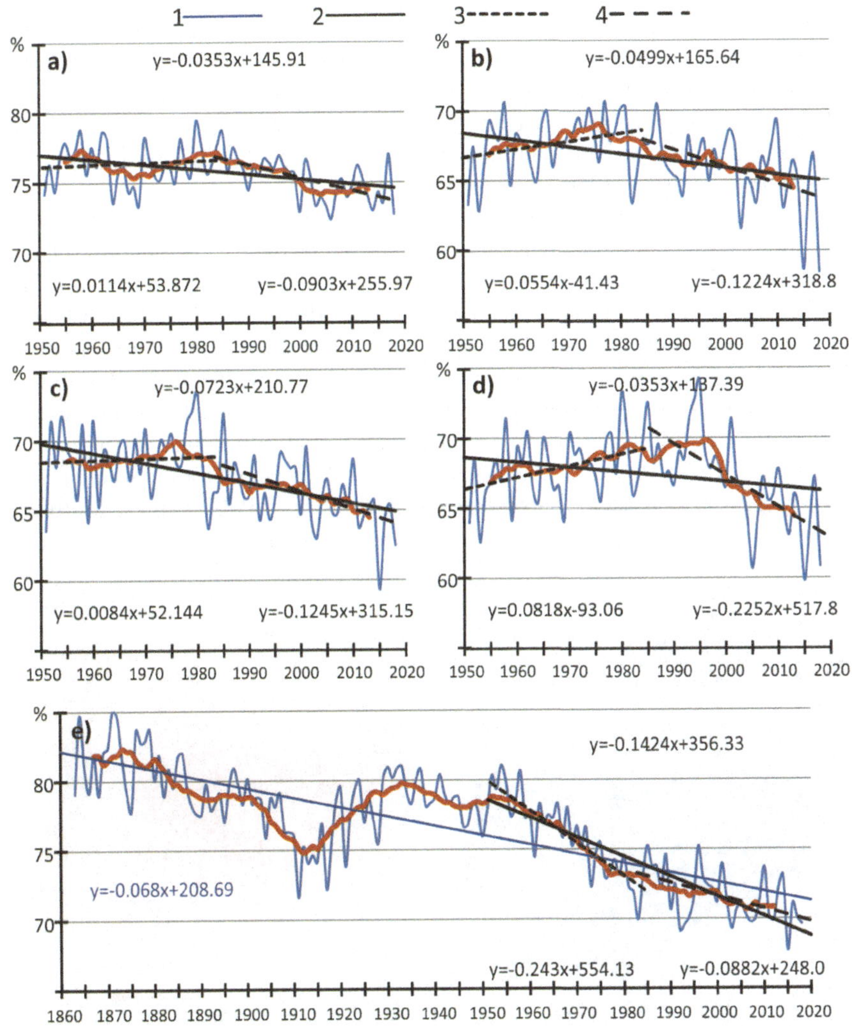

Fig. 12.3 Long-term variability of relative humidity at selected meteorological stations: **a** Łeba, **b** Słubice, **c** Warszawa, **d** Rzeszów, **e** Kraków (historical station)—mean annual values (blue) smoothed by 10-year consecutive average (red); straight lines—linear trends: 1—1863–2018, 2—1951–2018, 3—1951–1984, 4—1985–2018 (equations located regarding the periods)

(central PL), Rzeszów (southeastern PL) and Kraków (southern Poland, non-standard historical data series), clearly indicates the impact of geographical location on the relative humidity course. Long-term RH variability in Warszawa (Fig. 12.3b), Słubice (Fig. 12.3c) and Rzeszów (Fig. 12.3d), manifest an increase in relative humidity in the years 1951–1984. With the exception of Warszawa, where the trend value reaches almost 3%, the change is statistically insignificant. An extreme decrease in RH (up to 2% per decade and 7.5% in 1985–2018 period in Warszawa) can be noticed in the second half of the analysed period and considerably influences the long-term trend.

The only exception is Kraków, where the systematical drying of the air was noticed in the years 1951–2018 (more than 9% within the 68-year period). It has been observed since the beginning of the constant instrumental measurements of RH, i.e. 1863, with the short period of less than three decades between the years 1910 and 1940 when a significant increase in relative humidity in Kraków occurred, bringing the final drop of RH about 10.5% in 156 years.

Long-term variability demonstrates not only spatial but also seasonal differentiation. Spring and summer in Poland are characterised by the lowest values of relative humidity (Figs. 12.4a and 12.5a). The areal mean of the whole country does not exceed 65%, with central region as the driest one (RH < 60%). The warm half of the year has been affected also by the most intensive changes in moisture conditions (Figs. 12.4, 12.5, 12.6, and 12.7). They are statistically significant for the central-western and southern regions in spring, while in summer the whole country (except for two stations, Fig. 12.5a) has experienced substantial shift. It has run up to −1% per 10 years (Figs. 12.4a and 12.5a), whereas the relative trend for the whole period is the biggest (9% and 8%, respectively) in the centre of Poland (Figs. 12.4b and 12.5b).

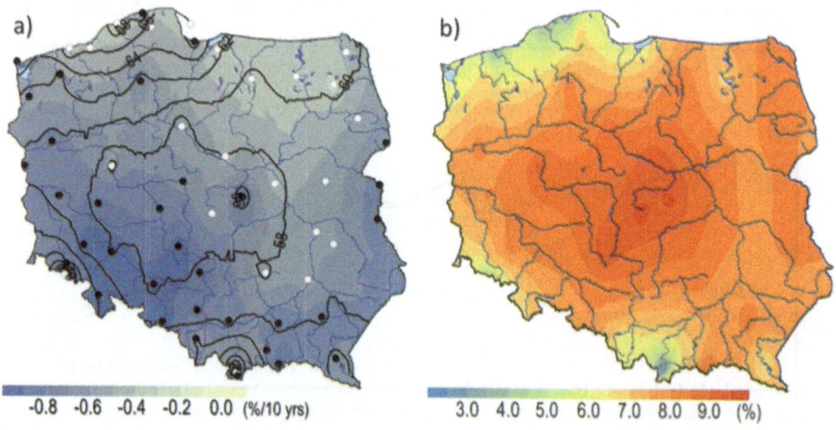

Fig. 12.4 Relative humidity variability and change in spring (MAM): **a** mean values (isolines, %) and absolute trend for the whole period 1951–2018 (%/10 years), **b** relative trend the whole period 1951–2018 (%/10 years). Black dots—statistically significant trends (α = 0.05)

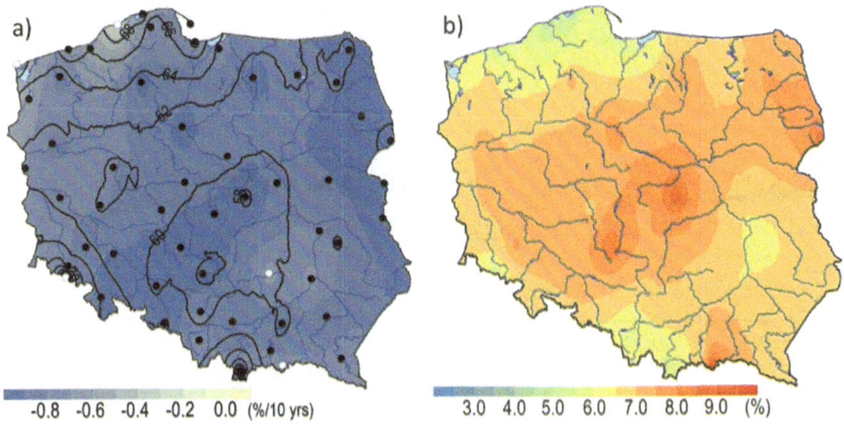

Fig. 12.5 Relative humidity variability and change in summer (JJA): **a** mean values (isolines, %) and absolute trend for the whole period 1951–2018 (%/10 years), **a** relative trend the whole period 1951–2018 (%/10 years). Black dots—statistically significant trends ($\alpha = 0.05$)

For particular locations, Rzeszów—situated in the southeastern part of the country—has experienced an almost 7% drop in RH in spring and a little less in summer (0.8% per 10 years). The more intensive change was noticed only in Kraków, where it reached −13% in spring and −19% in summer in the years 1951–2018, when the decrease in the first half of the period was more intense than at the turn of the century (Figs. 12.6e and 12.7e). A similar change pattern, i.e. a steady drop, also characterises the coastal region (Łeba). Nevertheless, the magnitude of the change (despite its significance) is inconsiderable. The other stations were afflicted with the previously mentioned model, where the first half-period displayed a positive trend. However, for the warm half of the year, this increase is almost imperceptible and statistically insignificant, and the last decades demonstrated an essential moisture decrease.

The cold half of the year does not bring any fundamental contribution to the variability and change in moisture conditions in Poland (Figs. 12.8, 12.9, 12.10, and 12.11). Although some trends can be noticed, especially in winter, for many locations, they are insignificant (Fig. 12.8a). For southern Poland, the relative winter trend reaches negative 5% and confirms tropospheric drying in this season as well (Fig. 12.8b). What was surprising was the significant change in both half-periods in autumn in Warszawa (Fig. 12.11c). The opposite sign of the trends (changes of +5.8% and −7.2%, respectively) resulted statistically as without any significant change. Nevertheless, the case is worth mentioning as a consequence of some modification in local conditions.

The same should be indicated regarding Kraków, where almost all the trends are statistically significant and, considering RH variability within a 156-year period, confirm the meaningful drying of the lower troposphere above the city (Figs. 12.9e

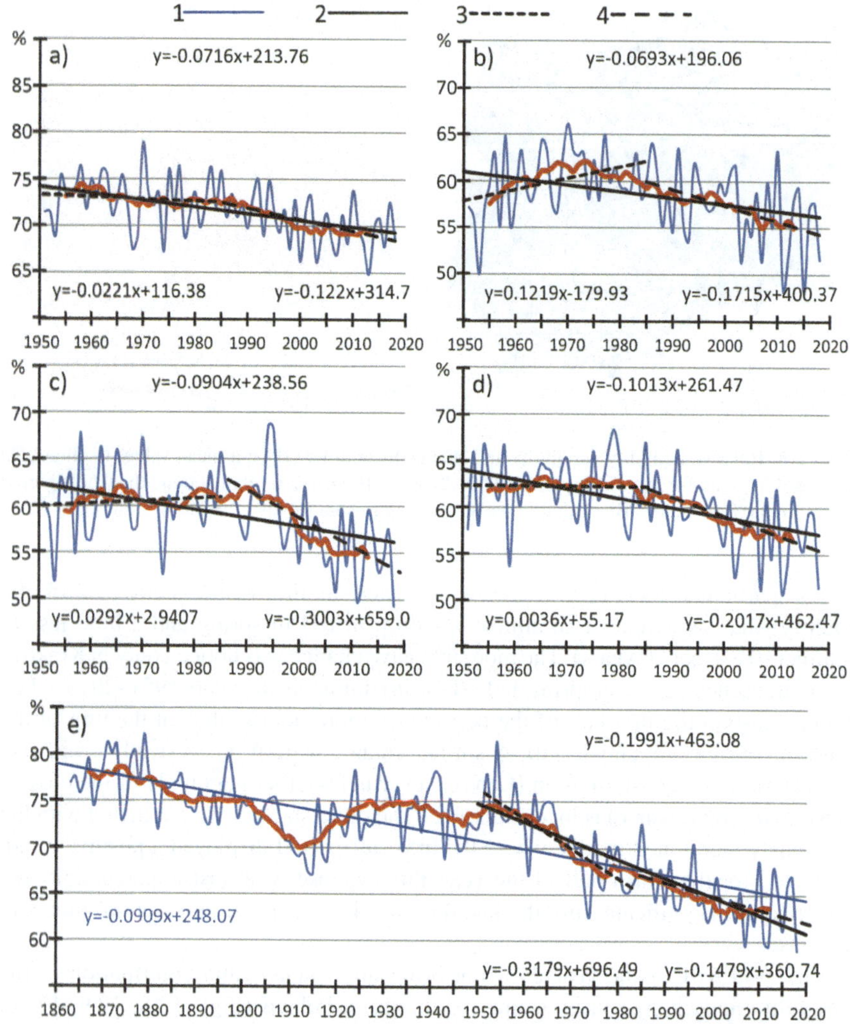

Fig. 12.6 Long-term variability of relative humidity at selected meteorological stations: **a** Łeba, **b** Słubice, **c** Warszawa, **d** Rzeszów, **e** Kraków (historical station)—mean spring (MAM) values (blue) smoothed by 10-year consecutive average (red); straight lines—linear trends: 1—1863–2018, 2—1951–2018, 3—1951–1984, 4—1985–2018 (equations located regarding the periods)

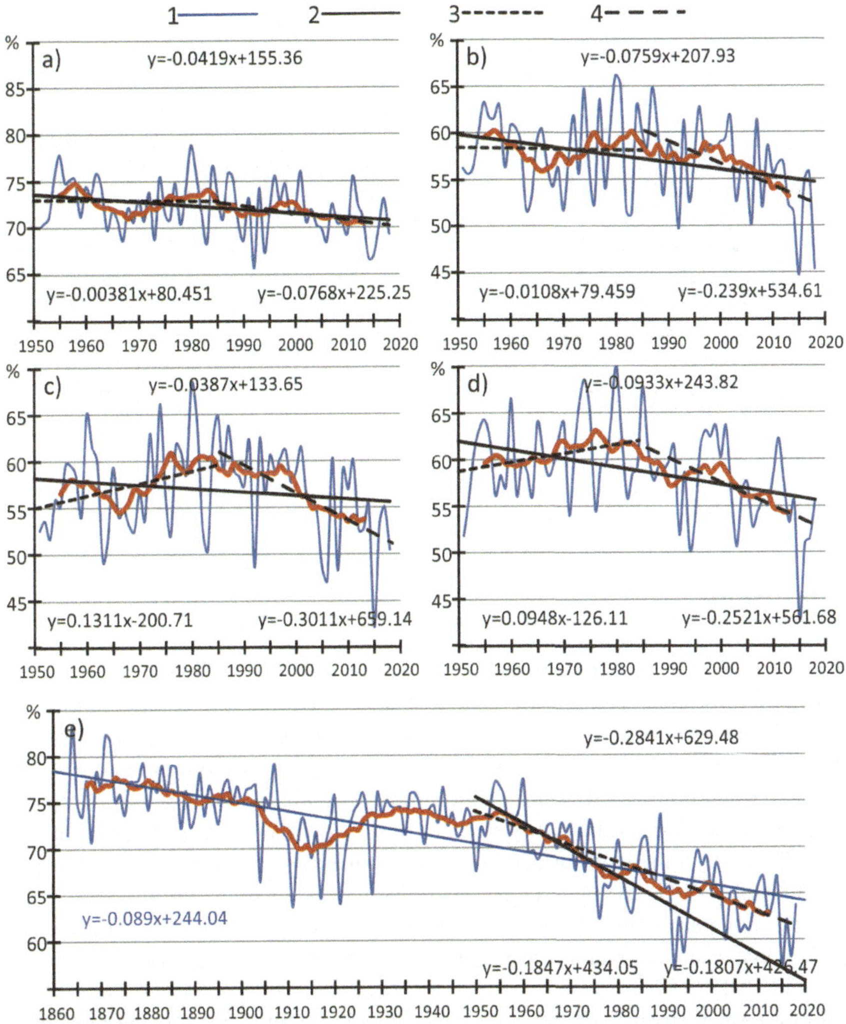

Fig. 12.7 Long-term variability of relative humidity at selected meteorological stations: **a** Łeba, **b** Słubice, **c** Warszawa, **d** Rzeszów, **e** Kraków (historical station)—mean summer (JJA) values (blue) smoothed by 10-year consecutive average (red); straight lines—linear trends: 1—1863–2018, 2—1951–2018, 3—1951–1984, 4—1985–2018 (equations located regarding the periods)

and 12.11e). Although the historical station has always been located in the same place, the surroundings have changed significantly.

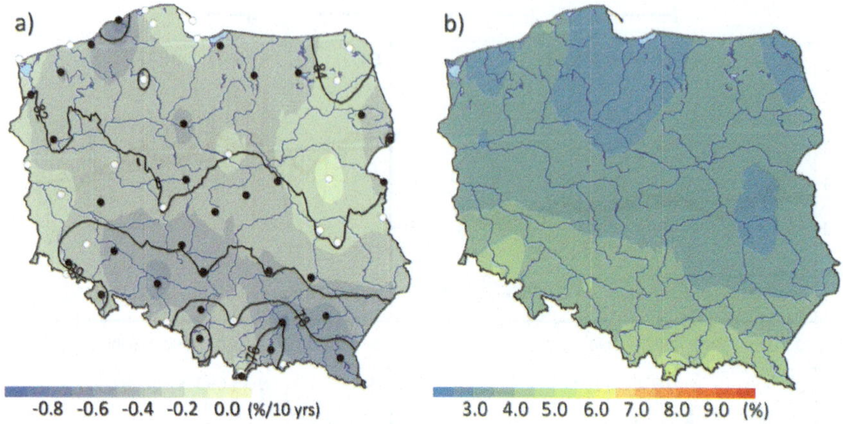

Fig. 12.8 Relative humidity variability and change in winter (DJF): **a** mean values (isolines, %) and absolute trend for the whole period 1951–2018 (%/10 years), **b** relative trend the whole period 1951–2018 (%/10 years). Black dots—statistically significant trends ($\alpha = 0.05$)

Conclusions

Tropospheric moisture content varies with time and space. Basic processes that control water vapour content are evaporation, condensation and precipitation. In the areas of moisture deficit, water vapour transport plays also an important role via advection and convection processes. Nevertheless, the maximum amount of water vapour in the atmosphere depends on the air temperature. With increasing air temperature, increasing specific humidity has been observed. Nevertheless, the trend magnitude is lower over the land surface, where the moisture sources are limited (Wypych and Bochenek 2018; Wypych et al. 2018). Moreover, relative humidity describing the saturation percentage, decreases because with raising air temperature the amount of available water vapour is insufficient to execute increasing moisture capacity (Wypych 2018).

Long-term variability of moisture conditions in Poland confirms the tendencies proved worldwide. What is more, regional or even local studies have emphasised that surface humidity may be affected by factors other than rising temperatures. These include changes in land-use, including irrigation and reservoirs, but especially enlarged range of built-up areas. Their importance in heat balance affects also essentially the water cycle, especially saturation deficit with a fundamental regional or local consequences. It has been proved, especially for spring (March–May) and summer (June–August), when the moisture shortage is the most evident, with the most intensive decreasing trends all around the country, particularly in urban areas. Overall, the situation of perceptible moisture shortage, if not changed, will have broad scale environmental consequences as well as effects on human health and the economy.

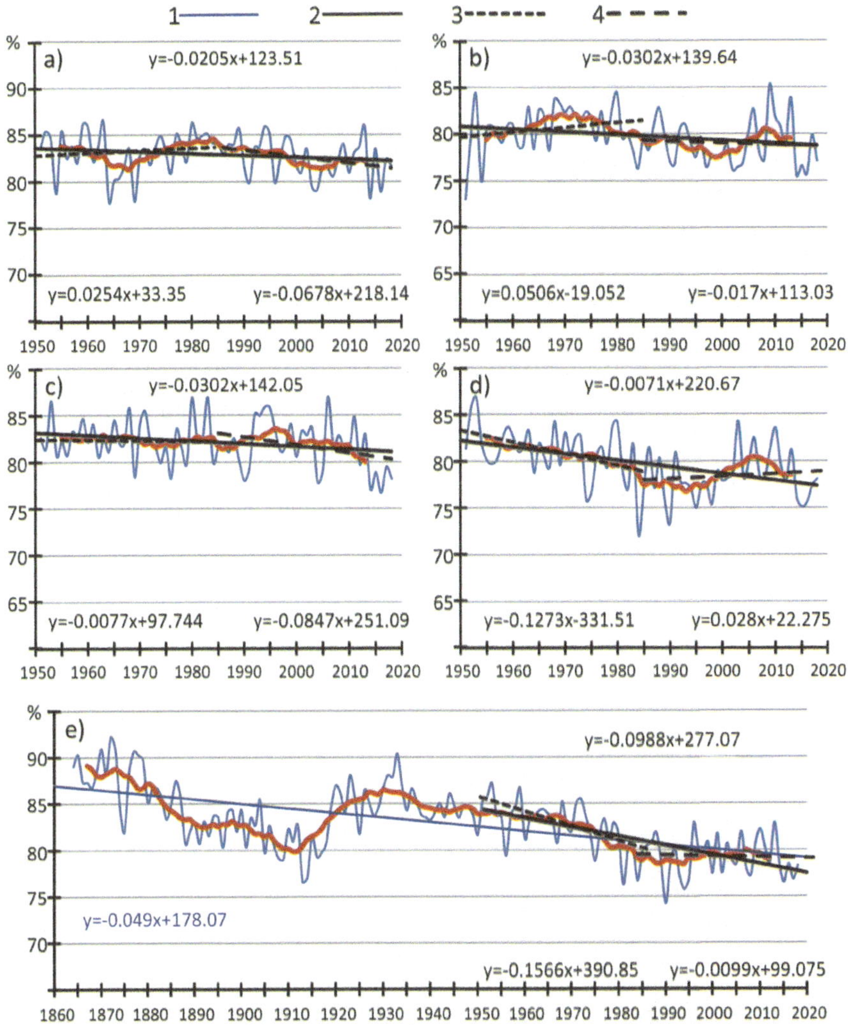

Fig. 12.9 Long-term variability of relative humidity at selected meteorological stations: **a** Łeba, **b** Słubice, **c** Warszawa, **d** Rzeszów, **e** Kraków (historical station)—mean winter (DJF) values (blue) smoothed by 10-year consecutive average (red); straight lines—linear trends: 1—1863–2018, 2—1951–2018, 3—1951–1984, 4—1985–2018 (equations located regarding the periods)

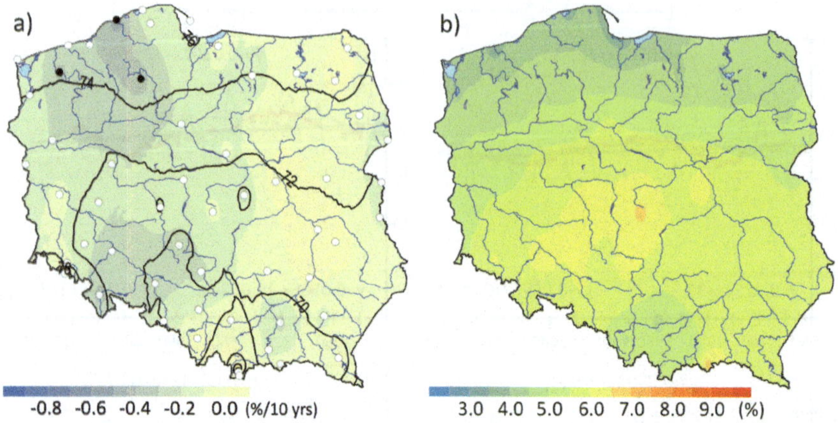

Fig. 12.10 Relative humidity variability and change in autumn (SON): **a** mean values (isolines, %) and absolute trend for the whole period 1951–2018 (%/10 years), **b** relative trend the whole period 1951–2018 (%/10 years). Black dots—statistically significant trends ($\alpha = 0.05$)

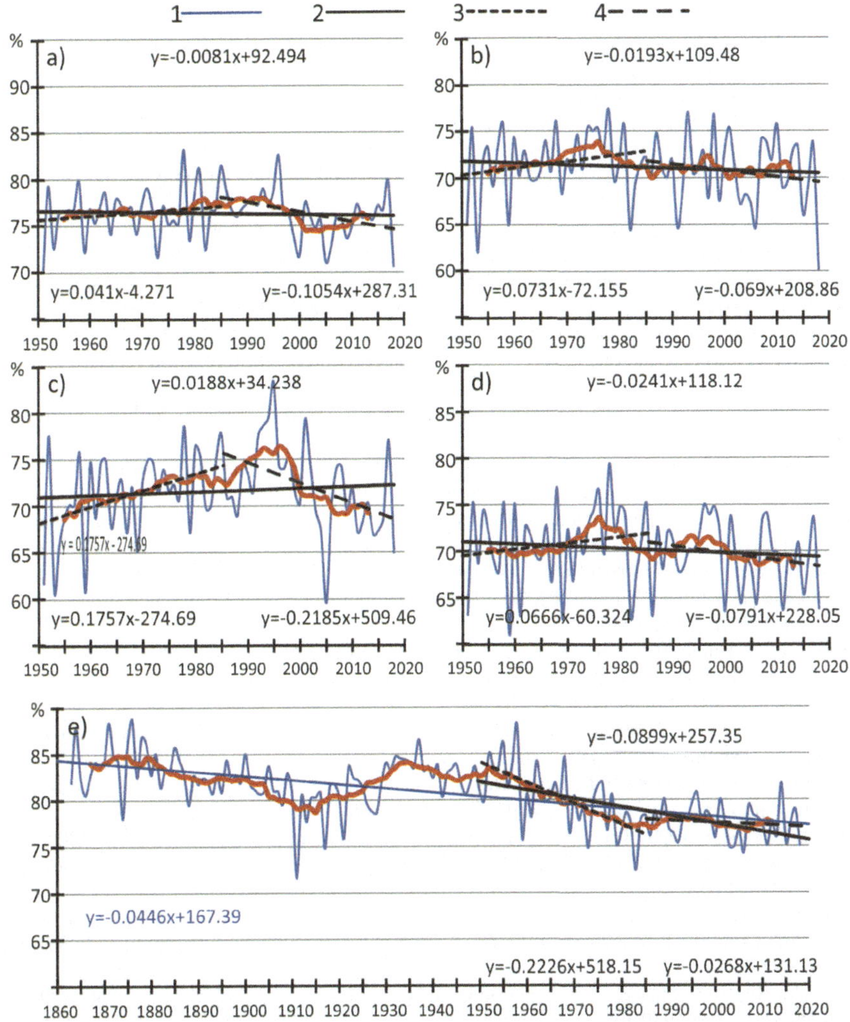

Fig. 12.11 Long-term variability of relative humidity at selected meteorological stations: **a** Łeba, **b** Słubice, **c** Warszawa, **d** Rzeszów, **e** Kraków (historical station)—mean autumn (SON) values (blue) smoothed by 10-year consecutive average (red); straight lines—linear trends: 1—1863–2018, 2—1951–2018, 3—1951–1984, 4—1985–2018 (equations located regarding the periods)

References

Allan RP, Ringer MA, Slingo A (2003) Evaluation of moisture in the Hadley Centre Climate Model using simulations of HIRS water vapor channel radiances. Q J R Meteorol Soc 129:3371–3389. https://doi.org/10.1256/qj.02.217

Allan RP, Zveryaev II (2011) Variability in the summer season hydrological cycle over the Atlantic-Europe region 1979–2007. Inter J Climatol 31:337–348. https://doi.org/10.1002/joc.2070

Brázdil R, Budikowá M (1994) Climate variation in Central Europe—facts or inhomogenities? In: Heino R (ed) Climate variations in Central Europe, Proceedings of the European Workshop on Climate Variations, Finland, pp 9–29
Bryś T (2003) Variability of saturation deficit in Wroclaw in the 20th century and some of its conditions. In: Pyka JL, Dubicka M, Szczepankiewicz-Szmyrka A, Sobik M, Błaś M (eds) Man and climate in the 20th century. Studia Geograficzne, Uniwersytet Wrocławski, 75, pp 203–215
Dai A (2006) Recent climatology, variability, and trends in global surface humidity. J Clim 19:3589–3606
Dubaniewicz H (1977) Wpływ miasta na kształtowanie się wilgotności powietrza obszarów przyległych na przykładzie Łodzi. Studia Regionalne 1(6):95–103 (in Polish)
Dubicka M, Rosiński D, Szymanowski M (2003) The influence of urban environment on air humidity in Wrocław. In: Pyka JL, Dubicka M, Szczepankiewicz-Szmyrka A, Sobik M, Błaś M (eds) Man and climate in the 20th century. Studia Geograficzne, Uniwersytet Wrocławski, 75, pp 504–527
Elliott WP (1995) On detecting long-term changes in atmospheric moisture. Clim Change 31:349–367
Gaffen DJ, Ross RJ (1999) Climatology and trends of U.S. surface humidity and temperature. J Clim 12:811–828
Głowicki B (1970a) Frekwencja klas wilgotności względnej we Wrocławiu w 1961r., Acta Univ Wratisl Prace Obserw Meteor Wrocł, 130(16):115–120 (in Polish)
Głowicki B (1970b) Przebieg niedosytu wilgotności we Wrocławiu w 1961r., Acta Univ Wratisl Prace Obserw Meteor Wrocł, 130(16):121–127 (in Polish)
Gluza A, Kaszewski B (1984) Zróżnicowanie temperatury i wilgotności względnej powietrza w Lublinie. In: Klimat i Bioklimat Miast, Materiały I Ogólnopolskiej Konferencji, Łódź, 22–24 listopada 1984, pp 107–114
Groisman PY, Knight RW, Karl TR, Easterling DR, Sun BM, Lawrimore JH (2004) Contemporary changes of the hydrological cycle over the contiguous United States: trends derived from in situ observations. J Hydrometeorology 5:64–85
Gumiński R (1927) Wilgotność powietrza w Polsce (wahania roczne i rozkład geograficzny). Prace Meteorol. i Hydrol 3:1–71 (in Polish)
HadCRUH: Hadley Centre and Climate Research Unit global surface humidity dataset from 1973 to 2003. http://hadobs.metoffice.com/hadcruh/ (12 April 2011)
Hall A, Manabe S (1999) The role of water vapor feedback in unperturbed climate variability and global warming. J Clim 12:2327–2346. https://doi.org/10.1175/1520-0442(1999)012%3c2327:Trowvf%3e2.0.Co;2
Heino R (1994) Changes of relative air humidity in Finland. In: Heino R (ed) Climate variations in Central Europe, Proceedings of the European Workshop on Climate Variations, Finland, pp 314–320
Hohendorf E (1955) Miesięczne i roczne sumy niedosytów wilgotności powietrza w Polsce. Prace PIHM 41:1–79 (in Polish)
Hohendorf E (1960) Rozmieszczenie średnich wieloletnich niedosytów wilgotności powietrza w głównych porach roku w Polsce. Wiadomości Instytutu Melior 1(3):8–23 (in Polish)
Hohendorf E (1967) Opracowanie statystyczne ciągów obserwacyjnych niedosytu wilgotności powietrza, Prace i Studia Kom. Inż., i Gosp. Wodnej 8:263–299 (in Polish)
Hohendorf E (1969) Niedosyty wilgotności powietrza w Polsce w latach 1881–1939, Roczniki Nauk Roln, Ser F, 73(3):413–483 (in Polish)
Ingram W (2010) A very simple model for the water vapor feedback on climate change. Q J R Meteorol Soc 136:30–40. https://doi.org/10.1002/qj.546
Kahn BH, Teixeira J, Fetzer EJ, Gettelman A, Hristova-Veleva SM, Huang X, Kochanski AK, Köhler M, Krueger SK, Wood R, Zhao M (2011) Temperature and water vapor variance scaling in global models: Comparisons to satellite and aircraft data. J Atmos Sci 68:2156–2168. https://doi.org/10.1175/2011JAS3737.1

Kłysik K (1985) Wpływ struktury termiczno-wilgotnościowej przyziemnych warstw powietrza na klimat lokalny w wybranych warunkach terenowych. Acta Geographica Lodziensia 49:1–119 (in Polish)

Kosiba A (1952) Zagadnienie bilansu wilgoci na Śląsku w świetle zmian wilgotności względnej. Przegląd Meteorologiczny i Hydrologiczny 5(3/4):58–69 (in Polish)

Kotońska B (1974) Wilgotność powietrza w Poznaniu, Badania Fizjogr. nad Polską Zach., Ser. A – Geogr. Fizyczna, XXVII:71–87 (in Polish)

Kveták Š (1985) Premenlivost' relativnej vlhkosti vzduchu v Hurbanove. In: Zbornik prac SHMÚ. Bratislava, 26, pp 175–241

Kveták Š (1993) Vlhkostný character obdobia 1961–1990 v Hurbanove. Meteorologické zprávy 46:48–53

Lewińska J (2000) Klimat miasta. Zasoby, zagrożenia, kształtowanie, IGPiK, Kraków (in Polish)

Manabe S, Wetherald RT (1967) Thermal equilibrium of the atmosphere with a given distribution of relative humidity. J Atmos Sci 24:241–259. https://doi.org/10.1175/1520-0469(1967)024%3c0241:TEOTAW%3e2.0.CO;2

Mattar C, Sobrino JA, Julien Y, Morales L (2011) Trends in column integrated water vapor over Europe from 1973 to 2003. Inter J Climatol 31:1749–1757. https://doi.org/10.1002/joc.2186

Michna E (1972) O wilgotności względnej powietrza w Polsce. Przegląd Geofizyczny 17(1):3–14 (in Polish)

Mieruch S, Nöel S, Bovensmann H, Burrows JP (2008) Analysis of global water vapor trends from satellite measurements in the visible spectral range. Atmos Chem Phys 8:491–504. https://doi.org/10.5194/acp-8-491-2008

Młostek E, Sobik M (1984) Usłovija vłažnosti vozducha w gorode v izbrannych sinopticzeskich situacjach na primiere goroda Wrocława. In: Mikroklimat i mezoklimat gorodskich aglomeracji. Brno, pp 36–43

Morland J, Collaud Coen M, Hocke K, Jeannet P, Mätzler C (2009) Tropospheric water vapor above Switzerland over the last 12 years. Atmos Chem Phys 9:5975–5988. https://doi.org/10.5194/acp-9-5975-2009

Niedźwiedź T (1973) Temperatura i wilgotność powietrza w warunkach rzeźby pogórskiej Karpat (na przykładzie doliny Raby koło Gaika-Brzezowej). Zeszyty Naukowe UJ, Prace Geograficzne 32:7–88 (in Polish)

Ning T, Elgered G, Willén U, Johansson JM (2013) Evaluation of the atmospheric water vapor content in a regional climate model using ground-based GPS measurements. J Geophys Res Atmos 118:329–339. https://doi.org/10.1029/2012JD018053

Obrębska-Starklowa B, Grzyborowska A (1997) Tendencje zmian wilgotności względnej powietrza w pogórskim odcinku doliny Raby w latach 1971–1992. Roczniki AR w Poznaniu 291(17):15–38 (in Polish)

Obrębska-Starklowa B, Kowanetz L, Zając J (1986a) Ciśnienie pary wodnej w Beskidach i na Podhalu, Zesz. Nauk. UJ, Prace Geogr 67:45–67 (in Polish)

Obrębska-Starklowa B, Kowanetz L, Zając J (1986b) Wilgotność względna powietrza w Beskidach i na Podhalu, Zesz. Nauk. UJ, Prace Geogr 64:21–47 (in Polish)

Ortiz de Galisteo JP, Bennouna Y, Toledano C, Cachorro V, Romero P, Andrés MI, Torres B (2014) Analysis of the annual cycle of the precipitable water vapor over Spain from 10-year homogenized series of GPS data. Q J R Meteorol Soc 140:397–406. https://doi.org/10.1002/qj.2146

Peixoto JP, Oort AH (1996) The climatology of relative humidity in the atmosphere. J Clim 9:3443–3463

Ross RJ, Elliott WP (1996) Tropospheric water vapor climatology and trends over North America: 1973–93. J Clim 9:3561–3574

Satke W (1904) Względna wilgotność w Tarnopolu. Rozpr AU Wydział Mat Przyr 43:417–434 (in Polish)

Schneider EK, Kirtman BP, Lindzen RS (1999) Tropospheric Water Vapor and Climate Sensitivity. J Atmos Sci 56:1649–1658. https://doi.org/10.1175/1520-0469(1999)056%3c1649:TWVACS%3e2.0.CO;2

Schröder M, Lockhoff M, Forsythe JM, Cronk HQ, Vonder Haar TH, Bennartz R (2016) The GEWEX Water Vapor Assessment: Results from Intercomparison, Trend, and Homogeneity Analysis of Total Column Water Vapor. J Appl Meteorol Climatol 55:1633–1649. https://doi.org/10.1175/JAMC-D-15-0304.1

Tarajkowska M (1974) Wpływ miasta na warunki wilgotnościowe powietrza na przykładzie Częstochowy, Zeszyty Naukowe Uniwersytetu Łódzkiego. Nauki Matematyczno- Przyrodnicze, Seria II 63:43–56 (in Polish)

Trybowska E, Trybowski C (1967) Wilgotność powietrza w Rabce. Prace PIHM 91:41–54 (in Polish)

Vincent LA, van Wijngaarden WA, Hopkinson R (2007) Temperature and humidity trends in Canada for 1953–2005. J Clim 20:5100–5113

Wierzbicki D (1878) Ruch dzienny prężności pary i wilgotności powietrza w Krakowie, Pamiętniki AU Wydz. Mat Przyr 4:153–173 (in Polish)

Wierzbicki Z (1959a) Niedosyt wilgotności powietrza w Polsce. Acta Geoph Pol 7(3/4):430–440 (in Polish)

Wierzbicki Z (1959b) O niedosycie wilgotności powietrza w Polsce, Annales UMCS, sec. B, XIV(1):1–45 (in Polish)

Wierzbicki Z (1960) Niedosyt wilgotności powietrza w Polsce w latach 1949–58. Prace PIHM 57:24–48 (in Polish)

Willett KM (2007) Creation and analysis of HadCRUH: A new global surface humidity dataset, Climatic Research Unit, School of Environmental Sciences, University of East Anglia, 1–174

Willett KM, Dunn RJH, Thorne PW, Bell S, de Podesta M, Parker DE, Jones PD, Williams CN (2014) HadISDH land surface multi-variable humidity and temperature record for climate monitoring. Clim Past 10:1983–2006. https://doi.org/10.5194/cp-10-1983-2014

Willett KM, Jones PD, Gillett NP, Thorne PW (2008) Recent changes in surface humidity: development of the HadCRUH dataset. J Clim 21:5364–5383. https://doi.org/10.1175/2008JCLI2274.1

Willett KM, Williams CN Jr, Dunn RJH, Thorne PW, Bell S, de Podesta M, Jones PD, Parker DE (2013) HadISDH: an updateable land surface specific humidity product for climate monitoring. Clim Past 9:657–677. https://doi.org/10.5194/cp-9-657-2013

Wypych A (2008) Niedosyt wilgotności powietrza miarą zmienności warunków wilgotnościowych w Krakowie. Acta Agrophysica 12(1):277–288 (in Polish)

Wypych A (2010) Twentieth century variability of surface humidity as the climate change indicator in Kraków (Southern Poland). Theor Appl Climatol 101:475–482

Wypych A (2018) Para wodna w troposferze nad Europą, Instytut Geografii i Gospodarki Przestrzennej UJ, 200 pp (in Polish)

Wypych A, Bochenek B (2018) Vertical Structure of Moisture Content over Europe, Advances in Meteorology, ID: 3940503

Wypych A, Bochenek B, Różycki M (2018) Atmospheric Moisture Content over Europe and the Northern Atlantic. Atmosphere 9:18. https://doi.org/10.3390/atmos9010018

Ye A, Fetzer EJ (2010) Atmospheric moisture content associated with surface air temperatures over northern Eurasia. Inter J Climatol 30:1463–1471. https://doi.org/10.1002/joc.1991

Chapter 13
Precipitation Change

Ewa Łupikasza and Łukasz Małarzewski

Abstract This chapter focuses on trends in annual, seasonal and monthly precipitation totals and frequency based on 52 series of daily precipitation covering the period 1951–2018. Trends in precipitation indices between 1951 and 2018 were rarely statistically significant. The temporal course of precipitation characteristics was rather dominated by fluctuation and altering dry and wet decades which occurred more or less simultaneously in spring and autumn— the wet periods included the 1960s, 1970s and the second half of 1990s and were separated by dry period in the 1980s and in the early 1990s. Spatial distribution of trend direction in precipitation totals was not entirely consistent with the distribution of trends in the corresponding precipitation frequency. In spring, precipitation totals were increasing faster than its frequency, indicating some increase in precipitation intensity in northern Poland. In southwestern Poland, spring precipitation totals were lowering. In summer, the spatial pattern of trends was complicated, which may result from more frequent condition for free convection compared to other seasons. In autumn, both characteristics increased at most stations. In winter, precipitation totals were significantly increasing in northern Poland and decreasing in southern Poland. On monthly scale, March had the highest number of significant trends.

Precipitation Change

Precipitation plays a crucial role in the Earth system triggering many geographical processes and affecting life. Detection of precipitation trends is problematic due to its great variability, which is an inherent feature of this climate element. The comprehensive revision of the studies on precipitation in the historical context can be found in Przybylak et al. (2010). The longest chronological series of precipitation are available for Wrocław from 1799 (Pyka 2003), Warsaw from 1813 (Gorczyński 1912; Marciniak and Kożuchowski 1990), and Cracow from August

E. Łupikasza (✉) · Ł. Małarzewski
Faculty of Natural Sciences, Institute of Earth Sciences, University of Silesia in Katowice, Katowice, Poland
e-mail: ewa.lupikasza@us.edu.pl

M. Falarz (ed.), *Climate Change in Poland*, Springer Climate,
https://doi.org/10.1007/978-3-030-70328-8_13

1894 (Trepińska 1997; Twardosz 1997; Twardosz and Cebulska 2009). The observations of precipitation types in Cracow started earlier in 1792 (Twardosz 1999). Initial research on precipitation in all of Poland focused on its intra-annual course (Kosińska-Bartnicka 1927) and probability of the occurrence of dry and wet years in 1900–1950 (Kaczorowska 1962). Many studies on precipitation changes concerned small areas or even single locations (Gorczyński 1912; Trepińska 1969; Hohendorf 1970; Kożuchowski 1985b; Twardosz 1999, 2007; Filipiak 2007). Data from Cracow shows variability in annual precipitation totals with no significant trends in the nineteenth century—the first part of the nineteenth century was found to be wet while the second part was rather dry (Twardosz 2007). One of the first studies of precipitation trends in the entire country showed a significant increase in precipitation totals in February and May and a decrease in October and March between the periods of 1891 and 1930 (Kołodziej 1965; Kożuchowski 1983), and in 1948–1963 over a large territory of Poland (Kołodziej 1965). The increasing tendency was also a dominant feature of precipitation in Poland in the period 1931–1980 (Kożuchowski 1982b; Brazdil and Kożuchowski 1986). As reported in many later papers, the increasing trend in precipitation characteristics in Poland was mostly insignificant (e.g. Żmudzka 2002; Czarnecka and Nidzgorska-Lencewicz 2012; Skowera et al. 2014; Malinowska and Jakusik 2015). As trends in precipitation strongly depend on the period under consideration, some papers reported spatial variability in trend directions (Zinkiewicz 1970; Zawora and Ziernicka 2003; Kożuchowski 2013; Skowera et al. 2014), decreasing trends in some seasonal precipitation indices or fluctuations and oscillation as a dominant feature in long-term series (Kożuchowski 1982a, 1985a; Przedpełska 1993; Kirschenstein 2005; Skowera et al. 2014). The oscillations in seasonal and annual precipitation series in Poland were of short-term nature, i.e. between 2 and 4 years (Kożuchowski 1985a; Miętus 1996; Kożuchowski and Żmudzka 2003). Moreover, no significant trends were found in the long-term course of extreme precipitation indices. It is, however, worth mentioning that increases dominated among these insignificant trends (Pinskwar et al. 2019). Nevertheless, some significant changes in precipitation extremes were reported for short-term trends—calculated for moving 30-year periods or when the origin of precipitation was accounted for. These changes mainly included decreases in summer extreme precipitation. Persistent increasing trends in moving 30-year periods were recorded sporadically. Spatial patterns of trend directions in spring showed the most complex pattern of all seasons (Łupikasza 2010; Łupikasza et al. 2011). Degirmendžić and Kożuchowski (2017) performed a detailed study on the occurrence and changes in precipitation of Mediterranean origin in Poland (MCP), which are classified as extreme events. Although occurring rarely (2% probability of occurrence a year), precipitation of Mediterranean origin constitutes almost 10% of the annual total in Poland. In the period 1958–2008, the declining share of MCP in total precipitation was found in Poland. Mean annual and seasonal precipitation of Mediterranean origin was concluded to decrease significantly in mountainous areas. In the lowlands, these changes proved to be insignificant, except for winter. Although insignificant, decreasing trends in the frequency of MCP in combination with no changes in its intensity indicates the possible increase in MCP efficiency (Degirmendžić and Kożuchowski 2017).

Data and Methods

In this chapter, the focus is on trends in basic annual, seasonal and monthly characteristics of precipitation, including its totals and frequency (number of days with precipitation) based on data covering the period 1951–2018. The maps of precipitation characteristics found in this chapter present a highly generalised picture resulting from only 52 stations used in this study. The real distributions of both precipitation totals and frequency are more complicated due to its dependence on many factors of various scales with the relief playing a significant role in topographically variable areas. Since precipitation is not normally distributed, the Mann-Kendall method was used to test statistical significance of trends while the rate of trends was calculated with Sen's slop method. The maps in this chapter show the statistical significance of trends (using the threshold $\alpha \leq 0.05$ for significant trends and $\alpha \leq 0.1$ for weakly significant trends) and relative trends expressed as the percentage of the average (1951–2018) index value for each station. Trends were calculated as change per decade.

Total Precipitation

Annually, the territory of Poland receives c.a. 640 mm of precipitation on average (1951–2018, averaged over all stations). Its spatial distribution strongly depends on the relief, and thus the driest are the lowlands, running through the central part of the country from west to east, and the wettest are the mountainous areas located on the southern fringes (Fig. 13.1). The lowest annual precipitation is measured in the Great Poland Lowlands and Kujawy region (e.g. 503 mm in Kalisz), where, in the driest years, the annual total may drop below 300 mm (Chomicz 1977; Paszyński and Niedźwiedź 1991). Low precipitation totals were explained by the location of the Great Poland Lowlands in the shadow of the Pomeranian Lakeland with respect to westerlies or the smoothness of the ground surface resulting from the lack of forests (Paszyński 1955) or gradual transformation into steppe (Lambor 1954). However, it occurred that last of these theories failed to be true. In mountainous areas, the relation between relief and precipitation is the strongest. Therefore, the highest totals are measured at the Kasprowy Wierch station (1760 mm).

Long-term variability in annual precipitation totals shown in Fig. 13.2 indicated the wetter conditions in 1960s and 1970s, when the highest annual totals exceeded 750 mm (1966, 1970, 1974), and since the mid-1990s. When the mountain stations (Śnieżka, Kasprowy Wierch) and Zakopane were excluded, the annual total lowered to 597.5 mm. The average annual precipitation total has not been changing considerably in Poland as it is indicated by the values delivered by other authors, e.g. 605($\pm$8) mm for 1901–1980 (Jokiel and Kożuchowski 1989), 601 mm for 1891–1990 and 606 mm for 1951–1970 (Zawora et al. 2000/2001), 590 mm for a 50-year period (Kożuchowski and Żmudzka 2003).

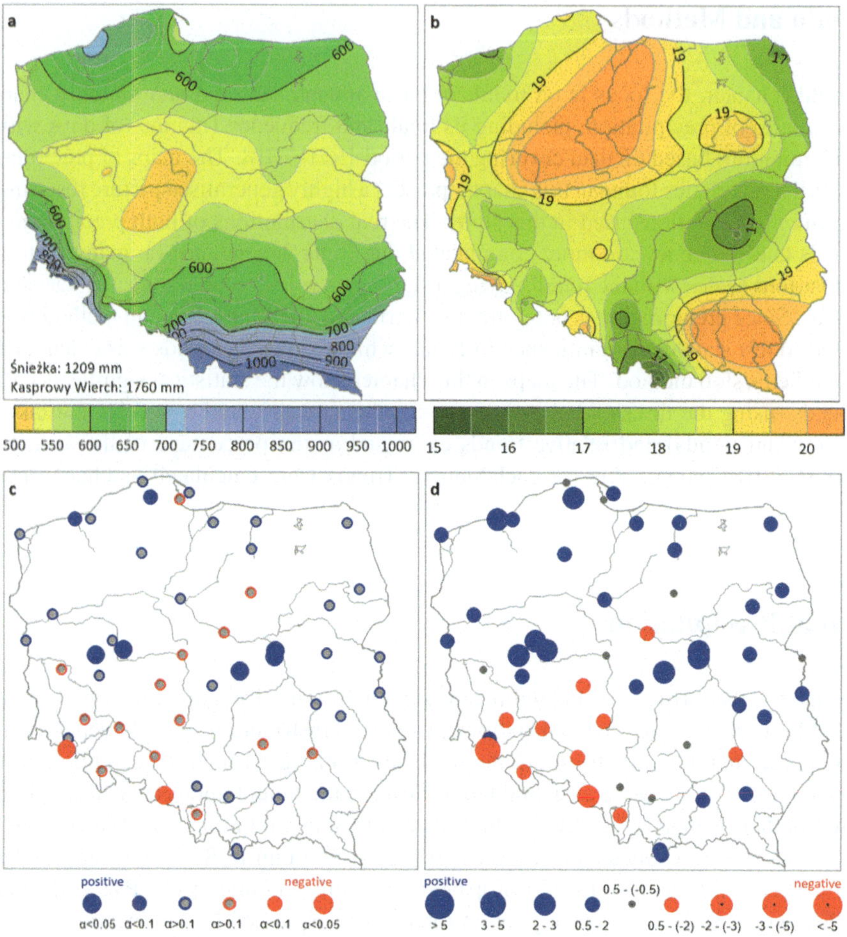

Fig. 13.1 Annual precipitation totals in Poland in the period 1951–2018: **a** spatial distribution of precipitation totals [mm], **b** coefficient of variability [%], **c** statistical significance and direction of trends, **d** relative trends [%]

The record precipitation in the country series was in 2010 (849 mm). In contrast, dry years were those in 1980s and 1990s (Fig. 13.2). Considering station data, the annual precipitation varied between 259 mm in 2015 (Kalisz) and 2600 mm in 2001 (Kasprowy Wierch). The coefficient of variability, ranging between 15 and 22%, was usually the highest in the driest parts of Poland and in southeastern region.

Although spatial distribution of the Mann-Kendall trends showed quite clear dominance of precipitation increases in the northern part and decreases in the southern part of Poland (Fig. 13.1d), most of these trends were statistically insignificant; except for single stations in central (5 stations), northwestern (2 stations) and southwestern (2 stations) Poland (Fig. 13.1c). The rate of significant positive trends varied

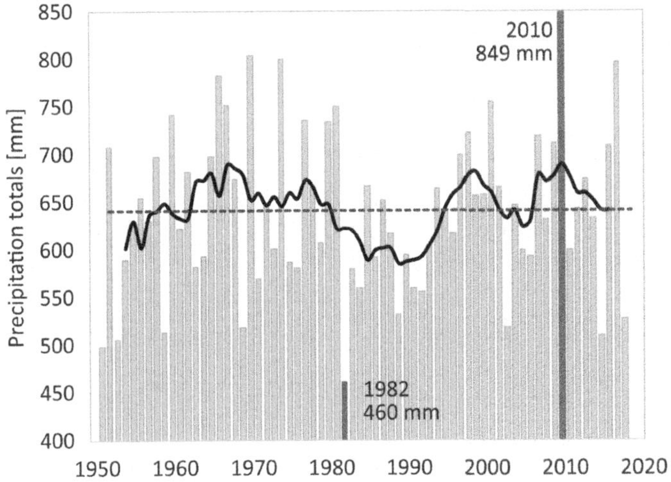

Fig. 13.2 Long-term variability in annual precipitation totals in Poland calculated as arithmetical average from station series, 1951–2018

between +12.9 and +17.7 mm per decade and negative between –59.0 and –17.6 mm per decade. According to Kożuchowski (2013), the current climate warming may contribute to moisture deficits in the already driest part of Central Poland, where, except for downward trends in precipitation, also increases in air temperature are present, which results in increased evapotranspiration.

In Poland, there is clear seasonality in precipitation distribution throughout the year. Figure 13.3 shows box-plots for seasonal and monthly totals and its contribution to the annual precipitation total. The summer maximum and winter minimum of precipitation indicate the domination of the continental influences on the precipitation regime. However, the prevalence of autumn over spring precipitation is a manifestation of the maritime influences (Fig. 13.3). On average in summer, Poland receives c.a. 239 mm that constitute c.a. 40% of annual total. In the lowlands, summer totals are lower than 220 mm, and they slightly increase to the north and much faster to the south. The maximum summer precipitation at mountain stations is as high as 379 mm at Śnieżka and 621 mm at Kasprowy Wierch. Summer precipitation twice or three times exceeds winter precipitation (DJF) that constitutes at most 20% of annual total (Fig. 13.3c). The month with the lowest monthly totals is February (Fig. 13.3d).

The country series of summer precipitation shows no clear tendency, however some features are possible to be identified, i.e. the dry periods in the 1980s and in the first half of the 1990s, a wider range of variability before the mid-1980s with the maximum in 1960 (351 mm), and a smaller range of variability, mostly due to higher totals in dry years since the second half of the 1990s, except for 2015 when the lowest total of 131 mm occurred (Fig. 13.6b).

Summer precipitation trends reveal the lowest spatial order of all seasons. Less coherent pattern of summer precipitation trends may be due to more frequent free

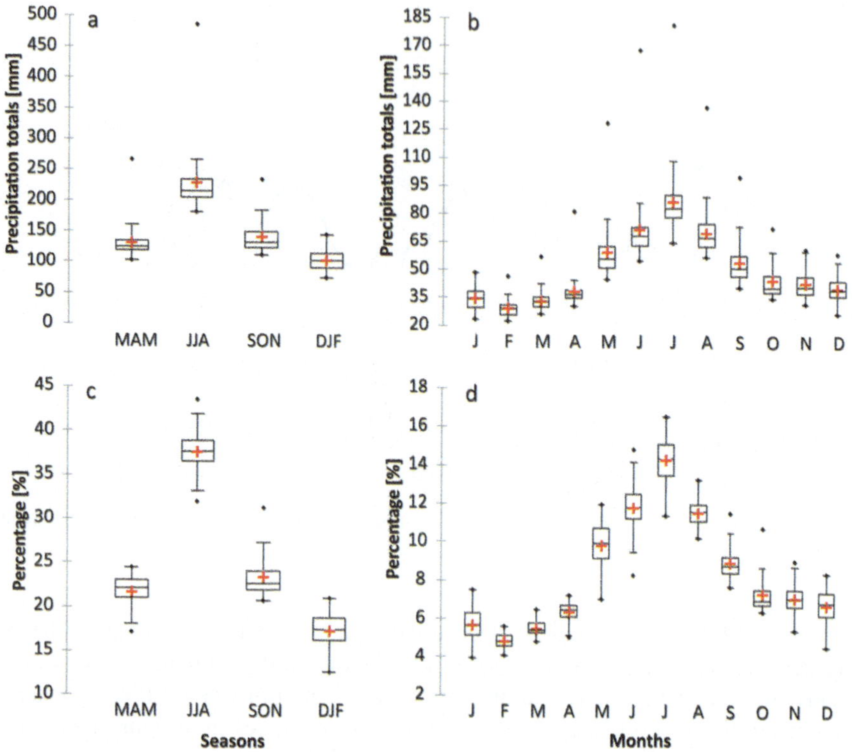

Fig. 13.3 Variability in annual course of precipitation totals in Poland, **a** seasonal precipitation totals, **b** monthly precipitation totals, **c** percentage of precipitation totals falling in seasons, **d** percentage of precipitation totals falling in particular months, red cross—arithmetical average, line in box—median, box—the quartiles, whiskers—outliers limit

convection that is of a local nature. However, southwestern Poland may be pointed out as an area with uniform but insignificant decreases in precipitation totals. The relative magnitudes of these negative trends equalled at most 2% of the long-term average at a majority of the stations. In the eastern part of Poland, relative trends were even smaller, varying between −0.5 and +0.5% of the long-term average. In principle, no station experienced a significant change in summer precipitation totals. At two stations, trends were weakly significant (Fig. 13.4c, d). At other stations, no changes or increasing trends dominated.

In spring, most of Poland received between more than 100 and 130 mm. The lowest summer totals are usually measured in the Great Poland Lowlands. At mountain stations in spring, precipitation was lower by 103 mm (Śnieżka) and 197 mm (Kasprowy Wierch) compared to summer totals (Figs. 13.5a and 13.4a). The highest coefficient of variability covered the central part of the country (Fig. 13.5b). The long-term course of country series of spring precipitation totals in Fig. 13.6a indicates 2010 as the wettest and 1982 as the driest years in the series. In 1960s and

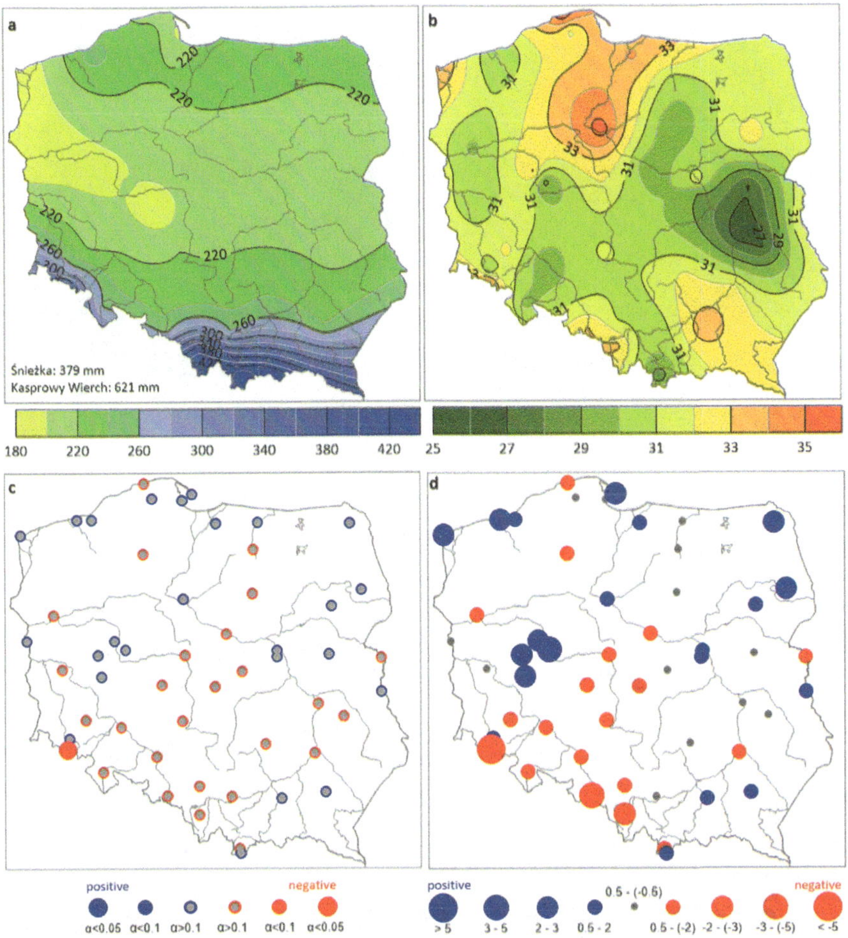

Fig. 13.4 Summer precipitation totals in Poland in the period 1951–2018: **a** spatial distribution of precipitation totals [mm], **b** coefficient of variability [%], **c** statistical significance and direction of trends, **d** relative trends [%]

1990s, spring precipitation was clearly higher than the long-term average. The dry period with totals lower than average lasted much longer compared to other seasons, i.e. from the early 1970s to the early 1990s. Taking station data into consideration, the extreme spring precipitation totals involved the Warszawa-Bielany station in 1953 (35 mm) and Kasprowy Wierch in 1978 (721 mm). In the starting two decades, year-to-year variability in spring precipitation was greater than in succeeding years (Fig. 13.6a). There is a clear order in spatial distribution of trend directions that indicates southwestern Poland as an area of decreasing spring precipitation; however, these changes are mostly insignificant. In the other regions, spring precipitation was increasing with several significant trends, mainly in eastern Poland (Fig. 13.5c). The

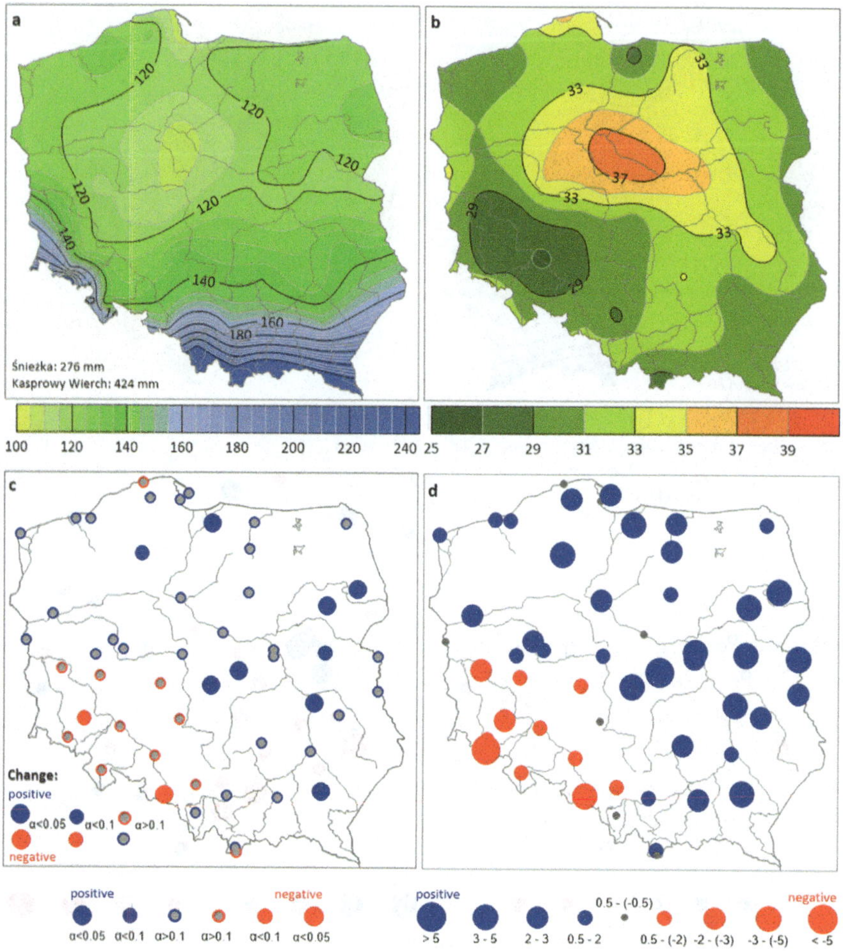

Fig. 13.5 Spring precipitation totals in Poland in the period 1951–2018: **a** spatial distribution of precipitation totals [mm], **b** coefficient of variability [%], **c** statistical significance and direction of trends, **d** relative trends [%]

rate of the significant increasing trends varied between +4.6 mm and +7.6 mm per decade.

In autumn and winter, the contrast between precipitation totals in northwestern and central Poland was bigger than in spring and summer. This contrast is related to low-pressure systems moving from the Atlantic Ocean more frequently in cold part of the year, crossing the Baltic Sea and influencing weather conditions of northern, particularly northwestern, Poland. However, in autumn, the range of spatial variability in precipitation totals was lower than in spring and summer. Like in other seasons, the country series showed two periods of increased and one short period of decreased precipitation with respect to long-term average. The wet periods included the 1960s,

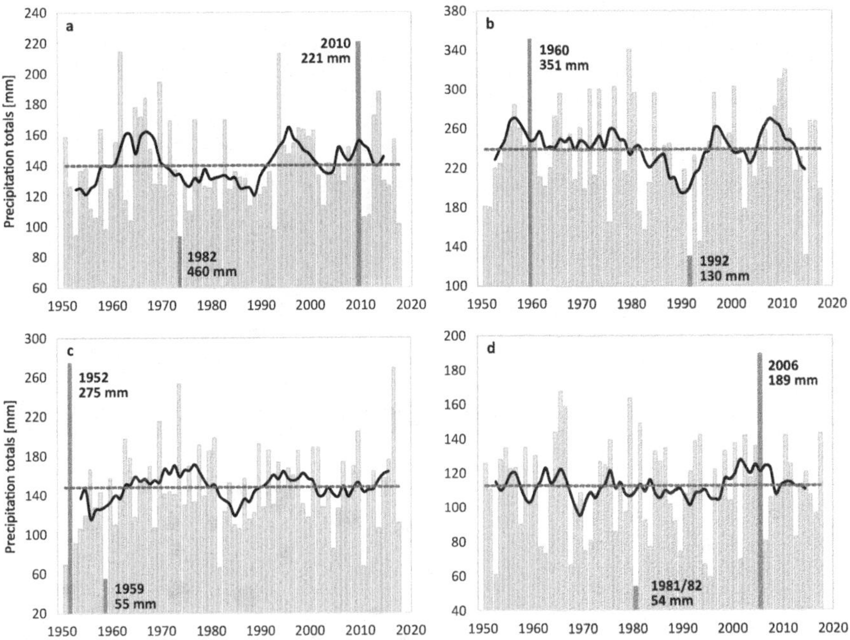

Fig. 13.6 Long-term variability in seasonal precipitation totals in Poland (**a** spring, **b** summer, **c** autumn, **d** winter) calculated as arithmetical averages from station series, 1951–2018

1970s and 1990s, while the dry period included the 1980s. Since the beginning of the 2000s, autumn precipitation has been close to the average conditions (Fig. 13.6c). Interestingly, the extreme years both occurred at the beginning of the research period in the 1950s: 1959 with 55 mm and 1952 with 275 mm. According to the station data, the wetter conditions occurred in 1964 (748 mm, Kasprowy Wierch) and the driest in 2011 (17 mm, Warszawa Okęcie). A majority of trends in autumn precipitation were positive but insignificant. The rate of relative trends was the fastest in southeastern Poland, but a significant increase at the rate of +7.7 to +21.0 mm per decade only occurred at a few stations in the central part of southern Poland. Positive trends were rare and spread in between negative ones with no clear order (Fig. 13.7c, d).

In winter, the range of spatial variability in precipitation totals was almost half as much as in other seasons when excluding mountain stations (from 74 mm in Kłodzko to 141 mm in Bielsko-Biała). The country series of winter precipitation totals revealed no clear wet and dry periods except for the pentad at the beginning of the twenty-first century that was slightly more abundant in precipitation compared to the average conditions (Fig. 13.6d). This period included the wettest season in the series—2006/2007 with country average of 189 mm. The driest season was that in 1981/1982 with precipitation as low as 54 mm. Based on station data, the extremes occurred in 1964 (643 mm, Kasprowy Wierch) and 1996 (23 mm, Siedlce). The stations in central Poland revealed the highest variability (Fig. 13.8b). Winter is the

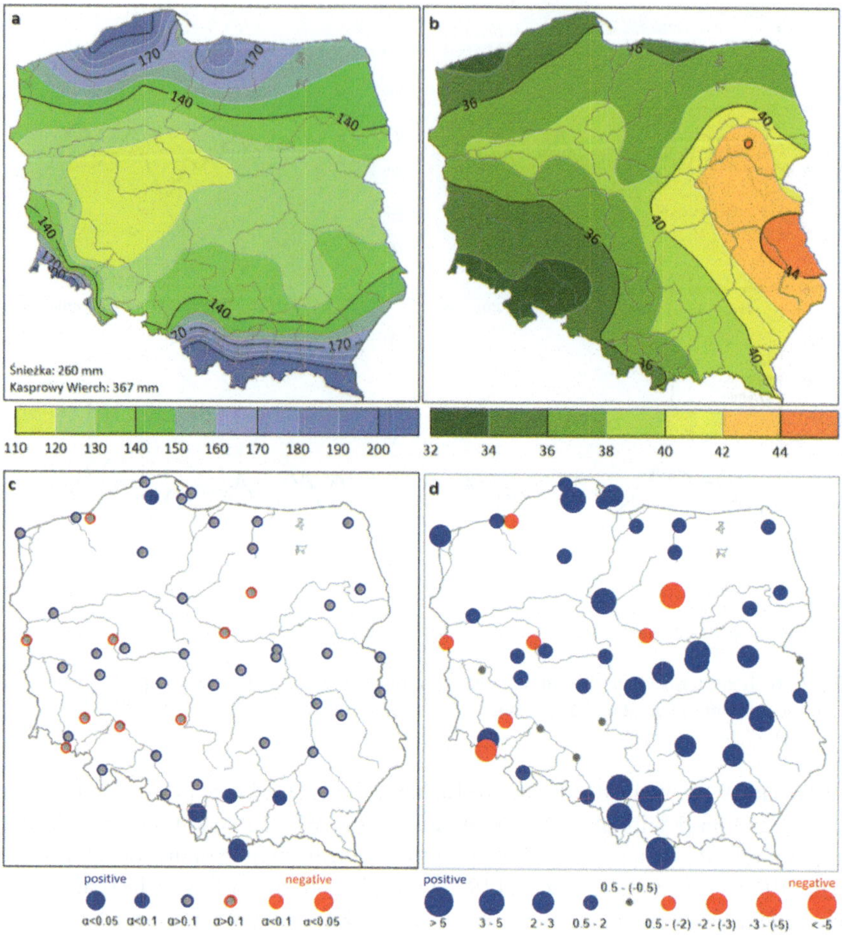

Fig. 13.7 Autumn precipitation totals in Poland in the period 1951–2018: **a** spatial distribution of precipitation totals [mm], **b** coefficient of variability [%], **c** statistical significance and direction of trends, **d** relative trends [%]

season with the highest number of significant trends in precipitation totals, mainly at stations in the northern part of Poland (Fig. 13.8c). The rate of the significant trends varied between +3.4 mm and +6.3 mm per decade. The relative trends showed that at many stations increases were as big as >5% of the long-term average of winter precipitation (Fig. 13.8d). The uniform decrease in winter precipitation totals covered the south of Poland with only single significant trends (Fig. 13.8c, d).

Furthermore, monthly trends in precipitation totals were also rarely significant. The number of stations with significant positive trends was almost twice as high as negative, but no clear pattern throughout the year was possible to identify (Fig. 13.9). The highest number of positive trends in winter precipitation was confirmed by the

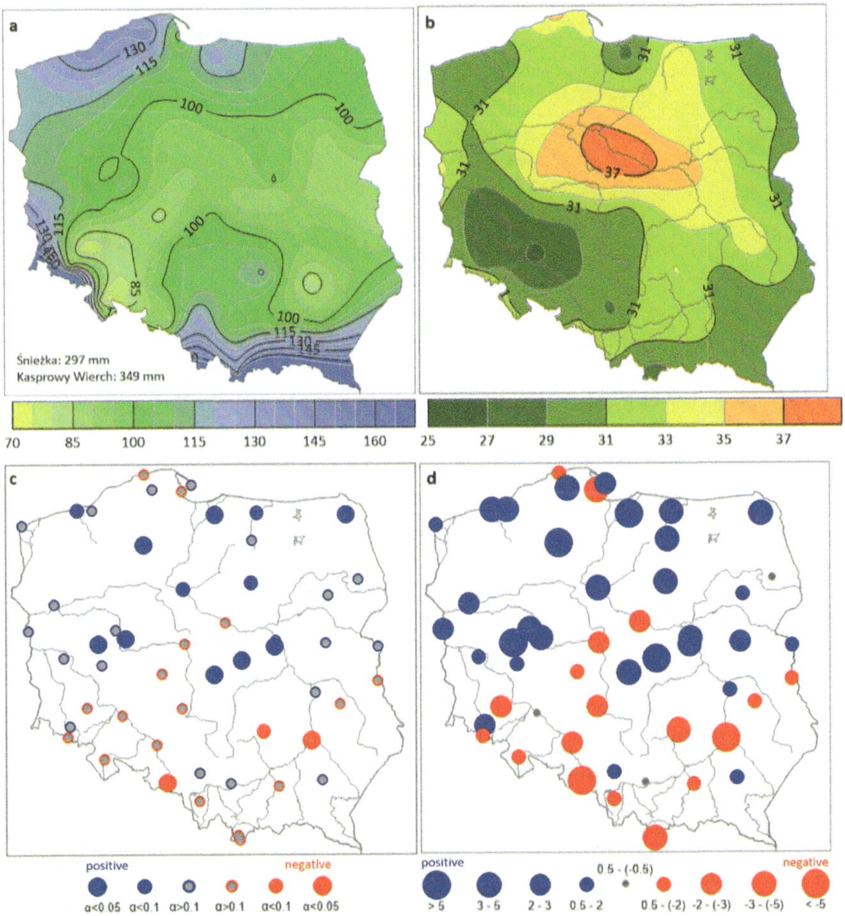

Fig. 13.8 Winter precipitation totals in Poland in the period 1951–2018: **a** spatial distribution of precipitation totals [mm], **b** coefficient of variability [%], **c** statistical significance and direction of trends, **d** relative trends [%]

number of significant increases in December and January, which were higher than in other months. However, the maximum frequency was found in March (21 meteorological stations with significant trends); which was also concluded by Degirmendžić et al. (2004). In most cases, the rate of these trends varied between 2–3 mm per decade and occurred in northwestern and central Poland (Fig. 13.9a, b). In addition, significant negative monthly trends were extremely rare with the maximum of five stations in April.

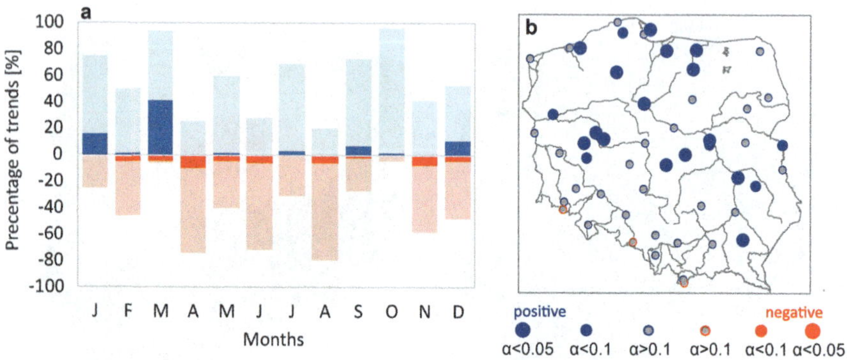

Fig. 13.9 Frequency of trends in monthly precipitation totals in Poland 1951–2018, red—negative trends, blue—positive trends, light—insignificant trends, dark—significant trends (**a**) and distribution of the significance of trends in March precipitation totals (**b**)

Number of Days with Precipitation

Spatial distribution of the frequency of precipitation in Poland in general followed that of precipitation totals. On an annual scale, the lowest number of days with precipitation occurred in the belt stretching from west to east in central Poland (less than 160 days), with a minimum of 145 days at Szepietowo station located in the northeast of this belt (Fig. 13.10a). These numbers increased to the north and south of Poland, reaching the maximum of 241 and 227 days at the mountain stations of Śnieżka and Kasprowy Wierch (Fig. 13.10a). High values of variability coefficient in annual number of days with precipitation cover large swaths of western Poland and its northeastern counterpart. The country series of annual number of days with precipitation showed no significant changes. Both the highest and the lowest values of the index occurred in the first part of the research period, i.e. in 1959 (128 days) and in 1970 (200 days) (Fig. 13.11). During the first two decades, there was a slight tendency rather than significant trend to increase in the frequency of annual precipitation, but in the following years, the moving average (solid dark line) fluctuated around the long-term average. Since 2000, a slight decrease can be noticed. Considering the station series, the highest and lowest number of days with precipitation occurred, respectively, in 1970 (282 days, Śnieżka) and 1959 (89 days, Opole). Like in case of precipitation totals, most of the trends in annual number of days with precipitation were statistically insignificant. The significant ones, if any, were only positive trends at single stations in central and southwestern parts of eastern Poland. In the central part of southern Poland, the relative magnitude of trends was as small as ±0.5 day per decade (Fig. 13.10c, d).

Seasonal and monthly distributions of the precipitation frequency demonstrate different patterns compared to precipitation totals. Although winter had the lowest precipitation totals, it was the most highest with respect to the precipitation frequency, constituting on average almost 29% of annual number of days with precipitation

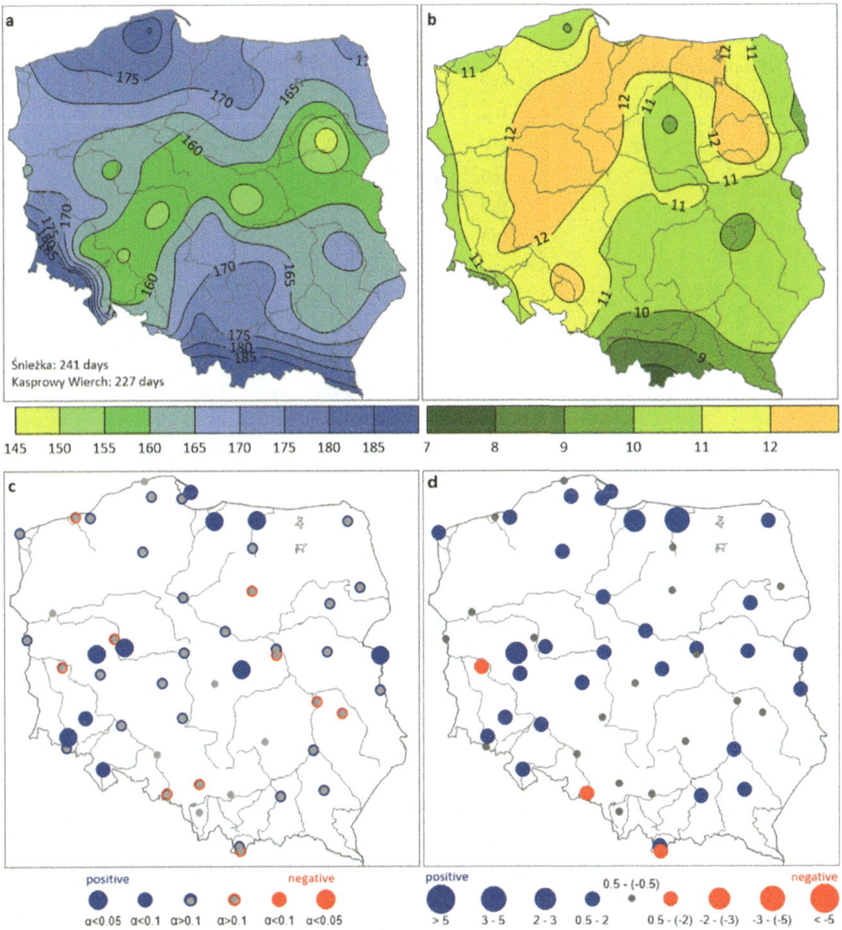

Fig. 13.10 Annual number of days with precipitation in Poland in the period 1951–2018: **a** spatial distribution of precipitation totals [mm], **b** coefficient of variability [%], **c** statistical significance and direction of trends, grey dots in the map "c" are for 0 trend rates, **d** relative trends [%]

(Fig. 13.12a, c). This means that winter precipitation was the least efficient of all seasons. Of all winter months, December was the most rainy; in January the number of days with precipitation was slightly lower than in December. In summer and spring, the number of days with precipitation was almost the same (c.a. 40 days) but the range of spatial variability in spring was clearly higher than in summer. The annual course of monthly number of days with precipitation was bimodal, with the first maximum in December and the secondary in July; the minimums were noted in April and September (Fig. 13.12b, d).

Spatial distribution of the seasonal precipitation frequency (number of days with precipitation, NoDPrc), its coefficient of variability and trends are presented in

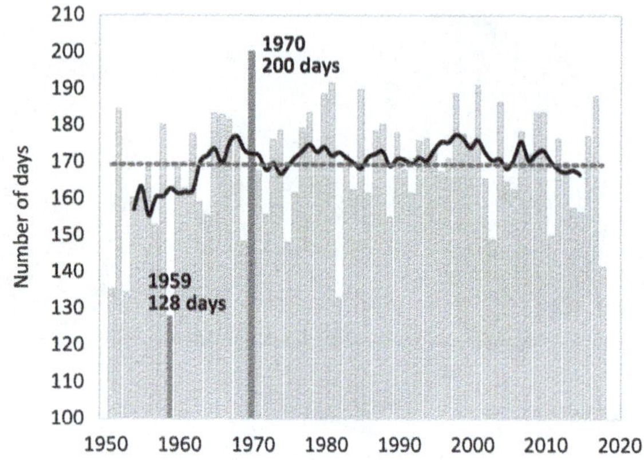

Fig. 13.11 Long-term variability in annual number of days with precipitation in Poland calculated as arithmetical averages from station series, 1951–2018

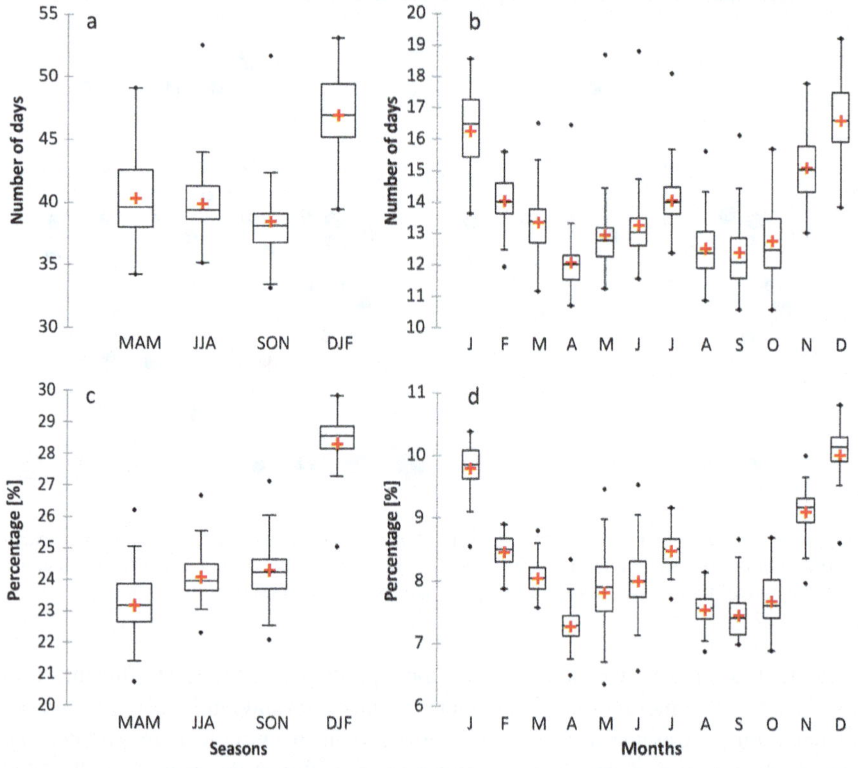

Fig. 13.12 Variability in annual course of precipitation frequency in Poland, **a** seasonal number of days with precipitation, **b** monthly number of days with precipitation, **c** percentage of annual number of days with precipitation in seasons, **d** percentage of annual number of days with precipitation in particular months, red cross—arithmetical average, line in box—median, box—the quartiles, whiskers—outliers limit

Figs. 13.13, 13.14, 13.15, 13.16, and 13.17. In spring, the precipitation frequency over large swaths of northern and central Poland did not change significantly, ranging between 34 and 38 days (Fig. 13.13a). In southern Poland, it increased up to 46 days and reached 59 days at Śnieżka and 61 days at Kasprowy Wierch. The coefficient of variability followed the distribution of the number of days with precipitation (Fig. 13.13b). The long-term course of country series depicted an increase in the NoDPrc in the first two decades of the research period, due to which in the 1960s and the first half of the 1970s NoDPrc was higher than the long-term average. Furthermore, the second period of increased precipitation frequency happened in the 1990s. Although the moving average did not demonstrate a clear change, since the mid-1990s

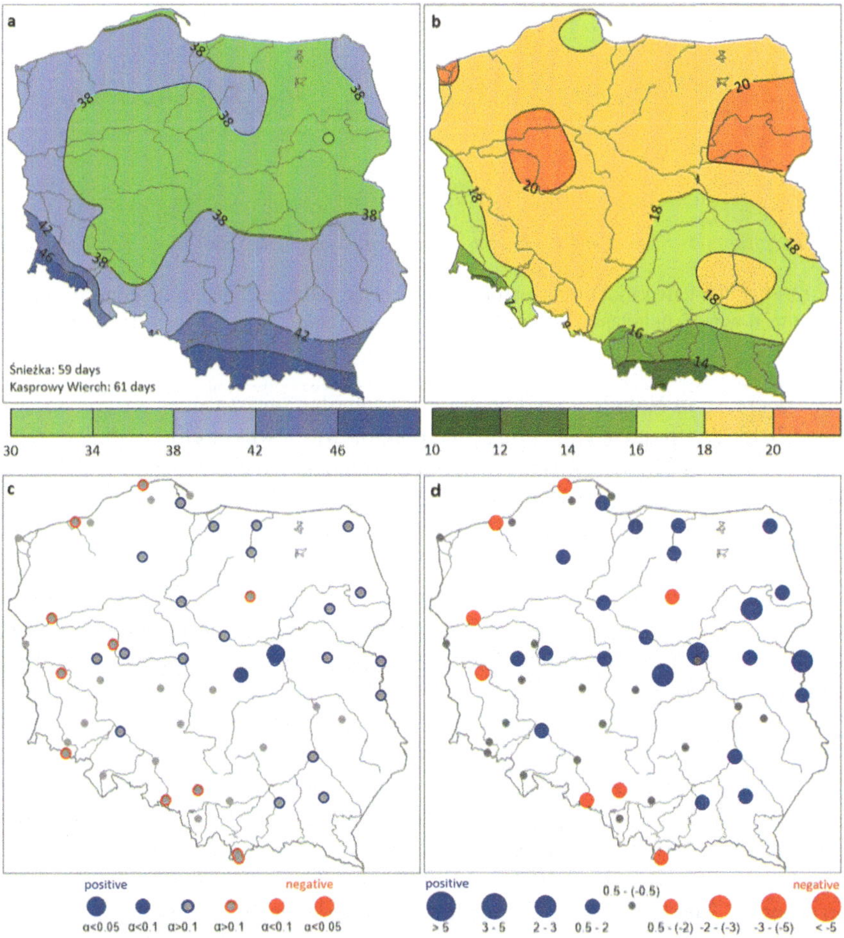

Fig. 13.13 Number of days with spring precipitation in Poland in the period 1951–2018: **a** spatial distribution of precipitation totals [mm], **b** coefficient of variability [%], **c** statistical significance and direction of trends, grey dots in the map "c" are for 0 trend rates, **d** relative trends [%]

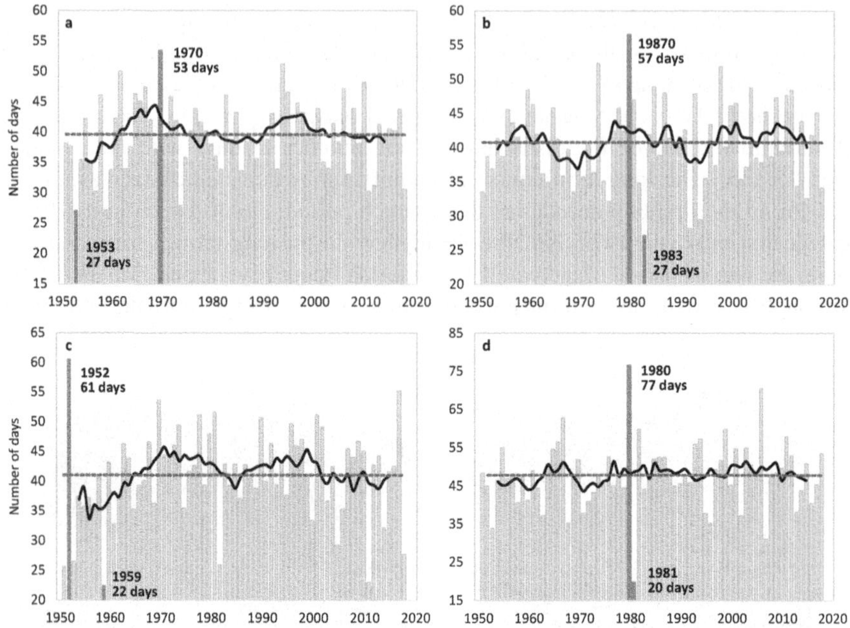

Fig. 13.14 Long-term variability in seasonal number of days with precipitation (**a** spring, **b** summer, **c** autumn, **d** winter) in Poland calculated as arithmetical averages from station series, 1951–2018

some decrease in NoDPrc could be noticed when the lowest values are considered. Since 2013, the frequency of spring precipitation was lower or close to the average. The lowest and the highest spring precipitation frequency was noted in 1953 (27 days) and 1970 (53 days on average in Poland), respectively (Fig. 13.14a).

Trends in the spring number of days with precipitation were practically insignificant, except for two stations in central Poland. Despite insignificant, their spatial distribution showed a rather ordered pattern that includes the majority of positive trends in northeast Poland, positive trends distributed along the western margins of the country and a concentration of trends of a very low rate in southern part of the country (Fig. 13.13c, b).

In summer, the average number of days with precipitation reached 40 days over vast areas of Poland. The precipitation frequency was lowest in the central part of eastern Poland (Fig. 13.15a). The coefficient of variability found its lowest values in northern Poland, and it was relatively low in the east (Fig. 13.15b). The country series indicated three periods with elevated NoDPrc, i.e. between the mid-1950s and 1960s, between the second parts of 1970s and 1980s, and in the first decade of the twenty-first century. The precipitation frequency was lower than average in the second half of the 1960s, the first half of the 70s, the first half of the 90s. The country series also revealed temporal changes in the range of variability in the number of days with summer precipitation. At the end of the 1960s, this range equalled 14 days, and it was as high as 30 days from the beginning of the 1970s until the end of the twentieth

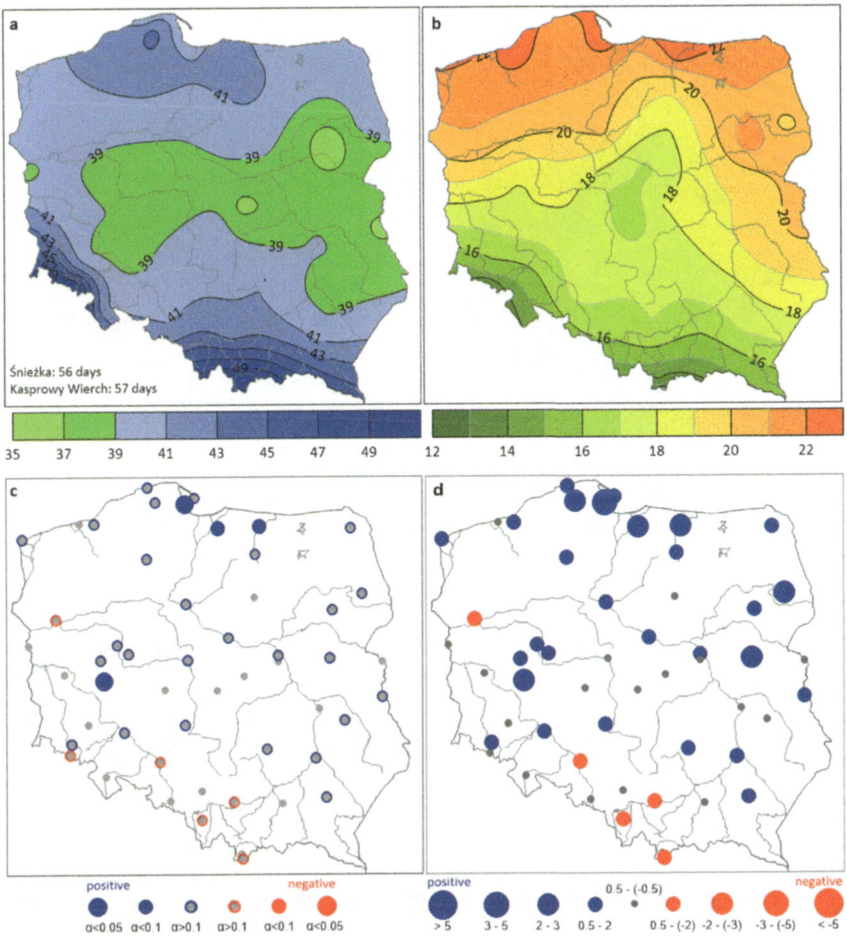

Fig. 13.15 Number of days with summer precipitation in Poland in the period 1951–2018: **a** spatial distribution of precipitation totals [mm], **b** coefficient of variability [%], **c** statistical significance and direction of trends, grey dots in the map "c" are for 0 trend rates, **d** relative trends [%]

century. It again diminished to 16 days at the beginning of the twentieth century. The highest and the lowest number of days with precipitation based on country series occurred in 1980 (57 days) and in 1983 (27 days) (Fig. 13.14b). In summer, like in other seasons, significant changes in the NoDPrc were limited to single stations, most of which were located in the central part of northern Poland (Fig. 13.15c). The relative rate of these trends exceeded 3% of the average index value (Fig. 13.15d). Spatial distribution of trends in the summer precipitation frequency shared some features of the annual distribution, i.e. the domination of the positive trends over negative mostly in the northeastern part of Poland. In summer, the occurrence of

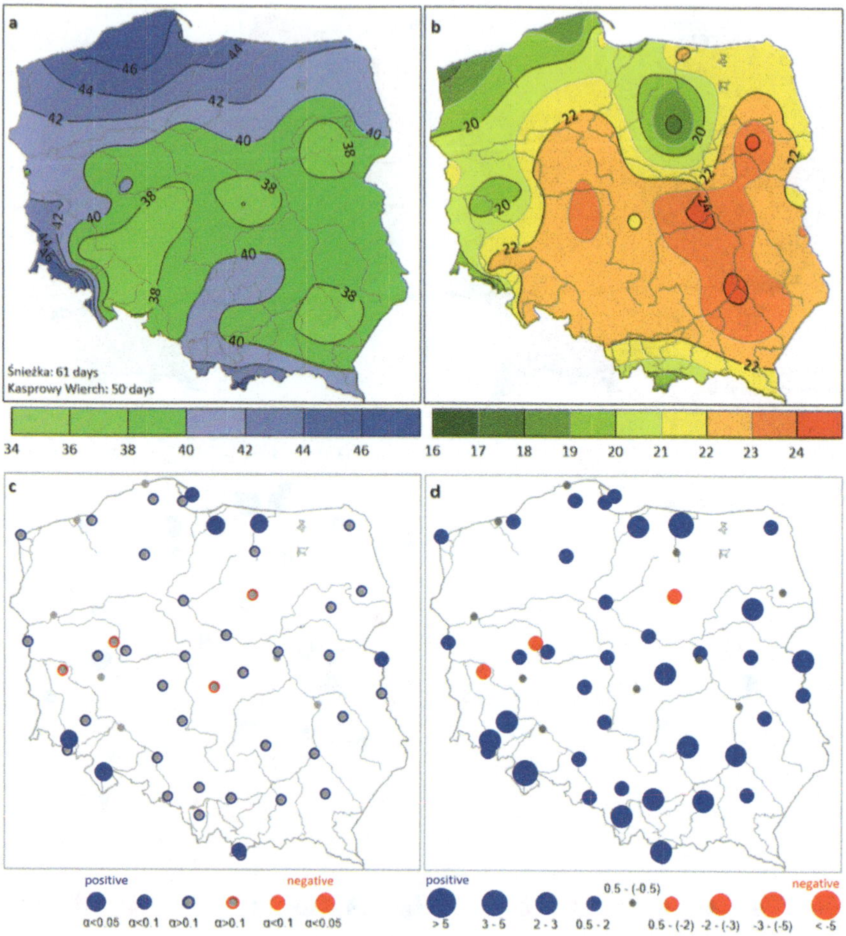

Fig. 13.16 Number of days with autumn precipitation in Poland in the period 1951–2018: **a** spatial distribution of precipitation totals [mm], **b** coefficient of variability [%], **c** statistical significance and direction of trends, grey dots in the map "c" are for 0 trend rates, **d** relative trends [%]

negative trends was limited to southern margins of Poland; the weakest trends were spread in the central and southern part of the country (Fig. 13.15d).

In autumn, the area with a relatively low number of days with precipitation was as large as in summer, but it was slightly shifted to the south. In this season, the number of days with precipitation in the north, particularly in northeast, was comparable with the number in the southern part of the country, excluding the mountain stations, where unvaryingly the maximums were noted (61 days—Śnieżka, 50 days—Kasprowy Wierch) (Fig. 13.16a). In this season, the highest coefficients of variability were characteristic of the central and southern parts of eastern Poland (Fig. 13.16b). The country series of autumn precipitation frequency revealed the most clear temporal

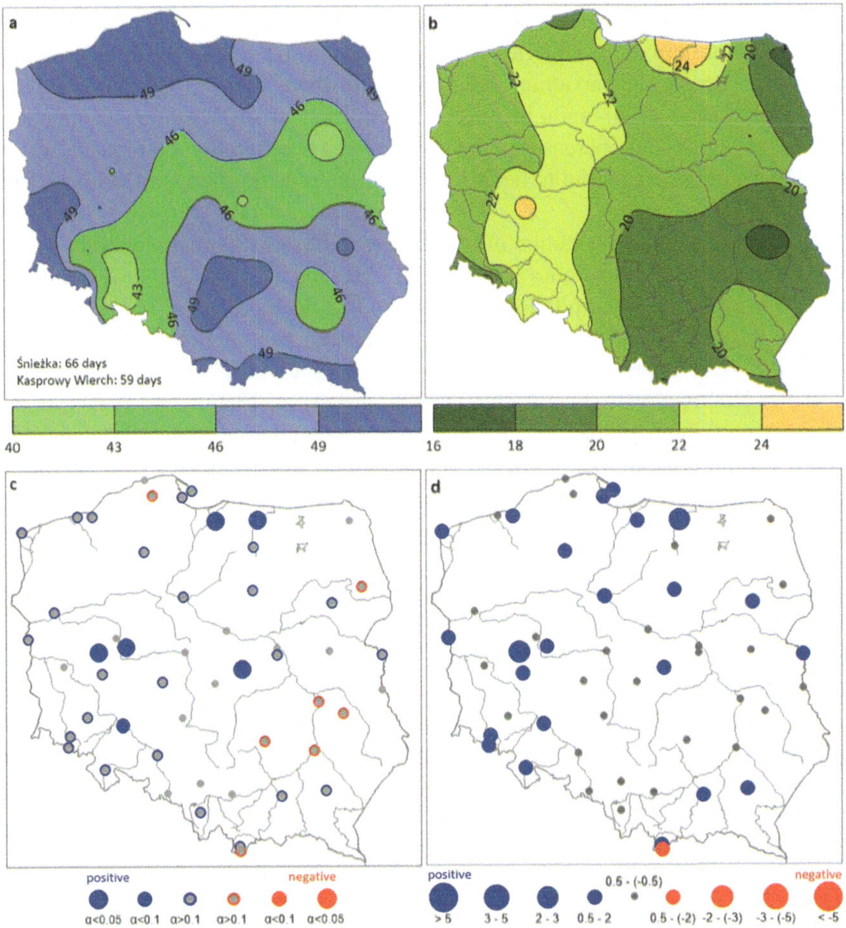

Fig. 13.17 Number of days with winter precipitation in Poland in the period 1951–2018: **a** spatial distribution of precipitation totals [mm], **b** coefficient of variability [%], **c** statistical significance and direction of trends, grey dots in the map "c" are for 0 trend rates, **d** relative trends [%]

short-term changes. During the initial two decades (the 1950s and 1960s), there was an apparent increase in the NoDPrec. From the beginning of the 1970s until the end of the twentieth-century precipitation was noted more frequently than on average, however, in the middle of this period (1982–1989), there was a sequence of years with relatively lower frequency of autumn precipitation. Since the beginning of the twenty-first century, the moving average has showed no noticeable trend, however, both the highest and the lowest seasonal index values suggest a decrease in the frequency of autumn precipitation. It is also important to note that since the mid-1990s relatively low index values have been occurring every 4–5 years (Fig. 13.14c). The lowest and the highest number of days in the country series were those in 1959

(22 days) and in 1952 (61 days). In autumn significant trends were only noted at single stations in northern and southern Poland. The rate of the significant trends reached 2 days per decade in maximum cases, which is more than 5% of the long-term average, but such a maximum increase occurred only at single stations. Generally, although insignificant, increases in the frequency of precipitation dominated, which in southern Poland occurred to be faster compared to its northern part (Fig. 13.16c, d).

In general, spatial distribution of the number of days with precipitation in winter was similar to that of the entire year (Figs. 13.10a and 13.17a). The lowest number of days with winter precipitation (less than 43 to 46 days) were recorded at the stations located within the belt stretching in eastern part of Poland from E to W and in the western part in SW-NE direction. Additionally, the lowest average values of winter NoDPrec varied between 40 and 66 days (Fig. 13.17a). According to the country series, the frequency of precipitation in winter underwent the least long-term changes of all seasons. The moving averages (solid line in Fig. 13.14d) oscillated around the long-term average. It is interesting that the maximum and minimum values of the NoDPrec for winter were noted year by year, i.e. in 1980 and 1981, respectively (Fig. 13.14d). When considering the number of days with winter precipitation at particular stations and in particular years, the extreme values ranged from 58 to 102 days (the lowest at stations) in 1980 and from 14 to 31 days (the highest at stations) in 1981. Although the number of significant trends in winter precipitation frequency was like in other seasons, at many stations the relative trends were very small, falling within the range of ±0.5% of average index value. These trends were particularly frequent in the central and eastern part of southern Poland. In the northern and eastern part of the country, positive trends dominated. The relative rate of significant trends varied between 1 and 3 days per decade, which in extreme cases constitute more than 5% of the long-term average index value (Fig. 13.17d).

Intra-annual pattern of monthly trends in the frequency of precipitation showed that the significant ones, if there were any, were the positive trends (Fig. 13.18). The maximum number of these trends was found in June and January (15 stations,

Fig. 13.18 Frequency of trends in monthly number of days with precipitation in Poland 1951–2018, red—negative trends, blue—positive trends, light—insignificant trends, dark—significant trends

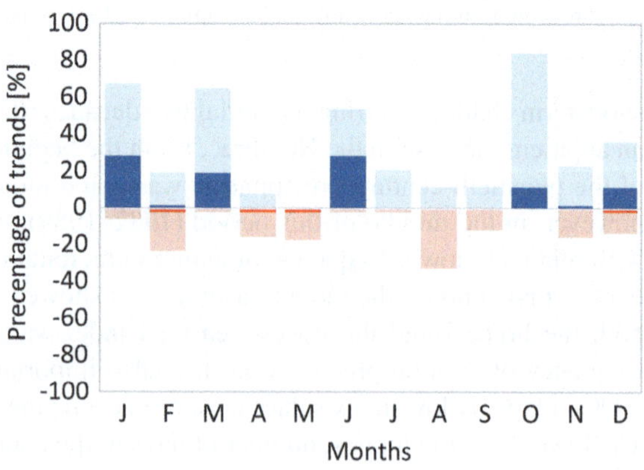

29% of the stations). Significant increasing trends were also found in March (19% of stations), October and November (12% of stations in each month). Negative trends were rare and practically insignificant. On the whole, months with no single significant trends were July, August and September (Fig. 13.18).

Discussion

The results of trend analysis in precipitation totals and frequency did not reveal significant changes between 1951 and 2018 on both the annual and seasonal scales. The temporal course of precipitation characteristics is rather dominated by short-term, i.e. several year-long tendencies or fluctuation and the occurrence of dry and wet decades, which was also indicated in many previous studies (e.g. Kożuchowski 1982a, 1985a; Przedpełska 1993; Żmudzka 2002; Kirschenstein 2005; Czarnecka and Nidzgorska-Lencewicz 2012; Skowera et al. 2014; Malinowska and Jakusik 2015; Mager et al. 2009). Such dry and wet periods occurred more or less simultaneously in spring and autumn in the case of both indices—the wet periods included the 1960s, 1970s and the second half of 1990s and were separated by dry period in the 1980s and in the early 1990s. Similar periods were distinguished in the course of annual precipitation totals and on the regional scale (e.g. Kożuchowski 1990; Kirschenstein 2005; Twardosz 2007; Ziernicka-Wojtaszek 2006). In summer, the wet and dry periods were less clear and not synchronised. The prevalence of fluctuation with no clear wet and dry periods in the country series of winter precipitation indices may rise from opposite trend directions in the indices in northern and southern Poland. However, the domination of fluctuations in winter precipitation was also found in other studies on a local scale (e.g. Twardosz and Cebulska 2009).

Overall, trends in precipitation characteristics are rarely significant compared to other climate elements, however, patterns in insignificant precipitation trends are also worthy of attention (Łupikasza 2017). Spatial distribution of trend direction in precipitation totals was not entirely consistent with the distribution of trends in the corresponding precipitation frequency, nonetheless, some common features in each season were possible to identify. These features included an increase in annual precipitation totals and frequency in northern and central Poland on an annual scale. In southern Poland, positive trends in the precipitation totals dominated, while no trends in the precipitation frequency were found at most stations. In spring, precipitation totals were increasing much faster than the precipitation frequency, which may indicate some increase in precipitation intensity, particularly in northern part of Poland, but with the exception of southwestern Poland, where spring precipitation totals were lowering. In summer, the spatial pattern of trends in precipitation characteristics was complicated with respect to direction, i.e. was characterised by the least spatial order, which may result from more frequent conditions for free convection compared to other seasons. In autumn, the differences in spatial distribution of trends in precipitation totals and frequency were the least—both characteristics increased at most stations. In winter, when the number of significant trends was the greatest, precipitation totals were significantly increasing in northern Poland and decreasing in southern Poland. Considering monthly precipitation totals—the highest number of significant changes was found in March, which confirms the results by Szwed (2019). The increase in the winter precipitation frequency in northern Poland was insignificant and of a very slow rate, while in southern Poland the precipitation frequency demonstrated neither significant nor insignificant trends. In short, this may indicate an increase in the intensity of winter precipitation in northern Poland and a decrease in its intensity in southern part of the country.

Generally, the updated series of precipitation characteristics showed that recent years did not influence trends in precipitation characteristics in Poland. This study confirms that long-term trends in precipitation in Poland continue to be insignificant (e.g. Kożuchowski 1985a; Miętus 1996; Kożuchowski and Żmudzka 2003; Mager et al. 2009; Czarnecka and Nidzgorska-Lencewicz 2012). As it stems from literature, trends in extreme precipitation were also insignificant and of various direction (Łupikasza 2017; Degirmendžić and Kożuchowski 2016, 2017; Pinskwar et al. 2019).

Conclusions

In summer, spring and on an annual scale, precipitation totals were decreasing in southwestern Poland, while in the other parts of Poland there was either a tendency to increase or no trends were found. In autumn, increasing trends dominated, while in winter opposite trends were found in northern (increases) and southern Poland (decreases). Winter was found to have the highest number of significant trends, but on a monthly scale, trends were most numerous in March.

In terms of the precipitation frequency, positive trends dominated on an annual scale and in spring and summer, particularly in northern part of the country. Stations with no trends were often located in southern Poland. In autumn, the precipitation frequency increased at a great majority of stations. Lastly, the winter precipitation frequency was increasing in northern and western Poland but with the slowest rate of all seasons, while most stations in southern Poland experienced no trends in this precipitation characteristic.

References

Braździl R, Kożuchowski K (1986) Some aspects of precipitation variability in Poland in the period of 1991–1980. Scripta Fac Sci Nat Univ Purk Brun 16(8):365–382

Chomicz K (1977) Materiały do poznania agroklimatu Polski. PWN Warszawa, 296 pp

Czarnecka M, Nidzgorska-Lencewicz J (2012) Wieloletnia zmienność sezonowych opadów w Polsce. Woda-Środowisko-Obszary Wiejskie 12, 2(38):45–60

Degirmendžić J, Kożuchowski K (2016) Niże Śródziemnomorskie jako czynnik klimatu Polski. Wydawnictwo Uniwersytetu Łódzkiego, 165 pp

Degirmendžić J, Kożuchowski K (2017) Mediterranean cyclones, the atmospheric moisture content and precipitation in Poland. Geographia Polonica 90(1):5–20. https://doi.org/10.7163/GPol.0075

Degirmendžić J, Kożuchowski K, Żmudzka E (2004) Changes of air temperature and precipitation in Poland in the period 1951–2000 and their relationship to atmospheric circulation. International Journal of Climatology 24(3):291–310. https://doi.org/10.1002/joc.1010

Filipiak J (2007) Obserwacje i pomiary opadów atmosferycznych w Gdańsku w XVIII wieku. In: Piotrowicz K, Twardosz R (eds) Wahania klimatu w różnych skalach przestrzennych i czasowych, Kraków, pp 365–373

Gorczyński W (1912) Materyały do poznania opadów w Królestwie Polskim. Część VI: Opady w Warszawie (1803–1910). Nakładem Towarzystwa Naukowego Warszawskiego, Warszawa, pp 133–151

Hohendorf E (1970) Zmienność opadów atmosferycznych w ostatnim stuleciu w Bydgoszczy 1861–1960. Prace i Studia Kom Gosp Wod i Surow 10:221–237

Jokiel P, Kozuchowski K, (1989) Zmiany wybranych charakterytyk hydroklimatycznych Polski w bieżącym stuleciu. [Changes in selected hydro-climatic characteristics in Polans]. Dokum. Geogr. IGZ PAN

Kaczorowska Z (1962) Opady w Polsce w przekroju wieloletnim. [Precipitation in Poland from the long-term perspective]. Prace geograficzne nr 33. Wyd. Geologiczne, Warszawa, 112 pp

Kirschenstein M (2005) Wieloletnie zmiany sum opadów atmosferycznych na wybranych stacjach północno-zachodniej Polski. Słupskie Prace Geograficzne 2:199–214

Kołodziej J (1965) Opady atmosferyczne w Polsce w latach 1948–1963 w porównaniu ze średnimi z okresu 1891–1930. Przegląd Geofizyczny X(XVIII) 3–4:291–295

Kosińska-Bartnicka S (1927) Opady w Polsce (wysokość częstość i charakter klimatyczny) [Precipitation in Poland (totals, frequency and climatological character)]. Prace Meteorologiczne i Hydrologiczne V:93

Kożuchowski K (1982a) Periodicity of variations of the annual precipitation totals in Poland. Acta Geophysica XXX(3):287–242

Kożuchowski K (1982b) Współczesne tendencje zmian rocznych sum opadów w Polsce. Przegląd Geograficzny XXVII(3–4):251–259

Kożuchowski K (1983) Wiekowe zmiany rocznych sum opadów w Polsce. Czasopismo Geograficzne LIV(2):231–237

Kożuchowski K (1985a) Zmienność opadów atmosferycznych w Polsce w stuleciu 1881–1980. Acta Geographica Lodziensia 48:158

Kożuchowski K (1985b) Zmienność opadów atmosferycznych w Łodzi w okresie 1904–1979. Acta UL, Folia Geogr 4:3–17

Kożuchowski K (ed) (1990) Materiały do poznania historii klimatu w okresie obserwacji instrumentalnych. Wyd Uniwersytetu Łódzkiego, Łódź, p 55

Kożuchowski KM (2013) Ocena higroklimatycznych warunków wegetacji w Polsce (Assessment of hygroclimatic conditions of vegetation in Poland). Monitoring Środowiska Przyrodniczego 14:103–111

Kożuchowski K, Żmudzka E (2003) 100-year series of areally averaged temperatures and precipitation totals in Poland. Acta Univ Wratisl 2542, Studia Geogr 75:116–122

Lambor J (1954) Stepowienia środkowych obszarów Polski. Prace PIHM 34, Warszawa, 48 pp

Łupikasza E (2010) Spatial and temporal variability of extreme precipitation in Poland in the period 1951–2006. Int J Climatol 30:991–1007. https://doi.org/10.1002/joc.1950

Łupikasza EB, Hansel S, Matschullat J (2011) Regional and seasonal variability of extreme precipitation trends in southern Poland and central-eastern Germany 1951–2006. Int J Climatol 31:2249–2271. https://doi.org/10.1002/joc.2229

Łupikasza EB (2017) Seasonal patterns and consistency of extreme precipitation trends in Europe, December 1950 to February 2008. "Climate Research", 72, pp 217–237. https://doi.org/10.3354/cr01467

Mager P, Kasprowicz T, Farat R (2009) Change of air temperature and precipitation in Poland in 1966–2006. W: Climate change and agriculture in Poland—impacts, mitigation and adaptation measures. Pr zbior Red J Leśny Acta Agrophysica Rozprawy i Monografie 169(1):19–38

Malinowska M, Jakusik E (2015) Charakterystyka opadów atmosferycznych w centralnej części Pojezierza Kaszubskiego w latach 1971–2010. [Characteristics of precipitation in central part of Kashubian Lake District, 1971–2010]. Sci Rev Eng Env Sci 69:273–285

Marciniak K, Kożuchowski K (1990) Aneks. In: Kożuchowski K (ed) Materiały do poznania historii klimatu w okresie obserwacji instrumentalnych. Wyd Uniwersytetu Łódzkiego, Łódź

Miętus M (1996) Zmienność temperatury i opadów w rejonie polskiego wybrzeża Morza Bałtyckiego i jej spodziewany przebieg do roku 2030. Mat Bad, Ser Meteorologia 26:1–72

Paszyński J (1955) Opady atmosferyczne dorzecza Odry i ich związek z hipsometrią i zalesieniem. Prace Geograficzne IG PAN, 4, 90 pp

Paszyński J, Niedźwiedź T (1991) Klimat [w] Starkel L (ed.) Geografia Polski. Środowisko Przyrodnicze. Wydawnictwo Naukowe PWN, pp 296–355

Pińskwar I, Choryński A, Graczyk D, Kundzewicz ZW (2019) Observed changes in extreme precipitation in Poland: 1991–2015 versus 1961–1990. Theor Appl Climatol 135:773–787

Przedpełska W (1993) Zmienność sum opadów atmosferycznych uśrednionych dla obszaru Polski. [Variability of precipitation averaged over Poland].Wiadomości IMGW XXXVII(1):17–31

Przybylak R, Majorowicz J, Brázdil R, Kejna M (eds) (2010) The Polish climate in the European context: an historical overview. Springer, Berlin, Heidelberg, and New York, p 535

Pyka J (2003) Meteorological observations and measurements in Wrocław. Acta Univ Wratisl Studia Geogr 75:11–22

Skowera B, Kopcińska J, Kopeć B (2014) Changes in thermal and precipitation conditions in Poland in 1971–2010. Annals of Warsaw University of Life Sciences—SGGW. Land Reclam 46(2):153–162

Szwed I (2019) Variability of precipitation in Poland under climate change. Theor Appl Climatol 135:1003–1015. https://doi.org/10.1007/s00704-018-2408-6

Trepińska J (1969) Ogólna charakterystyka opadów w Krakowie 1916–1965. Folia Geogr, ser Geogr Phys 3:117–138

Trepińska J (1997) Wstęp. In: Trepińska J (ed) Wahania klimatu w Krakowie. Instytut Geografii UJ, Kraków, pp 1792–1995

Twardosz R (1997) Pomiary opadów atmosferycznych na Stacji Meteorologicznej w Krakowie. In: Trepińska J (ed) Wahania klimatu w Krakowie. Instytut Geografii UJ, Kraków, pp 1792–1995

Twardosz R (1999) Warunki pluwialne w Krakowie w latach 1792–1998. Czasopismo Geogr 2:221–234

Twardosz R (2007) Opady atmosferyczne. In: Matuszko D (ed) Klimat Krakowa w XX wieku. Instytut Geografii UJ, Kraków

Twardosz R, Cebulska M (2009) Observations and measurements of precipitation in the Polish Province of Galicia in the nineteenth century. In: Przybylak R, Majorowicz J, Brázdil R, Kejna M (eds) The Polish climate in the European context: an historical overview. Springer, Berlin, Heidelberg, and New York, Chapter 23, pp 457–471

Zawora T, Janur E, Olszańska A, Skowera B, (2000/2001) Porównanie norm opadów atmosferycznych na obszarze Polski. [Comparison of precipitation norms in Poland]Ann. UMCS Sect. B t. 40/41 46 s. 391–398

Zawora T, Ziernicka A (2003) Precipitation variability in time in Poland in the light of multi-annual mean values (1991–2000). Acta Universitatis Wratislavensis, 2542, Studia Geograficzne 75:123–128

Ziernicka-Wojtaszek A (2006) Zmienność opadów atmosferycznych na obszarze Polski w latach 1971–2000.W: Klimatyczne aspekty środowiska geograficznego. Pr zbior Red J Trepińska, Z Olecki Kraków Inst Geogr i Gosp Przestrz UJ, 139–148

Zinkiewicz A (1970) Opady atmosferyczne w okresie 1951–1965 w stosunku do wieloletnich na obszarze Polski, Ann UMCS, B XXV(2):109–151

Żmudzka E (2002) O zmienności opadów atmosferycznych na obszarze Polski nizinnej w drugiej połowie XX wieku (About the variability of precipitation in lowland Poland in the second mid-twentieth century). Wiadomości IMGW 25(46)4:23–38

Chapter 14
Snow Cover Change

Małgorzata Falarz and Ewa Bednorz

Abstract The chapter analyses snow cover data for 60 weather stations in Poland: 52 series for the period 1950/51–2017/18 (68 winter seasons) and eight longer series of 80 to 104 winter seasons. Two basic characteristics of snow cover were examined: snow cover duration and seasonal maximum depth of snow cover. The coefficient of variability and absolute and relative trends per 10 years were investigated. The most important results of the study are as follows: (1) year-to-year variability of snow cover duration and its seasonal maximum depth is the largest in regions of poorest snow cover; (2) the number of days with snow cover during the 68 winter seasons has a negative time trend throughout Poland (with a minimum of −4 to −5 days for 10 years in north-eastern Poland); this tendency is statistically significant in most area of Poland except for the highlands and some parts of the mountainous areas; (3) the relative changes in the number of days with snow cover are the most significant in regions with a short duration of snow cover (western Poland; −8 to −10% per 10 years); (4) the maximum depth of snow cover throughout the considered period revealed a negative trend in most area of Poland, statistically significant in the highlands and mountainous areas (−2 to −6%/10y); only in north-eastern Poland is the trend positive, statistically insignificant, (5) in longer periods (80–104 winter seasons), the snow cover duration and maximum snow cover depth are characterised by a slight negative or near zero time trend at most weather stations; (6) in the winter seasons 1969/70 and 1995/96 the longest snow cover duration in most areas of Poland was recorded, while in the 1939/40 and 1962/63 seasons, the maximum depth of the snow cover was the highest. The trend values of snow cover duration in Poland are comparable to similar values averaged over the Northern Hemisphere, where the duration of the snow season has declined by 5 days per 10 years since the

M. Falarz (✉)
Faculty of Natural Sciences, Institute of Earth Sciences, University of Silesia in Katowice, Katowice, Poland
e-mail: malgorzata.falarz@us.edu.pl

E. Bednorz
Faculty of Geographical and Geological Sciences, Institute of Physical Geography and Environmental Planning, Department of Meteorology and Climatology, Adam Mickiewicz University, Poznań, Poland
e-mail: ewabedno@amu.edu.pl

M. Falarz (ed.), *Climate Change in Poland*, Springer Climate,
https://doi.org/10.1007/978-3-030-70328-8_14

winter of 1972/1973. On the other hand, the negative trend of maximum snow cover depth is not as significant in Poland as the European average, which is −11.4% per decade.

Introduction

Snow cover is an important climatic variable which strongly influences environmental conditions on a local and global scale. Owing to its physical features—mainly high albedo, being one of the highest of all natural surfaces, and low conductivity—snow strongly modifies the surface–atmosphere energy fluxes (Armstrong and Brun 2008). Thus, it causes significant changes in the diabatic heating on the Earth's surface and strongly affects the air temperature (Cohen 1994). Besides its climatic impact, snow cover influences the water balance by modifying winter retention, and prepossesses several branches of human activity, causing different kinds of hazards.

On the other hand, snow cover is a phenomenon strongly determined by climatic conditions. High sensitivity to changes in temperature and precipitation makes snow cover a good indicator of climate change. The nature of the impact, however, is dependent on geographic location, latitude and elevation (Dong and Menzel 2019; Brown and Robinson 2011). It was ascertained that the European snow cover extent is primarily temperature dependent and the increasing air temperatures have decreased snow accumulation and shortened snow cover period in mid-latitudes and at mid- and low elevations (e.g. Brown and Robinson 2011; Bulygina et al. 2009; Henderson and Leathers 2010; Ye and Lau 2017). The relation between precipitation variability and change and the snowpack is less obvious, mostly because temperature largely controls the snowfall/precipitation ratio (Feng and Hu 2007). Large-scale studies concerning snow cover duration have noted mostly declining trends in the Northern Hemisphere (e.g. Brown 2000; Choi et al. 2010; Bulygina et al. 2009). Besides shortening the snow period, decreases in snow depth were found over Europe as well, except for in the coldest regions (Fontrodona Bach et al. 2018).

In the mid-latitudes long-term snow cover changes are partly modulated by the influence of short-term fluctuations associated with variability in atmospheric circulation, which influence winter temperature and precipitation patterns (Ye and Lau 2017; Brown 2019). The role of the predominant winter teleconnection pattern over the Euro-Atlantic region, i.e. North Atlantic Oscillation (NAO), as a driving factor is emphasised (i.e. Clark et al. 1999; Bednorz 2002, 2004; Falarz 2007; Ye and Lau 2017). Therefore, although a declining tendency of snow cover is expected in Poland under the warming climate, a large inter-winter variability makes the clear identification of snow cover trends difficult. In most studies concerning the second half of the twentieth century only a slight decreasing trend of snow cover characteristics in Poland was detected (Falarz 2002, 2004, 2008; Nowosad and Bartoszek 2007; Czarnecka 2012; Szwed et al. 2017). At the same time, the large variability between individual seasons and the increasing frequency of mild and snowless winters occurring in a moderate climate zone was emphasised (e.g. Jaagus 1997; Niedźwiecki

1998). An occurring slight discrepancy between the current general ideas concerning snow cover tendency, namely, the expectations of a strongly declining trend, and the results of previous studies make the issue of contemporary changes in snow cover depth and duration in Poland most relevant.

Data and Methods

The chapter analyses snow cover data for 60 weather stations of the Institute of Meteorology and Water Management—National Research Institute, i.e. 52 series for the period 1950/51–2017/18 (68 winter seasons) and eight longer series of 80 (1938/39–2017/18 at Kasprowy Wierch) to 104 (1914/15–2017/18 in Zakopane) winter seasons. The data originate from everyday snow cover observations at 6 UTC. A day with snow cover is a day with snow of at least 1 cm depth, covering at least 50% of the observed area. The quality of data was controlled and corrected. The incomplete series (see Table 3.3 in Chap. 3 *Data and Method of Investigation*) was filled in by methods of regression analysis or ratios using data for one to three neighbouring stations, located in areas of similar snow conditions.

Examination and correction of the homogeneity of the snow cover series are described in Chap. 4. *Homogeneity of Climate Series.* Data series for 19 out of 60 weather stations (i.e. 32% of the series) were considered heterogeneous. All inhomogeneous series were corrected.

Two basic characteristics of snow cover were examined: (1) snow cover duration (i.e. number of days with snow cover in the winter season) and (2) seasonal maximum depth of snow cover (in cm). The period from 1 August to 31 July of the following year was considered a "snow year" at the high mountain stations (Kasprowy Wierch, Śnieżka, Hala Gąsieniowa).

The statistics described in Chap. 3 *Data and Method of Investigation* (i.e. the coefficient of variability, absolute and relative trend per 10 years) were used to study the long-term changes and variability of snow cover. The analysis was carried out for both the whole multi-year period and its first (1950/51–1983/84) and second half (1984/85–2017/18).

Number of Days with Snow Cover

The number of days with snow cover in the winter season averaged for the period 1950/51–2017/18 is the lowest in the west of Poland and is below 40 days (with a minimum of 34 days in Szczecin and Słubice) and rises as it moves east and northeast to more than 80 days (89 days in Suwałki; Fig. 14.1a). Much longer periods of snow cover occur in the mountainous areas of southern Poland, where it is observed on the highest peaks of the Sudeten for more than half the year (187 days on Śnieżka), and on the peaks of the Tatra Mountains—more than 7 months of the year (220 days

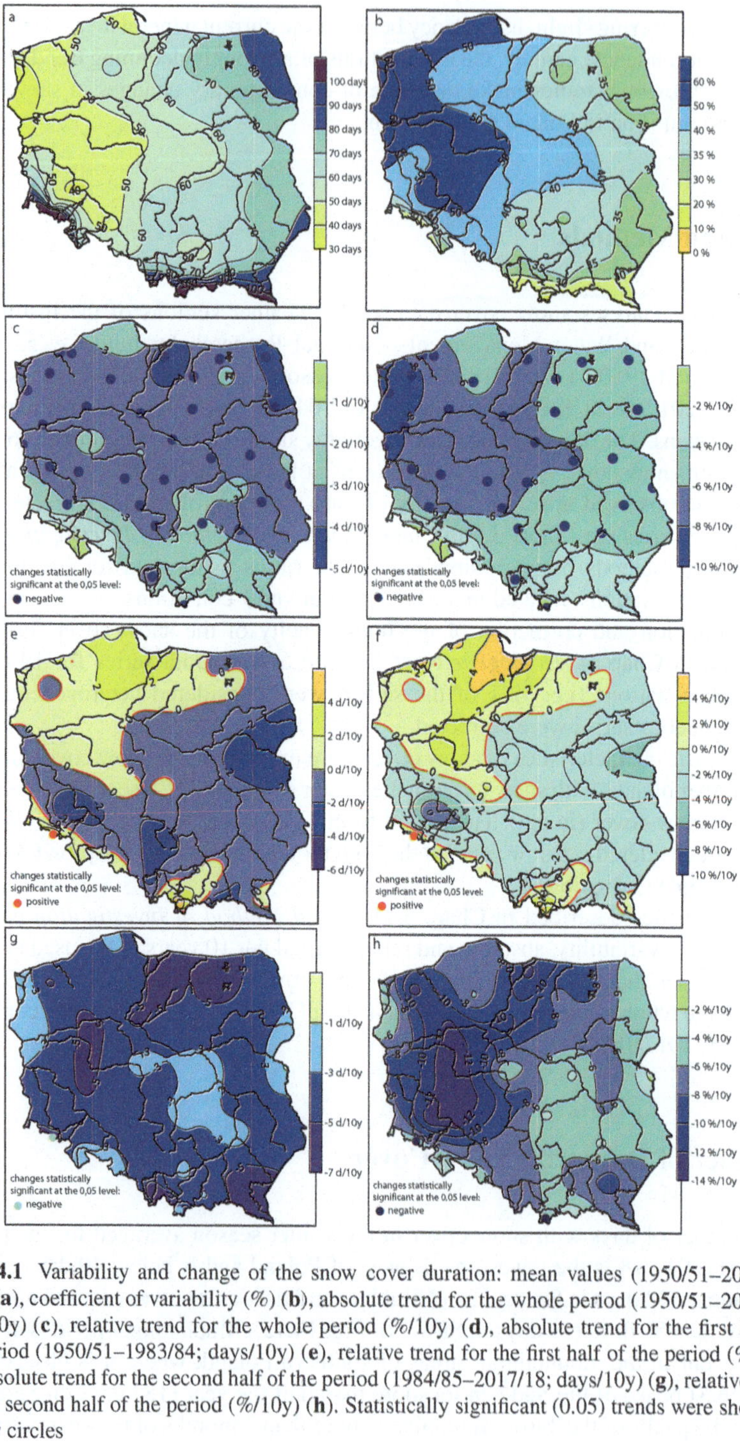

Fig. 14.1 Variability and change of the snow cover duration: mean values (1950/51–2017/18; days) (**a**), coefficient of variability (%) (**b**), absolute trend for the whole period (1950/51–2017/18; days/10y) (**c**), relative trend for the whole period (%/10y) (**d**), absolute trend for the first half of the period (1950/51–1983/84; days/10y) (**e**), relative trend for the first half of the period (%/10y) (**f**), absolute trend for the second half of the period (1984/85–2017/18; days/10y) (**g**), relative trend for the second half of the period (%/10y) (**h**). Statistically significant (0.05) trends were shown in **c**–**h** by circles

on Kasprowy Wierch). The year-to-year variability of snow cover expressed by the coefficient of variability is the smallest in the mountains, where at the highest peaks it does not exceed 10% (8% in Kasprowy Wierch), and the largest in the west of Poland—with values of more than 60% (with a maximum of 67% in Szczecin and Słubice; Fig. 14.1b). At the eastern borders of Poland, the coefficient of variability of the number of days with snow cover is 30–35%. The year-to-year snow cover variability is therefore inversely proportional to its duration in the winter season.

The absolute trend of the number of days with snow cover for the whole investigated period (1950/51–2017/18) is negative throughout Poland, statistically significant at 33 of the 60 meteorological stations (Fig. 14.1c). An insignificant trend is observed on the eastern coast, in the highlands and most of the mountainous areas. In the rest of Poland, the snow cover duration revealed a significant reduction in the winter season, with a minimum of about −4 to −5 days per 10 years in Mazury (north-eastern Poland). In the highest mountain ranges of Poland, extremely varying trends are observed: significant negative changes in the higher parts of the Tatra Mountains (Hala Gasienicowa: −2.5 days per 10 years, Kasprowy Wierch: −2.3 days per 10 years) and close to zero in Karkonosze (−0.32 days per 10 years on Sniezka summit). Changes in snow cover duration expressed in percentages (as a relative trend per 10 years) give a slightly different view (Fig. 14.1d). In all mountainous areas they are very small (−0.2 to −1.4% per 10 years in the highest parts of the Sudeten and Tatra Mountains), moderate in north-eastern Poland (−4 to −6% per 10 years), and significant in areas with a short snow cover duration (western Poland): −8 to −10% per 10 years (−10%/10 y in Słubice on the western border of Poland).

A look at the trend value of the number of days with snow cover in the two subperiods of the multi-year period considered leads to the conclusion that the trend described above was not a constant value over time. In the first half of the period (i.e. in the 34-year period 1950/51–1983/84) the snow cover duration was marked by a positive trend of change in north-western Poland (the coast, Pomorze, Wielkopolska and western Mazury) and in mountainous areas (Tatra Mountains, Sudeten; Figs. 14.1e, f). This trend proved to be statistically significant at the Śnieżka summit (+8 days per 10 years, i.e. 4.2% per 10 years). In the rest of Poland, this characteristic feature was a slight statistically insignificant negative trend (with a minimum of −8.6% per 10 years in Legnica and −4.2 days per 10 years in Terespol). During the second half of the period considered (i.e. in the 34-year period 1984/85–2017/18), the number of days with snow cover revealed a negative tendency throughout Poland (Figs. 14.1g, h). This trend was statistically significant at the peak of Śnieżka in Karkonosze (−6.6 days per 10 years, i.e. −3.5% per 10 years) and on Hala Gąsienicowa in the Tatras (−6.9 days per 10 years, i.e. −3.8% per 10 years). Despite the lack of statistical significance of trends at other meteorological stations, the values of the decrease in the number of days with snow cover are rather large: in some regions they reach −6 and −7 days per 10 years. The relative trend values are also significant during this period: in Wielkopolska, Nizina Śląska and parts of Pomorze they are below −10% per 10 years (with a minimum of −14% per 10 years in Poznań).

Maximum Seasonal Snow Cover Depth

Maximum seasonal snow cover depth averaged for 67 winter seasons varies spatially in Poland (Fig. 14.2a). It reaches 204 cm at the highest elevation, i.e. Kasprowy Wierch (1987 m a.s.l.) in the Tatra Mountains and 151 cm on Śnieżka (1603 m a.s.l.) in the Sudety. Much smaller values are recorded in lowland Poland, where maximum snow cover depth averaged for the entire study period ranges from 12 cm in the west to more than 30 cm in the northeast. However, in most of the country, it does not exceed 25 cm. The amount of snow is changeable from year to year and the coefficient of variability of maximum seasonal snow cover depth exceeds 50% at majority in most of the study areas (Fig. 14.2b). The coefficient of variability is the highest in the least snowy lowland regions of western (>70%) and central (>60%) Poland and takes much lower values in the mountains (<40%). Owing to inter-annual variability, snow depth during extremely snowy winters on Kasprowy Wierch exceeds 300 cm, as in seasons 1954/1955, 1961/1962, 1966/1967, 1994/1995 (355 cm), 1999/2000 and 2008/2009, but in some seasons it hardly reaches 100 cm, as in the winters of 1983/1984 and 2010/2011. On Śnieżka, extreme snow depth exceeding 250 cm occurred in the winters of 1951/1952 and 1955/1956, while an extremely snow-poor winter came in 1989/1990, with the maximum snow depth amounting to 38 cm. In lowland Poland, extreme values of snow cover depth can exceed 80 cm, particularly in the eastern part. In the exceptionally snowy winters of 1969/1070 and 1978/1979, snow depth exceeded 50 cm at more than 20 stations, including number of lowland locations. Since the 1990s, only one winter has occurred, when more than 50 cm of snow was observed at more than 10 stations (season 2004/2005).

During the entire study period a decreasing trend in the maximum snow cover depth was observed in a majority of the study areas (Fig. 14.2c, d). However, at some stations (10 out of 60) the trend was positive; most of these stations were located in the western part of Mazury lake district, in north-eastern Poland. A negative trend reached the highest rate of −8 cm/10 years on Śnieżka. However, in most of Poland it did not exceed −1 cm/10 years. This rate makes the speed of changes by −2 to −6%/10yrs and in several stations, located mostly in the south, the negative trend of changes is statistically significant at $p = 0.05$.

Dividing the period since 1950 into two subperiods, indicated a substantial difference in trends of snow cover depth in Poland during the two over 30-year-long intervals. In the first 34 years (1950/51–1983/84) almost half of the stations (29 out of 60), mostly located in lowland Poland, experienced an increasing trend of maximum snow depth, exceptionally exceeding 3 cm/10yrs (Figs. 14.2e, f). Negative trends in southern Poland were statistically significant only in the mountains and they reached a rate of more than −20 cm per decade. While positive trends in lowland Poland reached >12% of snow depth per decade in single locations, negative changes in the south hardly exceeded −10%/10yrs.

During the succeeding sub-period 1984/85–2017/18, negative trends in maximum seasonal snow cover depth were observed over most of Poland (at 43 out of 60 stations). However, they were statistically significant in only two locations in the

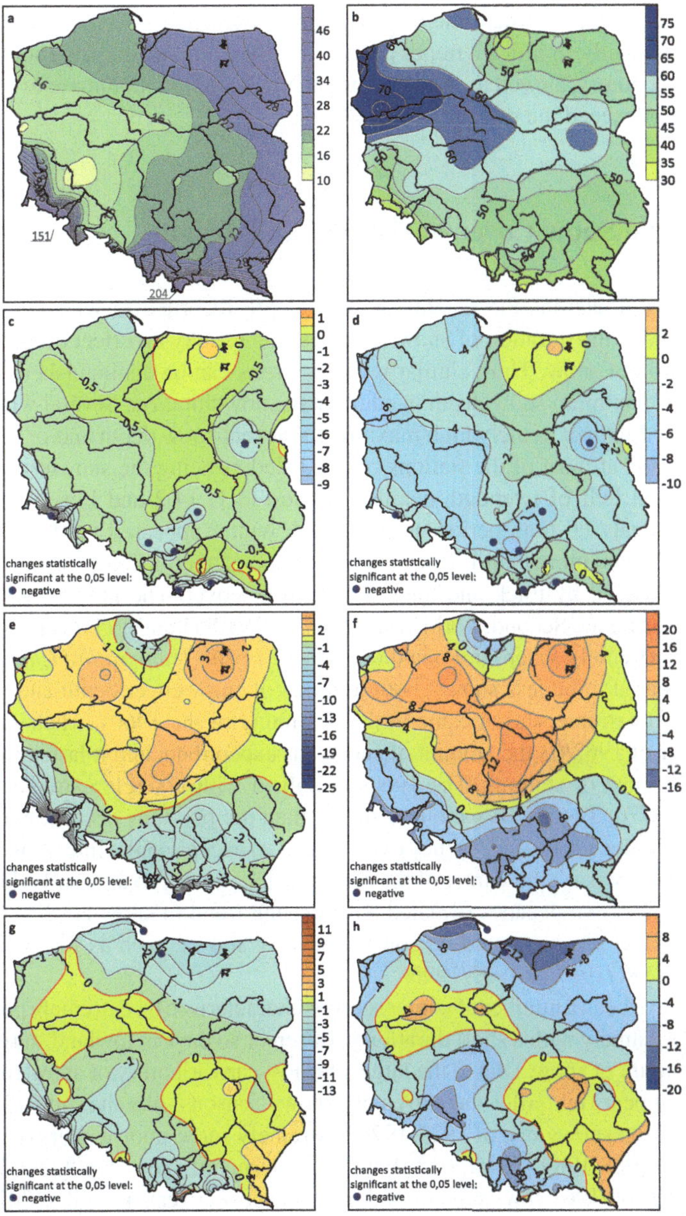

Fig. 14.2 Variability and change of the maximum seasonal snow cover depths: mean values (1950/51–2017/18; cm) (**a**), coefficient of variability (%) (**b**), absolute trend for the whole period (1950/51–2017/18; cm/10y) (**c**), relative trend for the whole period (%/10y) (**d**), absolute trend for the first half of the period (1950/51–1983/84; cm/10y) (**e**), relative trend for the first half of the period (%/10y) (**f**), absolute trend for the second half of the period (1984/85–2017/18; cm/10y) (**g**), relative trend for the second half of the period (%/10y) (**h**). Statistically significant (0.05) trends were shown in **c**–**h** by circles

north (Figs. 14.2g, h), where the rate of changes exceeded −3 cm/10yrs. Positive, but not statistically significant trends appeared in the north-western and southeastern parts of the country, as well as in single mountainous and submountainous locations. They did not exceed the rate of 8% per 10 years.

Long-Term Snow Cover Variability and Trends

In the lowland stations the maximum snow depth recorded in winter is strongly correlated with the duration of snow cover, as it is usually an effect of long-lasting accumulation of snow. Thus, similar winter seasons can be distinguished as snowy in terms of maximum snow depth in addition to snow duration. Several winters that characterised with extremely high maximum seasonal snow depth and/or long snow cover duration at most of the stations can be listed as being the snowiest in Poland since the first half of twentieth century (Tables 14.1, 14.2 and Figs. 14.3, 14.4). Within the analysed long period, the first such winter appeared in 1923/24, when snow cover lasted more than 100 days in most stations, then 1928/29 followed which was characterised by thick and long-lasting snow cover. The next snowy winters appeared during the Second World War (1939/40, 1940/41 and 1941/41) which were known as extremely severe in Poland (see Chap. 11 *Air Temperature Change*). The 1950s appeared to be less snowy, both in terms of snow cover depth and duration, and the next extreme season came in the beginning of the 1960s, when during the cold winter of 1962/63 the amount of snow was exceptional and it lasted more than 90 days at most of stations. The 1960s dissolved with a single extremely snowy season 1969/70. The 1970s were less snowy than the previous decade, however, at the end of the decade (1978/79) they experienced an exceptional winter season which came to be known as "the winter of the century", mostly due to unusually deep snow cover. The next two decades revealed a decreasing trend of snow cover occurrence and only single seasons could be considered very snowy, such as for example the winters of 1986/1987 or 1995/1996.

There have been a number of winter seasons characterised by extremely poor-snow conditions, namely with a short duration (a dozen or so days in lowland stations) and a small amount of snow. Due to the large year-to-year variability of snow conditions they may appear right after the extremely snowy seasons, like the mild 1924/1925 winter after the snowy one of 1923/1924. Extremely snow-poor winters occurred in 1974/1975, 1988/1989, 2014/2015 (both in the terms of snow depth and duration) and also in 1960/1961 (low number of days with snow cover), as well as 2015/2016. It is worth noting that during the last several years of the study period only mild or average winters occurred in Poland and in most locations a declining trend in snow duration and depth has been observed since 2010/2011.

Snow cover trends in eight stations with a long dataset, i.e. covering from 80 (Kasprowy Wierch) to 104 (Zakopane) winter seasons, were analyzed (Figs. 14.3 and 14.4). The two stations (Kasprowy Wierch and Zakopane) represent mountainous or submountainous climatic conditions, and the remaining six represent the lowland

Table 14.1 List of the 10 most snowy and 10 least snowy winters in terms of number of days with snow cover. Each season is marked by year of the first winter months, e.g. 1978 means winter 1978/1979; number of days with snow cover is given in brackets

Gdynia	Toruń	Poznań	Kalisz	Puławy	Kraków	Zakopane	Kasprowy Wierch
Most snowy winters							
1969 (116)	1969 (129)	1969 (121)	1969 (123)	1995 (123)	1995 (133)	1931 (170)	1972 (258)
1923 (99)	2005 (110)	1995 (96)	1923 (117)	1969 (117)	1969 (111)	1941 (169)	1954 (255)
1930 (99)	1995 (102)	1928 (91)	2005 (107)	2012 (112)	2005 (110)	1995 (164)	1952 (253)
1928 (96)	1928 (100)	1923 (86)	1962 (97)	1946 (112)	1923 (106)	1943 (159)	1990 (246)
1986 (95)	1962 (95)	1962 (84)	1928 (89)	2005 (110)	1928 (102)	1981 (154)	2007 (241)
1995 (94)	1941 (94)	1941 (83)	1941 (89)	1928 (109)	1962 (102)	1966 (153)	1996 (240)
1941 (89)	1978 (94)	1978 (82)	1995 (87)	1941 (104)	1978 (99)	1952 (151)	1956 (240)
1946 (87)	1923 (89)	2012 (80)	2012 (84)	1927 (104)	1963 (96)	1937 (150)	1973 (240)
1962 (84)	1930 (88)	1939 (80)	1952 (82)	1940 (103)	1930 (93)	1999 (149)	1941 (239)
1984 (83)	1952 (88)	1946 (79)	1930 (81)	1923 (101)	1961 (89)	1921 (147)	1979 (239)
Least snowy winters							
2014 (15)	2017 (23)	1991 (14)	2017 (14)	1949 (27)	1960 (32)	1917 (94)	1962 (199)
1929 (14)	1997 (22)	2016 (14)	1991(12)	2015 (26)	2014 (30)	1958 (93)	1948 (199)
2015 (13)	2015 (22)	2017 (13)	1924 (12)	2006 (25)	1993 (29)	1960 (92)	1976 (196)
1924 (13)	2014 (21)	1951 (12)	1974 (12)	2007 (22)	2015 (26)	1953 (89)	1968 (194)
1956 (13)	1972 (21)	1972 (11)	1988 (12)	1947 (21)	1989 (23)	2006 (89)	1982 (193)
2007 (13)	1974 (19)	1929 (11)	1943 (10)	1924 (21)	1988 (23)	1990 (85)	1949 (192)
1991 (12)	2006 (17)	1924 (10)	2013 (10)	1989 (19)	2006 (21)	1924 (84)	1967 (186)
1974 (10)	1924 (17)	2014 (9)	1929 (10)	1988 (17)	2013 (18)	2000 (82)	2012 (185)
1960 (8)	1988 (13)	1974 (8)	2014 (9)	2014 (15)	1974 (14)	2015 (68)	2011 (184)
1972 (3)	1929 (8)	1988 (2)	1949 (7)	1974 (14)	1924 (10)	2013 (61)	2000 (182)

Table 14.2 List of the 10 most snowy and 10 least snowy winters in terms of maximum snow cover depth. Each season is marked by year of the first winter months, e.g. 1978 means winter 1978/1979; maximum snow cover depth (cm) is given in brackets

Gdynia	Toruń	Poznań	Kalisz	Puławy	Kraków	Zakopane	Kasprowy Wierch
Most snowy winters							
1978 (60)	1939 (55)	1939 (46)	1962 (50)	2009 (48)	1962 (65)	1962 (88)	1943 (439)
1939 (45)	1964 (55)	1969 (46)	1939 (47)	1946 (46)	1951 (52)	1951 (104)	1938 (388)
1955 (43)	1969 (50)	1964 (35)	1978 (44)	1939 (44)	1939 (46)	1939 (72)	1994 (355)
2010 (41)	1978 (43)	1940 (34)	1969 (40)	1969 (43)	1969 (46)	1969 (85)	1947 (340)
1930 (38)	2009 (40)	2009 (33)	1940 (38)	1928 (38)	1940 (45)	1940 (42)	2008 (335)
1946 (38)	1928 (38)	1978 (29)	1981 (38)	1981 (36)	1946 (41)	1946 (74)	1948 (335)
1964 (38)	1962 (37)	1962 (28)	1937 (36)	1978 (35)	2009 (40)	2009 (59)	1999 (325)
1986 (36)	2010 (35)	2010 (28)	1928 (34)	2010 (34)	1986 (39)	1986 (73)	1966 (320)
1962 (35)	2000 (31)	1928 (27)	1933 (31)	1999 (34)	1928 (36)	1928 (68)	1954 (312)
1984 (34)	1940 (29)	1982 (25)	1986 (30)	2012 (32)	1965 (36)	1965 (41)	1961 (305)
Least snowy winters							
2008 (7)	1942 (7)	2017 (4)	1925 (6)	2007 (8)	1925 (9)	2017 (33)	1942 (130)
1974 (6)	2015 (6)	1951 (4)	1926 (6)	1956 (8)	2000 (9)	2015 (32)	1985 (125)
1996 (6)	2008 (6)	1991 (4)	1983 (6)	1958 (7)	1934 (9)	1988 (62)	1993 (125)
1956 (5)	1921 (6)	1972 (4)	1996 (4)	1997 (7)	1931 (9)	1920 (65)	1977 (120)
1971 (5)	2016 (5)	1927 (4)	1988 (4)	1960 (6)	1960 (7)	1914 (50)	1990 (107)
1983 (4)	2014 (5)	2014 (4)	2013 (3)	1988 (6)	1993 (6)	1915 (89)	2013 (106)
1988 (4)	1974 (5)	1924 (3)	1943 (3)	1993 (5)	1924 (5)	1916 (50)	1989 (103)
1991 (4)	1920 (5)	1960 (3)	1991 (3)	1974 (5)	2017 (5)	1917 (48)	1940 (100)
1994 (4)	1960 (4)	1974 (2)	2014 (3)	1996 (3)	2015 (5)	1918 (45)	2010 (94)
1972 (3)	1988 (4)	1988 (1)	1974 (2)	2014 (3)	1988 (5)	1919 (41)	1983 (92)

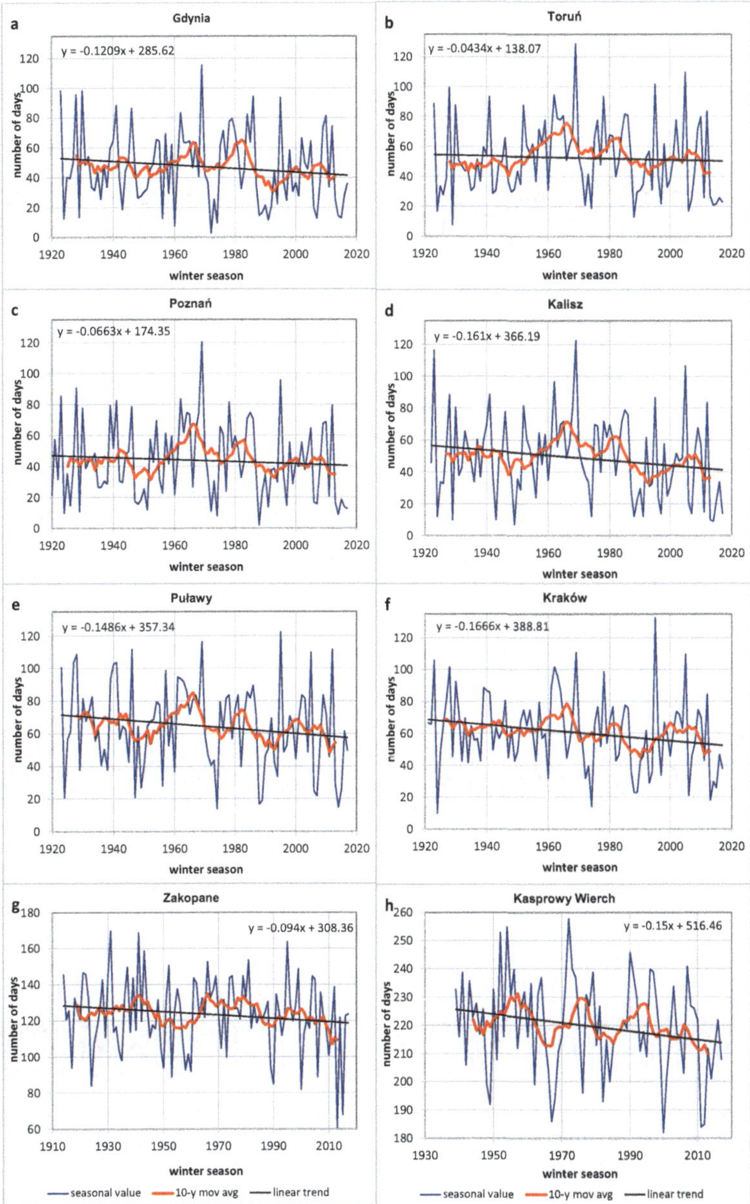

Fig. 14.3 Long-term variability of the seasonal number of days with snow cover in stations ordered from the north to the south of Poland: **a** Gdynia (sea-side station), period 1924–2017; **b** Toruń (central Poland); period 1920–2017; **c** Poznań (western-central Poland), period 1920–2017; **d** Kalisz (central Poland), period 1922–2017; **e** Puławy (central-eastern Poland), period 1921–2017; **f** Kraków (southern Poland), period 1921–2017; **g** Zakopane (submountainous station, southern Poland), period 1914–2017; **h** Kasprowy Wierch (mountainous station), period 1938–2017. Mean annual value (blue), 10-year consecutive average (red) and line trend (black) are shown. Only in Kraków the trend is statistically significant at $p = 0.05$

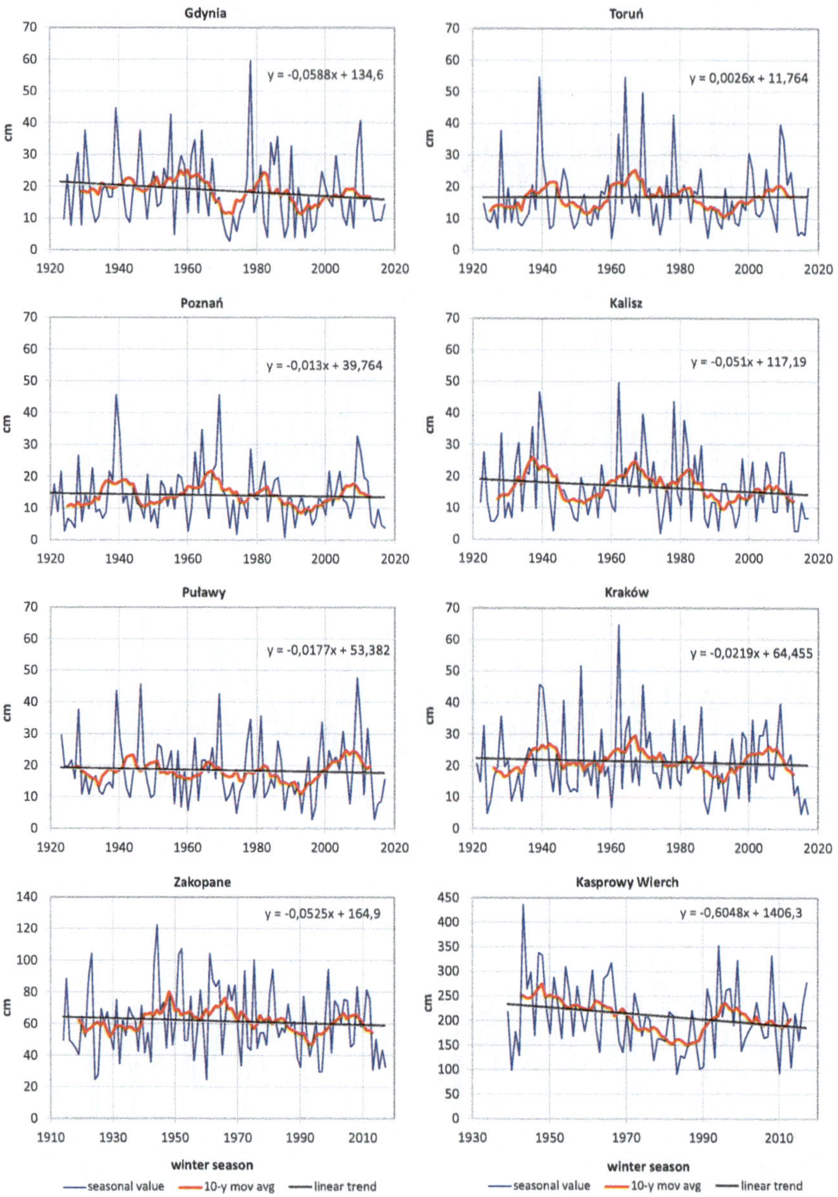

Fig. 14.4 Long-term variability of the maximum seasonal snow cover depth (cm) in stations ordered from the north to the south of Poland: **a** Gdynia (sea-side station), period 1924–2017; **b** Toruń (central Poland); period 1920–2017; **c** Poznań (western-central Poland), period 1920–2017; **d** Kalisz (central Poland), period 1922–2017; **e** Puławy (central-eastern Poland), period 1921–2017; **f** Kraków (southern Poland), period 1921–2017; **g** Zakopane (submountainous station, southern Poland), period 1914–2017; **h** Kasprowy Wierch (mountainous station), period 1938–2017. Mean annual value (blue), 10-year consecutive average (red) and line trend (black) are shown. Trends are not statistically significant

part of Poland. In every location negative trends of the seasonal number of days with snow cover were detected. However, they were statistically significant at $p = 0.05$ only in Kraków and achieved a rate of −1.7 day/10yrs. The insignificant rate of long-term decline in the number of days with snow cover in the remaining stations ranged from −0.4 day/10yrs in Toruń to −1.6 day/10yrs in Kalisz.

Long-term tendencies in maximum snow cover depth appeared to be positive (but very weak) in Toruń (changes close to zero per 10 yrs) and in other stations they ranged from approximately −0.6 cm/10yrs in Gdynia to −0.13 cm/10yrs in Poznań. In a snowy mountainous station i.e. Kasprowy Wierch maximum seasonal snow cover depth decreased by approximately −6 cm/10yrs.

Discussion and Conclusion

The chapter analyses the variability and trends of two basic snow cover characteristics: the number of days with snow cover and its maximum depth in the winter season. The most important results of the study are as follows:

- the year-to-year variability of the snow cover duration and its seasonal maximum depth is the largest in regions of poorest snow cover (in western Poland the coefficient of variation is above 60%);
- the number of days with snow cover during the 68 winter seasons (1950/51–2017/18) has a negative time trend throughout Poland (with a minimum of −4 to −5 days for 10 years in north-eastern Poland); this tendency is statistically significant in most areas of Poland except for the highlands and some parts of the mountainous areas;
- the relative changes in the number of days with snow cover are the most significant in regions with a short duration of snow cover (western Poland; −8 to −10% per 10 years);
- the maximum depth of snow cover throughout the considered period revealed a negative trend in a majority of Poland, statistically significant in the highlands and in mountainous areas (−2 to −6%/10y); only in north-eastern Poland is the trend positive, statistically insignificant;
- in longer periods (80 to 104 winter seasons), the snow cover duration and maximum depth of snow cover are characterised by a slight negative or near zero time trend at most weather stations; only in Kraków the number of days with snow cover reveals statistically significant negative tendency (−1,7 days per 10 years);
- in the winter seasons of 1969/70 and 1995/96 the longest snow cover duration in most areas of Poland was recorded, while in the 1939/40 and 1962/63 seasons, the maximum depth of the snow cover was the highest.

Considering data from 60 weather stations for the entire winter season refined the results of the long-term trends of snow cover that were obtained earlier for data from 43 stations for December to February by Szwed et al. (2017) and present them spatially using isolines.

It is significant that the two snow cover characteristics analysed show different and even opposing tendencies during certain periods in the same area. For example, in the period 1950/51–2017/18 in north-eastern Poland (Mazury) there is a negative trend of the snow cover duration, whereas a positive trend of the maximum snow cover depth. The depth of the snow cover is shaped mainly by the amount of snowfall, while the snow cover duration in the winter season is largely due to thermal conditions. All in all, this is the reason for the significant differences in the trends of the two characteristics analysed in different regions of Poland. This feature of snow cover variability is significant for the entire area of Northern Eurasia, where significant increases in winter snow accumulation but a shorter snowmelt season was detected (Bulygina et al. 2009).

The reason for the statistically significant negative trend of snow cover duration in Kraków-Obs. UJ in the long-term period (1921/22–2017/18) should be found in the increasing influence of the urban heat island on the analysed element. The station was founded in 1792 outside the city limits. The development of Krakow has now made it so that this place is located in the city centre and under the impact of dense buildings and increased traffic.

Statistically significant changes in the number of days with the snow cover and its maximum depth are recorded throughout the period considered also in the areas of the highest mountain ranges of Poland (Tatra Mountains, Karkonosze), i.e. in areas where the local (regional) impact of anthropogenic factors is at its minimum.

The negative trend of snow cover found in the investigation based on the data until 2018 presented in this chapter is much more noticeable than in the results previously obtained in Poland: until 2013 (Szwed et al. in 2017 show a trend of snow cover duration, statistically significant at 0.05 at only a few stations), and even more so until 1998 (Falarz in 2004 presents a statistically insignificant trend of snow cover duration throughout Poland, with the exception of mountainous areas). This indicates a progressive negative changes of snow conditions in Poland. Moreover, the projected seasonal snow cover depth in Poland for both the near (2021–2050) and far (2071–2100) future is much lower than today (Szwed et al. 2019).

Overall the values of the snow cover duration trends in Poland are comparable to analogue values averaged over the Northern Hemisphere, where the duration of the snow season has declined by 5 days per decade since the winter of 1972/1973 (Choi et al. 2010). On the other hand, the negative trend of maximum snow cover depth is not as significant in Poland as the European average reaching −11.4%/decade (Fontrodona Bach et al. 2018). However, similarly to other areas of Europe, the acceleration of the downward trend has been noted in Poland since the 1980s. Far greater changes have been recorded in the Arctic, where in the spring snow cover extend decreased by 46% (Brown et al. 2010).

References

Armstrong RL, Brun E (eds) (2008) Snow and climate: physical processes, surface energy exchange and modeling. Cambridge University Press, 256 p

Bednorz E (2002) Snow cover in western Poland and macroscale circulation conditions. Inter J Climatol J R Meteorol Soc 22(5):533–541

Bednorz E (2004) Snow cover in eastern Europe in relation to temperature, precipitation and circulation. Inter J Climatol J R Meteorol Soc 24(5):591–601

Brown RD (2000) Northern Hemisphere snow cover variability and change, 1915–97. J Clim 13:2339–2355

Brown I (2019) Snow cover duration and extent for Great Britain in a changing climate: Altitudinal variations and synoptic scale influences. Inter J Climatol 39(12):4611–4626. https://doi.org/10.1002/joc.6090

Brown RD, Robinson DA (2011) Northern Hemisphere spring snow cover variability and change over 1922–2010 including an assessment of uncertainty. Cryosphere 5:219–229. https://doi.org/10.5194/tc-5-219-2011

Brown R, Derksen C, Wang LB (2010) A multi-data set analysis of variability and change in Arctic spring snow cover extent, 1967–2008. J Geophys Res-Atmos 115:D16111

Bulygina ON, Razuvaev VN, Korshunova NN (2009) Changes in snow cover over Northern Eurasia in the last few decades. Environ Res Lett 4:045026. https://doi.org/10.1088/1748-9326/4/4/045026

Choi G, Robinson DA, Kang S (2010) Changing Northern Hemisphere snow seasons. J Clim 23:5305–5310

Clark MP, Serreze MC, Robinson DA (1999) Atmospheric controls on Eurasian snow extent. Int J Climatol 19:27–40. https://doi.org/10.1002/(SICI)1097-0088(199901)19:1%3c27:AID-JOC346%3e3.0.CO;2-N

Cohen J (1994) Snow cover and climate. Weather 49:150–156

Czarnecka M (2012) Częstość występowania i grubość pokrywy śnieżnej w Polsce. Acta Agrophys 19(3):501–514

Dong C, Menzel L (2019) Recent snow cover changes over central European low mountain ranges. Hydrol Process, 1–17. https://doi.org/10.1002/hyp.13586

Falarz M (2002) Long-term variability in reconstructed and observed snow cover over the last 100 winter seasons in Cracow and Zakopane (southern Poland). Clim Res 19(3):247–256

Falarz M (2004) Variability and trends in the duration and depth of snow cover in Poland in the 20th century. Int J Climatol 24:1713–1727

Falarz M (2007) Snow cover variability in Poland in relation to the macro- and mesoscale atmospheric circulation in the 20th century. Inter J Climatol 27:2069–2081. https://doi.org/10.1002/joc.1505

Falarz M (2008) Changes of extreme nival conditions in Poland during the second half of the 20th century. Meteorol Z 17(3):339–444

Feng S, Hu Q (2007) Changes in winter snowfall/precipitation ratio in the contiguous United States. J Geophys Res Atmos 112(D15). https://doi.org/10.1029/2007JD008397

Fontrodona Bach A, van der Schrier G, Melsen LA, Klein Tank AMG, Teuling AJ (2018) Widespread and accelerated decrease of observed mean and extreme snow depth over Europe. Geophys Res Lett 45:312–319

Henderson GR, Leathers DJ (2010) European snow cover extent variability and associations with atmospheric forcings. Inter J Climatol 30:1440–1451

Jaagus J (1997) The impact of climate change on the snow cover pattern in Estonia. Clim Change 36:65–77

Niedźwiecki M (1998) Charakterystyka pokrywy śnieżnej w Łodzi w latach 1950–1989. Acta Univ Lodziensis Folia Geograph Phys 3:265–277

Nowosad M, Bartoszek K (2007) Wieloletnia zmienność grubości pokrywy śnieżnej w okolicy Lublina. In: Piotrowicz K, Twardosz R (red) Wahania klimatu w rożnych skalach przestrzennych i czasowych. Instytut Geografii i Gospodarki Przestrzennej UJ, Kraków, pp 411–421
Szwed M, Dobler A, Mezghani A, Saloranta TM (2019) Change of maximum snow cover depth in Poland—trends and projections. Q J Hung Meteorol Serv 123(4):487–500
Szwed M, Pińskwar I, Kundzewicz ZW, Graczyk D, Mezghani A (2017) Changes of snow cover in Poland. Acta Geophys 65:65–76. https://doi.org/10.1007/s11600-017-0007-z
Ye K, Lau NC (2017) Influences of surface air temperature and atmospheric circulation on winter snow cover variability over Europe. Int J Climatol 37:2606–2619. https://doi.org/10.1002/joc.4868

Chapter 15
Change of Wind

Joanna Wibig

Abstract The wind was analyzed on the basis of speed and direction data from three observation times at 41 meteorological stations in Poland from 1966–2018. The average annual and seasonal wind speeds and the incidence of atmospheric calms were calculated. Wind frequencies from different directions were determined using a 16-sector wind rose. Particular attention was paid to winds exceeding 8 ms^{-1}. The seasonal and annual trends of strong winds were counted. The average wind speed was 3.6 ms^{-1} and ranged from 1.4 ms^{-1} in Zakopane to 12 ms^{-1} on Śnieżka. On the daily scale, the highest wind speed appeared at noon. In the annual cycle, the average wind speed at noon was the highest in spring and in the other times in winter, the lowest was always in summer. Windless weather occurred most often in the morning (12.3% of observations), less often in the evening (8.3%) and the least often at noon (3.0%). The long-term average of the highest annual wind speed oscillated between 8.3 ms^{-1} in Tarnów and 42.5 ms^{-1} on Śnieżka. Strong winds were most often recorded from the west, west southwest and west northwest directions. In southern Poland it was also the southern direction. The decreasing trends in annual and seasonal series of wind speed were observed in Poland as well as in other European countries. A number of authors attribute this trend to an increase in surface roughness.

Introduction

Wind is one of the basic meteorological and climatological elements determining the conditions in a given area. In addition to air temperature and humidity, it is also one of the parameters determining biometeorological conditions. It affects thermal sensations by increasing discomfort during cold periods or by improving biometeorological conditions during hot periods. It affects the rate of evaporation, thanks to

J. Wibig (✉)
Department of Meteorology and Climatology, Faculty of Geographical Sciences, University of Lodz, Lodz, Poland
e-mail: joanna.wibig@geo.uni.lodz.pl

M. Falarz (ed.), *Climate Change in Poland*, Springer Climate,
https://doi.org/10.1007/978-3-030-70328-8_15

which it significantly affects the local water balance. Despite such a significant importance of wind, there are relatively few items devoted to it in Polish climatological literature.

The list of classic publications opens with an article by Bartnicki (1930) about bottom air currents in Poland based on a 25-year data series from 1886 to 1910. Further works were created after World War II. Gumiński (1952) presented the first results of research on the spatial distribution of wind speed and directions based on measurements from selected meteorological stations in Poland. The analyzed data series came from before 1945. In the same year, Piasecki's work (1952) appeared on the maximum wind speeds in Poland. It was the first publication dealing with the wind from the point of view of threats to human life and various branches of the economy, but it only covered the observation period of 11 years from 1928 to 1938. Parczewski (1960) proposed classification of wind speed in relation to turbulence generation (Table 15.1). This procedure would allow connection of wind study results with such processes as evaporation, vertical transport of water vapour, formation of fogs, dew, hoar-frost and rime.

The wind was also one of the elements included in the Polish Climate Atlas (1971, 1973) and in Atlas Rzeczpospolitej Polskiej (Niedźwiedź et al. 1994). It presents the monthly and annual wind speed percentage distribution for 8 main directions divided into speed classes based on 30-year data series (1931–1960) for 12 stations, as well as for the decade of 1951–1960 for another 120 stations. The spatial differentiation of strong and very strong winds and their frequency dependent on individual types of atmospheric circulation were analysed by Adamczyk (1996). She used data from 16 meteorological stations from the period 1956–1965 and from three meteorological times. It was shown that strong and very strong winds occur most frequently in winter at cyclonal and in spring at anticyclonal types of circulation. The frequency of moderate, strong and very strong winds was also presented in the Climate Atlas of elements and phenomena harmful to agriculture (1990). The frequency of wind with a speed greater than 8 ms^{-1} was analyzed by Krawczyk (1994) and frequency of strong and very strong wind was investigated by Stopa-Boryczka (1989) as well as

Table 15.1 Wind speed classification proposed by Parczewski (1960)

Wind speed interval (ms^{-1})	Wind speed subinterval (ms^{-1})	Word mark of the wind speed interval	General characteristic of turbulence
0	0	Calm	Absent
>0 – 4	>0 – 2	Very weak	Weak
	>2 – 4	Weak	
>4 – 10	>4 – 7	Moderate	Moderate
	>7 – 10	Quite strong	
>10 – 20	>10 – 15	Strong	Strong
	>15 – 20	Very strong	
>20	>20	Gale	Very strong

Paszyński and Niedźwiedź (1991). Żurański and Jaśpińska (1996) analysed extreme wind speeds in relation to wind direction for the whole country basing on data from 23 stations in Poland from the period 1966–1990. The authors considered 12 sectors, each of 30 degrees and showed that the strongest winds blow from the western sectors. Wind velocities for the sectors from northern through eastern to southern were much weaker than those from the western sector. Spatial distribution of wind speed in Poland was also analysed by Wierzbicki (1968) and Wierzbicki and Bratkowski (1970). Recently, a detailed analysis of winds with maximum speeds in Poland was presented by Lorenc (2012). She distinguished three genetically different meteorological situations that threaten the occurrence of maximum wind velocities at gusts exceeding 17 ms^{-1}. These are winds associated with general atmospheric circulation, regional foehn winds and small-scale whirlwinds and squalls. Calms in Poland were investigated by Parczewski (1974) and Kowalewski (2005).

A number of papers were published on wind conditions in the Carpathian Mountains. Even before World War II, Sokołowski's work (1927) about winds in the Tatras and Milata's paper (1936) about the frequency of lower winds and mountain winds in the Carpathians appeared. Wierczek (1958) used Milata's data from 1891 to 1910 to depict wind roses in the Carpathian region west of the Biała Dunajec River basin. Hess (1965) presented wind relations in the Western Carpathians based on a 10-year observation series (1952–1961) from 89 stations. A special chapter in the monograph on the climate of the Tatra Mountains was devoted to the wind (Otruba and Wiszniewski 1974). Kożuchowski (1977) wrote about the altitudinal zonation of anemological relations in the Tatra Mountains. A detailed analysis of wind relations throughout the Carpathians was made by Niedźwiedź et al. (1985). They used data from 84 stations from the decade of 1961–1970. Their analysis covered the entire Polish part of the Carpathians and concerned the incidence of major wind directions, average speeds, calms and strong winds, maximum wind speeds and daily average wind speed.

Foehn winds were investigated in various mountain areas. Kwiatkowski (1972, 1975 and 1979) studied them in the Sudetes, Malicki and Michna (1966) in the Bieszczady, Lewińska (1958) near Rytro and Rymanów, Milata (1936), Stachlewski (1974) and Ustrnul (1992) in the Polish Carpathians.

Anemometric conditions in northern Poland were studied by Taranowska (1957), Trzeciak (2001) and Tarnowska (2011). The first one described the occurrence of wind on the coast, the second one concentrated on storm winds. In the third paper, an analysis of strong and very strong winds on the Polish Baltic Sea Coast has been presented. The average wind speed and direction data of eight observations during the day from the years 1971–1990 from 4 stations was analysed. It showed a greater share of strong and very strong winds in winter and early spring—mostly from land directions with a little variation in the frequency of strong wind during the day. Anemometric conditions in the Warmian-Masurian Lake District were studied by Grabowski (1996) in 1971–1990 and Pożarska and Grabowski (2015) in 1991–2010.

Wind gusts were studied by Lorenc (1968 and 1996) for the period 1961–1975 and Kolendowicz et al. (2016) in the period 2001–2015. In the latter paper, wind gust cases were divided into convective and non-convective ones. The authors showed that the highest threat in terms of frequency and strength of peak wind gusts due to convection occurs in July. Peak wind gusts in a non-convective days are the highest and the most frequent in January, and are at a minimum during summer.

Several authors investigated wind energy resources in Poland. Lorenc (1992 and 1996) showed, based on data from 55 station in Poland from the period 1966–1990, that more than 40% of the area of Poland has winds suitable for wind power generation. The most favored in this respect are: the Slovenian seashore from Świnoujście to Reda, the Suwałki Region, most of Mazovia, the Silesian and the Beskid Żywiecki Mts. and the Dynowskie Foothills.

Data

The study uses data from 1951 to 2018 from 41 meteorological stations in Poland from three observation times: 6, 12 and 18. Data were taken from the archives of the Polish Institute of Meteorology and Water Management–National Research Institute (PIMWM-NRI). These are the values of the average wind speed and its direction. According to the World Meteorological Organization guidelines, the measurements should be performed using an anemometer placed 10 m above the ground, and the speed and direction should be 10-minute averages. The wind gauge should be placed so, that terrain obstacles do not disturb the observation. Table 15.2 shows the location of anemometers used in this analysis. In addition to the geographical coordinates and the altitude of the station above sea level, it also includes the height of the anemometer above the ground. It is usually different from 10 m. In addition, in many cases, this value changed during the measurement period, therefore the heights given should be considered as average.

Average Wind Speed

The average annual wind speed in Poland was 3.6 ms^{-1} and ranged from 1.4 ms^{-1} in Zakopane and 1.6 ms^{-1} in Nowy Sącz to 12.1 ms^{-1} on Śnieżka Mt. (Fig. 15.1). While Śnieżka Mt. is really the windiest place in Poland, the low wind speed in Zakopane and Nowy Sącz are probably due to specific location of anemometers. There are also the places with the highest frequency of calms. Lorenc (1996) reported that the location of anemometers in Zakopane is incorrect. However, Niedźwiedź et al. (1985) reported that a low average wind speed and a high frequency of calms are typical for stations located in the valleys like Zakopane and Nowy Sącz. So it was decided not to exclude these stations from further analysis.

The spatial distribution of the average wind speed indicates that the largest values, exceeding the value 4 ms^{-1} are observed on the coast (Łeba 5.0 ms^{-1}, Ustka 4.5 ms^{-1}, Hel 4.3 ms^{-1}) as well as in Warszawa (4.0 ms^{-1}), on Kasprowy Wierch Mt. (6.7 ms^{-1}) and the aforementioned Śnieżka Mt. (12.1 ms^{-1}). The average wind speed on Śnieżka

Table 15.2 Location of meteorological stations together with their geographical position, altitude and height above the ground

Station	Latitude	Longitude	Altitude (m a.s.l.)	Height (m a.g.l.)
Chojnice	53°43'	17°32'	165	13
Elbląg	54°10'	19°26'	40	19
Koło	52°12'	18°40'	115	14
Koszalin	54°12'	16°09'	32	13
Łeba	54°45'	17°32'	2	14
Lębork	54°33'	17°45'	38	12
Olsztyn	53°46'	20°25'	133	13
Suwałki	54°08'	22°57'	184	10
Świnoujście	53°55'	14°14'	6	11
Szczecin	53°24'	14°37'	1	24
Toruń	53°02'	18°35'	69	12
Ustka	54°35'	16°52'	6	20
Białystok	53°06'	23°10'	148	12
Hel	54°36'	18°49'	1	23
Bielsko Biała	49°48'	19°00'	398	20
Gorzów	52°45'	15°49'	72	13
Jelenia Góra	50°54'	15°48'	342	16
Kalisz	51°47'	18°05'	138	15
Kasprowy Wierch Mt.	49°14'	19°59'	1991	15
Kielce	50°49'	20°42'	260	15
Kłodzko	50°26'	16°37'	356	14
Kołobrzeg	54°11'	15°35'	3	18
Krakow	50°05'	19°48'	237	11
Legnica	51°12'	16°12'	122	16
Lesko	49°28'	22°21'	420	14
Łódź	51°44'	19°24'	187	12
Lublin	51°13'	22°34'	238	12
Nowy Sącz	49°37'	20°42'	292	16
Opole	50°38'	17°58'	165	14
Poznań	52°25'	16°51'	83	10
Rzeszów	50°06'	22°03'	200	16
Siedlce	52°11'	22°15'	152	12
Słubice	52°21'	14°36'	21	17
Śnieżka Mt.	50°44'	15°44'	1603	6
Tarnów	50°02'	20°59'	209	14

(continued)

Table 15.2 (continued)

Station	Latitude	Longitude	Altitude (m a.s.l.)	Height (m a.g.l.)
Warszawa	52°10'	20°58'	105	12
Wieluń	51°13'	18°34'	200	15
Włodawa	51°33'	23°32'	177	12
Wrocław	51°06'	16°53'	120	17
Zakopane	49°18'	19°56'	844	18
Zielona Góra	51°56'	15°32'	192	13

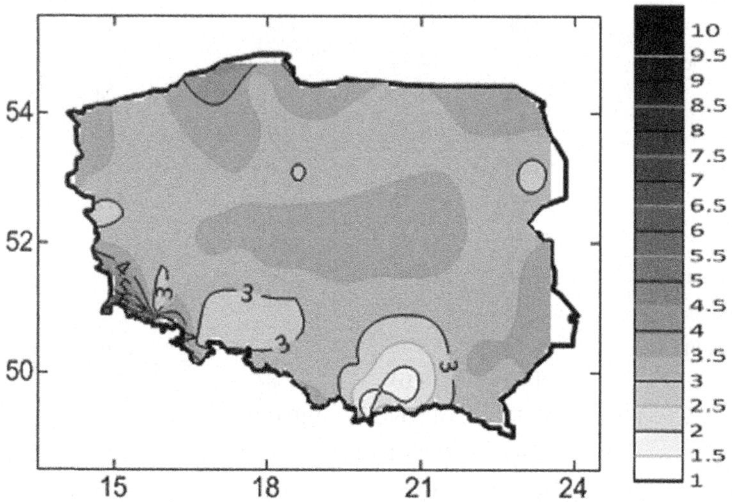

Fig. 15.1 Average annual wind speed (ms^{-1}) in the period 1966–2018

Mt. is almost twice as high as on Kasprowy Wierch Mt. although Śnieżka Mt. is significantly lower. This is due to different topographic conditions. Average wind speed exceeding 3.5 ms^{-1} are observed in northern, central and south-eastern Poland. Winds with an average speed of less than 3 ms^{-1} occurs in the foothills of the Sudetes and the Carpathians. Such low average values are probably due to a high frequency of calms caused by the terrain around the station. In the remaining area the average wind speed is in the range from 3.0 to 3.5 ms^{-1}.

The average wind speed clearly changes both in the annual and daily cycle (Fig. 15.2). In the daily cycle the highest average speed, at noon, reaches 4.5 ms^{-1} on average in Poland. Slightly lower speeds occur in winter and autumn, respectively 4.4 and 4.2 ms^{-1}. In summer, the average wind speed is definitely lower and reaches only 3.8 ms^{-1}. In the morning and evening, the wind speed is definitely lower than at noon, and the differences between the morning and the evening are insignificant. The annual cycle is slightly different. The highest average wind speeds are observed

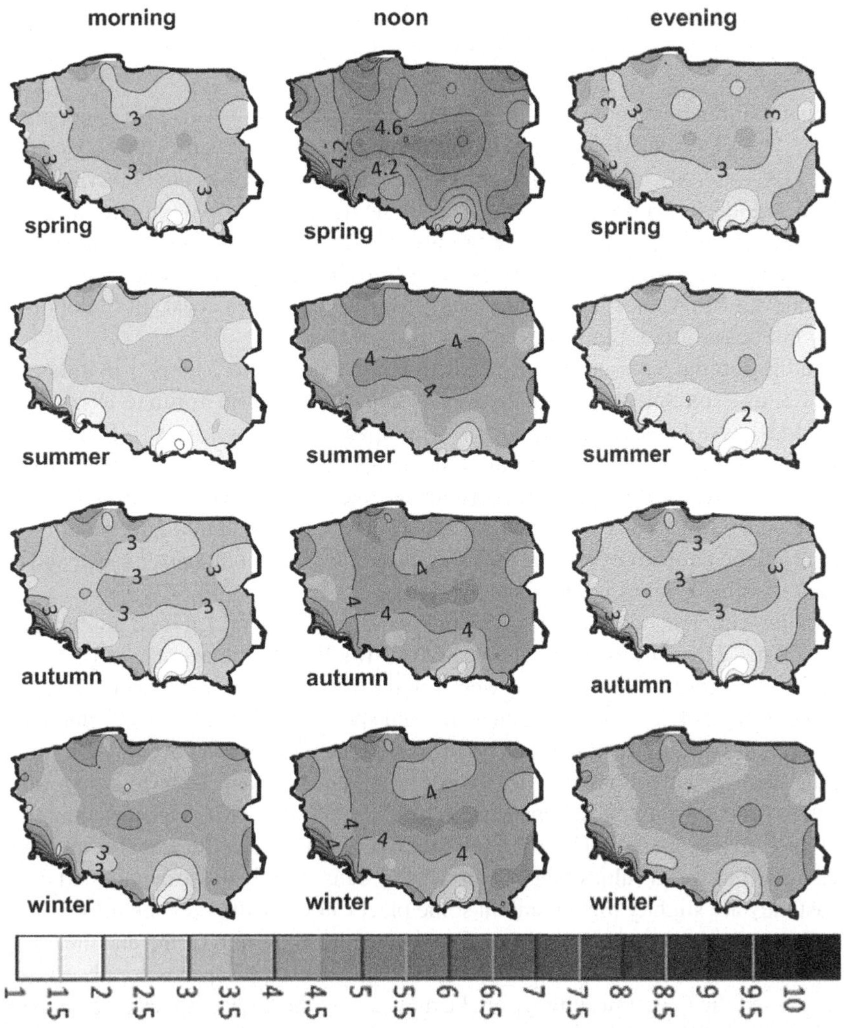

Fig. 15.2 Average wind speed in seasons and observation times in the period 1966–2018

in winter (respectively 3.9 and 4.0 ms^{-1} in the morning and evening), slightly lower in spring and autumn 3.3 ms^{-1} in both seasons and both in the morning and evening. The lowest average wind speeds occur in the morning and evening in summer and reach 2.7 ms^{-1} for both.

In the spring, at noon wind speeds in Poland vary from 2.5 ms^{-1} in Zakopane to 10.4 ms^{-1} on Śnieżka Mt. In most of the country the average values exceed 4.0 ms^{-1}, only in southern Poland they are lower than this value. In the morning and evening, the spread of the average values is greater. In the morning changes are from 1.1 ms^{-1} in Tarnów to 12.3 ms^{-1} on Śnieżka Mt., in the evening from 1.4 ms^{-1} in Zakopane to

11.9 ms^{-1} on Śnieżka Mt. Values lower than 3 ms^{-1} occur in southern and western Poland and in the lower reaches of the Vistula. In Małopolska, they fall below 2 ms^{-1} in places.

In the summer, at noon, wind speeds in Poland vary from 1.9 ms^{-1} in Zakopane to 7.8 ms^{-1} on Śnieżka Mt. For most of the country, the average values are between 3.0 and 4.0 ms^{-1}, only in the north and in the center of Poland they are higher than this value and in Małopolska lower than this. In the morning and evening, the spread of the average values is greater. In the morning, it ranges from 0.9 ms^{-1} in Nowy Sącz to 9.9 ms^{-1} on Śnieżka Mt., in the evening from 0.9 ms^{-1} in Zakopane to 9.4 ms^{-1} on Śnieżka Mt. Values higher than 3 ms^{-1} occur on the sea coast. In Małopolska, they fall below 2 ms^{-1} in places.

In the autumn, at noon, wind speeds in Poland vary from 2.0 ms^{-1} in Zakopane to 11.3 ms^{-1} on Śnieżka Mt. In the west, south, the belt from Toruń to Olsztyn and around Białystok, the average values of wind speed fall below 4.0 ms^{-1}. In the rest of the country they are higher than this value. In the morning and evening the spread of the average values is greater. The morning and evening values change from 1.0 ms^{-1} in Zakopane to 13.1 ms^{-1} on Śnieżka Mt. The spatial distribution is very similar to that from the noon observations, however, the values are about 1 ms^{-1} lower and the 4 ms^{-1} isoline is replaced by a 3 ms^{-1} isoline.

In the winter, at noon, wind speeds in Poland vary from od 1.9 ms^{-1} in Zakopane to 14.8 ms^{-1} on Śnieżka Mt. The spatial distribution of the average wind is very similar to that of autumn; in the west, south, the belt from Toruń to Olsztyn and around Białystok, the average values of wind speed fall below 4.0 ms^{-1}, in the rest of the country they are higher than this value. In the morning and evening the spread of the average values is greater, like in the case of other seasons. In the morning values change from 1.3 ms^{-1} in Zakopane to 15.6 ms^{-1} on Śnieżka Mt., in the evening from 1.4 ms^{-1} in Zakopane to 15.5 ms^{-1} on Śnieżka Mt. In almost the whole country, the average wind speed values range from 3.0 to 4.0 ms^{-1}, only in the central part of the coast they are slightly higher, and in some places in the south they are lower.

To sum up, the average wind speed at noon is the highest in spring and the lowest in summer. In the morning and evening, the highest average speeds are observed in winter, and the lowest in summer. In the daily cycle, the highest speeds are observed at the noon, while in the morning and evening the spread of average speeds in Poland is greater than at the noon. The highest average wind speeds are observed on Śnieżka Mt., followed by Kasprowy Wierch Mt. Apart from the high-mountain region, the highest wind speeds occur in the central part of the coast (Ustka and Łeba) and central and north-eastern Poland.

Frequency of Calms

Windless weather is relatively rare in Poland (Table 15.3). Calms are most often observed in the morning (during 11.26% of observations on average), less often in the evening (during 8.33% of observations on average) and least often appear at

Table 15.3 Percentage of calms in the morning, noon and evening measurements in the period 1966–2018

Station	Morning	Noon	Evening
Chojnice	5.80	2.31	4.56
Elbląg	8.29	1.66	3.54
Koło	5.28	1.91	4.35
Koszalin	5.49	1.54	6.24
Łeba	3.67	0.66	3.06
Lębork	17.14	2.99	9.47
Olsztyn	9.13	1.76	6.67
Suwałki	7.03	1.86	5.18
Świnoujście	3.38	0.98	3.55
Szczecin	5.24	1.05	3.12
Toruń	5.56	1.08	5.83
Ustka	0.52	0.21	0.72
Białystok	10.69	2.37	14.1
Hel	1.85	0.82	1.35
Bielsko Biała	7.19	4.53	7.17
Gorzów	6.53	1.84	4.33
Jelenia Góra	34.65	8.33	22.14
Kalisz	5.91	2.51	5.40
Kasprowy Wierch Mt.	3.41	3.12	2.89
Kielce	19.67	2.64	11.12
Kłodzko	31.59	8.38	16.34
Kołobrzeg	3.98	0.95	4.92
Krakow	23.87	5.32	18.60
Legnica	10.35	2.82	4.53
Lesko	17.60	4.78	9.44
Łódź	6.72	1.06	4.87
Lublin	5.58	1.31	7.14
Nowy Sącz	52.92	18.28	38.62
Opole	9.24	3.38	6.34
Poznań	6.79	1.55	5.95
Rzeszów	12.94	3.12	10.90
Siedlce	7.90	1.72	6.79
Słubice	17.35	2.16	11.73
Śnieżka Mt.	1.15	1.38	0.77
Tarnów	22.28	4.95	19.30
Warszawa	4.25	0.89	3.27
Wieluń	3.08	0.79	3.02

(continued)

Table 15.3 (continued)

Station	Morning	Noon	Evening
Włodawa	3.66	1.50	3.46
Wrocław	8.86	2.19	7.10
Zakopane	43.53	10.84	31.84
Zielona Góra	1.75	0.80	1.63
mean	11.26	2.98	8.33
minimum	0.52	0.21	0.72
maximum	52.92	18.28	38.62

noon (on average during 2.98% of observations). In the spatial distribution calms are the least frequent at seaside stations. At noon, they appear there in less than 1% of observations (0.21% at Ustka, 0.66% at Łeba, 0.82% at Hel, 0.95% at Kołobrzeg and 0.98% at Świnoujście); in the evening they occur in 0.72–4.92% of observations, and in the morning in 0.52–3.98% of observations. Most calms were observed at the stations in Nowy Sącz (52.92, 18.28 and 38.62% in the morning, at noon and in the evening, respectively), Zakopane (43.53, 10.84 and 31.84% respectively) and Kłodzko (31.59, 8.38 and 16.34% respectively). These values significantly exceed those at other stations, because the frequency of calms is strongly dependent on local topographic conditions. Parczewski (1974) and Niedźwiedź et al. (1985) noted their increased number in the valleys of the Sudetes and Carpathians. Parczewski (1974) estimated that the frequency of "atmospheric silence" (calms and winds at <2 ms^{-1}) in Zakopane exceeds 80%. Niedźwiedź et al. (1985) estimated the incidence of calms in Nowy Sącz and Zakopane in 1961–1970 at 48% and 47%, respectively.

Frequency of Wind in Different Speed Classes

The incidence of wind in different speed ranges was also studied. According to Parczewski's division (Parczewski 1960), four classes were taken into account: weak wind with a speed not exceeding 4 ms^{-1}, moderate wind with a speed of above 4 ms^{-1} but not exceeding 10 ms^{-1}, strong wind with a speed of above 10 ms^{-1} but not exceeding 20 ms^{-1} and very strong wind with a speed of above 20 ms^{-1} (Table 15.1). Wind frequencies in the first three speed ranges at individual observation stations are shown in Fig. 15.3. In Poland, a weak wind not exceeding speed 4 ms^{-1} is the most frequent. Its average frequency in Poland is 59% at noon, and in the morning and evening 65.5 and 68.8%, respectively. It appears least frequently on Śnieżka Mt. (13.5% in the morning, 22.1% at noon and 12.5% in the evening), and then on Kasprowy Wierch Mt. (32.2%, 39.5% and 31.1%, respectively). At other stations, the frequency of low winds is much higher and ranges from 41.7% to 83.9% in the morning, from 38.0% to 83.6% at noon and from 52.6% to 84.4% in the evening. Moderate winds with speeds of 4 to 10 ms^{-1} are much rarer. Their average frequency

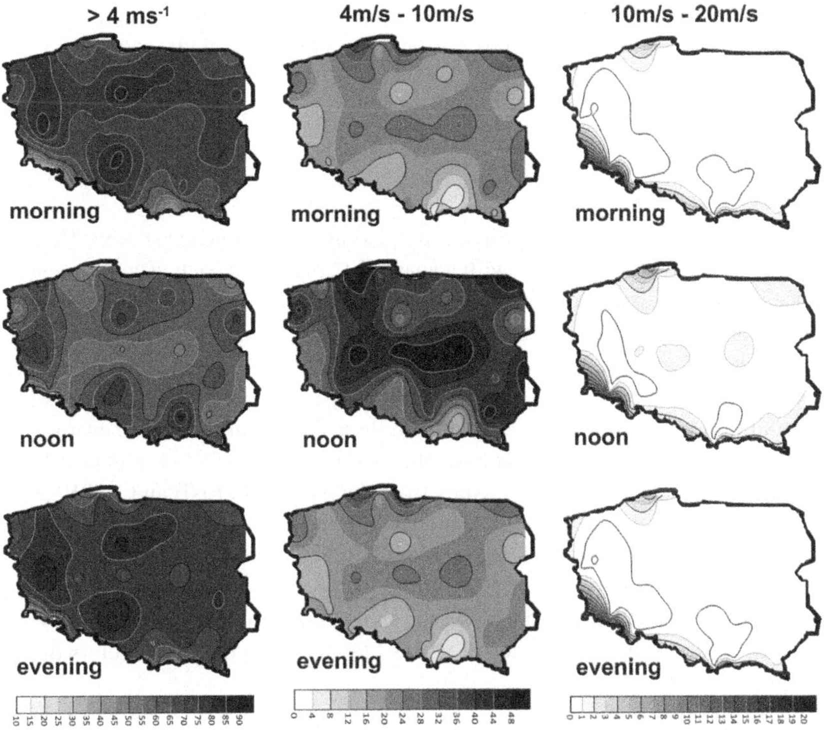

Fig. 15.3 Frequencies of wind in classes defined by Parczewski (1960) in the morning, at noon and in the evening observations

in Poland is 35.5% at noon, 20.8% in the morning and only 20.4% in the evening. Most often, they appear on Kasprowy Wierch Mt. (45.4% in the morning, 40.4% at noon and 46.7% in the evening), and then on Śnieżka Mt. (29.7%, 30.2% and 30.5%, respectively). At other stations, the frequency of moderate winds is much lower and ranges in the morning from 3.0% in Zakopane to 38.4% in Ustka, at noon from 6.7% in Zakopane to 53.9% in Łeba and in the evening 3.0% in Zakopane to 35.9% at Hel. Such low values in Zakopane result from a very high frequency of calms and probably from the incorrect location of the anemometer. Strong winds with speeds of 10 to 20 ms^{-1} are very rare. Their average frequency in Poland is 2.0% in the morning, 2.2% at noon and 2.1% in the evening. Therefore, it is difficult to talk about the daily cycle of their occurrence. Most often they appear on Śnieżka Mt. (39.6% in the morning, 34.6% at noon and 35.5% in the evening). Besides Śnieżka Mt. a 10% frequency of strong winds is only exceeded by the station on Kasprowy Wierch Mt., where strong wind appears in 18.4% of cases in the morning, 16.2% at noon and 18.4% in the evening. At other stations, the frequency of strong winds varies from 0.1 to 7.4%. Very strong winds are extremely rare in Poland. Only on Śnieżka Mt. there frequency is higher than 10% (15.4% in the morning, 11.7% at noon and

14.7% in the evening). At other stations, their frequency does not exceed 1%, and at three-fourths of stations such winds have not been observed at all.

Wind Direction

Wind frequency from individual directions was calculated for each station. The wind rose was divided into 16 sectors with a span of 22.5 degrees each. The frequency of silence in most cases did not exceed a few percent, so the average wind frequency of each sector is just over 6%. Figure 15.4 presents the frequency distribution of wind from individual sectors at all analyzed stations. In Poland, wind from the western sector is the most frequent. The median frequency from this direction exceeds 11%, the maximum value is close to 17%, which is three times higher than the average. For at least half of the stations, winds from the WSW, S, E and SW sectors also have a frequency of over 6%. In Poland, winds from the NNE, NE, ENE and NNW sectors are the least frequent. Out of 41 analyzed stations, at 31 wind most often blows from the W sector, on four from the S sector. On two stations wind most often blows from WNW and at the next two from SW. There is one station where wind most often blows from sector N and another with the most common wind from sector WSW. Figure 15.5 shows the frequencies of winds blowing from different sectors divided into speed classes.

Extreme Winds

The long-term average of the highest annual wind speed oscillated in the examined period between 8.3 ms^{-1} in Tarnów and 42.5 ms^{-1} on Śnieżka Mt. (Fig. 15.6). At only two stations, the long-term average of the highest annual wind speed was lower than 10 ms^{-1}, i.e., in Tarnów and Białystok. At the same time, at only seven stations the speed exceeded 15 ms^{-1}. It was in Łeba, Suwałki, Świnoujście, Ustka, Bielsko-Biała, on Kasprowy Wierch Mt. and Śnieżka Mt.. The maximum wind speed recorded throughout the period considered varied from 13 ms^{-1} in Zielona Góra and Tarnów, to 60 ms^{-1} on Śnieżka Mt. The value 15 ms^{-1} was not exceeded at only three stations during the period considered: in Białystok, Tarnów and Zielona Góra. At eight stations the maximum wind speed exceeded 25 ms^{-1}. It was in Łeba, Słubice, Kalisz, Lesko, Jelenia Góra, Bielsko Biała, Kasprowy Wierch Mt. and Śnieżka Mt..

On the seasonal scale, the lowest maximum wind speed was recorded in summer (Fig. 15.7). The long-term average of the highest annual wind speed oscillated in the examined period in summer between 5.7 ms^{-1} in Zakopane and 30.3 ms^{-1} on Śnieżka Mt. At 22 stations, i.e. over 50% of all stations, the long-term average of the highest summer wind speed was lower than 10 ms^{-1}. At the same time, this speed exceeded 15 ms^{-1} only at two stations. It was on Kasprowy Wierch Mt. and Śnieżka Mt. The maximum wind speed recorded throughout the studied period varied from

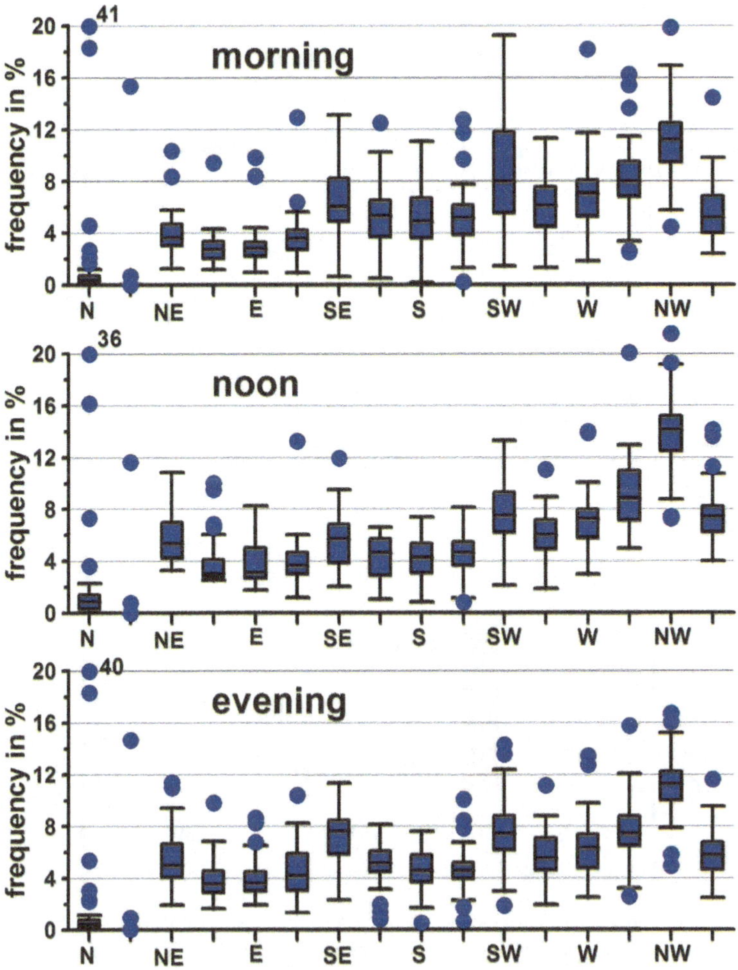

Fig. 15.4 Box and whiskers graphs showing the distribution of frequencies of winds from 16 main directions in the period 1966–2018

9 ms^{-1} in Tarnów to 45 ms^{-1} on Śnieżka Mt. (Fig. 15.8). The value 15 ms^{-1} was not exceeded at 28 stations during the period under consideration. At three stations, the maximum wind speed exceeded 25 ms^{-1}. It was in Słubice, Kasprowy Wierch Mt. and Śnieżka Mt.

In the remaining seasons, the maximum wind speeds were similar. In the winter, the long-term average of the highest annual wind speed fluctuated between 7.5 ms^{-1} in Tarnów and 40.0 ms^{-1} on Śnieżka Mt. At seven stations, the long-term average of the highest annual wind speed was below 10 ms^{-1}. It was in Olsztyn, Białystok, Zielona Góra, Toruń, Tarnów, Nowy Sącz and Zakopane. At the same time, at four stations: in Łeba, Bielsko Biała and on Kasprowy Wierch Mt. and Śnieżka Mt., this

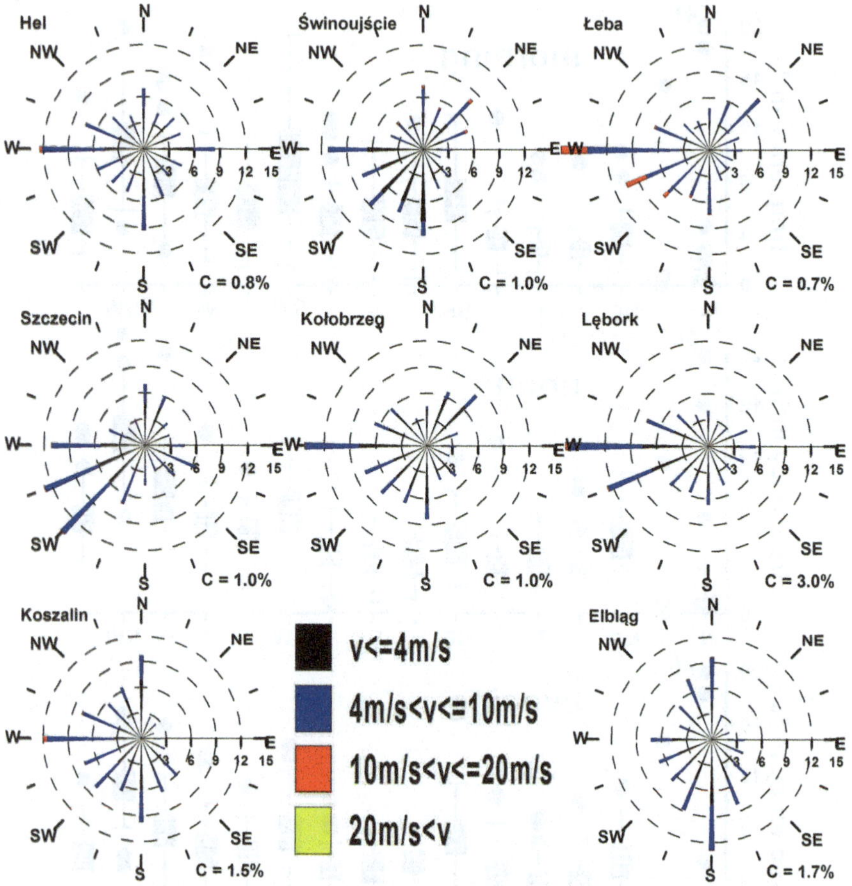

Fig. 15.5 The frequencies of winds from different sectors divided into speed classes at the noon. Percentage of calms is given in the low right corner of each graph

speed exceeded 15 ms^{-1}. The maximum wind speed recorded throughout the period varied from 13 ms^{-1} in Zielona Góra and Tarnów to 50 ms^{-1} on Śnieżka Mt. At five stations, in Zielona Góra, Tarnów, Kołobrzeg, Białystok and Toruń, the value 15 ms^{-1} was not exceeded. At six stations, the maximum wind speed exceeded 25 ms^{-1}. It was in Łeba, Słubice, Bielsko Biała, Jelenia Góra, Kasprowy Wierch Mt. and Śnieżka Mt.

In the spring, the long-term average of the highest annual wind speed oscillated in the period between 7.5 ms^{-1} in Tarnów and 37.0 ms^{-1} on Śnieżka Mt. At 10 stations, the long-term average of the highest annual wind speed was below 10 ms^{-1}: in Kołobrzeg, Białystok, Zielona Góra, Gorzów Wielkopolski, Toruń, Siedlce, Wieluń, Tarnów, Nowy Sącz and Zakopane. At the same time, at three stations: in Łeba, on Kasprowy Wierch Mt. and Śnieżka Mt., this speed exceeded 15 ms^{-1}. The maximum wind speed recorded throughout the period varied from 11 ms^{-1} in Zielona Góra to

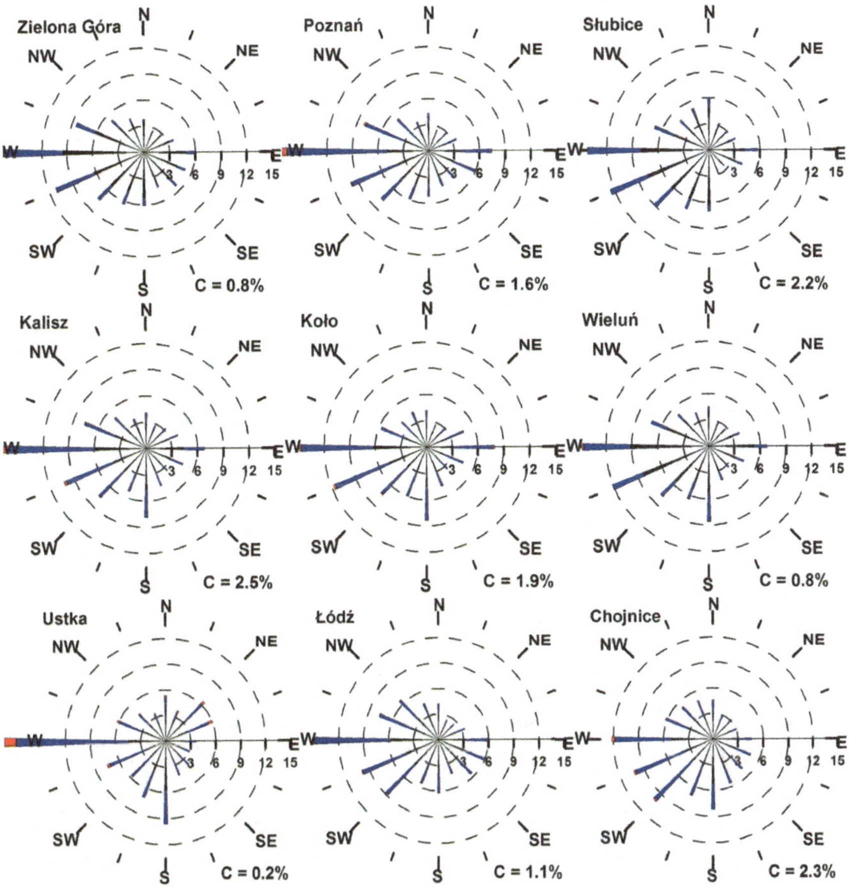

Fig. 15.5 (continued)

60 ms^{-1} on Śnieżka Mt.. At four stations, in Białystok, Zielona Góra, Nowy Sącz and Tarnów, the value did not exceeded 15 ms^{-1}. The maximum wind speed exceeded 25 ms^{-1} only on Kasprowy Wierch Mt. and Śnieżka Mt.

In the autumn, the long-term average of the highest annual wind speed oscillated between 7.2 ms^{-1} in Tarnów and 36.0 ms^{-1} on Śnieżka Mt.. At 11 stations, the long-term average of the highest annual wind speed was below 10 ms^{-1}. Those were Kołobrzeg, Olsztyn, Białystok, Zielona Góra, Gorzów Wielkopolski, Wieluń, Opole, Kielce, Tarnów, Nowy Sącz and Zakopane. At the same time, at five stations: in Łeba, Ustka, Bielsko Biała, on Kasprowy Wierch Mt. and Śnieżka Mt., the speed exceeded 15 ms^{-1}. The maximum wind speed recorded throughout the period varied from 13 ms^{-1} in Tarnów to 50 ms^{-1} on Śnieżka Mt.. At six stations, in Olsztyn, Białystok, Zielona Góra, Kielce, Rzeszów and Tarnów, the value did not exceeded 15 ms^{-1}.

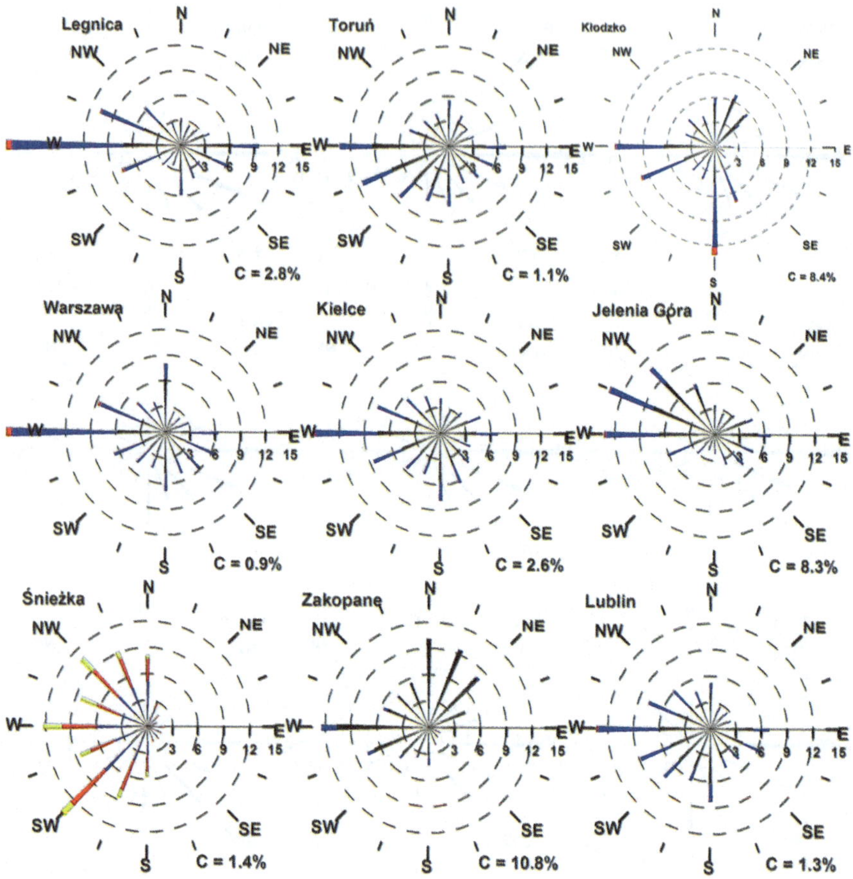

Fig. 15.5 (continued)

The maximum wind speed exceeded 25 ms^{-1} at five stations: in Słubice, Bielsko Biała, Lesko, on Kasprowy Wierch Mt. and Śnieżka Mt.

The geographical location of Poland favors the inflow of air from various directions. The frequency of strong winds is conditioned by the pressure situation, i.e. also the direction of the air mass inflow. In this part of the work, 8 ms^{-1} was adopted as the lower limit of strong wind speed. This is a slightly lower value than 10 ms^{-1} proposed by Parczewski (1960). It was lowered compared to previous considerations, because in a large area of Poland the frequency of winds with a speed exceeding 10 ms^{-1} was rare, and in Gorzów Wielkopolski they did not occur at all. The occurrence of strong winds was analyzed in all three observation times together, therefore the maximum number of such occurrences during a year could reach 1095 (1098 in a leap year). Of course, this value did not occur at any station; however, on Śnieżka Mt., there were 728 cases per year on average, which means that on average wind above 8 ms^{-1} was recorded during two of the three observations per day. At the station on Kasprowy

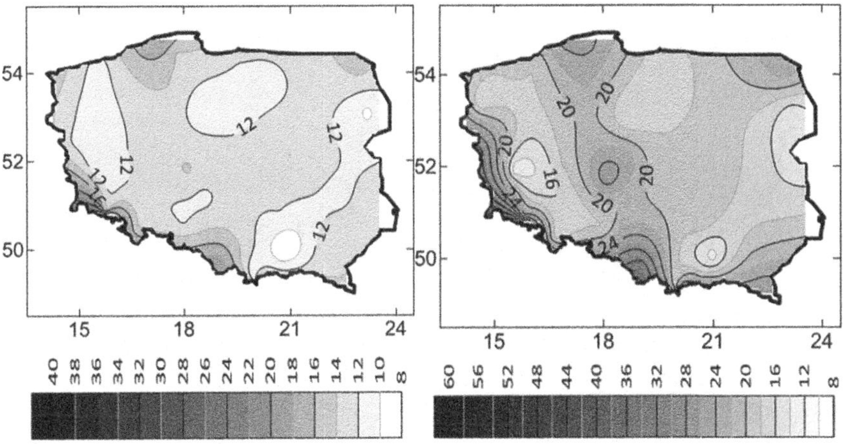

Fig. 15.6 The long-term average (on the left) and maximum (on the right) of yearly maximum wind speed in Poland in the period 1966–2018 in ms^{-1}

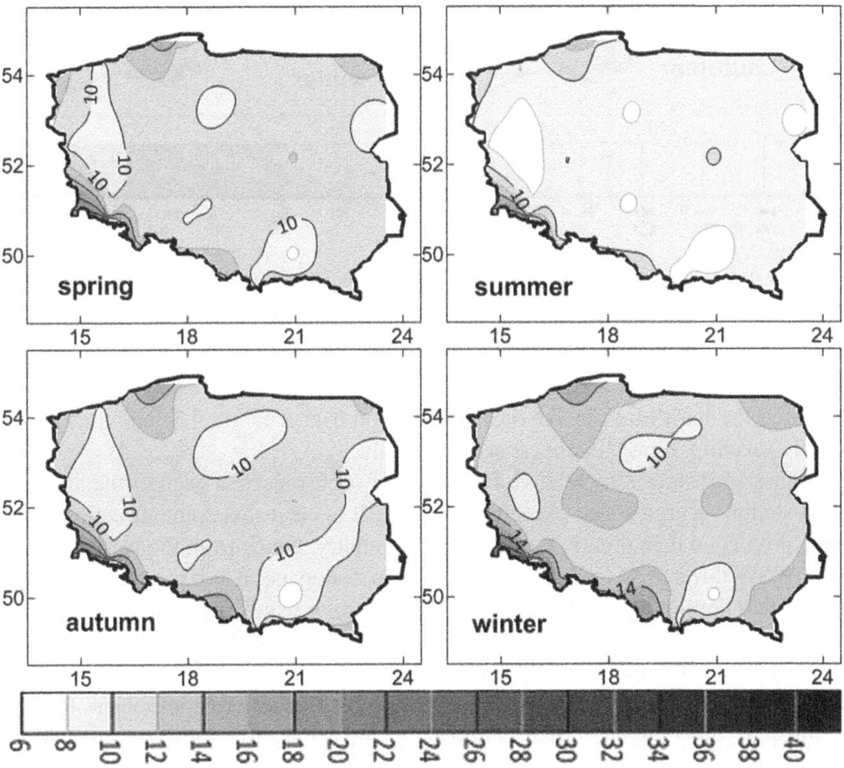

Fig. 15.7 The long-term average of seasonal maximum wind speed in Poland in the period 1966–2018 in ms^{-1}

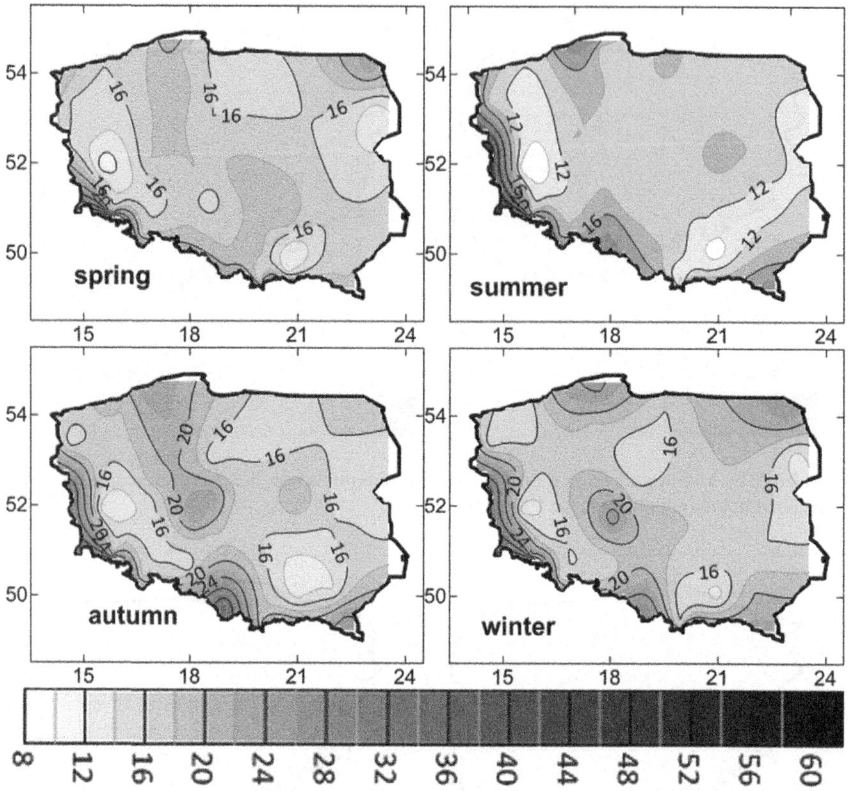

Fig. 15.8 The absolute maximum of seasonal maximum wind speed in Poland in the period 1966–2018 in ms^{-1}

Wierch Mt., second in the row, there were only 390 such cases in a year. Strong winds appeared the least often in Zakopane, where on average only 4 wind cases with a speed exceeding 8 ms^{-1} were reported annually.

Figure 15.9 shows the relative frequency of occurrences at each of the analyzed wind stations with a speed exceeding 8 ms^{-1}. It is clear that some directions were more privileged than others. For most of the country, winds from the western sector from WSW through W to WNW dominated. In Zakopane, Kłodzko and Lesko there was a clear dominance of the southern sector. Strong winds from this sector constitute over 30% of all cases of strong wind at these stations. On Śnieżka Mt., where the frequency of winds was highest, the distribution of directions was more even; such winds were recorded from all directions with the western component. In order to illustrate the frequency distribution of strong wind directions, all cases where the frequency of strong winds exceeded 10% were counted and taken on a graph (Fig. 15.10). In addition, colors indicate cases in which winds were more frequent than 20, 30 and 40% from a given direction. There is a clear dominance of three sectors: western W, where at 35 out of 41 stations the frequency of strong winds

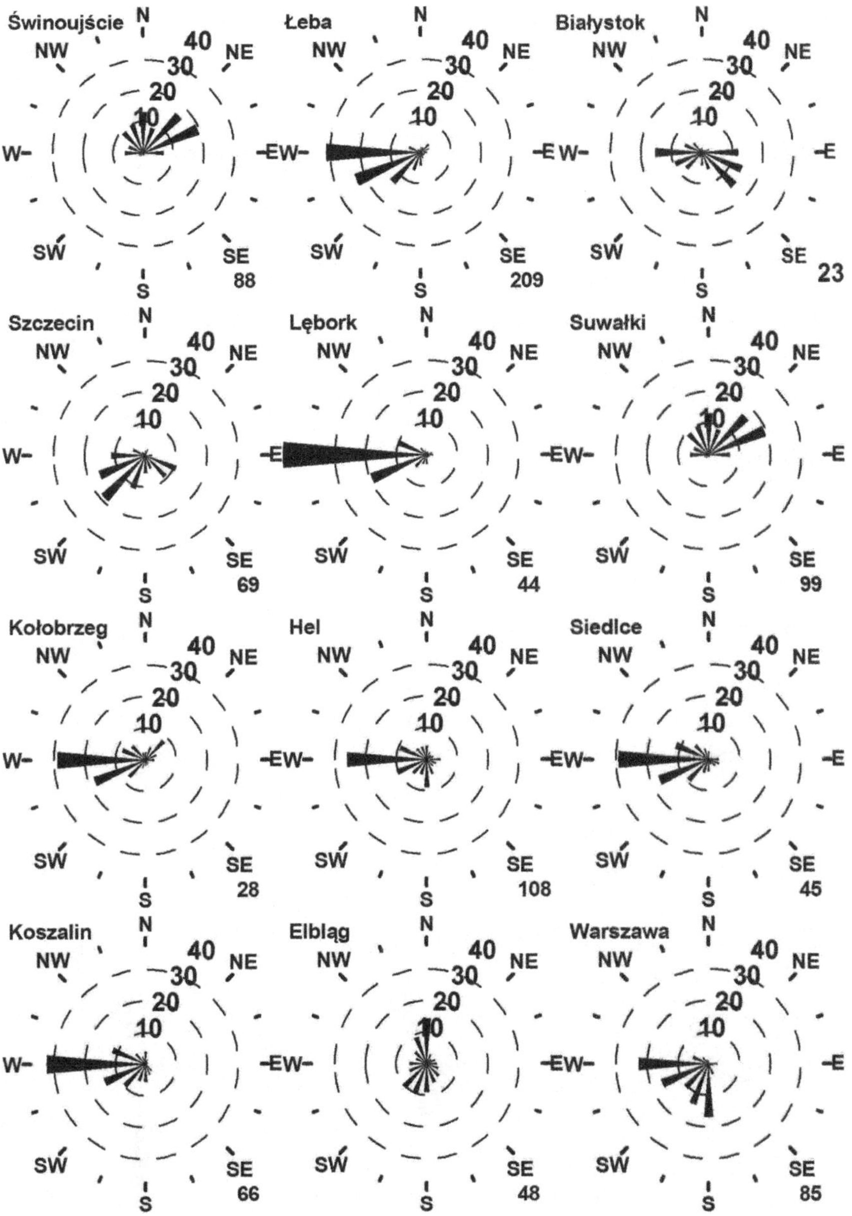

Fig. 15.9 The relative frequencies of strong winds (>8 ms^{-1}) in relation to their directions. The mean annual number of cases of such winds is given in the low right corner of each graph

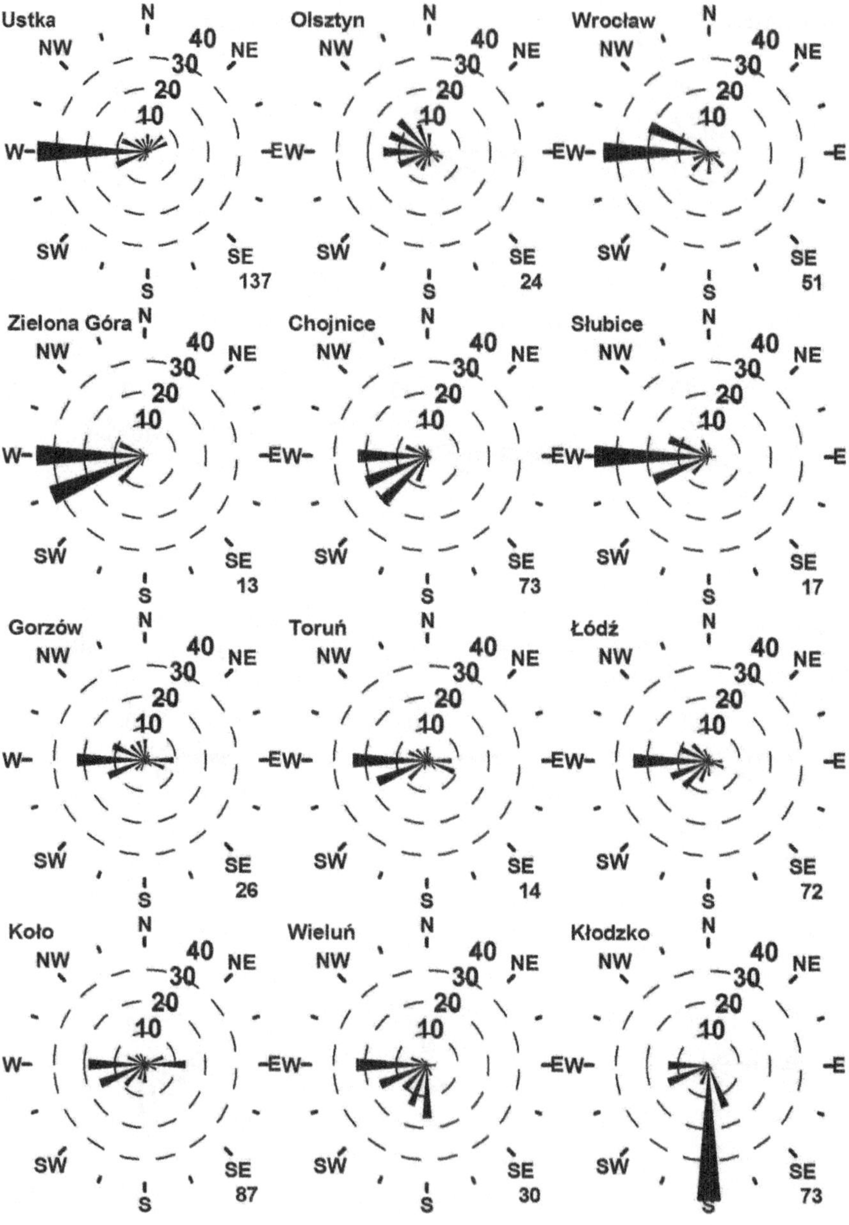

Fig. 15.9 (continued)

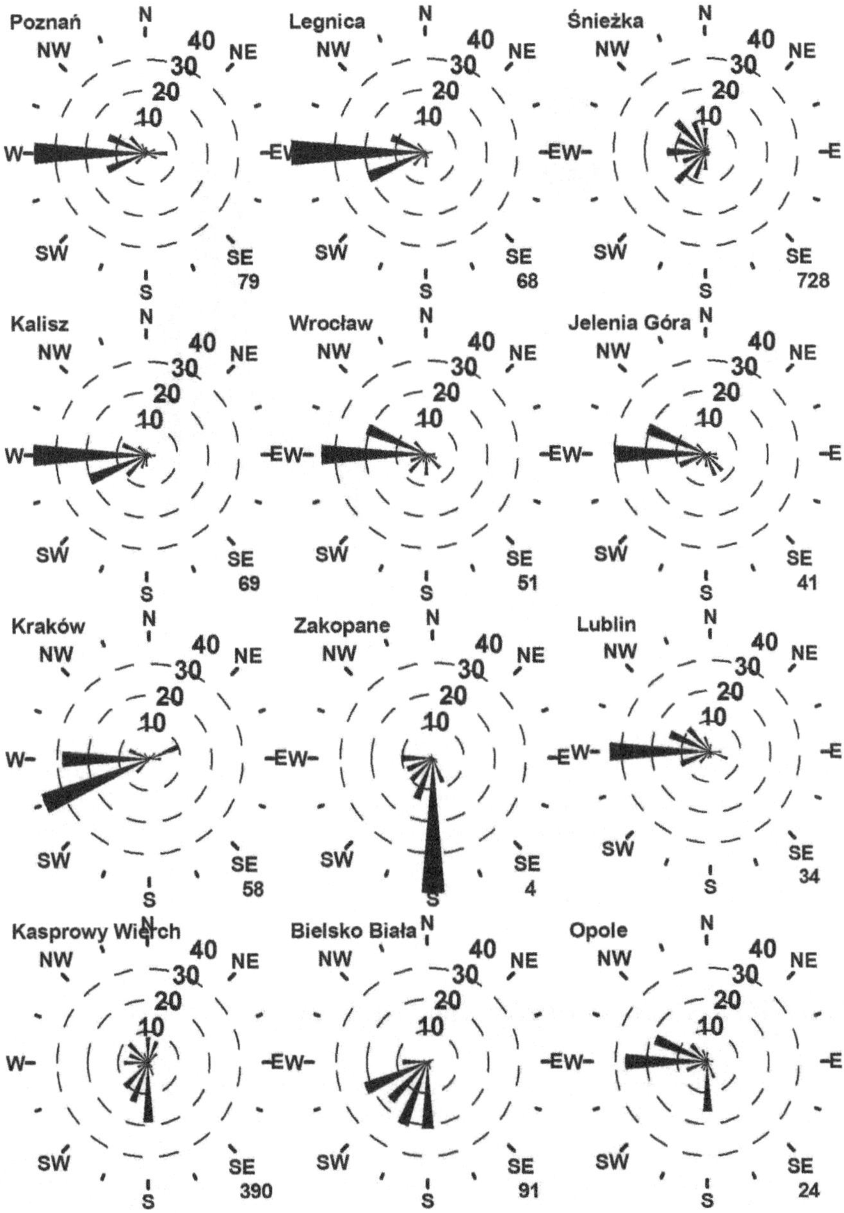

Fig. 15.9 (continued)

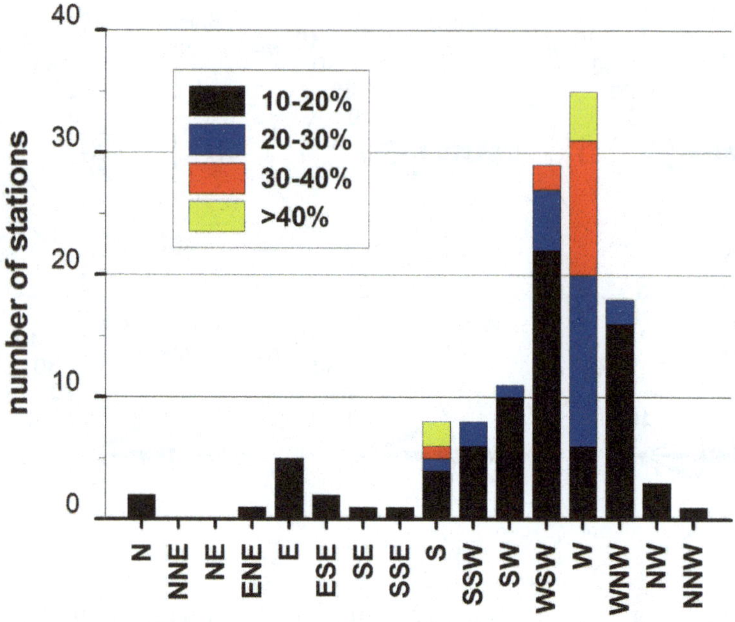

Fig. 15.10 Number of stations with at least 10% of strong wind cases from a given sector (>8 ms^{-1})

exceeded 10% (including 29 with more than 20%, 15 with above 30% and 4 with above 40%), WSW and WNW (at 29 and 18 stations, respectively). At no station did the frequency of strong winds exceed 10% for directions NNE and NE. The southern direction also requires attention. Although only at 8 stations the frequency of strong winds from this direction exceeded 10%, but at three of them it was as much as 30%, and at two even more than 40%.

Long-Term Changes of Extreme Wind Speed

An important element of climate change analysis is the search for long-term changes. In this study, variability of the 75th and 95th percentiles and maximum annual and seasonal wind speed values were examined. The coefficients of the linear trend equation were determined using the least squares method, and analyzed using the Student's t-test at a significance level of $p \leq 0.05$.

For most of Poland, the 75th and 95th percentile values and maximum values are decreasing (Fig. 15.11). The decreases are small; they do not exceed 0.4 ms^{-1} per decade in the case of the 75th percentile, 0.7 ms^{-1} per decade in the case of the 95th percentile and 1.0 ms^{-1} per decade for maximum values in a year. The trends are statistically significant at 29, 32 and 23 stations out of 41 (75th, 95th percentile and maximum values in a year, respectively). Increases are recorded on the central coast

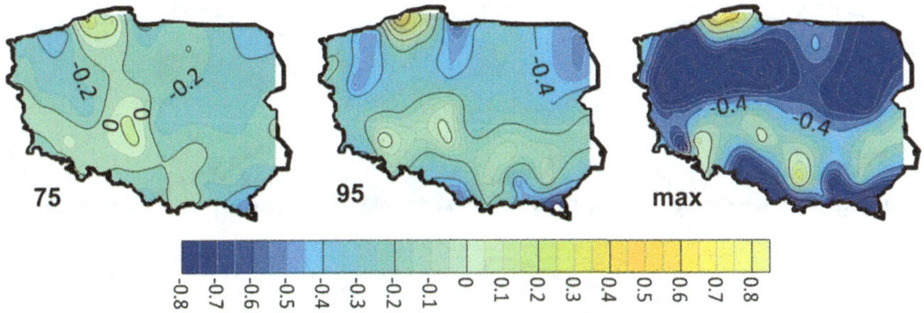

Fig. 15.11 The trend coefficients in ms^{-1} per decade of yearly 75th percentile, 95th percentile and yearly maximum wind speed in Poland in the period 1966–2018

(Łeba, Ustka, Lębork), at individual stations in the center (Wieluń, Kalisz) and in the south (Kraków), but they do not reach the level of statistical significance.

The trends are the weakest in summer as compared to other seasons (Fig. 15.12). The downward trends apply to almost the entire area of Poland; they are of the same size as those per year and do not exceed 0.4 ms^{-1} per decade in the case of the 75th percentile, 0.7 ms^{-1} per decade in the case of the 95th percentile and 1.0 ms^{-1} per decade for maximum values during summer. They are statistically significant at more than half of the analyzed stations. Increases are recorded on the coast and in Ustka, where they are statistically significant.

In the remaining seasons, the downward trends also dominate and are statistically significant in more than half of Poland (Fig. 15.12). The maximum values are of similar order to those in summer and throughout the year; however, the area covered by stronger trends is significantly larger. In all seasons, an increase in wind speed is observed on the central coast (Łeba, Ustka, Lębork). In all analyzed periods, the weakest trends occur in the belt extending longitudinally from Ustka through Kalisz to Opole, and in the belt extending latitudinally from Wrocław through Opole, Wieluń, Kielce, Kraków and Rzeszów.

Decreasing trends in annual and seasonal series of wind speed have also been observed in other countries in Europe. This surprising result has also been presented in a number of papers from Europe. Brazdil et al. (2017) and Zahradníček et al. (2019) show it for the Czech Republic; Péliné Németh et al. (2011) for Hungary; Azorin-Molina et al. (2014, 2016, 2017) for Portugal and Spain; Romanić et al. (2015) for the Balkans; Minola et al. (2016) for Sweden; Laapas and Venäläinen (2017) for Finland and Kohler et al. (2018) for Germany. Worldwide wind-stilling has been described by Roderick et al. (2007) and McVicar et al. (2012). Many authors see the reason for the increase in surface roughness. This thesis can be confirmed by the fact that, simultaneously, based on NCEP/NCAR data on zonal and meridional wind velocity at a height of 10 m above ground level from 35 grids from the period 1951–2005, Central Europe has witnessed increases in wind speed, as well as in number of days with strong wind (>8 ms^{-1}), both statistically significant (Araźny et al. 2007).

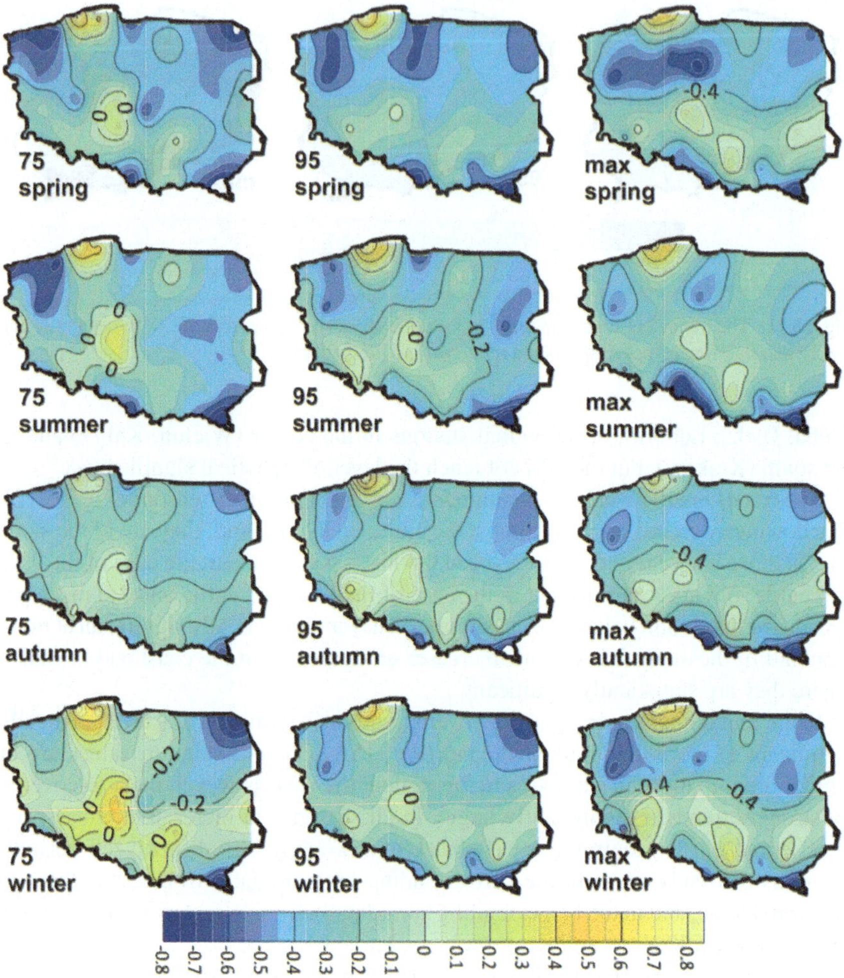

Fig. 15.12 The trend in ms^{-1} per decade of seasonal 75th percentile, 95th percentile and seasonal maximum wind speed in Poland in the period 1966–2018

Long-Term Changes of Wind Direction

Wind is characterized by two variables: speed and direction. In order to examine how the wind direction changed in the analyzed period, the frequency of wind appearing from each of 16 sectors and the frequency of calms were calculated in each year and season on each station. The number of wind observations in a given sector was divided by the number of all observations in the season (or year, as appropriate). The least squares regression coefficients of these frequency series were then calculated. Time was an independent variable. The statistical significance of trend coefficients was

Table 15.4 Results of trend analysis in frequencies of wind from different sectors: number of stations with statistically significant increase in frequency/number of stations with statistically significant decrease in frequency

Wind direction	Winter	Spring	Summer	Autumn	Year
N	7/4	**2/14**	6/5	1/4	6/9
NNE	**11/0**	6/6	**15/1**	5/1	**13/2**
NE	**12/0**	8/7	**11/0**	4/3	**21/2**
ENE	4/1	6/5	8/0	5/2	**13/1**
E	1/6	**4/12**	3/4	3/5	4/8
ESE	2/9	3/4	7/4	2/2	**12/4**
SE	4/9	11/7	5/2	1/1	**21/5**
SSE	5/7	11/5	9/2	**11/2**	**20/2**
S	**0/10**	4/11	3/6	0/3	7/14
SSW	**16/0**	10/9	8/5	5/1	**14/2**
SW	**24/2**	12/7	9/6	8/1	**18/4**
WSW	**18/1**	12/8	5/3	**10/2**	**15/6**
W	1/3	7/10	**2/16**	2/5	**1/18**
WNW	4/2	**14/4**	6/9	5/1	**15/6**
NW	5/1	**12/4**	5/4	5/0	**14/4**
NNW	6/1	6/8	4/1	3/0	**15/1**
Calms	**0/30**	**1/19**	**1/24**	**1/28**	**0/31**

analyzed using the Student's *t*-test, with $p \leq 0.05$ considered significant. Table 15.4 presents the number of statistically significant cases of wind increase/decrease in individual sectors. It turns out that a significant number of significant changes were observed. Amazingly, there are more increases. This is a little surprising, because the number of all observations in each season is constant. It turns out that these increases are the result of a widespread fall in calms in all seasons. However, a cursory analysis of these decreases, presented on the example of the station in Kalisz in Fig. 15.13, indicates that this is the result of multiple changes of instruments, rather than changes in climatic conditions. The sensitivity of subsequent instruments increased, and the number of calms observed decreased with this increase.

The second surprising observation is the decrease in the frequency of types E, S and W, in favor of the frequency of sectors directly adjacent to them. It seems to be a psychological effect. The observer observed the wind meter indicator fluctuations for 2 or 10 min, and then rated the average direction "*by eye*," subconsciously more often choosing the main direction, rather than secondary ones. Currently, wind directions are averaged automatically; hence, the greater balance between wind frequencies from neighboring directions.

In this light it is difficult to talk about clear trends of changes in the surface wind direction. However, it is worth realizing how subjective assessment can affect observation results.

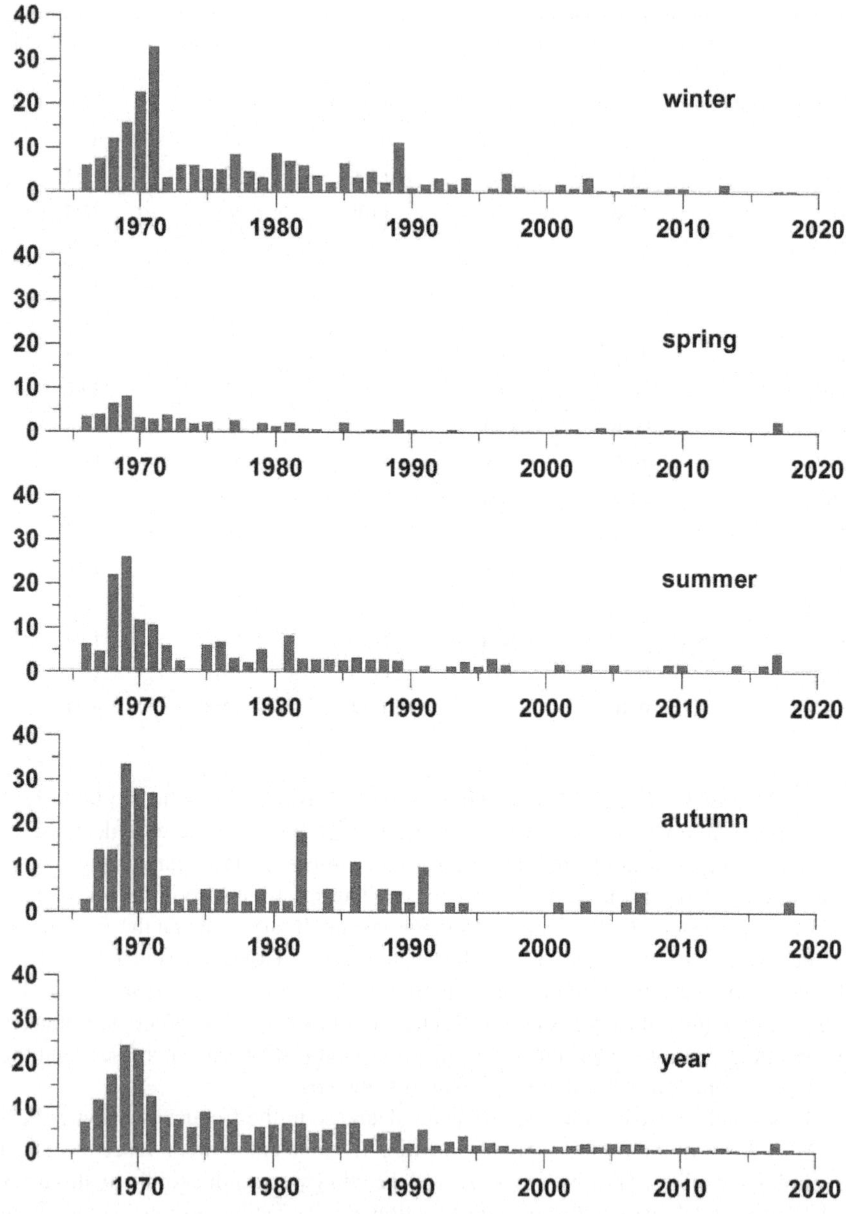

Fig. 15.13 Long-term course of relative frequencies (%) of calms at Kalisz

Summary

The wind was analyzed on the basis of speed and direction data from three observation times at 41 meteorological stations in Poland from 1966 to 2018. Due to a large change in the homogeneity of the series caused by the transition from Wild' anemometers to cup anemometers, the data from 1951 to 1965 were abandoned. The average annual and seasonal wind speeds were calculated taking into account three observation times. The incidence of atmospheric calms was investigated. Wind frequencies from different directions were determined, taking into account the 16-sector wind rose. Particular attention was paid to winds exceeding 8 ms^{-1}, the frequency of their occurrence and directions of air masses inflow during strong winds were examined. The seasonal and annual trends of strong winds were counted.

The average wind speed in Poland was 3.6 ms^{-1} and ranged from 1.4 ms^{-1} in Zakopane to 12 ms^{-1} on Śnieżka Mt. In the annual cycle the average wind speed at noon was the highest in spring and lowest in summer. In the case of the morning and evening observations, the highest wind speed was observed in winter and the lowest in summer. In the daily scale, the highest wind appeared at noon. There was no significant differences between the average wind speed at the morning and evening observations.

Windless weather was relatively rare in Poland. It appeared the most often in the morning (12.3% of observations), less often in the evening (8.3%) and the least often at noon (3.0%). They were extremely rare at the coast (less than 1% of all observations), and appeared the most often at stations located in mountain valleys: Zakopane, Kłodzko and Nowy Sącz.

During the analyzed period, winds from the western (W) sector were the most often observed besides those from the west southwestern (WSW) and southern (S) sectors. The least frequent wind in Poland was from the north north-east (NNE) and north-east (NE) directions.

The long-term average of the highest annual wind speed oscillated between 8.3 ms^{-1} in Tarnów and 42.5 ms^{-1} on Śnieżka Mt. On the seasonal scale, the lowest maximum wind speeds were recorded in summer and the highest in spring. Strong winds were most often recorded from the west (W), west southwest (WSW) and west northwest (WNW) directions. At several stations in southern Poland it was also the southern (S) direction.

The decreasing trends in annual and seasonal series of wind speed were observed in Poland as well as in other European countries. However, a number of authors attribute this trend to an increase in surface roughness. This thesis can be confirmed by the fact that simultaneously, based on NCEP/NCAR data on zonal and meridional wind velocity at a height of 10 m above ground level from 35 grids from the period 1951–2005, Central Europe witnessed increases in wind speed as well as in the number of days with strong wind (>8 ms^{-1}), both statistically significant (Araźny et al. 2007).

References

Adamczyk AB (1996) Charakterystyka wiatrów silnych i bardzo silnych w Polsce. Zesz IGiPZ PAN 37:5–42

Araźny A, Przybylak R, Vizi Z et al (2007) Mean and Extreme wind velocities in Central Europe 1951–2005. Geogr Pol 80(2):69–78

Azorin-Molina C, Guijarro JA, McVicar TR et al (2016) Trends of daily peak wind gusts in Spain and Portugal, 1961–2014. J Geophys Res 121:1059–1078. https://doi.org/10.1002/2015JD024485

Azorin-Molina C, Vicente-Serrano SM, McVicar TR et al (2014) Homogenization and assessment of observed near-surface wind speed trends over Spain and Portugal, 1961–2011. J Clim 27:3692–3712. https://doi.org/10.1175/JCLI-D-13-00652.1

Azorin-Molina C, Vicente-Serrano SM, McVicar TR et al (2017) Assessing the impact of measurement time interval when calculating wind speed means and trends under the stilling phenomenon. Int J Climatol 37:480–492. https://doi.org/10.1002/joc.4720

Bartnicki L (1930) Prądy powietrzne dolne w Polsce. Prace Geofiz 3(3):2–98

Brázdil R, Valík A, Zahradníček P, Řezníčková L et al (2017) Wind-stilling in the light of wind speed measurements: the Czech experience. Clim Res 75:131–143

Grabowski J (1996) Analiza warunków anemometrycznych i zachmurzenia w wybranych miejscowościach Pojezierza Warmińsko-Mazurskiego w latach 1971–1990. Acta Academiae Agriculturae Ac Technicae Olstenensis. Agricultura 64:3–12

Gumiński R (1952) Rozkład kierunków i prędkości wiatru na niektórych stacjach meteorologicznych Polski. Wiad Sł Hydr i Met t 3(2a):45–64

Hess M (1965) Piętra klimatyczne w polskich Karpatach Zachodnich. Zesz Nauk UJ, Prace Geogr 11:1–258

Kohler M, Metzger J, Kalthoff N (2018) Trends in temperature and windspeed from 40 years of observations at a 200-m high meteorological tower in Southwest Germany. Int J Climatol 38:23–34. https://doi.org/10.1002/joc.5157

Kolendowicz L, Taszarek M, Czernecki B (2016) Convective and non-convective wind gusts in Poland, 2001–2015. Met Hydr Water Manag 4(2):15–21

Kowalewski M (2005) Cisze atmosferyczne w Polsce. In: Bogdanowicz E, Kossowska Cezak U, Szkutnicki J (eds) Ekstremalne zdarzenia hydrologiczne i meteorologiczne. IMGW, Warszawa, pp 366–370

Koźmiński C (ed) (1990) Climate atlas of elements and phenomena harmful to agriculture (Atlas klimatyczny elementów i zjawisk szkodliwych dla rolnictwawPolsce). IUNG, Puławy

Kożuchowski K (1977) Piętrowy układ stosunków anemologicznych w Tatrach. Acta Univ Lodz Ser 2: Folia Geogr 5:3–11

Krawczyk B (1994) Średnia liczba dni z wiatrem o prędkości v > 8 ms^{-1}. In: Atlas zasobów, walorów i zagrożeń środowiska geograficznego Polski. IGiPZ PAN, Warszawa

Kwiatkowski J (1972) Feny w Kotlinie Jeleniogórskiej. Acta UnivWratisl 173:3–46

Kwiatkowski J (1975) Zasięg fenów sudeckich i ich wpływ na mezoklimat regionów południowo-zachodniej i środkowej Polski. Prz Geofiz 20:15–30

Kwiatkowski J (1979) Zjawiska fenowe w Sudetach i na przedpolu Sudetów. Probl Zagosp Ziem Gór 20:243–280

Laapas M, Venäläinen A (2017) Homogenization and trend analysis of monthly mean and maximum wind speed time series in Finland, 1959–2015. Int J Climatol 37:4803–4813. https://doi.org/10.1002/joc.5124

Lewińska J (1958) Wiatry ryterskie i rymanowskie. Prz Geofiz 11(1):17–26

Lorenc H (1968) O porywistości wiatru w Warszawie. Wiad Służby Hydr Met 3–4

Lorenc H (1992) Zasoby wiatru w Polsce. Mat Bad IMGW Meteorologia 18, IMGW, Warszawa

Lorenc H (1996) Struktura i zasoby energetyczne wiatru w Polsce. Mat Bad IMGW Meteorologia 25

Lorenc H (2012) Maksymalne prędkości wiatru w Polsce. IMGW PIB, Warszawa

Malicki A, Michna E (1966) O występowaniu wiatrów halnych w Bieszczadach Zachodnich. Ann UMCS, Ser B 21:133–142
McVicar TR, Roderick ML, Donohue RJ et al (2012) Global review and synthesis of trends in observed terrestrial near-surface wind speeds: implications for evaporation. J Hydrol 416–417:182–205
Milata W (1936) Wiatry halne w Karpatach. Wiad Geogr 14(5–7):53–55
Minola L, Azorin-Molina C, Chen DL (2016) Homogenization and assessment of observed near-surface wind speed trends across Sweden, 1956–2013. J Climate 29:7397–7415. https://doi.org/10.1175/JCLI-D-15-0636.1
Niedżwiedż T, Orlicz M, Orliczowa J (1985) Wiatr w Karpatach Polskich. Zesz IGiPZ PAN 6
Niedźwiedź T, Paszyński J, Czekierda D (1994) Wiatry. In: Atlas Rzeczypospolitej Polskiej. IGiPZ PAN, Warszawa
Otruba J, Wiszniewski W (1974) Vietor. Klima Tatier. SAV, Bratislava, pp 233–245
Parczewski W (1960) Klasyfykacja przedziałów predkości wiatru w zastosowaniu do opracowań klimatycznych i bioklimatycznych. Prz Geofiz 5(2):117–122
Parczewski W (1974) Całodobowe cisze atmosferyczne i słabe wiatry dolne w Polsce. Prace IMGW 1:9–38
Paszyński J, Niedźwiedź T (1991) Klimat. In: Starkel L (ed) Geografia Polski środowisko przyrodnicze. PWN, Warszawa
Péliné Németh C, Radics K, Bartholy J (2011) Seasonal variability of Hungarian wind climate. Acta Silvatica Lignaria Hungarica 7:39–48
Piasecki D (1952) Wiatry o maksymalnych prędkościach na obszarze Polski w latach 1928–1938. Wiad Sł Hydr i Met 3(2a):65–101
Polish Climate Atlas (Atlas klimatyczny Polski) (1971) Część tabelaryczna, z. 1, Wiatr, PIHM, Warszawa
Polish Climate Atlas (Atlas klimatyczny Polski) (1973) red. W. Wiszniewski, IMGW, PPWK, Warszawa
Pożarska K, Grabowski J (2015) Zmienność warunków anemometrycznych w województwie warmińsko-mazurskim w latach 1991–2010. Woda-Środowisko-Obszary Wiejski 15, 4(52):29–38
Roderick ML, Rotstayn LD, Farquhar GD et al (2007) On the attribution of changing pan evaporation. Geophys Res Lett 34:L17403. https://doi.org/10.1029/2007GL031166
Romanić D, Curić M, Jovičić I et al (2015) Long-term trends of the 'Koshava' wind during the period 1949–2010. Int J Climatol 35:288–302. https://doi.org/10.1002/joc.3981
Sokołowski M (1927) Wiatry w Tatrach. Wierch Mt. 5:36–41
Stachlewski W (1974) Wiatry fenowe w wybranych miejscowosciach wzdłuż profilu Kasprowy Wierch Mt. – Kraków. Prz Geofiz 27(1):81–94
Stopa-Boryczka M (ed) (1989) Atlas współzależności parametrów meteorologicznych i geograficznych w Polsce, 5. Wyd. UW, Warszawa
Taranowska S (1957) Występowanie wiatrów na wybrzeżu. Wiad Sł Hydr Met 4(1):3–10
Tarnowska K (2011) Wiatry silne na polskim wybrzeżu Morza Bałtyckiego. Prace i Studia Geograficzne 47:197–204
Trzeciak S (2001) Wiatry sztormowe na polskim wybrzeżu Bałtyku. Wyższa Szkoła Morska w Szczecinie, Szczecin
Ustrnul Z (1992) Potencjalne warunki występowania wiatrów fenowych w Karpatach Polskich. Zesz Nauk UJ, Prace Geogr 90:97–112
Wierczek E (1958) Częstotliwość występowania poszczególnych kierunków wiatrów dolnych w Karpatach zachodnich (na obszarze województwa krakowskiego). Prz Geofiz 11(3–4):229–234
Wierzbicki Z (1968) Rozkład prędkości wiatrów w Polsce na wysokości 10 i 25 m nad gruntem. Prace PIHM 93
Wierzbicki Z, Bartkowski Z (1970) Charakterystyka klimatologiczna pola dolnych wiatrów w Polsce. PIHM, Wyd Kom i Łączn, Warszawa

Zahradníček P, Brázdil R, Štěpánek P et al (2019) Differences in wind speeds according to measured and homogenized series in the Czech Republic, 1961–2015. Int J Climatol 39:235–250. https://doi.org/10.1002/joc.5800

Żurański JA, Jaśpińska B (1996) Directional analysis of extreme wind speeds in Poland. J Wind Eng Ind Aerod 65:13–20

Chapter 16
Change of Thunderstorms and Tornadoes

Zuzanna Bielec-Bąkowska, Mateusz Taszarek, and Leszek Kolendowicz

Abstract In this chapter, on the basis of meteorological observation series from 47 meteorological stations for the years 1951–2018 and data from the Scientific Station of the Climatology Department of the Jagiellonian University in Kraków for the years 1901–2018, the analysis of the long-term variability and annual course of the thunderstorms occurrence in Poland was undertaken. The results of research on the occurrence of tornadoes in Poland for the years 1810–2018 were also presented, and the data used in the analysis come from the European Severe Weather Database (ESWD) and historical sources. Additionally, long-term changes in meteorological conditions leading to the occurrence of thunderstorms and tornadoes were identified. For this purpose data from the ERA5 reanalysis were used. The results obtained revealed that the period of thunderstorm activity during the year became longer than at the beginning of the periods under consideration and shifts towards the first months of the year. It was also found that in the period 1951–2018 in the eastern part of the country can observe signals of an increase in the frequency of days with thunderstorms, while in the western part trends are rather downward. However, long-term variability of the thunderstorms occurrence during the period of over 100 years does not indicate a clear tendency of changes and only in the cool season the trend is increasing and statistically significant. It was also determine that tornadoes are really rare phenomena, but always occurring in Poland. Each year, on average 5 weak tornadoes, 1–2 significant tornadoes (up to F4 in Fujita scale) and 4 waterspouts are reported. At the same time, it has been shown that with the ongoing climate

Z. Bielec-Bąkowska (✉)
Faculty of Natural Sciences, Institute of Earth Sciences, University of Silesia in Katowice, Katowice, Poland
e-mail: zuzanna.bielec-bakowska@us.edu.pl

M. Taszarek · L. Kolendowicz
Faculty of Geographical and Geological Sciences, Institute of Physical Geography and Environmental Planning, Department of Meteorology and Climatology, Adam Mickiewicz University, Poznań, Poland
e-mail: mateusz.taszarek@amu.edu.pl

L. Kolendowicz
e-mail: leszko@amu.edu.pl

M. Falarz (ed.), *Climate Change in Poland*, Springer Climate,
https://doi.org/10.1007/978-3-030-70328-8_16

change, environmental conditions are becoming consistently more conducive for severe convective phenomena occurrence in Poland.
clearpage

Introduction

As in most regions of the world, thunderstorms and tornadoes are considered to be among the most violent meteorological phenomena occurring in Poland. Their occurrence is often accompanied by significant damage to the natural environment and human activities. This damage is caused mainly by thunderstorms associated with lightning, strong gusty winds, heavy rainfall, large hail and tornadoes. However, until the 1990s, there was little comprehensive research using long series of observations on the occurrence of thunderstorms and tornadoes in Poland. This gap was filled by studies conducted after 1990. They focused primarily on the temporal and spatial variability of thunderstorms (Kolendowicz 1997; Bielec-Bąkowska 2003, 2013; Taszarek et al. 2015) and tornadoes (Taszarek and Brooks 2015), as well as their synoptic conditions (Kolendowicz 1998, 2005; Bielec-Bąkowska 2002; Kolendowicz et al. 2017). At the same time, awareness of progressive climate change and related changes in the occurrence of extreme events increased (IPCC 2013). This awareness prompted more detailed analyses of the occurrence of thunderstorms and tornadoes. It was especially important to check whether, in the light of climate change, conditions conducive to the formation of convective phenomena are more common (Allen et al. 2014; Seeley and Romps 2015; Allen 2018) or whether the number of thunderstorms and tornadoes is changing over time (Groenemeijer and Kühne 2014; Finney et al. 2018; Taszarek et al. 2019). It should be noted that both phenomena significantly differ in their incidence, as well as in the sources of data providing information about their occurrence. For this reason, the analysis of temporal and spatial variability of the occurrence of tornadoes presented in this paper differs slightly from the climatological aspects of thunderstorms occurrence.

Data and Methods

The analysis of spatial and temporal variability of **thunderstorms** occurrence in Poland was based on meteorological observations from 47 synoptic stations, which are part of the meteorological station network of IMGW–PIB (Polish Institute of Meteorology and Water Management—National Research Institute). The data span the period from 1951 to 2018 (Table 3.3, Fig. 16.1). The data included information on the occurrence of thunderstorm on a given calendar day. If the thunderstorm occurred at the turn of days, it was included in both (however, such cases occurred sporadically). The research does not take into account the division of thunderstorms into

Fig. 16.1 Location of meteorological stations

close and distant events, or their duration. Particular attention was paid to thunderstorms occurring in the cool half of the year. In the case of several stations, a 2–4-year gap in observations was present, however, they usually concerned the beginning or end of the observational period and were related to the transition into automatic observations or relocation. In such cases, the data from the shorter observational series were taken into account. The long-term variability was also presented on the basis of over a 100-year-old Kraków observational series from 1901 to 2018. These data come from the Scientific Station of the Climatology Department of the Jagiellonian University in Kraków (Kraków Observatory), which has the longest and the most complete series of meteorological observations and measurements in Poland.

Research on the occurrence of **tornadoes** in Poland was conducted using information about their occurrence in the period from 1810 to 2018 derived from the European Severe Weather Database (ESWD) and historical sources from the nineteenth and twentieth centuries. Press reports and social media were also taken into account in the analysis and verification of tornado reports.

The study also identified **long-term changes in meteorological conditions leading to the occurrence of thunderstorms and tornadoes**. For this purpose, data from the European Centre for Medium-Range Weather Forecasts (ECMWF) ERA5 reanalysis (Copernicus Climate Change Service 2017) were used. The data with a 0.25° horizontal grid spacing, 137 vertical sigma levels and 1-hour temporal resolution provide the opportunity to model thunderstorm climatology. In this study we use ERA5 data covering the period 1979–2018. Based on the previous estimates performed by Taszarek et al. (2019), a proxy for thunderstorm situation can be defined when mixed-layer CAPE exceeds 150 J kg^{-1} and convective precipitation occurs. For severe thunderstorms, an additional proxy of WMAXSHEAR (for further details see Taszarek et al. 2019) exceeding 400 m^2/s^2 is included. We use an additional threshold value of Significant Tornado Parameter (STP; Thompson et al. 2003) exceeding 1 to define situations with environmental conditions supporting occurrence of tornadoes. The usefulness of STP in studying climatological aspects of tornadoes has been previously shown by Gensini and Brooks (2018).

Number of Days with Thunderstorms and Tornadoes

Days with Thunderstorms

The number of days with thunderstorms was determined on the basis of data gathered from 47 meteorological stations in Poland belonging to the IMGW–PIB network (The Polish Institute of Meteorology and Water Management, National Research Institute) from the period 1951–2018 (Fig. 16.1).

Considering a long-term variability of thunderstorms, days with thunderstorms were most frequently observed in the southeastern part of Poland (over 29 days) and less frequently on the coast (below 15 days; Fig. 16.2). In particular years, the number of days ranged from 4 days in Ustka in 1976 to 54 days on Kasprowy Wierch

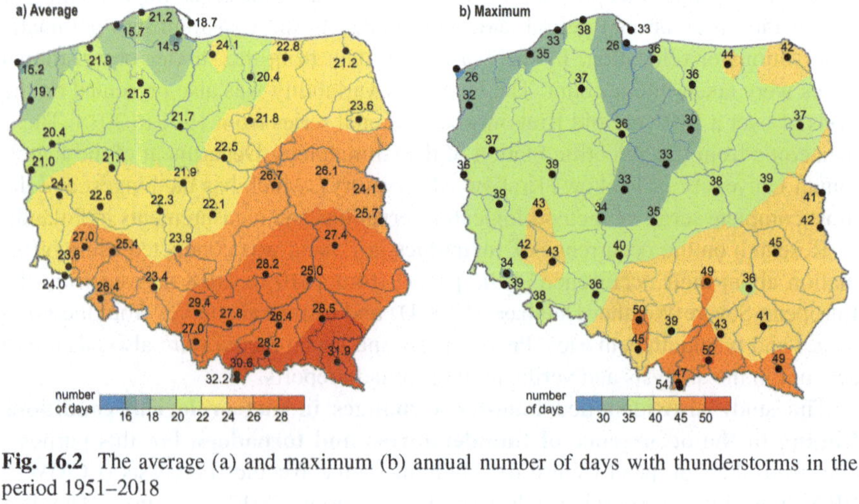

Fig. 16.2 The average (a) and maximum (b) annual number of days with thunderstorms in the period 1951–2018

in 1963. At 22 stations, the lowest annual number of days with thunderstorms was lower than or equal to 10. Such low numbers of annual days with thunderstorms were usually recorded in the northern part of Poland at the beginning of the studied period, i.e. the 1950s, and in the 2000s. In the southern part of the country, there were usually no less than 12–18 days with thunderstorms annually. In this region and in the eastern part of Poland, the largest annual number of days analyzed were also recorded (above 40 days), while on the coast of the Baltic Sea the highest annual number of days with thunderstorm reached 26 (Fig. 16.2).

Analysing thunderstorm activity from a seasonal point of view, it was found that the largest number of days with thunderstorms is observed in summer, with a maximum in July (on average from 3.7 days in Ustka and Gdańsk to 7.9 days on Kasprowy Wierch). The frequency of the analysed phenomenon in spring is higher than in autumn, and it is sporadic in winter (Fig. 16.3). It is worth noting, however, that at stations located in the northern part of Poland the number of days with thunderstorms is distinctly lower, but thunderstorms are more often recorded in the cold half of the year (October–March), accounting for about 7–8% of all cases at coastal stations.

An analysis of the multiannual data indicates that there is no uniform trend in Poland regarding the frequency of the analysed phenomenon. Most of the stations that recorded a decrease in the number of days with thunderstorms were found northwest of the line connecting Elbląg in the north with the town of Bielsko-Biała in the south (Fig. 16.4a). An increase in the number of days with thunderstorms was observed southeast of this line. The greatest changes (statistically significant at the confidence level of 0.05) were recorded in Terespol (1.9 days per 10 years), Włodawa (1.5 days) and Kielce (1.7 days). The greatest decrease (statistically significant at $p < 0.05$) was noted on Śnieżka (−1.2 days), Kasprowy Wierch (−1.0 days) and in Siedlce (−1.0 days).

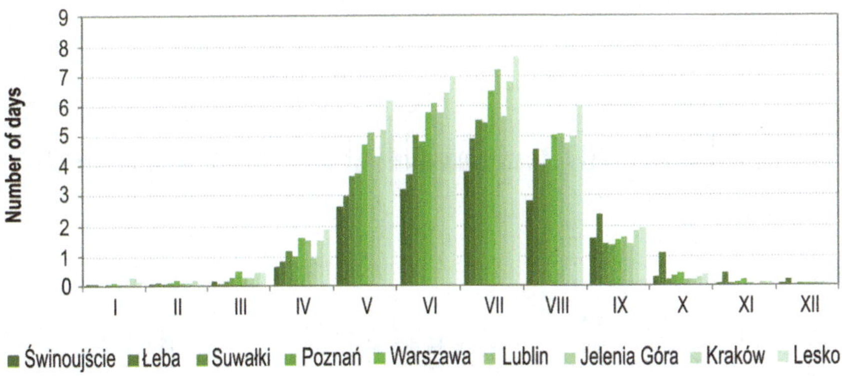

Fig. 16.3 The average monthly number of days with thunderstorms at selected stations in Poland in the period 1951–2018

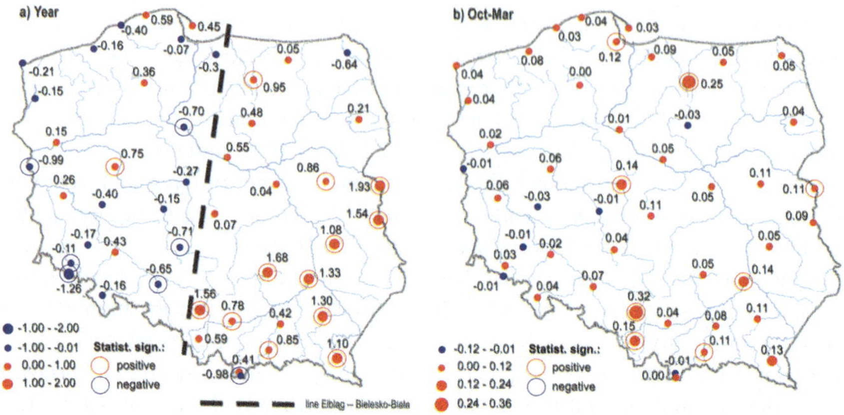

Fig. 16.4 Long-term changes in the number of days with thunderstorm in Poland in the period 1951–2018: (a) annual value [day/10 years] and (b) from October to March [day/10 years]

The long-term variability of the number of days with thunderstorms at selected stations over the years 1951–2018 is shown in Fig. 16.5. This confirms the regularities described above in the value and direction of changes in the frequency of thunderstorms in Poland. Most of them should only be considered as indications of an increase or decrease in the number of days with thunderstorms. More pronounced

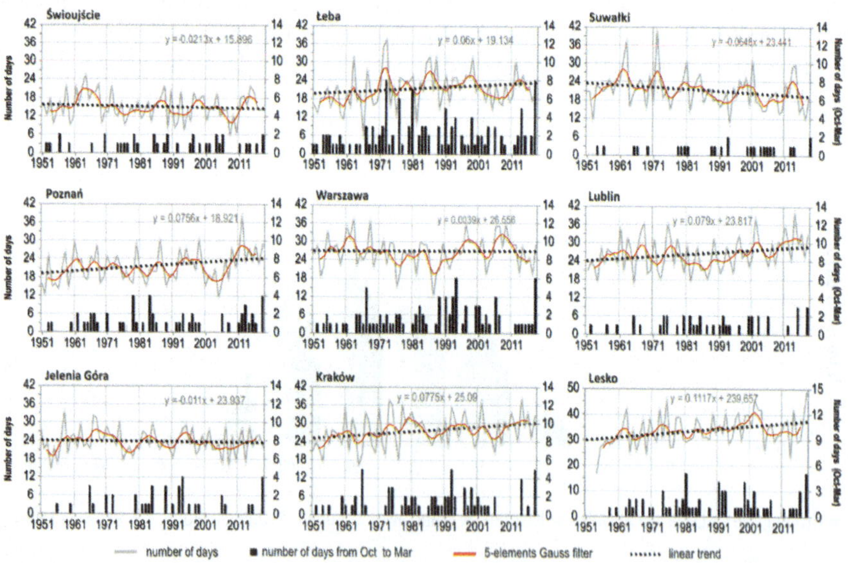

Fig. 16.5 Long-term variability of days with thunderstorms at the selected meteorological stations in the period 1951–2018 (number of days e.g. 2015 for the period Oct–Mar means sum of days from October–December 2015 and from January–March 2016)

changes mean an increase in the number of thunderstorm days from 0.8 to 1.0, and even almost to 2.0 days in 10 years, or a decrease at a rate of about 1 day per decade. It should be noted, however, that the average number of days with thunderstorms on a country scale is around 24. Although this means a change of approximately 5–10% of the total annual number of days with thunderstorms, its impact on the environment is low.

Another characteristic feature of the long-term variability of thunderstorms in Poland is the increased variability in the number of days with thunderstorms in the north and also in the east of the country. The coefficient of variation in these regions over the years 1951–2018 fluctuated around 27–36% (about 31% in Olsztyn and Świnoujście to 35% in Ustka and Gdańsk and from 27% in Nowy Sącz to 36% in Terespol). In the central and southern parts of Poland, these values were half these percentages, reaching 17.1% in Łódź, 18.5% in Jelenia Góra and Rzeszów, and 19.7% in Warszawa. However, it should be emphasized that the occurrence of thunderstorms is characterized by significant regional diversity. Their frequency is affected by the different circulation conditions in particular regions of the country and the strong influence of the natural features of the particular region. Due to this, it is difficult to identify longer or shorter periods with a similar number of days with thunderstorms, even at stations situated close to each other.

As previously mentioned, thunderstorms occur mainly in the warm half of the year. However, in recent decades there has been a slight increase in their frequency in the cold period. This is probably related to the change in atmospheric circulation and the accompanying climate changes, which cause an increase in the frequency of instability in the atmosphere (see also Sect "Data and Methods"). The average number of days with thunderstorms in the months from October to March in the studied period ranged from 0.4 to 1.8 days per decade (Fig. 16.6). The best conditions for thunderstorms in the cold half of the year occur at the coast and in the south of the

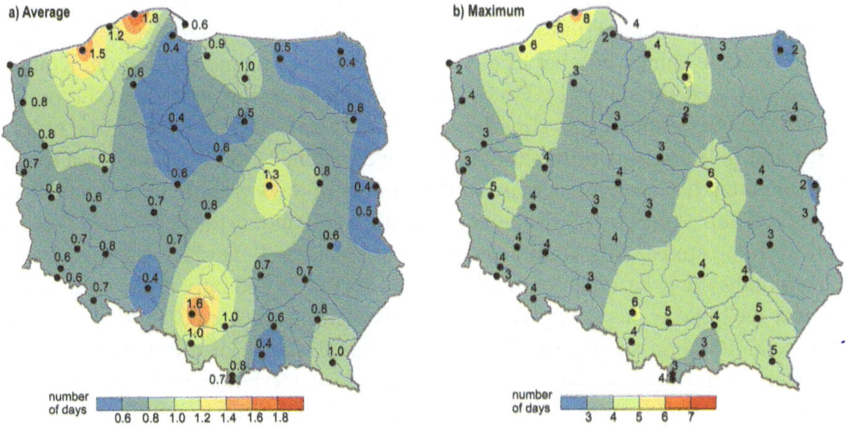

Fig. 16.6 The average (a) and maximum (b) number of days with thunderstorms from October to March in the period 1951–2018

country. At the coast, it is the result of the warming influence of the Baltic Sea in the cold half of the year, whereas in the south of the country impact of the mountains and highlands as well as the inflow of warmer air from the south through the Moravian Gate are the most significant. This is also confirmed by the maximum number of days with thunderstorms in the cold season in particular years. The largest increase of such days was recorded in Łeba. Eight such days were recorded in this city in 1973, and then 7 days in 1981 (Fig. 16.6). Among the remaining stations, 5 or more days with thunderstorms were also recorded in Koszalin, Ustka, Mława, Warszawa, Terespol, Katowice, Kraków and Lesko. This usually occurred 1 or 2 times over a period of many years. Such a large number of investigated days only occurred in Łeba and Katowice, 5 and 4 days, respectively. Due to the fact that thunderstorms in the cold half of the year are rare and their number is small, the increase in the number of days is small. Apart from northern Poland, an increased number of analysed days occurred mainly in the years 1980–2007. In most of Poland, the tendency of the changes described above is not statistically significant (Fig. 16.4b). Exceptions to this are Gdańsk, Olsztyn, Koło, Terespol, Katowice, Bielsko-Biała and Nowy Sącz, where there was a slight and statistically significant ($p < 0.05$) increase in the number of days with thunderstorms (from about 0.1 to 0.3 days per 10 years). It is worth noting, however, that there have been indications of an increase in the number of days with thunderstorms almost all over the country.

The changes described above in the number of days with thunderstorms relate to a period of 68 years. However, when considering data from other research periods, including much longer ones, it has been noted that the number of days with thunderstorms clearly varies depending on the chosen period and region of Poland. However, analysing changes in the number of days with thunderstorms from long observational series (1885–2000) it was found that these changes are very small and in most cases they do not show statistically significant trends (Bielec-Bąkowska 2003). An example of such changes is the occurrence of thunderstorms at the Scientific Station of the Climatology Department of the Jagiellonian University in Krakow (the Krakow Observatory) from the period 1901–2018 (Fig. 16.7).

The average annual number of days with thunderstorms recorded at the Kraków station was 27.4 (Table 16.1). However, for example, in the 1940s and the period 1969–1995, this number was slightly more than 24 days a year, while at the turn of 1930s and in the period 1996–2007 this number reached over 30 days. As a result, although there has been an increase in the number of days with thunderstorms from the second half of the twentieth century in southern Poland, in the period 1901–2018, there were no statistically significant trends of changes. It can be even seen that in some months the occurrence of thunderstorms decreases and only in the cold half of the year the trend is increasing and statistically significant ($p < 0.05$).

Special attention should be drawn to the annual course of the number of days with thunderstorms. The data shows that not only is the number of thunderstorms occurring in spring greater than in autumn, but also the largest monthly number of days with thunderstorms was recorded several times in spring (Table 16.2). It was also found that from around the mid-1970s the period of thunderstorm activity during

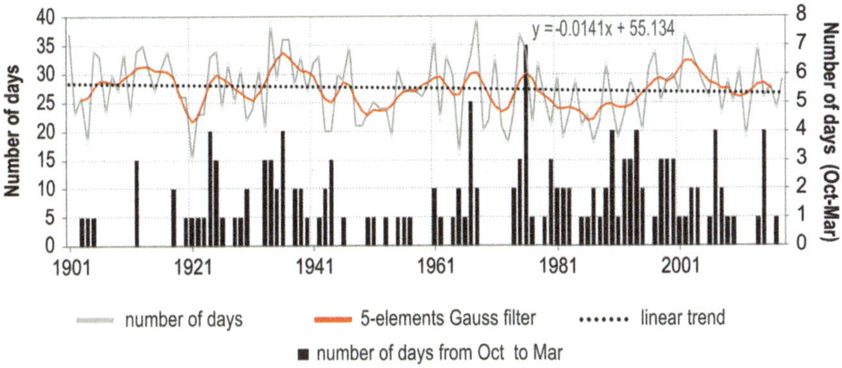

Fig. 16.7 Long-term variability of days with thunderstorms at Kraków Observatory meteorological stations in the period 1901–2018 (number of days, e.g. 2015 for the period October–March means sum of days from October–December 2015 and from January–March 2016)

the year is often longer than in previous years and is shifting towards the first months of the year.

Days with Tornadoes

The reporting of phenomena such as tornadoes share a number of problems associated with the lack of witnesses, the lack of evidence (photographs, videos), the lack of a system to archive events and finally the accuracy of reporting. As pointed out by Groenemeijer and van Delden (2007), some convective events in Europe are described as tornadoes instead of wind gusts thanks to a desire to experience a tornado. In Poland, tornadoes were regarded for a long time by society as a strange and rare phenomena reserved mainly for the United States (Taszarek and Brooks 2015; Taszarek and Gromadzki 2017). Doswell (2003) described this situation as a self-fulfilling prophecy in which denying the existence of tornadoes resulted in a lack of record keeping of such events.

Tornado reports presented in this analysis are derived from the European Severe Weather Database (ESWD) and cover the period 1810–2018. In total, 361 reports are included in the analysis (Table 16.3). Among them, about 26% are significant tornadoes (rated F2–F4 on the Fujita scale), about 60% are weak tornadoes (rated F0–F1 and unrated due to the lack of information on the damage caused) and about 14% are waterspouts (tornadoes occurring over the water surface along the coast of the Baltic Sea). An analysis of these tornado categories in the period 2007–2018 when the credibility of tornado reporting significantly increased shows a higher percentage of waterspouts (37%) and lower percentages of significant tornadoes (16%) and weak tornadoes (47%).

Table 16.1 Number of days with thunderstorms at Kraków Observatory meteorological stations in the period 1901–2018

Index/ Number of days	Months												Year	X–III
	I	II	III	IV	V	VI	VII	VIII	IX	X	XI	XII		
The lowest	0	0	0	0	1	1	1	0	0	0	0	0	15	0
Average	1.3	1.4	1.2	2.2	5.3	6.1	6.6	5.1	2.2	1.3	1.1	1.0	27.4	1.2
The highest	4	3	3	7	11	12	12	13	7	2	2	1	40	7
S.D.	0.9	0.7	0.5	1.5	2.3	2.5	2.6	2.4	1.3	0.4	0.3	0	5.5	1.3
V(%)	65.2	51.0	39.3	69.6	44.3	42.1	40.2	46.6	59.7	34.6	25.5	0	20.2	106.8
Trend [days/10 years]	0.03	0.01	0.03	**−0.09**	−0.10	0.08	−0.07	−0.05	**−0.07**	0.01	−0.01	0.00	−0.14	**0.08**

S.D.—standard deviation (hPa), V—coefficient of variability (%), **0.08**—statistically significant at $p < 0.05$

Table 16.2 Long-term variability of the number of days with thunderstorms at Kraków Observatory meteorological stations in the period 1901–2018

Year	Months												Sum	X-III
	I	II	III	IV	V	VI	VII	VIII	IX	X	XI	XII		
1901				5	9	6	9	8					37	
1902					3	2	7	6	4			1	23	1
1903				2	6	4	6	4	3	1			26	1
1904				2	3	4	4	3	1			1	18	1
1905				2	4	8	12	6	2				34	
1906				3	11	5	5	6	3				33	
1907					8	4	6	4	1				23	
1908					8	4	11	7					30	
1909					2	9	5	6	5				27	
1910				2	8	12	10	2					34	
1911				2	8	4	3	4	2				23	3
1912		1	2	1	9	10	8	3					34	
1913				3	8	10	4	5	5				35	
1914				4	4	5	10	5	3				31	
1915					10	3	6	7	1				27	
1916					8	7	7	8	1				31	
1917					7	9	7	8	2	1			34	2
1918	1			4	5	4	8	6	2				30	
1919					5	2	12	4	2	1			26	1
1920				4	4	2	10	4	2				26	1
1921	1				3	1	4	4	1	1			15	1
1922				1	4	5	5	6	2				23	1
1923			1	1	5	3	3	6	2	1		1	23	4
1924		2		1	11	6	6	4	1	1	1		33	3
1925			1		9	4	12	7		1			34	1
1926				2	5	7	5	3	2				24	
1927				2	5	8	12	5					32	1
1928		1		1	5	6	3	6	2	1			25	1
1929					10	4	5	10	1			1	31	2
1930			1	4	5	4	2	3	3				22	
1931				1	7	8	2	5	1				24	
1932					6	5	9	5	4	2			31	3
1933			1			7	7	2	2	1			20	3
1934		2		6	4	8	9	8		2			39	2
1935				6	3	8	6	2	2	2			29	4
1936	1		1		9	7	9	4	5				36	
1937				2	6	4	8	13	2		1		36	2
1938			1		2	7	8	7	1	2			28	2
1939				1	5	9	8	9	1				33	1
1940			1		5	6	6	3	6				27	
1941					7	7	8	8	1			1	32	1
1942				2	4	6	10	7	2	1	1		33	2
1943				5	1	5	4	2	2	1			20	3

(continued)

Table 16.2 (continued)

1944	1		1		1	3	8	4	2				20	
1945					7	9	10	3	1				30	1
1946		1			5	10	6	7					29	
1947				6	2	7	8	9	3				35	
1948				1	8	4	4	2	2				21	
1949				1	3	4	5	5	3				21	1
1950			1		3	5	6	6	1	1			23	1
1951				1	2	5	7	6	4				25	
1952				2	1	6	5	7	2	1			24	1
1953				1	3	7	10	1	2				24	
1954					6	5	3	3	2				19	1
1955			1	1	2	2	11	11	2		1		31	1
1956					3	9	4	6	5				27	1
1957			1	3	2	6	11	4					27	
1958					6	4	8	6	3				27	
1959					4	7	10	5					26	
1960				1	4	7	7	6	2	1			28	2
1961			1	2	9	10	7	4	3				36	1
1962		1		3	4	3	8	3					22	
1963				4	10	5	5	3	6				33	1
1964		1		2	5	9	6	4	1		2		30	2
1965				1	5	5	2	2	1				16	1
1966			1	2	5	6	8	4	2	1			29	5
1967		3	1		4	5	10	7	3				33	2
1968			2	4	5	5	5	12	7				40	
1969					5	7	5	3					20	
1970					6	7	1	5	3				22	
1971					9	12	8	3	1				33	
1972				2	5	5	4	4	1				21	
1973					4	3	3	5	1		1	1	18	2
1974					5	7	7	3				1	23	3
1975			2	1	6	11	3	9	3	2			37	7
1976	4		1		2	3	10	9	4		1		34	1
1977				1	3	5	4	5	3				21	
1978				1	5	11	8	5	1	1			32	1
1979				1	7	9	5	2	1			1	26	3
1980		1	1	3	2	6	6	2					21	2
1981	1		1		5	5	8	6	2	2			30	2
1982					3	4	4	5	3				19	2
1983	1		1	1	6	3	5	5	1				23	
1984				4	7	5	5	6	2				29	1
1985		1		2	6	4	4	3	1				21	1
1986	1			1	3	6	5	6			1		23	2
1987	1				2	4	6	2	3				18	1
1988			1	1	9	3	3	2	1		1	1	22	2

(continued)

Table 16.2 (continued)

1989				7	7	9	6	2		1			32	4
1990		1	2	3	5	7	1	5	1				25	1
1991			1	1	2	4	2	5	3	1			19	3
1992		1	1	2	2	4	7	5	1				23	3
1993	3			1	7	9	7	1	1				29	4
1994	1		3	1	7	1	3	6	3	1			26	2
1995			1	1	3	5	4	5	1				20	
1996					6	7	7	9	1		1		31	1
1997				2	4	9	10	4	4	1			34	3
1998	1		1	1	5	11	3	2	1				25	3
1999		3		1	5	7	9	2	2				29	3
2000	1		2	1	5	5	5	10		1			30	1
2001					7	2	10	6	2				27	1
2002		1		2	9	9	7	7	1	1			37	2
2003			1	2	7	7	10	5	1	1			34	2
2004			1	1	6	6	9	7	1				31	
2005				1	5	7	11	2	2				28	1
2006			1	1	3	6	4	7	1	2	1		26	4
2007	1				9	9	6	7	1	1			34	2
2008	1				5	4	5	5	2		1		23	1
2009				1	8	12	8						29	1
2010			1		4	4	8	8					25	
2011				5	5	7	8	6					31	
2012				1	3	5	7	2	1				19	
2013				1	7	12	3	2	1			1	27	1
2014				5	4	2	11	9	2	2			35	4
2015			2		5	5	7	4	3				26	
2016				1	6	10	7	2	2				28	1
2017			1	1	3	5	8	4	2				24	
2018				3	3	7	6	7	2			1	29	

Table 16.3 Tornado reports used in the analysis

Category		Since 1810	Percentage (%)		Since 2007	Percentage (%)	
Waterspout		51	14.1		48	36.9	
Unrated	Weak	154	42.7	59.8	28	21.5	46.9
F0		7	1.9		6	4.6	
F1		55	15.2		27	20.	
F2	Significant	65	18.0	26.0	15	11.5	16.2
F3		25	6.9		6	4.6	
F4		4	1.1		0	0.0	

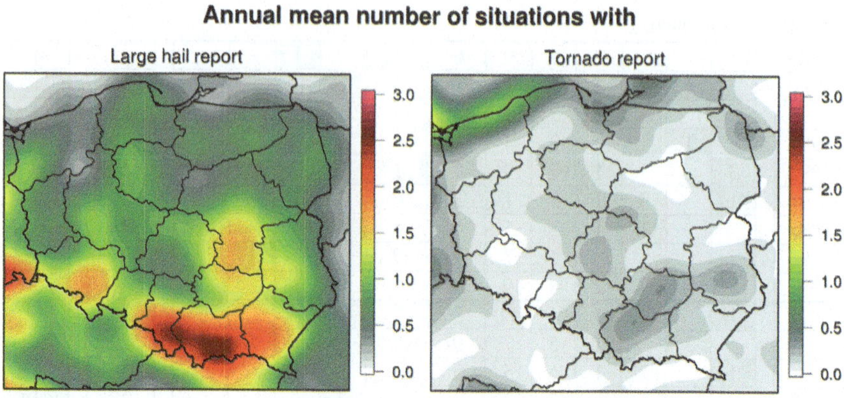

Fig. 16.8 Spatial distribution of 2 cm+ large hail (left) and tornado (right) reports in 100 × 100 km^2 boxes (with applied interpolation smoothing), based on ESWD reports between 2007 and 2018

The spatial distribution of tornado occurrence is estimated by taking the number of tornado situations (a "situation" is considered as a unique hour with a tornado report) in 100 × 100 km grid boxes over the period 2007–2018 (which is then smoothed by interpolation). The results indicate that tornadoes are most frequent along the coast of the Baltic Sea, where an average of 1 tornado (all waterspouts) per 10 km^2 is reported each year (Fig. 16.8). Tornadoes occurring over land surface are less frequent with the reporting density varying from close to zero to one tornado every two years. Increased tornado frequency can be seen in southeastern Poland.

For comparison, large hail situations (more than 2 cm in diameter) from the ESWD covering the same period are also presented (Fig. 16.8). It is worth highlighting that both tornadoes and large hail are predominantly produced by supercell thunderstorms. It can be seen that the density of large hail events is on average 4–7 times higher than tornadoes. Similarly as with tornadoes, the southeastern part of the country has the highest density of hail events. Overall, these results are in agreement with the climatological aspects of thunderstorms in Poland pointing to a peak frequency in southeastern Poland (Bielec-Bąkowska 2003, 2013; Taszarek et al. 2015).

Based on the records from the twenty-first century (Fig. 16.9), it can be estimated that on average 5 weak tornadoes, 1–2 significant tornadoes and 4 waterspouts are reported each year (approximately 10 events per year), which is in line with estimates from Taszarek and Brooks (2015). However, comparing the years 2008 (17 tornadoes including 7 significant) and 2015 (4 tornadoes including 2 waterspouts) it can be seen that the year-to-year variability is relatively large. Comparing tornado reports from recent years with data from the nineteenth and twentieth centuries, it can be seen that significant underestimations were made before the year 2000. Approximately 15 tornado reports including a few significant reports occurred in each decade. According to estimates from Taszarek and Gromadzki (2017), these tornadoes resulted in an average of 5 fatalities per decade indicating that tornadoes

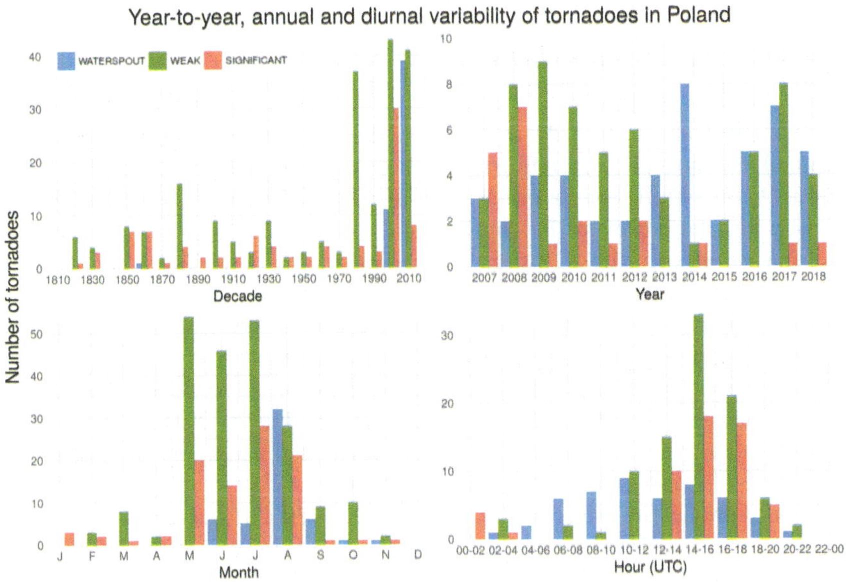

Fig. 16.9 Decadal (top left), year to year (top right), annual (bottom left) and diurnal (bottom right) variability of tornadoes in Poland, based on tornado reports between 1810 and 2018

in Poland are rather a rare phenomenon and pose a low risk to society compared with, for example, the United States (Ashley 2007; Sutter and Simmons 2010). An analysis of the annual cycle indicates that the period with increased tornado activity lasts from May until August which coincides with the period of cloud-to-ground lightning activity in Poland (Taszarek et al. 2015). Tornadoes in Poland also occur from September to April, but these are very rare events. A tornado has never been reported in December. Considering a period of 209 years, the highest number of weak tornadoes occurred in May while significant tornadoes were most frequently reported in June. Waterspouts have a clearly defined peak in August (Fig. 16.9).

The diurnal cycle of tornado occurrence coincides well with surface-based convection. Peak tornado occurrence is between 14:00 and 16:00 UTC which is between 16:00 and 18:00 local time in Poland. The highest frequency of significant tornadoes also occurs between 16:00 and 18:00 UTC. Tornadoes are highly unlikely to occur over land during the night and in the morning. Waterspouts do not have such a well-defined diurnal cycle, but they are more likely to form in the morning hours. Their peak frequency is observed between 10:00 and 12:00 UTC.

Atmospheric Conditions Leading to the Occurrence of Thunderstorms and Tornadoes

As mentioned above, tornado reporting in Poland features a strong spatial and temporal inhomogeneity, therefore identifying long-term changes in their frequency may not be credible. However, instead of studying tornado reports alone, it is possible to study long-term changes in atmospheric conditions leading to their occurrence (Gensini and Brooks 2018). This can be done with the use of reanalysis data. Reanalysis products are generated by the assimilation of almost all available observational data (*e.g.* surface observations, satellite information, radiosondes) over a given period of time. The primary goal of reanalysis is to provide a numerical snapshot of atmospheric conditions that are as close as possible to reality (Thorne and Vose 2010). In this study, we use ERA5 data covering the period 1979–2018 (Hersbach and Dee 2016) and three proxies describing environmental conditions supporting the occurrence of thunderstorms and tornadoes (Table 16.4).

Defined proxies are used to estimate a 40-year (1979–2018) spatial and temporal pattern in climatology and long-term changes. As can be seen in Fig. 16.10, environmental conditions leading to the formation of thunderstorms and severe thunderstorms are the most frequent in the southeastern part of Poland, which is in agreement with climatological patterns in lightning and severe weather reports (Taszarek et al. 2019). Estimates of tornado environments show a very low potential in Poland with only a few hours per year, and therefore no specific spatial pattern can be established. It is noteworthy that all threats have a long-term positive trend, indicating that convective events are becoming more frequent with a globally warming climate, particularly due to increases in low-level moisture and temperature.

A deeper insight into long-term changes in particular locations is presented in Fig. 16.11. Statistically significant ($p < 0.05$) increasing trends of thunderstorm situations can be seen for Kołobrzeg, Gorzów Wielkopolski, Poznań, Łódź, Wrocław and Jelenia Góra. The steepest slope occurs in Łódź (more than 10 h per decade). In addition, a high year-to-year variability of thunderstorm situations can be seen. Statistically significant increasing trends of severe thunderstorm situations can be seen for Elbląg, Gorzów Wielkopolski (including the steepest slope with 3.5 h per decade), Bydgoszcz and Łódź. The environmental conditions supporting tornadic situations are very rare and usually feature 1 situation per 1–2 years, which confirms

Table 16.4 Proxies for defining situations with environmental conditions supporting the occurrence of thunderstorm, severe thunderstorm and tornadic thunderstorm in ERA5 reanalysis

Category	Shortcut	Threshold values of parameters from ERA5
Thunderstorm	TSTM	ML CAPE > 150 J/kg, conv. precp > 0.25 mm/h
Severe thunderstorm	SVRT	ML CAPE > 150 J/kg, conv. precp > 0.25 mm/h, WMAXSHEAR > 400 m2/s2
Tornadic thunderstorm	TORN	ML CAPE > 150 J/kg, conv. precp > 0.25 mm/h, Significant Tornado Parameter > 1

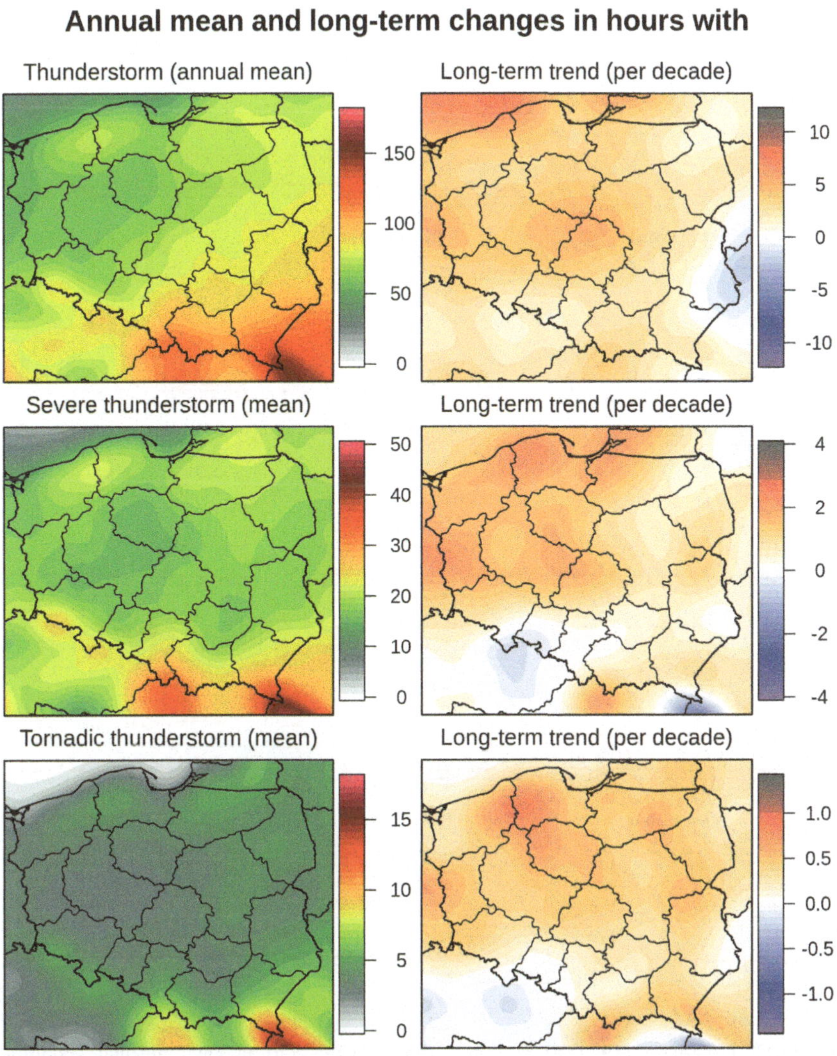

Fig. 16.10 Annual mean and long-term trends (linear regression) in environmental conditions supporting development of thunderstorm, severe thunderstorm and tornadic thunderstorm for selected locations in Poland in the period 1979–2018, based on ERA5 reanalysis

the estimates obtained from an analysis of the tornado reports. Statistically significant increasing trends can only be seen for Gorzów Wielkopolski and Bydgoszcz. Overall, no large changes in tornadic environmental conditions can be defined.

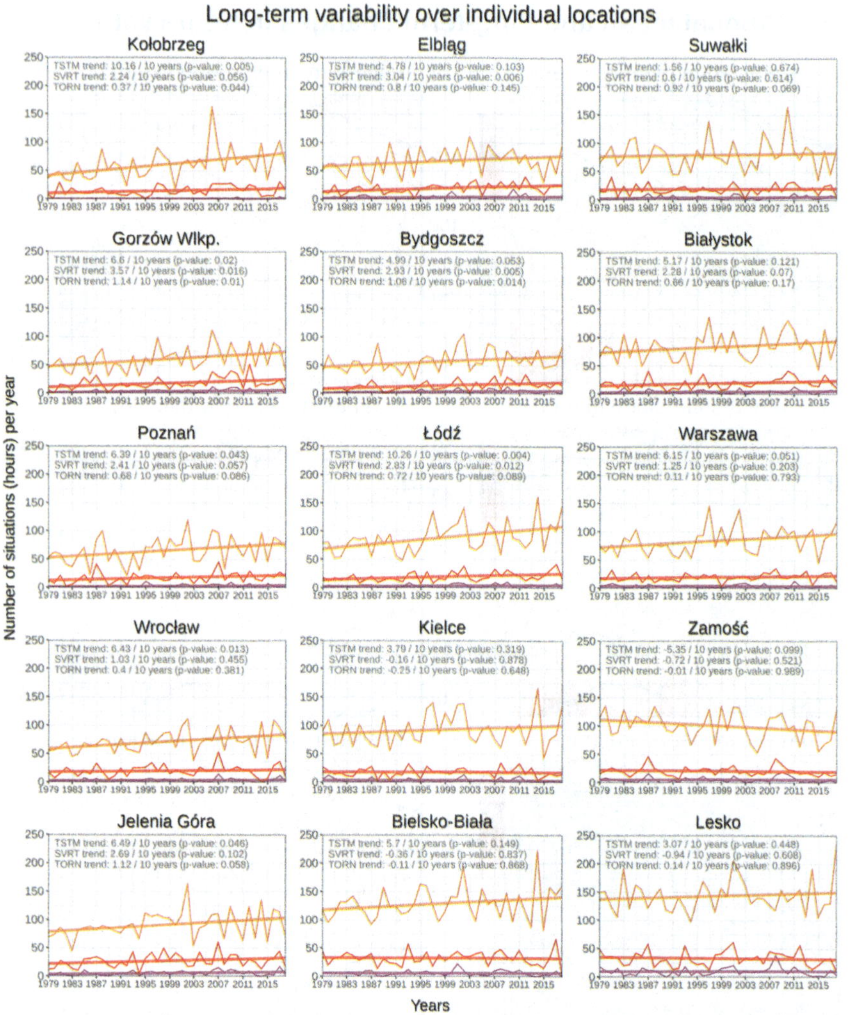

Fig. 16.11 Long-term (1979–2018) trends (linear regression) in environmental conditions supporting development of thunderstorm (orange), severe thunderstorm (red) and tornadic thunderstorm (magenta) for selected locations in Poland, based on ERA5 reanalysis

Concluding Remarks

This research indicates that, despite the observed climate changes, the spatial differentiation of the number of days with thunderstorms and its annual cycle does not differ significantly from similar characteristics recorded in earlier research periods.

The least number of days with thunderstorms occurs from October to March (usually less often than once a season). After this period, a rapid increase in the number of days with thunderstorms is observed in May, and then stabilizes between May and August, until it quickly drops in October. The largest number of days with thunderstorms is observed in summer, with a maximum in July (from 3.7 to 7.9 days). It should also be noted that the frequency of the analysed phenomenon in spring is higher than in autumn. It was also found that over the last 40–50 years the period of thunderstorm activity during the year was often longer than in previous years and shifts towards the beginning of the convective season.

These regularities are specific for most regions of Europe. Enno et al. (2013) found this same period to be the maximum for the Baltic countries, Sonnadara et al. (2006) for Sweden, Schultz et al. (2005) for Austria, Novak and Kyznarova (2011) for the Czech Republic and Antonescu and Burcea (2010) for Romania. The latest studies based on lightning detection data also pointed to the period from June to August as the most prone to thunderstorm occurrence in Europe (Enno et al. 2020; Taszarek et al. 2019; Poelman et al. 2016).

More pronounced changes were found in the long-term variability of thunderstorms occurrence in Poland. Long-term thunderstorm activity tendencies in Poland (1951–2018) indicate that an increase in the frequency of days with thunderstorms can be observed in the eastern part of the country, and for some stations the changes are statistically significant. In the western part of Poland, trends of thunderstorm activity are rather downward. However, such tendencies are rather local and may result from year-to-year fluctuations of favourable thunderstorm synoptic patterns and changes in convective-related indices such as lapse rates, boundary layer humidity and wind shear (Brooks 2013; Seeley and Romps 2015). This is confirmed by the results of the analysis of the long-term variability of thunderstorm occurrence in Poland in 1885–2000 (Bielec-Bąkowska 2003) and the occurrence of thunderstorms in Kraków in 1901–2018. They indicate that, in the long-term periods mentioned above, there were no statistically significant tendencies of changes and only in the colder half of the year is the trend increasing and statistically significant.

Studying tornadoes in Poland has now become possible thanks to changes that took place over the last 15 years. These included the significant increase in the exchange of weather information (*e.g.* social media, cameras in mobile phones, widespread access to the Internet), the increase in severe weather monitoring (formation of the POLRAD radar network), and more systematic efforts to collect severe weather reports (formation of the European Severe Weather Database). Although with each year the European tornado database expands, it is still too short a timeframe to derive reliable conclusions regarding long-term trends and even spatial patterns (given how rare these events are). Estimates based on reanalysis indicate that together with the globally warming climate, environmental conditions are becoming consistently more conducive for severe convective thunderstorms in Poland, including tornadoes.

Based on the observational records from the twenty-first century, it can be estimated that on average 5 weak tornadoes, 1–2 significant tornadoes and 4 waterspouts are reported each year in Poland (approximately 10 events per year). A similar potential is observed in northeastern United States (*e.g.* Michigan and Pennsylvania;

Farney and Dixon 2015), but as confirmed with environmental estimates, Poland has a relatively low threat of tornadoes. An important addition to our knowledge of the occurrence of the phenomena in question was the discovery of dozens of historical tornado cases that took place over the last 200 years in Poland. This finding contradicts the popular statement that *"tornadoes in Poland are a new thing and have become more frequent due to the changing climate"*. Historical records indicate that this phenomenon is not new, but Poland has been and is vulnerable to the occurrence of devastating tornadoes even up to F4 on the Fujita scale. Although such disasters are very rare, they are possible as evidenced over the last 200 years.

References

Allen JT (2018) Climate change and severe thunderstorms. Oxford Research Encyclopedia of Climate Science, Oxford University Press, Oxford

Allen JT, Karoly DJ, Walsh KJ (2014) Future Australian severe thunderstorm environments. Part II: the influence of a strongly warming climate on convective environments. J Clim 27:3848–3868

Antonescu B, Bucea S (2010) A cloud-to-ground lighting climatology for Romania. Mon Weather Rev 138:579–591

Ashley WS (2007) Spatial and temporal analysis of tornado fatalities in the United States: 1880–2005. Weather Forecast 22(6):1214–1228

Bielec-Bąkowska Z (2002) Zróżnicowanie przestrzenne i zmienność wieloletnia występowania burz w Polsce. Wydawnictwo Uniwersytetu Śląskiego, Katowice

Bielec-Bąkowska Z (2003) Long-term variability of thunderstorms occurrence in Poland in the 20th century. Atmos Res 67–68:35–52

Bielec-Bąkowska Z (2013) Burze i grady w Polsce. Prace Geograficzne Instytutu Geografii i Gospodarki Przestrzennej UJ 132:99–132

Brooks HE (2013) Severe thunderstorms and climate change. Atmos Res 123:129–138

Copernicus Climate Change Service (C3S) (2017) ERA5: Fifth generation of ECMWF atmospheric reanalyses of the global climate. Copernicus Climate Change Service Climate Data Store (CDS), date of access: April 2019. https://cds.climate.copernicus.eu/cdsapp#!/home

Doswell CA III (2003) Societal impacts of severe thunderstorms and tornadoes: lessons learned and implications for Europe. Atmos Res 67–68:135–152

Enno SE, Briede A, Valiukas D (2013) Climatology of thunderstorms in the Baltic countries, 1951–2000. Theor Appl Climatol 111:309–325

Enno SE, Sugier J, Alber R, Seltzer M (2020) Lightning flash density in Europe based on 10 years of ATDnet data. Atmos Res 235:104769

Farney TJ, Dixon PG (2015) Variability of tornado climatology across the continental United States. Int J Climatol 35(10):2993–3006

Finney DL, Doherty RM, Wild O, Stevenson DS, MacKenzie IA, Blyth AM (2018) A projected decrease in lightning under climate change. Nature Clim Change 8:210–213

Gensini VA, Brooks HE (2018) Spatial trends in United States tornado frequency. npj Clim Atmos Sci 1(1):38

Groenemeijer P, Kühne T (2014) A climatology of tornadoes in Europe: results from the European Severe Weather Database. Mon Weather Rev 142:4775–4790

Groenemeijer P, van Delden A (2007) Sounding-derived parameters associated with large hail and tornadoes in the Netherlands. Atmos Res 83:473–487

Hersbach H, Dee D (2016) ERA5 reanalysis is in production. ECMWF newsletter 147(7):5–6

IPCC (2013) In: climate change 2013: The physical science basis. Contribution of Working Group I to the Fifth Assessment Report of the Intergovernmental Panel on Climate Change [Stocker

TF, Qin D, Plattner G-K, Tignor M, Allen SK, Boschung J, Nauels A, Xia Y, Bex V, Midgley PM (eds)]. Cambridge University Press, Cambridge, United Kingdom and New York, NY, USA, 1535 pp

Kolendowicz L (1997) Dni z burzą na obszarze Polski Północno-Zachodniej w latach 1951–1990. Badania Fizjograficzne nad Polską Zachodnią. Seria A—Geografia Fizyczna 48:57–82

Kolendowicz L (1998) Zjawiska burzowe w Polsce a warunki synoptyczne. Bogucki Wydawnictwo Naukowe

Kolendowicz L (2005) Wpływ cyrkulacji atmosferycznej oraz temperatury i wilgotności powietrza na występowanie dni z burzą na obszarze Polski. Wydawnictwo Naukowe UAM, Poznań

Kolendowicz L, Taszarek M, Czernecki B (2017) Atmospheric circulation and sounding-derived parameters associated with thunderstorm occurrence in Central Europe. Atmos Res 191:101–114

Novak P, Kyznarova H (2011) Climatology of lightning in the Czech Republic. Atmos Res 100:318–333

Poelman DR, Schulz W, Diendorfer G, Bernardi M (2016) The European lightning location system EUCLID—Part 2: observations. Nat Hazards Earth Syst Sci 16:607–616

Schulz W, Cummins K, Diendorfer G, Dorninger M (2005) Cloud-to-ground lightning in Austria: a 10-year study using data from a lightning location system. J Geophys Res 110(D9)

Seeley JT, Romps DM (2015) The effect of global warming on severe thunderstorms in the United States. J Clim 28:2443–2458

Sonnadara U, Cooray V, Götschl T (2006) Characteristics of cloud-to-ground lightning flashes over Sweden. Phys Scr 74(5):541–548

Sutter D, Simmons KM (2010) Tornado fatalities and mobile homes in the United States. Nat Hazards 53(1):125–137

Taszarek M, Allen J, Púčik T, Groenemeijer P, Czernecki B, Kolendowicz L, Lagouvardos K, Kotroni V, Schulz W (2019) A climatology of thunderstorms across Europe from a synthesis of multiple data sources. J Clim 32:1813–1837

Taszarek M, Brooks HE (2015) Tornado climatology of Poland. Mon Weather Rev 143(3):702–717

Taszarek M, Czernecki B, Kozioł A (2015) A cloud-to-ground lightning climatology for Poland. Mon Weather Rev 143(11):4285–4304

Taszarek M, Gromadzki J (2017) Deadly tornadoes in Poland from 1820 to 2015. Mon Weather Rev 145(4):1221–1243

Thompson RL, Edwards R, Hart JA, Elmore KL, Markowski P (2003) Close proximity soundings within supercell environments obtained from the rapid update cycle. Weather and Forecast 18(6):1243–1261

Thorne PW, Vose RS (2010) Reanalyses suitable for characterizing long-term trends. Bull Am Meteorol Soc 91:353–361

Chapter 17
Change of Hail Frequency

Zuzanna Bielec-Bąkowska

Abstract Data on the number of days with hail from 40 meteorological stations for the years 1966–2018 and from the Scientific Station of the Climatology Department of the Jagiellonian University in Kraków for the years 1901–2018 were used in this study. As a result, it was determined that during analysed periods, from April to September, the average number of days with hail ranges from 0.3 to 1.7, and exceed 10 days in mountainous areas. The largest number of days with hail mostly is recorded in the spring months, and also in the autumn by the Baltic Sea. The spatial differentiation of the occurrence of hail in Poland and their annual course largely depend on the impact of local conditions and the meteorological conditions prevailing over a given area (types of atmospheric circulation, air masses and atmospheric fronts). Except for a few stations, no significant changes in long-term variability of the number of days with hail were observed in the period under investigation (1966–2018). Also in Krakow, from 1901 to present day, no significant changes in the number of days with hail have been recorded. However, it is worth paying attention to the significant changes in the number of days examined from year-to-year.

Introduction

Hailfall is classified as an extreme meteorological phenomenon not only on account of its very rare occurrence, but also due to the often serious nature of material losses it causes. However, due to the low availability of information about the occurrence, volume and intensity of hail, researching this phenomenon is a particularly difficult task. For this reason, there is little published research on hail in Poland. Most of the research is either dedicated to exceptional hailfall occurrences or covers short periods (Twardosz et al. 2010; Bielec-Bąkowska 2013). This situation began to change after the year 2000 when the number of papers on hail occurrence significantly increased.

Z. Bielec-Bąkowska (✉)
Faculty of Natural Sciences, Institute of Earth Sciences, University of Silesia in Katowice, Katowice, Poland
e-mail: zuzanna.bielec-bakowska@us.edu.pl

M. Falarz (ed.), *Climate Change in Poland*, Springer Climate,
https://doi.org/10.1007/978-3-030-70328-8_17

First of all, studies covered long-term and annual variability of hail occurrence (Twardosz et al. 2011; Suwała 2011; Bielec-Bąkowska 2013) as well as the circulation conditions favouring hail formation (Bielec-Bąkowska 2010; Twardosz 2005, 2010; Twardosz et al. 2010). In recent years, research has also been undertaken into the occurrence of particularly high hailfall and the meteorological conditions with which they are associated (Kłokowska and Lorenc 2012; Pilorz 2015; Taszarek and Suwała 2015). Unfortunately, given the nature of hail, its low frequency of occurrence, and the high dependence on local conditions, a number of issues related to hailfall remain to be investigated.

Data

Data on the number of days with hail from 40 meteorological stations belonging to the IMGW-PIB network (The Polish Institute of Meteorology and Water Management—National Research Institute) from the period 1966–2018 were used in this study (Fig. 17.1). Hail occurs mainly in the warm half of the year, therefore, changes in the occurrence of hail from April to September were considered in the analysis.

Fig. 17.1 Location of meteorological stations

This decision was based on the fact that, in most of Europe, this phenomenon occurs mainly in the warm season (Fraile et al. 2003; Simeonov and Georgiev 2003; Chromá et al. 2005; Taszarek et al. 2019) and due to frequent classification of all types of ice pellets in Polish databases under the same category as hail precipitation. This caused numerous errors in the database for the cold half of the year, especially visible at mountain stations. At the same time, research on the occurrence of hail at the Scientific Station in Kraków showed that hail events occurring in the period from October to March constitute to about 7–10% of all cases of such precipitation (Bielec 1996; Twardosz et al. 2010). Therefore, it seems that the adopted assumptions reflect very well the most important features of hail variability.

Additionally, hail data from the Scientific Station of the Climatology Department of the Jagiellonian University in Kraków (Kraków Observatory) from the period 1901–2018 were taken into consideration.

Number of Days with Hail

Hail is a very rare phenomenon in Poland and its spatial variability in most of the country is relatively low. A slightly higher frequency of hailfall is noticeable in areas with varied topography, mainly in mountainous and foothill areas, and at selected sites in northern Poland. In northern Poland this may be attributable to exceptionally favourable environmental conditions and the warming effect of the Baltic Sea.

The average number of days with hail from April to September in Poland in the period 1966–2018 ranged from 0.3 days in Terespol to 1.7 days in Elbląg. The only places where it was higher than this range were the Śnieżka and Kasprowy Wierch mountain stations, where the number of days was 7.0 and 11.5, respectively (Fig. 17.2). However, it must be noted that these values are probably overstated as a result of the incorrect classification of all ice precipitation as hailfall in the database until 1980–1985 (Bielec-Bąkowska 2013). From 1985, the average values of hailfall were 3.5 days at Śnieżka and 9.5 days at Kasprowy Wierch.

The maximum number of days with hailfall in particular years ranged from 2 to 8, with the highest values recorded in the south and the west of Poland. The number of days of hail reached over 20 at the Śnieżka and Kasprowy Wierch mountain stations, however, after 1985, the number of days was 9 and 20, respectively. Except for the mountain stations, the number of years with no hail ranged from 23% to 26% in Siedlce, Koszalin, Lesko and Zakopane (12–14 years) to approximately 70% in Mława and Terespol (37–38 years) (Fig. 17.2).

As mentioned above over 90% of hail occurs from April to September, and the highest frequency is recorded in the spring months (Fig. 17.3), mainly in April and May (0.3 days on average in Poland; a maximum of 0.4–0.5 days at particular stations). In the warmest months, this phenomenon occurs sporadically and the number of days with hail tends to increase again in September, especially on the coast (*e.g.* in Łeba and Hel). At the stations located in the east of Poland, a gradual

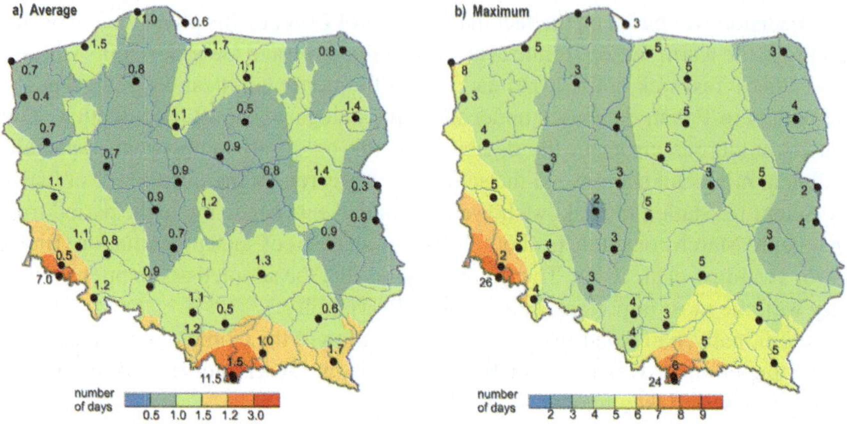

Fig. 17.2 The average (a) and maximum (b) annual number of days with hail in the period 1966–2018

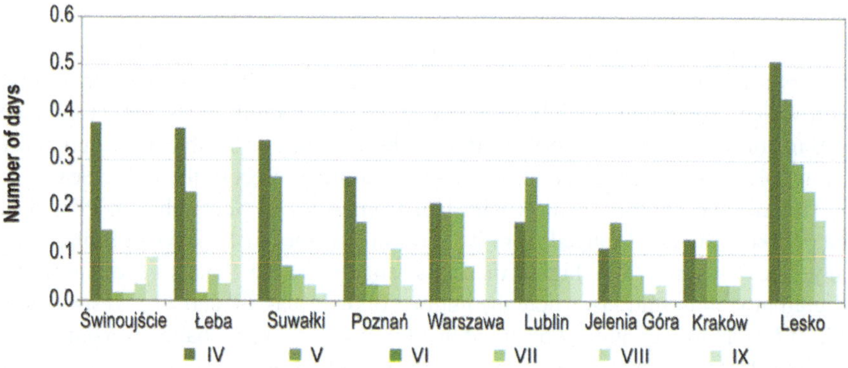

Fig. 17.3 The average monthly number of days with hail at selected stations in Poland in the period 1966–2018

decrease in the number of days of hail from spring to autumn can be observed (*e.g.* in Suwałki, Lublin, Lesko).

The long-term variations of hail occurrence in Poland are strongly related to the location of the station. However, it should be noted that the number of cases studied was very small and there was no hailfall at all in many years. Over a large area of Poland, the changes are very small and not statistically significant ($p < 0.05$). Nevertheless, it can be seen that most of Poland have an increase in hailfall (Figs. 17.4 and 17.5). A statistically significant decrease in the number of days with hail was only recorded at three mountain or foothill stations and at two stations located on the Baltic coast. However, in the case of the Śnieżka and Kasprowy Wierch mountain stations, the significant decrease in the number of days with hailfall (by −2.5 and −1.5

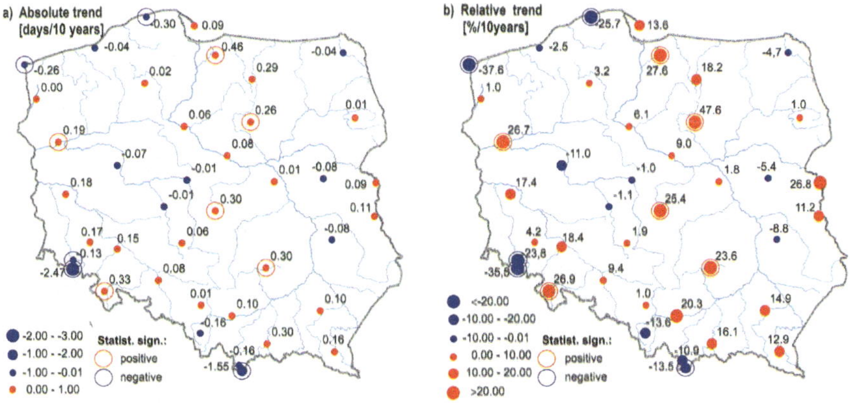

Fig. 17.4 Long-term changes in the number of days with hail in Poland from April to September in the period 1966–2018

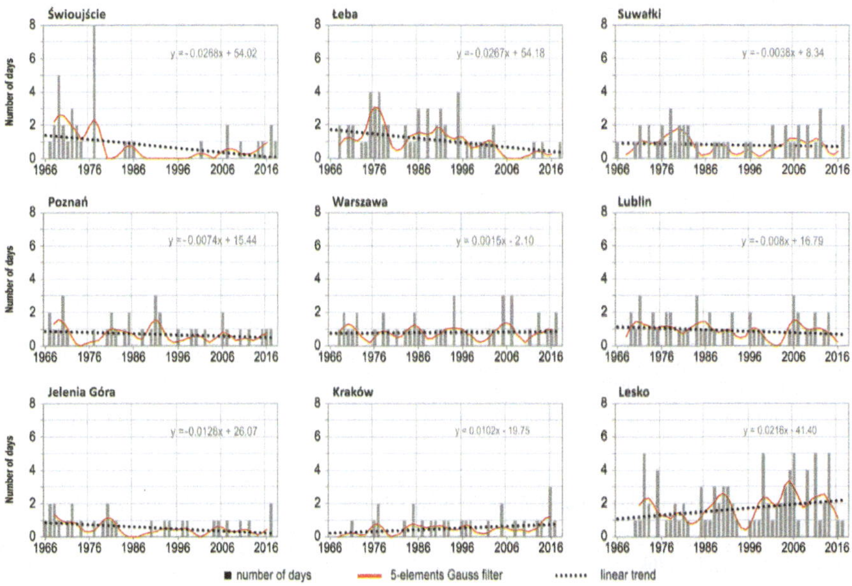

Fig. 17.5 Long-term variability of days with hail at the selected meteorological stations in the period 1966–2018 (from April to September)

days per decade, respectively) is probably due to incorrect classification in the databases. For the shorter study period (from 1986), these values indicate a weak and statistically insignificant increase of about 0.5 days per decade.

The above changes in the occurrence of hail in Poland were compared with the trend observed from 1901 at the station Kraków Observatory. Based on this comparison, the occurrence of hail was not found to follow a clear trend over the long term

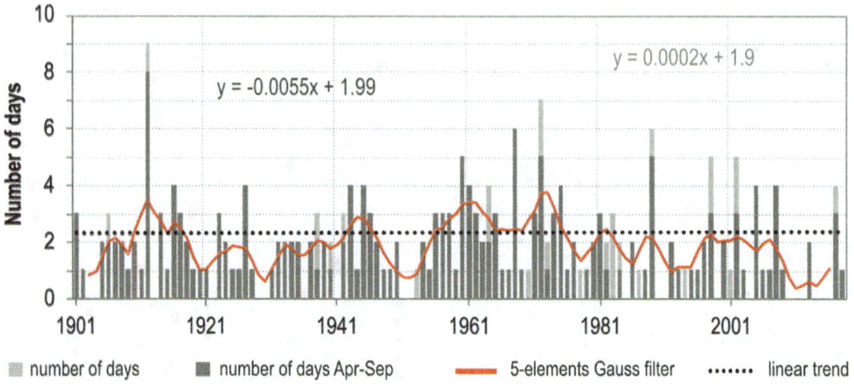

Fig. 17.6 Long-term variability of days with hail at Kraków Obserwatorium meteorological stations in the period 1901–2018

of the study period. A significant increase or decrease in the frequency of hailfall usually lasts for several or several dozen years (Fig. 17.6). However, it is worth noting that hail was very rarely reported in Kraków in the last two to three decades of the study period. The average number of days with hail in the period 2000–2018 was 2.2, while the number of days with hail was 2.3 for the entire study period (Table 17.1). At the same time, after the year 2000 no hailfall at all was recorded in as many as 7 years compared with 21 such years from 1901–2018.

With regard to the annual course of hail occurrence, the absence of hail in January and February and the increased frequency of hail in the period from October to December in the years 1964–2002 are worth noting (Fig. 17.7). In the mentioned period (1964–2002), from October to December, as many as 14 days with hail were recorded (about 6% of all of the cases in the study period), while in the remaining 79 years there were only 6 such days. It would appear that the above changes in the annual and seasonal number of days with hail is associated with the impact of

Table 17.1 Number of days with hail at Kraków Obserwatorium meteorological stations in the period 1901–2018

Index/ Number of days	Months												Year
	I	II	III	IV	V	VI	VII	VIII	IX	X	XI	XII	
Average	–	–	1.3	1.3	1.3	1.2	1.1	1.1	1.1	1.3	1.3	1.0	2.3
The highest	–	–	3	3	4	4	2	2	2	2	2	1	9
S.D.	–	–	0.7	0.5	0.7	0.6	0.3	0.3	0.2	0.5	0.4	0.0	1.5
V(%)	–	–	55.9	42.0	51.5	49.5	29.4	29.4	22.8	35.4	34.6	0.0	64.7
Trend [days/10 years]	–	–	0.00	**0.06**	−0.01	0.02	**0.00**	**−0.03**	**−0.02**	**0.04**	**−0.18**	0.00	0.00

S.D.—standard deviation (hPa), V—coefficient of variability (%), **0.08**—statistically significant at $p < 0.05$

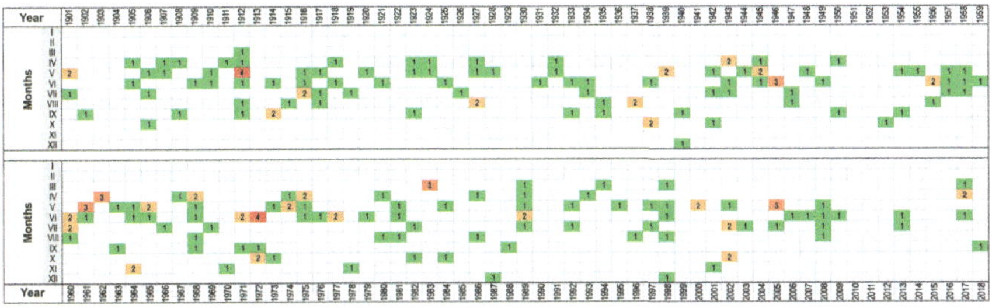

Fig. 17.7 Number of days with hail in Kraków in the period 1901–2018

the city on the formation of convection phenomena (Changnon et al. 1979; Punge and Kunz 2016), which mainly concerns the formation of hail linked inseparably to the buildup of *Cumulonimbus* clouds. These clouds have been observed ever more frequently in Kraków, also in the cold half of the year. However, the number of days with thunderstorms that often accompany *Cumulonimbus* clouds is not increasing as fast (Matuszko 2014). As can be seen, a similar tendency is observable in the case of hailfall, which has been less frequent in recent years than before 1980. Perhaps this can be explained by the thermal and humidity conditions of the air over Kraków. On the one hand, the warming effect of the city drives convection and, on the other hand, decreased humidity inhibits the development of clouds and precipitation (Knight and Knight 2001). The increased number of days with hail in the cool half of the year in the period 1964–2002 may have been associated with a distinct increase in the concentrations of particulate matter caused by the strong industrialization of the city (an increase in the number of condensation nuclei), and the significant reduction in air pollution in Krakow over last two decades probably also affected the frequency of convective phenomena occurrence.

Conclusions and Discussion

The frequency of hail occurrence in Poland is similar to that observed in Central Europe, although it is much lower than in southern Europe and in the areas of the world with the highest frequency of convection phenomena (Punge and Kunz 2016; Prein and Holland 2018). The average number of days with hail from April to September ranges from 0.3 to 1.7, and can even exceed 10 days in mountainous areas. Such considerable variance in the number of days with hail is associated with the impact of local conditions. In the north of Poland, in the coastal area (Koszalin, Łeba) and in the lakelands (Elbląg, Olsztyn, Białystok), the inflow of significant amounts of moisture (Zinkiewicz and Michna 1955; Tuovinen et al. 2009) and the warming effect of the Baltic Sea drive the higher frequency of hail occurrence (up to 1.5–1.7 days on average per year or up to 8 days with hail in Świnoujście in 1977). In lowland areas, which prevail in Poland, days with hail are recorded once in several years.

However, as in the case of other highland and mountainous regions of the world (Kunz et al. 2009; Punge and Kunz 2016), the south of Poland is characterized by the most favourable conditions for hail occurrence, notably the movement of low-pressure systems over the area, accompanied by the advection of polar air with a westerly or southerly component (Twardosz et al. 2010; Suwała and Bednorz 2013). In such meteorological conditions, the diversified topography is conducive to the intensification of upward air currents and the formation of *Cumulonimbus* cloud complexes. The conditions described above are also behind the characteristic annual course of hail occurrence, where the largest number of days with hail is recorded in the spring months, and also in the autumn by the Baltic Sea. Contrary to the long-term variability of the number of days with a thunderstorm (see Chapter 16), no significant changes in the number of days with hail were observed in the period under investigation (1966–2018), except for a few stations. However, even in these few cases, the changes are very small and do not have any significant impact on the environment. A similar lack or slight change in the occurrence of hail has been observed since 1901 in Kraków. However, it is worth paying attention to large year-to-year changes and to the periods with higher hail frequency as well as to the periods with almost no hail. Explanations for these changes should probably be sought in the circulation conditions occurring in particular regions of the country in particular years.

However, it must be emphasized that this study analysed only the number of days with hail without taking into account its diameter, which determines the amount of damage to the environment, property, etc. The most dangerous hailfall occurs in the south and east of Poland, which is mainly attributable to the above-mentioned orographic and circulation conditions (Pilorz 2015; Taszarek and Suwała 2015).

References

Bielec Z (1996) Występowanie burz w Krakowie w latach 1896–1995. Master's thesis, archive of the Climatology Department of the Jagiellonian University in Kraków

Bielec-Bąkowska Z (2010) Występowanie gradów w Polsce w świetle cyrkulacji atmosfery (1966–2006). In: Ciupa T, Suligowski R (eds) Woda w badaniach geograficznych. Uniwersytet Jana Kochanowskiego, Kielce, pp 359–374

Bielec-Bąkowska Z (2013) Burze i grady w Polsce. Prace Geograficzne Instytutu Geografii i Gospodarki Przestrzennej UJ 132:99–132

Changnon SAJ, Gatz D, Bertnes J, Sonka S, Bartlett J, Hassett J (1979) Studies of impacts of urban-related weather and climate changes at Chicago and St. Louis. Final Report on NSF Grant ENV77-15375, Causes and Impacts of Urban and Lake Influences on Precipitation 3, http://www.isws.illinois.edu/pubdoc/CR/ISWSCR-217.pdf

Chromá K, Brázdil R, Tolasz R (2005) Spatio-temporal variability of halistorms for Moravia and Silesia in the summer half-year of the period 1961–2000. Meteorologický časopis 8:65–74

Fraile R, Berthet C, Dessent J, Sánchez JL (2003) Return period of severe hailfalls computed from hailpad data. Atmos Res 67–68:189–202

Kłokowska K, Lorenc H (2012) Ryzyko występowania gradu w Polsce. In: Lorenc H (ed) Klęski żywiołowe a bezpieczeństwo wewnętrzne kraju. Instytut Meteorologii i Gospodarki Wodnej–Państwowy Instytut Badawczy, Warszawa, pp 80–97 (in Polish)

Knight CA, Knight NC (2001) Hailstorms. In: Doswell III ChA (ed) Severe convective storms. American Meteorological Society, AMS Monograph 28(50):223–254

Kunz M, Sander J, Kottmeier Ch (2009) Recent trends of thunderstorm and hailstorm frequency and their relation to atmospheric characteristics in southwest Germany. Int J Climatol 29:2283–2297

Matuszko D (2014) Zachmurzenie i burze w aspekcie ekstremalnych zdarzeń atmosferycznych. Prace Geograficzne Instytutu Geografii i Gospodarki Przestrzennej UJ 139:79–92

Pilorz W (2015) Very large hail occurrence in Poland from 2007 to 2015. Contemp Trends in Geosci 4(1):45–55

Prein AF, Holland GJ (2018) Global estimates of damaging hail hazard. Weather and Clim Extremes 22:10–23

Punge HJ, Kunz M (2016) Hail observations and hailstorm characteristics in Europe: a rev. Atmos Res 176–177:159–184

Simeonov P, Georgiev ChG (2003) Severe wind/hail storms over Bulgaria in 1999–2001 period: synoptic- and meso-scale factors for generation. Atmos Res 67–68:629–643

Suwała K (2011) Hail occurrence in Poland. Quaestiones Geographicae 30(3):115–126

Suwała K, Bednorz E (2013) Climatology of hail in Central Europe. Quaestiones Geographicae 32(3):99–110

Taszarek M, Allen J, Púčik T, Groenemeijer P, Czernecki B, Kolendowicz L, Lagouvardos K, Kotroni V, Schulz W (2019) A climatology of thunderstorms across Europe from a synthesis of multiple data sources. J Clim 32:1813–1837

Taszarek M, Suwała K (2015) Large hail in Poland in 2012. Quaestiones Geographicae 34(1):75–84

Tuovinen JP, Punkka AJ, Rauhala J, Hohti H, Schultz DM (2009) Climatology of severe hail in Finland: 1930–2006. Mon Weather Rev 137(7):2238–2249

Twardosz R (2005) Dobowy przebieg opadów atmosferycznych w ujęciu synoptycznym i probabilistycznym na przykładzie Krakowa (1886–2002). Instytut Geografii i Gospodarki Przestrzennej UJ, Kraków

Twardosz R (2010) A synoptic analysis of the diurnal cycle of thunderstorm precipitation in Kraków (Southern Poland). Int J Climatol 30:1008–1013

Twardosz R, Niedźwiedź T, Łupikasza E (2010) Burze w Krakowie i ich uwarunkowania cyrkulacyjne. In: Ciupa T, Suligowski R (eds) Woda w badaniach geograficznych. Uniwersytet Jana Kochanowskiego, Kielce, pp 303–313

Twardosz R, Niedźwiedź T, Łupikasza E (2011) Zmienność i uwarunkowania cyrkulacyjne występowania postaci i typów opadów atmosferycznych na przykładzie Krakowa. Wydawnictwo Uniwersytetu Jagiellońskiego, Kraków

Zinkiewicz W, Michna E (1955) Częstość występowania gradu w woj. lubelskim w zależności od warunków fizjograficznych. Annales Universitatis Marie Curie-Skłodowska Lublin – Polonia, Sectio B, 10(5):224–267

Chapter 18
Change of Fog Frequency

Anita Bokwa, Agnieszka Wypych, and Monika J. Hajto

Abstract Data on fog occurrence from 26 Polish meteorological stations were analysed for the period 1966–2018. Annual and seasonal number of days with fog, number of days with fog of the duration <6 h, 6-12 h and >12 h, and number of days with dense, moderate and thin fog were studied in terms of spatial and temporal variability. In the high mountains, the fog frequency is the highest (about 300 days per year), no statistically significant multi-annual changes of most of the indices can be defined and the inter-annual variability is very low. At the seaside, the number of days with fog is the lowest (about 28 days per year on average) and the inter-annual variability is high. Therefore, no statistically significant trends can be determined for most of the indices. In the lowland areas of northern and central Poland, and in the uplands of southern Poland, high variability in fog occurrence and long-term trends can be observed in each region. For stations with the highest fog frequency, statistically significant increasing trends were obtained while for the others the trends were either decreasing or not significant statistically. As fog is a very localised phenomenon, downscaling methods of satellite image processing have to be developed in order to improve the spatial approach in long-term analyses.

A. Bokwa (✉) · A. Wypych · M. J. Hajto
Faculty of Geography and Geology, Institute of Geography and Spatial Management, Jagiellonian University, Kraków, Poland
e-mail: anita.bokwa@uj.edu.pl

A. Wypych
e-mail: agnieszka.wypych@uj.edu.pl

M. J. Hajto
e-mail: monika.hajto@imgw.pl

A. Wypych · M. J. Hajto
Institute of Meteorology and Water Management—National Research Institute, Kraków, Poland

M. Falarz (ed.), *Climate Change in Poland*, Springer Climate,
https://doi.org/10.1007/978-3-030-70328-8_18

Introduction

Fog is a meteorological element which can become a hazardous phenomenon. It can cause problems with traffic (flight delays, automobile and marine accidents) or be associated with critical conditions in air pollution, resulting from air pollutants becoming trapped in the fog droplets and reaching high concentrations, causing the formation of smog or in some cases acid fog. In addition, the fog has an important role in maintaining radiation balance and, as a result, long-term changes in the frequency of fog can play an important role in the accuracy of climate model predictions. Fog is a very localized phenomenon that can form as a result of advection, radiative cooling or a weather front moving over an area. Its frequency and spatial distribution is closely related to orography and proximity to the sea (Avotniece et al. 2015).

Studies concerning fog typically concentrate on various aspects of the phenomena, e.g. spatial distribution, long-term changes, chemical properties, origin of the phenomenon, interaction with land use/land cover. This paper deals with long-term changes in fog frequency in Poland, including fog duration and intensity. The results presented below should be interpreted in the context of similar studies undertaken in Poland and in other European countries.

Bendix (1994) studied fog spatial distribution with the application of satellite imagery in the Po valley (Italy) and in Germany (Bendix 2001, 2002), but no long-term trends were established. However, Giulianelli et al. (2014), found a decreasing trend since 1980s in the Po valley. Cermak et al. (2009) used satellite-derived fog and low stratus maps of Europe for the period 2004–2008 and showed a large difference between southern Europe with frequencies around 15–20% and northern Europe with values reaching 50–60%. Egli et al. (2017) completed similar maps for the period 2006–2015; inter-annual trends were found most pronounced in winter, with vast areas of Central Europe showing a distinct decrease. For the Swiss Plateau, Scherrer and Appenzeller (2014) analysed fog and low stratus trends in the period 1901–2012 and found no significant trends. However, for the subperiod 1984–2012, a significant decrease can be seen. Avotniece et al. (2015) studied fog spatial distribution and trends in Latvia from 1960 to 2012 and found a strong decreasing trend all over the country, resulting from gradual decrease in industrial activities and the resultant improvements in air quality and the observed increase in air temperature. The warming was the most significant in the winter and this might have triggered a decrease in the formation of advective fog, which in winter usually forms when warm and moist air flows over a cool or snow-covered surface. Those results confirm the findings from Europe for the period 1976–2006, published by Vautard et al. (2009). One reason for the general decline in fog and low stratus in winter may be the rising temperatures in the lower atmosphere due to climatological changes. Also, improvements in air quality lead to a reduction in aerosols that act as condensation nuclei in supersaturated conditions. This leads to further reduction in fog and low stratus formation as well (Klemm and Lin 2016). Oldenborgh et al. (2010) extended the analyses completed by Vautard et al. (2009) and focused on dense fog occurrence. In large parts of Europe the decreases correspond to half the number of foggy and misty days. The decrease in number of foggy and misty days is spatially and temporally correlated with the decrease in SO_2 emissions for all fog ranges. The spatial correlation between trends in fog and urbanization is lower, but also positive for all ranges studied. For inter-annual variability, the effects of circulation dominate, but for this trend other factors are more important. The study that does not follow the general trend is the one by Veljović et al. (2015), who analysed fog occurrence in Belgrade, Serbia, for the period from 1973 to 2005 and found an increase in the annual number of fog days. Fog occurrence has been studied in other parts of the world as well, e.g. in the United States (Witiw and LaDochy 2008), South Asia (Syed et al. 2012; Gautam and Singh 2018); China (Niu et al. 2010; Quan et al. 2011; Fu et al. 2014). The results obtained so far for various parts of Europe show a general decreasing tendency of fog frequency in the last decades, but, on the other hand, a large spatial variability of the phenomenon can be seen, especially in mountainous areas.

In Poland, fog occurrence variability was determined for various regions. Trzeciak (1992) studied fog frequency for the western part of the Polish seaside, for the period 1961–1985, but temporal variability was not analysed. In the case of urban areas, Wypych (2003) found a decreasing trend for fog in Kraków in the period 1961–2000. Łupikasza and Niedźwiedź (2016) analysed fog occurrence at three stations in Southern Poland representing the Silesian Upland (Katowice), the Carpathian Foothills (Bielsko-Biała) and the basins of the Carpathian Foredeep (Kraków), in the years 1966–2015. A decreasing trend was found only for Kraków, while at the two other stations a high variability of the element studied was observed.

Fog occurs in mountainous areas much more often than in lower locations. Hess (1965) studied fog occurrence in the Polish Carpathians for the period 1952–1961 and found that at the mountain peaks, fog is noted 10 times more often than at the mountain foot. Moreover, in lower locations fog is mainly of radiative origin, while at the peaks advective processes are what determine the occurrence of the phenomenon. Żarnowiecki (2000) analysed fog occurrence in the Holy Cross Mts. and their foothills for the period 1955–1965. Even though the station representing the highest parts of the mountains is located at an altitude of 587.7 m a.s.l., mean annual fog frequency reaches 37.7% there, while at the foothills it is 8–10%. The spatial distribution of fog in the Sudety Mts. was studied by Błaś et al. (2002). Fog is the most frequently observed atmospheric phenomenon there, being present on average 45% of the time, with 250–300 days of fog per year. Błaś and Sobik (2000, 2004) compared fog frequency and vertical distribution in the European mountains and found an increase in the phenomenon with altitude, together with its high spatial variability. Lange et al. (2003) reported 240 days with fog annually on average in the Erzgebirge Mts. between 1971 and 1997. The study for the Fichtelgebirge Mts., completed by Wrzesinsky and Klemm (2000), covered only one year (1997). It was dedicated to fog chemistry, but additionally showed a much higher fog frequency in the cold half of the year than in the warm one.

Data and Methods

Data concerning fog occurrence were collected for 35 stations for the period 1966–2018. The data included:

- daily number of hours with fog,
- data on visibility in metres observed at 06, 12 and 18 UTC.

After the homogenization procedure, 9 stations were excluded from further study (see Chapter 4 *Homogeneity of Climate Series*).

The indices obtained from the data listed above were the following:

- number of days with fog; a day with fog is defined as a day when fog was observed at the station, regardless the fog's duration or intensity,
- number of days with fog of a given duration: <6 h, 6–12 h, >12 h,
- number of days with fog of a given intensity, separately for 06, 12 and 18 UTC; the intensity was defined with the visibility data: (1) dense fog: visibility <200 m; (2) moderate fog: visibility <500 m; (3) thin fog: visibility <1000 m.

The indices were calculated at annual and seasonal resolution. Only two seasons were distinguished: cold (October–March) and warm (April–September) half-year, due to the phenomenon's general frequency observed for most stations. The mean annual and seasonal values were calculated for each index, together with the coefficient of variability (i.e. standard deviation divided by the mean value). The multi-annual course of each index was studied with linear regression analysis (if the series was complete enough to allow such calculation). In case, where the trend obtained was statistically significant ($\alpha = 0.05$), the relative trend per 10 year (in percent) was calculated.

Number of Days with Fog

For the period 1966–2018, **mean annual number of days with fog** varied from 67.6 days in Kraków and 67.7 days in Mława to 27.7 days in Hel (Fig. 18.1). The stations analysed can be grouped in a few regions and in each region large differences between the stations can be seen. For example, in the north-western region, including the seaside (stations: Koszalin, Łeba, Hel, Świnoujście and Szczecin), the values vary from 56.7 days in Koszalin to the already mentioned 27.7 days in Hel. The proximity to the seashore, resulting in the exposure to frequent strong winds is the factor that largely determines fog frequency.

In the north-eastern region (stations: Mikołajki, Toruń, Mława and Białystok) the values range is from 45.5 days in Toruń to the previously mentioned 67.7 days in Mława and in this case the determining factor is the location of Mława, in an area with vast marshy areas, and the impact of urban structures on fog in case of Toruń.

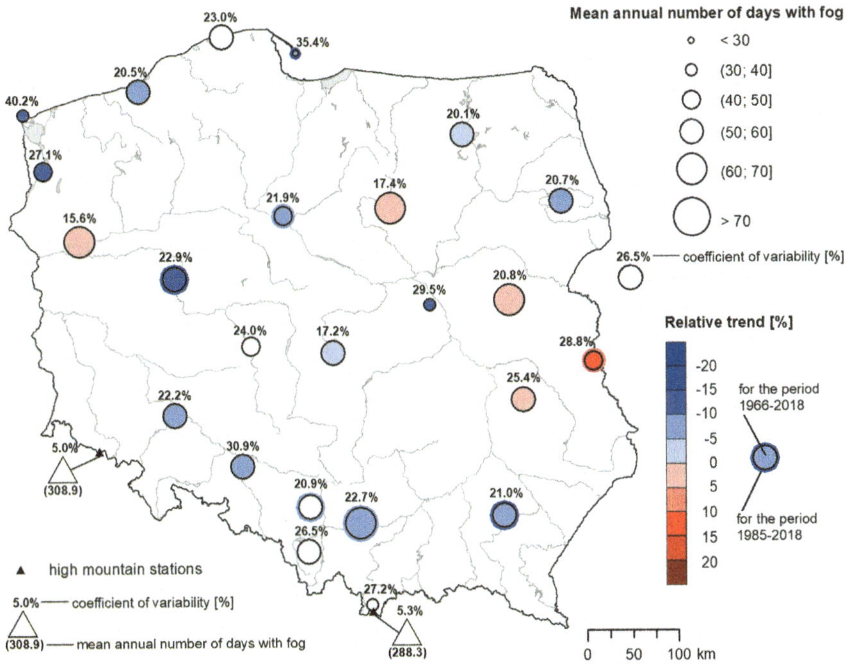

Fig. 18.1 The mean annual number of days with fog in the period 1966–2018, the coefficient of variability (%) at the stations studied and relative trends (%/10 years) for the periods 1966–2018 and 1985–2018. Explanations: only those relative trends which are statistically significant are marked. Data for two high mountain stations: Kasprowy Wierch and Śnieżka are presented separately. Trends which are statistically insignificant are indicated with colourless circles

In the central region (stations: Gorzów Wielkopolski, Poznań, Warsaw, Siedlce, Kalisz and Łódź), the lowest number of days with fog is noted for Warsaw (36.4), while the highest is for Gorzów Wielkopolski (63.0). The impact of urban structures can be seen again in the case of Warsaw, while, in the case of Gorzów, the location in the western part of the country and direct exposure to humid Atlantic air masses seems to be a determining factor.

In the southern region (stations: Wrocław, Lublin, Włodawa, Opole, Katowice, Kraków, Rzeszów, Bielsko-Biała), the largest value is assigned to Kraków (67.6 days) and the lowest to Włodawa (44.5 days). The southern region has a very diversified relief, with uplands and basins located along mountain belts. That factor is decisive in the case of Kraków, a large city located in a concave landform where cold air pools and air temperature inversions often form. Włodawa is located in the eastern part of the country where climatic conditions are more continental than in the western part.

Two high mountain stations: Kasprowy Wierch and Śnieżka had to be analysed separately due to the fact that fog frequency is much higher there than at stations placed in lower locations. The high mountain stations experience fog most of the year. The mean annual numbers of days with fog are 288.3 for Kasprowy Wierch in the Carpathians and 308.9 for Śnieżka in the Sudety Mts. An interesting feature is that for Zakopane, located directly at the foothills of the Tatra Mts., where the Kasprowy Wierch Mt. is located, the value is 39.3 days only. That shows a very high spatial variability of fog frequency in the mountainous areas.

Figure 18.2 shows the annual number of days with fog in the period 1966–2018 for stations representing the regions mentioned above. In all cases, a large inter-annual variability of the index can be seen. The basic differences among the regions include much higher values in the mountains than in the lower locations. No clear tendency of multi-annual changes can be observed for the mountains and the seashore regions. For the rest of the country, a slight decreasing tendency can be observed, with the exception of Gorzów Wielkopolski. The statistical properties of the data series in a national scale are presented in the further sections.

For the period 1985–2018, the lowest and highest values (mountain stations excluded) were noted at the same stations as in the period 1966–2018: 26.7 days in Hel and 70.2 days in Mława (for an explanation of subperiods used see Chapter 3 *Data and Methods of Investigation*). The spatial variability remained the same as described above. The coefficient of variability for the period 1966–2018 shows that for most of the stations the annual number of days with fog does not change much from year to year. The values exceeded 25% for 9 stations only and the highest value was 40% in Świnoujście. In the period 1985–2018, the index was even less variable

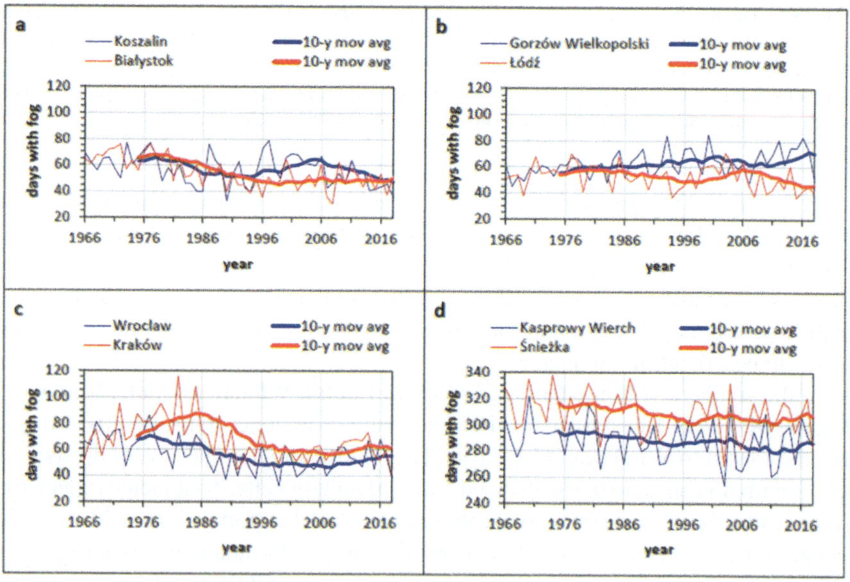

Fig. 18.2 Annual numbers of days with fog in the period 1966–2018 for stations representing: (a)—Baltic seashore (Koszalin) and the north-eastern region (Białystok), (b)—the western and central region (Gorzów Wielkopolski and Łódź), (c)—the southern region (Wrocław and Kraków), (d)—the two high mountain stations: Kasprowy Wierch and Śnieżka. Explanations: *10-y mov avg* means 10-year moving average of number of days with fog

as only in four cases the coefficient exceeded 25% and the highest value reached 37% in Świnoujście. In the case of both high mountain stations, the coefficient was as low as 5% for both periods. For 19 stations, the linear trends were statistically significant for the period 1966–2018 (14 decreasing [maximum value: −5.3 days/10 years in Szczecin and Poznań] and 5 increasing [max. value: 4.9 days/10 years in Włodawa]), while for the years 1985–2018 they were significant for seven stations only (6 decreasing and 1 increasing). Relative trends were calculated and for the period 1966–2018 they ranged from +11%/10 years in Włodawa to −14.3%/10 years in Świnoujście. For the period 1985–2018, the values were from +8.4%/10 years in Włodawa to −18.3%/10 years in Hel. From the six stations for which the trends were significant for both periods, it is important to note that for Hel and Rzeszów, the values for 1985–2018 were much larger than for the whole period 1966–2018. For Hel, the value for 1966–2018 was −6.9%/10 years, while for Rzeszów the values reached: 1966–2018: −6.2%/10 years, 1985–2018: −14.2%/10 years. For high mountain stations the trends were statistically insignificant.

Most of the days with fog are observed during the **cold half of the year (October–March)**. The highest mean value for the period 1966–2018 was noted for Mława and Gorzów Wielkopolski (48.0 days), the lowest for Hel (16.4) (Fig. 18.3). In the case of high mountain stations, the values reached 144.7 for Kasprowy Wierch and 160.4 for Śnieżka. For the period 1985–2018, the highest and lowest values were very similar: 50.0 for Gorzów, 16.2 for Hel, 144.9 for Kasprowy Wierch and 160.3 for Śnieżka. The spatial variability of the values was similar to that for annual values described above. The coefficient of variability for the period 1966–2018 shows that for most of the stations the seasonal number of days with fog changes significantly from year to year. The values exceeded 25% for 15 stations and the highest value was 39% in Świnoujście. In the period 1985–2018, the index was less variable as in 12 cases the coefficient exceeded 25% and the highest value reached 39% in Świnoujście and Hel. In the case of both high mountain stations, the coefficient was as low as 5–7% for both periods. For 13 stations, the linear trends were statistically significant for the period 1966–2018 (9 decreasing [max. value: −3.9 days/10 years in Poznań] and 4 increasing [max. value: 3.5 days/10 years in Włodawa]), while for the years 1985–2018, they were significant for three stations only (2 decreasing and 1 increasing). Relative trends were calculated and for the period 1966–2018 they ranged from +11.7%/10 years in Włodawa to −12.0%/10 years in Świnoujście. For the period 1985–2018, the values ranged from +12.5/10 years in Włodawa to −18.5%/10 years in Hel. For high mountain stations the trends were statistically insignificant.

In the case of **warm half of the year (April–September)**, the highest mean value for the period 1966–2018 was noted for Łeba (32.1 days), while the lowest was for Świnoujście (8.4) (Fig. 18.4). In the case of high mountain stations, the values reached 143.6 for Kasprowy Wierch and 148.6 for Śnieżka. For the period 1985–2018, the highest and lowest values were very similar: 31.7 for Łeba, 6.1 for Świnoujście, 140.9 for Kasprowy Wierch and 145.9 for Śnieżka. The spatial variability of the values was similar to that described above for annual values. The coefficient of variability for the period 1966–2018 shows that at all the stations the seasonal number of days with fog changes significantly from year to year. The values exceeded 25% for all

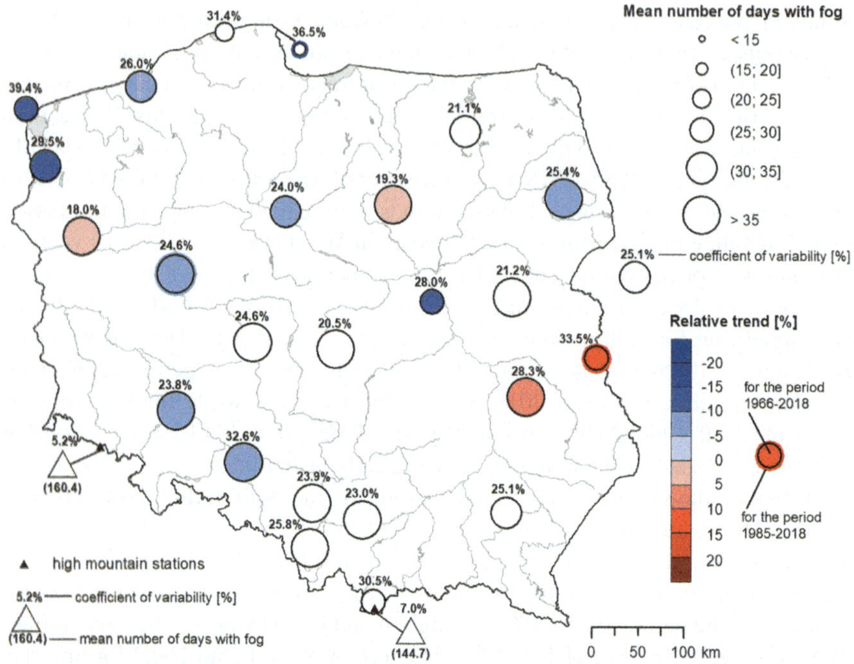

Fig. 18.3 The mean number of days with fog in the cold half of the year in the period 1966–2018, the coefficient of variability (%) at the stations studied and relative trends (%/10 years) for the periods 1966–2018 and 1985–2018. Explanations: only those relative trends which are statistically significant are marked. Data for two high mountain stations: Kasprowy Wierch and Śnieżka are presented separately. Trends which are statistically insignificant are indicated with colourless circles

stations and the highest value was 72% in Świnoujście. In the period 1985–2018, the index was equally variable and the highest value reached 78% in Świnoujście. In the case of both high mountain stations, the coefficient was as low as 8–9% for both periods. For 22 stations, the linear trends were statistically significant for the period 1966–2018, including the two high mountain stations (19 decreasing [max. value: –2.9 days/10 years in Śnieżka Mt.] and 3 increasing [max. value: 2.3 days/10 years in Siedlce]), while for the years 1985–2018 they were significant for 10 stations (all of them decreasing). Relative trends were calculated and for the period 1966–2018they ranged from +9.4%/10 years in Siedlce to –22.3%/10 years in Warsaw. For the period 1985–2018, the values ranged from –11.5%/10 years in Łódź to –23.4%/10 years in Rzeszów, and for nine stations the decreasing trend was much larger in the years 1985–2018 than in the whole period 1966–2018. For high mountain stations the trends were statistically insignificant.

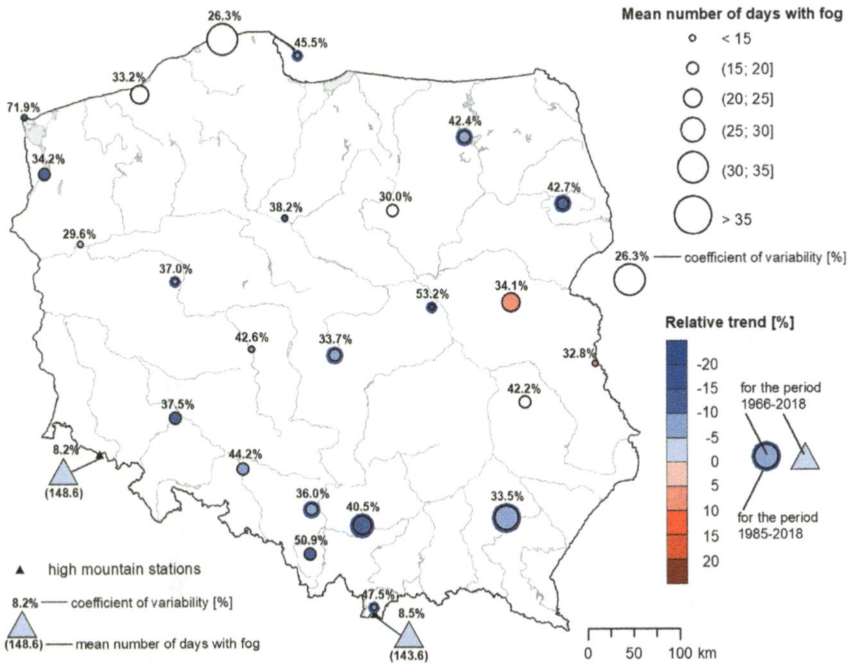

Fig. 18.4 The mean number of days with fog in the warm half of the year in the period 1966–2018, the coefficient of variability (%) at the stations studied and relative trends (%/10 years) for the periods 1966–2018 and 1985–2018. Explanations: only those relative trends which are statistically significant are marked. Data for two high mountain stations: Kasprowy Wierch and Śnieżka are presented separately. Trends which are statistically insignificant are indicated with colourless circles

Number of Days with Fog with a Given Duration

For the period 1966–2018, the **mean annual number of days with fog with a duration of <6 h** varied from 43.7 days in Mława to 18.4 days in Hel (Fig. 18.5). In the case of high mountain stations, the values reached 63.0 for Kasprowy Wierch and 48.4 for Śnieżka. For the period 1985–2018, the highest and lowest values were very similar: 50.3 days in Mława, 18.1 for Hel, 64.6 for Kasprowy Wierch and 49.2 for Śnieżka. Again, in each region, a large spatial variability can be observed. The coefficient of variability for the period 1966–2018 shows that for some of the stations the seasonal number of days with fog changes significantly from year to year, as for 10 stations the values exceeded 25% and the highest value was 37% in Opole. In the period 1985–2018, the index was less variable as in 5 cases the coefficient exceeded 25% and the highest value reached 37% in Opole. In the case of both high mountain stations, the coefficient was as low as 12–17% for both periods. For 12 stations, the linear trends were statistically significant for the period 1966–2018 (7 decreasing [max. value: −3.6 days/10 years in Warsaw] and 5 increasing [max. value: 4.7 days/10 years in Włodawa]), while for the years 1985–2018 they

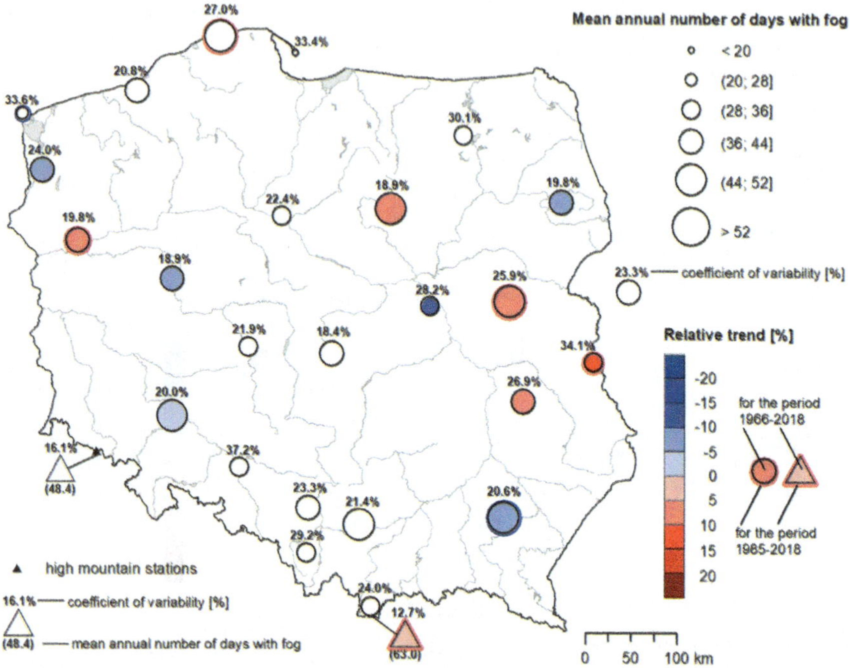

Fig. 18.5 The mean annual number of days with fog with a duration of <6 hours in the period 1966–2018, the coefficient of variability (%) at the stations studied and relative trends (%/10 years) for the periods 1966–2018 and 1985–2018. Explanations: only those relative trends which are statistically significant are marked. Data for two high mountain stations: Kasprowy Wierch and Śnieżka are presented separately. Trends which are statistically insignificant are indicated with colourless circles

were significant for 7 stations only (1 decreasing and 6 increasing). Relative trends were calculated and for the period 1966–2018 they ranged from +13.9%/10 years in Włodawa to −12.5%/10 years in Warsaw. For the period 1985–2018, the values ranged from +12.9/10 years in Świnoujście to −14.1%/10 years in Rzeszów, and in the case of two stations significant changes in the trends intensity can be seen. The increasing trend in Włodawa was much smaller in 1985–2018: 9.3%/10 years, while in Rzeszów the value for 1966–2018 was smaller: −5.1%/10 years. For high mountain stations the trends were statistically significant only for Kasprowy Wierch (1966–2018: +3.8%/10 years; 1985–2018: +7.3%/10 years).

During the **cold half of the year (October–March)**, the mean seasonal number of days with fog with a duration of <6 h varied from 29.0 days in Mława to 10.2 days in Hel (Fig. 18.6). In the case of high mountain stations the values reached 24.5 for Kasprowy Wierch and 16.3 for Śnieżka. For the period 1985–2018, the highest and lowest values were very similar: 31.1 days in Mława, 10.2 for Hel, 24.7 for Kasprowy Wierch and 16.6 for Śnieżka. Again, in each region a large spatial variability can be observed. The coefficient of variability for the period 1966–2018 shows that for most of the stations the seasonal number of days with fog changes significantly from

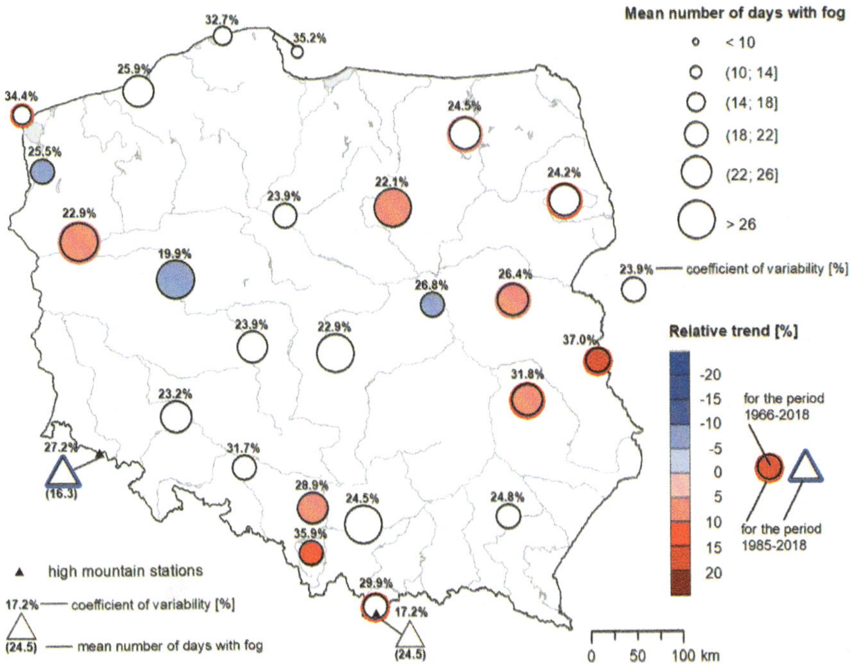

Fig. 18.6 The mean number of days with fog with a duration of <6 hours in the cold half of the year in the period 1966–2018, the coefficient of variability (%) at the stations studied and relative trends (%/10 years) for the periods 1966–2018 and 1985–2018. Explanations: only those relative trends which are statistically significant are marked. Data for two high mountain stations: Kasprowy Wierch and Śnieżka are presented separately. Trends which are statistically insignificant are indicated with colourless circles

year to year, as for 15 stations the values exceeded 25% and the highest value was 37% in Włodawa. In the period 1985–2018 the situation is very similar, i.e. the values exceeded 25% for 15 stations as well, but the highest value of 37% was noted at Łeba and Hel. In the case of high mountain stations, the coefficient was as low as 17% for Kasprowy Wierch for both periods but for Śnieżka it reached 27–29%. For 10 stations, the linear trends were statistically significant for the period 1966–2018 (3 decreasing [max. value: −1.5 days/10 years in Poznań] and 7 increasing [max. value: 3.3 days/10 years in Włodawa]), while for the years 1985–2018 they were significant for 9 stations only (1 decreasing and 8 increasing). Relative trends were calculated and for the period 1966–2018, they ranged from +15.9%/10 years in Włodawa to −7.3%/10 years in Warsaw. For the period 1985–2018, the values ranged from +15.1/10 years in Zakopane to +7.4%/10 years in Gorzów Wielkopolski. For high mountain stations the trends were statistically significant only for Śnieżka and reached −10.2%/10 years.

During the **warm half of the year (April–September)**, the mean seasonal number of days with fog with a duration of <6 h varied from 27.6 days in Łeba to 6.5 in

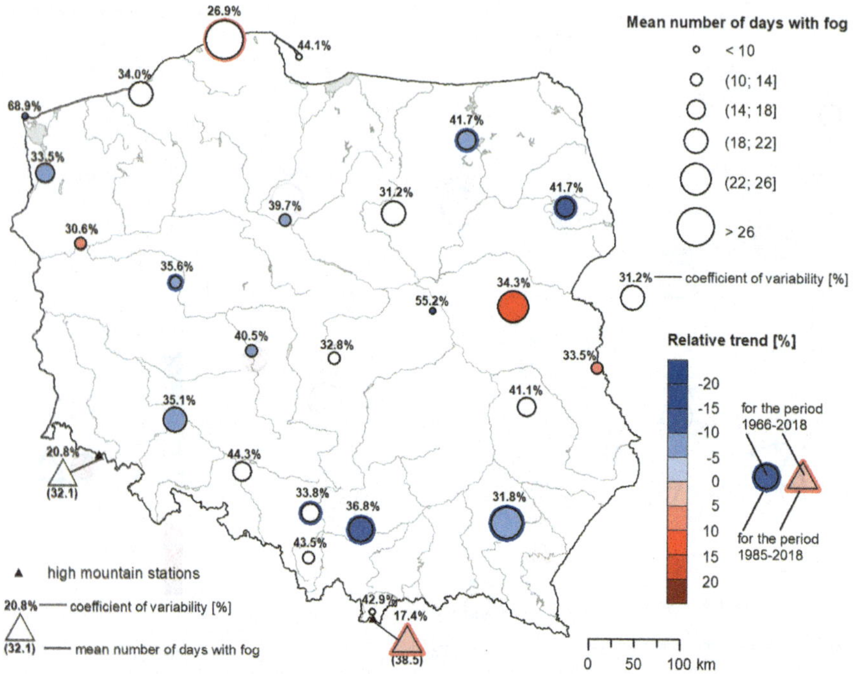

Fig. 18.7 The mean number of days with fog with a duration of <6 hours in the warm half of the year in the period 1966–2018, the coefficient of variability (%) at the stations studied and relative trends (%/10 years) for the periods 1966–2018 and 1985–2018. Explanations: only those relative trends which are statistically significant are marked. Data for two high mountain stations: Kasprowy Wierch and Śnieżka are presented separately. Trends which are statistically insignificant are indicated with colourless circles

Świnoujście (Fig. 18.7). In the case of high mountain stations the values reached 38.5 for Kasprowy Wierch and 32.1 for Śnieżka. For the period 1985–2018, the highest and lowest values were very similar: 27.9 days in Łeba, 5.1 for Świnoujście, 24.7 for Kasprowy Wierch and 16.6 for Śnieżka. Again, in each region a large spatial variability can be observed. The coefficient of variability for the period 1966–2018 shows that for all of the stations the seasonal number of days with fog changes significantly from year to year, all values exceeded 25%, with the highest one 69% for Świnoujście. In the period 1985–2018, the situation is very similar, and again the highest value of 69% was obtained for Świnoujście. In the case of high mountain stations, the coefficient reached 17–21% for both periods. For 15 stations, the linear trends were statistically significant for the period 1966–2018 (11 decreasing [max. value: −2.1 days/10 years in Warsaw] and 4 increasing [max. value: 2.4 days/10 years in Siedlce]), while for the years 1985–2018 they were significant for 9 stations (7 decreasing and 2 increasing). Relative trends were calculated and for the period 1966–2018 they ranged from +10.6%/10 years in Siedlce to −21.9%/10 years in Warsaw. For the period 1985–2018, the values were from +8.6/10 years in Łeba to −21.3%/10 years in Rzeszów, and for five stations the decreasing trends were intensified. For high mountain stations the trends were statistically significant for Kasprowy Wierch only and reached +9.8%/10 years.

In case of **days with fog with a duration of 6–12 h**, the analyses could be performed for the whole year and for the cold half of the year only, as in the warm half such days occurred very irregularly. The **mean annual** number of days with fog with a duration of 6–12 h varied from 15.0 days in Kraków to 5.9 days in Warsaw (Fig. 18.8). In the case of high mountain stations the values reached 61.5 for Kasprowy Wierch and 50.8 for Śnieżka. For the period 1985–2018, the highest and lowest values were very similar: 13.7 days in Mława, 4.8 for Warsaw, 61.9 for Kasprowy Wierch and 50.9 for Śnieżka. Again, in each region a large spatial variability can be observed. The coefficient of variability for the period 1966–2018 shows that for all of the stations the seasonal number of days with fog changes significantly from year to year, as for all of them the values exceeded 25% and the highest value was 67% in Świnoujście. In the period 1985–2018, the index was equally high and the highest value reached 74% in Świnoujście. In the case of both

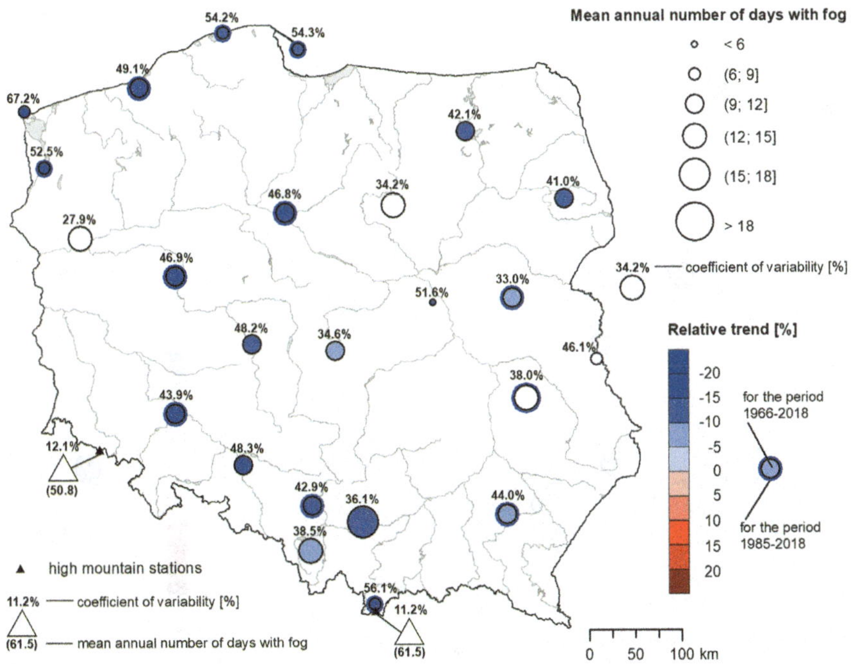

Fig. 18.8 The mean annual number of days with fog with a duration of 6–12 hours in the period 1966–2018, the coefficient of variability (%) at the stations studied, and relative trends (%/10 years) for the periods 1966–2018 and 1985–2018. Explanations: only those relative trends which are statistically significant are marked. Data for two high mountain stations: Kasprowy Wierch and Śnieżka are presented separately. Trends which are statistically insignificant are indicated with colourless circles

high mountain stations, the coefficient was as low as 10–12% for both periods. For 20 stations, the linear trends were statistically significant for the period 1966–2018 (all of them decreasing [max. value: −2.2 days/10 years in Świnoujście]), while for the years 1985–2018 they were significant for 12 stations only (all of them decreasing). Relative trends were calculated and for the period 1966–2018 they ranged from −5.6%/10 years in Siedlce to −28.6%/10 years in Świnoujście. For the period 1985–2018, the values were from −13.9%/10 years in Katowice to −37.8%/10 years in Hel, and for eight stations the decreasing trends were intensified. For high mountain stations the trends were statistically insignificant.

In the **cold half of the year**, the mean seasonal number of days with fog with a duration of 6–12 h varied from 12.5 days in Kraków to 3.9 days in Hel (Fig. 18.9). In the case of high mountain stations the values reached 29.3 for Kasprowy Wierch and 20.5 for Śnieżka. For the period 1985–2018, the highest and lowest values were very similar: 12.5 days in Mława, 4.0 for Hel and Łeba, 29.6 for Kasprowy Wierch and 20.9 for Śnieżka. Again, in each region a large spatial variability can be observed. The coefficient of variability for the period 1966–2018 shows that for all the stations the

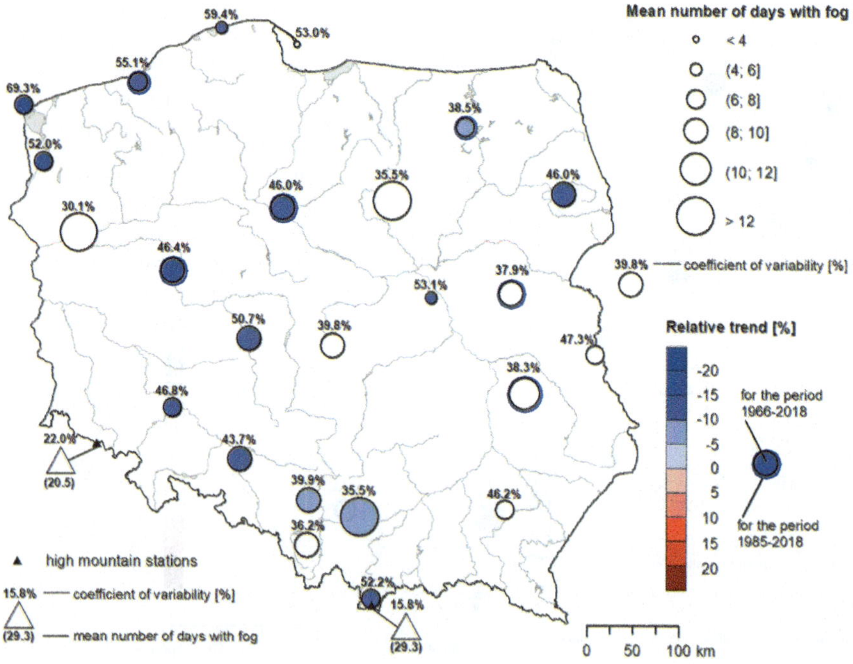

Fig. 18.9 The mean number of days with fog with a duration of 6–12 hours in the cold half of the year in the period 1966–2018, the coefficient of variability (%) at the stations studied, and relative trends (%/10 years) for the periods 1966–2018 and 1985–2018. Explanations: only those relative trends which are statistically significant are marked. Data for two high mountain stations: Kasprowy Wierch and Śnieżka are presented separately. Trends which are statistically insignificant are indicated with colourless circles

seasonal number of days with fog changes significantly from year to year, as for all of them the values exceeded 25% and the highest value was 69% in Świnoujście. In the period 1985–2018, the situation is very similar, i.e. the values exceeded 25% for all stations as well, but the highest value of 76% was noted at Świnoujście. In the case of high mountain stations, the coefficient was as low as 15–22% in both periods. For 15 stations, the linear trends were statistically significant for the period 1966–2018 (all of them decreasing (max. value: −1.6 days/10 years in Świnoujście)), while for the years 1985–2018 they were significant for 7 stations only (all of them decreasing). Relative trends were calculated and for the period 1966–2018 they ranged from −6.4%/10 years in Kraków to −25.7%/10 years in Świnoujście. For the period 1985–2018, the values ranged from −12.2%/10 years in Lublin to −21.1%/10 years in Poznań. and for four stations the decreasing trends were intensified. For high mountain stations the trends were statistically insignificant.

Days with **fog with a duration of >12 h** occur very rarely in most of the territory of Poland, except the mountains. The mean annual number of such days varied from 8.0 days in Bielsko-Biała to 1.8 days in Warsaw in the years 1966–2018. In the case of high mountain stations the values reached 163.7 for Kasprowy Wierch and 209.8 for Śnieżka. For the period 1985–2018, the values were the following: 7.1 days in Bielsko-Biała, 1.4 days in Warsaw and Łeba, 159.3 days at Kasprowy Wierch and 206.1 days at Śnieżka. The coefficient of variability could be calculated for high mountain stations only and for both periods it was as low as 8–10%. No statistically significant trends were found for those two stations.

Number of Days with Fog of a Given Intensity

Three categories of fog intensity were considered: (1) dense fog: visibility <200 m, (2) moderate fog: visibility <500 m, (3) thin fog: visibility <1000 m. For each combination of the factors mentioned, only the mean values of the series could be calculated, except high mountain stations where the trends can also be established. The highest **mean annual** numbers of days with fog of a given intensity are the following:

- at 6 UTC: dense fog: 1966–2018: 9.3 days, 1985–2018: 9.4 days (Gorzów Wielkopolski); moderate fog: 1966–2018: 9.7 days, 1985–2018: 9.6 days (Kraków); thin fog: 1966–2018: 11.4 days, 1985–2018: 11.5 days (Lublin),
- at 12 UTC: dense fog: 1966–2018: 2.2 days, 1985–2018: 2.3 days (Gorzów Wielkopolski); moderate fog: 1966–2018: 2.7 days, 1985–2018: 2.5 days (Mława); thin fog: 1966–2018: 6.7 days, 1985–2018: 6.2 days (Bielsko-Biała),
- at 18 UTC: dense fog: 1966–2018: 3.3 days, 1985–2018: 2.9 days (Bielsko-Biała); moderate fog: 1966–2018: 2.6 days (Bielsko-Biała), 1985–2018: 2.6 days (Lublin); thin fog: 1966–2018: 5.7 days, 1985–2018: 5.8 days (Bielsko-Biała).

For high mountain stations, the **mean annual** number of days with dense fog is very high for all measurement terms, while for other intensities the values are much lower. The linear trends could be calculated for dense fog series only. For

Kasprowy Wierch, the mean annual number of days with dense fog in the years 1966–2018 reached 147 at 6 UTC, 148 at 12 UTC and 154 at 18 UTC. For Śnieżka, the values are 173, 149 and 162, respectively. For the period 1985–2018, the values for Kasprowy Wierch are: 154, 158 and 160, while for Śnieżka they reach: 176, 153 and 165. The coefficient of variability for all the series mentioned varied in the range of 9–15%. For the period 1966–2018, linear trends were statistically significant for Kasprowy Wierch only, and for all terms they presented an increase of the index (relative trends in the range 2.8–5.6%/10 years). In the years 1985–2018, significant trends were obtained for 6 (both stations) and 18 (Kasprowy Wierch) UTC, only. Unlike for the longer period, all trends are negative (relative trends from −3.4 to −5.5%/10 years). Most of the days with dense fog were noted in the **cold half of the year**: 53–64% of the annual values. In the period 1966–2018, linear trends were statistically significant for all terms for Kasprowy Wierch and in the case of Śnieżka for 12 and 18 UTC. They indicated an increasing tendency (relative trends in the range 4.0–7.4%/10 years). However, for the period 1985–2018, none of the seasonal trends was found to be statistically significant.

Discussion and Conclusion

Through the analyses presented above it was possible to determine a few basic features of fog occurrence in Poland. First of all, mountain areas show a distinctively different regime of fog frequency than the rest of the country. Most days of the year (around 80%) are foggy, and on about 60% of those days, the fog lasts over 12 h. For each measurement term, on over 50% of days a dense fog is noted. Those values are linked to the fact that the basis of low-level clouds is often located below mountain peaks, and, additionally, there are many weather phenomena specific to mountainous climates which can cause water vapour condensation. No significant multi-annual changes in fog occurrence can be defined for the mountainous areas and the inter-annual variability is very low. The only exception could be the frequency of dense fog but the trends are opposite for 1966–2018 and 1985–2018, and furthermore, they are not significant for many cases. Those results are in accordance with similar studies for other mountains in Europe, as mentioned in the introductory section.

Another area that shows a distinctively different regime of fog frequency than the rest of the country is the seaside area, located northernmost and exposed to the impact of the open sea, represented by the station in Hel. Frequent strong winds contribute to the fact that the number of days with fog is the lowest there (in comparison to the rest of the country) and the inter-annual variability is high. Therefore, no statistically significant trends can be determined.

In the lowland areas of northern and central Poland, and in the uplands of southern Poland, specific local conditions seem to be the factors determining the fog frequency. In each region, a high variability in fog occurrence can be observed among the stations located there. In terms of long-term trends, there is no unification. For those stations where fog occurs most often, statistically significant increasing trends were obtained, while for the others the trends are either decreasing or not significant statistically. The only area which shows some uniformity is Central-Eastern Poland, represented by the stations in Siedlce, Lublin and Włodawa. Strong increasing trends in annual number of days with fog and days with fog lasting <6 h can be observed at all three stations. Similar features can be found for two stations located at similar latitude but further to the west: Mława and Gorzów Wielkopolski. However, at other neighbouring stations, opposite tendencies can be observed, so no other region with uniform tendencies can be identified. Those results are verified by the study published by Vautard et al. in 2009 (data for the years 1977–2007), where the whole territory of Poland is included in a zone with a decrease in the days with low visibility. For most series analysed in the present paper, high values of the coefficient of variability were found and that feature seems to be the most important one concerning the application of the results obtained. In terms of seasonal variability, fog occurs mainly in the cold half of the year, but, other than mountainous areas, it occurs very irregularly and can constitute a potentially dangerous weather factor for many human activities. As far as fog duration is concerned, for those stations where a statistically significant increasing trend in the number of days with fog (annual and in the cold half of the year) was found, a significant increase in days with fog <6 h was found as well. For longer durations, only decreasing or no trends were obtained which might suggest that we can expect similar conditions in the future. Bendix (1994, 2001, 2002) showed that satellite images can be very useful in fog research. Taking under consideration that fog is a very localized phenomenon, it seems necessary to develop downscaling methods of satellite image processing in order to improve fog prediction on a local scale as well as the spatial approach in long-term analyses.

References

Avotniece Z, Klavins M, Lizuma L (2015) Fog climatology in Latvia. Theor Appl Clim 122:97–109

Bendix J (1994) Fog climatology of the Po Valley. Rivista di Meteorologia Aeronautica 3–4:25–36

Bendix J (2001) A 10-years fog climatology of Germany and the Alpine region based on satellite data—Preliminary results. In: abstracts of the Second International Conference on Fog and Fog Collection, St. John's, Canada, 15–20 July, 2001:357–360

Bendix J (2002) A satellite-based climatology of fog and low-level stratus in Germany and adjacent areas. Atmos Res 64:3–18

Błaś M, Sobik M (2000) Mgła w Karkonoszach i wybranych masywach górskich Europy (Fog in Giant Mts. and selected European massifs). Opera Corcontica 37:35–46

Błaś M, Sobik M (2004) The distribution of fog frequency in the Carpathians. Geogr Polon 77:19–34

Błaś M, Sobik M, Quiel F, Netzel P (2002) Temporal and spatial variations of fog in the Western Sudety Mts., Poland. Atmos Res 64:19–28

Cermak J, Eastman RM, Bendix J, Warren SG (2009) European climatology of fog and low stratus based on geostationary satellite observations. Q J R Meteorol Soc 135:2125–2130

Egli S, Thies B, Drönner J, Cermak J, Bendix J (2017) A 10 year fog and low stratus climatology for Europe based on Meteosat second generation data. Q J R Meteorol Soc 143:530–541

Fu GQ, Xu WY, Yang RF, Li JB, Zhao CS (2014) The distribution and trends of fog and haze in the North China Plain over the past 30 years. Atmos Chem Phys 14:11949–11958

Gautam R, Singh MK (2018) Urban heat island over Delhi punches holes in widespread fog in the Indo-Gangetic Plains. Geoph Res Let 45:1114–1121

Giulianelli L, Gilardoni S, Tarozzi L, Rinaldi M, Decesari S, Carbone C, Facchini MC, Fuzzi S (2014) Fog occurrence and chemical composition in the Po valley over the last twenty years. Atmos Environ 98:394–401

Hess M (1965) Mgła (Fog). In: Piętra klimatyczne w Polskich Karpatach Zachodnich (Vertical climatic zones in the Polish Western Carpathians). Zesz Nauk UJ, Pr Geogr 11:195–198

Klemm O, Lin NH (2016) What causes observed fog trends: air quality or climate change. Aero Air Qual Res 16:1131–1142

Lange CA, Matschullat J, Zimmermann F, Sterzik G, Wienhaus O (2003) Fog frequency and chemical composition of fog water—a relevant contribution to atmospheric deposition in the eastern Erzgebirge, Germany. Atmos Environ 37:3731–3739

Łupikasza E, Niedźwiedź T (2016) Synoptic climatology of fog in selected locations of southern Poland (1966–2015). Bull Geogr, Phys Geogr Ser 11:5–15

Niu S, Lu C, Yu H, Zhao L, Lü J (2010) Fog research in China: an overview. Adv Atmos Sci 27(3):639–661

Quan J, Zhang Q, He H, Liu J, Huang M, Jin H (2011) Analysis of the formation of fog and haze in North China Plain (NCP). Atmos Chem Phys 11:8205–8214

Scherrer SC, Appenzeller C (2014) Fog and low stratus over the Swiss Plateau—a climatological study. Int J Climatol 34:678–686

Syed FS, Körnich H, Tjernström M (2012) On the fog variability over south Asia. Clim Dyn 39:2993–3005

Trzeciak S (1992) Mgły i zjawiska im towarzyszące w zachodniej części polskiego wybrzeża Bałtyku (Fogs and phenomena accompanying them in the western part of the Polish Baltic coast). Szczecińskie Roczniki Naukowe (Annales Scientiarum Stetinenses) 7(1):67–85

van Oldenborgh GJ, Yiou P, Vautard R (2010) On the roles of circulation and aerosols in the decline of mist and dense fog in Europe over the last 30 years. Atmos Chem Phys 10:4597–4609

Vautard R, Yiou P, van Oldenborough GJ (2009) Decline of fog, mist and haze in Europe over the past 30 years. Nat Geosci 2:115–119

Veljović K, Vujović D, Lazić L, Vučković V (2015) An analysis of fog events at Belgrade International Airport. Theor Appl Climatol 119:13–24

Witiw MR, LaDochy S (2008) Trends in fog frequencies in the Los Angeles Basin. Atmos Res 87:293–300

Wrzesinsky T, Klemm O (2000) Summertime fog chemistry at a mountainous site in central Europe. Atmos Environ 34:1487–1496

Wypych A (2003) Air humidity and fogs in Cracow in the period 1961–2000 in relation to synoptic situations In: Górka Z (ed) Changes of geographical environment and those of socio-economic phenomena. Zesz Nauk UJ, Pr Geogr 112:105–114

Żarnowiecki G (2000) Występowanie dni z mgłą na stacjach meteorologicznych Święty Krzyż, Kielce i Nowa Słupia (The occurrence of days with fog in Święty Krzyż, Kielce and Nowa Słupia). Rocz Świętokrz Ser B 27:265–279

Chapter 19
Changes in Bioclimatic Indices

Magdalena Kuchcik, Krzysztof Błażejczyk, and Agnieszka Halaś

Abstract This paper assesses changes in bioclimatic conditions in Poland using two indices: the Universal Thermal Climate Index (UTCI) and Physiological Subjective Temperature (PST). The indices were calculated using the BioKlima 2.6 software package on the basis of several meteorological parameters (temperature, humidity, cloudiness, wind speed), measured daily at 12:00 UTC, from the period 1951–2018 at 24 stations representing different biometeorological regions. First, the spatial distribution of UTCI and PST values and the frequency of different thermal stress and thermal sensation categories are presented, with the various thermal stress and sensation categories grouped into 3 for each index. Then, the absolute values and relative percentage trends (compared to 1951) are calculated for ten-year periods and days with specific characteristics. The analysis shows a significant increase in the minimum and mean values of the applied indices during the period under investigation over all areas of Poland, including the mountains, especially in the north-east (where UTCI minimum values rose up to 1.8 °C per decade) and Tarnów in the south-east (up to 2.2 °C per decade). The rise of UTCI summer maximum values was significant only in the north-east and north-west of Poland (0.3–0.5 °C per decade) and in the mountain foothills (0.3–0.4 °C per decade). The changes in PST maximum values were insignificant. On average, in every decade, the number of cold days reduced by 2.7 in the case of PST and 3.2 on the UTCI. However, in the north-east, the reduction was much greater, at 7 days per 10 years (UTCI) and 6 days when considering PST. Regionally, the greatest increase in the mean and minimum values of the indices used and the decrease in the frequency of cold days were observed in north-eastern, eastern and western Poland, i.e. the regions most exposed to frequent changes in air circulation, which is hypothesised to play the main role in shaping bioclimatic conditions in Poland.

M. Kuchcik (✉) · K. Błażejczyk
Polish Academy of Sciences, Climate Impacts Laboratory, Institute of Geography and Spatial Organization, Warsaw, Poland
e-mail: mkuchcik@twarda.pan.pl

A. Halaś
Polish Academy of Sciences, Past Landscape Dynamic Laboratory, Institute of Geography and Spatial Organization, Warsaw, Poland

M. Falarz (ed.), *Climate Change in Poland*, Springer Climate,
https://doi.org/10.1007/978-3-030-70328-8_19

Introduction

Human beings are permanently under the influence of atmospheric stimuli. They impact particular systems and organs of our bodies. Special attention must be paid to the so-called "thermal environment", which comprises both the atmospheric heat exchange conditions (stress) and the physiological response (strain) (Jendritzky et al. 2012). The magnitude of thermal impacts depends not only on air temperature but also on the intensity of solar radiation, air humidity and wind speed, which together create man's bio-thermal environment (Gasparini et al. 2015; Köppe et al. 2004). Human organisms react to ambient stimuli and endeavour to balance the heat budget and to keep the thermal equilibrium of the body core.

Balancing the human heat budget in variable atmospheric conditions is achieved by the autonomous thermoregulatory system, additionally supported by behavioural adaptation (International… 2003; Havenith 2001; Kenney and Munce 2003; Parsons 2003). The type and magnitude of adaptation depend on the thermal environment. In hot conditions, heat equilibrium within the body is mainly regulated through increased bodily sweating and consequent evaporation-induced cooling (Elizondo and Bullard 1971; Givoni and Goldman 1973; Kenney 1985; Hajat et al. 2007; Cheshire 2016). In a cold environment, heat balance is regulated by reduction of heat loss from the body (through vasoconstriction) and by production of heat by shivering (le Blanc 1975; Clark and Edholm 1985; Holmér 1988; Guyton and Hall 2006).

Several biometeorological indicators of heat exposure have been developed to incorporate physical properties of the ambient environment. In general, thermal stress indices can be divided into 3 groups (NIOSH 1986; Parsons 2003): 1) Indices that are based on calculations involving the heat balance equation ("rational indices"), 2) Indices that are based on objective and subjective strain ("empirical indices") and 3) indices based on direct measurements of environmental variables ("direct indices"). Indices of the first two groups depend on many variables (both meteorological and physiological). However, the third group needs special, unique measuring devices. The main advantage of rational and direct indices is to assess how ambient conditions influence the human organism. Recently, reviews of biometeorological indices have been presented by Epstein and Moran (2006) and Błażejczyk et al. (2012), as well as by de Freitas and Grigorieva (2017).

To assess changes in bioclimatic conditions in Poland, the authors have used two indices which represent different relationships between the atmosphere and the human organism. The Universal Thermal Climate Index (UTCI, Błażejczyk et al. 2012; Bröede et al. 2012) shows the physiological reactions of an organism to atmospheric stimuli and the intensity of such reactions, defined by objective measures of heat stress and cold stress. However, Physiological Subjective Temperature (PST, Błażejczyk and Matzarakis 2007; Błażejczyk and Kunert 2011) tells us how the Central European population feels thermal stimuli of ambient conditions: PST indicates subjective thermal sensations in man.

Methods

To calculate daily values of UTCI and PST, data of air temperature, relative humidity, total cloud cover and wind speed at 12:00 UTC were applied. The analysed period comprises 68 years, from 1951 to 2018. Because of the complexity of bioclimatic indices, only 24 stations have been chosen.

Calculations of bioclimatic indices were made using the BioKlima 2.6 software package. We consider minimum, mean and maximum values and frequencies of selected categories of thermal stress (UTCI) and thermal sensations (PST). Trends and their statistical significance were verified with Statgraphics Centurion XVI, version 16.2.04. The relative trends of the percentage rise/fall of particular days were calculated using 1951 as the baseline.

The UTCI is defined as the air temperature (Ta) of the reference condition, causing the same model response as the actual condition. Thus, UTCI is the air temperature which would produce, under reference conditions, the same thermal strain as in the actual thermal environment. The offset, i.e. the deviation of UTCI from air temperature, depends on the actual values of air temperature (Ta) and mean radiant temperature (Tmrt), wind speed (va) and humidity, expressed as water vapour pressure (vp) or relative humidity (RH). Table 19.1 presents the labelled stress categories. In the analysis, all stress categories according to UTCI ≤ -13.0 °C were grouped as cold stress, while UTCI > 32 °C was defined as heat stress.

The Physiological Subjective Temperature (PST) is derived from the Man-ENvironment heat EXchange model, first published in 1994 (Błażejczyk 1994). In the most recent modifications (Błażejczyk 2005; Błażejczyk and Kunert 2011), the MENEX_2005 model proposes several thermo-physiological indices, including Physiological Subjective Temperature (PST). PST represents the subjective feeling of the thermal environment that is caused by signals of cold and/or warm receptors in the skin, and in the nervous system. Thus, PST is defined as the temperature that is formed around the skin surface, under clothing, after 15–20 min of adaptation to

Table 19.1 UTCI equivalent temperature categorised in terms of thermal stress

UTCI (°C)	Objective thermal stress	UTCI categories considered in research
< −40	extreme cold stress	Cold stress
(−40, −27]	very strong cold stress	
(−27, −13]	strong cold stress	
(−13, 0]	moderate cold stress	*Not considered*
(0, 9]	slight cold stress	
(9, 26]	no thermal stress	
(26, 32]	moderate heat stress	
(32, 38]	strong heat stress	Heat stress
(38, 46]	very strong heat stress	
> 46	extreme heat stress	

Table 19.2 Thermal sensations according to PST

PST (°C)	Subjective thermal sensation	PST categories considered in research
< −36	Frosty	Cold
(−36, −16]	Very cold	
(−16, 4]	Cold	
(4, 14]	Cool	Comfort
(14, 24]	Comfortable	
(24, 34]	Warm	
(34, 44]	Hot	Hot
(44, 54]	Very hot	
>54	Sweltering	

maintain homeothermy. In the analysis, particular thermal sensation categories were grouped into 3: cold (PST ≤ 4.0 °C), hot (PST > 34.0 °C) and comfort, which includes cool and warm sensations (Table 19.2). As a result, the PST analysis comprised all the days of the year, though UTCI represented only some days, the most "extreme" ones.

Spatial Distribution of Bioclimatic Conditions in Poland

The bioclimatic conditions of Poland are spatially differentiated. According to specific patterns of geographical environment, the monitoring stations can be divided into three groups: coastal (which are influenced by the Baltic Sea), mountainous (where vertical changes of meteorological parameters create the climate and bioclimate of the area) and lowland (which are strongly influenced by oceanic and continental air masses traversing Poland, mainly from the west and east).

However, considering the spatial distribution of absolute maximum, minimum and mean values of UTCI and PST and the number of days with specific conditions, this simple climatic regionalisation will get complicated. The highest UTCI values are noted in the north-west part of Poland and in Tarnów, situated in the south in the Sandomierski Basin, close to the Carpathian foothills (UTCI exceeds 42 °C) (Fig. 19.1a). The spatial differentiation of yearly mean UTCI values follows diagonal belts running from north-west to south-east: the lowest values are noted in the north-east, the highest in the south, except in the mountains (Fig. 19.1b). Stations located in north-eastern and eastern Poland are influenced by frequent impacts of arctic and polar continental air masses. This results in great seasonal differentiation in the bioclimate, with severe winters and hot summers. On average, according to UTCI, 11.9–20.8% of days in the year (43–76 days) represent cold stress and only 0.8–1.8% (3–5 days) show heat stress (Fig. 19.2a, c).

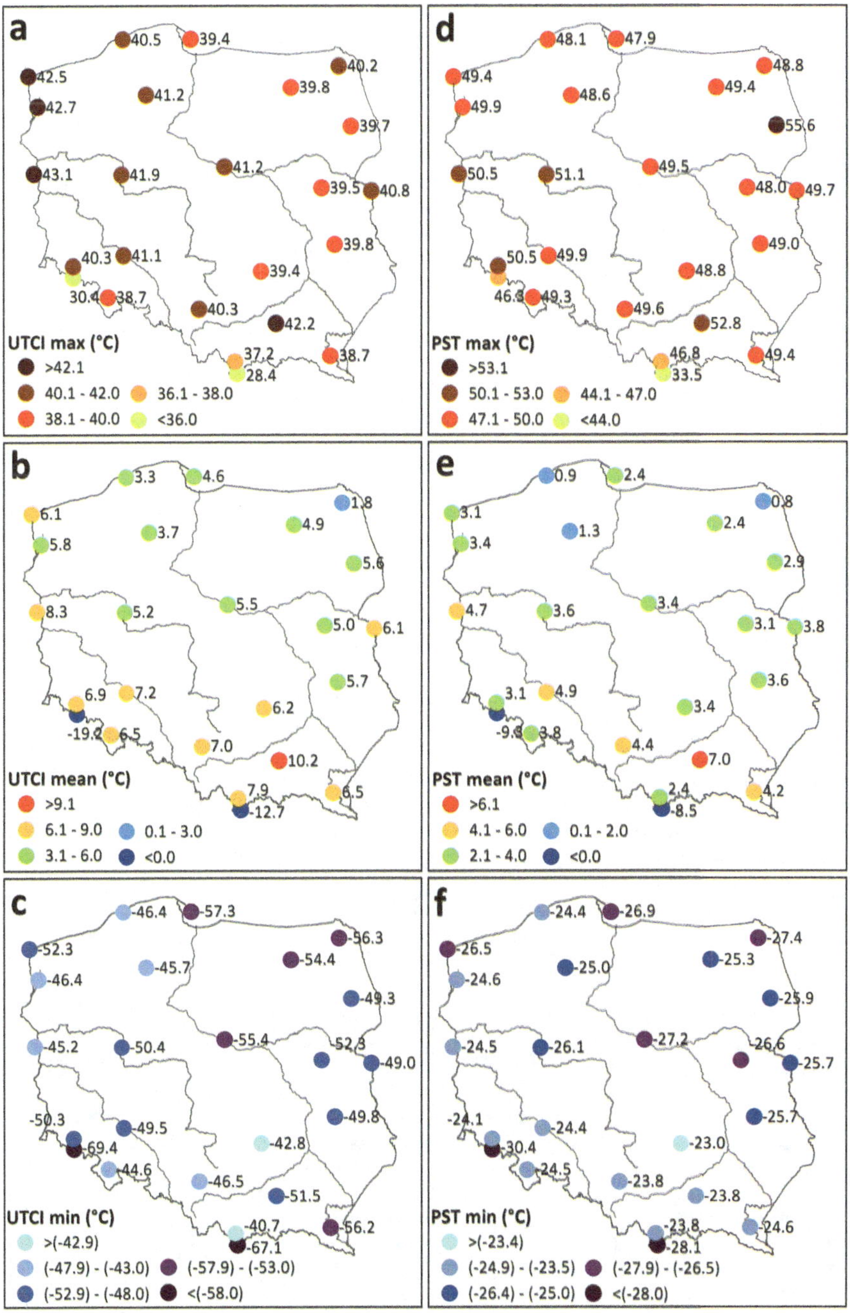

Fig. 19.1 Yearly mean, the absolute high (max) and the absolute low (min) values of UTCI (a, b, c) and PST (d, e, f), 1951–2018

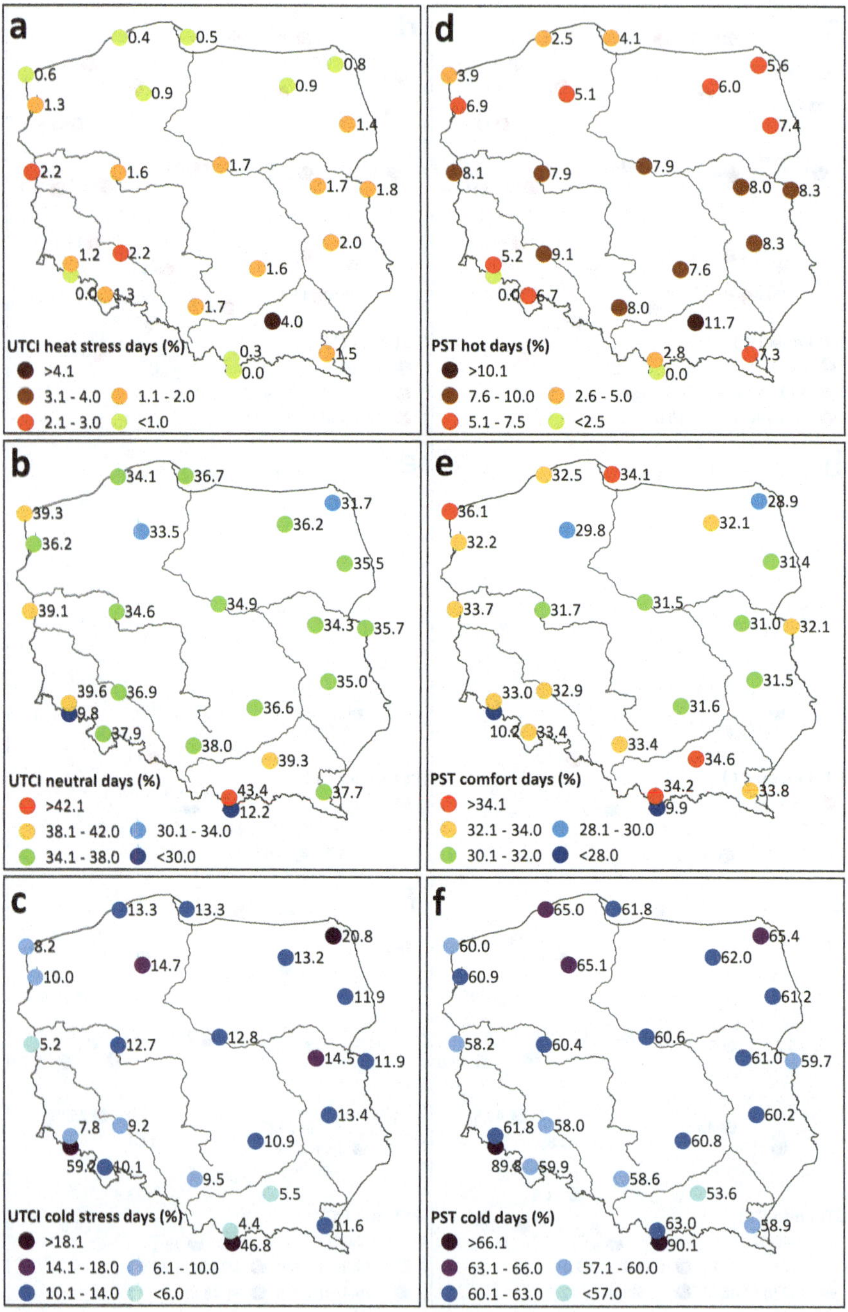

Fig. 19.2 Mean frequency of selected UTCI (a, b, c) and PST categories (d, e, f), 1951–2018

Because Fig. 19.2 presents the frequency of selected UTCI and PST categories (%), the authors decided not to duplicate this data in the text but describe the corresponding yearly number of days.

At the mountain stations, very low values of bioclimatic indices are registered (Fig. 19.1c, f). These regions are characterised by increased frequencies of cold stress and cold thermal sensations, which, according to PST, occur almost 330 days a year (Fig. 19.2f). Heat stress days and hot sensations are almost never observed there (Fig. 19.2a, d).

High absolute UCTI or PST values on the coast are not reflected in the frequency of heat-stress days. Coastal locations are characterised by lower frequency of heat stress (0.4–0.6% of the year, meaning just 1–2 days) and hot thermal sensations (9–15 days) (Fig. 19.2a, d).

Within lowland stations, bioclimatic characteristics vary significantly. In western locations, the impact of oceanic (mostly polar) air masses is observed, giving, on average, only 19 days with cold stress (UTCI) and 8 days with heat stress. The most southerly parts of the lowland areas (Wrocław) can be considered the warmest in Poland, with significantly high frequency of heat stress (8 days) and 33 days with hot thermal sensations. The highest number of heat-stress days is noted in Tarnów (4%, which means 15 per year on average, but 35 days in 2012 and 2015) (Fig. 19.2a).

Changes of Bioclimatic Conditions in Poland, 1951–2018

Multiannual changes of bioclimatic conditions in Poland were analysed for 24 stations, representing different regions. We considered mean, maximum and minimum annual values (°C) of UTCI and PST. The changes in the frequency (number of days per year) of selected UTCI and PST categories were also studied. To compare changes in bioclimatic characteristics, we used 10-year trends of the variables used. The relative trends of the rise/fall of particular days in percentage terms were calculated in relation to 1951.

Changes of Universal Thermal Climate Index (UTCI)

Over such a long period (1951–2018), the changes in UTCI values were noticeable and mostly statistically significant at the 5% level. The average trend of the annual maximum values of UTCI (UTCImax)—which represent the summer (June–August)—is the smallest (+0.2 °C per 10 years), though it reaches +0.5 °C/10 years in the north-east of Poland. 11 of the analysed stations are characterised by small, positive and statistically significant trends from +0.2 °C to +0.5 °C per 10 years and for 2 stations at specific locations (Kasprowy Wierch—the highest station in Poland and Hel—located on the outthrust Baltic Sea peninsula), the trend of UTCImax is negative (Fig. 19.3a).

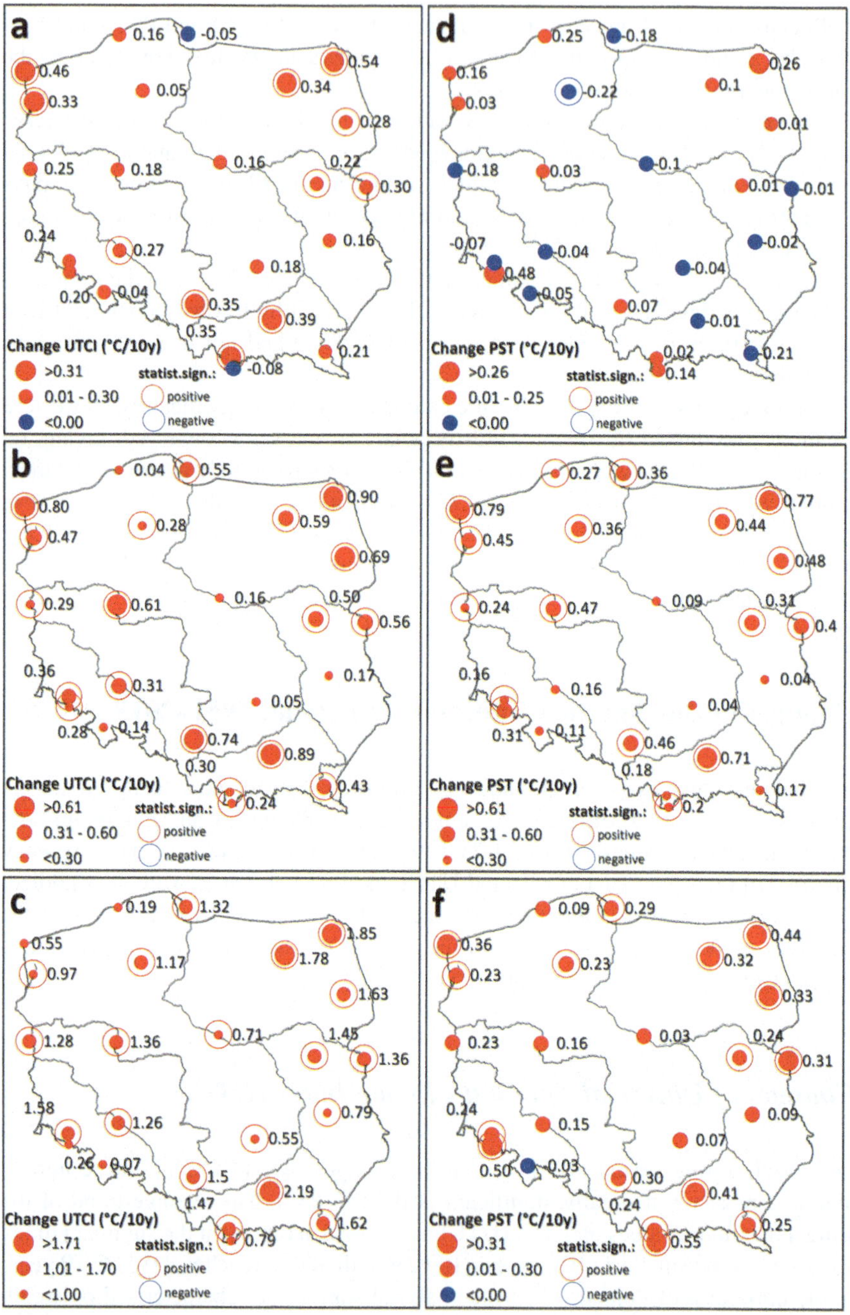

Fig. 19.3 10-years trends of UTCI maximum (a), mean (b) and minimum (c) values and PST maximum (d), mean (e) and minimum (f) values, 1951–2018. Statistical significance at the 5% level

On average, mean UTCI (calculated as the average of particular trends) has increased by +0.4 °C per 10 years, but the trends differ between cities and regions. The biggest rise characterises the coolest (Suwałki +0.9 °C/10 years) and the warmest areas of Poland (Tarnów +0.9 °C/10 years, Świnoujście +0.8 °C/10 years). A much lower rising trend occurs in the central part of Poland (from +0.3 °C to +0.6 °C per 10 years), while even lower trends characterise the mountains (from +0.2 °C to + 0.4 °C per 10 years). However, the positive rising trends are statistically significant for all but 5 of the 24 analysed stations (Fig. 19.3b).

The average trend of the lowest (annual minimum) UTCI values (UTCImin)—which represent the winter—was +1.1 °C/10 years, which was much higher than the trend of mean UTCI. The highest trend was observed in Tarnów (the warmest station), where it reached +2.2 °C per 10 years, but in the north-east of Poland, trends are also very high and equal +1.8 °C/10 years in Suwałki, and only slightly less in Mikołajki and Białystok. At stations located in the central part of Poland, the trends of UTCImin do not show significant regularities (Fig. 19.3c).

Not only trends but also year-to-year changes are important. Figures 19.4, 19.5, 19.7 and 19.8 show examples of UTCI and PST multi-year course in 4 chosen stations. They represent the coastal, the westernmost area (Świnoujście) and mountain regions (Śnieżka), as well as the coldest (Suwałki) and the warmest (Tarnów) areas of Poland.

There are clear rising trends of UTCI at the coldest and the warmest stations. However, in Suwałki, throughout the period, there are sub-periods of clearly lower UTCI values (1970–1986) and of elevated UTCI values 1992–2002, with one, distinct peak in 1994 (Fig. 19.4a, b). In Tarnów, the growth of UTCI is more stable, and the line of 10-year averages lies much closer to the trend line (Fig. 19.4d, e).

When analysing year-to-year changes of UTCI values, it should be emphasised that the fluctuations of UTCImin are 3 times bigger than of UTCImax, especially at stations located in northern and eastern Poland, e.g. Świnoujście, Mikołajki, Białystok, Lesko and Tarnów (Fig. 19.4). The smallest year-to-year changes of UTCImin are typically for summit stations (Kasprowy Wierch and Śnieżka), but even more stable conditions in this context occur at the foot of the mountains (Zakopane, Jelenia Góra). The changes of UTCImax are the highest on the western, northern and north-eastern edges of Poland, the areas frequently reached by fresh air masses (Świnoujście, Szczecin, Ustka, Suwałki). The smallest year-to-year UTCImax changes characterise southern regions of Poland (Katowice, Lesko).

In humans, physiological reactions to heat and cold stress or the perception of the environment as neutral according to UTCI, are more important than UTCI values themselves: this is why the changes in the frequency of thermal stress categories are analysed. In this study, we considered these three categories of thermal stress (Table 19.1): heat, cold and no thermal stress.

The obvious rise of UTCI values results in the increase of heat-stress days (most spectacularly in Tarnów and Suwałki, but high also in Wrocław or Słubice) and neutral days (especially high in Świnoujście). The lowest changes in the number of heat-stress days are observed on the sea coast. The decrease of cold-stress days is high and stable (as in Tarnów) or outstanding but of great variability, as in Suwałki, characterised by periods of high frequency of cold-stress days (1970–1984, up to

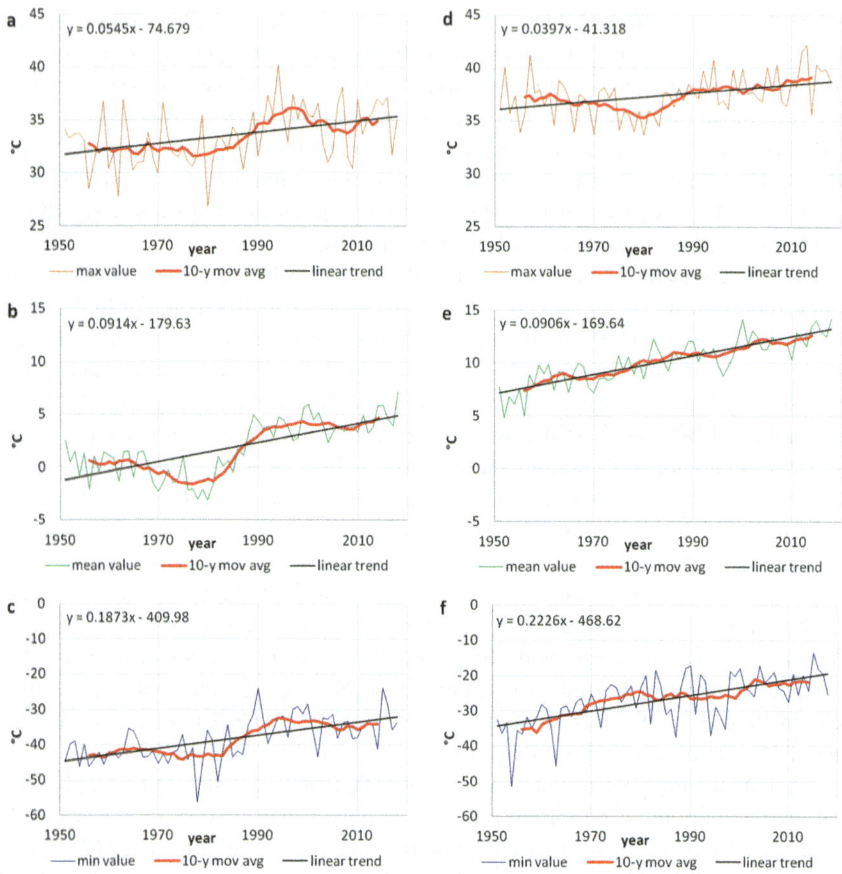

Fig. 19.4 Changes in maximum (max), mean and minimum (min) annual values of UTCI in Suwałki (a, b, c) and Tarnów (d, e, f), 1951–2018

115 days in 1980), followed by low frequency years (48 days in 1990 and only 32 in 2015) (Fig. 19.5).

Primarily, at all stations, a strong, negative trend in the annual number of cold-stress days is observed. Its average value is −3.2 days/10 years, which gives 6% per decade compared to 1951. The greatest reduction in cold-stress days is observed in Suwałki (−6.9 days/10 years; relative trend: − 7% per 10 years); Hel (−6.2 days/10 years; relative trend: −8.9%) and Białystok (−5.8 days/10 years; relative trend: −9.2%). At 18 of the 24 analysed stations, the decrease in the number of cold-stress days is statistically significant (Fig. 19.6c, f).

At most of the stations which experienced a significant drop in cold stress, a significant rise in neutral thermal conditions was noted. The average increasing trend of no-thermal-stress days equals 1.7 days/10 years (relative trend: 1.5%/10 years)

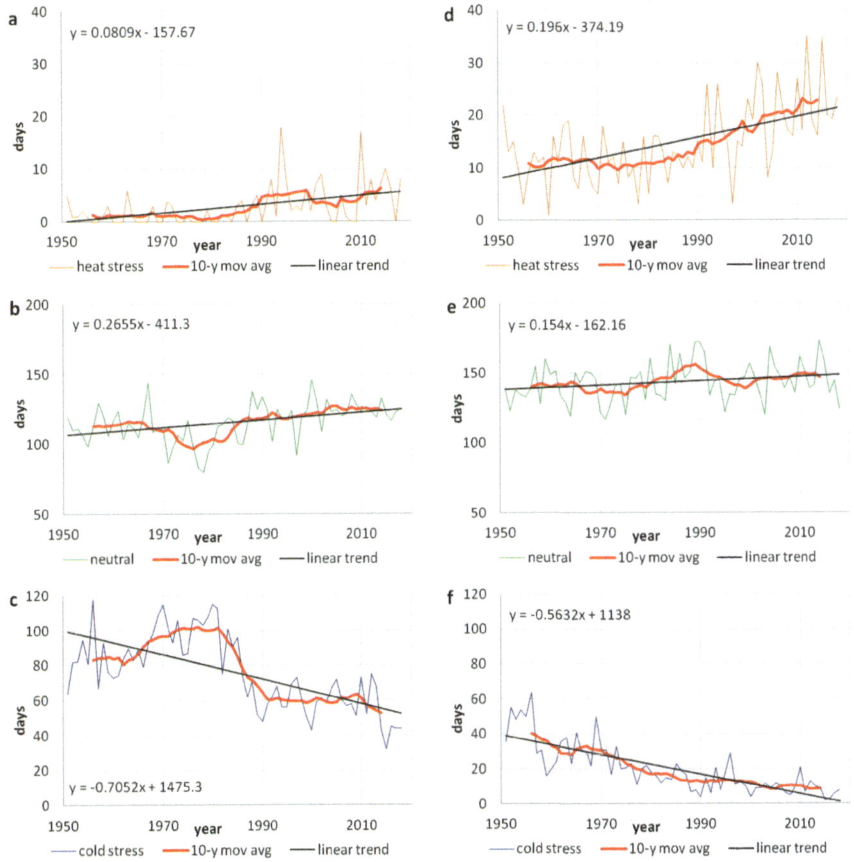

Fig. 19.5 Changes in annual frequency of days with different thermal stress categories according to UTCI (heat, neutral, cold stress) in Suwałki (a, b, c), Tarnów (d, e, f), Świnoujście (g, h, i) and Śnieżka (j, k), 1951–2018

but it ranges from +5.3 days/10 years in Świnoujście, through 1.3–2.9 days/10 years in central Poland to 0.8–1.0 days/10 years in Lesko and Zakopane (Fig. 19.6b, d).

The rise of heat stress is much lower than the reduction in cold-stress days, at an average of 0.5 days/10 years, but it gives a 49% relative trend per decade as compared to 1951. In the south, it reaches 1.9 days/10 years in Tarnów and 0.6–1.1 days/10 years in the east of Poland (Terespol, Suwałki, Białystok, Lublin). However, most stations are characterised by small, statistically insignificant rises in heat stress days (Fig. 19.6a). Although the absolute value of the trend in Suwałki was only 0.8 days/10 years, the relative trend showed almost 700% rise in heat-stress days in the analysed period as compared to 1951, since the observed frequency of those days has risen from 1.4 in the decade 1951–1960 to 6.3 in the last 10 years (Fig. 19.6d).

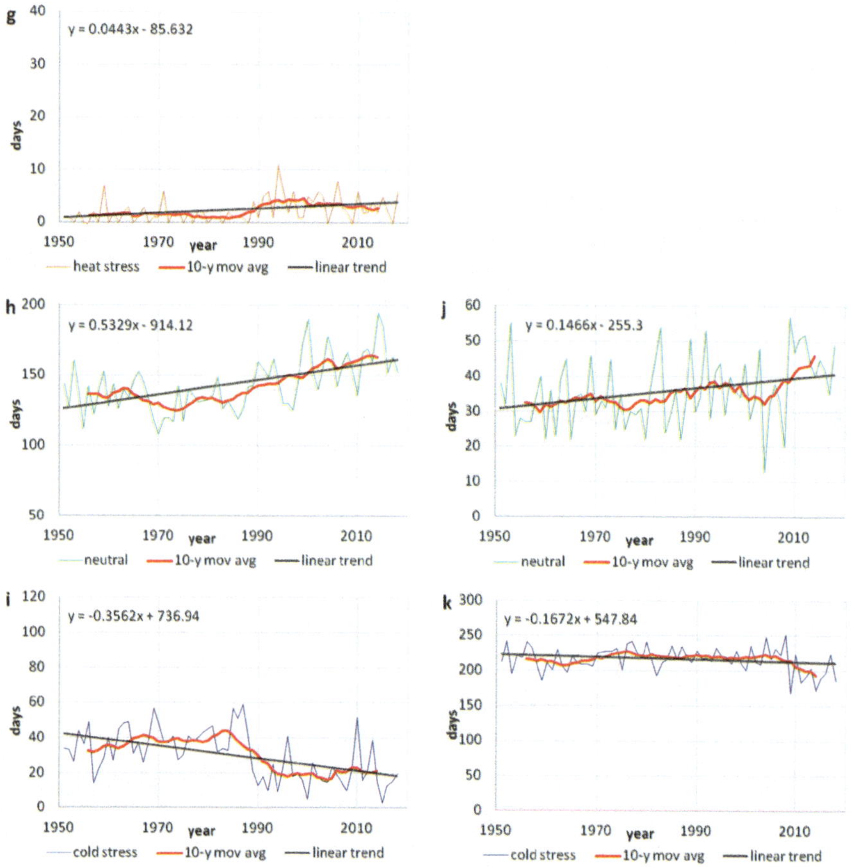

Fig. 19.5 (continued)

Changes of Physiological Subjective Temperature (PST)

At the majority of stations, changes in mean annual PST values are statistically significant at the 5% level. The average trend is +0.33 °C per 10 years. However, they vary spatially. The highest trends were found for the coldest and the warmest regions (Suwałki +0.8 °C, Świnoujście +0.8 °C and Tarnów +0.7 °C per 10 years). At stations located in the central part of Poland (Wrocław, Płock, Kielce, Lublin), trends are, for the most part, low and statistically insignificant. Comparing the mountain regions, a higher trend occurs in the Sudetes (+0.3 °C/10 years at Śnieżka) than in the Carpathians (0.2 °C/10 years at Kasprowy Wierch) (Fig. 19.3e).

Changes in the minimum annual values of PST are also statistically significant. These represent the cold season (December–March). The average trend is + 0.2 °C/10 years and, again, high values (about +0.4 °C/10 years) were found for Świnoujście, Suwałki and Tarnów and the lowest for central Poland. The highest

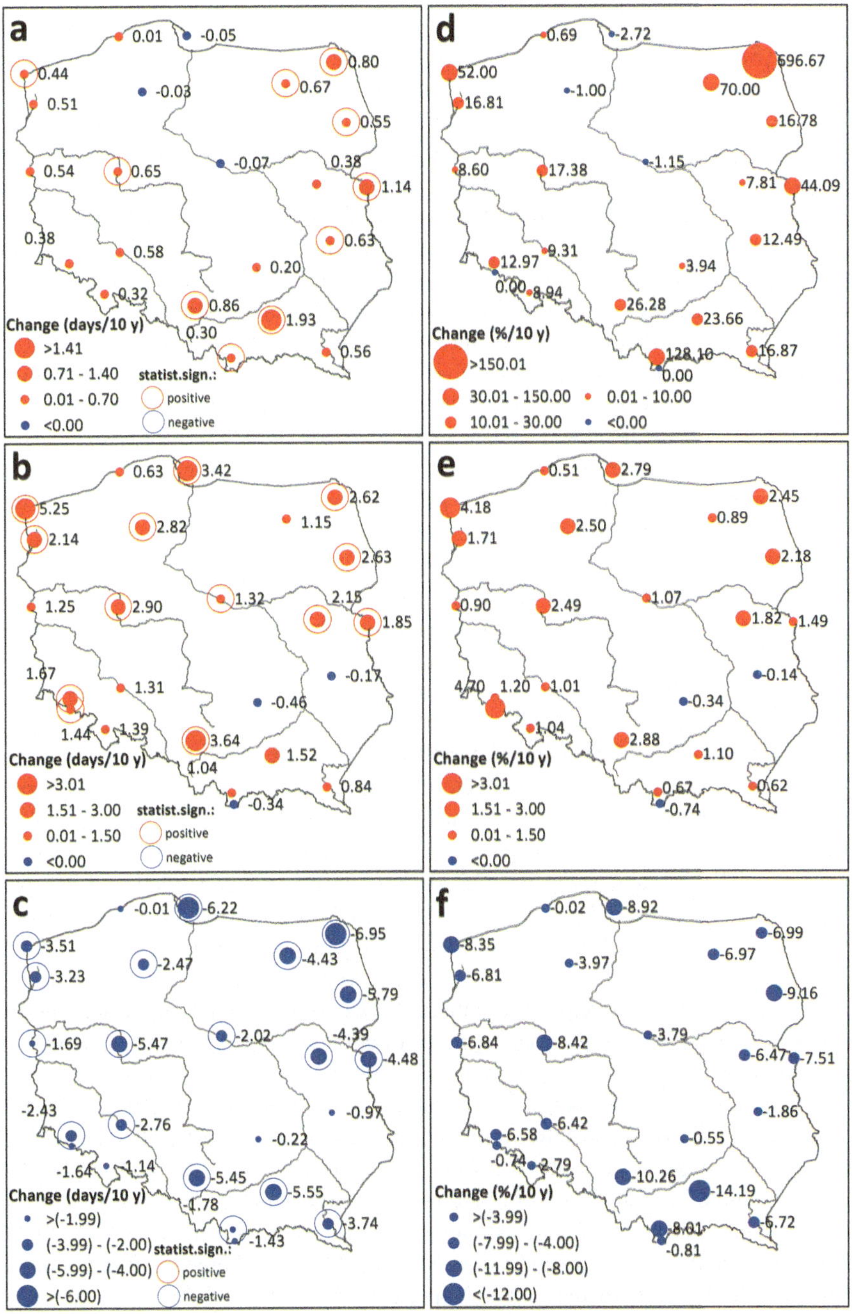

Fig. 19.6 10-years trends of days with different thermal stress categories according to UTCI: absolute value of the trend of heat-stress days (a), neutral days (b), cold-stress days (c), and relative value of the trend, compared to 1951, of heat-stress days (d), neutral days (e), cold-stress days (f), 1951–2018. Statistical significance at the 5% level

trends of PSTmin (+0.5 °C, +0.6 °C/10 years) are also observed at elevated mountain stations (Śnieżka, Kasprowy Wierch) (Fig. 19.3f).

However, annual maximum figures for PST—which represent the warm season (June–August)—did not change significantly during the period studied; the average trend is only +0.02 °C per 10 years. Only one station, Chojnice, recorded a trend that is statistically significant and negative (Fig. 19.3d).

When considering year-to-year changes in PST values, it can be emphasised that the greatest fluctuations occur at PSTmax, especially at stations located in cold areas (mountains, north-eastern and eastern Poland). In the case of PSTmin, higher year-to-year variability was observed at coastal and lowland stations.

Physiological Subjective Temperature doesn't show such strong growth as UTCI. The highest rise concerns the yearly mean PST value. As with UTCI, this growth is steadier in Tarnów than in Suwałki, where periods of lower values appeared between 1976 and 1982 (Fig. 19.7).

In bioclimatic research, assessment of the frequency of different categories of the indicators used is very important. As mentioned, in the present research, we considered three categories of thermal sensations: cold, comfortable and warm (see Table 19.2).

When analysing the course of the frequency of different thermal sensation categories, we can see great year-to-year variability, especially in the case of cold and comfortable sensations (Fig. 19.8). The most stable conditions were found at mountain stations, though the increase in days of hot sensations in Tarnów is noticeably greater than in Suwałki, where, in turn, the decrease in cold sensation days is higher. In the north, at both western and eastern stations, clear periods of lower frequency of comfort days and higher occurrence of cold days are noted (between 1969 and 1986), while, in the south, there is no such clear pattern.

At all stations, we observe negative trends in the annual number of cold sensation days. The average value is −2.7 days/10 years (relative trend: −1.2% per 10 years). The greatest reduction in cold sensation days was observed in Świnoujście (−6.8 days; −2.8% per decade relatively to 1951), Suwałki (−6.1 days; −2.3% relative trend/10 years), Tarnów (−4.7 days; −2.2%/10 years) and Katowice (−4.7 days). For 17 stations, these decreases are statistically significant. At mountain stations, as well as in central Poland, the change trends of cold days are small and insignificant (Fig. 19.9c, f).

An increase in the number of comfortable and hot-sensation days was observed in the studied period, especially in the north-east and north-west of Poland, where average values are +1.6 and +1.2 days per 10 years, respectively. The greatest increase in comfortable days was found for Świnoujście (+5.1), and of hot days for Tarnów (+3.8). However, at many stations, such trends are insignificant (Fig. 19.9). Świnoujście and Suwałki are characterised by the highest relative trend of hot days, compared to 1951, with over 21% increase in hot-sensation days per 10 years (Fig. 19.9d).

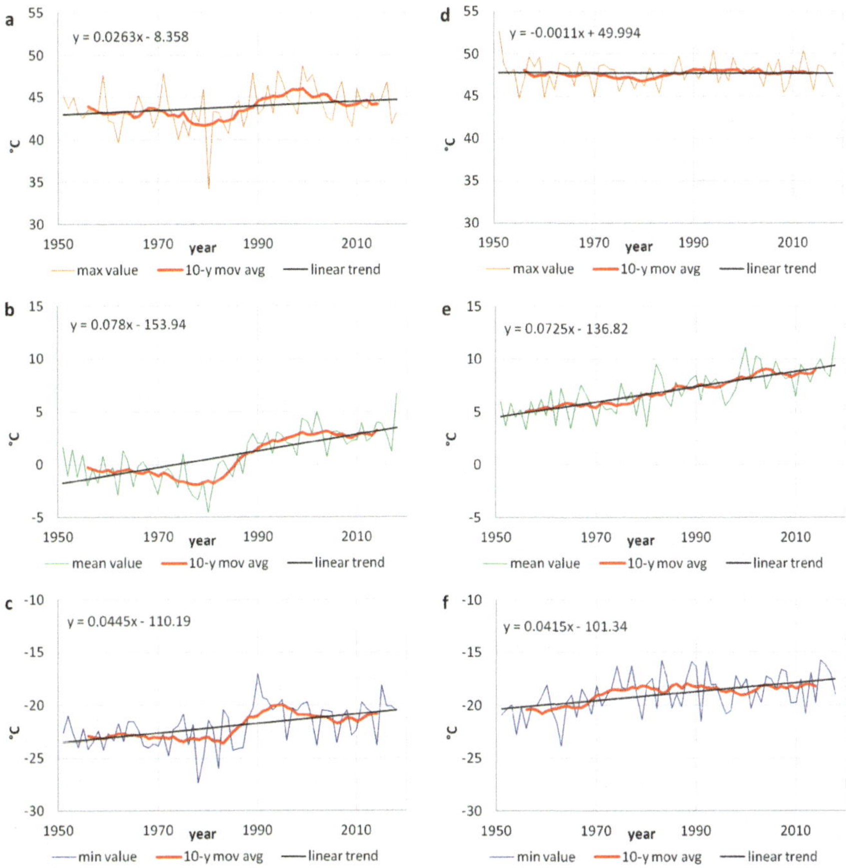

Fig. 19.7 Changes in maximum (max), mean and minimum (min) annual values of PST in Suwałki (a, b, c) and Tarnów (d, e, f), 1951–2018

General Assessment of Bioclimatic Conditions' Changes

The analysis of the changes in bioclimatic conditions has shown significant increases in minimum and mean values of the applied indices during the period studied. Positive changes in maximum UTCI values (which occur in the summer) were significant in the north-east and north-west of Poland (0.3–0.5 °C per 10 years), as well as in the foothills of the Sudetes and Carpathian Mountains (0.3–0.4 °C per 10 years), though the changes of PST maximum values were insignificant. Such data suggest that the general increase in subjective temperatures and thermal stress in Poland depends mostly on changes in climate variables (temperature, humidity, cloudiness and wind speed) in the cold months. This is confirmed by trends in the frequency of thermal sensations and thermal stress categories related to cold environments. Both UTCI and PST minimum values showed significant increases in the whole area of Poland,

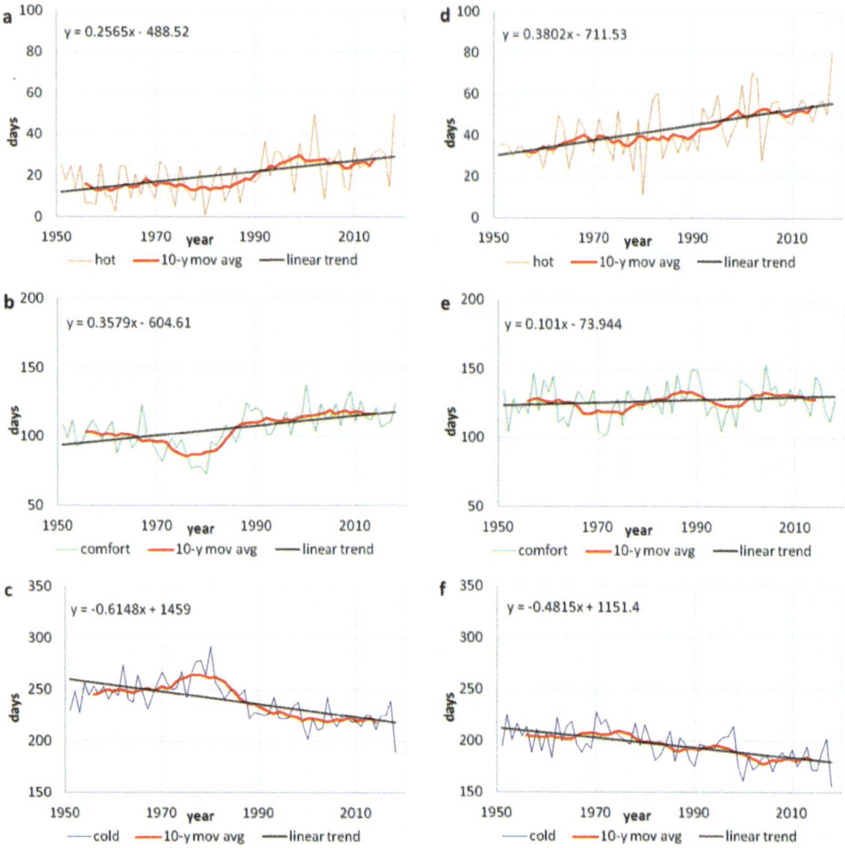

Fig. 19.8 Changes in annual frequency of days with different sensations categories according to PST (hot, comfort, cold) in Suwałki (a, b, c), Tarnów (d, e, f), Świnoujście (g, h, i) and Śnieżka (j, k), 1951–2018

including the mountains, but especially in the north-east, where UTCImin growth was up to 1.8 °C per 10 years, and Tarnów in the south-east (up to 2.2 °C per 10 years). On average, in every decade, the number of cold days fell by 2.7 days in the case of PST (average relative trend: −1.2%/10 years), and −3.2 days per 10 years on the UTCI (average relative trend: −6.0%/10 years), but in the north-east the reduction was much greater, at 7 days per 10 years (UTCI, relative trend: −7.0%/10 years) and 6 days when considering PST (relative trend: −2.3%/10 years).

Regionally, the greatest increase in mean and minimum values of the indices used and decrease in the frequency of cold days were observed in north-eastern, eastern and western Poland, i.e. in the regions most exposed to frequent changes in air circulation. To these regions, Tarnów has to be added because, after a huge rise in UTCI, it has become the warmest city among those analysed in terms of UTCI. In the central region, the changes were smaller. This, however, does not reflect the

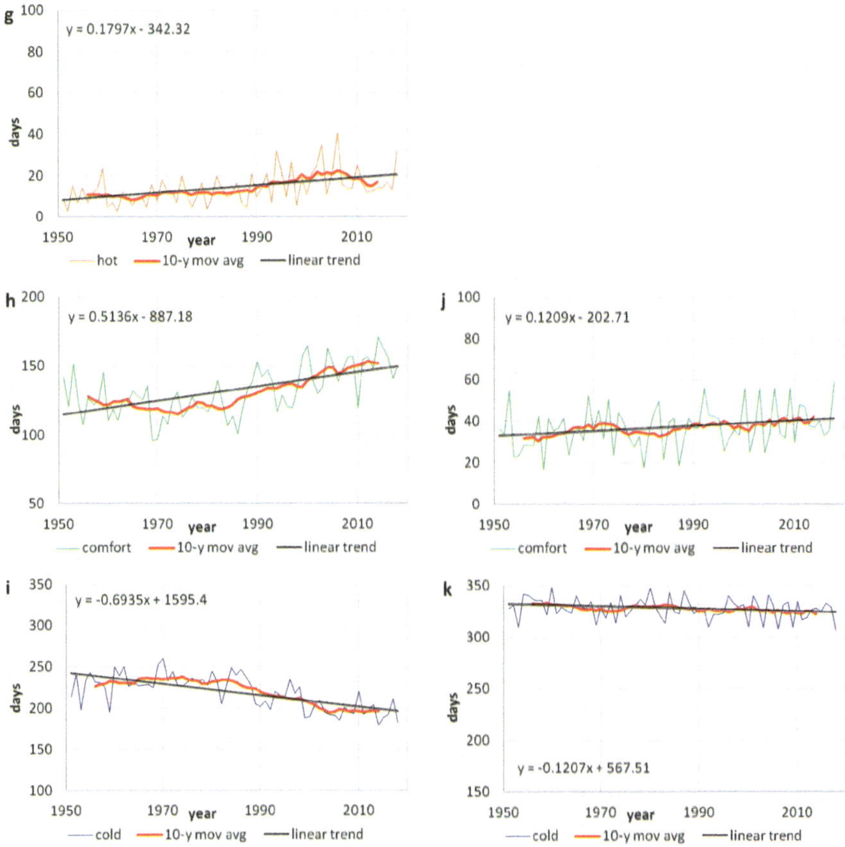

Fig. 19.8 (continued)

pattern of growing sunshine throughout Poland and even more strongly suggests an important role for air circulation in forming bioclimatic conditions in Poland.

Discussion and Conclusions

In previous research dealing with long-term changes in Poland's bioclimate, different indicators were used. Błażejczyk and Twardosz (2002, 2010) analysed changes in bioclimatic conditions in Krakow in the nineteenth and twentieth centuries, using many indexes, but also physiological subjective temperature. In the years 1826–2006, PST grew significantly in January by 0.95 °C per 100 years. In April, PST increased from about 1 °C at the beginning to 4 °C at the end of the studied period. PST was used in a few Polish descriptions of the bioclimate of health resorts in the years 1971–1990 (Kozłowska-Szczęsna et al. 2002; Błażejczyk and Kunert 2011) or the

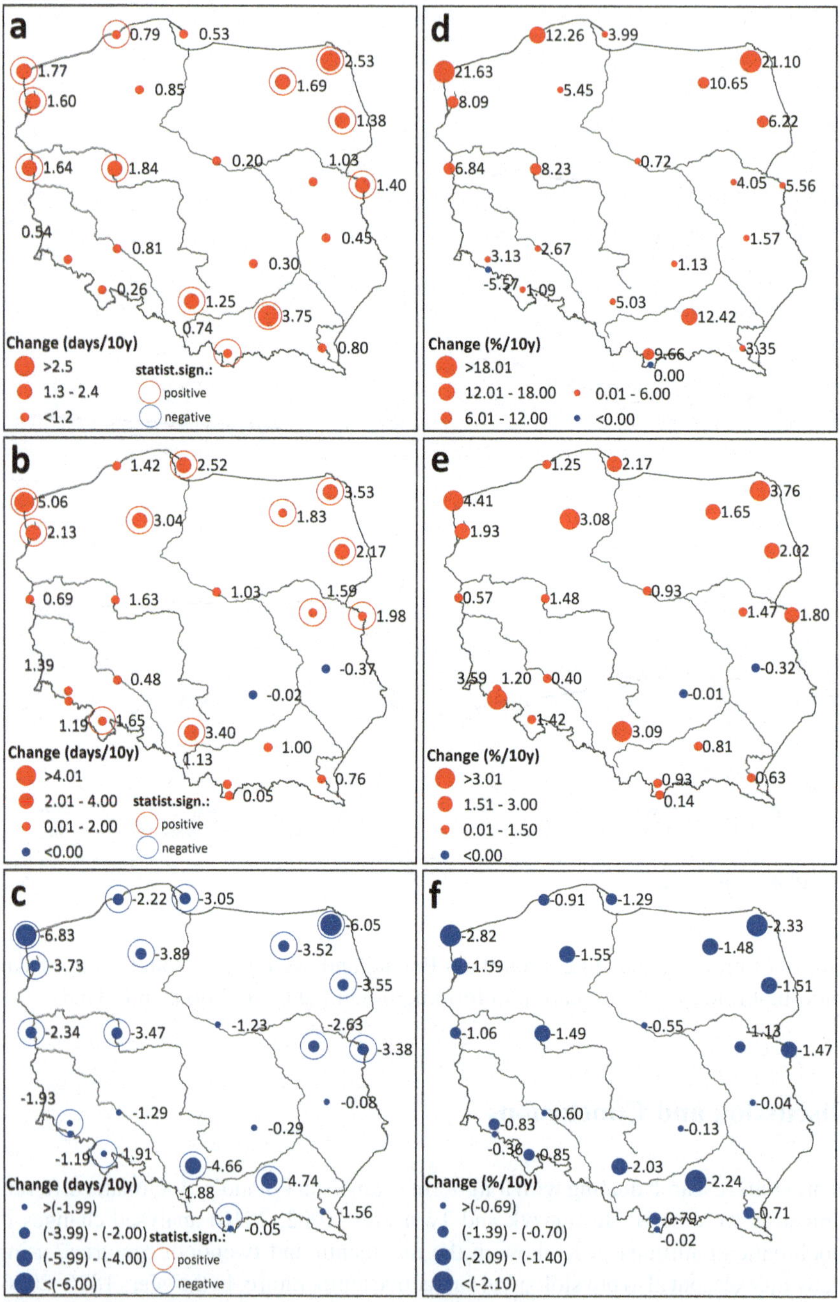

Fig. 19.9 10-years trends of days with different thermal sensations categories according to PST: absolute trend value of hot days (a), comfort days (b), cold days (c), and relative trend value, compared to 1951, of hot days (d), comfort days (e), cold days (f), 1951–2018. Statistical significance at the 5% level

therapeutic potential of the Polish climate (1991–2000, Kuchcik et al. 2013), but no trends were calculated then, so the comparison is impossible.

The Universal Thermal Climate Index, since its final development in 2009, has gained worldwide recognition and it was used in numerous different climate-human studies, which it was mainly created for (e.g. Kuchcik et al. 2013; Morabito et al. 2014; Nastos and Matzarakis 2012; Urban and Kyselý 2014; Błażejczyk et al. 2015, 2018; Gao et al. 2018), but also in bioclimate-change studies, e.g. in Hungary in the years 1971–2000 (Németh 2011) or in China in the years 1981–2010 (Chi et al. 2018), the whole of Europe (di Napoli et al. 2018) and across the world, including Poland. Some papers do not analyse the trends, observing only the values or the frequency of heat-stress days, as in Lublin in the years 1952–2010 (Dobek and Krzyżewska 2015) or in Warsaw in the years 1998–2015 (Rozbicka and Rozbicki 2018).

Only a few studies have analysed the trends and the changes in UTCI values. The analysis of UTCI trends in the years 1975–2014 in the biggest Polish cities confirms the results obtained in this paper: the highest and statistically significant fall in the number of cold-stress days (reaching 3 days/10 years from 1975 to 1989) and rise in heat-stress days was noticed in north-east Poland, with only a significant fall in cold-stress days on the coast (in Gdańsk, close to Hel) and no rise in heat-stress days, etc. (Kuchcik 2017). Another long-term analysis of UTCI (1966–2015) in the whole area of Poland showed very similar results, e.g. a rise in heat-stress days of 1.3 days/10 years in north-east Poland (Suwałki) or of 2.0 days/10 years in the south (Rzeszów, close to Tarnów) (Tomczyk and Owczarek 2019).

The multiannual pattern of days with different thermal sensations and thermal stress categories also shows an increase in hot and neutral days and a decrease in the number of cold days. The observed changes in bioclimatic indices do not exactly reflect the pattern of growing sunshine duration throughout Poland, it correlates more with increases of mean cloudiness especially in winter in the years 1985–2018, which is described in Chap. 10 *Change of cloudiness*.

References

Błażejczyk K (1994) New climatological- and -physiological model of the human heat balance outdoor (MENEX) and its applications in bioclimatological studies in different scales. Zeszyty IGiPZ PAN 28:27–58

Błażejczyk K (2005) New indices to assess thermal risks outdoors. In: Holmér I, Kuklane K, Gao Ch (eds) Environmental ergonomics XI, Proc. Of the 11th International Conference, 22–26 May, Ystat, Sweden, pp 222–225

Błażejczyk K, Baranowski J, Błażejczyk A (2015) Wpływ klimatu na stan zdrowia w Polsce: stan aktualny oraz prognoza do 2100 roku. Wyd. Akademickie SEDNO, Warsaw

Błażejczyk A, Błażejczyk K, Baranowski J, Kuchcik M (2018) Heat stress mortality and desired responses of healthcare system in Poland. Int J Biometeorol 62:307–318. https://doi.org/10.1007/s00484-017-1423-0

Błażejczyk K, Epstein Y, Jendritzky G, Staiger H, Tinz B (2012) Comparison of UTCI to selected thermal indices. Int J Biometeorol 56(3):515–535. https://doi.org/10.1007/s00484-011-0453-2

Błażejczyk K, Kunert A (2011) Bioklimatyczne uwarunkowania rekreacji i turystyki w Polsce, 2nd edn, Monografie IGiPZ PAN, p 13
Błażejczyk K, Matzarakis A (2007) Assessment of bioclimatic differentiation of Poland based on the human heat balance. Geographia Polonica 80(1):63–82
Błażejczyk K, Twardosz R (2002) La variabilité des charges thermiques de l'homme a Cracovie au XXe siecle. Publications de l'AIC 14:413–419
Błażejczyk K, Twardosz R (2010) Long-term changes of bioclimatic conditions in Cracow (Poland). In: Przybylak R, Majorowicz R, Brázdil J, Kejna M (eds) The Polish climate in the European context: an historical overview. Springer, Science+Business Media B.V, pp 235–246
Bröde P, Fiala D, Błażejczyk K, Holmér I, Jendritzky G, Kampmann B, Tinz B, Havenith G (2012) Deriving the operational procedure for the Universal Thermal Climate Index (UTCI). Int J Biometeorol 56(3):481–494. https://doi.org/10.1007/s00484-011-0454-1
Cheshire WP (2016) Thermoregulatory disorders and illness related to heat and cold stress. Auton Neurosci: Basic Clin 196:91–104. https://doi.org/10.1016/j.autneu.2016.01.001
Chi X, Cubasch U, Cao W (2018) The thermal comfort and its changes in the 31 provincial capital cities of mainland China in the past 30 years. Theor Appl Climatol 132(1–2):599–619. https://doi.org/10.1007/s00704-017-2099-4
Clark RP, Edholm OG (1985) Man and his thermal environment. E. Arnold Ltd., London
de Freitas CR, Grigorieva AE (2017) A comparison and appraisal of a comprehensive range of human thermal climate indices. Int J Biometeorol 61(3):487–512. https://doi.org/10.1007/s00484-016-1228-6
Dobek M, Krzyżewska A (2015) Wybrane zagadnienia z bioklimatu Lublina. Annales Universitatis Mariae Curie-Skłodowska Lublin-Polonia, Sectio B 70(2):117–129. https://doi.org/10.17951/b.2015.70.2.117.
Di Napoli C, Pappenberger F, Cloke HL (2018) Assessing heat-related health risk in Europe via the Universal Thermal Climate Index (UTCI). Int J Biometeorol 62(7):1155–1165. https://doi.org/10.1007/s00484-018-1518-2
Elizondo RS, Bullard RW (1971) Local determination of sweating and the assessment of the "set point". Int J Biometeorol 15(2–4):273–280
Epstein Y, Moran DS (2006) Thermal comfort and heat stress indices. Indust Health 44:388–398
Gao Ch, Kuklane K, Östergren P, Kjellstrom T (2018) Occupational heat stress assessment and protective strategies in the context of climate change. Int J Biometeorol 6:359–371. https://doi.org/10.1007/s00484-017-1352-y
Gasparrini A, Guo Y, Hashizume M, Lavigne E, Zanobetti A, Schwartz J, Tobias A, Tong S, Rocklöv J, Forsberg B, Leone M, de Sario M, Bell ML, Guo Y-LL, Wu C-fu, Kan H, Yi S-M, de Sousa Zanotti Stagliorio Coelho M, Saldiva PHN, Honda Y, Kim H, Armstrong B (2015) Mortality risk attributable to high and low ambient temperature: a multicountry observational study. Lancet 386:369–375. https://doi.org/10.1016/S0140-6736(14)62114-0
Givoni B, Goldman RF (1973) Predicting effects of heat acclimatization on heart rate and rectal temperature. J Appl Physiol 35(6):875–879
Guyton AC, Hall JE (2006) Medical physiology, 11th edn. Elsevier Saunders, Philadelphia
Hajat S, Kovats RS, Lachowycz K (2007) Heat-related and cold-related deaths in England and Wales: who is at risk? Occup Environ Med 64(2):93–100
Havenith G (2001) An individual model of human thermoregulation for the simulation of heat stress response. J Appl Physiol 90:1943–1954
Holmér I (1988) Assessment of cold stress in terms of required clothing insulation—IREQ. Int J Ind Ergon 3:159–166
International Union of Physiological Sciences—Commission for Thermal Physiology (2003) Glossary of terms for thermal physiology. J Thermal Biol 28:75–106
Jendritzky G, de Dear R, Havenith G (2012) UTCI—why another thermal index? Int J Biometeorol 56(3):421–428. https://doi.org/10.1007/s00484-011-0513-7
Kenney WL (1985) A review of comparative responses of men and women to heat stress. Environ Res 37(1):1–11

Kenney WL, Munce TA (2003) Aging and human temperature regulation. J Appl Physiol 95(6):2598–2603
Köppe C, Kovats S, Jendritzky G, Menne B (2004) Heat-waves: risks and responses. Health and Global Environmental Change Series, No. 2, WHO Europe, Copenhagen, Denmark
Kozłowska-Szczęsna T, Błażejczyk K, Krawczyk B, Limanówka D (2002) Bioklimat uzdrowisk polskich i możliwości jego wykorzystania w lecznictwie, IGiPZ PAN Mo-nografie, 3, Warsaw
Kuchcik M (2017) Zmiany warunków termicznych w Polsce na przełomie XX i XXI wieku i ich wpływ na umieralność. Prace Geograficzne, IGiPZ PAN, Warsaw
Kuchcik M, Błażejczyk K, Szmyd J, Milewski P, Błażejczyk A, Baranowski J (2013) Potencjał leczniczy klimatu Polski. IGiPZ PAN, Wyd. Akademickie SEDNO, Warsaw
le Blanc J (1975) Man in the cold. ChC Thomas Publ, Springfield
Morabito M, Crisci A, Messeri A, Capecchi V, Modesti PA, Gensini GF, Orlandini S (2014) Environmental temperature and thermal indices: what is the most effective predictor of heat-related mortality in different geographical contexts? Hindawi Publishing Corporation, The Sci World J 2014, Article ID 961750. https://doi.org/10.1155/2014/961750
Nastos PT, Matzarakis A (2012) The effect of air temperature and human thermal indices on mortality in Athens, Greece. Theor Appl Climatol 108:591–599
Németh Á (2011) Changing thermal bioclimate in some Hungarian cities, Acta climatologica et chorologica, Universitatis Szegediensis 44–45:93–101
NIOSH (1986) Criteria for a recommended standard: occupational exposure to hot environment, National Institute for Occupational Safety and Health, 101–110, DHHS (NIOSH) Publication No 86–113, Washington, DC
Parsons KC (2003) Human thermal environments: the effects of hot, moderate, and cold environments on human health, comfort and performance. Taylor & Francis, London and New York
Rozbicka K, Rozbicki T (2018) Variability of UTCI index in South Warsaw depending on atmospheric circulation. Theor Appl Climatol 133:511–520. https://doi.org/10.1007/s00704-017-2201-y
Tomczyk AM, Owczarek M (2019) Occurrence of strong and very strong heat stress in Poland and its circulation conditions. Theor Appl Climatol. https://doi.org/10.1007/s00704-019-02998-3
Urban A, Kyselý J (2014) Comparison of UTCI with other thermal indices in the assess-ment of heat and cold effects on cardiovascular mortality in the Czech Republic. Int J Environ Res Public Health 11:952–967. https://doi.org/10.3390/ijerph110100952

Chapter 20
Change of Weather Types

Katarzyna Piotrowicz and Dominika Ciaranek

Abstract This chapter's overarching goal is to evaluate the change tendencies of weather types identified on the basis of three meteorological elements: air temperature, cloudiness and precipitation as well as by the latest Woś classification (Woś 2010). It was determined that the number of days with very warm weather (3--), especially cloudy without precipitation and with precipitation (310 and 311) increased, while days which are fairly frost (9--) and/or have ground-frost (5-- and 6--) decreased. It was found that there is a decreasing tendency in the number of weather types in a year (a statistically significant trend in 75% of the surveyed stations) in Poland. This indicates that the weather throughout the year was more stable.

Introduction

In order to learn more about the climatic conditions of a given city or region, especially their long-term variability, it is essential to analyse the frequency of weather types. This concept, in accordance with the assumptions of complex climatology, is understood to mean the daily course of values of individual meteorological elements with similar characteristics, but analysed together. According to many climatologists, characterising the climate by analysing individual meteorological elements does not give a full picture of its features (Kossowski 1968; Woś 1999). Such an analysis omits a very important fact—their interdependence.

In Poland and internationally, there are few studies analysing the climate of a given place in terms of a complex climatology, especially taking into account the long-term series of meteorological measurements and observations (Maheras 1984, 1985, 1988; Lotko-Łozińska 1992; Michailidou et al. 2009a, b). Existing weather classifications most often belong to bioclimatic or synoptic classifications, which were developed for specific purposes (Błażejczyk 2004; Durło 2005; Kaszewski 1992).

K. Piotrowicz (✉) · D. Ciaranek
Faculty of Geography and Geology, Institute of Geography and Spatial Management, Jagiellonian University, Kraków, Poland
e-mail: k.piotrowicz@uj.edu.pl

M. Falarz (ed.), *Climate Change in Poland*, Springer Climate,
https://doi.org/10.1007/978-3-030-70328-8_20

The most important contribution to understanding the diversity of weather types in Poland was made by Woś (1999, 2010). In the course of his many years of research (Woś 1968, 2010), the author has modified his classification several times. This work uses its latest version (Woś 2010), which was, compared to the previous one (Woś 1999), limited to 8 (out of 11) thermal types of weather. Methodological details of this classification are presented in the next Sect. ("Data and Methods").

The weather classifications of Woś have been used by many authors, including Kaszewski (1992), Kożuchowski (1996), Lotko-Łozińska (1994), Nagórska (1998), Durło (2003), Więcław (2004), Piotrowicz (2010), Ciaranek and Piotrowicz (2014), and Piotrowicz and Ciaranek (2020). It was also modified by introducing additional meteorological elements (e.g. wind speed or sunshine duration) and slightly different ranges of values (Marsz 1992; Lotko-Łozińska 1994; Ferdynus 1997; Ferdynus and Marsz 2000; Ferdynus 2004; Piotrowicz 2010; Dobrowolska 2018). This was the result of, among other things, the adaptation of Woś' weather classification to the analysis of weather types' structure and annual climate seasonality in polar areas or various Asian climate zones. A classification similar to Woś' assumptions but developed for the needs of phytoclimatology was introduced by Durło (2005).

Weather types distinguished according to Woś classification (or its modifications) were also analysed in connection with various types of atmosphere circulation (e.g. Bogucki and Woś 1994; Nagórska 1998; Więcław 2004; Piotrowicz et al. 2016). They were also used to determine the seasonal climate structure (including season distinction) and to Polish climate regionalisation (Woś 1999, 2010; Piotrowicz 2010; Kożuchowski 2011). Until now, there have been few papers analysing changes and variability of weather types over many years (Piotrowicz 2010). This chapter is therefore the first study that presents the trends in changing weather types in several dozen Polish cities in the period 1951–2018.

Data and Methods

The meteorological data (1951–2019) from 24 stations representative of various climate regions in Poland (Woś 2010; Kożuchowski 2011) as well as the latest Woś classification (2010) were used in the paper. The author, dividing the weather of each day into specific types, took into account three meteorological elements: air temperature (daily average, maximum and minimum), average daily cloud cover and daily total precipitation (Table 20.1). Their values were classified into specific ranges: eight for temperature, three for cloud cover and two for rainfall. As a result, Woś (2010) distinguished 48 types of weather, which were identified by a number and description (Table 20.1).

Woś (1999) also makes it possible to analyse weather types taking into account only air temperature (thermal weather types), cloudiness and precipitation (weather subtypes). These elements may also be combined into three weather units: (1) warm, starting with codes 3, 2 and 1, i.e. very warm, warm and cool weather types; (2) frosts, i.e. ground-frost cold (5) and ground-frost very cold (6); (3) frosty, i.e. starting with

Table 20.1 Weather types classification by Woś (2010)

No	Partition		Name of the days
Air temperature (Thermal weather types)			
3--	T_{mean} >15.0°C	T_{max}>0.0°C	very warm
2--	T_{mean} 5.1-15.0°C		warm
1--	T_{mean} 0.1-5.0°C		cool
5--	T_{mean} >0.0°C	T_{min}≤0.0°C T_{max}>0.0°C	ground-frost, cold
6--	T_{mean} ≤0.0°C		ground-frost, very cold
8--	T_{mean} 0.0÷-5.0°C	T_{max}≤0.0°C	moderately frosty
9--	T_{mean} -5.1÷-15.0°C		fairly frosty
0--	T_{mean} <-15.0°C		very frosty
Cloudiness (C) and Precipitation (P)			
-00	C ≤20% and P <0.1 mm		sunny without precipitation
-01	C ≤20% and P ≥0.1 mm		sunny with precipitation
-10	C 21-79% and P <0.1 mm		cloudy without precipitation
-11	C 21-79% and P ≥0.1 mm		cloudy with precipitation
-20	C ≥80% and P <0.1 mm		very cloudy without precipitation
-21	C ≥80% and P ≥0.1 mm		very cloudy with precipitation

Eg. 210 – warm, cloudy without precipitation

code 8, 9 and 0, which corresponds to types moderately frosty, fairly frosty and very frosty (Table 20.1). The absence of any element in the highlighted code (symbol) is replaced by a dash (-). For example, the code 300 means very warm and sunny day without precipitation, while 811—moderately frosty, cloudy with precipitation (see Table 20.1).

This chapter briefly presents the average annual number of days of distinguished types. Their more specific characteristics can be found in the work of Woś (2010), in which the data from 46 stations (1951–2000) were used. The provided values give the opportunity to compare changes that have occurred in the average number of individual types within the last 18 years. Special attention was paid to the long-term variability of the analysed types of weather. Their trends were determined on the linear regression equation basis.

In addition, the meteorological measurements and observations from the climatological station in Kraków (Jagiellonian University Observatory) from 1901 to 2018 (118 years) were used in the chapter.

Mean Annual Number of Days with Particular Weather Types

Depending on the region, the most common types of weather in Poland are:

- warm, cloudy without precipitation (210), especially in the northern part of the country (Łeba, Szczecin),
- warm, very cloudy with precipitation (221), especially in the south (Bielsko-Biała, Zakopane),

- very warm, cloudy without precipitation (310), especially in the central part (Warszawa, Poznań).

Each of these three types occurred on average from about 30 to just over 50 in a year (Tables 20.2–20.4). At stations representing mountain areas (Kasprowy Wierch, Śnieżka) the dominant type of weather is fairly frosty, very cloudy with precipitation (921), on average about 49 days a year (Table 20.4).

There are slightly less frequent days in Poland:

- warm, cloudy with precipitation (211),
- warm, very cloudy without precipitation (220),
- very warm, cloudy with precipitation (311),
- cool, very cloudy with precipitation (121).

In total, the seven types of weather mentioned above constitute over 50% of the days of the year at most of the analysed stations. Only in Zakopane (49.1%), Śnieżka (35.7%) and Kasprowy Wierch (29.5%), i.e. stations located above 850 m a.s.l. the incidence was lower. In addition to the aforementioned 921 weather type, moderately frosty, very cloudy with precipitation (821) were also very frequent (Tables 20.2–20.4). As noted by Kożuchowski (2011) and what was confirmed in this study, a greater concentration of the frequency of weather types occurs on the coast of the

Table 20.2 Mean annual number of days with very warm (3--), warm (2--) and cool (1--) weather types in selected stations in Poland in the period 1951–2018

Station	very warm weather types						warm weather types						cool weather types					
	300	301	310	311	320	321	200	201	210	211	220	221	100	101	110	111	120	121
Łeba	10.0	0.4	30.3	18.0	2.9	9.1	11.8	0.4	49.9	34.8	12.6	39.4	0.8	0.0	10.3	8.0	8.6	23.3
Suwałki	11.3	0.5	33.3	18.0	3.2	8.5	8.9	0.2	39.8	24.4	13.4	35.3	0.2	0.0	5.4	4.1	8.1	19.0
Szczecin	14.3	0.5	39.6	22.7	4.1	11.7	8.9	0.2	46.5	33.6	15.7	38.6	0.4	0.1	8.3	7.8	7.7	16.5
Chojnice	10.7	0.3	33.0	15.7	4.4	10.4	7.8	0.2	43.2	28.9	15.1	41.4	0.4	0.0	6.1	5.1	7.7	20.9
Toruń	12.5	0.5	41.8	20.6	5.1	12.1	7.7	0.3	40.6	24.5	15.5	36.2	0.3	0.1	6.5	5.3	8.1	16.8
Olsztyn*	13.4	0.6	33.0	19.3	3.1	8.7	9.6	0.3	40.1	29.5	12.5	35.5	0.4	-	6.0	5.3	7.3	18.6
Białystok	12.0	0.5	37.8	20.2	3.1	9.3	8.8	0.3	39.6	22.7	13.0	35.0	0.3	0.1	4.9	3.3	7.9	17.2
Gorzów Wlk.	13.1	0.4	41.7	19.4	6.4	13.3	8.5	0.2	42.3	26.7	16.9	43.0	0.4	0.0	6.7	6.1	7.4	17.3
Poznań	13.9	0.4	45.2	22.4	4.0	10.4	8.6	0.2	45.8	26.3	14.1	36.3	0.4	0.0	7.4	5.6	7.1	16.1
Warszawa	13.6	0.5	48.1	24.0	2.9	8.1	8.5	0.3	45.3	26.1	11.1	32.9	0.2	-	6.5	4.5	7.9	15.6
Zielona Góra	14.5	0.4	42.4	21.2	5.1	11.2	9.1	0.3	45.4	26.8	15.5	41.6	0.6	0.0	6.6	6.3	6.0	17.1
Wrocław	14.6	0.6	42.5	21.9	5.7	13.4	8.0	0.2	44.2	22.0	17.0	38.7	0.1	0.0	5.8	4.0	6.1	13.9
Łódź	14.6	0.6	40.9	21.0	3.7	10.7	8.8	0.3	43.5	24.7	12.9	38.4	0.4	0.0	6.1	4.6	6.4	17.3
Lublin	15.8	0.5	39.5	21.2	3.5	10.7	10.9	0.3	40.3	22.6	11.8	37.1	0.4	0.0	6.2	4.4	6.2	16.7
Włodawa	14.4	0.5	42.8	20.4	5.2	11.4	8.8	0.2	38.8	20.8	13.8	33.4	0.2	0.0	5.1	3.5	6.5	14.6
Śnieżka	1.0	0.1	2.0	1.4	0.0	0.1	8.3	0.3	36.1	22.8	6.5	35.0	2.6	0.1	9.2	6.5	3.9	26.7
Katowice	13.1	0.7	39.6	23.1	4.1	11.9	8.6	0.4	40.5	23.7	13.3	43.0	0.3	-	4.5	3.9	5.1	16.1
Kraków	16.4	0.9	38.7	24.8	3.4	10.7	9.8	0.3	39.3	24.6	11.5	38.7	0.1	-	4.7	4.0	5.1	15.3
Kielce	13.1	0.7	36.1	20.9	4.3	11.3	8.3	0.5	39.3	23.0	12.6	38.8	0.2	-	4.3	3.5	5.3	16.1
Rzeszów**	16.3	0.6	41.9	23.8	4.0	11.6	10.4	0.3	41.3	22.9	13.2	35.9	0.4	0.0	5.8	4.2	6.3	15.6
Bielsko Biała	14.8	0.8	35.5	23.1	3.0	12.1	11.6	0.4	42.5	22.6	12.6	50.2	0.3	0.1	5.3	3.1	4.6	16.2
Zakopane	4.9	0.4	17.1	14.8	1.4	5.1	9.3	0.4	42.5	28.9	10.5	50.5	0.1	-	3.5	2.4	3.0	15.0
Kasprowy Wierch	0.4	0.0	0.6	0.5	-	0.0	8.3	0.3	32.3	23.9	4.9	23.9	2.3	0.1	10.1	6.5	4.2	21.7
Lesko***	1.1	12.2	27.7	27.3	11.9	1.4	1.1	8.9	35.1	33.6	50.9	7.3	0.0	0.2	4.0	4.2	17.8	2.5
Kraków Obs.****	17.4	1.1	38.8	22.4	8.8	20.3	10.3	0.6	34.0	19.4	16.5	44.5	0.7	0.1	6.3	4.6	7.4	19.1

* – no data for years 2014-2017, ** – data from 1952, *** – data from 1955, **** – data from 1901-2018

Colour scale: 1.0 4.0 10.0 30.0 40.0 days

Table 20.3 Mean annual number of days with ground-frost—cold (5--) and very cold (6--) weather types in selected stations in Poland in the period 1951–2018

Station	ground-frost, cold weather types						ground-frost, very cold weather types					
	500	501	510	511	520	521	600	601	610	611	620	621
Łeba	4.0	0.2	13.4	8.1	5.3	13.2	2.6	0.0	7.4	4.2	3.8	7.1
Suwałki	4.8	0.1	11.7	6.1	6.1	13.6	2.5	0.1	7.5	3.5	5.8	11.1
Szczecin	4.8	0.1	13.4	6.7	6.0	10.1	3.1	0.1	6.9	3.8	4.0	4.5
Chojnice	3.6	0.1	12.0	6.1	6.7	14.4	2.2	0.1	7.7	4.1	5.2	9.0
Toruń	5.1	0.2	15.9	7.1	6.9	12.4	2.9	0.1	8.7	4.0	5.4	6.9
Olsztyn*	5.6	0.1	12.9	6.5	5.3	12.9	2.7	0.1	7.2	4.8	4.8	8.8
Białystok	5.8	0.1	14.6	6.7	6.3	14.6	3.0	0.1	7.6	3.9	5.5	10.1
Gorzów Wlk.	3.9	0.1	11.9	6.5	5.4	13.3	3.1	0.1	7.0	4.1	4.7	6.0
Poznań	5.0	0.1	14.9	6.7	5.6	11.7	3.4	0.1	8.1	3.8	4.2	6.2
Warszawa	4.7	0.1	13.9	6.5	5.7	11.5	2.6	0.1	8.1	4.1	4.8	7.6
Zielona Góra	3.0	0.2	10.0	6.1	4.3	13.2	2.6	0.1	6.9	4.0	3.6	8.0
Wrocław	6.7	0.2	19.4	8.7	6.6	12.0	3.7	0.0	9.5	3.8	4.2	6.4
Łódź	4.5	0.1	13.2	6.4	5.9	13.1	2.8	0.1	7.5	4.2	4.6	9.2
Lublin	5.4	0.1	12.6	6.1	4.8	13.4	2.6	0.1	7.1	4.1	4.2	10.3
Włodawa	5.4	0.1	15.1	6.7	6.6	14.0	2.4	0.1	8.2	3.7	5.4	9.2
Śnieżka	3.4	0.1	9.6	4.8	2.6	15.5	2.4	0.2	7.0	6.0	2.4	19.6
Katowice	5.5	0.2	16.1	7.1	5.4	14.4	3.6	0.1	8.1	3.9	3.8	10.1
Kraków	6.4	0.2	17.4	9.1	5.6	12.6	3.2	0.0	8.9	4.8	3.7	9.0
Kielce	6.2	0.2	16.7	8.3	5.7	15.3	3.3	0.1	8.9	4.7	5.0	10.8
Rzeszów**	6.8	0.1	16.4	7.0	5.2	11.3	3.1	0.0	8.7	3.5	4.4	8.8
Bielsko Biała	5.3	0.2	13.0	5.4	4.1	12.8	4.0	0.1	7.1	3.6	3.1	10.9
Zakopane	7.2	0.1	19.5	9.4	5.1	16.2	6.7	0.1	13.6	6.5	3.6	15.9
Kasprowy Wierch	4.3	0.1	11.2	5.4	3.1	14.5	3.7	0.1	9.8	5.6	2.9	18.6
Lesko***	0.3	6.5	9.0	12.4	15.7	2.3	0.2	4.5	5.7	6.5	12.8	1.3
Kraków Obs.****	4.0	0.2	9.5	4.5	5.8	12.1	2.2	0.1	6.1	3.0	4.6	7.5

* – no data for years 2014-2017, ** – data from 1952, *** – data from 1955, **** – data from 1901-2018

1.0 4.0 10.0 15.0 18.0 days

Baltic Sea, and the dispersion of the frequency of individual types increases to the southeast. Of course, mountain areas have a fundamentally different weather pattern. This region is represented by stations on Śnieżka and Kasprowy Wierch.

Not all of 48 distinguished weather types occurred at each station in the analysed years. At least one of the weather types, very frosty (0--), did not occur in 58% of the stations. At 33% of the stations there were no moderately frosty, sunny with precipitation (801) weather types, and 25% were cool, sunny with precipitation (101). Some types occur sporadically, once every few years, hence their average number per year is 0.0 (Tables 20.2–20.4).

Seasonal weather variation can be defined using the presence of frost (Tmax > 0 °C and Tmin ≤ 0 °C) and the thermal seasons (winter T < 0 °C, spring and autumn T = 0–5 °C, summer T > 15 °C) as defined by the Woś classification (2010). Table 20.5 shows the potential period of occurrence of individual thermal weather types at selected stations. It presents that very warm weather types (3--) can occur in Poland (except in mountain areas, where this period is definitely shorter) between March and November, and warm (2--) throughout the year. Other types of weather appear at the turn of the year. Compared to the years 1951–2018 in Kraków, the potential period of individual weather types was only slightly different in the Kraków Observatory in the years 1901–2018 (Table 20.5). The occurrence period of 3--, 8--, 9-- and 0-- types has been extended. The reason for this was the increase in the frequency of occurrence

Table 20.4 Mean annual number of days with moderately frosty (8--), fairly frosty (9--) and very frosty (0--) weather types in selected stations in Poland in the period 1951–2018

Station	moderately frosty weather types						fairly frosty weather types						very frosty weather types					
	800	801	810	811	820	821	900	901	910	911	920	921	000	001	010	011	020	021
Łeba	0.5	0.0	2.6	1.8	3.5	5.7	2.2	0.0	3.6	1.8	1.2	2.4	0.1	-	0.0	0.1	-	-
Suwałki	0.3	-	3.2	2.0	7.0	10.6	3.7	0.1	10.0	4.3	5.8	7.7	1.8	0.1	1.4	0.6	0.1	0.3
Szczecin	0.3	0.0	2.9	1.1	4.1	3.9	2.5	0.1	4.5	1.6	1.5	1.6	0.3	0.0	0.2	0.1	-	0.0
Chojnice	0.5	0.0	3.7	2.0	7.5	7.7	3.2	0.1	6.4	3.0	3.3	4.1	0.4	0.1	0.4	0.1	0.0	0.1
Toruń	0.4	0.0	2.6	1.5	5.0	5.3	3.0	0.0	6.9	2.5	2.8	3.7	0.8	0.0	0.4	0.2	0.0	0.1
Olsztyn*	0.4	0.0	3.1	2.4	6.0	7.5	3.6	0.2	7.7	4.0	3.5	4.9	1.1	0.0	0.6	0.3	0.0	0.1
Białystok	0.3	0.0	2.6	1.5	6.9	9.0	3.6	0.2	8.3	3.7	4.6	6.8	1.4	0.1	1.2	0.6	0.0	0.3
Gorzów Wlk.	0.5	0.0	2.9	1.4	4.7	5.4	2.5	0.1	4.9	1.8	2.1	2.5	0.2	0.0	0.2	0.1	-	0.1
Poznań	0.5	0.0	3.0	1.4	4.9	5.1	2.9	0.1	5.4	2.0	2.0	2.9	0.4	-	0.3	0.1	0.0	0.1
Warszawa	0.4	-	3.3	1.6	5.6	6.5	3.0	0.1	7.5	2.6	3.4	4.0	0.8	0.0	0.7	0.1	0.0	0.1
Zielona Góra	0.5	0.0	2.9	1.9	4.7	7.6	2.5	0.1	4.3	2.2	2.2	3.6	0.3	-	0.2	0.1	-	0.0
Wrocław	0.2	-	1.9	0.8	3.8	4.1	2.5	0.1	5.4	1.4	2.1	2.3	0.3	0.0	0.5	0.1	-	0.1
Łódź	0.5	-	3.1	1.7	4.9	7.1	3.3	0.1	6.4	2.5	2.8	4.9	0.5	0.1	0.5	0.3	-	0.1
Lublin	0.4	0.0	3.0	2.0	5.3	9.3	3.5	0.2	7.4	3.3	3.5	6.4	0.9	0.0	0.7	0.4	0.1	0.2
Włodawa	0.2	-	3.2	1.6	6.4	8.3	3.5	0.1	8.2	3.1	4.6	5.6	1.0	0.1	1.1	0.5	0.1	0.2
Śnieżka	2.1	0.1	6.1	5.4	3.0	30.2	4.1	0.4	8.9	13.3	2.1	49.4	0.3	0.0	0.5	1.3	0.0	1.8
Katowice	0.2	0.0	2.0	1.5	4.2	8.0	2.9	0.1	5.2	2.8	2.0	5.2	0.4	0.0	0.3	0.2	0.0	0.1
Kraków	0.3	0.0	2.2	1.2	4.5	6.7	2.8	0.1	6.3	3.2	2.4	4.9	0.6	0.1	0.5	0.3	0.0	0.2
Kielce	0.2	-	2.1	1.6	4.9	8.3	3.2	0.1	6.9	3.6	3.1	5.7	0.8	0.0	0.7	0.3	0.0	0.1
Rzeszów**	0.3	0.0	2.8	1.6	4.7	7.6	3.2	0.1	7.0	2.8	3.1	5.5	0.9	0.0	0.9	0.4	0.0	0.1
Bielsko Biała	0.5	-	2.5	1.0	3.1	9.1	3.4	0.1	4.9	2.4	2.0	6.4	0.5	-	0.3	0.2	-	0.2
Zakopane	0.1	-	1.4	1.8	2.1	12.0	4.3	0.3	8.9	5.2	2.1	11.6	0.7	0.0	0.5	0.3	0.0	0.3
Kasprowy Wierch	3.5	0.1	7.4	5.6	3.8	25.3	7.1	0.5	13.3	17.7	3.1	49.3	1.0	0.1	1.2	3.2	0.1	3.8
Lesko***	0.0	0.3	2.6	1.5	14.6	1.1	0.3	2.6	5.6	3.6	9.6	0.5	0.1	0.4	0.5	0.2	0.3	0.0
Kraków Obs.****	0.4	0.0	2.6	1.1	4.6	5.9	2.4	0.1	5.2	1.7	3.3	4.9	0.4	0.0	0.5	0.1	0.1	0.2

* – no data for years 2014-2017, ** – data from 1952, *** – data from 1955, **** – data from 1901-2018

1.0 4.0 10.0 30.0 40.0 days

Table 20.5 Potential period of occurrence of thermal weather types in selected stations in Poland in the period 1951–2018

Weather types	Łeba	Suwałki	Poznań	Warszawa	Kraków	Kraków Obs. (1901-2018)
3--	03.04-23.10	07.04-12.10	21.03-30.10	26.03-31.10	26.03-17.11	21.03-21.11
2--	01.01-31.12	01.01-31.12	01.01-31.12	01.01-31.12	01.01-31.12	01-01-31.12
1--	03.10-22.05	21.09-24.05	01.10-19.05	26.09-22.05	26.09-30.05	25.09-30.05
5--	24.09-09.06	15.09-11.06	19.09-29.05	27.09-25.05	19.09-01.06	19.09-01.06
6--	26.10-16.04	10.10-27.04	18.10-18.04	13.10-19.04	17.10-18.04	17.10-19.04
8--	01.11-31.03	23.10-18.04	31.10-10.04	31.10-08.04	31.10-07.04	20.10-07.04
9--	15.11-21.03	24.10-30.03	16.11-24.03	31.10-27.03	09.11-28.03	30.10-28.03
0--	31.12-06.02	21.11-08.03	17.12-22.02	14.12-28.02	17.12-05.03	04.12-05.03

in recent years (years) of days with temperatures above 15 °C, characteristic for the summer period (summer), and a greater number of frosty days in the first half of the twentieth century, especially in 1920–1945 (Fig. 20.1). Details on the long-term variability and change in trends of individual types of weather are presented in the next section.

The potential period of occurrence of weather subtypes, determined on the basis of cloudiness and precipitation, covers the whole year, from January 1st to December 31st. Thus, the occurrence of weather subtypes during the year depends on the air temperature (thermal weather types).

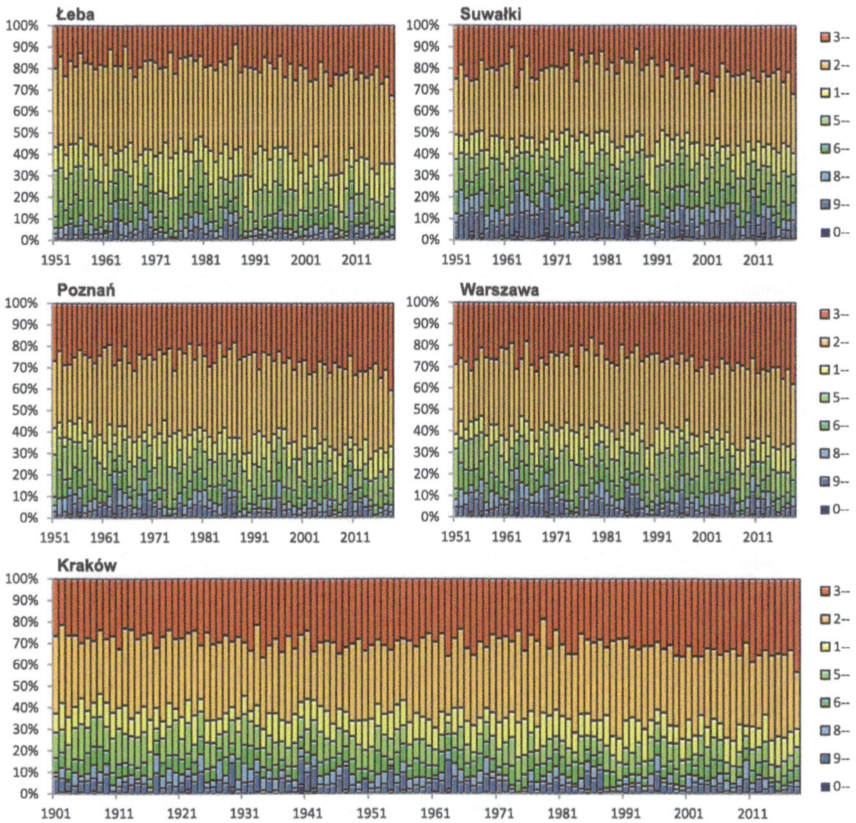

Fig. 20.1 Frequency of thermal weather types in selected stations in the period 1951–2018 and in Kraków in the period 1901–2018

Long-Term Variability of Days with Particular Weather Types

Based on the frequency of the occurrence of eight thermal types and six weather subtypes, it is possible to determine the long-term variability of weather conditions. It also allows for analysis of extreme events, which, in the case of data from the Kraków Observatory (1901–2018), could occur with the frequency of even only "once in 100 years".

Figures 20.1 and 20.2 show the frequency of thermal types and weather subtypes at selected stations representing various Polish climate regions. The multi-year course of the analysed days is quite similar at each station, although their frequency varies. In Łeba, which represents the Baltic Sea coast, since the beginning of the twenty-first century, the number of very warm days (3--) has increased significantly, while ground-frost (5-- and 6--) has decreased (Table 20.6). Very frosty days (0--) on

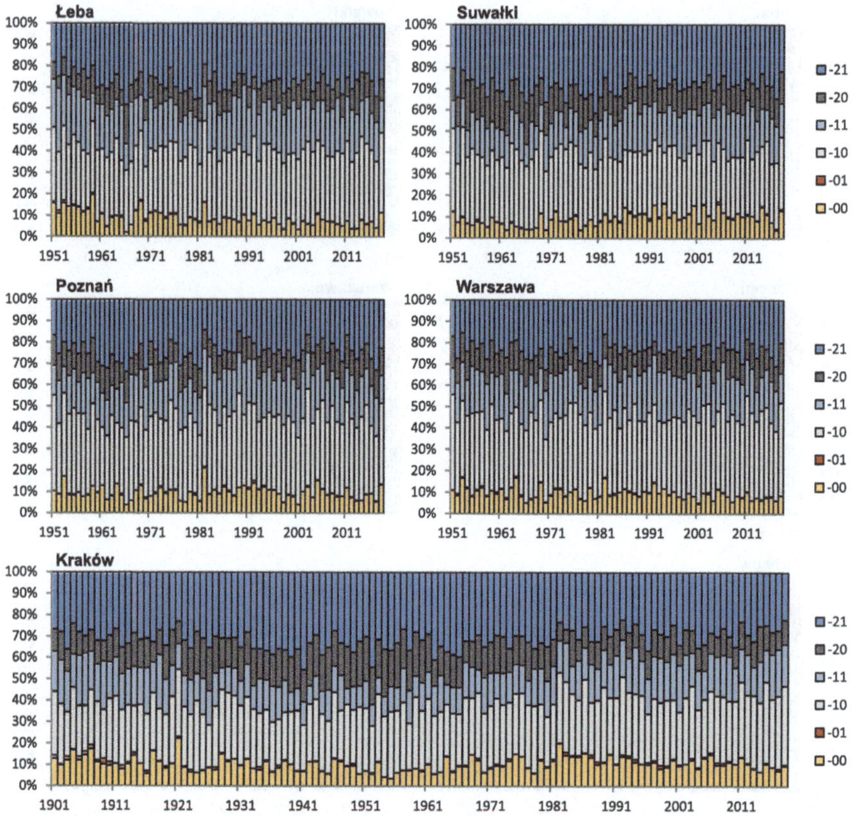

Fig. 20.2 Frequency of weather subtypes in selected stations in the period 1951–2018 and in Kraków in the period 1901–2018

the coast occur sporadically. There were only 12 of them in Łeba throughout the entire period. There were certainly more such days in Suwałki (on average 4.2 days a year; 0.8%), although they did not appear in each winter in the analysed years (Fig. 20.1). In Poznań, Warszawa and Kraków, i.e. cities that well represent the central and southern parts of the country (except for mountain areas), in the multi-annual course, the dominant incidence of days with very warm and warm weather (3-- and 2--) can be seen. Also at these stations, there was a marked increase in the frequency of days 3-- (Fig. 20.1, Table 20.6). The calculated linear trend of these days for all considered stations ranged from 0.48 days/10 years at Kasprowy Wierch and 0.99 day/10 years at Śnieżka, for 3–4 days/10 years at stations in most parts of the country, to 5.24 days/10 years in Zakopane. Only at two stations (Białystok, Lublin; Table 20.6) were the changes not statistically significant at 0.05. It can therefore be concluded that, with the exception of the central-eastern part of the country, the increase in the incidence of very warm days will most often affect the incidence of other thermal weather types. At most stations, the increase in the frequency of days

Table 20.6 Trends of changes in thermal weather types and subtypes of weather (days/10 years) in Poland in the period 1951–2018

Station	Thermal weather types								Subtypes of weather					
	3--	2--	1--	5--	6--	8--	9--	0--	-00	-01	-10	-11	-20	-21
Łeba	4.16	+	+	-2.30	-1.62	-	-	-	-3.62	-0.26	2.62	-	+	+
Suwałki	2.12	+	+	+	-	-0.91	-2.43	-	2.64	0.13	-2.17	2.84	-2.28	-
Szczecin	3.67	+	-	-1.29	-1.30	-0.89	-	-	=	-	=	3.83	-1.83	-2.16
Chojnice	3.86	=	+	-	-1.00	-1.16	-1.86	-	-1.79	-0.18	1.84	2.41	-2.53	=
Toruń	3.72	+	+	-2.48	-	-0.98	-1.67	-	-	-0.25	+	3.62	-2.85	-
Olsztyn*	2.01	-	+	-2.30	-1.15	-	-1.70	-	-2.08	-0.18	-	-	=	=
Białystok	+	+	+	+	-	-	-1.99	-	-	-0.32	+	1.58	-1.25	-
Gorzów Wlk.	4.20	+	-	-1.30	-	-	-	-	-1.33	-	+	-	+	+
Poznań	4.04	+	=	-2.48	-	-0.92	-	-	-	=	+	+	-	-
Warszawa	3.63	+	+	-1.53	-1.08	-	-1.40	-	-1.64	-0.19	1.97	=	-	+
Zielona Góra	3.87	=	=	-	-1.02	-	-	-	-1.45	-	2.21	=	+	-
Wrocław	4.67	+	-	-1.48	-1.37	-	-	-0.29	-2.54	-	=	-	+	2.55
Łódź	3.20	+	+	-1.56	-1.00	=	-1.56	-	-1.47	-	=	=	+	+
Lublin	+	2.17	=	-2.13	-	=	-	-	-1.39	-0.17	-	-1.67	1.64	1.99
Włodawa	1.77	2.02	2.33	-2.64	-1.16	-	-1.39	-	-	-0.19	=	+	-	+
Śnieżka	0.99	2.93	+	-	=	-	-2.47	-	-2.12	-	+	-1.83	+	3.22
Katowice	4.01	=	-	-	-1.16	-	-1.33	-	-2.64	-0.23	2.18	-	+	+
Kraków	3.68	=	=	-	-	-	-	-	-	-	+	=	+	=
Kielce	2.45	=	+	=	-	-	-1.52	-	-2.97	-	+	-1.73	2.08	2.23
Rzeszów**	4.15	+	+	=	-	-	-1.09	-	-	=	+	=	+	2.13
Bielsko Biała	4.50	=	-	-	-1.00	-	-	-	-	-	=	1.40	=	-
Zakopane	5.24	-	-	-	-	-	-	-	-2.53	-0.20	+	+	+	+
Kasprowy Wierch	0.48	3.21	-	=	=	-	-2.11	-0.77	-2.60	-0.20	2.31	-2.75	1.80	+
Lesko***	3.35	+	+	0.68	=	+	=	=	=	-1.65	+	2.34	4.89	-1.33
Kraków Obs.****	2.30	+	+	-1.71	-0.39	-0.41	-0.71	-	=	-	1.78	+	-1.11	-1.14

* – no data for the years 2014-2017, ** – data from 1952, *** – data from 1955, **** – data from 1901-2018; = no trend, +/- – positive / negative trends, but not statistically significant at 0.05, values (positive / negative) – trend statistically significant at the level 0.05

3-- was due to a decrease in frosty days (5-- and 6--), both or one of them, and/or moderately frosty (8--) and fairly frosty (9--). It was such a tendency to change that the thermal types of weather in the Kraków Observatory were in the years 1901–2018 (Fig. 20.1, Table 20.6). The increase in the number of days 3-- was 2.30 days/10 years, while the decrease in frosty days −1.71 (5--) and −0.39 days/10 years (6--) and frosty days −0.41 (8--) and −0.71 days/10 years (9--) (Table 20.6).

In the case of weather subtypes (Fig. 20.2, Table 20.6), the largest changes in the analysed years occurred on days with no precipitation (--0). At 58% of the stations there was a statistically significant decrease in sunny without precipitation (-00) days, from 1.33/10 years in Gorzów Wielkopolski to −3.62/10 years in Łeba, most often caused by an increase in cloudy and/or very cloudy without days precipitation (-10 and -20). Only in Suwałki was an increase in the number of these days found, amounting to 2.64 days/10 years, also at the cost of a decrease in cloudy and very cloudy days without precipitation (-10 and -20), amounting to −2.17 and 2.28 days/10 years, respectively (Fig. 20.2, Table 20.6). Among the subtypes, days with precipitation and sunny (-01) appear rarely. They were slightly more frequent in 1901–1920 and 1981–2000 in Kraków (Fig. 20.2). In northern Poland, there was a slight increase in the number of cloudy with precipitation days (-11; from −1.58 to 3.88 days/10 years; Table 20.6), while in southern Poland, a slight increase in very cloudy days with precipitation (-21; to 3.22 days/10 years on Śnieżka). In 1901–2018 in Kraków (Fig. 20.2), statistically significant changes in weather subtypes occurred

Table 20.7 Trends of changes (days/10 years) of the most frequently occurring weather types in Poland in the period 1951–2018

Station	310	311	210	211	220	221	510	521	820	821	900	910
Łeba	2.27	1.30	0.94	-	0.68	1.17	-	-	=	-	-0.27	-
Suwałki	=	1.94	-	0.94	-0.83	=	=	=	-	-	-	-0.83
Szczecin	1.18	2.28	+	1.34	-	=	-	-0.59	-0.38	-0.44	-	-0.51
Chojnice	2.32	1.76	+	+	-0.56	+	+	-0.81	-0.59	-	-0.39	-0.62
Toruń	2.22	2.62	+	0.86	-	+	-	-1.05	-0.48	-	-0.27	-0.56
Olsztyn*	+	1.19	-	-	+	+	-	-1.02	=	=	-0.50	-0.56
Białystok	+	1.25	+	+	-	+	0.67	-	+	-	-0.54	-
Gorzów Wlk.	2.57	+	=	-	+	+	-	-	-	-	-	-
Poznań	1.98	1.86	+	=	+	+	-	-1.11	-0.38	-	-	-
Warszawa	2.79	1.20	+	-	+	+	=	-0.58	-	=	-0.32	-
Zielona Góra	3.02	1.05	=	+	+	=	-	-	=	-	-	-
Wrocław	2.84	1.11	-	-	0.98	1.60	-	+	-	-	-	-
Łódź	1.91	1.19	=	-	+	1.30	-	-0.86	=	+	-	-0.50
Lublin	1.09	+	+	-	0.67	1.75	-0.96	-	+	+	-	-
Włodawa	1.37	+	+	+	+	1.40	-	-1.38	-	=	-0.35	-
Śnieżka	0.43	0.35	1.88	+	0.39	1.39	-0.72	+	-	+	-0.56	-
Katowice	3.09	0.89	=	-	=	+	+	-	+	-	-0.30	-
Kraków	2.47	0.82	-	=	+	+	=	-	=	=	-	-0.45
Kielce	1.75	0.70	-	-1.17	0.73	1.64	+	=	+	-	-	-
Rzeszów**	2.35	0.69	-	-	+	+	-	=	-	=	-	-0.52
Bielsko Biała	2.10	1.62	-	=	+	=	-	-	+	=	-	-
Zakopane	2.55	1.78	=	-1.06	+	1.33	-	=	=	=	-	-
Kasprowy Wierch	0.30	0.18	1.72	+	0.53	1.60	+	+	0.35	=	-	=
Lesko***	1.51	1.59	0.32	-0.68	1.90	-0.44	=	-0.37	0.69	-0.07	0.08	-0.18
Kraków Obs.****	1.72	0.58	0.90	=	=	=	-0.38	-0.66	-0.26	=	-0.04	-0.24

* – no data for the years 2014-2017, ** – data from 1952, *** – data from 1955, **** – data from 1901-2018, ; = no trend, +/- – positive / negative trends, but not statistically significant at 0.05, values (positive / negative) – trend statistically significant at the level 0.05

only in case of cloudy (-10; rise) and very cloudy without precipitation (-20; fall), as well as very cloudy with precipitation subtypes (-21; fall) (Table 20.6). The trends of changes of the most common weather types at the analysed stations are presented in Table 20.7. At most stations, a statistically significant increase occurred in the days marked as 310 and 311 (Fig. 20.3). In some regions of Poland, the number of days 210, 211, 220 and 221 also increased, while 521, 820, and especially 900 and 910 decreased (Fig. 20.3).

Number of Weather Types in a Year

While analysing the long-term variability in the number of weather types in a year, it is also possible to evaluate weather stability/instability. Fewer types of weather in a year indicate that they occur for 365/366 days a year in longer sequences. Similarly, the more there are in a year, the more they form short strings or occur individually. Therefore, during the year, the weather variability increases day by day, i.e. the weather is unstable in such a year.

The most important characteristics of the number of weather types at the studied stations are presented in Table 20.8. In Poland, on average, there are 32–35 weather types per year. The largest number of them occurs in the eastern, especially in the

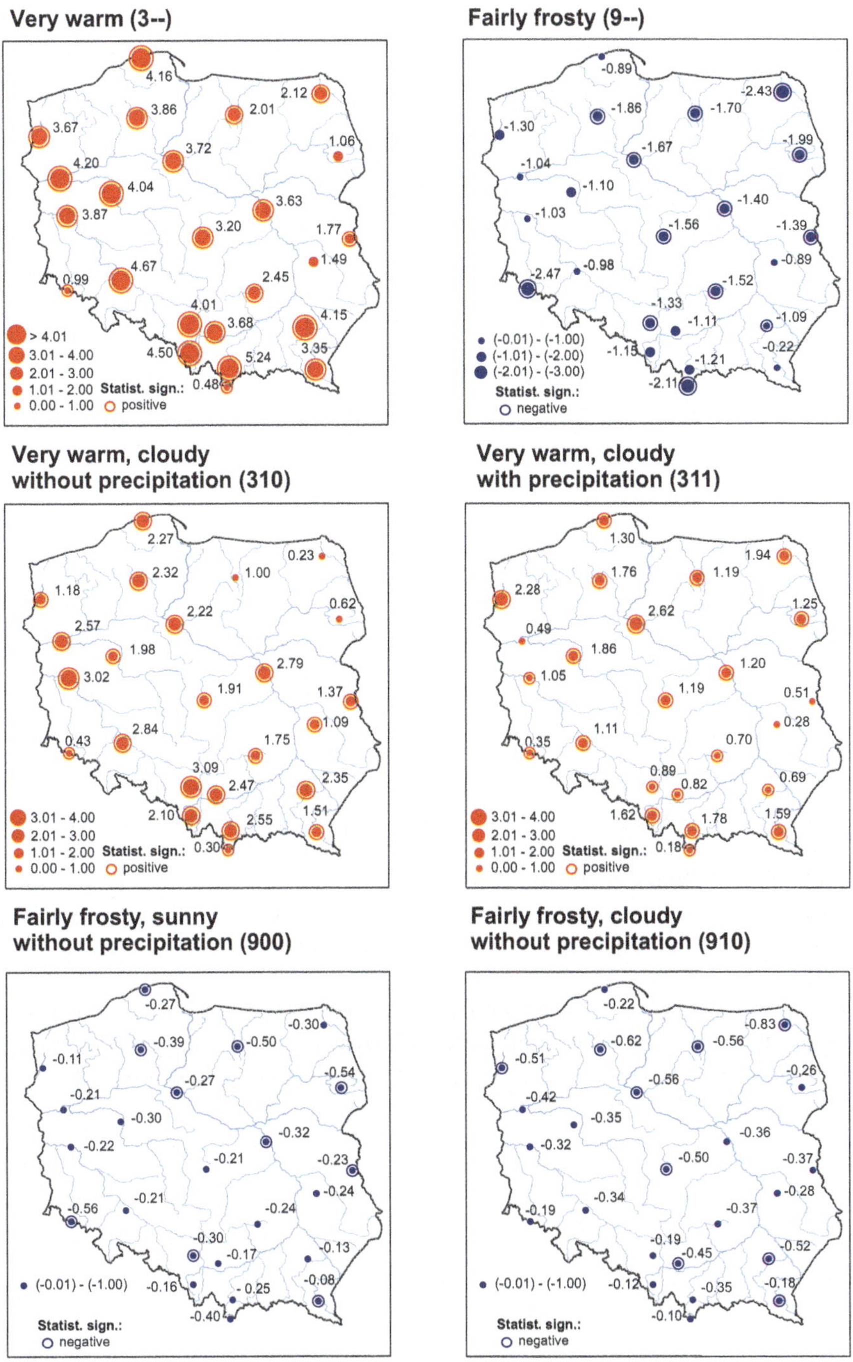

Fig. 20.3 Trends of changes (days/10 years) of selected weather types in Poland in the period 1951–2018

Table 20.8 Characteristics of the number of weather types in a year in Poland in the period 1951–2018

Station	Mean	Max (year)		Min (year)		S.D.	V(%)	Trend weather types/10 year
Łeba	32.3	37	(1953, 1969)	26	(2016)	2.7	8.3	-0.55
Suwałki	35.0	41	(1956)	30	(2015)	2.1	5.9	-
Szczecin	32.1	38	(1954)	27	(1977, 2008)	2.6	8.1	-0.40
Chojnice	33.8	39	(1963)	27	(2000)	2.3	6.7	-0.55
Toruń	34.0	41	(1963)	27	(1992)	2.4	7.1	-0.42
Olsztyn*	34.2	39	(1956)	29	(2008)	2.7	7.9	-0.42
Białystok	35.0	40	(1952)	30	(2017)	2.1	5.9	-0.32
Gorzów Wlk.	32.9	38	(1987)	26	(1974)	2.2	6.7	-0.31
Poznań	33.1	40	(1987)	28	(2015)	2.4	7.1	-0.31
Warszawa	33.8	39	(1956, 1987)	27	(2008)	2.3	6.7	-0.32
Zielona Góra	33.2	40	(1956)	26	(2015)	2.7	8.0	-0.33
Wrocław	32.3	39	(1956)	26	(2015)	2.4	7.4	-0.45
Łódź	34.4	40	(1956)	28	(2008)	2.2	6.5	-0.28
Lublin	34.8	41	(1956)	31	(2007, 2015)	2.3	6.5	-0.33
Włodawa	34.5	39	(1956, 1963)	28	(1990)	2.1	6.0	-0.29
Śnieżka	33.8	40	(1964)	29	(1997)	2.0	6.0	-
Katowice	33.7	39	(1963, 1987)	28	(2014)	2.2	6.5	-0.49
Kraków	33.8	40	(1969)	29	(1990)	2.1	6.4	-0.34
Kielce	34.2	39	(1982)	29	(1966)	2.0	5.8	-0.23
Rzeszów**	34.5	39	(1986)	29	(2014)	2.0	5.9	-0.26
Bielsko Biała	33.9	39	(1963, 1996)	28	(2008, 2014)	2.4	7.2	-
Zakopane	33.4	40	(1955)	29	(2007)	2.2	6.5	-
Kasprowy Wierch	34.4	40	(1954)	31	(1990,2008,2014)	2.0	5.8	-
Lesko***	33.6	42	(1963)	27	(1967, 2000)	7.4	22.1	-
Kraków Obs.****	34.0	39	(1907,1929,1969)	29	(1975, 1990)	2.4	7.0	-0.15

* – no data for the years 2014-2017, ** – data from 1952, *** – data from 1955, **** – data from 1901-2018, S.D. – standard deviation, V – coefficient of variability, - – negative trends, but not statistically significant at 0.05, values – trend statistically significant at the level 0.05

north-eastern part of the country (Suwałki), while the smallest in the west, and especially in the north-west (Szczecin). The spatial distribution of the number of weather types in a year therefore refers to the coefficient of thermal continentalism (Wypych 2010; Witek et al. 2015).

Over the years, the most variable weather was in the years 1950 and 1960, while the most stable was at the beginning of the twentieth century (1901–1920) and since 1990, especially in the twenty-first century (2001–2018) (Fig. 20.4). Trends in the change in the number of weather types are negative at all stations, mostly statistically significant at 0.05 (Table 20.8) and indicate increasing stability, i.e. less day-to-day weather variability. Therefore, some types of weather occur in longer periods.

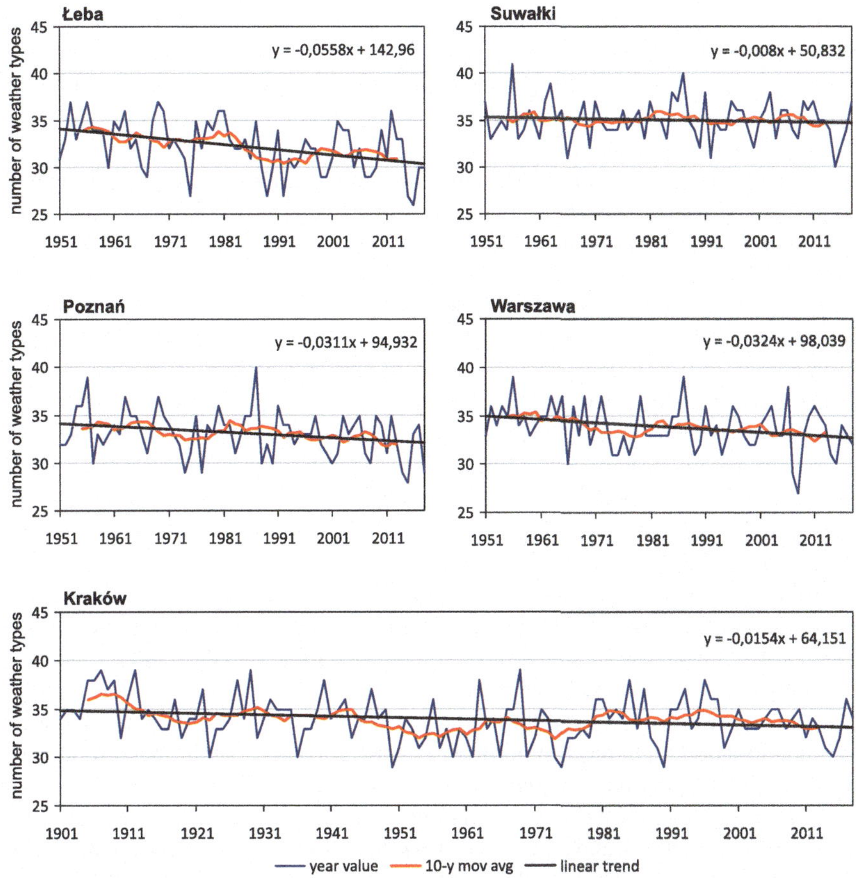

Fig. 20.4 Long-term variability of number of weather types in a year (1951–2018) and in Kraków (1901–2018)

Conclusions

The most common types of weather in Poland are: warm, cloudy without precipitation (210), warm, very cloudy with precipitation (221) and very warm, cloudy without precipitation (310). Their number, depending on the station, usually fluctuates between 30 and 50 days a year. At mountain stations (Kasprowy Wierch, Śnieżka) the dominant type of weather is fairly frosty, very cloudy with precipitation (921), on average about 50 days a year. In general, the fewest types of weather in the year occur in western Poland and the north-west, while the most in eastern, especially in north-eastern, Poland. This distribution of weather type variability refers to the continental indicator, which is a measure of the continent/ocean impact on climate elements (Kożuchowski 2011) and the spatial diversity of the average annual air temperature amplitude in Poland (Woś 2010).

Weather types are most differentiated by air temperature. There was a clear increase in days with very warm weather (3--), especially those marked with the symbol 310 and 311, even at high mountain stations, while the decline concerned ground-frost (5-- and 6--; especially 510 and 521) and fairly frosty (9--; especially 900 and 910) weather types.

These results are therefore consistent with the research of Michalska (2011) and Wójcik and Miętus (2014). The author stated that a particularly large increase in air temperature occurred in Poland in the twenty-first century in 2001–2008 (Michalska 2011). Wójcik and Miętus (2014), based on research from 1951 to 2010, indicate not only a statistically significant increase in the annual average temperature in the entire country of slightly over 0.2 °C/10 years, but also springs (0.36 °C/10 years), summers (up to 0.2 °C/10 years) and autumns, while deceleration or even a slight decrease in temperature in winter. According to these authors, the fastest temperature increase occurs in the northern part of the country (the Baltic Sea coast and the Lake District) and in the Carpathians, while the slowest in the highlands and in the Sudety (Wójcik and Miętus 2014). Also, the decrease in the number of days with frost was analysed in detail, by Graczyk and Kundzewicz (2016), Wypych et al. (2017) and Bielec-Bąkowska et al. (2018).

In the case of weather subtypes (distinguished on the basis of cloudiness and precipitation), the largest changes over the years occurred on days with no precipitation (--0), a decrease in sunny days without precipitation (-00) and an increase in cloudy and/or very cloudy without precipitation (-10 and -20). In the weather subtypes of precipitation (--1), there was a slight increase in the number of cloudy with precipitation (-11) days in northern Poland, whereas a slight increase in very cloudy with precipitation (-21) days in southern Poland.

In the case of long-term variability of rainfall sums in Poland (in the year, half-year and seasons), Czarnecka and Nidzgorska-Lencewicz (2012) did not find a statistically significant trend. According to the authors, however, there was a small tendency of rainfall to increase in the spring and autumn seasons, and a decreasing share of summer rainfall in the annual total was noted in most parts of the country (Czarnecka, Nidzgorska-Lencewicz 2012). Marosz et al. (2011) found an increase in rainfall sums in the northern part of the country (in Pomerania and the coast) but a decrease in the central and southern parts of Poland. Statistically significant changes in the amount of cloudiness in 1951–2000 and 1971–2000 did not cover the whole country and even their direction of change was different (Żmudzka 2003; Filipiak and Miętus 2009). Nevertheless, in Kraków and Łódź, the total cloudiness fell in the second half of the twentieth century (Matuszko 2007; Wibig 2008), which may correspond with the increase in the number of days with the cloudy without precipitation subtype (-10).

On the whole, not all of the 48 identified weather types occur every year and at each station. In the years 1951–2018, their number ranged from 26 to 42, and on average it was 32–35 types. Slightly more of them were in the 1950s and 1960s, while fewer after 2001. There is a tendency for the number of weather types in a year to decrease (a statistically significant trend in 75% of the surveyed stations) in Poland. This indicates that the weather varies less in the year, i.e. more stable. Thus, particular types of weather have been grouping since 1951 (and in Kraków since 1901) in slightly longer sequences of days. Francis et al. (2018) also point to an increased persistence of North American Weather Regimes. The authors hypothesis that the reason for this is rapid Arctic warming. They suggest that further warming may favour persistent weather patterns that can lead to weather extremes and increased frequency of long-duration events (LDEs).

In short, this may be associated with longer heat waves in the summer and warmer periods in the cool half of the year, i.e. refer to a higher frequency of days with the thermal type of very warm, warm and cool weather (3--, 2-- and 1--).

References

Bielec-Bąkowska Z, Piotrowicz K, Krępa-Adolf E (2018) Trends in the frost-free season with parallel circulation and air mass statistics in Poland. Időjárás 122(4):375–392. https://doi.org/10.28974/idojaras.2018.4.2

Błażejczyk K (2004) Bioklimatyczne uwarunkowania rekreacji i turystyki w Polsce. Prace Geograficzne, IGiPZ PAN 192:1–291

Bogucki J, Woś A (1994) Typy pogody a typy cyrkulacji atmosferycznej w rejonie Poznania. Badania Fizjograficzne nad Polską Zachodnią 45, Seria A:25–42

Ciaranek D, Piotrowicz K (2014) Stability of thermal conditions in Central Europe. In: Kanakidou M, Mihalopoulos N, Nastos P (eds) COMECAP, 12 International Conference of Meteorology, Climatology and Physics of the Atmosphere. 28–31 May 2014, Heraklion, University of Crete, Greece 1:198–203

Czarnecka M, Nidzgorska-Lencewicz J (2012) Wieloletnia zmienność sezonowych opadów w Polsce. Woda-Środowisko-Obszary Wiejskie 12(2):45–60

Dobrowolska K (2018) Typy pogody w wybranych strefach klimatycznych Azji. Praca Doktorska. Wydział Nauk o Ziemi Uniwersytet Śląski

Durło G (2003) Bioclimatic calendar a new form of climatological elaboration Analele Universitatii din Oradea. Seria Geografie 13:99–104

Durło G (2005) Typologia warunków pogodowych na potrzeby fitoklimatologii. Woda-Środowisko-Obszary Wiejskie 5(14):129–136

Ferdynus J (1997) Główne cechy klimatu morskiego strefy subpolarnej północnego Atlantyku w świetle struktury stanów pogód. Wyższa Szkoła Morska, Wydział Nawigacyjny, Gdynia

Ferdynus J (2004), Roczna struktura stanów pogody w Hornsundzie (SW Spitsbergen). Polish Polar Studies, 30 International Polar Sympsium, Gdynia, 81–94

Ferdynus J, Marsz AA (2000), Struktura stanów pogód i sezonowość pogodowa. In: Marsz AA, Styszyńska A (eds) Główne cechy klimatu rejonu Polskiej Stacji Antarktycznej im. H. Arctowskiego (Antarktyka Zachodnia, Szetlandy Południowe, Wyspa Króla Jerzego), Wyższa Szkoła Morska, Gdynia, pp 143–162

Filipiak J, Miętus M (2009) Spatial and temporal variability of cloudiness in Poland, 1971–2000. Int J Climatol 29:1294–1311. https://doi.org/10.1002/joc.1777

Francis JA, Skific N, Vavrus SJ (2018) North American weather regimes are becoming more persistent: is Arctic amplification a factor? Geophys Res Letters 45:11,414–11,422. https://doi.org/10.1029/2018gl080252

Graczyk D, Kundzewicz ZW (2016) Changes of temperature-related agroclimatic indices in Poland. Theor Appl Climatol 124:401–410. https://doi.org/10.1007/s00704-015-1429-7

Kaszewski BM (1992) Typy cyrkulacji a typy pogody w Polsce. Rozprawy Habilitacyjne Wydziału Biologii i Nauk o Ziemi UMCS, 42, Lublin

Kossowski J (1968) 1968, O częstości głównych typów pogody w Polsce. Przegląd Geofizyczny 13(3):283–292

Kożuchowski K (1996) Typy pogody w Szczecinie. In: Kożuchowski K (ed) Współczesne zmiany klimatyczne. Klimat Szczecina i współczesne zmiany klimatyczne w rejonie Morza Bałtyckiego. Rozprawy i Studia Uniwersytetu Szczecińskiego 224:29–32

Kożuchowski K (2011) Klimat Polski. Nowe spojrzenie, Wydawnictwo Naukowe PWN, Warszawa

Lotko-Łozińska A (1992) Zagadnienia typologii pogód w polskiej literaturze klimatologicznej na tle wybranych prac literatury światowej. Badania Fizjograficzne nad Polską Zachodnią 44, Seria A:141–151

Lotko-Łozińska A (1994) Założenia i kryteria klasyfikacji stanów pogody występujących w Poznaniu w latach 1920–1990. Badania Fizjograficzne nad Polską Zachodnią 45, Seria A:68–80

Maheras P (1984) Weather-type classification by factor analysis in the Thessaloniki area. J Climatol 4(4):437–443

Maheras P (1985) Correspondences between types of circulation and weather types. Application to the Area of Thessaloniki. Zeitschrift für Meteorologie 35(1):26–35

Maheras P (1988) The synoptic weather types and objective delimitation on the winter period in Greece. Weather 43:40–45

Marosz M, Wójcik R, Biernacik D, Jakusik E, Pilarski M, Owczarek M, Miętus M (2011) Zmienność klimatu polski od połowy xx wieku. Rezultaty projektu klimat. Prace i Studia Geograficzne 47:51–66

Marsz AA (1992) Struktura pogód i roczna sezonowość klimatu Stacji Arctowskiego. Problemy Klimatologii Polarnej 2:30–49

Matuszko D (2007) Zachmurzenie. In: Matuszko D (ed) Klimat Krakowa w XX wieku. IGiGP UJ, Kraków, pp 75–86

Michailidou C, Maheras P, Arseni-Papadimititriou A, Kolyva-Machera F, Anagnostopoulou C (2009a) A study of weather types at Athens and Thessaloniki and their relationship to circulation types for the cold-wet period, Part I: two-step cluster analysis. Theor Appl Climatol 97(1–2):163–177

Michailidou C, Maheras P, Arseni-Papadimititriou A, Kolyva-Machera F, Anagnostopoulou C (2009b) A study of weather types at Athens and Thessaloniki and their relationship to circulation types for the cold-wet period, Part II: discriminant analysis. Theor Appl Climatol 97(1–2):179–194

Michalska B (2011) Tendencje zmian temperatury powietrza w Polsce. Prace i Studia Geograficzne 47:67–75

Nagórska D (1998) Termiczne typy pogody w Koszalinie a typy cyrkulacji atmosferycznej. Badania Fizjograficzne nad Polską Zachodnią 49, Seria A:163–169

Piotrowicz (2010) Sezonowa i wieloletnia zmienność typów pogody w Krakowie. IGiGP UJ, Kraków

Piotrowicz K, Bielec-Bąkowska Z, Domonkos P (2016) High atmospheric pressure and accompanying cold season weather types in Poland (1951–2010). Clim Res 67:165–177. https://doi.org/10.3354/cr01364

Piotrowicz K, Ciaranek D (2020) A selection of weather type classification systems and examples of their application. Theor Appl Climatol 140:719–730. https://doi.org/10.1007/s00704-020-03118-2

Wibig J (2008) Cloudiness variations in Łódź in the second half of the 20th century. Int J Climatol 28:479–491. https://doi.org/10.1002/joc.1544

Więcław M (2004) Masy powietrza nad Polską i ich wpływ na typy pogody. Wydawnictwo Akademii Bydgoskiej, Bydgoszcz

Witek M, Bednorz E, Forycka-Ławniczek H (2015) Kontynentalizm termiczny w Europie. Badania Fizjograficzne nad Polską Zachodnią 6, Seria A:171–182. https://doi.org/10.14746/bfg.2015.6.13

Wójcik R, Miętus M (2014) Niektóre cechy wieloletniej zmienności temperatury powietrza w Polsce (1951–2010). Przegląd Geograficzny 86(3):339–364

Woś A (1968) Zarys klimatu Poznania w pogodach. Poznańskie Towarzystwo Przyjaciół Nauk, Wydział Matematyczno-Przyrodniczy 9(1):1–67

Woś A (1999) Klimat Polski. PWN, Warszawa

Woś A (2010) Klimat Polski w drugiej połowie XX wieku. Wydawnictwo Naukowe UAM, Poznań

Wypych A (2010) Variability of the European climate on the basis of differentiation of indicators of continentalism. In: Przybylak R, Majorowicz J, Brázdil R, Kejna M (eds) The Polish climate in the European context: an historical overview, Springer, Berlin, Heidelberg, and New York, pp 473–484

Wypych A, Sulikowska A, Ustrnul Z, Czekierda D (2017) Variability of growing degree days in Poland in response to ongoing climate changes in Europe. Int J Biometeorol 61:49–59. https://doi.org/10.1007/s00484-016-1190-3

Żmudzka E (2003) Wielkość zachmurzenia w Polsce w drugiej połowie XX wieku. Przegląd Geofizyczny 34(1):159–185

Part IV
Future Climate Change

Małgorzata Szwed

Climate is dynamic. Thus, climate change is nothing new. Climate has changed in the past, it is changing now, and it will continue changing in the future. What distinguishes the present climate change from previous ones is, apart from natural factors, greater human impact on climate.

The globally averaged combined land and ocean surface temperature data, as calculated by a linear trend, show a warming of 0.85 (0.65–1.06) °C, over the period from 1880 to 2012, for which multiple independently produced datasets exist (cf. IPCC, 2013). Each of the 19 years of the 21st century, 2001–2019, belongs to a set of the 20 warmest years on record globally. The noticeable warming has been widespread across the globe, including Poland. An increase of 0.8 °C was calculated for the area of Poland in the second half of the 20th century (Fortuniak et al., 2001; Kożuchowski and Żmudzka, 2001). Furthermore, even stronger warming has been projected for the future. However, the pace of the further warming depends on such uncertain factors as the population increase and socio-economic development, the development of technology, and changes in energy strategy and the land use, i.e. sectors primarily responsible for emission and sequestration of carbon dioxide. A global policy of climate change mitigation and its effectiveness are also imperative.

Global climate change does not only concern noticeable growth of mean temperature. Apart from average values, extreme values—minima and maxima—have also undergone considerable changes. Changes of temperature translate, among others, into changes of characteristics and occurrence of thermal seasons or the length of the growing season. However, observed climate change extends far beyond changes of temperature. Changes concerning intense winds, heavy precipitation events, and periods without precipitation, changes of snow cover, etc., have already been observed. Observation and climate model simulations indicate acceleration of the water cycle. Therefore, global warming will also result also in changes in all elements of the water balance. What is more, climate change is followed by substantial changes of many other physical, as well as biological and socio-economic characteristics.

Some impacts are projected to be advantageous, and others—disadvantageous or neutral.

Projection of the future climate presents a number of difficulties. Climatic models are not perfect, and thus they are subject to considerable uncertainty. However, they were built based on the laws of physics, and experience shows that these simulations basically reflect climate change in the past. Older projections made with the help of climate models in the beginning of the 21st century and earlier, as well as simulations of the impact of different factors on the climate, have been confirmed by observations. Climate forecasting models are constantly being improved. However, the complicated and non-linear nature of climate means that climate models will always require refinement and improvement.

Part IV will present considerations on future climate changes in Poland based on the latest generation of climate model simulations, i.e. the European domain of the Coordinated Downscaling Experiment Initiative (EURO-CORDEX), and they will only apply to basic climate parameters—temperature and precipitation, as well as their derivatives regarding some economic indexes.

This is not a place for discussion of whether, or rather to what extent, anthropogenic climate change will be slowed down, therefore the results of model simulations for two climate change scenarios, mild and more severe, will be considered.

Climate change in terms of future temperature and precipitation changes has been described using many indexes. Their selection, however subjective, was inspired by "indexes of extreme" recommended within the European Climate Assessment & Dataset project (ECA&D), with some of the ECA indexes being faithfully implemented, and the formula (e.g. threshold values) modified for a few.

References

Fortuniak K, Kożuchowski K, Żmudzka E (2001) Trendy i okresowość zmian temperatury powietrza w Polsce w drugiej połowie XX wieku. Przegl Geofiz XLVI(4):283–303.

IPCC (2013) Climate change 2013, The physical science basis. Contribution of working group I to the fifth assessment report of the intergovernmental panel on climate change. In: Stocker TF, Qin D, Plattner G.-K, Tignor M, Allen SK, Boschung J, Nauels A, Xia Y, Bex V, Midgley PM (eds). Cambridge University Press, Cambridge, United Kingdom and New York, NY, USA, 1535 pp

Kożuchowski K, Żmudzka E (2001) Ocieplenie w Polsce, skala i rozkład sezonowy zmian temperatury powietrza w drugiej połowie XX wieku. Przegl Geofiz XLVI(1-2):81–90.

Chapter 21
Projections of Temperature Changes in Poland

Małgorzata Szwed

Abstract The present paper examines projected future changes in the mean and extreme thermal conditions for Poland. The study uses a set of eight regional climate models from EURO-CORDEX experiment for two representative concentration pathways (RCP4.5 and RCP8.5) and two future time horizons (2021–2050 and 2071–2100), with reference to the period of 1971–2000. In this study, the mean value of temperature and its absolute extreme (maximum and minimum) values, as well as indexes related to specific thermal conditions, were considered. According to this study, all models have shown an agreement on a systematic upward trend in the mean temperature, both for the near and far future and for both RCPs. It is projected that in the future the extreme hot conditions not only will be hotter, but they will also last longer and/or appear more frequently. In contrast, frosty periods and their frequency will decrease. As a rule, the thermal changes in the future are more pronounced for the far future (2071–2100) than for the near future (2021–2050) and more for the RCP8.5 scenario than for RCP4.5.

Introduction

There are many papers in the scientific literature concerning the projection of climate change in Poland. Polish scientists quickly joined the new trend of climate studies—research on climate change. The first research studies presenting temperature projections for Poland date back to 1990s, e.g. Liszewska and Osuch (1999). Along with the development of subsequent modifications of models or appearance of the new generation of climate models, more studies came into existence. In these studies, changes in various thermal indexes were considered. The average and extreme values for different time intervals were analysed from daily values, through monthly and seasonal, to annual or average values for the entire future period. It is enough to quote: Szwed et al. (2007) based on first simulations of HadCM3-PRECISE regional climate model developed by the Hadley Centre, Szwed et al. (2010), Szwed and Graczyk

M. Szwed (✉)
Institute for Agricultural and Forest Environment, Polish Academy of Sciences, Poznań, Poland
e-mail: mszwed@man.poznan.pl

M. Falarz (ed.), *Climate Change in Poland*, Springer Climate,
https://doi.org/10.1007/978-3-030-70328-8_21

(2006) using results of RCM projections originating from the ENSEMBLES project, or finally, numerous research studies analysing climate projections stemming from the latest generation of EURO-CORDEX models, such as: Romanowicz et al. (2016), Mezghani et al. (2017), Piniewski et al. (2017), and Brzóska and Jaczewski (2017), etc.

Data and Methodology

Future temperature changes in Poland were considered based on the newest generation of climate model simulations, i.e. the European domain of the Coordinated Downscaling Experiment Initiative (EURO-CORDEX), the European branch of the global CORDEX framework (Jacobs et al. 2014).

The climate projection framework within CORDEX is based on the set of new global model simulations related to the IPCC Fifth Assessment Report. This generation of scenario simulations is based on so-called representative concentration pathways (RCPs), i.e. prescribed greenhouse gas concentration pathways throughout the twenty-first century, corresponding to different radiative forcing stabilisation levels by the year 2100. This approach differs from the earlier scenario runs employed in the fourth IPCC assessment cycle, which were based on the SRES GHG emission scenarios (IPCC 2000).

A set of eight regional climate model simulations (Table 21.1) from the EURO-CORDEX experiment was selected. These projections can be seen as an update of old scenarios. To produce reliable high-resolution climate projections of precipitation and temperature for Poland, the output of the projections was bias-adjusted and downscaled to 5 km. More detailed information on these bias-adjusted climate projections is given in Mezghani et al. (2017). The data is available in the public domain at http://dx.doi.org/10.4121/uuid:e940ec1a-71a0-449e-bbe3-29217f2ba31d.

In order to produce unbiased projections across Poland, two experiments corresponding to the two targeted radiative forcing values of +4.5 Wm^{-2} and +8.5 Wm^{-2} in 2100 (respectively: RCP4.5 and RCP8.5), relative to pre-industrial values were

Table 21.1 List of available GCM-run-RCM combinations composing the multi-model ensemble, based on EURO-CORDEX, used in the determination of projections

GCM	RCM
CNRM-CERFACS-CNRM-CM5	CLMcom-CCLM4-8-17
ICHEC-EC-EARTH	CLMcom-CCLM4-8-17
ICHEC-EC-EARTH	SMHI-RCA4
ICHEC-EC-EARTH	KNMI-RACMO22E
ICHEC-EC-EARTH	DMI-HIRHAM5
IPSL-IPSL-CM5A-MR	SMHI-RCA4
MPI-M-MPI-ESM-LR	CLMcom-CCLM4-8-17
MPI-M-MPI-ESM-LR	SMHI-RCA4

carried out. The RCP4.5 is an intermediate pathway in which radiative forcing is stabilised after 2100 and global warming is estimated to be around 2.5 °C towards the end of the century. The RCP8.5 is the highest emission scenario, also referred to as "business as usual", and implies a global warming of more than 4° in 2100 (Clarke et al. 2007).

Future temperature changes were described using many indexes. Their selection, however subjective, was inspired by "indexes of extreme" recommended within the European Climate Assessment & Dataset project (ECA&D). Most of the ECA indexes were implemented, while for a few the formula was modified (e.g. by adaptation of threshold values). More details can be found on https://www.ecad.eu/.

Except for average annual value of temperature, the absolute extreme (maximum and minimum) values as well as indexes related to specific thermal condition were considered. As for the thermal conditions, such indexes as the largest number of consecutive days with maximum temperature above 30 °C, mean number of days with minimum temperature above 20 °C, mean number of days with both maximum temperature above 30 °C and minimum temperature above 20 °C, mean number of days with minimum temperature below 0 °C and mean number of days with minimum temperature below −15 °C were studied.

Projected indexes/characteristics of temperature were simulated for two future time horizons, i.e. for the near (2021–2050) and far future (2071–2100) as well as for one historical (reference) period of 1971–2000.

This chapter discusses only the projected changes in thermal conditions in the future, understood as differences in the value of indexes in the future in relation to the values from the reference period. Thus, only such differences are described in the next subsection 3, without reference to future index values.

Data were processed with the usage of the Climate Data Operators (CDO) (Schulzweida 2019). The ArcMap GIS platform was used to describe spatial variability of analysed thermal characteristics.

Results

Changes in Mean Annual Temperature

The mean annual temperature is projected to increase in the future for the whole territory of Poland. The higher increases are projected for north-eastern and eastern areas (Fig. 21.1). In the near future (NF), the increases are projected to be in the range of 1–1.5 °C, for RCP4.5 and RCP8.5, respectively, with the highest values in the north-east. As for the far future (FF) and RCP4.5, the mean annual temperature will likely grow by about 2 °C for most of the area of Poland. The highest increases of temperature, as expected, are projected for FF and RCP8.5. The projected changes in temperature vary from 3 °C in the West to almost 4° in north-east and in the high

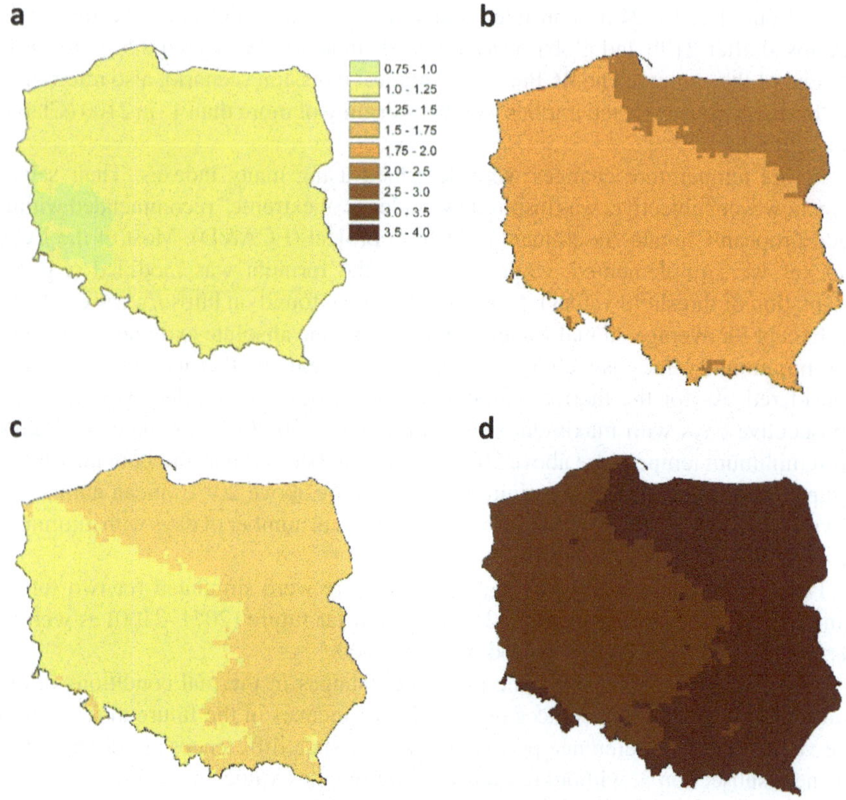

Fig. 21.1 Changes in the mean annual temperature in °C. Difference (increase) in the future compared to the reference period 1971–2000. Upper row: RCP4.5; **a** for period 2021–2050; **b** for period 2071–2100; lower row RCP8.5; **c** for period 2021–2050; **d** for period 2071–2100

mountains. The isoline of the 3.5° increase in temperature runs almost along the Vistula River valley.

Changes in Extreme Temperature

Changes in the Absolute Maximum Temperature

Projected increases in absolute maximum temperature reach much higher values. In the period of 2021–2050, changes in the absolute maximum temperature will be probably on average higher by 2–3°, while on the coast could reach 4–5 °C more as compared to the reference period of 1971–2000 (Fig. 21.2). The lowest changes are projected for Lublin Highland. For FF and RPC4.5, simulations show similar

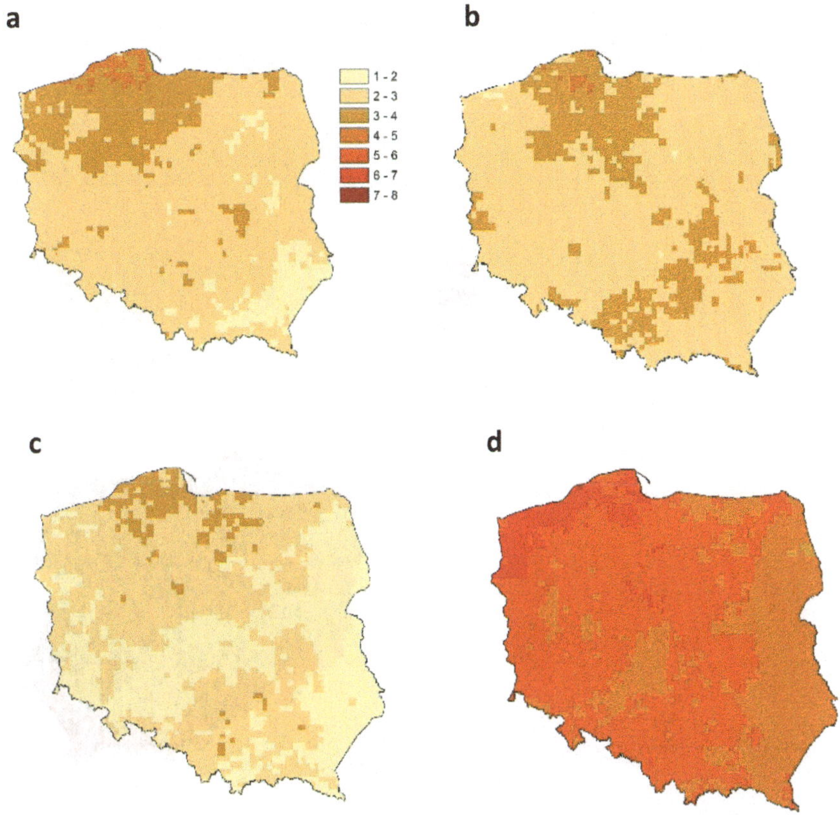

Fig. 21.2 Changes of the absolute maximum temperature in °C. Difference (increase) in the future compared to the reference period 1971–2000. Upper row: RCP4.5; **a** for period 2021–2050; **b** for period 2071–2100; lower row RCP8.5; **c** for period 2021–2050; **d** for period 2071–2100

increases as for NF (up to 4–5° in the northern part, with 2–3° for most of the area). The warming rate is accelerating under the RCP8.5 emission scenario and for the far future horizon. In this case the increases are significantly higher: from 3–4° in the East to more than 7° on the coast.

Changes in the Absolute Minimum Temperature

When it comes to the changes in the absolute minimum temperature, the simulations for NF are almost identical for both analysed RCPs. They vary by about 1 to 3–4° in Western Pomerania (Fig. 21.3) with the value of 2–3 for most of the area. For the FF, there are projected greater differences between both RCPs. For the RCP4.5, the simulations predict increases in the absolute minimum temperature mostly by 4–6°,

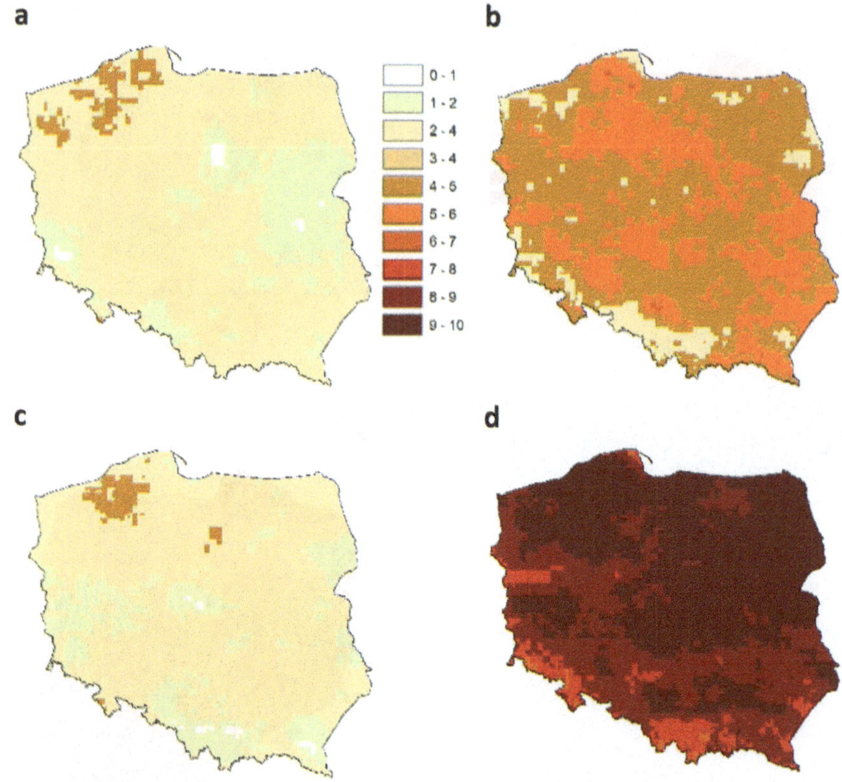

Fig. 21.3 Changes in the absolute minimum temperature in °C. Difference (increase) in the future compared to the reference period 1971–2000. Upper row: RCP4.5; **a** for period 2021–2050; **b** for period 2071–2100; lower row RCP8.5; **c** for period 2021–2050; **d** for period 2071–2100

while for the RCP8.5 they start from 5–6° in the mountains and 6–7 in the South and reach more than 10° in the north-eastern parts of Poland.

Changes in Number of Days with Specific Thermal Conditions

Changes in the Largest Number of Consecutive Days with Maximum Temperature Above 30 °C

It is projected that in the future the extreme thermal conditions not only will be stronger but also they will last longer and appear more often. The number of consecutive days with maximum temperature above 30 °C points at really extreme weather conditions. In the near future, the longest such period may be extended by another few days (3–4 on average), regardless of the adopted RCP. The longest extension is

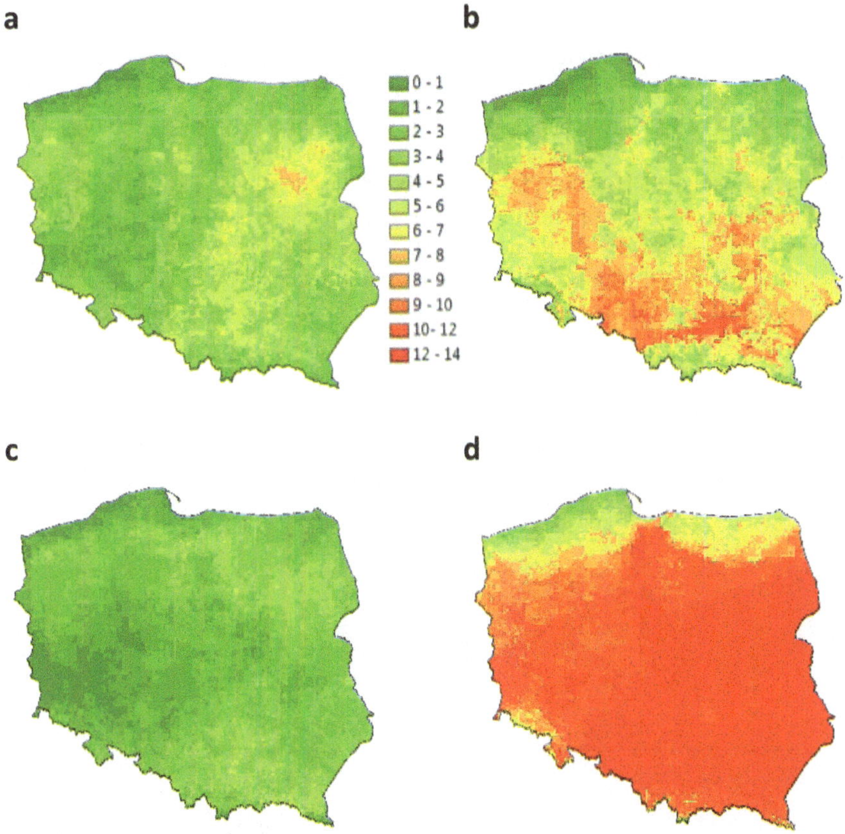

Fig. 21.4 Changes in the largest number of consecutive days with maximum temperature above 30 °C. Difference (increase) in the future compared to the reference period 1971–2000. Upper row: RCP4.5; **a** for period 2021–2050; **b** for period 2071–2100; lower row RCP8.5; **c** for period 2021–2050; **d** for period 2071–2100

projected in the Narew River catchment (Fig. 21.4). Based on model simulations, a prolonged period of extremely high temperatures will significantly increase in FF. While for RCP4.5 the extension will be almost 2 weeks only in the larger river valleys in southern and western parts, in RCP8.5 it will cover almost the entire country, except the coast. However, at the northern outskirts of Poland, the extension of such a period will be almost a week.

Changes in the Mean Number of Days with Minimum Temperature Above 20 °C

In addition to extending the period with very high temperatures during the day and an increase in the number of such days during the year, the number of hot nights will also rise. For the near future and both RCPs, almost no increase in the mean number of days with minimum temperature above 20 °C for the whole territory of Poland (Fig. 21.5) is projected. For the far future and RCP4.5, the increase in the number of such days is still small. For most of the area, it is 1–2 days (mainly in the central part of Poland), but for lake districts, the South, highlands and mountains, no changes are expected. However, for FF and RCP8.5, according to climate projections, a significant increase in the number of warm nights with a minimum temperature above 20 °C is expected. While no changes are projected in the mountains, for the rest of the country, the number of such days could increase by no less than 5 days, and in the valleys of large rivers, there may be even 10 more such days per year.

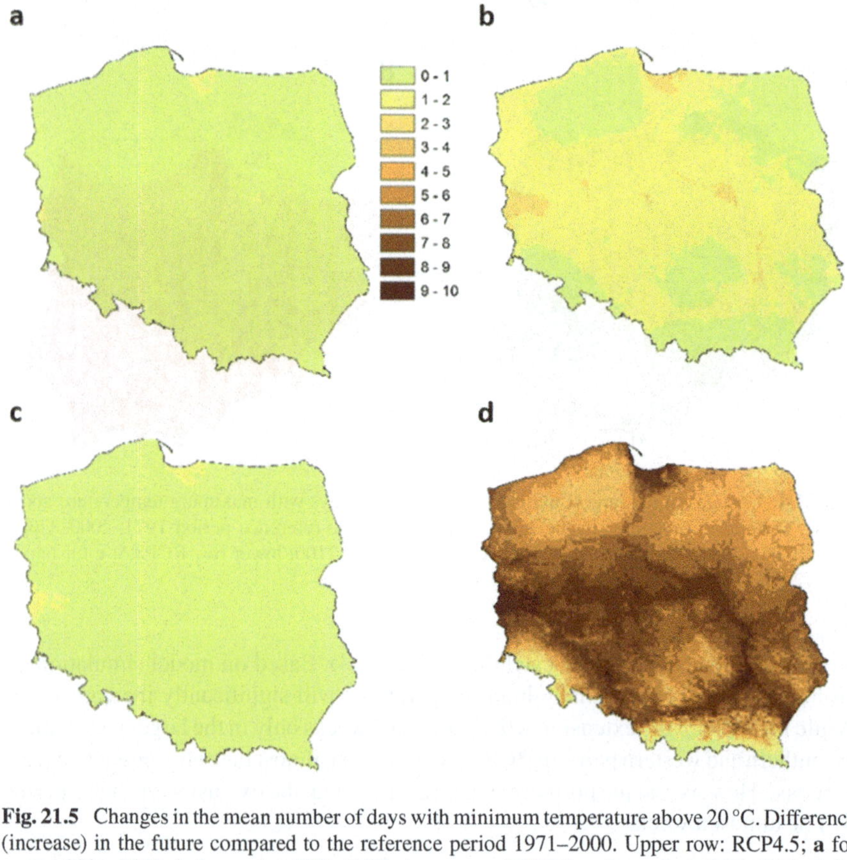

Fig. 21.5 Changes in the mean number of days with minimum temperature above 20 °C. Difference (increase) in the future compared to the reference period 1971–2000. Upper row: RCP4.5; **a** for period 2021–2050; **b** for period 2071–2100; lower row RCP8.5; **c** for period 2021–2050; **d** for period 2071–2100

Changes in the Mean Number of Days with Both Maximum Temperature Above 30 °C and Minimum Temperature Above 20 °C

Another index from the group of extreme/specific thermal conditions is the mean number of days with both maximum temperature above 30 °C and minimum temperature above 20 °C in a year. Projection for NF and both RCPs point to minor changes in the number of such days for the whole country. Increases will become higher at the end of the twenty-first century; however, for RCP4.5 they will mostly be at the level of 1–2 days in a year (Fig. 21.6). For FF and RCP8.5, the simulations project significant changes. Thus, along the river valleys, the largest increases are projected, i.e. up to 10 days more. In the lowlands and highlands, there will likely be 5–8 more

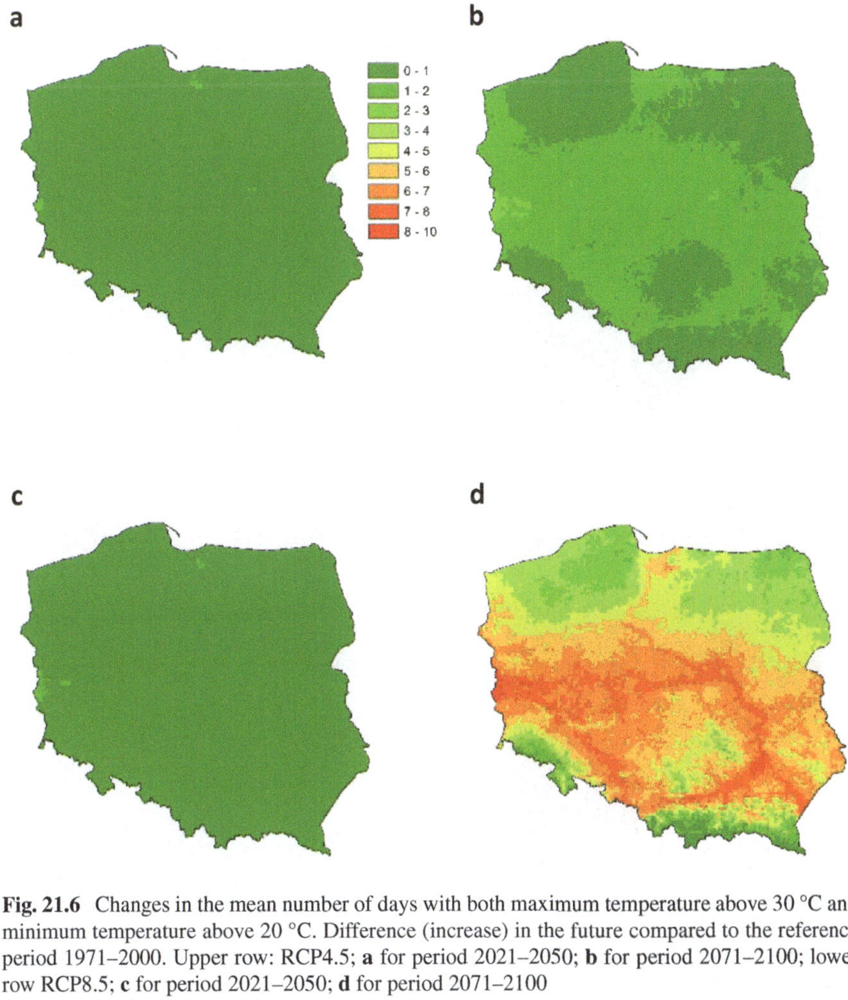

Fig. 21.6 Changes in the mean number of days with both maximum temperature above 30 °C and minimum temperature above 20 °C. Difference (increase) in the future compared to the reference period 1971–2000. Upper row: RCP4.5; **a** for period 2021–2050; **b** for period 2071–2100; lower row RCP8.5; **c** for period 2021–2050; **d** for period 2071–2100

such days, while in lake districts about 2–4 days on average. The mountainous areas will remain almost unchanged.

Changes in the Mean Number of Days with Minimum Temperature Below 0 °C and Below −15 °C

Finally, two indexes regarding the appearance of negative temperatures in the future were analysed. Intuitively, it seems that in a warmer climate, on average, there should be fewer incidences of frosts and icy days. However, this does not mean that in the future a cold winter cannot take place. As to the number of days with a minimum temperature below 0 °C, model-based projections indicate large decreases (Fig. 21.7).

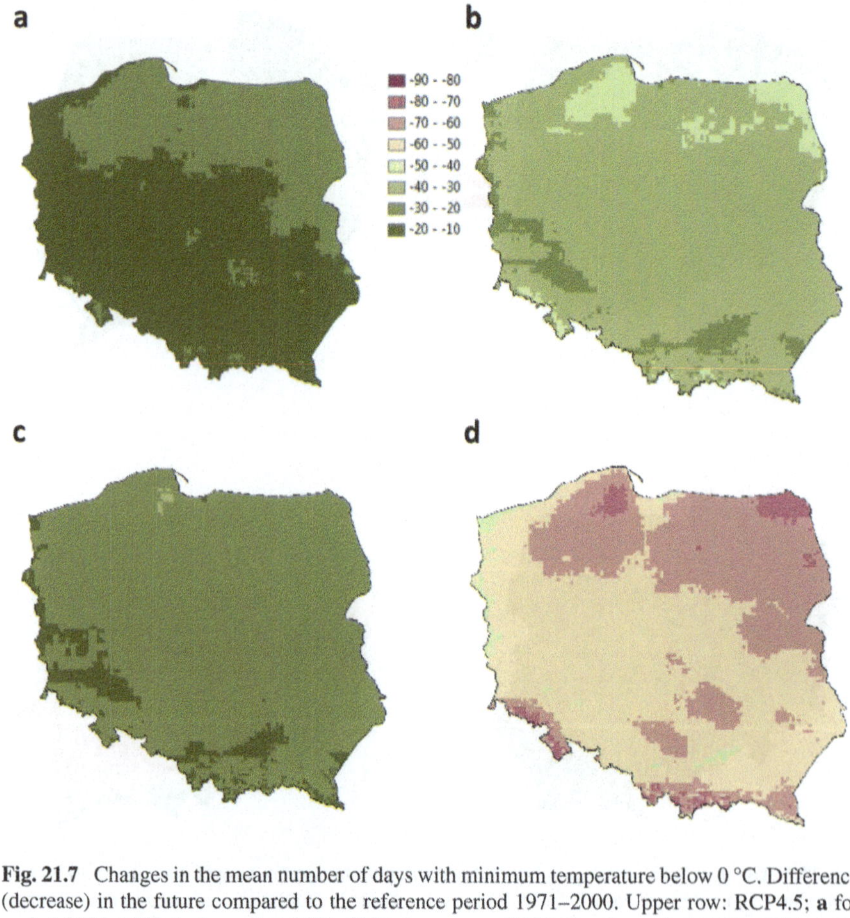

Fig. 21.7 Changes in the mean number of days with minimum temperature below 0 °C. Difference (decrease) in the future compared to the reference period 1971–2000. Upper row: RCP4.5; **a** for period 2021–2050; **b** for period 2071–2100; lower row RCP8.5; **c** for period 2021–2050; **d** for period 2071–2100

For the NF and RCP4.5, the territory of Poland is divided into two parts, where in the South the number is expected to decrease by 10–20 such days and in the North by 20–30 days. For the RCP8.5, the expected decreases will likely be on the level of 20–30 days for most areas in Poland. For FF, at the end of twenty-first century, under RCP4.5, the decreases reached 30–40 days. Some areas in the South are projected to have smaller decreases (by 20–30 days) and some areas in the North with even greater decreases (40–50 days). Projections for FF and RCP8.5 differ significantly from the others. The number of days with temperatures below 0 °C in the areas from the Northwest via Polish lowlands and highlands to the Southeast, will probably decrease by 50–60 a year. In the Lake District, even bigger decreases are expected, up to 60–70 days. In the mountainous areas, they reach the value of 80–90 days.

Projections for the future also indicate decreases in the number of days with a minimum temperature below −15 °C for the whole territory of Poland. For both future horizons and both RCPs, there are similar spatial patterns of changes—the gradient of values grows from West to the East (Fig. 21.8). For the NF and both RCPs, projections are almost identical, i.e. there are expected decreases in the number of such days by 1–2 in the West and 4–5 in the East. For the FF and RCP4.5, decreases vary from 1–2 to 6–7 days, while for RCP8.5 to even more than 9 days.

Discussion and Conclusion

This study makes note of projected future changes in the mean and extreme thermal conditions and indices for Poland. The multi-model ensemble mean of eight regional model simulations from the EURO-CORDEX experiment for two representative concentration pathways: RCP4.5 and RCP8.5 and two time horizons, near future (2021–2050) and far future (2071–2100) was used. According to this study, all models have shown an agreement on a systematic upward trend in mean temperature in Poland, for both the near and the far future and for both RCPs.

The mean annual temperature is projected to increase in the future for the whole territory of Poland. In general, higher increases are projected for north-eastern and eastern areas. The increases are projected to range from 1 °C for NF and RCP4.5 to almost 4° for FF and RCP8.5. As global warming progresses, the model simulations have begun to predict an increase in the mean temperature. Results of the set of global climate simulations from DDC (scenario GS, i.e. Greenhouse Gas plus Sulphate Integration) for Central Europe and Poland analysed by Liszewska and Osuch (1999) indicated an increase in the mean temperature, the highest in winter. Parry (2000) reported future increases in the mean seasonal (summer and winter) temperature for Poland based on results obtained within the ACACIA project for four IPCC SRES scenarios: B1, B2, A1 and A2, and for time horizon of the 2080s, as compared to the control period 1961–1990.

Results of multi-model climate projections originating from the ENSEMBLES project (Van der Linden and Mitchell 2009) under A1B scenario, showed a significant upward trend in temperature across Europe. For the territory of Poland as a whole,

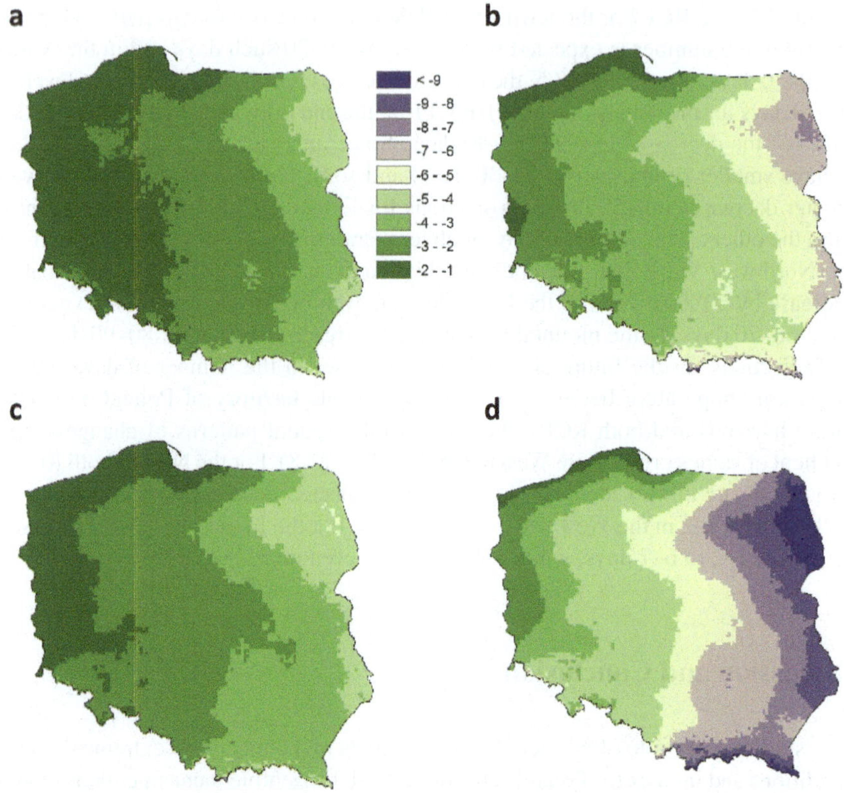

Fig. 21.8 Changes in the mean number of days with minimum temperature below −15 °C. Difference (decrease) in the future to the reference period 1971–2000. Upper row: RCP4.5; **a** for period 2021–2050; **b** for period 2071–2100; lower row RCP8.5; **c** for period 2021–2050; **d** for period 2071–2100

the model-based climate projections indicate an increase in mean annual temperature by 3–3.5 °C by the end of twenty-first century, relative to 1961–1990, depending on the emission scenario and the climate model.

Jacobs et al. (2014) projected future changes in temperature for Europe, based on new high-resolution climate projections generated in EURO-CORDEX project. For the far future (2071–2100) they project robust and statistically significant warming in Poland, in the range of 2–2.5 °C for RCP4.5 and of 3.5–4 °C for RCP8.5. Using nine GCM-RCM combinations from EURO-CORDEX, Mezghani et al. (2017) estimated that the annual mean temperature across Poland is expected to increase by about 1 °C by NF and 2 °C by FF following the RCP4.5, with very low spatial variability but a clear north-east to southwest gradient: Even higher increases were projected for the RCP8.5 emission scenario. Romanowicz et al. (2016), who also based their study on climate simulation originating from EURO-CORDEX project, delivered similar results for selected catchments in Poland. It can be found, in this study, that changes in mean annual air temperature indicate an increase of about 1 °C for the near future (2021–2050) and 2 °C for the far future (2071–2100) periods in comparison to the reference period of 1971–2000.

Furthermore, projected increases in absolute maximum and minimum temperature reach much higher values than the mean values. Studies by Kjellström et al. (2011) and Fischer and Schär (2009) also emphasised that the simulated changes in both cold and warm temperature extremes are larger than the corresponding changes in the mean. In the present study we found that, in the period of 2021–2050, changes in the absolute maximum temperature could, on average, be higher by 2–3 ° across Poland, up to 4–5 °C on the coast (RCP4.5). For the far future horizon of 2071–2100, the increases are significantly higher. For RCP8.5, they range from 3–4 ° in the East to more than 7 ° on the coast. The projections of the changes in the absolute minimum temperature mostly indicate increases by 2–3 °C, for NF and both RCPs. For the FF, there are projected greater differences between RCPs. For RCP8.5, the projected increases may reach more than 10 ° in the north-eastern parts of Poland in comparison to the reference period.

It is interesting to see the results of this research in the context of other studies related to projections of maximum and minimum temperature in Poland. According to the projection for the future with the simulation model HadRM3-PRECIS and SRES A1B scenario, Szwed et al. (2007) stated that the absolute maximum daily temperature is projected to increase in the future (2071–2100) by 2–6 °C in the north-east. In absolute terms, the highest increases of the modelled maximum are projected on the Baltic coast, Mazurian Lake District and the Suwałki District, whereas the smallest in the Wielkopolska Region. Nikulin et al. (2011) analysed maximum and minimum temperatures across Europe based on an ensemble of RCA3 climate model simulation under the SRES A1B emission scenario. By the end of the twenty-first century, the simulated high-temperature extremes will intensify all over Europe with an increase in maximum temperature above 20 °C by about 2–4 °C over Northern Europe, including Poland. Next, minimum temperatures above 20 °C are projected to increase by 4–8 °C by the end of the twenty-first century across Central Europe, including Poland. Based on EURO-CORDEX model simulations, Mezghani et al. (2017) analysed the absolute changes in annual means of minimum and maximum temperature. Under the RCP4.5 scenario, they projected a change in minimum temperature varying between 0.8 °C and 1.6 °C. Changes in maximum temperature are expected to have a similar magnitude and they are less pronounced than those for minimum temperature. Piniewski et al. (2017), also employing EURO-CORDEX model simulations, found a robust increase in the annual mean of minimum and maximum temperatures over the Vistula and Odra River basins (area similar to the territory of Poland) for each combination of future horizons and RCPs.

In the future, it is projected that the extreme hot conditions will not only be hotter but will also last longer and appear more often. The number of consecutive days with maximum temperature above 30 °C indicates very extreme weather conditions. In the near future, the longest such period may be extended by another few days (3–4, on average), regardless of the adopted RCP. Even longer extensions are projected for the end of twenty-first century—almost 2 weeks in the big river valleys in southern and western parts of Poland under RCP4.5 and in nearly the entire country, except coasts, under RCP8.5.

For the near future and both RCPs, no increase in the mean number of days with minimum temperature above 20 °C is expected for the whole territory of Poland. In fact, the increases are projected only for FF and RCP8.5. The number of such days could mostly increase by no less than 5 days across Poland, but along the valleys of large rivers, it is projected to increase by 10 days per year. In that case, projections for NF and both RCPs show minor changes in the number of days with both maximum temperature above 30 °C and minimum temperature above 20 °C for the whole country. Increases will become higher at the end of the twenty-first century. For FF and RCP8.5, the largest increases by up to 10 days are projected. In the lowlands and highlands, it is probable that the increases will mostly be in the range of 5–8 days, and in lake districts of 2–4 days, on average.

Based on the HadRM3-PRECIS regional model simulation with SRES A1B scenario, in the future horizon, both the number of days with maximum temperature equal to or greater than 35 °C and the number of nights with minimum temperature equal to or greater than 20 °C are projected to increase by 5–30 days, with the highest increases in the Southeast. These thermal indices will not change along the Baltic coast. In addition, the number of hot nights will see an even greater increase. The maximum increases in relation to the reference period 1961–1990 are predicted for the Sandomierz Valley and Southeastern Poland. In this area, the number of extremely hot days followed by extremely hot nights can likely increase in the future by over 10 days in the average year (Szwed and Graczyk 2006; Szwed et al. 2007). This coincides also with the results of projections from the RegCM (REGional Climate Model) under SRES B1, A1B and A2 scenarios for number of summer days (with maximum temperature above 25 °C) presented by Jaczewski et al. (2014).

Based on climate projection, it can be expected that, in the future, frost will gradually retreat from the area of Poland. As to the number of days with a minimum temperature below 0 °C, model simulations used in this paper project large decreases. For the NF and RCP4.5, it is expected to decrease by about 20 such days, on average and for FF and RCP8.5—by 40–50 days across Poland, except for the Lake District (up to 60–70 days) and in the mountainous areas (80–90 days). Projections for the future also point towards decreases in the number of days with minimum temperature below −15 °C for the whole territory of Poland. For the future horizon of 2071–2100 and RCP4.5, decreases will vary from 1 to 6–7 days, while for RCP8.5 to even more than 9 days.

With the results of simulation from the HadRM3-PRECIS regional model and SRES A1B emission scenario, Szwed et al. (2007) affirmed that absolute minimum daily temperature is projected to be higher by 6 (for the central part of Poland and the Baltic coast) to even 12 °C (in the East), with relation to the present. It was also predicted that, in an average future year, frosty days with a minimum temperature below −15 °C will appear sporadically in 2071–2100. Jaczewski et al. (2014) compared changes of selected thermal indices for the time period 2011–2030 relative to 1971–1990, based on bias-corrected projections from the RegCM (REGional Climate Model) under SRES B1, A1B and A2 scenarios and reported a decrease in the number of frosty and icy days. Brzóska and Jaczewski (2017) used the results of three RCM simulations from the EURO-CORDEX branch for RCP4.5 and RCP8.5 scenarios. Their results show an increase in the global temperature, resulting in a decrease in the number of frosty and icy days in Poland.

To conclude, the climate will continue to change in the future. All climate models have shown an agreement on a systematic upward trend in the mean temperature in Poland. In the future, the extreme hot thermal conditions will become even hotter, last longer and appear more frequently, while the amplitude of frosty periods and their frequency will decrease. Nonetheless, climate change extends far beyond changes in temperature.

Acknowledgements Support of the project CHASE-PL (Climate change impact assessment for selected sectors in Poland) of the Polish–Norwegian Research Programme operated by the National Centre for Research and Development (NCBiR) under the Norwegian Financial Mechanism 2009–2014 in the frame of Project Contract No. Pol Nor/200799/90/2014 is gratefully acknowledged.

Author thanks to Iwona Pińskwar for technical support using Climate Data Operator (CDO) software.

References

Brzóska B, Jaczewski A (2017) Przyszłe zmiany wybranych wskaźników klimatycznych dla Polski na podstawie wyników dynamicznego downscalingu (Future changes of selected climate indices in Poland based on the results of dynamical downscaling). Prace Geograficzne UJ 149:7–14. https://doi.org/10.4467/20833113pg.17.007.6923

Clarke LE, Edmonds JA, Jacoby HD, Pitcher HM, Reilly JM, Richels RG (2007) Scenarios of greenhouse gas emissions and atmospheric concentrations, Sub-report 2.1a, U.S. Climate Change Science Program. Synthesis and Assessment Product 2.1a, CCSP, Washington, 154 pp

Fischer EM, Schär C (2009) Future changes in daily summer temperature variability: driving processes and role for temperature extremes. Clim Dyn 32. https://doi.org/10.1007/s00382-008-0473-8

IPCC (2000) Special report on emission scenarios In: N Nakicenovic et al.(eds). Cambridge University Press, New York, 599 pp

Jacobs D, Petersen J, Eggert B, Alias A, Christensen OB, Bouwer LM, Braun A, Colette A, Déqué M, Georgievski G, Georgopoulou E (2014) EURO-CORDEX: new high-resolution climate change projections for European impact research. Reg Environ Change 4(2):563–578

Jaczewski A, Brzoska B, Wibig J (2014) Comparison of temperature indices for three IPCC SRES scenarios based on RegCM simulations for Poland in 2011–2030 period. Meteorol Z 24:99–106. https://doi.org/10.1127/metz/2014/0457

Kjellström E, Nikulin G, Hansson U, Strandberg G, Ullerstig A (2011) 21st century changes in the European climate: uncertainties derived from an ensemble of regional climate model simulations. Tellus 63A:24–40

Liszewska M, Osuch M (1999) Analiza wyników globalnych modeli klimatu dla Europy Środkowej i Polski (Analysis of the results of global climate models for Central Europe and Poland). In: Zmiany i zmienność klimatu Polski (Change and variability of Polish climate) Polish Conference, Łódź, 4–6 November, 1999, pp 129–142 (in Polish)

Mezghani A, Dobler A, Haugen JE, Benestad RE, Parding KM, Piniewski M, Kardel I, Kundzewicz ZW (2017) CHASE-PL Climate Projection dataset over Poland-Bias adjustment of EURO-CORDEX simulations. Sci Data Discuss, Earth Syst. https://doi.org/10.5194/essd-2017-51

Nikulin G, Kjellstro EM, Hansson U, Strandberg G, Ullerstig A (2011) Evaluation and future projections of temperature, precipitation and wind extremes over Europe in an ensemble of regional climate simulations. Tellus A DynMeteorol Oceanogr 63(1):41–55. https://doi.org/10.1111/j.1600-0870.2010.00466.x

Parry ML (ed) (2000) Assessment of potential effects and adaptations for climate change in Europe: The ACACIA Project. Jackson Environment Institute, University of East Anglia, Norwich, UK, 320 pp

Piniewski M, Mezghani A, Szcześniak M, Kundzewicz ZW (2017) Regional projections of temperature and precipitation changes: Robustness and uncertainty aspects. Meteorol Z 26(2):223–234. https://doi.org/10.1127/metz/2017/0813

Romanowicz RJ, Bogdanowicz E, Debele SE, Doroszkiewicz J, Hisdal H, Lawrence D, Meresa HK, Napiórkowski JJ, Osuch M, Strupczewski WG, Wilson D, Wong WK (2016) Climate change impact on hydrological extremes: preliminary results from the Polish-Norwegian project. Acta Geophys 64(2):477–509. https://doi.org/10.1515/acgeo-2016-0009

Schulzweida U (2019) CDO User Guide (Version 1.9.6) http://doi.org/10.5281/zenodo.2558193

Szwed M, Graczyk D (2006) Thermal seasons in Poland—the present and the future based on HadRM3-PRECIS results. Papers on Global Change 13:59–77

Szwed M, Graczyk D, Pińskwar I, Kundzewicz ZW (2007) Projections of climate extremes in Poland. Geographia Polonica 80(2):191–202

Szwed M, Karg G, Pińskwar I, Radziejewski M, Graczyk D, Kędziora A, Kundzewicz ZW (2010) Climate change and its effect on agriculture, water resources and human health sectors in Poland. Nat Hazards Earth Syst Sci 10:1725–1737. https://doi.org/10.5194/nhess-10-1725-2010

Van der Linden P, Mitchell JFB (eds) (2009) ENSEMBLES: climate change and its impacts: summary of research and results from the ENSEMBLES project. Met Office Hadley Centre, FitzRoy Road, Exeter EX1 3 PB, UK. 160 pp

Chapter 22
Projections of Precipitation Changes in Poland

Iwona Pińskwar and Adam Choryński

Abstract In a warmer climate, the consequences of future precipitation changes could be much more severe than nowadays. Therefore, assessment of these future changes is an issue worthy of consideration, facilitating proper adaptation to water-related problems. In this study, ensembles of eight regional climate models for two representative concentration pathways (RCP4.5 and RCP8.5) and two-time horizons (2021–2050 and 2071–2100) were used to gain insight into changes in precipitation relating to the reference period of 1971–2000. Analyses were made of various indices, such as annual totals, maximum 24 h total, 3-day total and the number of three-day periods with precipitation totals greater than 50 mm; consecutive wet days (CWD) and the number of wet periods longer than 5 days; simple daily intensity index (SDII); consecutive dry days; number of dry and hot days and also number of days with intense precipitation equal or above the thresholds of 20 mm per day. It was found that most indices based on the models under examination will increase and this applies to both mean annual precipitation totals as well as in intense precipitation and scarcity of water. Future changes will be more severe for higher RCP and the far time horizon.

Introduction

The assessment of future changes in precipitation is very important and greatly needed. A rise in magnitude, frequency and duration of extreme precipitation may increase climate-related flood risk. Moreover, changes in annual totals and dry periods have an impact on several sectors in Poland, from agriculture, forestry, through the energy sector to water supply. This study presents projections of changes in annual sum as well as nine other precipitation indices for Poland for the near time horizon 2021–2050 and far time horizon: 2071–2100; under two representative concentration pathways: RCP4.5 and RCP8.5.

I. Pińskwar (✉) · A. Choryński
Institute for Agricultural and Forest Environment, Polish Academy of Sciences, Poznań, Poland
e-mail: iwona.pinskwar@isrl.poznan.pl

M. Falarz (ed.), *Climate Change in Poland*, Springer Climate,
https://doi.org/10.1007/978-3-030-70328-8_22

Climate models are compatible and predict an increase in mean, as well as in minimum and maximum temperatures. Changes in precipitation are more difficult to project because of spatial diversity, high natural variability, seasonality and the complexity of the phenomenon involving the sub-grid scale features such as topography or land use, which makes it difficult to properly assess the intensity of precipitation on a smaller scale. Thus, downscaling methods are considered to be useful in this case. The simulations in this study have been bias-corrected using the quantile mapping method (Mezghani et al. 2016).

Simulations of past projections from the end of the twentieth century showed an increase in total precipitation for Northern Europe and drier conditions for Southern Europe. Poland, located in mid-latitudes, is placed between these pronounced changes. The first IPCC Report (1992) stated that for Southern Europe (35°–50°N, 10°W-45°E) there will be an increase in precipitation in winter, but summer precipitation will decrease by 5–15% (IPCC "Business-as-Usual" scenario; changes from pre-industrial). Results of global climate simulations for Central Europe and Poland analysed by Liszewska and Osuch (1999) indicated an increase in precipitation. Perry (2000) also presented future changes in summer and winter precipitation for Poland for three time horizons and four IPCC SRES scenarios: B1, B2, A1 and A2 (Nakicenovic et al. 2000). Six models indicated an increase in winter rainfall and one a decrease, but for summer this uncertainty of model projections was higher: five models indicated a decrease and two—an increase. The more time distant the projections, the more visible the changes in precipitation. According to Christensen and Christensen (2007), for Eastern Europe at the end of the twenty-first century, for the A2 scenario, there should be a precipitation increase during winter, which will be lower during spring, but decrease for summer and autumn, which means less precipitation than nowadays. The mentioned results for winter and summer were comparable to those obtained for the A1B scenario (IPCC 2007). Projections of changes in annual precipitation total (HadRM3, scenarios A2 and B2, period 2071–2100) showed a decrease for Poland, despite an increase in precipitation during winter for A2; for B2 simulations of the model showed also a decrease during winter (Pińskwar 2010).

As global warming progresses, the model simulations have begun to predict an increase in annual precipitation, like multi-model ensemble climate projections obtained within the ENSEMBLES Project (van der Linden and Mitchell 2009) under the A1B scenario. According to this study, for the near future 2021–2050, an increase was shown for winter and autumn precipitation and almost no changes for spring and summer. More visible changes were predicted for the period 2071–2100: more precipitation during winter, spring and autumn and less during summer. The latest climate model simulations for the RCP4.5 and RCP8.5 scenarios projected an increase in annual precipitation all across the studied country for the periods 2021–2050 and 2071–2100. Although multi-model ensemble mean of models are consistent on the overall positive changes for winter and spring, the changes for autumn and summer are more moderate, almost positive, but for southern Poland precipitation are expected to decrease though (Mezghani et al. 2016).

Also, changes in extreme precipitation for Poland were more vague than for other regions. According to the projection of changes corresponding to the A2 scenario for several models (Tuomenvirta et al. 2006), the maximum daily precipitation will increase during the winter throughout nearly all of Europe. It will decrease only in the southern part of Spain, southern Italy and the island of Sicily. In the summer, the maximum daily rainfall will increase in Belarus, Germany, Poland and parts of Northern Europe. Bates et al. (2008) stated that in the future intensive rainfall will increase, especially in high and medium latitudes, particularly where average rainfall increases. Kamiguchi et al. (2006) used model MRI-CGCM2.3 (A1B scenario for the period 2080–2099) and showed that intensive rainfall indicators (5-day rainfall sum, number of days with precipitation above 10 mm, average wet day intensity and share total annual precipitation above the 95th percentile) will increase almost worldwide. Fewer days with precipitation above 10 mm will be expected in Latin America, Chile, southern Great Britain and Spain, partly in Madagascar and southern Australia. In general, indices of extreme precipitations may increase where they are already high. Pińskwar (2010) presented projections of the future changes for scenario A2, which indicate in most cases an increase in intense rainfall and for the B2 scenario, for which increases may be smaller, in a number of cases intensive rainfall may be lower, specifically in the southwestern part of the country.

The latest projections of models show a likely further increase in extreme precipitation. Maximum 1-day and maximum 5-day precipitation totals are projected to increase in almost all regions around the globe, including Central and Eastern Europe (Hay et al. 2016). The use of a new generation of climate models strengthens the results obtained in previous studies with the help of RCMs for the SRES A1B scenario. Lehtonen et al. (2014) studied the multi-model-mean results for the A1B forcing at the end of the twenty-first century and projected a maximum 24 h precipitation total to increase by 10–30% across most of Europe in winter and about 10–20% in summer in the north, but a small decrease in the south of the continent. Jacob et al. (2014) compared a multi-model ensemble for the SRES A1B scenario to the new regional EURO-CORDEX data set for RCP8.5 and RCP4.5 for the period 2071–2100 with respect to 1971–2000. The projected seasonal mean changes in heavy precipitation (95th percentile) for these three scenarios are relatively similar, with some regional differences. In particular, the amplitude of change is stronger for RCP8.5 than for A1B in several regions. For winter, the RCP8.5 scenario projects the most dramatic increases in heavy precipitation (up to 35%) in Poland, whereas A1B projects less pronounced changes (up to 25%). Similarly to this study, projected changes for the RCP4.5 scenario, compared to RCP8.5, show convergent seasonal patterns of changes, but the amplitude of changes is much smaller. The magnitude of changes for SRES A1B mostly lies in-between the two RCPs. Rajczak and Schär (2017) studied projections of the 99th percentile of all-day precipitation

(P99) over the European continent with a multi-model ensemble of 12 and 50 km resolution EURO-CORDEX RCMs (EU-COR) forced by RCP4.5 and RCP8.5 and compared those to ENSEMBLES RCMs (A1B) results. For Central and Eastern Europe, including Poland, they projected an increase for the period 2070–2099 with respect to 1981–2010 for all seasons with the highest increase taking place in winter. Some 90% of the ensemble members agreed on the direction of change, except for summer. The range of projections of ENSEMBLES RCMs are close to those for EU-COR, except for summer (changes in projections of ENSEMBLES RCMs are much smaller).

Also indices related to water shortages show a likely further increase. Kamiguchi et al. (2006; the A1B scenario for 2080–2099) presented the value of the CDD index—the longest dry period (with daily precipitation below 1 mm), which will increase in most areas of the Southern Hemisphere, and will decrease in the high latitudes of the Northern Hemisphere, in the Arabian Peninsula and China. Bates et al. (2008) showed that in areas where the average rainfall is currently decreasing, the longest dry period CDD (with daily precipitation below 1 mm) may increase in the future. According to Pińskwar (2010), for both SRES A2 and B2 scenarios changes in warm and dry periods will be more severe in the future. Osuch et al. (2016) investigated the SPI value for six different climate model runs under the scenario A1B for the period 1971–2099. They showed better water availability during the winter months and a decrease in SPI in the summer period, which means increased water shortage. The authors also indicated considerable inter-model variability on regional and local scales.

Changes in precipitation may lead to problems associated with excess water, dramatic and high-impact floods, but also droughts, which nowadays are severe and may get even worse in the future. With the use of bias-corrected simulations from the EURO-CORDEX project, this research analyses future changes in the characteristics of mean and intense precipitation and also dry and hot spells across Poland, comparing the past time periods of 1971–2000 with two future time horizons: 2021–2050 and 2071–2100.

Data and Methodology

This study has been performed with the use of the multi-model ensemble mean of eight regional model simulations for two representative concentration pathways: RCP4.5 and RCP8.5 and two time horizons, near future (NF): 2021–2050 and far future (FF): 2071–2100, stemming from the EURO-CORDEX project (see Chapter 21). Simulations of these models are available also for the past interval, 1971–2000, so indices were calculated with reference to this past period. Analysing simulations from an ensemble of various models reduces related uncertainties.

Several indices were used in this study to recognise future changes in precipitation. Among them are annual total precipitation, indices related to wet periods and mean intensity, such as the number of consecutive wet days (CWD) and the number of wet spells longer than 5 days, as well as simple daily intensity index (SDII). In addition, indices to assess future changes in intense precipitation were used, including maximum 24 h total, 3-day total and the number of 3-day periods with precipitation totals greater than 50 mm, and also number of days with intense precipitation equal to or greater than the threshold of 20 mm per day and indices related to dry periods, such as consecutive dry days, i.e. days with daily precipitation amount below 1 mm and also the number of dry and hot days (with daily precipitation amount below 1 mm and with the maximum temperature equal to or greater than 30 °C). Data were processed with the usage of the Climate Data Operators (CDO) (Schulzweida et al. 2012). Some indices were calculated on the basis of ECA indices, which are implemented in CDO (CWD, SDII, maximum 24 h total, number of days with intense precipitation equal to or above the threshold of 20 mm per day, CDD) and other indices were calculated with the use of CDO operators.

Results

Generally, changes in precipitation in the future are more pronounced for the far future (2071–2100) than for the near future (2021–2050) and for RCP8.5 than for RCP4.5. The changes in extreme precipitation are likely to be more pronounced than in the mean precipitation in the future. When temperature increases, the water holding capacity of the atmosphere rises, and thus the possibility of intense precipitation also grows. On the other hand, higher temperatures, especially hot spells during summer, are conducive to long dry periods and a higher probability of drought occurrence.

Changes in Mean Precipitation and Wet Periods

The annual mean precipitation totals are projected to increase for the whole country, with higher increases in southern and northern areas (Fig. 22.1). For the NF, more precipitation than nowadays, at about 25–50 mm, is predicted for RCP4.5. Only in northern and southern parts of the country the increase of annual totals may exceed 50 mm, while for RCP8.5 more areas could experience an increase exceeding 50 mm. For the FF, simulations show more areas with precipitation exceeding 50 mm and exceeding 75 mm in the southern, northern and eastern parts of Poland for RCP4.5. For RCP8.5, the increase is the highest: in the west up to 100 mm, in the east up to 125 mm, and also higher values for the northern and southern parts of country.

Changes in the largest number of consecutive wet days (CWD) do not show regular patterns for RCPs and future time horizons, with increases and decreases up to 5 days (Fig. 22.2). It is projected that that longest spell with precipitation above 1 mm will be shortened to about 2.5 days in south-eastern part of Poland and for the FF (both RCPs) for the upper Odra River basin. The longest wet spell will be extended about 5 days in the eastern part for the NF and RCP4.5 and in the far future for RCP8.5.

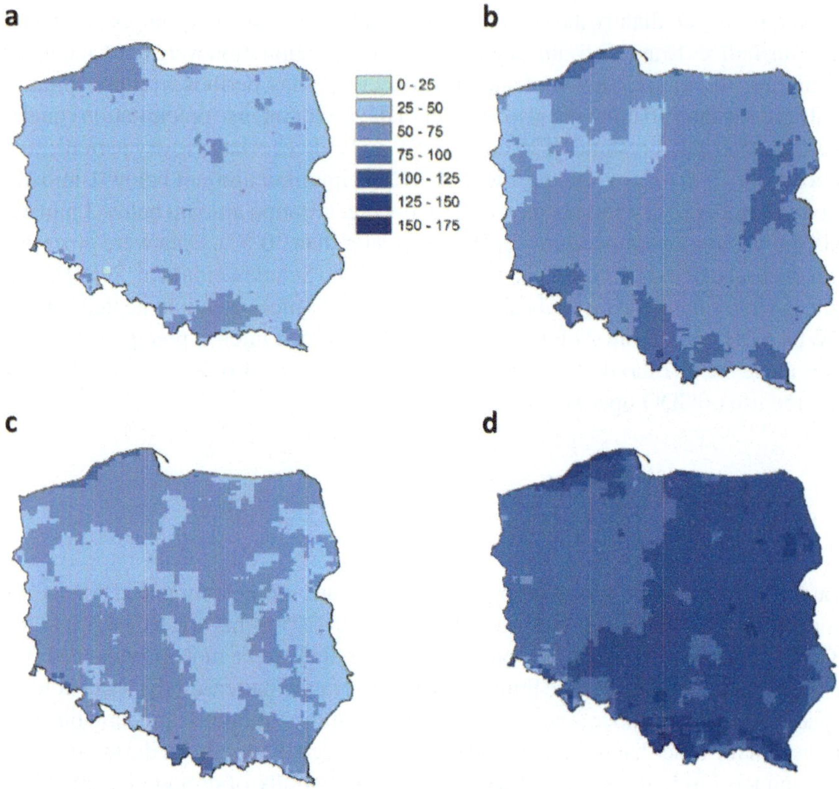

Fig. 22.1 Changes in mean annual precipitation totals in mm. Difference in reference to the period 1971–2000. Upper row: RCP4.5; **a** for period 2021–2050; **b** for period 2071–2100; lower row RCP8.5; **c** for period 2021–2050; **d** for period 2071–2100

In the case of the number of wet periods (CWD) lasting more than five days calculated for the entire 30-year period, changes are more regular (Fig. 22.3). In the future, there will be more wet spells for most of the area of Poland for both RCPs and time horizons, up to 10 and up to 20 periods especially in the eastern part of the country. Less wet spells (up to 10 days) are projected to occur in the west for both RCPs and in the south-eastern parts of the country for the near future for RCP4.5 and both time horizons for RCP8.5.

In the future, also the mean precipitation amount on wet days, SDII (i.e. above 1 mm), is projected to increase. For the NF and both RCPs (Fig. 22.4), mean daily precipitation could be more intense by 0.25–0.5 mm, while for the FF by about 0.5 mm for RCP4.5 and up to 0.75 mm for most of the area of Poland, as well as up to 1 mm for the southern part of the country for RCP8.5.

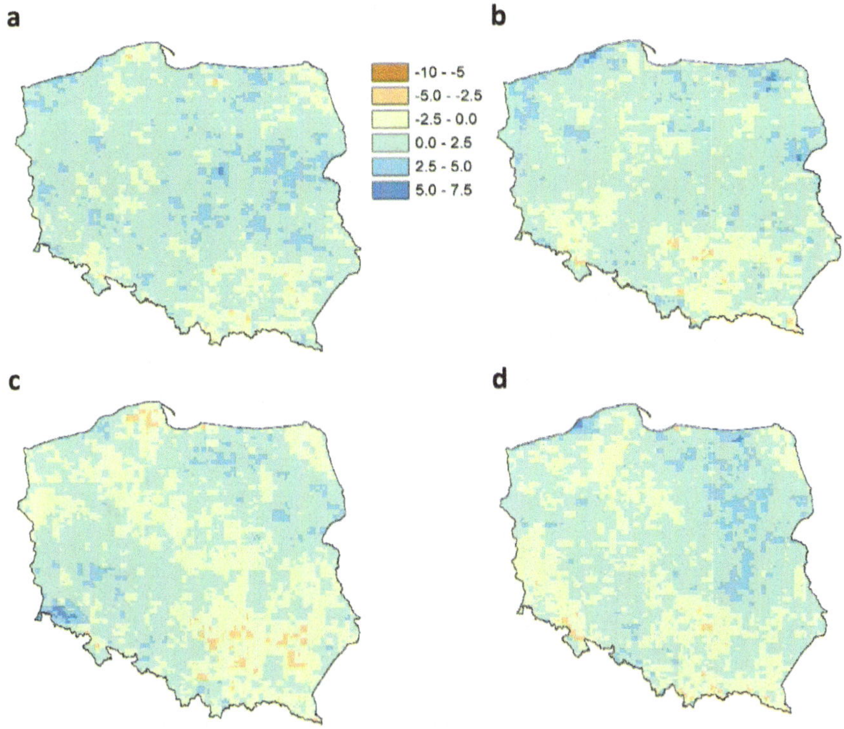

Fig. 22.2 Changes in the largest number of consecutive wet days (CWD). Difference in reference to the period 1971–2000. Upper row: RCP4.5; **a** for period 2021–2050; **b** for period 2071–2100; lower row RCP8.5; **c** for period 2021–2050; **d** for period 2071–2100

Changes in Extreme Precipitation

Changes in extreme precipitation show a rather consistent pattern. For nearly the whole country, simulations of models predict an increase in values of indices and this increase will become higher at the end of the twenty-first century. In the near future, one can expect that the maximum daily precipitation (Fig. 22.5) will be higher by 20–40 mm than nowadays, and for the far future even by 60–80 mm.

Similarly, simulation models show changes for maximum 3-day precipitation totals (Fig. 22.6). This 3-day maximum may decrease only in a small area and increase up to 50 mm for both RCPs and period 2021–2050, and is to be much higher (up to 75–100 mm) for RCP8.5 in the end of the century. Also, in the future, there will be a higher number of 3-day periods with precipitation totals greater than 50 mm for the studied periods (Fig. 22.7); up to 25 spells in nearly the whole country and up to 50 in the south of Poland for the near future and both RCPs and the far future and RCP4.5. A greater number of these very wet periods can be expected for RCP8.5 and the period 2071–2100; in the southern part even up to 75 and 100.

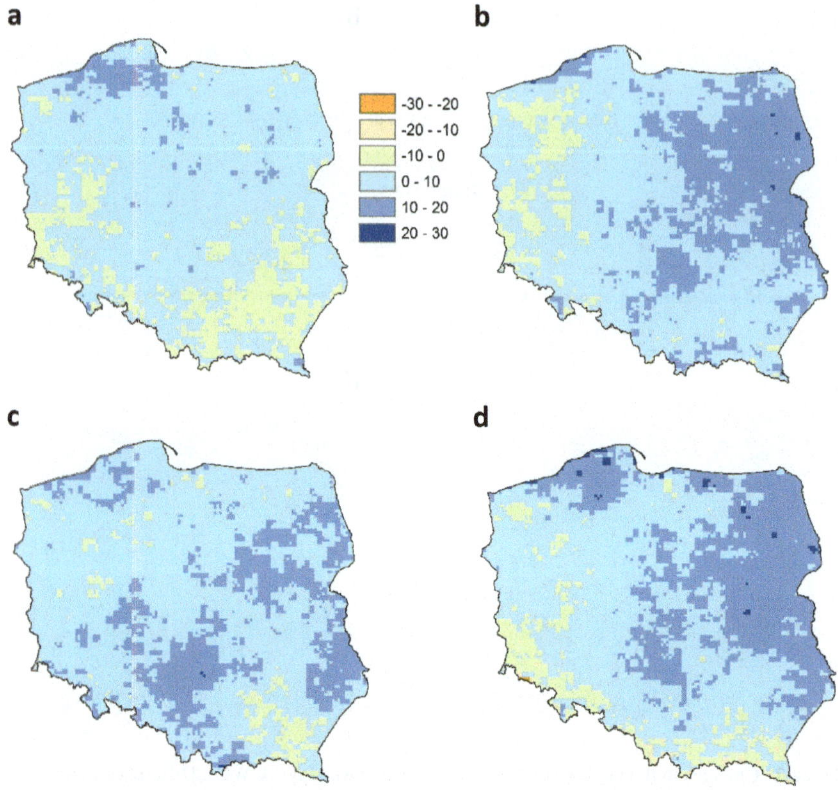

Fig. 22.3 Changes in the number of wet periods (CWD) of more than 5 days for 30 years. Difference in reference to the period 1971–2000. Upper row: RCP4.5; **a** for period 2021–2050; **b** for period 2071–2100; lower row RCP8.5; **c** for period 2021–2050; **d** for period 2071–2100

In the future, there will be more days with daily precipitation amounts equal to or greater than 20 mm (Fig. 22.8). For the NF and both RCPs, the number of days with extreme precipitation may increase by 0.5–1.0. For the time horizon 2071–2100, this increase will be higher: by 2 days for the southern part of Poland for RCP4.5 and even up to 3–4 days in this area for RCP8.5.

Changes in Dry and Dry and Hot Spells

In both the near and far future and both RCPs, the longest dry period (CDD, Fig. 22.9) will be even longer, on average by 5 days, especially in the eastern part of Poland and shorter on average by 7 days in the western part. However, changes in the largest number of consecutive dry days (CDD) are unevenly distributed. In the case of the longest dry and hot period (Fig. 22.10), the models are more comparable. In the

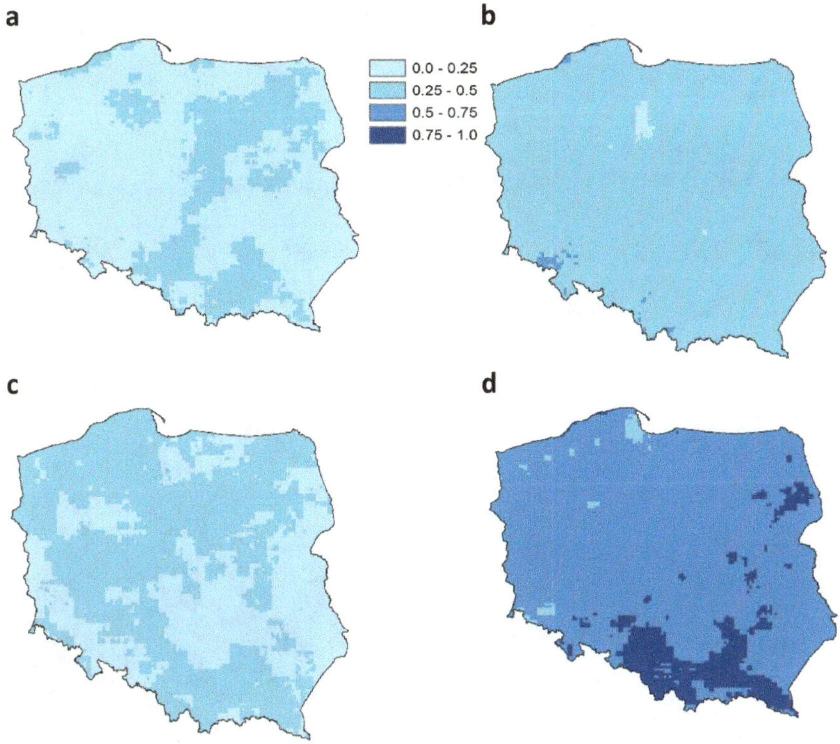

Fig. 22.4 Changes in the simple daily intensity index (SDII—the mean precipitation amount of wet days, i.e. above 1 mm). Difference in reference to the period 1971-2000. Upper row: RCP4.5; **a** for period 2021–2050; **b** for period 2071–2100; lower row RCP8.5; **c** for period 2021–2050; **d** for period 2071–2100

future, this period will be longer by 3–7 days and even up to 12 days in the eastern part of the country for RCP8.5 in the end of the twenty-first century. This means that it is very likely there will be a significant extension of dry and hot waves.

Conclusions

This study brings to light projected future changes in annual precipitation total as well as in indices of intense and extreme precipitation and dry periods for Poland. The multi-model ensemble mean of eight regional model simulations has been used for two representative concentration pathways: RCP4.5 and RCP8.5 and two time horizons, near future: 2021–2050 and far future: 2071–2100. Apart from the two indices, namely CWD and CDD, the model simulations are rather consistent as to the direction of changes in future precipitation. This information about changes in

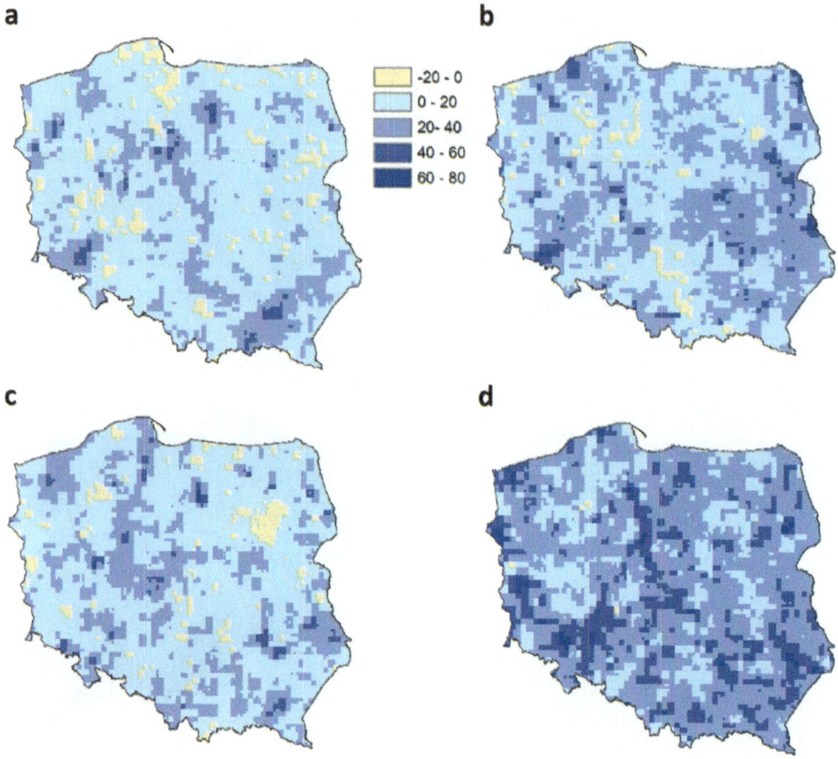

Fig. 22.5 Changes in maximum 24 h precipitation totals. Difference in reference to the period 1971–2000. Upper row: RCP4.5; **a** for period 2021–2050; **b** for period 2071–2100; lower row RCP8.5; **c** for period 2021–2050; **d** for period 2071–2100

precipitation are vital for climate change adaptation and flood risk reduction (Zhang et al. 2013). Every increase in extreme precipitation can contribute to more severe floods in the future, while increases in dry periods lead to extended droughts. Recent studies indicate that the projection of precipitation extremes is associated with various uncertainties related to emission scenarios of greenhouse gases, GCMs, RCMs and statistical downscaling methods as well as by natural variability of climate. However, the direction of projected changes in mean, heavy precipitation and dry periods across Poland are quite consistent and correspond with continental-scale results. Projections also show that the highest values of indices may not increase for the whole country (CWD and CDD), but their frequency may increase, as shown in the case of CWD. Hence, the results can be regarded as fairly robust.

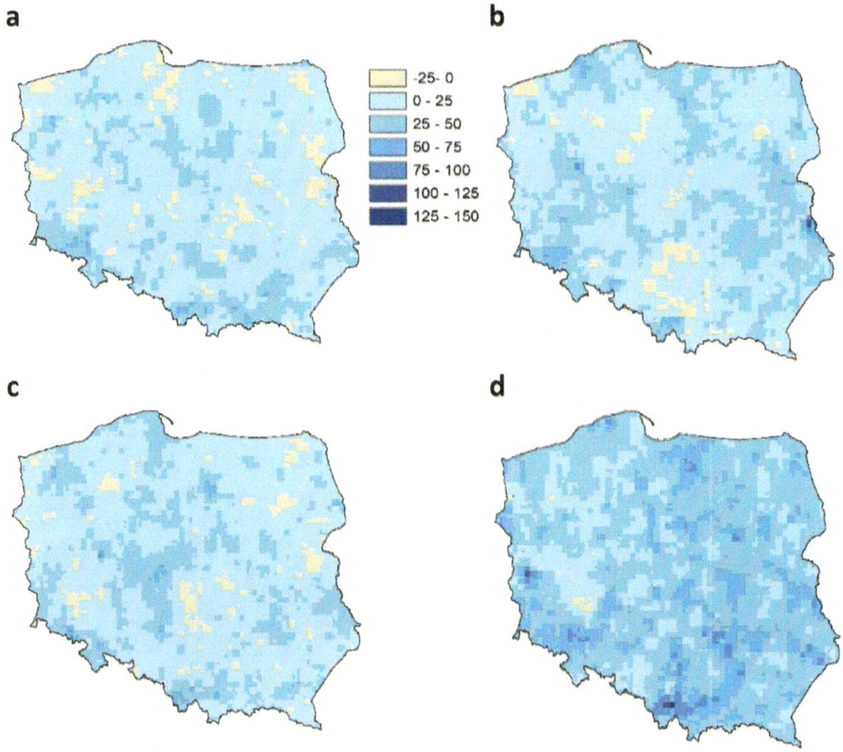

Fig. 22.6 Changes in maximum 3-day precipitation totals. Difference in reference to the period 1971–2000. Upper row: RCP4.5; **a** for period 2021–2050; **b** for period 2071–2100; lower row RCP8.5; **c** for period 2021–2050; **d** for period 2071–2100

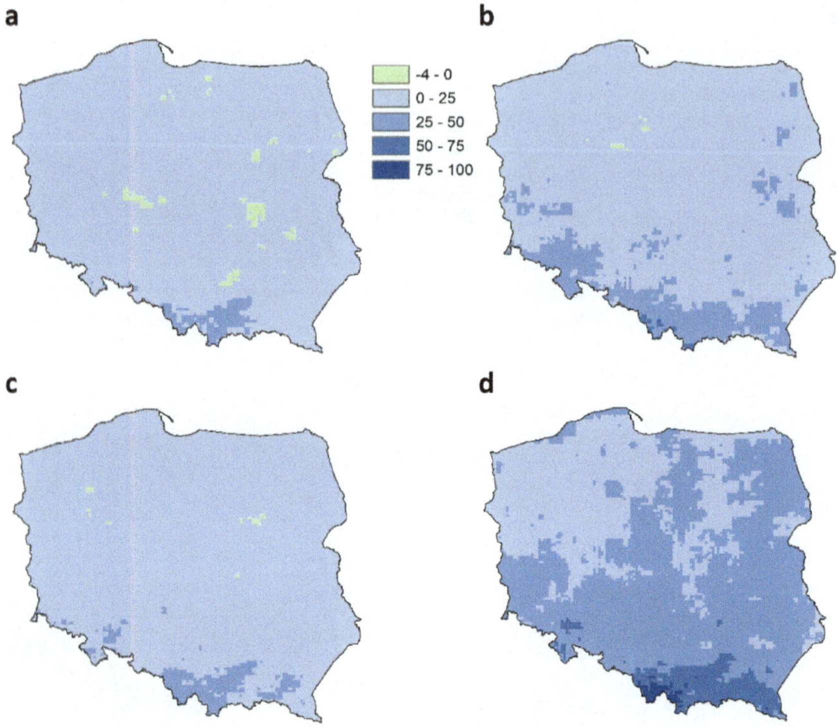

Fig. 22.7 Changes in the number of 3-day periods with precipitation totals greater than 50 mm for 30 years. Difference in reference to the period 1971–2000. Upper row: RCP4.5; **a** for period 2021–2050; **b** for period 2071–2100; lower row RCP8.5; **c** for period 2021–2050; **d** for period 2071–2100

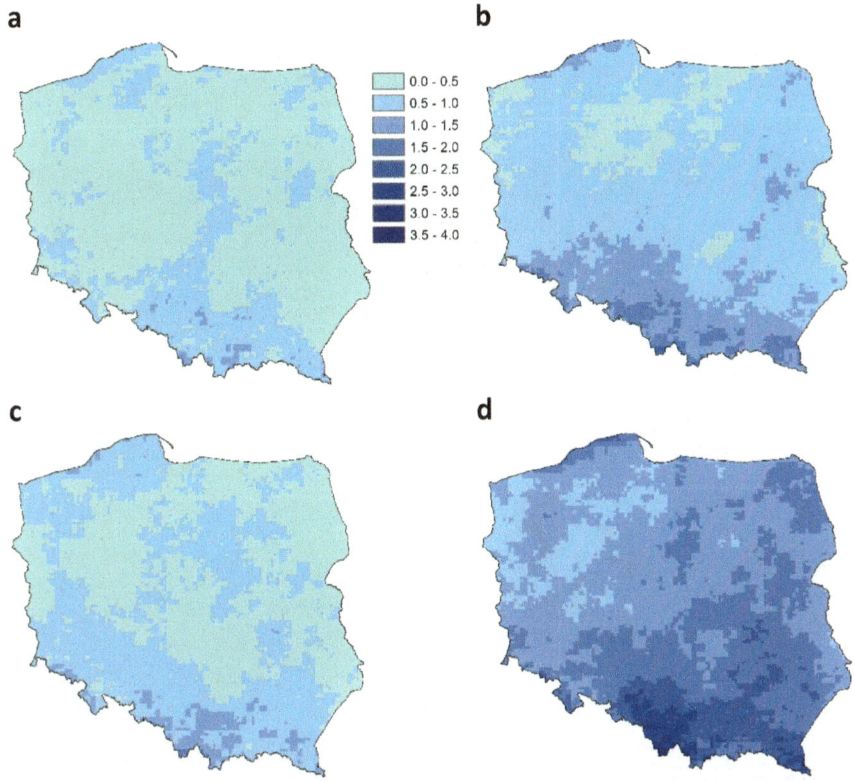

Fig. 22.8 Changes in the number of days with daily precipitation amounts equal to or greater than 20 mm. Difference in reference to the period 1971–2000. Upper row: RCP4.5; **a** for period 2021–2050; **b** for period 2071–2100; lower row RCP8.5; **c** for period 2021–2050; **d** for period 2071–2100

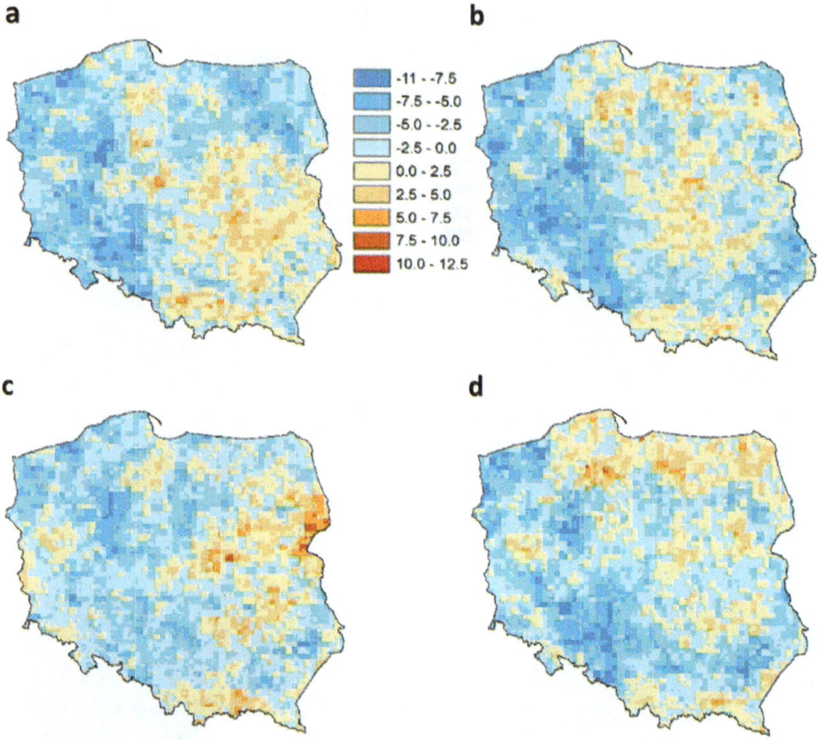

Fig. 22.9 Changes in the largest number of consecutive dry days (CDD) with daily precipitation amounts less than 1 mm. Difference in reference to the period 1971–2000. Upper row: RCP4.5; **a** for period 2021–2050; **b** for period 2071–2100; lower row RCP8.5; **c** for period 2021–2050; **d** for period 2071–2100

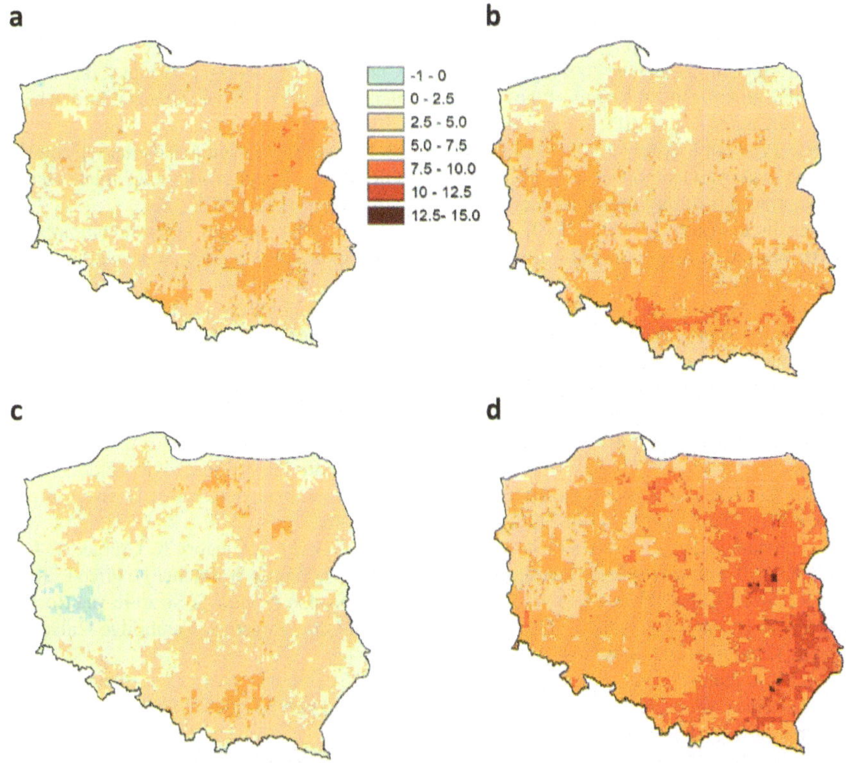

Fig. 22.10 Changes in the longest period with daily precipitation amounts less than 1 mm and maximum daily temperature equal or greater than 30°C. Difference in reference to the period 1971–2000. Upper row: RCP4.5; **a** for period 2021–2050; **b** for period 2071–2100; lower row RCP8.5; **c** for period 2021–2050; **d** for period 2071–2100

Acknowledgements Support of the project CHASE-PL (Climate change impact assessment for selected sectors in Poland) of the Polish–Norwegian Research Programme operated by the National Centre for Research and Development (NCBiR) under the Norwegian Financial Mechanism 2009–2014 in the frame of Project Contract No. Pol Nor/200799/90/2014 is gratefully acknowledged.

References

Bates BC, Kundzewicz ZW, Wu S, Palutikof JP (eds.) (2008) Climate change and water. Technical Paper of the Intergovernmental Panel on Climate Change, IPCC Secretariat, Geneva, Switzerland, p 210

Christensen JH, Christensen OB (2007) A summary of the PRUDENCE model projections of changes in European climate by the end of this century. Clim Change 81:7–30

Hay JE, Easterling D, Ebi KL, Kitoh A, Parry M (2016) Conclusion to the special issue: Observed and projected changes in weather and climate extremes. Weather and Climate Extremes 11:103–105

IPCC (1992) Climate change: The 1990 and 1992 IPCC Assessments Report. J.T. Houghton (ed.), Cambridge, UK, 212 pp

IPCC (2007) Climate change 2007: synthesis report. Contribution of Working Groups I, II and III to the Fourth Assessment Report of the Intergovernmental Panel on Climate Change. In: Pachauri RK, Reisinger A. (eds) IPCC, Geneva, 104 pp

Jacob D, Petersen J, Eggert B, Alias A, Christensen OB, Bouwer LM, Braun A, Colette A, Déqué M, Georgievski G, Georgopoulou E (2014) EURO-CORDEX: new high-resolution climate change projections for European impact research. Reg Environ Change 4(2):563–578

Kamiguchi K, Kitoh A, Uchiyama T, Mizuta R, Noda A (2006) Changes in precipitation-based extremes indices due to global warming projected by a global 20-km-mesh atmospheric model. Sola 2:064–067

Lehtonen I, Ruosteenoja K, Jylhä K (2014) Projected changes in European extreme precipitation indices on the basis of global and regional climate model ensembles. Int J Climatol 34:1208–1222

Liszewska, M, Osuch M (1999) Analiza wyników globalnych modeli klimatu dla Europy Środkowej i Polski. (Analysis of the results of global climate models for Central Europe and Poland). In: Zmiany i zmienność klimatu Polski (Change and variability of Polish climate) Polish Conference, Łódź, 4–6 November, 1999:129–142 (in Polish)

Mezghani A, Dobler A, Haugen JE (2016) CHASEPL Climate Projections: 5-km Gridded Daily Precipitation & Temperature Dataset (CPLCP-GDPT5). Norwegian Meteor. Inst. Dataset., https://doi.org/10.4121/uuid:e940ec1a-71a0-449e-bbe3-29217f2ba31d

Nakicenovic N, Alcamo J, Davis G, de Vries B, Fenhann J, Gaffin S, Gregory K, Grübler A, Jung TY, Kram T, La Rovere EL, Michaelis L, Mori S, Morita T, Pepper W, Pitcher H, Price L, Riahi K, Roehrl A, Rogner H-H, Sankovski A, Schlesinger M, Shukla P, Smith S, Swart R, van Rooijen S, Victor N, Dadi Z (2000) IPCC special report on emission scenarios. Cambridge University Press, Cambridge and New York, p 599

Osuch M, Romanowicz RJ, Lawrence D, Wong WK (2016) Trends in projections of standardized precipitation indices in a future climate in Poland. Hydrol Earth Syst Sci, 20, 1947–1969, www.hydrol-earth-syst-sci.net/20/1947/2016/

Parry ML (ed) (2000) Assessment of potential effects and adaptations for climate change in Europe: The ACACIA Project. University of East Anglia, Norwich, UK, Jackson Environment Institute, p 320

Pińskwar I (2010) Projekcje zmian w ekstremach opadowych w Polsce. (Projections of changes in extreme precipitation in Poland) Monografia KGW PAN, 34 (in Polish)

Rajczak J, Schär C (2017) Projections of future precipitation extremes over Europe: a multimodel assessment of climate simulations. J Geophys Res Atmos 122:10,773-10, 800 https://doi.org/10.1002/2017jd027176

Schulzweida U, Kornblueh L, Quast R (2012) CDO User's Guide (Version 1.5.9)

Tuomenvirta H, Jylhä K, Kilpeläinen T (2006) Extreme precipitation events in the Baltic Sea catchment—observations and future projections. Abstract. International Conference: Extreme hydrometeorological events in Poland and their impacts—European context. Warsaw, 7–9 December, 82–84

van der Linden P, Mitchell JFB (eds.) (2009) ENSEMBLES: climate change and its impacts: summary of research and results from the ENSEMBLES project. Met Office Hadley Centre, FitzRoy Road, Exeter EX1 3 PB, UK. 160 pp

Zhang X, Hegerl GC, Seneviratne SI, Stewart R, Zwiers FW, Alexander LV (2013) WCRP grand challenge: understanding and predicting weather and climate extremes http://www.wcrpclimate.org/gc-extremes-documents

Chapter 23
Projected Changes in Thermal Indices Related to the Agriculture and Energy Sectors

Dariusz Graczyk, Iwona Pińskwar, and Adam Choryński

Abstract The increase in air temperature by 2100 projected by climate models will affect the value of thermal characteristics that are important from the point of view of the agriculture and energy sectors. Noticeable changes for each of the nine calculated indices are already visible in the so-called near future (2021–2050). At the end of the current century, they are projected to increase so dramatically that they will significantly affect the analysed sectors. The growing season in projections based on the RCP 8.5 scenario may be extended in some regions of Poland by up to 2 months and the frost-free period in almost the entire territory of Poland by at least 40 days. The annual number of growing degree days for both temperature thresholds (5 °C and 10 °C) will also increase significantly. From the energy consumption point of view, the positive effect of warming will be a significant reduction in the demand for energy used to heat buildings (heating degree days). This positive impact can, however, could be reduced by a large increase in energy demand due to refrigeration and air conditioning (cooling degree days). Even in warmer climates, there will still be some days with a high demand for energy for heating. An increase in the demand for energy for cooling during the warmest days may overload energy production and transmission systems.

Introduction

Climate model projections of future thermal conditions in Central Europe and in Poland are in consensus that temperatures will continue to increase. The range of possible warming depends mainly on the greenhouse gas emission scenario (IPCC SRES) in the previous generation of models and scenarios of concentration (RCP) in more recent approaches. The time horizon of analysis is also important. For different models, the increase in the average annual air temperature in the near future (before 2050) is noticeable and in line with current trends. Its value is estimated at 1–1.4 °C compared to the reference period of 1971–2000. The projections for the end of this

D. Graczyk (✉) · I. Pińskwar · A. Choryński
Institute for Agricultural and Forest Environment, Polish Academy of Sciences, Poznań, Poland
e-mail: darekgraczyk@wp.pl

M. Falarz (ed.), *Climate Change in Poland*, Springer Climate,
https://doi.org/10.1007/978-3-030-70328-8_23

century predict a temperature increase of 3.8 °C for RCP 8.5 scenario and 1.9 °C for RCP 4.5, assuming limitations on greenhouse gas emissions (Piniewski et al. 2017).

Projections of average temperature values due to high inter-annual, seasonal or monthly variability of this parameter do not sufficiently describe the thermal conditions in the analysed time and space. Much more precise information on future thermal conditions is provided by analyses based on the so-called characteristic days when the air temperature reaches or exceeds certain threshold values. The results of climate models containing information about the future values of different meteorological characteristics also aid in calculating more complex indices. Thus, it can be estimated whether and to what extent climate change affects various sectors of the economy and human activity. Useful information on their vulnerability to the effects of global warming can be provided by analysis of currently occurring extreme climatic phenomena such as heat waves or prolonged droughts. Some analyses of climate models, e.g. Beniston et al. (2004) and Kysely (2009) found that heat waves that are currently considered to be very severe, and at the end of the twenty-first century may be much more frequent, may become a yearly phenomenon.

The impact of climate change on some sectors of the economy has previously been studied for Poland as well with the use of the results of climate model projections. The impact on public health and mortality associated with high temperatures in the future has been studied by Błażejczyk et al. (2017, 2018) and Kuchcik (2013). Projections of agrometeorological characteristics were analysed by Graczyk and Kundzewicz (2016). The impact on health, water management and agriculture has also been considered in the literature (Szwed et al. 2010).

The purpose of this research is to examine the future values of nine indices related to two important sectors of the Polish economy—agriculture and energy.

Data and Methodology

Projections of eight climate models created for two emission scenarios RCP 4.5 and RCP 8.5 as parts of the EUROCORDEX project have been used for calculation of indices. More detailed information on the models can be found in Chapter 21. Model simulations were used to calculate the value of nine indices vital to both the agriculture and energy sectors. Each index has been analysed for two time horizons:

- near future (NF) covering the years 2021–2050;
- far future (FF) covering the years 2050–2100.

The interval 1971–2000 was selected as the reference period.

List of indices and short description of calculation methods:

1. Growing season length (GSL) calculated as the period between the occurrence of the first 6-day period with Tavg > 5 °C in the first half of the year and the first occurrence of the 6-day period with Tavg < 5 °C in the second half of the year.

2. Beginning of growing season (GSB) calculated as the day of the year that begins the first 6-day period of the year with Tavg > 5 °C.
3. Frost-free period length (FFP) calculated as the number of days between the occurrence of the last spring day with Tmin < 0 °C and the first autumn day with Tmin < 0 °C.
4. Growing degree days for threshold values 5 °C (GDD5) calculated as the annual sum of positive deviations of daily mean air temperature above the threshold 5 °C.
5. Growing degree days for threshold values 10 °C (GDD10) calculated as the annual sum of positive deviations of daily mean air temperature above the threshold 10 °C.
6. Heating degree days (HDD) calculated as the annual sum of negative deviations of daily mean air temperature below the threshold 17 °C.
7. Maximum daily value of heating degree days (HDDmax) calculated as the maximum daily value of negative deviations of daily mean air temperature below the threshold 17 °C.
8. Cooling degree days (CDD) calculated as the annual sum of positive deviations of daily mean air temperature above the threshold 20 °C.
9. Maximum daily value of cooling degree days (CDDmax) calculated as the maximum value of positive deviations of daily mean air temperature above the threshold 20 °C.

Calculations were performed using CDO Command Line Operators (CDO) (Schulzweida et al. 2012). Most indices were calculated on the basis of ECA indices (implemented in CDO). The indices CDD and GDD10 were calculated with the use of modified CDO operators (e.g. changing temperature thresholds).

Results

Indices Important for Agriculture

Growing Season Length

The average from eight climate models (Fig. 23.1) predicts the extension of the growing season as early as in the period of 2021–2050. Compared to the reference period 1971–2000, the growing season will be extended by as many as 15–30 days for the whole country. In the case of simulations based on the RCP 4.5 concentration scenario, there are areas in the south and centre of the country where the increase of length of the growing season will be less than 15 days. For the RCP 8.5 scenario, the extension of the vegetation period of less than 15 days occurs only in a very small area in southeastern Poland. At the end of the twenty-first century, the growing season will be extended even further. For the RCP 4.5 scenario, it will take 30–45 days longer in most of the country. In the simulations based on the RCP 8.5 scenario, the difference

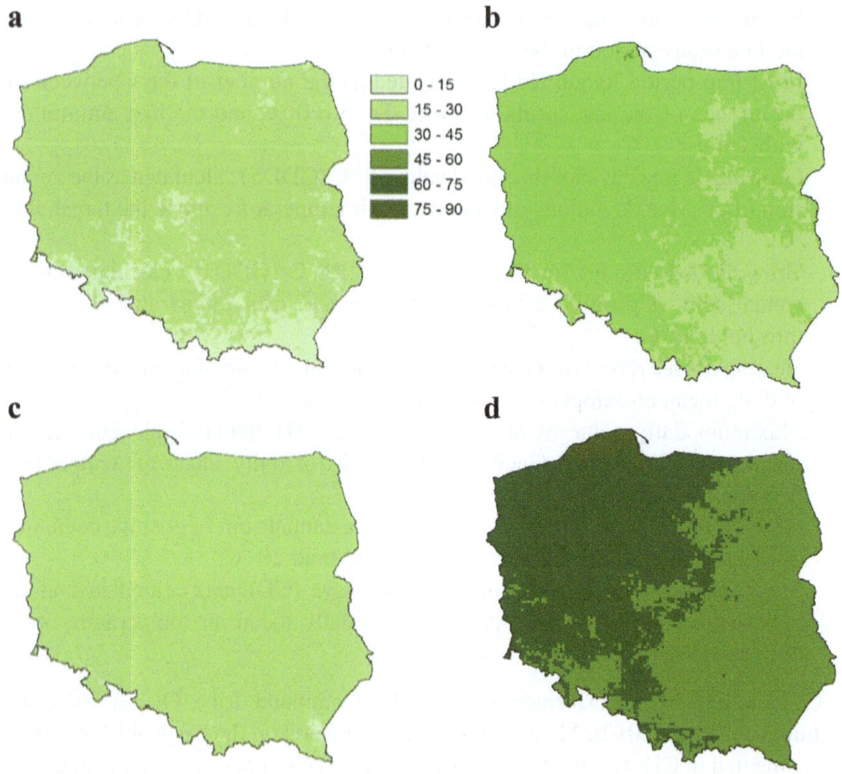

Fig. 23.1 Changes in mean annual growing season length. Difference in reference to the period 1971–2000. Upper row: RCP4.5; **a** for period 2021–2050; **b** for period 2071–2100; lower row RCP8.5; **c** for period 2021–2050; **d** for period 2071–2100

between the reference period and far future in the north and west of Poland may exceed 2 months. The extension of the growing season in the near future will occur to a large extent due to the acceleration of its onset. For both concentration scenarios, the growing season will start 7–14 days earlier than in the reference period. In the far future, the beginning of the growing season for the RCP 4.5 emission scenario will be accelerated in most parts of the country by up to 3 weeks and on the Baltic coast by more than a month. Simulations based on the RCP 8.5 scenario assume that, in the FF, the growing season in most of the country will start 5–6 weeks earlier than in 1971–2000 and in the eastern part of the coastal area by as many as 50 days (Fig. 23.2).

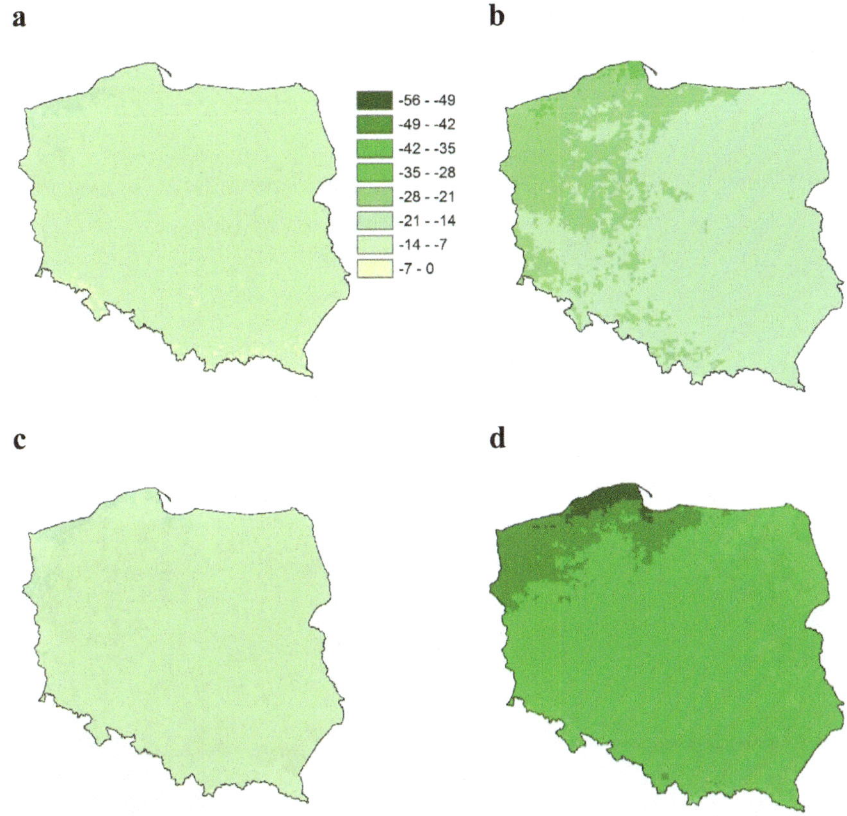

Fig. 23.2 Changes in the beginning of growing season. Difference in reference to the period 1971–2000. Upper row: RCP4.5; **a** for period 2021–2050; **b** for period 2071–2100; lower row RCP8.5; **c** for period 2021–2050; **d** for period 2071–2100

Frost-Free Period Length

The extension of the growing season will be accompanied by an increase in the length of the frost-free period (Fig. 23.3). In the near future (2021–2050), for both concentration scenarios, the increase in the length of the frost-free season for most area is in the range of 10 to 20 days. In 2071–2100, simulations for the RCP4.5 scenario provide for an extension of the frost-free period by further 10–20 days for southeastern, central and large parts of Western Poland. Even higher values may occur in a large area of Poland. Compared to the period 1971–2000, the no-frost period will be extended by up to 40 days and in areas near Gdańsk Pomerania by as many as 50. For projections based on the RCP 8.5 scenario in the far future, the no-frost period will be at least 40 days longer in the entire country, although in places in the north and in the foothills, the extension will exceed 2 months.

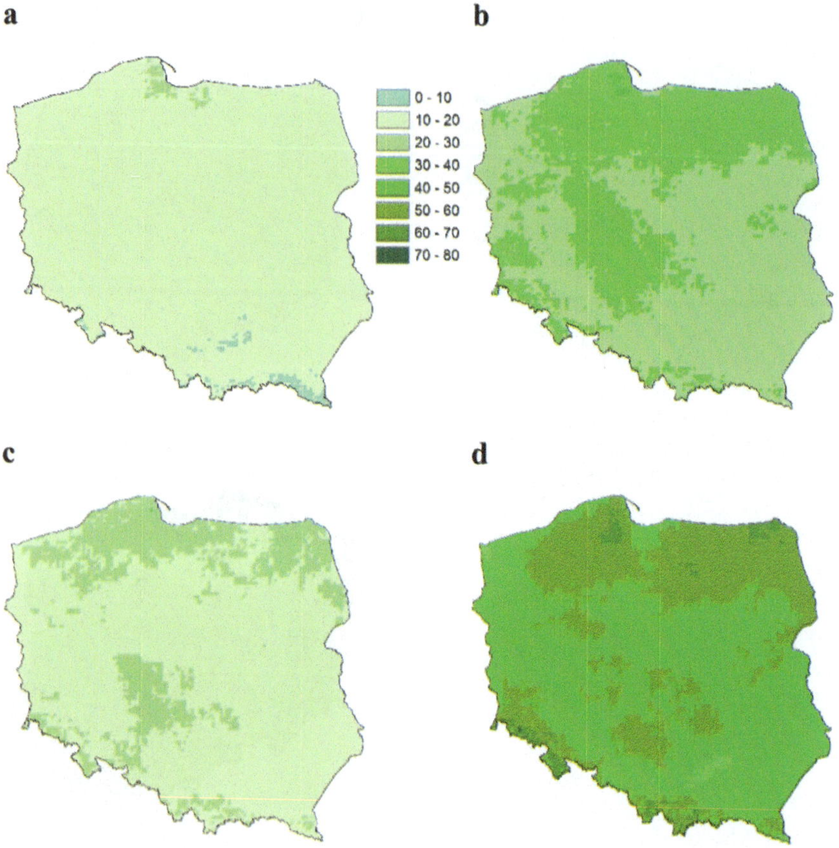

Fig. 23.3 Changes in mean annual length of frost-free period. Difference in reference to the period 1971–2000. Upper row: RCP4.5; **a** for period 2021–2050; **b** for period 2071–2100; lower row RCP8.5; **c** for period 2021–2050; **d** for period 2071–2100

Growing Degree Days

The extension of the vegetation period and the increase in the average air temperature in individual months of the year will increase the annual sum of growing degree days. For the 5 °C threshold temperature, for both RCP scenarios, the increase will be for nearly the entire country from 200 to 300 degree days, and in the case of the RCP 8.5 scenario, up to 400 degree days in the coastal area (Fig. 23.4). In the years 2071–2100, annual sums of GDD will increase in most of Poland by 400–500 degree days (RCP4.5). Furthermore, even higher values are predicted by simulations based on the RCP 8.5 scenario. The average of these ranges from 700 degree days in the north and northeast to 900 degree days in the south and west of the country.

Annual sums of growing degree days for plants with higher thermal requirements, for which the threshold value is 10 °C, in 2021–2050, according to the simulation

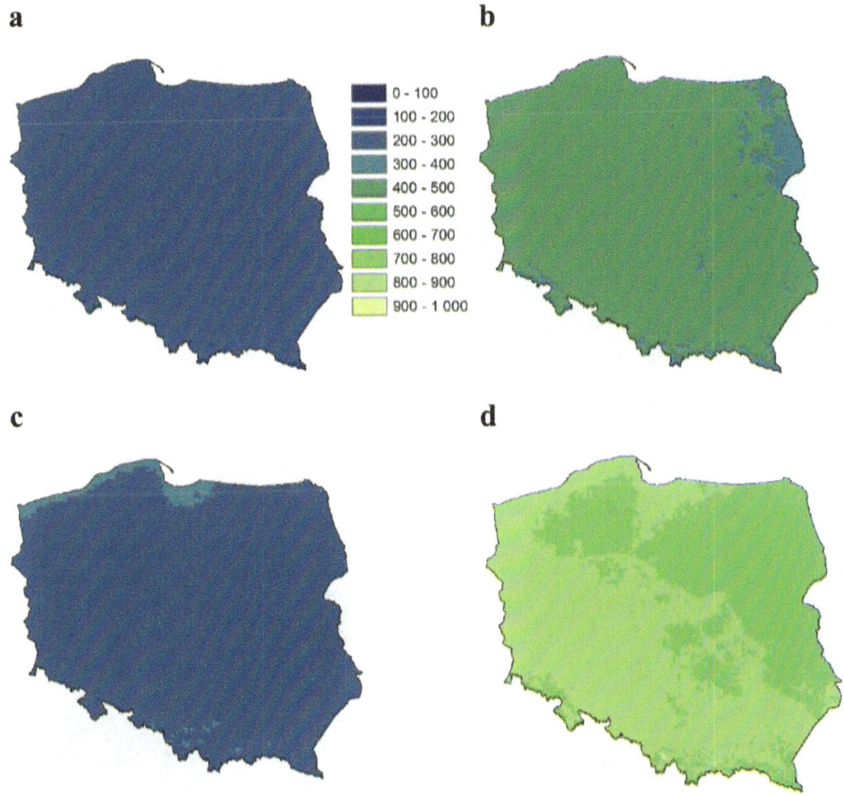

Fig. 23.4 Changes in annual sums of growing degree days for the threshold 5 °C. Difference in reference to the period 1971–2000. Upper row: RCP4.5; **a** for period 2021–2050; **b** for period 2071–2100; lower row RCP8.5; **c** for period 2021–2050; **d** for period 2071–2100

for the RCP 4.5 emission scenario, will increase in Poland on average from 100 to 200 degree days (Fig. 23.5). Simulations for RCP 8.5 concentration scenarios in the south of Poland (but not in the mountainous areas) and on the Baltic coast predict a greater increase. In the far future, throughout almost the entire country, the average annual sums of growing degree days according to the simulation for the RCP 4.5 scenario will be higher than at reference period from 200 to 300 degree days and in the south and west up to 500 degree days. The RCP 8.5 scenario provides evidence for an increase of 500–600 degree days and in the south even up to 800 degree days.

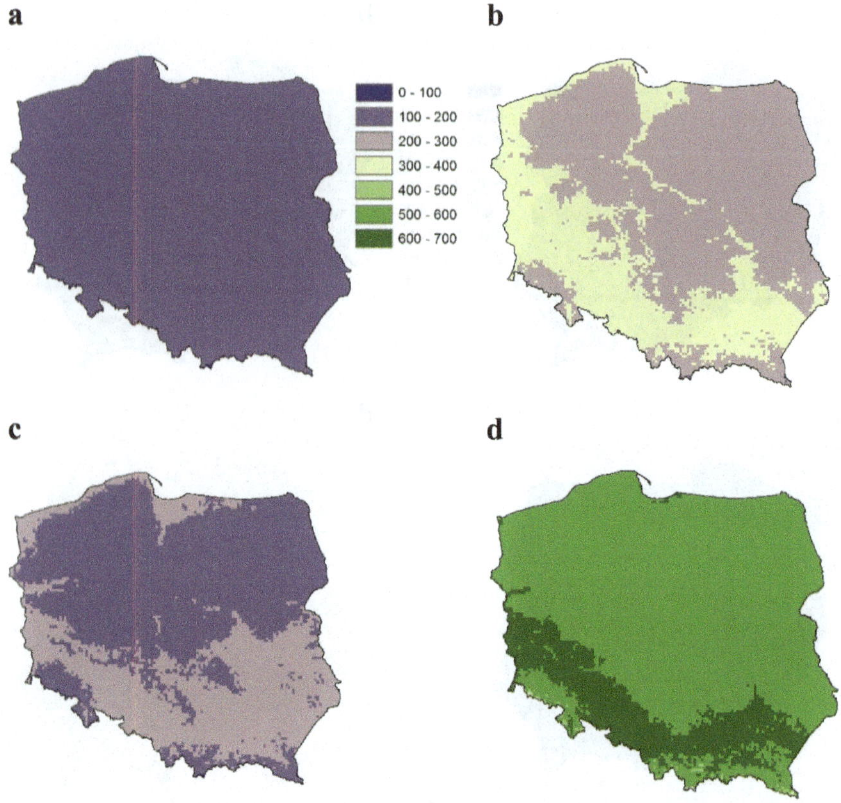

Fig. 23.5 Changes in annual sums of growing degree days for the threshold 10 °C. Difference in reference to the period 1971–2000. Upper row: RCP4.5; **a** for period 2021–2050; **b** for period 2071–2100; lower row RCP8.5; **c** for period 2021–2050; **d** for period 2071–2100

Indices Important for Energy Demands

Heating Degree Days

Temperature increases in the cool part of the year will lead to a decrease in the amount of energy used for heating purposes. The magnitude of this decrease for the near future and for the last decades of the twenty-first century is shown in Fig. 23.6. For nearly the entire area of Poland, in the RCP 4.5 scenario, the sum of annual heating degree days (HDD) drops by 200–400 degree days. Similar values are predicted by models for the RCP 8.5 scenario for southwest and west Poland, central Poland and locations in the Podkarpacie region. In the northeast and in the foothills, a decrease of 400 to 600 HDD is projected. The simulation results for the last decades of the twenty-first century indicate that the decrease in the value of this indicator will have a similar geographical distribution for both scenarios. In the RCP 4.5 scenario, even

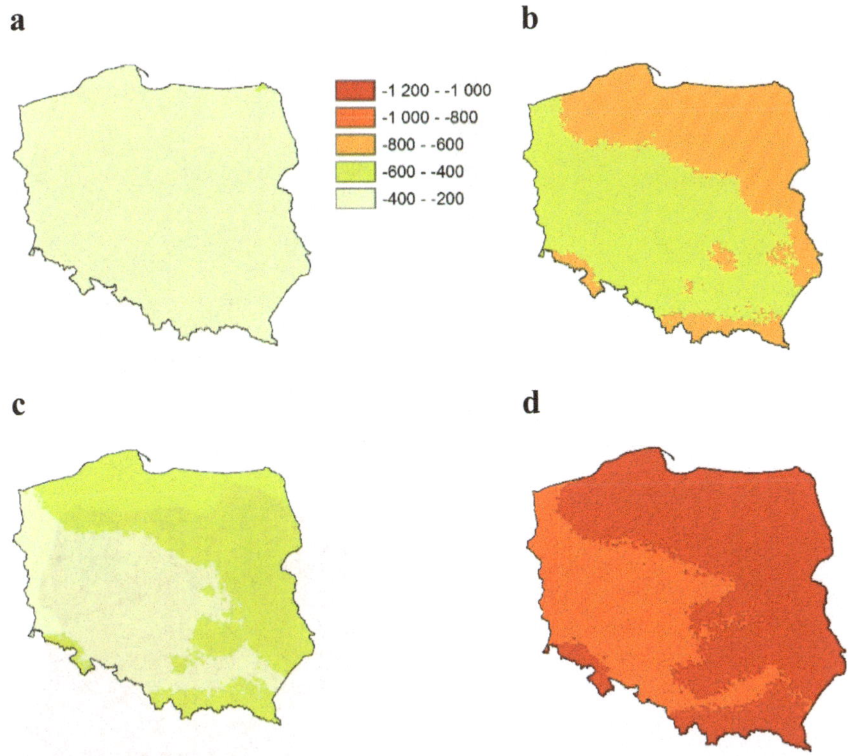

Fig. 23.6 Changes in annual sums of heating degree days. Difference in reference to the period 1971–2000. Upper row: RCP4.5; **a** for period 2021–2050; **b** for period 2071–2100; lower row RCP8.5; **c** for period 2021–2050; **d** for period 2071–2100

up to 800 degree days could decrease in the north, east and in the foothills. In the same areas, according to the RCP 8.5 scenario, the decrease is projected to be as high as 1200 degree days.

During the coldest days of exceptionally cold winters, the demand for energy to heat buildings increases rapidly. Figure 23.7 shows the difference in HDD during the coldest days that may occur in the future compared to those currently recorded. In the NF, these differences for both RCP scenarios do not exceed 5 degree days. Considering that currently, the largest daily HDD sums in large parts of Poland exceed 40 degree days, it can be concluded that also in the near future there may be days during which demand for heating energy will be very high. Although to a lesser extent, this will also be the case in the far future when the maximum daily HDD sums decrease in different places by more than 10 degree days.

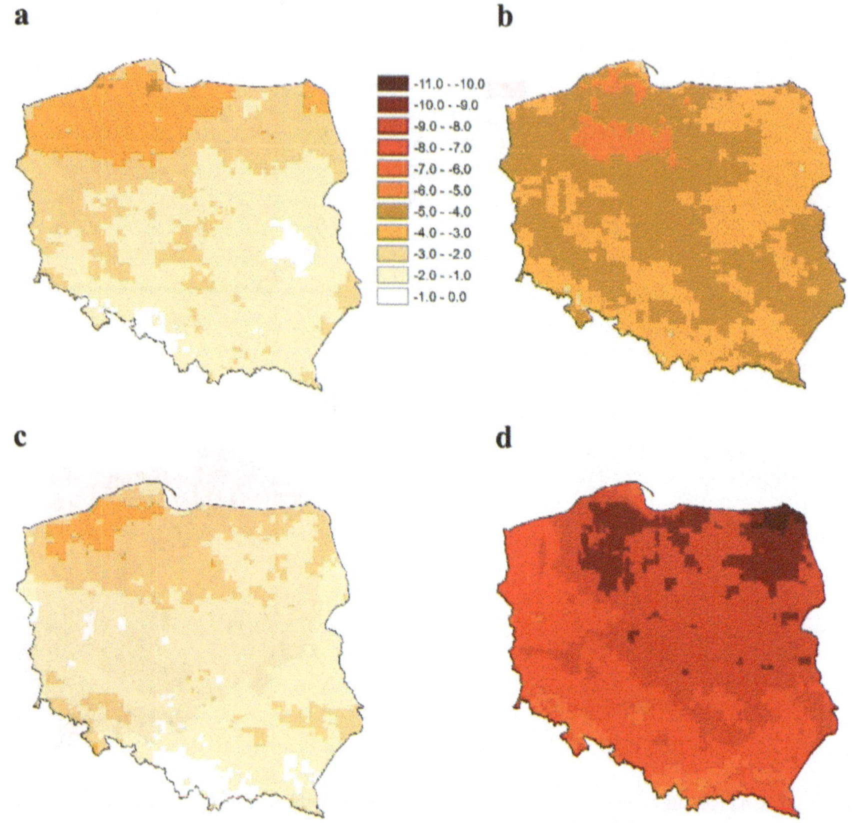

Fig. 23.7 Changes in the highest daily sum of heating degree days. Difference in reference to the period 1971–2000. Upper row: RCP4.5; **a** for period 2021–2050; **b** for period 2071–2100; lower row RCP8.5; **c** for period 2021–2050; **d** for period 2071–2100

Cooling Degree Days

The average annual cooling degree days (CDD) values will be higher in the future, regardless of the adopted time horizon and emission scenario (Fig. 23.8). In the years 2021–2050, the value of the CDD index will increase for both emission scenarios by 25 to 50 CDD almost throughout Poland. This is equivalent to the annual occurrence of one additional 7-day heat wave during which the average daily temperatures will be 3.6 to 7.2 °C higher than the average from the reference period. In some areas located in the north of Poland and in the foothills, CDD growth will be smaller—from 0 to 25 degree days. In the far future, large differences between scenarios are projected. For the RCP 4.5 emission scenario, for the most part of the country, the CDD is projected to increase by 75–100 degree days and the highest increases in the upper Odra valley and the upper Vistula basin may reach up to 125 degree days. For almost all of southern Poland (excluding mountainous areas), model simulations

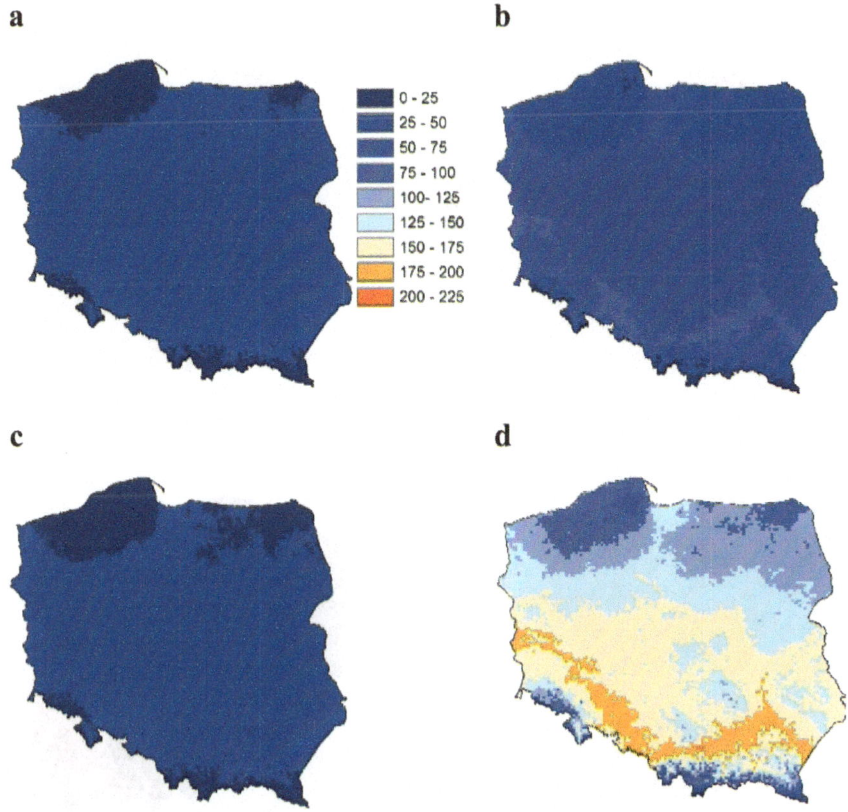

Fig. 23.8 Changes in annual sums of cooling degree days. Difference in reference to the period 1971–2000. Upper row: RCP4.5; **a** for period 2021–2050; **b** for period 2071–2100; lower row RCP8.5; **c** for period 2021–2050; **d** for period 2071–2100

based on the RCP 8.5 scenario predict an increase of CDD by 150–175 and in some places of the upper Odra and Vistula basins up to 200 degree days.

The hottest days currently recorded during heatwaves for most of Poland exceed the value of 8 degree days of the CDD index. In the NF, the values of this index may increase during the hottest days in almost the entire country by at least 2–3 degree days (Fig. 23.9). Although the nominal growth does not seem large compared to currently recorded values, it is projected to exceed 35%. During the hottest days recorded in the far future in a large area of the country, daily values may be 5–6 degree days (60–75%) higher than those recorded during the reference period.

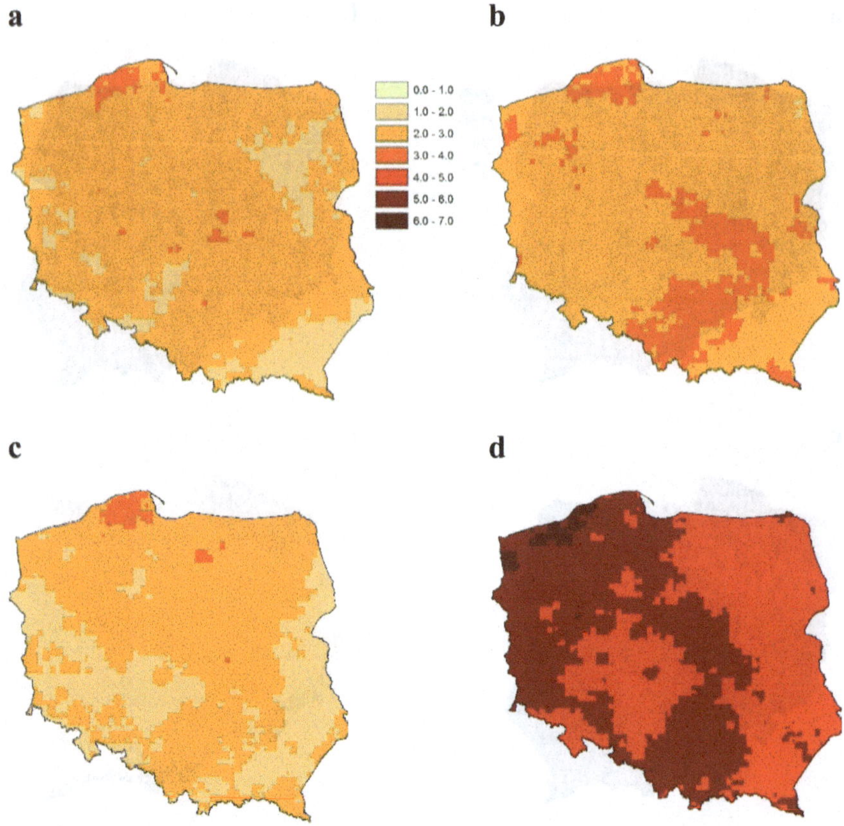

Fig. 23.9 Changes in the highest daily sum of cooling degree days. Difference in reference to the period 1971–2000. Upper row: RCP4.5; **a** for period 2021–2050; **b** for period 2071–2100; lower row RCP8.5; **c** for period 2021–2050; **d** for period 2071–2100

Summary and Discussion

The increase in air temperature visible in the projection of climate models will significantly affect the sectors analysed in this study. Changes in almost all of the examined indices will become evident already in the near future, 2021–2050. Furthermore, the growing season is projected to be longer in most parts of Poland by 15–30 days. At the end of the current century, according to models based on the RCP 8.5 scenario, the growing season may be up to 60 days longer than in the reference period. Similar values for central Poland were obtained by Nieróbca et al. (2013). For the HadCM-A2 model, it was an increase of 14 days to 2030 and 27 days in 2050. Projections of the ECHAM-A1B model, based on the scenario assuming lower greenhouse gas emissions, indicated an increase of 10 and 18 days, respectively. Graczyk and Kundzewicz (2016) analysed the period closer to the end of the century (2061–2090). For the projection by three models based on the SRES A2 scenario, the growing season may be longer by 40–60 days in the east to 70–90 days in the west of Poland. Projections are in line with the currently observed trends. A spatially diversified trend indicating an earlier beginning and later end of the growing season and, as a result, the extension of the growing season was reported e.g. in Tomczyk and Szyga-Pluta (2019). Extension in the first decade of the twenty-first century, according to Nieróbca et al. (2013), was found to vary from 8 to 16 days compared to the period of 1971–2000, depending on the region of Poland.

Considering only the thermal background, conditions of agricultural production will improve in the future. An increase in air temperature may effect with earlier occurrence of individual development phases by arable crops, e.g. Kozyra (2013). Increasing annual sums of growing degree days and extension of the frost-free period will favour crops with higher thermal requirements. This is already visible at the present time, e.g. Wypych et al. (2016). The beneficial effect may, however, be reduced partially or completely, for example by more frequent occurrence of intense drought and rainfall deficiencies in summer and the growing season, see Pinskwar (2010) and Szwed et al. (2010).

The increase in air temperature predicted by climate models will have a dual impact on energy demand. During the cold part of the year, the demand for heating energy will significantly decrease. This can bring savings in the near future. In contrast, demand for energy for refrigeration and air conditioning will increase. According to the projection by climate models, energy demand will increase until the end of the century and, in some parts of Poland, the growth may exceed even 75% of the currently recorded values. While air conditioning uses only electricity, heating depends on many sources of energy. Hence, it is difficult to estimate now whether profits will outweigh costs in the future. According to estimates by Damm et al. (2017), in 2036–2065 for the RCP 4.5 scenario, due to global warming, energy consumption will fall in almost all European countries. It is estimated at 1 to 2% for Poland.

Acknowledgements Support of the project CHASE-PL (Climate change impact assessment for selected sectors in Poland) of the Polish–Norwegian Research Programme operated by the National Centre for Research and Development (NCBiR) under the Norwegian Financial Mechanism 2009–2014 in the frame of Project Contract No. Pol Nor/200799/90/2014 is gratefully acknowledged.

References

Beniston M, Diaz HF (2004) The 2003 heat wave as an example of summers in a greenhouse climate? Observations and climate model simulations for Basel, Switzerland. Global Planet Change 44:73–81

Błażejczyk A, Błażejczyk K, Baranowski J, Kuchcik M (2017) Heat stress mortality and desired adaptation responses of healthcare system in Poland. Int J Biometeorol. https://doi.org/10.1007/s00484-017-1423-0

Błażejczyk K, Baranowski J, Błażejczyk A (2018) Climate related diseases. Current regional variability and projections to the year 2100. Quaestiones Geographicae 37(1), Bogucki Wydawnictwo Naukowe, Poznań, 23–36

Damm A, Köberl J, Prettenthaler F, Rogler N, Töglhofer C (2017) Impacts of +2°C global warming on electricity demand in Europe. Climate Services 7:12–30 https://doi.org/10.1016/j.cliser.2016.07.001

Graczyk D, Kundzewicz Z (2016) Changes of temperature-related agroclimatic indices in Poland. Theor Appl Climatol 124(1–2):401–410

Kozyra J (2013) Wpływ prognozowanych zmian temperatury powietrza na fenologię zbóż ozimych w Polsce (The Influence of forecasted air temperature changes on phenology of winter cereals in Poland). Monografie i Rozprawy Naukowe, IUNG-PIB Puławy, (40):ss114 (in Polish)

Kuchcik M (2013) The attempt to validate the applicability of two climate models for the evaluation of heat wave related mortality in Warszawa in the 21st century. Geogr Pol 86(4):295–311

Kysely J (2009) Recent severe heat waves in central Europe: how to view them in a long-term prospect? Int J Clim 30:89–109. https://doi.org/10.1002/joc.1874

Nieróbca A, Kozyra J, Mizak K, Wróblewska E (2013) Zmiana długości okresu wegetacyjnego w Polsce (Changing length of the growing season in Poland). Woda-Środowisko-Obszary Wiejskie 13, 2(42):81–94 (in Polish)

Piniewski M, Mezghani A, Szczesniak M, Kundzewicz ZW (2017) Regional projections of temperature and precipitation changes: Robustness and uncertainty aspects. Meteorologische Zeitschrift, 26(2):223–234

Pińskwar I (2010) Projections of changes in precipitation extremes in Poland. Monografia KGW PAN, 34 (in Polish)

Szwed M, Karg G, Pińskwar I, Radziejewski M, Graczyk D, Kędziora A, Kundzewicz ZW (2010) Climate change and its effect on agriculture, water resources and human health sectors in Poland. Nat Hazards Earth Syst Sci 10(8):1725–1737

Schulzweida U, Kornblueh L, Quast R (2012) CDO User's Guide (Version 1.5.9)

Tomczyk AM, Szyga-Pluta K (2019) Variability of thermal and precipitation conditions in the growing season in Poland in the years 1966–2015. Theor Appl Climatol 135:1517. https://doi.org/10.1007/s00704-018-2450-4

Wypych A, Sulikowska A, Ustrnul Z, Czekierda D (2017) Variability of growing degree days in Poland in response to ongoing climate changes in Europe. Int J Biometeorol 61:49. https://doi.org/10.1007/s00484-016-1190-3

Part V
Discussion and Conclusion

Chapter 24
Climate Change in Poland—Summary, Discussion and Conclusion

Małgorzata Falarz, Zuzanna Bielec-Bąkowska, Agnieszka Wypych, Dorota Matuszko, Tadeusz Niedźwiedź, Iwona Pińskwar, Katarzyna Piotrowicz, Zbigniew Ustrnul, Anita Bokwa, Janusz Filipiak, Dariusz Graczyk, Magdalena Kuchcik, Magdalena Opała-Owczarek, Małgorzata Szwed, and Joanna Wibig

Abstract The book presents the results of climate research throughout Poland in the pre-instrumental period (using proxy data), instrumental period (using mainly statistical methods, based on data from weather stations and grid data) and projected changes (using regional climate models). A total of 1100 years are covered, i.e. the

---

M. Falarz (✉) · Z. Bielec-Bąkowska · T. Niedźwiedź · M. Opała-Owczarek
Faculty of Natural Sciences, Institute of Earth Sciences, University of Silesia in Katowice, Katowice, Poland
e-mail: malgorzata.falarz@us.edu.pl

Z. Bielec-Bąkowska
e-mail: zuzanna.bielec-bakowska@us.edu.pl

T. Niedźwiedź
e-mail: tadeusz.niedzwiedz@us.edu.pl

M. Opała-Owczarek
e-mail: magdalena.opala@us.edu.pl

A. Wypych · D. Matuszko · K. Piotrowicz · Z. Ustrnul · A. Bokwa
Faculty of Geography and Geology, Institute of Geography and Spatial Management, Jagiellonian University in Kraków, Kraków, Poland
e-mail: agnieszka.wypych@uj.edu.pl

D. Matuszko
e-mail: d.matuszko@uj.edu.pl

K. Piotrowicz
e-mail: k.piotrowicz@uj.edu.pl

Z. Ustrnul
e-mail: zbigniew.ustrnul@uj.edu.pl

A. Bokwa
e-mail: anita.bokwa@uj.edu.pl

A. Wypych · Z. Ustrnul · J. Filipiak
Institute of Meteorology and Water Management—National Research Institute, Warsaw, Poland

I. Pińskwar · D. Graczyk · M. Szwed
Institute for Agricultural and Forest Environment, Polish Academy of Sciences, Poznań, Poland

M. Falarz (ed.), *Climate Change in Poland*, Springer Climate,
https://doi.org/10.1007/978-3-030-70328-8_24

period from about 1000 to 2100. The majority of examined climate elements, meteorological phenomena and indices show statistically significant changes at least in certain areas of Poland and at certain seasons of the year. Moreover, many elements demonstrate significant year-to-year variability and temporal fluctuations. Changes of particular climate elements are interrelated. The primary causative factors are both anthropogenic changes (greenhouse gas emissions resulting in increased greenhouse effect and global warming, local sources of air pollution) and natural changes: (1) circulation factors: changes in the intensity and location of atmospheric activity centres, changes in the frequency of advection from a specific sector, and the frequency of cyclonic and anticyclonic systems over Poland and (2) radiation factors (changes in values of global solar radiation, sunshine duration and cloudiness). These changes, especially visible after the 1980s, affect the trends of most climatic elements, meteorological phenomena and indices. The effects of these changes, both positive and negative, are evident in people's daily lives (e.g. decrease in bioclimatic cold stress, increase in bioclimatic heat stress, changes in conditions for recreation and sport) and economy (e.g. improvement of thermal agricultural conditions, changes in energy demand for heating buildings, air conditioning and refrigeration). A better understanding of the relationships of trends of the different climatic elements should be the aim of further research into the climate of Poland.

The book presents the results of climate research throughout Poland in the pre-instrumental period (using proxy data), instrumental period (using mainly statistical methods, based on data from weather stations and grid data) and projected changes (using regional climate models). A total of 1100 years are covered, i.e. the period

e-mail: iwona.pinskwar@isrl.poznan.pl

D. Graczyk
e-mail: darekgraczyk@wp.pl

M. Szwed
e-mail: mszwed@man.poznan.pl

J. Filipiak
Faculty of Oceanography and Geography, Institute of Geography,
Department of Meteorology and Climatology, University of Gdańsk, Gdańsk, Poland
e-mail: janusz.filipiak@ug.edu.pl

M. Kuchcik
Climate Impacts Laboratory, Institute of Geography and
Spatial Organization, Polish Academy of Sciences, Warsaw, Poland
e-mail: mkuchcik@twarda.pan.pl

J. Wibig
Faculty of Geographical Sciences, Department of Meteorology and Climatology,
University of Lodz, Łódź, Poland
e-mail: joanna.wibig@geo.uni.lodz.pl

from about 1000 to 2100. The most important results of these studies for specific periods, climate elements and indices are as follows:

Climate change before instrumental measurements:

- data for the Medieval Warm Period are of a high degree of uncertainty; however, the mean annual air temperature during that period might have been comparable to that of nowadays. Climate in this period had a great degree of oceanicity; during the Little Ice Age winters were on average about 2°C colder than today, while summers a little warmer. In this period the highest degree of climate continentality was noted;
- the highest precipitation totals in the entire historical period likely occurred mainly in the twelfth and thirteenth centuries; secondary maxima were noted in the first halves of the sixteenth and eighteenth centuries. On the contrary, during the second halves of the fourteenth and sixteenth centuries, low precipitation totals were observed;
- inconsistencies between proxy-based climate reconstructions remain in the details, which may be attributed to using different archives, uncertainties in chronologies, varied sensitivities of the proxy used in particular reconstructions, and finally to the not fully comparable meaning of the reconstruction results. As great potential for research still remains, further development of new data sources, sampling sites, parameters and more sophisticated methods of reconstruction could potentially increase our understanding of multicentennial past climate changes in Poland.

Change of atmospheric circulation:

- from the end of the nineteenth century to the present, large fluctuations and a general weak increasing trend of circulation from the westerly sector were found; according to the typology of J. Lityński and Grosswetterlagen an increase of western circulation in the 1980s and 1990s is noticeable, while at the turn of the 1960s and 1970s and during the last 10 years the weakening of this circulation pattern is marked; during the summer the weakening of western circulation was evident, particularly in the last several years;
- the circulation from the south had a slightly decreasing trend throughout the year except for the summer when the opposite trend was found;
- changes in the cyclonicity index expressing the frequency of low pressure systems over Poland have shown increasing trends in all seasons.

Air pressure change:

- the largest range of air pressure fluctuations occurred in the cool half-year, especially in the winter, and the smallest in the summer;
- days with a pressure of $\geq$ 1030 hPa (strong high) occur in Poland several times more often than days with a pressure of $\leq$ 990 hPa (deep low),

- no significant and statistically significant changes in both annual and seasonal, as well as monthly pressure values were found in their long-term course; the absence of such changes was also evident in the number of days with high and low pressure ($\geq$ 1030 i $\leq$ 990 hPa); similar regularities are also noticeable in the 118-year measurement series from the Scientific Station of the Climatology Department of the Jagiellonian University in Kraków (Kraków Observatory);
- the described features of the multi-annual and annual pressure courses indicate that changes in the studied element are characterised by the occurrence of alternating shorter periods of their clearly higher or lower values.

Solar radiation change:

- long-term trends of global solar radiation in Poland are, in most cases, statistically insignificant; a few significant tendencies were observed in southeastern and southwestern Poland (positive from 1990s, in summer, autumn and winter) and in southern Poland (negative for the whole twentieth century, in autumn, winter and year-round); no significant changes in the solar radiation in the high mountains have been demonstrated;
- in Kraków, for a 125-year series of global solar radiation values (partly reconstructed a periodicity of about 60 years is observed with three periods of relatively high values (1880–1900, 1940–1960, 1990–2018) separated by periods of low values: 1910–1930 and 1970–1990.

Sunshine change:

- there is a statistically significant growing trend in actual and relative sunshine duration in the period 1971–2018 throughout Poland, more strongly expressed in the western part of the country; these trends are the result of both circulation (an increase in the frequency of convective clouds in recent years, a decrease in the frequency of stratiform clouds) and anthropogenic factors (the improvement of air quality in Poland);
- the decrease in the number of days without sunshine was observed in the whole of Poland;
- long heliographic series confirmed the high multi-annual variability of sunshine duration in Poland and showed the periods of decreases and increases in sunshine duration also at the turn of the nineteenth and twentieth centuries, and in the first half of the twentieth century, although not as pronounced as the periods of "global dimming" and "global brightening".

Cloudiness change:

- in the majority of Poland the cloud cover did not change considerably in the period 1951–2018;
- in the south of the country there is a group of stations where a statistically significant rise in annual and seasonal mean cloud cover occurs; periods of smaller cloudiness were observed in the country during the first half of the 1970s and at the turn of the 1980s and 1990s;

- the maximum of the average annual total cloud cover in Poland was observed in 1966. The minimum annual cloudiness was recorded in Poland in 1982, and that year the low value of cloud cover occurred at the majority of stations in the country;
- strong changes are occasionally revealed in case of particular seasons in subperiods: 1951–1984 and 1985–2018; statistically significant increases in mean cloudiness were present in Poland in spring in the subperiod 1951–1984 and in winter in the years 1985–2018;
- considerable changes may be observed in case of the number of characteristic nephological days;
- the most important long-term changes are observed in the frequency of occurrence of different cloud genera: the low-level clouds excluding Stratus are becoming more frequent; this is also the case for Altocumulus and high-level clouds; the decrease in the occurrence of middle-level stratiform clouds also can be observed.

Air temperature change:

- there are observed increasing tendencies in average annual and seasonal air temperature, annual and seasonal maximum air temperature, annual and seasonal minimum air temperature, number of hot days and a decreasing trend in the number of frosty days;
- the trends were found irrespective of the length of the period considered; this is especially noticeable in the basic period under analysis (1951–2018), although it is the result of a significant increase in air temperature in the last 3 decades, confirmed independently by the results of analyses for various thermal indicators; these changes are in most cases highly statistically significant, however, particular attention should be paid to the trends in average and maximum air temperatures during the year and in the summer and winter.

Air humidity change:

- a slight increase in specific humidity in Poland is noted, significant particularly in summer (up to 0.2 $g{\cdot}kg^{-1}$ per 10 years) and autumn; the relative trend reaches approximately 3% of mean values;
- a statistically significant decrease in relative humidity as a result of increasing air temperature is observed starting in the second half of the twentieth century; the trend value of almost 1% per decade is observed in the southeastern part of the country and much less at the coast; the relative trend is the highest, about 5% of the mean, in central Poland;
- the most intensive decreasing trends in relative humidity all around the country have been found for spring (March–May) and summer (June–August).

Precipitation change:

- in summer, spring and on an annual scale, precipitation totals were decreasing in southwestern Poland, while in the other parts of Poland there was either a tendency to increase or no trends were found; in autumn, increasing trends dominated, while in winter opposite trends were found in northern (increases) and southern Poland (decreases); winter was found to have the highest number of significant trends, but on a monthly scale, trends were most numerous in March.

- in terms of the precipitation frequency (expressed by the number of days with precipitation), positive trends dominated on an annual scale and in spring and summer, particularly in the northern part of the country; stations with no trends were often located in southern Poland; in autumn, the precipitation frequency increased at a great majority of stations; lastly, the winter precipitation frequency was increasing in northern and western Poland but with the slowest rate of all seasons, while most stations in southern Poland experienced no trends in this precipitation characteristic.

Snow cover change:

- year-to-year variability of snow cover duration and its seasonal maximum depth is the largest in regions with the lowest values of snow cover;
- the number of days with snow cover during the 68 winter seasons has a negative time trend throughout Poland (with a minimum of −4 to −5 days per 10 years in northeastern Poland); this tendency is statistically significant in most area of Poland except for the highlands and some parts of the mountainous areas; the relative changes in the number of days with snow cover are the most significant in regions with a short duration of snow cover (western Poland; −8 to −10% per 10 years); the maximum depth of snow cover throughout the considered period revealed a negative trend in most area of Poland, statistically significant in the highlands and mountainous areas (−2 to −6%/10y); only in northeastern Poland is the trend positive, statistically insignificant;
- in longer periods (80–104 winter seasons), the snow cover duration and maximum snow cover depth are characterised by a slight negative or near zero time trend at most weather stations.

Change of wind:

- for most of Poland, the 75th and 95th percentile values and maximum values of extreme wind speed are decreasing; the decreases do not exceed 0.4 ms^{-1} per decade in the case of the 75th percentile, 0.7 ms^{-1} per decade in the case of the 95th percentile and 1.0 ms^{-1} per decade for maximum values in a year; the trends are statistically significant at 29, 32 and 23 stations from 41 (75th, 95th percentile and maximum values in a year, respectively); increases are recorded on the central coast, at individual stations in the centre and in the south, but they do not reach the level of statistical significance;
- the trends in extreme wind speed are the weakest in summer; in the remaining seasons, the downward trends dominate and are statistically significant in more than half of Poland; in all seasons, an increase in wind speed is observed on the central coast.

Change of thunderstorms and tornadoes:

- long-term thunderstorm activity tendencies in Poland (1951–2018) indicate that, in the eastern part of the country, signals of an increase in the frequency of days with thunderstorms can be observed, in the western part trends are rather downward; however, long-term variability of thunderstorm occurrence in Poland from 1885 to 2018 indicates that there was no statistically significant tendency of changes and only in the cool half of the year the trend is increasing and statistically significant;

- although with each year the European tornado database is expanding, it is still too short, a timeframe to derive reliable conclusions regarding long-term trends and even spatial patterns (given how rare these events are); based on the observational records from the twenty-first century, it may be estimated that, on average, 5 weak tornadoes, 1–2 significant tornadoes and 4 waterspouts are reported each year in Poland (approximately 10 cases per year);

Change of hail frequency:

- no significant changes in the number of days with hail were observed in the period under investigation (1966–2018); a similar lack of significant trend or slight change only has been observed since 1901 in Kraków; however, large year-to-year changes and periods with higher hail frequency as well as periods with almost no hail are noticeable; these changes are probably connected to circulation conditions in particular years.

Change of fog frequency:

- fog occurrence shows a high spatial and temporal variability in the territory of Poland in the study period (1966–2018);
- mountain areas show a distinctively different regime of fog frequency than the rest of the country; in the high mountains, the fog frequency is the highest (about 300 days per year), no statistically significant multi-annual changes of most of the indices can be defined and the inter-annual variability is very low.
- at the seaside, the number of days with fog is the lowest (about 28 days per year on average) and the inter-annual variability is high; no statistically significant trends can be determined for most of the indices.
- in the lowland areas of northern and central Poland, and in the uplands of southern Poland, a high variability in fog occurrence and long-term trends can be observed in each region; for areas with the highest fog frequency, i.e. in central-eastern Poland statistically significant increasing trends in annual number of days with fog and days with fog lasting < 6 h were obtained, while for the others the trends were either decreasing or not significant statistically.

Change of bioclimatic indices:

- the analysis shows a significant increase in minimum and mean values of Universal Thermal Climate Index (UTCI) and Physiological Subjective Temperature (PST) during the period 1951–2018;
- UTCI and PST minimum values (represented winter season) increased significantly over time in the whole area of Poland including the mountains, especially in the northeast (UTCImin growth was up to 1.8°C per 10 years) and Tarnów in the southeast (up to 2.2°C per 10 years).

- UTCI maximum values (summer season) were increasing significantly only in northeastern and northwestern Poland (0.3–0.5°C per 10 years) and at the foothills of the mountains (0.3–0.4°C per 10 years). The changes of maximum PST values were insignificant.
- on average, in every decade, the number of cold days was reduced to −2.7 days in the case of PST (average relative trend −1.2%/10 years) and −3.2 days per 10 years at UTCI (relatively −6.0%/10 years on average), but, in the northeast the reduction was much bigger, 7 days per 10 years (UTCI, relative trend −7.0%/10 years) and 6 days when considering PST (relatively −2.3%/10 years).

Change of weather types:

- the number of days with very warm weather (3–) increased in the period 1951–2018 (1901–2018); the trend of it at most stations was 3–5 days per 10 years; the rise was mainly due to a higher frequency of those days from March to November, especially those marked with the symbol 310 (very warm, cloudy without precipitation) and 311 (very warm, cloudy with precipitation), even at high mountain stations;
- there is a decreasing tendency in the number of days with frost (5– and 6–), especially cold, cloudy without precipitation (510) or very cloudy with precipitation (521) and fairly frosty (9–), especially sunny and/or cloudy without precipitation (900 and/or 910);
- there is a decreasing tendency in the number of weather types in a year (a statistically significant trend in 75% of the surveyed stations) in Poland; this indicates that the weather varies less in the year, i.e. it is more and more stable.

Future climate change: air temperature:

- all models have shown an agreement on a systematic upward trend in the mean temperature, both for the near and far future and for RCP4.5 and RCP8.5; in the future the extreme hot conditions not only will be hotter, but will also last longer and/or appear more frequently. In contrast, frosty periods and their frequency will decrease; the thermal changes in the future are more pronounced for the far future (FF: 2071–2100) than for the near future (NF: 2021–2050) and more for the RCP8.5 scenario than for RCP4.5.
- the mean annual temperature is projected to increase in the future for the whole territory of Poland; higher increases are projected for the northeastern and eastern areas. The increases are projected to range from 1°C for NF and RCP4.5 to almost 4°C for FF and RCP8.5.
- in the period of 2021–2050, changes in the absolute maximum temperature could, on average, be higher by 2–3°C across Poland, up to 4–5°C on the coast (RCP4.5); for the far future horizon of 2071–2100, the increases are significantly higher: for RCP8.5, they range from 3–4°C in the east to more than 7°C on the coast; the projections of the changes in the absolute minimum temperature mostly indicate increases by 2–3°C, for NF and both RCPs; for the FF, there are projected greater

differences between RCPs; for RCP8.5, the projected increases may reach more than 10°C in the northeastern parts of Poland in comparison to the reference period.
- for the near future and both RCPs, no increase in the mean number of days with minimum temperature above 20°C is expected for the whole territory of Poland; the increases are projected only for FF and RCP8.5.

Future climate change: precipitation:

- most precipitation indices based on the models under examination will increase and this applies to both the mean annual precipitation totals as well as in intense precipitation and scarcity of water; future changes will be more severe for higher RCP and the far time horizon.
- the changes in extreme precipitation are likely to be more pronounced than in the mean precipitation in the future.

Future climate change: thermal indices related to the agriculture and energy sectors:

- the growing season is projected to be longer in most parts of Poland by 15–30 days; at the end of the current century, according to models based on the RCP 8.5 scenario, the growing season may be up to 60 days longer than in the reference period;
- the frost-free period in almost the entire territory of Poland may by at least 40 days;
- for nearly the entire area of Poland, in the RCP 4.5 scenario, the sum of annual heating degree days (HDD) drops by 200–400; similar values are predicted by models for the RCP 8.5 scenario for southwest, west, central and southeast Poland; in the northeast and in the foothills, a decrease of 400 to 600 HDD is projected; the simulation for the end of the twenty-first century in the RCP 4.5 scenario indicates up to a 800 HDD decrease in the north, east and in the foothills; in the same areas, according to the RCP 8.5 scenario, the decrease is projected to be as high as 1200 degree days;
- in the period 2021–2050, the value of the cooling degree days index (CDD) will increase for both emission scenarios by 25 to 50 in majority of Poland; in the far future, large differences between scenarios are projected: increase by 75–100 CDD for the RCP 4.5 emission scenario, for most of the country and up to 125 in the upper Odra valley and by 150–175 for the RCP 8.5 scenario.

The **chrysophyte-based reconstruction** of winter severity showed similar features to data available from other European regions. Despite some periods of divergence, the **chironomid-inferred** mean August temperature reconstruction responded similarly to other archives. The **dendroclimatic reconstruction** of summer temperature variability for the Tatra Mountains compared with those obtained for the Alpine arc and Central Europe revealed similarities, especially during the cold periods associated with the Dalton and Late Maunder solar minima and sequences of major volcanic eruptions. Due to the relatively low precipitation signal present in the dendroclimatological records developed for the territory of Poland, there is a limited number of works with reconstructed precipitation in Poland in the pre-instrumental period based on tree rings in comparison to other parts of Europe.

Several research methods used for **atmospheric circulation** analysis produced rather consistent results, which are relatively convergent throughout Poland. The results are also not in contradiction with those of previously published works on circulation changes in the last 100 years in Europe. An increase in low pressure system frequency over all seasons in Poland explains, to some extent, the greater dynamism of weather conditions.

No significant changes in **air pressure** were also concluded at other European regions (Bielec-Bąkowska 2010, Sweeney 2000; Bielec-Bąkowska and Piotrowicz 2013) and in the North Atlantic region (Bärring and Fortuniak 2009). However, in the period 1961–2018 in the north and east of Poland, as well as at its southwestern end, there were signals of an increasing number of days with SLP $\geq$ 1030 hPa connected perhaps with an increase in the incidence of winter anticyclonal systems associated with the shift of the Asian High towards the northwest (Zhang et al. 2012).

The negative trend of annual **solar radiation** totals in Kraków over the entire 125-year period studied is accompanied by a positive trend in cloudiness in the city over the period 1826-2005, as proven by Lewik and co-authors (2010). Relatively high values of global solar radiation in southern Poland in the period 1990–2018 correspond with decreasing trend of cloudiness in 1985–2018 in summer and autumn (Filipiak 2021). The global solar radiation course in southern Poland largely corresponds to the periods of "global dimming" and "global brightening", described by researchers in different parts of the world (Liepert 2002; Alpert et al. 2005; Wild et al. 2005, 2007; Norris and Wild 2007; Ruckstuhl and Norris 2009; Wild 2009). There is an observed decrease in values until the end of the 1970s or 1980s, depending on the season of a year, and then an increase until the end of the twentieth century. The positive trend in solar radiation since the 1980s can be a significant factor affecting the positive trend of thermal conditions (Ustrnul et al. 2021) and the negative trend for snow cover (Falarz and Bednorz 2021).

The long-term variability of **sunshine duration** in Poland shows similar tendencies to the changes occurring in other parts of the world. This is confirmed both by the data from the multi-annual period of 1971–2018 from 31 stations in Poland and the long-term heliographic series from Warszawa, Puławy, Wrocław and Śnieżka. The periods of "global dimming" and "global brightening", well-known from the climatological literature are very clearly visible in the course of sunshine duration in Poland. The years 1971–1980 mark the end of "global dimming", which was manifested in very low values of actual and relative sunshine duration. "Global brightening" confirms a statistically significant growing trend in actual and relative sunshine duration, with its maximum in 2018 at all analysed stations in Poland. Similar tendencies in sunshine duration over a larger area indicate causes of global nature, only modified by local factors. Macro-scale conditions can be of both natural (related to circulation and volcanic eruptions) and anthropogenic origin caused by changes in the concentration of aerosols due to industrialisation and urbanisation, as well as measures to reduce their emissions.

The mean **cloud amount** in Poland does not differ significantly from the values recorded in neighbouring European countries and in the entire Baltic Sea basin due to the crucial role of atmospheric circulation in forming the nephologic conditions in Poland. An important role in this case is also played by processes of a local scale related to relief diversity and orography, land use and in particular factors connected with impact of the cities.

The demonstrated variability of **thermal conditions** in Poland is fully in line with the trends observed in recent decades in Europe and elsewhere in the world (IPCC 2013; Abram et al. 2019). This applies in particular to the courses of average air temperature values for which there are a number of key studies.

The maximum amount of water vapour in the atmosphere depends on the air temperature. The increasing temperature causes an increase in specific **humidity**. However, the trend magnitude is lower over land where moisture sources are limited (Wypych and Bochenek 2018; Wypych et al. 2018). In addition, relative humidity describing the saturation percentage decreases in conditions of increasing temperature as the amount of available water vapour is insufficient to execute increasing moisture capacity (Wypych 2018).

The temporal course of **precipitation** characteristics is rather dominated by short-term, several year-long fluctuations, and dry and wet decades. It was also indicated in a number of previous studies (Kożuchowski 1982, 1985; Żmudzka 2002; Kirschenstein 2005; Czarnecka and Nidzgorska-Lencewicz 2012; Skowera et al. 2014; Malinowska and Jakusik 2015; Mager et al. 2009). The area of Central Europe is on the maps published in the IPCC report (2013), shown as an area with positive but statistically insignificant trends in annual precipitation totals in the period 1979–2010 (according to CRU, GHCN, GPCC datasets) and a negative insignificant trend of frequency in the annual maximum number of consecutive dry days in the period 1951-2010 for grid data. The results presented in this book for the territory of Poland, based on data from meteorological stations, are fully in accordance with those of IPCC Report.

The trends of **snow cover** in Poland are mostly negative and are in line with those observed in Europe, Eurasia (Bednorz 2004; Takala et al. 2009; Fontrodona Bach et al. 2018; Henderson and Leathers 2010) and throughout the Northern Hemisphere (Fernandes et al. 2009; Brown and Robinson 2011; Flanner et al. 2011). Changes in snow cover in Poland are the result of changes in circulation, thermal and precipitation conditions, which have a great influence on snow cover (Bednorz 2004; Falarz et al. 2018). Changes in snow cover affect radiation balance, mainly due to spring snow cover-albedo feedback (Fernandes et al. 2009; Flanner et al. 2011). This is particularly important in areas such as Poland, where the ground is seasonally covered by snow (Barry and Gran 2011). Negative trends in the snow cover duration and its maximum depth have a negative impact on skiing and winter sports conditions (Marke et al. 2015). This is especially true in mountainous areas, where the ski season is the longest. It is therefore increasingly necessary to create artificial snow for the slopes provided a fulfilment of thermal criteria (Steiger and Mayer 2008). In the High Alpine Mountains, the categorisation of the difficulty has changed for a number of routes, and the beginning of the climbing season has shifted earlier (Mourey and Ravanel 2017).

The results of the study of snow cover changes in Poland are largely consistent with IPCC Reports (IPCC 2013, 2018; Abram et al. 2019), which highlight the fastest changes in snow cover extent in spring in the Northern Hemisphere (Déry and Brown 2007), especially in the Arctic (Brown et al. 2010). The average March and April Northern Hemisphere snow cover area decreased by 1.59% per decade over the 1967–2012 period (Déry and Brown 2007). A decline in Arctic sea ice extent has also been observed as well (Vaughan et al. 2013).

The decreasing trends in annual and seasonal series of **wind speed** were observed also in other countries in Europe, e.g. in Czech Republic (Brázdil et al. 2017, Zahradníček et al. 2019), in Hungary (Péliné Németh et al. 2011), in Germany (Kohler et al. 2018), in Portugal and Spain (Azorin-Molina et al. 2014; 2016; 2017), in the Balkans (Romanić et al. 2015), in Sweden (Minola et al. 2016) and in Finland (Laapas and Venäläinen 2017). A number of authors attribute this trend to an increase in surface roughness. This thesis can be confirmed by the increasing statistically significant tendency of wind speed and the number of days with strong wind (> 8 ms^{-1}) in Central Europe obtained in research based on zonal and meridional wind velocity at a height of 10 m above ground level from 35 grids in the period 1951–2005 (Araźny et al. 2007).

The spatial differentiation of the number of days with **thunderstorms** and its annual cycle does not differ significantly from similar characteristics recorded in earlier research periods. Over the last 40–50 years, the period of thunderstorm activity during the year was often longer than in previous years and shifts towards the beginning of the convective season. These regularities are specific for most regions of Europe, e.g. for the Baltic countries (Enno et al. 2013), for Sweden (Sonnadara et al. 2006), for Austria (Schulz et al. 2005), for the Czech Republic (Novak and Kyznarova 2011) and for Romania (Antonescu and Burcea 2010). The latest studies based on lightning detection data also pointed to the period from June to August as the most prone to thunderstorm occurrence in Europe (Enno et al. 2020; Taszarek et al. 2019; Poelman et al. 2016).

An important addition to our knowledge of the occurrence of **tornadoes** in question was the discovery of dozens of historical tornado cases that took place over the last 200 years in Poland. This finding contradicts the popular statement that *"tornadoes in Poland are a new thing and have become more frequent due to the changing climate"*. Historical records indicate that this phenomenon is not new, but Poland has been and is vulnerable to the occurrence of devastating tornadoes even up to F4 on the Fujita scale. Although such disasters are very rare, they are possible as evidenced over the last 200 years. Estimates based on the reanalysis indicate that along with the globally warming climate, environmental conditions are becoming consistently more conducive for severe convective thunderstorms in Poland, including tornadoes.

The number of days with **hail** in Poland is similar to that observed in Central Europe, although it is much lower than in Southern Europe and in the areas of the world with the highest frequency of convection phenomena (Punge and Kunz 2016;

Prein and Holland 2018). In the coastal area and the lakelands, the inflow of significant amounts of moisture (Zinkiewicz and Michna 1955; Tuovinen et al. 2009) and the warming effect of the Baltic Sea drive the higher frequency of hail occurrence. Southern Poland is the mountainous area of the most favourable conditions for hail occurrence: the diversified topography is conducive to the intensification of upward air currents and the formation of *Cumulonimbus* cloud complexes. The most dangerous hailfall occurs in the south and east of Poland, which is mainly attributable to orographic and circulation conditions (Pilorz 2015; Taszarek and Suwała 2015).

The regional feature of the amount of **fog** is not always correctly recognised in large-scale studies. For example, in the paper by van Oldenborgh et al. (2010), the data used present much better the spatial differences in the fog frequency in Poland than in case of the work by Vautard et al. (2009). The authors conclude that for inter-annual variability, the effects of circulation dominate, but, for the trend, other factors are more important, e.g. decreasing aerosol emissions over Europe, a larger fraction of urban land, moisture availability and heat generation at ground level.

The other analysis of UTCI trends in Poland confirms the results obtained in this study (Kuchcik 2017; Tomczyk and Owczarek 2020). The observed changes in **bioclimatic indices** (UTCI and PST) do not exactly reflect the pattern of increasing sunshine duration throughout Poland, however correlates more with increases of mean cloudiness especially in winter in the years 1985–2018. The multi-annual pattern of days with different thermal sensations and thermal stress categories also shows an increase in hot and neutral days, and a decrease in the number of cold days. The changes of the UTCI index value are bigger than the changes of the meteorological parameters analysed separately which confirms the legitimacy of using biometeorological indicators in the analyses of thermal stress.

The results of **weather types** investigation are consistent with the research of temperature, and cloudiness conditions (e.g. Wójcik and Miętus 2014; Wypych et al. 2017; Bielec-Bąkowska et al. 2018), precipitation (e.g. Czarnecka and Nidzgorska-Lencewicz 2012; Marosz et al. 2011) and cloudiness condition (e.g. Filipiak and Miętus 2009; Matuszko 2007, Wibig 2008). Particular types of weather have been grouping since 1951 (and in Kraków since 1901) in slightly longer sequences of days. Francis et al. (2018) also point to an increased persistence of North American Weather Regimes. The authors hypothesise that the reason for this is rapid Arctic warming. They suggest that further warming may favour persistent weather patterns that can lead to weather extremes and increased frequency of long-duration events (LDEs). In short, this may be associated with longer heat waves in the summer and warmer periods in the cool half of the year, i.e. refer to a higher frequency of days with the thermal type of very warm, warm and cool weather (3–, 2– and 1–).

The climate will continue to **change in the future**. All climate models have shown an agreement on a systematic upward trend in **the mean temperature** in Poland. In the future, the extreme hot thermal conditions will become even hotter, last longer and appear more frequently, while amplitude of frosty periods and their frequency will decrease. Nonetheless, climate change extends far beyond changes in temperature. The projected mean temperature change for Central Europe is about 1.5–2°C at 1.5°C global mean surface temperature (GMST) warming and 2–3°C at 2°C GMST warming compared to preindustrial time period (1861–1880) (IPCC 2018). The values are comparable to those obtained by RCPs.

Projected mean precipitation change for Central Europe is 0–5% at both 1.5°C and 2°C global mean surface temperature (GMST) warming compared to preindustrial time period (1861–1880) (IPCC 2018). Future increase in extreme precipitation can contribute to more severe floods, while increases in dry periods lead to extended droughts. Recent studies indicate that the projection of precipitation extremes is associated with various uncertainties related to emission scenarios of greenhouse gases, GCMs, RCMs and statistical downscaling methods as well as by natural variability of climate. However, the direction of projected changes in mean, heavy precipitation and dry periods across Poland are quite consistent and correspond with continental-scale results. Projections also show that the highest values of indices may not increase for the whole country.

Considering only the thermal background, conditions of **agricultural production** will improve **in the future**. An increase in air temperature may result in earlier occurrence of individual development phases by arable crops (Nieróbca et al. 2013). Increasing annual sums of growing degree days and extension of the frost-free period will favour crops with higher thermal requirements. The beneficial effect may, however, be reduced by more frequent occurrence of intense drought and rainfall deficiencies in summer and the growing season (Pinskwar 2010; Szwed et al. 2010).

The positive effect of **projected warming** will be a significant reduction in the **demand for energy** used to heat buildings. However, this positive impact could be reduced by a large increase in energy demand due to refrigeration and air conditioning. Even in warmer climates, there will still be some days with a high demand for energy for heating. An increase in the demand for energy for cooling during the warmest days may overload energy production and transmission systems. According to estimates by Damm et al. (2017), in 2036–2065 for the RCP 4.5 scenario, due to global warming, energy consumption will fall in almost all European countries. Such an estimation for Poland area is at 1 to 2%.

The majority of examined climate elements, meteorological phenomena and indices show statistically significant changes at least in certain areas of Poland and at certain seasons of the year. Moreover, many elements demonstrate significant year-to-year variability and temporal fluctuations. Changes of particular climate elements are interrelated. Schemes of current and future climate change connections in Poland based on the results presented in this book are shown in Figs. 24.1 and 24.2. The primary causative factors are both anthropogenic changes (greenhouse gas emissions resulting in increased greenhouse effect and global warming, local sources of air pollution) and natural changes: (1) circulation factors: changes in the intensity and location of atmospheric activity centres (Zhang et al. 2012; Falarz 2019), changes in the frequency of advection from a specific sector, and the frequency of cyclonic and anticyclonic systems over Poland and (2) radiation factors (changes in values of globar solar radiation, sunshine duration and cloudiness). These changes, especially visible after the 1980s, affect the trends of most climatic elements, meteorological phenomena and indices. The effects of these changes, both positive and negative, are

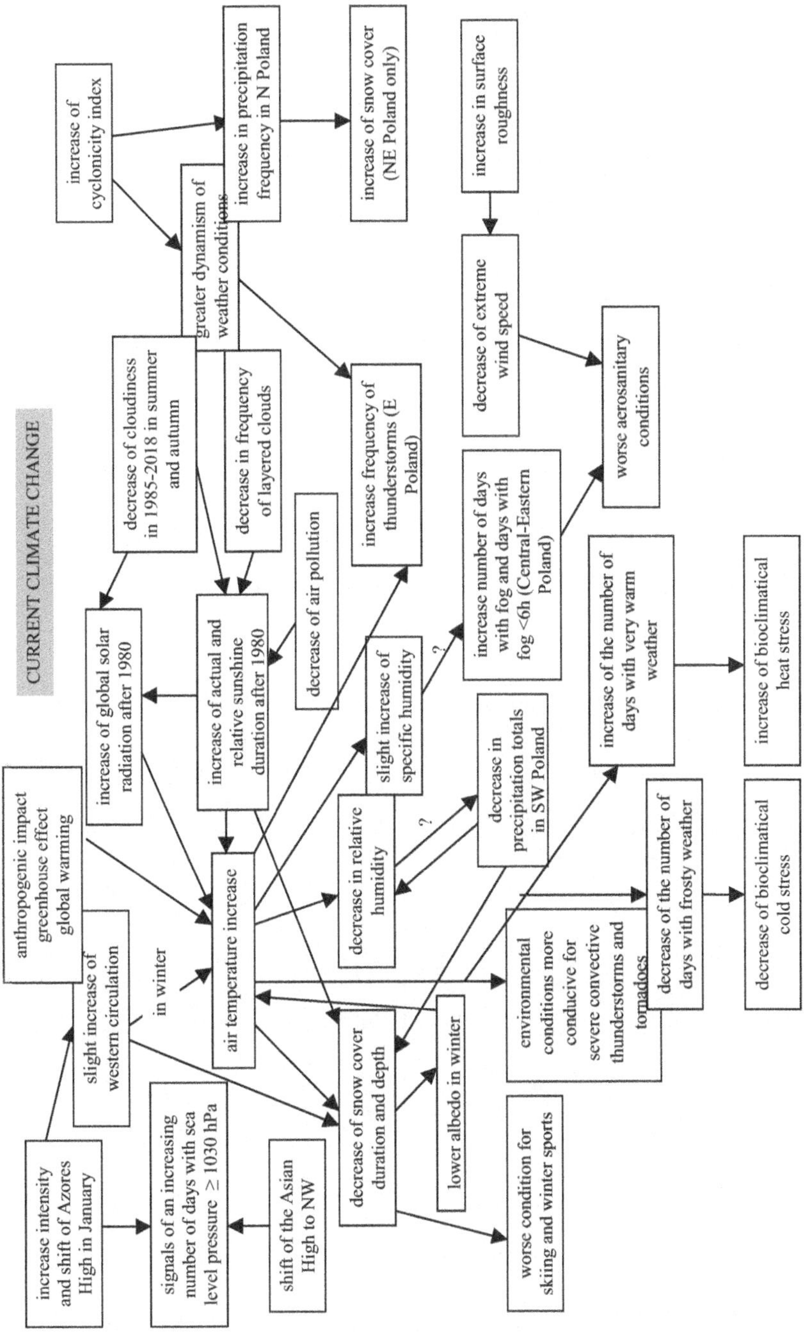

Fig. 24.1 Scheme of current climate change connections in Poland. Possible connections are marked by "?"

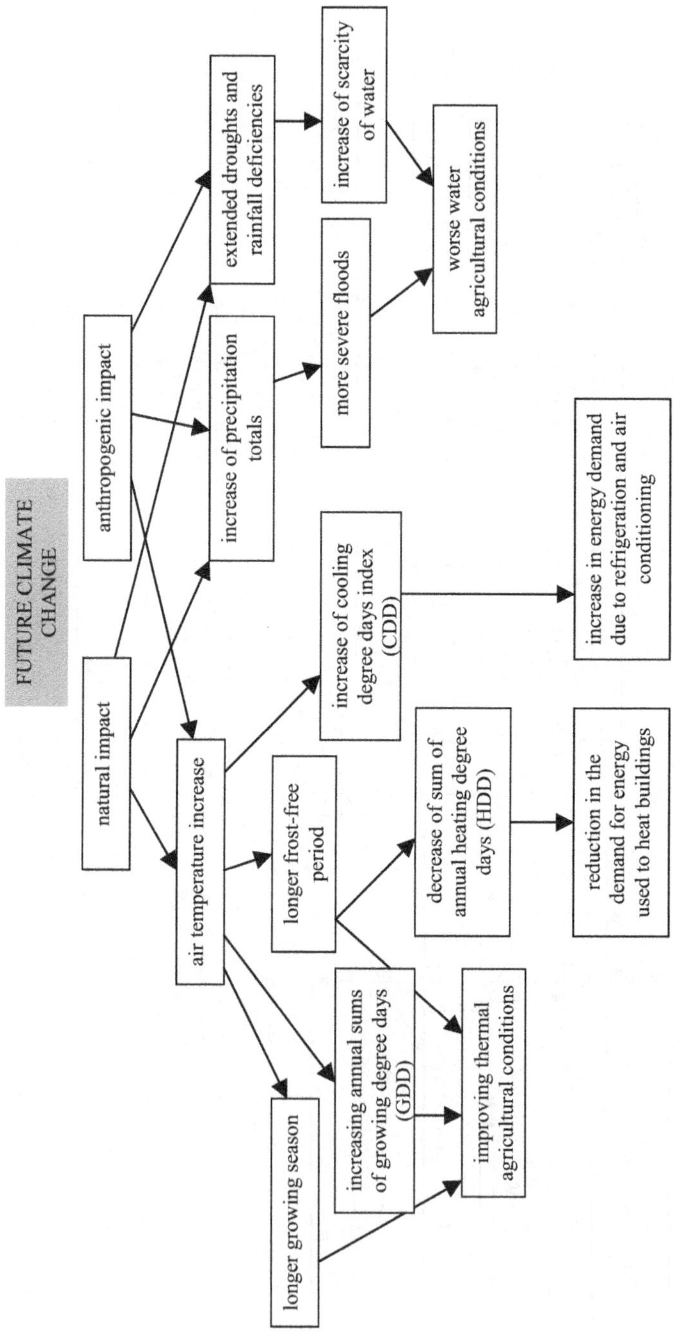

Fig. 24.2 Scheme of future climate change connections in Poland. Some economic effects were demonstrated

evident in people's daily lives (e.g. decrease in bioclimatic cold stress, increase in bioclimatic heat stress, changes in conditions for recreation and sport) and economy (e.g. improvement of thermal agricultural conditions, changes in energy demand for heating buildings, air conditioning and refrigeration).

A better understanding of the relationships of trends of the different climatic elements should be the aim of further research into the climate of Poland.

References

Abram N, Gattuso J-P, Prakash A, Cheng L, Chidichimo MP, Crate S, Enomoto H, Garschagen M, Gruber N, Harper S, Holland E, Kudela RM, Rice J, Steffen K, von Schuckmann K (2019) Framing and context of the report. In: IPCC Special Report on the Ocean and Cryosphere in a Changing Climate [H.-O. Pörtner, D.C. Roberts, V. Masson-Delmotte, P. Zhai, M. Tignor, E. Poloczanska, K. Mintenbeck, A. Alegría, M. Nicolai, A. Okem, J. Petzold, B. Rama, N.M. Weyer (eds.)]

Alpert P, Kishcha P, Kaufman YJ, Schwarzbard R (2005) Global dimming or local dim-ming? Effect of urbanization on sunlight availability. Geophys Res Lett 32:L17802. https://doi.org/10.1029/2005GL023320

Antonescu B, Burcea S (2010) A cloud-to-Ground lighting climatology for Romania. Mon Weather Rev 138:579–591

Araźny A, Przybylak R, Vizi Z et al (2007) Mean and extreme wind velocities in Central Europe 1951–2005. Geogr Pol 80(2):69–78

Azorin-Molina C, Guijarro JA, McVicar TR et al (2016) Trends of daily peak wind gusts in Spain and Portugal, 1961–2014. J Geophys Res 121:1059–1078. https://doi.org/10.1002/2015JD024485

Azorin-Molina C, Vicente-Serrano SM, McVicar TR et al (2014) Homogenization and assessment of observed near-surface wind speed trends over Spain and Portugal, 1961–2011. J Climate 27:3692–3712. https://doi.org/10.1175/JCLI-D-13-00652.1

Azorin-Molina C, Vicente-Serrano SM, McVicar TR et al (2017) Assessing the impact of measurement time interval when calculating wind speed means and trends under the stilling phenomenon. Int J Climatol 37:480–492. https://doi.org/10.1002/joc.4720

Bärring L, Fortuniak K (2009) Multi-indices analysis of southern Scandinavian storminess 1780-2005 and links to interdecadal variations in the NW Europe-North Sea region. Int J Climatol 29(3):373–384

Barry R, Gran TY (2011) The global cryosphere: Past, Present and Future. Cambridge University Press, Cambridge, p 498

Bednorz E (2004) Snow cover in eastern Europe in relation to temperature, precipitation and circulation. Int J Climatol: J R Meteorol Soc, 24(5), 591–601

Bielec-Bąkowska Z (2010) Silne wyże nad Europą W XX wieku, [w] Kolendowicz L. (red.), Klimat Polski na tle klimatu Europy. Warunki cyrkulacyjne i radiacyjne, Bogucki Wydawnictwo Naukowe, Seria Studia i Prace z Geografii i Geologii, 14:23–37

Bielec-Bąkowska Z, Piotrowicz K (2013) Long-term occurrence, variability and tracks of deep cyclones over Kraków (central Europe) during the period 1900–2010. Int J Climatol 33:677–689

Bielec-Bąkowska Z, Piotrowicz K, Krępa-Adolf E (2018) Trends in the frost-free season with parallel circulation and air mass statistics in Poland. Időjárás 122(4):375–392. https://doi.org/10.28974/idojaras.2018.4.2

Brázdil R, Valík A, Zahradníček P, Řezníčková L et al (2017) Wind-stilling in the light of wind speed measurements: the Czech experience. Clim Res 75:131–143

Brown RD, Robinson DA (2011) Northern Hemisphere spring snow cover variability and change over 1922–2010 including an assessment of uncertainty. The Cryosphere 5:219–229. https://doi.org/10.5194/tc-5-219-2011

Brown R, Derksen C, Wang LB (2010) A multi-data set analysis of variability and change in Arctic spring snow cover extent, 1967–2008. J. Geophys. Res.-Atmos. 115:D16111

Czarnecka M, Nidzgorska-Lencewicz J (2012) Wieloletnia zmienność sezonowych opadów w Polsce. Woda-Środowisko-Obszary Wiejskie 12(2):45–60

Damm A, Köberl J, Prettenthaler F, Rogler N, Töglhofer C (2017) Impacts of +2°C global warming on electricity demand in Europe. Clim Serv 7:12–30. https://doi.org/10.1016/j.cliser.2016.07.001

Déry SJ, Brown RD (2007) Recent Northern Hemisphere snow cover extent trends and implications for the snow-albedo feedback. Geophys Res Lett, 34, 6 (L22504)

Enno SE, Briede A, Valiukas D (2013) Climatology of thunderstorms in the Baltic countries, 1951–2000. Theor Appl Climatol 111:309–325

Enno SE, Sugier J, Alber R, Seltzer M (2020) Lightning flash density in Europe based on 10 years of ATDnet data. Atmos Res 235:104769

Falarz M (2019) Azores High and Hawaiian High: correlations, trends and shifts (1948–2018). Theor Appl Climatol 138:417–431

Falarz M, Nowosad M, Bednorz E, Rasmus S (2018) Review of Polish contribution to snow cover research (1880–2017). Quaest Geogr 37(1):7–22. https://doi.org/10.2478/quageo-2018-0002

Falarz M, Bednorz E (2021) Snow Cover Change. In: Falarz M. [ed.] Climate Change in Poland - Past, Present, Future, Springer Climate. https://doi.org/10.1007/978-3-030-70328-8_14

Fernandes R, Zhao HX, Wang XJ, Key J, Qu X, Hall A (2009) Controls on Northern Hemisphere snow albedo feedback quantified using satellite Earth observations. Geophys Res Lett 36:L21702

Filipiak J, Miętus M (2009) Spatial and temporal variability of cloudiness in Poland, 1971–2000. Int J Climatol 29:1294–1311. https://doi.org/10.1002/joc.1777

Filipiak J (2021) Change of Cloudiness. In: Falarz M. [ed.] Climate Change in Poland - Past, Present, Future, Springer Climate. https://doi.org/10.1007/978-3-030-70328-8_10

Flanner MG, Shell KM, Barlage M, Perovich DK, Tschudi MA (2011) Radiative forcing and albedo feedback from the Northern Hemisphere cryosphere between 1979 and 2008. Nat Geosci 4:151–155

Fontrodona Bach A, van der Schrier G, Melsen LA, Tank K, Teuling AJ (2018) Widespread and accelerated decrease of observed mean and extreme snow depth over Europe. Geophys Res Lett 45:312–319

Francis JA, Skific N, Vavrus SJ (2018) North American weather regimes are becoming more persistent: Is Arctic amplification a factor? Geophys Res Lett 45:11,414–11,422. https://doi.org/10.1029/2018gl080252

Henderson GR, Leathers DJ (2010) European snow cover extent variability and associations with atmospheric forcings. Int J Climatol 30:1440–1451

IPCC (2013) Climate Change 2013. The Physical Science Basis. Contribution of Working Group I to the Fifth Assessment Report of the Intergovernmental Panel on Climate Change [Stocker, T.F., D. Qin, G.-K. Platter, M. Tignor, S.K. Allen, J. Boschung, A. Nauels, Y. Xia, V. Bex and P.M. Midgley (eds.)]. Cambridge University Press, Cambridge, UK and New York, NY, USA

IPCC (2018) Global Warming of 1.5°C. An IPCC Special Report on the impacts of global warming of 1.5°C above pre-industrial levels and related global greenhouse gas emission pathways, in the context of strengthening the global response to the threat of climate change, sustainable development, and efforts to eradicate poverty [Masson-Delmotte, V., P. Zhai, H.-O. Pörtner, D. Roberts, J. Skea, P.R. Shukla, A. Pirani, W. Moufouma-Okia, C. Péan, R. Pidcock, S. Connors, J.B.R. Matthews, Y. Chen, X. Zhou, M.I. Gomis, E. Lonnoy, T. Maycock, M. Tignor, and T. Waterfield (eds.)]

Kirschenstein M (2005) Wieloletnie zmiany sum opadów atmosferycznych na wybranych stacjach północno-zachodniej Polski. Słupskie Prace Geograficzne 2:199–214

Kohler M, Metzger J, Kalthoff N (2018) Trends in temperature and windspeed from 40 years of observations at a 200-m high meteorological tower in Southwest Germany. Int J Climatol 38:23–34. https://doi.org/10.1002/joc.5157
Kożuchowski K (1982) Periodicity of variations of the annual precipitation totals in Poland. Acta Geophysica, XXX 3:242–287
Kożuchowski K (1985) Zmienność opadów atmosferycznych w Polsce w stuleciu 1881-1980. Acta Geographica Lodziensia 48:158
Kuchcik M (2017) Zmiany warunków termicznych w Polsce na przełomie XX i XXI wieku i ich wpływ na umieralność. Prace Geograficzne, IGiPZ PAN, Warsaw
Laapas M, Venäläinen A (2017) Homogenization and trend analysis of monthly mean and maximum wind speed time series in Finland, 1959–2015. Int J Climatol 37:4803–4813. https://doi.org/10.1002/
Lewik P, Matuszko D, Morawska-Horawska M (2010) Multi-annual variability of cloudiness and sunshine duration in Krakow between 1826 and 2005. In: The Polish Climate in the European Context: An Historical Overview. Advances in Global Change Research, Vol. XVIII, Przybylak R., Majorowicz J., Brazdil R., Kejna M. (eds). Springer, Dordrecht, The Netherlands, Heidelberg, Germany, London and New York, NY, pp 341–353
Liepert B (2002) Observed reductions of surface solar radiation at sites in the United States and worldwide from 1961 to 1990. Geophys Res Lett 29(10):1421. https://doi.org/10.1029/2002GL014910
Mager P, Kasprowicz T, Farat R (2009) Change of air temperature and precipitation in Poland in 1966–2006. W: Climate change and agriculture in Poland – impacts, mitigation and adaptation measures. Pr. zbior. Red. J. Leśny. Acta Agrophysica. Rozprawy i Monografie, 169(1):19–38
Malinowska M, Jakusik E (2015) Charakterystyka opadów atmosferycznych w centralnej części Pojezierza Kaszubskiego w latach 1971–2010. [Characteristics of precipitation in central part of Kashubian Lake District, 1971–2010]. Scientific Review—Engineering and Environmental Sciences No 69, 2015: 273–285 (Sci Rev Eng Env Sci 69, 2015)
Marke T, Strasser U, Hanzer F, Stotter J, Anna R, Wilcke I, Gobiet A (2015) Scenarios of Future Snow Conditions in Styria (Austrian Alps). J Hydrometeorol 16:261–277
Marosz M, Wójcik R, Biernacik D, Jakusik E, Pilarski M, Owczarek M, Miętus M (2011) Zmienność klimatu polski od połowy xx wieku. Rezultaty projektu klimat. Prace i Studia Geograficzne 47:51–66
Matuszko D (2007) Zachmurzenie. In: Matuszko D (ed) Klimat Krakowa w XX wieku. IGiGP UJ, Kraków, pp 75–86
Minola L, Azorin-Molina C, Chen DL (2016) Homogenization and assessment of observed near-surface wind speed trends across Sweden, 1956–2013. J Climate 29:7397–7415. https://doi.org/10.1175/JCLI-D-15-0636.1
Mourey J, Ravanel L (2017) Evolution of access routes to high mountain refuges of the Mer de Glace Basin (Mont Blanc Massif, France): An example of adapting to climate change effects in the Alpine High Mountains. J AlpE Res | Revue de géographie alpine, 105(4). https://doi.org/10.4000/rga.3790
Nieróbca A, Kozyra J, Mizak K, Wróblewska E. (2013) Zmiana długości okresu wegetacyjnego w Polsce (Changing length of the growing season in Poland). Woda-Środowisko-Obszary Wiejskie 13, 2(42):81–94
Norris JR, Wild M (2007) Trends in aerosolradiative effects over Europe inferred from observed cloud cover, solar 'dimming' and solar 'brightening'. J Geophys Res 112:D08214. https://doi.org/10.1029/2006JD0077941-13
Novak P, Kyznarova H (2011) Climatology of lightning in the Czech Republic. Atmos Res 100:318–333
Péliné Németh C, Radics K, Bartholy J (2011) Seasonal variability of Hungarian wind climate. Acta Silvatica & Lignaria Hungarica 7:39–48
Pilorz W (2015) Very large hail occurrence in Poland from 2007 to 2015. Contemporary Trends in Geoscience 4(1):45–55

Pińskwar I (2010) Projections of changes in precipitation extremes in Poland. Monografia KGW PAN, 34

Poelman DR, Schulz W, Diendorfer G, Bernardi M (2016) The European lightning location system EUCLID—Part 2: Observations. Nat Hazards Earth Syst Sci 16:607–616

Prein AF, Holland GJ (2018) Global estimates of damaging hail hazard. Weather Clim Extrem 22:10–23

Punge HJ, Kunz M (2016) Hail observations and hailstorm characteristics in Europe: A review. Atmos Res 176–177:159–184

Romanić D, Curić M, Jovičić I et al (2015) Long-term trends of the 'Koshava' wind during the period 1949–2010. Int J Climatol 35:288–302. https://doi.org/10.1002/joc.3981

Ruckstuhl C, Norris JR (2009) How do aerosol histories affect solar "dimming" and "brightening" over Europe? IPCC-AR4 models versus observations. J Geophys Res 114: D00D04 1029/2008JD011066:1–12

Schulz W, Cummins K, Diendorfer G, Dorninger M (2005) Cloud-to-ground lightning in Austria: a 10-year study using data from a lightning location system. J Geophys Res 110 (D9)

Skowera B, Kopcińska J, Kopeć B (2014) Changes in thermal and precipitation conditions in Poland in 1971–2010. Annals of Warsaw University of Life Sciences—SGGW. Land Reclamation 46(2):153–162

Sonnadara U, Cooray V, Götschl T (2006) Characteristics of cloud-to-ground lightning flashes over Sweden. Phys Scr 74(5):541–548

Steiger R, Mayer M (2008) Snowmaking and climate change. Futur Options Snow Prod Tyrolean Ski Resorts Mt Res Dev 28(3/4):292–298

Sweeney JC (2000) A three-century storm climatology for Dublin 1715–2000. Ir Geogr 33(1):1–14

Szwed M, Karg G, Pińskwar I, Radziejewski M, Graczyk D, Kędziora A, Kundzewicz ZW (2010) Climate change and its effect on agriculture, water resources and human health sectors in Poland. Nat Hazards Earth Syst Sci 10(8):1725–1737

Takala M, Pulliainen J, Metsamaki SJ, Koskinen JT (2009) Detection of snowmelt using spaceborne microwave radiometer data in Eurasia from 1979 to 2007. IEEE Trans Geosci Remote Sens 47:2996–3007

Taszarek M, Allen J, Púčik T, Groenemeijer P, Czernecki B, Kolendowicz L, Lagouvardos K, Kotroni V, Schulz W (2019) A climatology of thunderstorms across Europe from a synthesis of multiple data sources. J Clim 32:1813–1837

Taszarek M, Suwała K (2015) Large hail in Poland in 2012. Quaest Geogr 34(1):75–84

Tomczyk AM, Owczarek M (2020) Occurrence of strong and very strong heat stress in Poland and its circulation conditions. Theoret Appl Climatol 139:893–905

Tuovinen JP, Punkka AJ, Rauhala J, Hohti H, Schultz DM (2009) Climatology of severe hail in Finland: 1930–2006. Mon Weather Rev 137(7):2238–2249

Ustrnul Z, Wypych A, Czekierda D (2021) Air Temperature Change. In: Falarz M. [ed.] Climate Change in Poland - Past, Present, Future, Springer Climate. https://doi.org/10.1007/978-3-030-70328-8_11

van Oldenborgh GJ, Yiou P, Vautard R (2010) On the roles of circulation and aerosols in the decline of mist and dense fog in Europe over the last 30 years. Atmos Chem Phys 10:4597–4609

Vaughan, DG et al. (2013) Observations: Cryosphere. In: Climate change 2013: The physical science basis. Contribution of Working Group I to the Fifth Assessment Report of the Intergovernmental Panel on Climate Change [Stocker, T.F., D. Qin, G.-K. Plattner, M. Tignor, S.K. Allen, J. Boschung, A. Nauels, Y. Xia, V. Bex and P.M. Midgley (eds.)]. Cambridge University Press, Cambridge, UK and New York, NY, USA, pp 317–382

Vautard R, Yiou P, van Oldenborough GJ (2009) Decline of fog, mist and haze in Europe over the past 30 years. Nat Geosci 2:115–119

Wibig J (2008) Cloudiness variations in Łódź in the second half of the 20th century. Int J Climatol 28:479–491. https://doi.org/10.1002/joc.1544

Wild M. (2009) Global dimming and brightening: a review. J Geophys Res 114 (D10): D00D16. https://doi.org/10.1029/200 8JD011470

Wild M, Gilgen H, Roesch A, Ohmura A, Long CN, Dutton E, Forgan B, Kallis A, Russak V, Tsvetkov A (2005) From dimming to brightening: Decadal changes in solar radiation at earth's surface. Science 308(5723):847–850

Wild M, Ohmura A, Makowski K (2007) Impact of global dimming and brightening on global warming. Geophys Res Lett 34:L04702. https://doi.org/10.1029/2006GL028031

Wójcik R, Miętus M (2014) Niektóre cechy wieloletniej zmienności temperatury powietrza w Polsce (1951–2010). Przegląd Geograficzny 86(3):339–364

Wypych A, Sulikowska A, Ustrnul Z, Czekierda D (2017) Variability of growing degree days in Poland in response to ongoing climate changes in Europe. Int J Biometeorol 61:49–59. https://doi.org/10.1007/s00484-016-1190-3

Wypych A (2018) Para wodna w troposferze nad Europą, Instytut Geografii i Gospodarki Przestrzennej UJ, pp 200

Wypych A, Bochenek B (2018) Vertical Structure of Moisture Content over Europe, Advances in Meteorology, ID: 3940503

Wypych A, Bochenek B, Różycki M (2018) Atmospheric Moisture content over Europe and the Northern Atlantic, Atmosphere, 9(18). https://doi.org/10.3390/atmos9010018

Zahradníček P, Brázdil R, Štěpánek P et al (2019) Differences in wind speeds according to measured and homogenized series in the Czech Republic, 1961–2015. Int J Climatol 39:235–250. https://doi.org/10.1002/joc.5800

Zhang X, Lu Ch, Guan Z (2012) Weakened cyclones, intensified anticyclones and recent extreme cold winter weather events in Eurasia. Environ Res Lett 7:044044

Zinkiewicz W, Michna E (1955) Częstość występowania gradu w woj. lubelskim w zależności od warunków fizjograficznych (Hail falls frequency in Lubelskie region according to physiographic patterns). Annales Universitatis Marie Curie-Skłodowska Lublin – Polonia, Sectio B 10(5):224–267

Żmudzka E (2002) O zmienności opadów atmosferycznych na obszarze Polski nizinnej w drugiej połowie XX wieku (About the variability of precipitation in lowland Poland in the second mid-twentieth century). Wiadomości IMGW 25(4):23–38

Correction to: Climate Change in Poland

Małgorzata Falarz

Correction to:
M. Falarz (ed.),
Climate Change in Poland, Springer Climate,
https://doi.org/10.1007/978-3-030-70328-8

The original version of the book was inadvertently published with incorrect Figures 23 and 33 in Chapter 11 and also the affiliation city name "Katowice" of the volume editor is corrected in web version. The book has been updated with the changes.

The updated version of the book can be found at
https://doi.org/10.1007/978-3-030-70328-8_11
https://doi.org/10.1007/978-3-030-70328-8

M. Falarz (ed.), *Climate Change in Poland*, Springer Climate,
https://doi.org/10.1007/978-3-030-70328-8_25

The manufacturer's authorised representative in the EU is Springer Nature Customer Service Centre GmbH, Europaplatz 3, 69115 Heidelberg, Germany. If you have any concerns regarding our products, please contact ProductSafety@springernature.com

Printed and bound by CPI Group (UK) Ltd, Croydon, CR0 4YY
30/09/2026
02225320-0019